# Selected Titles in This Series

D0474088

# STOCHASTIC MODELS

*Canadian Mathematical Society*
*Société mathématique du Canada*

**CONFERENCE PROCEEDINGS • Volume 26**

# STOCHASTIC MODELS

Proceedings of the International Conference
on Stochastic Models in Honour of
Professor Donald A. Dawson
Ottawa, Canada
June 10–13, 1998

**Luis G. Gorostiza**
**B. Gail Ivanoff**
**Editors**

Published by the American Mathematical Society
for the Canadian Mathematical Society
PROVIDENCE, RHODE ISLAND

CMS CONFERENCE PROCEEDINGS
Series Editors
Anthony V. Geramita and Niky Kamran
Volume 26

Accepted for publication by the Canadian Mathematical Society
Published by the American Mathematical Society

Proceedings of the International Conference on Stochastic Models in Honour of Professor Donald A. Dawson held in Ottawa, Canada, June 10–13, 1998.

1991 *Mathematics Subject Classification*. Primary 60–06, 60G99.

---

**Library of Congress Cataloging-in-Publication Data**

International Conference on Stochastic Models in Honour of Professor Donald A. Dawson (1998 : Ottawa, Canada)

Stochastic models : proceedings of the International Conference on Stochastic Models in Honour of Professor Donald A. Dawson, Ottawa, Canada, June 10-13, 1998/Luis G. Gorostiza, B. Gail Ivanoff, editors.

p. cm. – (Conference proceedings / Canadian Mathematical Society, ISSN 0731–1036; v. 26)

Includes bibliographical references.

ISBN 0-8218-1063-4

1. Stochastic processes–Congresses. I. Dawson, Donald Andrew, 1937–. II. Gorostiza, L. G. (Luis G.), 1939–. III. Ivanoff, B. Gail. IV. Title. V. Conference proceedings (Canadian Mathematical Society); v. 26.

QA274.A1 I57 1998
519.2–dc21

00-022164
CIP

---

# Contents

# Preface

This book, dedicated to Professor Donald A. Dawson, comprises the refereed Proceedings of the International Conference on Stochastic Models (ICSM), which took place at Carleton University in Ottawa, Canada, June 10-13, 1998. Don's many students, postdoctoral fellows, and collaborators took great pleasure in celebrating the enormous personal and professional contributions Don has made throughout his career.

The ICSM was sponsored by the Laboratory for Research in Statistics and Probability (LRSP), Carleton University - University of Ottawa, and the School of Mathematics and Statistics, Carleton University. We gratefully acknowledge the Natural Sciences and Engineering Research Council of Canada (NSERC), whose financial support of the LRSP through a Major Facilities Access grant made this conference possible. As well, partial support for the work of the first editor (LG) was provided by the National Council for Science and Technology of Mexico and for the second editor (GI) by NSERC.

We also wish to express our sincere thanks to Mrs. Gillian Murray, the coordinator of the LRSP, for the superb way in which the local organization of the ICSM was carried out.

We are pleased to acknowledge the cooperation of the Canadian Mathematical Society in the preparation and publication of this volume and the assistance of Mrs. Debbie Iscoe in formatting the articles. Also, we are grateful to Ed Perkins for his help in the challenging task of summarizing Don's diverse mathematical achievements.

Most of all, we wish to thank Don for his leadership and friendship, and for his unfailing generosity. To paraphrase Ed Perkins, Don is one of the select group of "leaders in the field at the very highest level of scientific accomplishment (who) set high standards in character and leadership as well as science."

*Luis Gorostiza*
*Gail Ivanoff*

*November, 1999*

# Foreword

Donald A. Dawson is Director, The Fields Institute for Research in Mathematical Sciences, and Professor Emeritus and Distinguished Research Professor of Mathematics and Statistics, Carleton University. His current position as Director of the Fields Institute is the capstone of a long and distinguished career as one of the foremost probabilists of his generation.

Don obtained his B.Sc. in 1958 and his M.Sc. in 1959, both at McGill University. He subsequently completed his Ph.D. under the supervision of H.P. McKean, Jr. in 1963 at the Massachusetts Institute of Technology. He then returned to McGill as an Assistant (and subsequently Associate) Professor, where he remained from 1963 to 1970. In 1970, he moved to Carleton University, and was named Professor of Mathematics and Statistics in 1971. He has remained in this position until his recent retirement in June, 1999. Since 1996, he has been on leave during his tenure as Director of the Fields Institute in Toronto. Throughout his career, Don has accepted visiting positions around the world: Nagoya, Paris VI, Heidelberg, Wisconsin, Zurich, Università di Roma, Chapel Hill, the Weierstrass Institute, Saint-Flour, Cambridge, Montreal, Tokyo. He has supervised 24 Ph.D.'s and 29 postdoctoral and research fellows. He has published more than 100 refereed articles in addition to 7 research monographs and lecture notes, and as well has edited two books (a complete list of publications appears at the end of the Foreword). In addition to his term as Director of the Fields Institute, Don has tirelessly served the community by organizing numerous conferences, working on several editorial boards (including a term as co-editor of the Canadian Journal of Mathematics), chairing the Statistics Grant Selection Committee of NSERC, being Chairman of his department, and playing a central role in founding the Laboratory for Research in Statistics and Probability (Carleton University-University of Ottawa).

As is clear from this very brief synopsis of Don's career, he is one of the rare mathematicians who not only has attained the highest level of scientific achievement, but who has also contributed in every other way to the betterment of his discipline. He is a gifted teacher at both the undergraduate and graduate levels, and as well is a skilled administrator who has succeeded in significantly raising the profile of the mathematical community in Canada, improving the morale of mathematical researchers and promoting the discipline with governmental organizations. Only an individual with Don's scientific stature and personal integrity would have been capable of representing the mathematical community so well.

Don's remarkable achievements have been justly recognized with many honours: he was elected a Fellow of the Royal Society of Canada, Academy of Sciences, in 1987, and received the Gold Medal of the Statistical Society of Canada in 1991. He is a Fellow of the International Statistical Institute and of the Institute of Mathematical Statistics and was a Killam Senior Research Scholar from 1997 to 1999. He has been invited to give several extremely prestigious lectures, including the Jeffrey-Williams Lecture of the Canadian Mathematical Society (1994), an invited lecture at the International Congress of Mathematicians (Zurich, 1994), a Special Invited Lecture, Institute of Mathematical Statistics (1996), the SPA Lecture of the Fourth World Congress of the Bernoulli Society (Vienna, 1996), and the Davidson Dunton Research Lecture at Carleton University (1997). As well, Don has been recognized by awards for excellence in both teaching and research by Carleton University (a rare accomplishment, indeed!). Finally, he was the recipient of the Max Planck Award for International Cooperation (1997).

Don is the world leader in measure-valued processes. These processes include several classes of stochastic models for complex evolving systems which appear in many areas of science and engineering. One class is comprised of the "Dawson-Watanabe superprocesses" and their extensions, which arise as high-density limits of spatially distributed branching particle systems. Don and his collaborators are responsible for establishing many of the fundamental properties of measure-valued processes and for developing the basic techniques to study them. This area of research has attracted several of the best probabilists of our time, who have made use of and built on Don's work.

Don's paper in this volume gives a lucid overview of his motivations for proposing and investigating this class of models, and of some of the important results that have been obtained by him, his collaborators and others, and it also mentions open problems. To quote from the Introduction to his paper, the focus is "the potential role of infinite dimensional stochastic processes in developing mathematical models which can be used to explore the collective behaviour of a spatially distributed population of interacting individuals having an evolving internal structure over large space and time scales and the identification of universality classes of these processes".

In addition to the many areas of application in science and engineering that Don originally had in mind (including geographically distributed populations and chemical reactions, evolutionary biology, communications networks and other classes of spatially distributed systems), in the last few years measure-valued processes have arisen in applications that even he had not foreseen (percolation, lattice trees, scaling limits of many types of particle systems, Brownian excursions), and have become a central topic of study in probability and population genetics. All of these developments have confirmed Don's belief of the important role of these processes.

Don's major scientific contributions are not confined to measure-valued processes, for his interests are extremely broad and go far beyond this field. For example, his interest in phase transitions in statistical physics led to his work on large deviations for infinite dimensional processes. The breadth of his knowledge and interests is remarkable: queueing systems, spin glass models, mathematical finance, dynamical systems, population genetics, statistical inference, Dirichlet forms, interacting particle systems, large deviations. It is a list born out by the diversity of the interests of his students and collaborators. His students have gone on to use their training in a fundamental way in their chosen careers, some in academia and others in the private sector, and many have become leaders in their own right. The collective impact of his students, postdoctoral and research fellows on mathematics and statistics in Canada and other countries is huge. His numerous scientific collaborators (30 co-authors) from many parts of the world have been attracted to Don by his impressive knowledge of mathematics and science, the originality and depth of his ideas, and his generosity. The range of topics in the papers collected in this volume attests to the diversity of areas where Don has made and promoted significant contributions. The key to Don's scientific success may lie in the fact that he is always interested in learning new fields and is always eager to understand other people's ideas.

This abbreviated summary of Don's research contributions cannot do justice to the profound impact he has had on the discipline. He continues to be a leader in his fields of research and in probability in general, and will be a source of inspiration to mathematicians for a long time to come. We are privileged to know him as collaborator, teacher and friend.

*Luis Gorostiza*
*Gail Ivanoff*

# Publications of Donald A. Dawson

1. D.A. Dawson, *The construction of a class of diffusions*, Illinois Journal of Mathematics **8** (1964), 657–685.

2. D.A. Dawson, *The local diffusions of a harmonic sheaf*, Canadian Mathematical Bulletin **8** (1965), 307–316.

3. D.A. Dawson, *Potential theory and non-Markovian chains*, Z. Wahr. verw. Geb. **5** (1966), 118–138.

4. D.A. Dawson and D. Sankoff, *An inequality for probabilities*, Proceedings of the American Mathematical Society **18** (1967), 504–507.

5. D.A. Dawson and D. Sankoff, *A packing problem for measurable sets*, Canadian Journal of Mathematics **19** (1967), 749–756.

6. D.A. Dawson, *Equivalence of Markov processes*, Transactions of the American Mathematical Society **131** (1968), 1–31.

7. D.A. Dawson, *Generalized stochastic integrals and equations*, Transactions of the American Mathematical Society **147** (1970), 473–506.

8. D.A. Dawson, *Denumerable Markov Chains*, Canadian Mathematical Society Monograph No. 2, 1970, 100 pages.

9. D.A. Dawson, *Stochastic evolution equations*, Mathematical Biosciences **15** (1972), 287–316.

10. D.A. Dawson, *Information flow in discrete Markov systems*, Journal of Applied Probability **10** (1973), 63–83.

11. D.A. Dawson, *Information flow in some classes of Markov systems*, Journal of Applied Probability **11** (1974), 594–600.

12. D.A. Dawson, *Information flow in one–dimensional Markov systems*, Proceedings of the American Mathematical Society **43** (1974), 383–392.

13. D.A. Dawson, *Lecture Notes on Discrete Markov Systems*, Carleton Mathematical Notes No. **10** (1974), 144 pages.

14. D.A. Dawson, *Synchronous and a synchronous reversible Markov systems*, Canadian Mathematical Bulletin **17** (1974/75), 633–649.

15. D.A. Dawson, *Stochastic models for complex systems*, Selecta Statistica Canadiana **2** (1975), 46–64.

16. D.A. Dawson, *Stochastic evolution equations and related measure processes*, Journal of Multivariate Analysis **5** (1975), 1–52.

17. D.A. Dawson, *Gibbsian models and inference*, Proceedings of Symposium on Statistics and Related Topics, Carleton Mathematical Lecture Notes No. **12** (1975), 3.1–3.15.

18. D.A. Dawson, *Quasi–deterministic states for Markov systems*, Advances of Applied Probability **7** (1975), 231–232.

19. D.A. Dawson, *Information flow in graphs*, Stochastic Processes and their Applications **3** (1975), 137–151.

20. D.A. Dawson, *The nature and scope of applied probability*, Stochastic Processes and their Applications **3** (1975), 239–241.

21. D.A. Dawson, *The critical measure diffusion processes*, Z. Wahr. verw. Geb. **40** (1977), 125–145.

22. D.A. Dawson, *Stable states of probabilistic cellular automata*, Information and Control **34** (1977), 93–106.

23. D.A. Dawson, *Limit theorems for interaction free geostochastic systems*, Seria Matematica Societatis Janos Bolyai **24** (1978), 27–47.

24. D.A. Dawson, *Geostochastic calculus*, Canadian Journal of Statistics **6** (1978), 143–168.

25. D.A. Dawson and B.G. Ivanoff, *Branching diffusions and random measures*, Advances in Probability **5** (1978), A. Joffe and P. Ney, eds., 61–104.

26. A. Bose and D.A. Dawson, *A class of measure–valued Markov processes*, Lecture Notes in Mathematics **695** (1978), 15–125.

27. D.A. Dawson, *Sistemas de poblacion con distribuciones espaciales estocasticas*, in Proceedings Coloquio de Matematicas del CIEA-IPN, Conferencias Sobre Sistemas Estocasticos, Mexico, 1979, 1–48.

28. D.A. Dawson and K.J. Hochberg, *The carrying dimension of a measure diffusion process*, Annals of Probability **7** (1979), 693–703.

29. D.A. Dawson, *An infinite geostochastic system*, Multivariate Analysis **5** (1979), (ed. P.R. Krishnaiah), 119–136.

30. D.A. Dawson, *Critical behavior of the geostochastic logistic system*, C.R. Math. Rep. Acad. Sci. Canada **1** (1979), 79–82

31. D.A. Dawson, *Stochastic measure diffusion* Canadian Mathematical Bulletin **22** (1979), 129–138.

32. D.A. Dawson, *Invariant sets for spatially homogensous measure processes*, Seminaires Analyse et Controle de Systemes, Institut de Recherche d'informatique et d'automatique (France), 1979, 51–61.

33. D.A. Dawson, *Qualitative behavior of geostochastic systems*, Journal of Stochastic Processes and their Applications **10** (1980), 1–31.

34. D.A. Dawson and H. Salehi, *Spatially homogeneous random evolutions*, Journal of Multivariate Analysis **10** (1980), 1–31.

35. M. Csörgő, D.A. Dawson, J.N.K. Rao and A.K.M.E. Saleh (eds), *Statistics and Related Topics*, North Holland, Amsterdam, 1980.

36. D.A. Dawson, *An infinite geostochastic system*, Multivariate Analysis **5** (1980), 119–136.

37. D.A. Dawson, *Galerkin approximation of nonlinear Markov processes*, in Statistics and Related Topics, M. Csörgő, D.A. Dawson, J.N.K. Rao and A.K.Md.E. Saleh, eds., pages 317–339, North Holland, 1981.

38. D.A. Dawson, *Stochastic measure processes*, in Stochastic Nonlinear Systems in Physics, Chemistry and Biology, L. Arnold and R. Lefever, eds, 185–201, 1981.

39. D.A. Dawson and K.J. Hochberg, *Wandering random measures in the Fleming–Viot model*, Annals of Probability **10** (1982), 554–580.

40. D.A. Dawson and T.G. Kurtz, *Applications of duality to measure–valued processes*, Lecture Notes on Control and Optimization **42** (1982), 91–106, Springer-Verlag.

41. D.A. Dawson, *Critical dynamics and fluctuations for a mean–field model of cooperative behavior*, Journal of Statistical Physics **31** (1983), 29–85.

42. D.A. Dawson and K.J. Hochberg, *Qualitative behavior of a selectively neutral allelic model*, Theoretical Population Biology **23** (1983), 1–18.

43. D.A. Dawson and K. Fleischmann, *On spatially homogeneous branching processes in a random environment*, Math. Nachr. **113** (1983), 249–257.

44. D.A. Dawson and L.G. Gorostiza, *Limit theorems for supercritical branching random fields*, Math. Nachr. **119** (1984), 19–46.

45. D.A. Dawson and G.C. Papanicolaou, *A random wave process*, Applied Mathematics and Optimization **12** (1984), 97–114.

46. D.A. Dawson and K. Fleischmann, *Critical dimension for a model of branching in a random medium*, Zeit. fur Wahr. verw. Geb. **70** (1985), 315–334.

47. D.A. Dawson and K.J. Hochberg, *Function–valued duals for measure–valued processes with applications*, Contemporary Mathematics **41** (1985), 55–69.

48. D.A. Dawson, *Asymptotic analysis of multilevel stochastic systems*, Lecture Notes in Control and Optimization **69** (1985), 79–90, (M. Metivier and E. Pardoux, eds.), Springer-Verlag.

49. D.A. Dawson, *Stochastic ensembles and hierarchies*, Proceedings of the 15th Bernoulli Conference on Stochastic Processes and Applications, Nagoya, 1985, Lecture Notes in Mathematics **1203** (1986), 20–37, Springer–Verlag.

50. D.A. Dawson, *Measure–valued stochastic processes: construction, qualitative behavior and stochastic geometry*, Proc. Workshop on Spatial Stochastic Models, Lecture Notes in Mathematics **1212** (1986), 69–93, Springer–Verlag.

51. D.A. Dawson, K. Fleischmann, R.D. Foley and L.A. Peletier, *A critical measure–valued branching process with infinite mean*, Stochastic Analysis and Applications **4** (1986), 117–129.

52. D.A. Dawson and J. Gartner, *Large deviations and tunnelling for particle systems with mean–field interaction*, C.R. Math. Acad. Sci. Canada **8** (1986), 387–392.

53. D.A. Dawson and J. Gartner, *Large deviations from the McKean–Vlasov limit for weakly interacting diffusions*, Stochastics **20** (1987), 247–308.

54. D.A. Dawson and L.G. Gorostiza, *\*-solutions of evolution equations in Hilbert space*, Journal of Differential Equations **68** (1987), 299–319.

55. D.A. Dawson, *Stochastic models of parallel systems for global optimization*, in Hydrodynamic Behavior and Interacting Particle Systems and Applications, ed. G. Papanicolaou, The IMA Volumes in Mathematics and Its Applications, Vol. **9** (1987), 25–44, Springer-Verlag.

56. D.A. Dawson and J. Gärtner, *Large deviations and long time behaviour of stochastic particle systems*, Proc. 1st World Congress Bernoulli Society, Vol. **1** (1987), 651–654, VNU Science Press.

57. D.A. Dawson and J. Gärtner, *Long-time fluctuations of weakly interacting diffusions*, Proceedings 5th IFIP Conference on Stochastic Differential Equations, Eisenach, DDR, H.J. Engelbert and W. Schmidt, eds. *Lecture Notes on Control and Information Sciences* **96** (1987), 3–11, Springer-Verlag.

58. D.A. Dawson, I. Iscoe and E.A. Perkins, *Sample path properties of the support process of super–Brownian motion*, C.R. Math. Rep. Acad. Sci. Canada **10** (1988), 83–88.

59. D.A. Dawson and K.J. Fleischmann, *A stochastic explosive reaction system with sampling*, Journal of Applied Probability **26** (1989), 9–22.

60. D.A. Dawson and J. Gartner, *Long-time behavior of interacting diffusions*, in Stochastic Calculus in Applications, ed. J.R. Norris, 29–54, Pitman Research Notes in Mathematics **197** (1988).

61. D.A. Dawson and K. Fleischmann, *Strong Clumping of critical space-time branching models in subcritical dimensions*, Stochastic Processes and their Applications **30** (1988), 193–208.

62. C. Cutler and D.A. Dawson, *Estimation of dimension for spatially distributed data and related limit theorems*, Journal of Multivariate Analysis **28** (1989), 115–148.

63. D.A. Dawson, *Review of Stochastic Equations for Complex Systems by A.V. Skorohod*, Bulletin of the American Mathematical Society **20** (1989), 259–267.

64. B. Remillard and D.A. Dawson, *Laws of the Iterated Logarithm and Large Deviations for a Class of Diffusion Processes*, Canadian Journal of Statistics **17** (1989), 349–376.

65. D.A. Dawson, I. Iscoe and E.A. Perkins, *Super–Brownian motion and hitting probabilities*, Probability Theory and Related Fields **83** (1989), 135–205.

66. D.A. Dawson, K. Fleischmann and L.G. Gorostiza, *Stable hydrodynamic limit fluctuations of a critical branching particle system in a random medium*, Annals of Probability **17** (1989), 1083–1117.

67. D.A. Dawson and J. Gärtner, *Large deviations, free energy functional and quasipotential for a mean–field model of interacting diffusions*, Memoirs of the American Mathematical Society **398** (1989), 98 pages.

68. D.A. Dawson and L.G. Gorostiza, *Generalized solutions of stochastic evolution equations*, Proceedings of the Workshop on Stochastic Partial Differential Equations, Trento, Italy, Feb. 1988, Springer Lecture Notes **1390** (1989), 53–65.

69. D.A. Dawson and K.J. Hochberg and Y. Wu, *Multilevel branching systems*, in Proc. Bielefeld Encounters in Mathematics and Physics (1989), World Scientific (1990), 93–107.

70. C. Cutler and D.A. Dawson, *Nearest neighbour analysis of a family of fractal distributions*, Annals of Probability **18**, 256–271, 1990.

71. D.A. Dawson and L.G. Gorostiza, *Generalized solutions of a class of nuclear space valued stochastic evolution equations*, Applied Mathematics and Optimization **22** (1990), 241–263.

72. D.A. Dawson, L. Gorostiza and A. Wakolbinger, *Schrödinger processes and large deviations*, Journal of Mathematical Physics **31**(1990), 2385–2388.

73. D.A. Dawson and Z. Zheng, *Law of Large Numbers and Central Limit Theorem for Unbounded Jump Mean-Field Models*, Advances in Applied Mathematics **12** (1991), 293–326.

74. D.A. Dawson and K.J. Hochberg, *A multilevel branching model*, Advances in Applied Probability **23** (1991), 701–715.

75. D.A. Dawson and E.A. Perkins, *Historical Processes*, Memoirs of the American Mathematical Society **454** (1991), 179 pages.

76. D.A. Dawson and K. Fleischmann, *Critical branching in a highly fluctuating random medium*, Probability Theory and Related Fields **90** (1991), 241–274.

77. D.A. Dawson and B. Remillard, *A limit theorem for Brownian motion in a random scenery*, Canadian Mathematical Bulletin **34** (1991), 385–391.

78. D.A. Dawson, K. Fleischmann and S. Roelly, *Absolute continuity of the measure states in a branching model with catalysts*, Seminar on Stochastic Processes, 1990, 117–160, Birkhauser (1991).

79. D.A. Dawson and K. Fleischmann, *Diffusion and reaction caused by point catalysts*, SIAM J. Applied Math. **52** (1992), 163–180.

80. D.A. Dawson, *Infinitely divisible random measures and superprocesses*, in Stochastic Analysis and Related Topics, H. Korezlioglu and A.S. Ustunel, eds., Progress in Probability vol. **31**, Birkhauser 1992, 1–130.

81. D.A. Dawson and J. Gärtner, *Multilevel models of interacting diffusions and large deviations*, Proc. I.C.M.P. (1991), K. Schmüdgen, ed., Springer-Verlag 1992.

82. D.A. Dawson and K. Fleischmann, *Diffusion and reaction caused by point catalysts*, SIAM Journal of Applied Mathematics **52** (1992), 163–180.

83. D.A. Dawson and A. Greven, *Multiple time scale analysis of interacting diffusions*, Probability Theory and Related Fields **95** (1993), 467–508.

84. D.A. Dawson and A. Greven, *Multiple time scale analysis of hierarchically interacting systems*, in A Festschrift to honor G. Kallianpur, S. Cambanis, J.K. Ghosh, R.L. Karandikar and P.K. Sen, eds., Springer-Verlag, 1993, 41–50.

85. D.A. Dawson and A. Greven, *Hierarchical models of interacting diffusions: multiple time scale phenomena, phase transition and pattern of cluster formation*, Probability Theory and Related Fields, **96** (1993), 435–473.

86. D.A. Dawson, *Measure-valued Markov processes*, in École d'Été de Probabilités de Saint Flour XXI, Lecture Notes in Mathematics **1541**, Springer-Verlag 1993, 1–261.

87. D.A. Dawson and K. Fleischmann, *A super–Brownian motion with a single point catalyst*, Stochastic Processes and their Applications **49** (1994), 3–40.

88. D.A. Dawson and J. Gärtner, *Multilevel large deviations and interacting diffusions*, Probability Theory and Related Fields **98** (1994), 423–487.

89. D.A. Dawson and V. Vinogradov, *Almost-sure path properties of $(2, d, \beta)$-superprocesses*, Stochastic Processes and their Applications **51** (1994) 221–258.

90. D.A. Dawson and V. Vinogradov, *Mutual singularity of genealogical structures of Fleming–Viot and continuous branching processes*, in The Dynkin Festschrift, ed. M.I. Freidlin, Birkhäuser, 1994, 61–84.

91. D.A. Dawson, K.J. Hochberg and V. Vinogradov, *On path properties of super-2 processes I*, in Proceedings of the Conference and Workshop on Measure–valued Processes, Stochastic Partial Differential Equations and Interacting Systems, D.A. Dawson, ed., CRM Proceedings and Lecture Notes vol. **5** (1994), 69–82.

92. D.A. Dawson (ed.),*Proceedings of the Conference and Workshop on Measure-valued Processes, Stochastic Partial Differential Equations and Interacting Systems*, CRM Proceedings and Lecture Notes Volume **5** (1994), American Mathematical Society.

93. D.A. Dawson, K.J. Hochberg and V. Vinogradov, *On path properties of super-2 processes. II*, in Proc. Symposia in Pure Mathematics, vol. **57**, Stochastic Analysis, eds. M.C. Cranston and M.A. Pinsky, American Mathematical Society (1995), 385–404.

94. D.A. Dawson, K. Fleischmann and J.F. Le Gall, *Super–Brownian motion in catalytic media*, in C.C. Heyde, ed. Branching Processes: Proceedings of the First World Congress, Lecture Notes in Statistics **99** (1995), 122–134, Springer-Verlag.

95. D.A. Dawson and K. Fleischmann, *Super–Brownian motion in higher dimensions with absolutely continuous measure states*, Journal of Theoretical Probability **8** (1995), 179–206.

96. D.A. Dawson and J. Vaillancourt, *Stochastic McKean–Vlasov equations*, NoDEA **2** (1995), 199–229.

97. D.A. Dawson and P. March, *Resolvent estimates for Fleming–Viot operators and uniqueness of solutions to related martingale problems*, Journal of Functional Analysis **132** (1995), 417–472.

98. D.A. Dawson, K. Fleischmann, Y. Li and C. Mueller, *Singularity of super–Brownian local time at a point catalyst*, Annals of Probability **23** (1995), 37–55.

99. D.A. Dawson, A. Greven and J. Vaillancourt, *Equilibria and quasiequilibria for infinite collections of interacting Fleming–Viot processes*, Transactions of the American Mathematical Society **347** (1995), 2277–2360.

100. D.A. Dawson, *Interaction and hierarchy in measure–valued processes*, Proc. International Congress of Mathematicians, Zurich, 1994, Birkhäuser Verlag, Basel, 1995, 986–996.

101. D.A. Dawson and Y. Wu, *Multilevel multitype models of an information system*, in Classical and Modern Branching Processes, (eds. K.B. Athreya and P. Jagers), I.M.A. Volume **84**, 1996, Springer-Verlag, pages 57–72.

102. D.A. Dawson, Y. Li and C. Mueller (1996), *The support of measure–valued branching processes in a random environment*, Annals of Probability **23**, 1692–1718.

103. D.A. Dawson, K.J. Hochberg and V. Vinogradov, *High-density limits of hierarchically structured branching–diffusing populations*, Stochastic Processes and their Applications, **62** (1996), 191–222.

104. D.A. Dawson and A. Greven, *Multiple space–time scale analysis for interacting branching models*, Electronic Journal of Probability **1** (1996), paper 6, 1–84.

105. D.A. Dawson and E.A. Perkins, *Measure-valued processes and stochastic partial differential equations*, Canadian Mathematical Society 1945–1995, vol. **3** (1996), 19–112.

106. D.A. Dawson, *Hierarchical and mean–field stepping stone models*, in I.M.A. Volume 87 (1997), Progress in Population Genetics and Human Evolution, eds. P. Donnelly and S. Tavaré, 287–298.

107. D.A. Dawson, K.J. Hochberg and V. Vinogradov, *The Wiener process revisited*, Proc. of the seventh Japan-Russia Symposium on Probability Theory and Mathematical Statistics, ed. Prohorov, Shiryaev, Watanabe and Fukushima, 1996, World Scientific, 43–50.

108. D.A. Dawson, K.J. Hochberg and V. Vinogradov, *On weak convergence of branching particle systems undergoing spatial motion*, Israel Mathematical Conference Proceedings (1996).

109. D.A. Dawson and M.A. Kouritzin, *Invariance principles for parabolic equations with random coefficients*, Journal of Functional Analysis **149** (1997), 377–414.

110. D.A. Dawson and K. Fleischmann, *A continuous super–Brownian motion in a super–Brownian medium*, Journal of Theoretical Probability **10** (1997), 213–176.

111. D.A. Dawson and K. Fleischmann, *Longtime behaviour of a branching process controlled by branching catalysts*, Stochastic Processes and their Applications **71** (1997), 241–258.

112. D.A. Dawson and S. Feng, *Large deviations for the Fleming–Viot process with neutral mutation and selection*, Stochastic Processes and their Applications **77** (1998), 207–222.

113. D.A. Dawson, K. Fleischmann and G. Leduc, *Continuous dependence of a class of superprocesses on branching parameters and applications*, Annals of Probability **26**(1998), 652–601.

114. D.A. Dawson and E.A. Perkins, *Long–time behaviour and co–existence in a mutually catalytic branching model*, Ann. Probab. **26** (1998), 1088–1138.

115. D.A. Dawson and J. Gärtner, *Analytic aspects of multilevel large deviations*, in Asymptotic Methods in Probability and Statistics (ed. B. Szyszkowicz), Elsevier, Amsterdam (1998), 401–440.

116. D.A. Dawson and K. Fleischmann, *Catalytic and mutually catalytic branching*, in Proceedings of the Royal Netherlands Academy of Arts and Sciences Colloquium on Infinite Dimensional Stochastic Analysis, Amsterdam, February 1999.

117. D.A. Dawson and E.A. Perkins, *Measure–valued processes and Renormalization of Branching Particle Systems*, in Stochastic Partial Differential Equations: Six Perspectives, eds. R. Carmona and B. Rozovskii, American Mathematical Society Mathematical Surveys and Monographs vol. **64** (1999), pp. 45–106.

118. D.A. Dawson and A. Greven, *Hierarchically interacting Fleming–Viot processes with selection and mutation: multiple space time scale analysis and quasi–equilibria*, Electronic Journal of Probability vol. **4** (1999), Paper no. 14, pages 1–81.

119. D.A. Dawson, J. Vaillancourt and H. Wang, *Stochastic partial differential equations for a class of interacting measure–valued diffusions*, Annales de L'Institut Henri–Poincaré, to appear.

120. D.A. Dawson, K. Fleischmann and C. Mueller, *Finite time extinction of superprocesses with catalysts*, Annals of Probability, to appear.

# List of Invited Lectures

Particle representations for measure–valued processes
**Thomas G. Kurtz**, University of Wisconsin - Madison

Finite time extinction of super–Brownian motions with catalysts
**Klaus Fleischmann**, Weierstrass Institute for Applied Analysis and Stochastics

May clustering be strictly easier in free branching than in a stepping stone model?
**Anton Wakolbinger**, Universität Frankfurt

Distinctive behaviour of multilevel branching systems
**Kenneth J. Hochberg**, Bar–Ilan University

Longtime behaviour of spatial population growth models
**Andreas Greven**, Universität Erlangen–Nürnberg

Comparative genomics and evolutionary inference
**David Sankoff**, Université de Montréal

Stochasticity and determinism in time series
**Colleen D. Cutler**, University of Waterloo

Level crossing probabilities of the Ornstein–Uhlenbeck process
**Joseph Denes**, Agricultural University, Gödöllő

Asymptotic behaviour of occupation times
**Bruno Remillard**, Université du Québec à Trois Rivières

Path properties of processes of financial modeling
**Miklós Csörgő**, Carleton University

Model misspecification, Euler residuals and heavy tails
**Reg Kulperger**, University of Western Ontario

Rescaled voter models and super–Brownian motion
   **J. Ted Cox**, University of British Columbia

On de Giorgi's conjecture and related problems
   **Nassif Ghoussoub**, University of British Columbia

Branching random walk in a catalytic medium
   **Achim Klenke**, Universität Erlangen–Nürnberg

Mutually catalytic branching models
   **Edwin Perkins**, University of British Columbia

Weighted sequential quantile processes and their application to
change–point analysis
   **Barbara Szyszkowicz**, Carleton University

Minimax estimation of exponential family means over $\ell_p$ bodies
   **Brenda MacGibbon**, Université du Québec à Montréal

Intermittent behavior and spectral properties of the heat equation with
random potential
   **Jürgen Gärtner**, Technische Universität Berlin

Fourier analysis applied to super stable processes
   **Amit Bose**, Carleton University

On the conditioned exit measures of super Brownian motion
   **Tom Salisbury**, York University

The critical exponent for blow–up of an SPDE
   **Carl Mueller**, University of Rochester

Some models of coagulation and gelation
   **Peter March**, The Ohio State University

Steady state analysis with heavy traffic limits for semi–open networks
   **Raj Srinivasan**, University of Saskatchewan

From particles to mezoscopic and macroscopic equations
   **Peter Kotelenez**, Case Western Reserve University

Construction and stationarity for interacting biomolecular evolutions
in a correlated diffusive medium
   **Jean Vaillancourt**, Université de Sherbrooke

Ewens sampling formula and related topics
  **Shui Feng**, McMaster University

Regular variation of natural exponential families and convergence to Tweedie models
  **Vladimir Vinogradov**, University of Northern British Columbia and University of Toronto

Valuation of barrier options with alternative jump–diffusion underlying prices
  **Hao Wang**, Algorithmics Inc.

Convolution in exact filtering
  **Michael Kouritzin**, University of Alberta

Do foreign exchange markets have memory?
  **Donna M. Salopek**, York University and The Fields Institute

Some exceptional sets in configuration space
  **Byron Schmuland**, University of Alberta

Self–intersection of distribution–valued Gaussian processes
  **Luis Gorostiza**, Centro de Investigación y de Estudios Avanzados

Set–indexed Markov processes
  **Gail Ivanoff**

# List of Participants

Mathieu Blanchette
Department of Computer Science & Engineering
University of Washington
Seattle, WA 98195-2350 USA
blanchem@cs.washington.edu

Matthias Birkner
Fachbereich Mathematik
J. W. Goethe - Universität
D-60054 Frankfurt am Main, GERMANY
birkner@math.uni-frankfurt.de

Douglas Blount
Department of Mathematics and Statistics
Arizona State University
Tempe, AZ 85287 USA
blount@math.la.asu.edu

Amit Bose
Department of Mathematics and Statistics
Carleton University
1125 Colonel By Drive
Ottawa, ON K1S 5B6 CANADA
abose@math.carleton.ca

J. Theodore Cox
Department of Mathematics
Syracuse University
Syracuse, NY 13244-1150 USA
jtcox@gumby.syr.edu

Miklós Csörgő
Department of Mathematics and Statistics
Carleton University
1125 Colonel By Drive
Ottawa, ON K1S 5B6 CANADA
mcsorgo@math.carleton.ca

Colleen Cutler
Department of Statistics
University of Waterloo
Waterloo, ON N2L 3G1 CANADA
cdcutler@math.uwaterloo.ca

Donald A. Dawson
The Fields Institute for Research in Mathematical Sciences
222 College Street
Toronto, Ontario M5T 3J1 CANADA
ddawson@fields.utoronto.ca

Joseph Denes
Department of Mathematics and Statistics
Carleton University
1125 Colonel By Drive
Ottawa, ON K1S 5B6 CANADA
jdenes@math.carleton.ca

Steven N. Evans
Department of Statistics
University of California at Berkeley
367 Evans Hall
Berkeley, CA 94720-3860 USA
evans@stat.berkeley.edu

Shui Feng
Department of Mathematics and Statistics
McMaster University
1280 Main Street West
Hamilton, ON L8S 4K1 CANADA
shuifeng@mcmail.cis.mcmaster.ca

Klaus Fleischmann
Institut of Applied Analysis and Stochastics
Mohrenstrasse 39
D-10117 Berlin, GERMANY
fleischmann@wias-berlin.de

Jürgen Gärtner
Institut für Angewandte Analysis and Stochastik
Hausvogteiplatz 5-7
D-0 1086 Berlin, GERMANY
jg@math.tu-berlin.de

Luis G. Gorostiza
Department of Mathematics
Centro de Investigación y de Estudios Avanzados
A.P. 14-740
México 07000 D.F. MEXICO
gortega@servidor.unam.mx

Eric Gourdin
GERAD and Département de Mathématiques Appliqées École Polytechnique
38-40, Rue du General Leclerc
F-92794 Issy-Les-Moulineaux FRANCE
eric.gourdin@cnet.francetelecom.fr

Andreas Greven
Institut für Mathematik
Humboldt Universitat zu Berlin
Unter den Linden 6
D 10099 Berlin GERMANY
greven@mi.uni-erlangen.de

Kenneth J. Hochberg
Department of Mathematics and Computer Science
Bar-Ilan University
52900 Ramat-Gan
Israel & College of Judea and Samaria
44837 Ariel, ISRAEL
hochberg@macs.biu.ac.il

Gail Ivanoff
Department of Mathematics and Statistics
University of Ottawa
P.O. Box 450, Stn A
Ottawa, ON K1N 6N5 CANADA
givanoff@praxis.cc.uottawa.ca

Brigitte Jaumard
Ecole Polytechnique de Montréal et GERAD
Departement de Mathématiques et de Genie Industriel
Succursale centre-ville, Case Postale 6079
Montréal (Québec) H3C 3A7 CANADA
brigitt@crt.umontreal.ca

Intae Jeon
Department of Mathematics
The Ohio State University
231 W 18th Ave.
Columbus, Ohio 43212 USA
jeon@math.ohio-state.edu

Peter J. Kempthorne
Kempthorne Analytics, Inc.
63 Wharf Street
Salem, MA 01970 USA
pjk@kemp.com

Achim Klenke
Mathematisches Institut
Universitaet Erlangen-Nuernberg
Bismarckstrasse 1 1/2
91054 Erlangen, GERMANY
klenke@mi.uni-erlangen.de

Michael Kouritzin
Department of Mathematical Sciences
University of Alberta
Edmonton, AB T6G 2G1 CANADA
mkouritz@math.ualberta.ca

Reg Kulperger
Dept. of Statistics and Actuarial Sciences
Room 3005 EMS Bldg.
University of Western Ontario
London, ON N6A 5B9 CANADA
rjk@fisher.stats.uwo.ca

Thomas G. Kurtz
Department of Mathematics and Statistics
University of Wisconsin
480 Lincoln Dr.
Madison, WI 53706-1388 USA
kurtz@math.wisc.edu

Zeng-Hu Li
Department of Mathematics
Beijing Normal University
Beijing 100875 P.R. CHINA
lizh@email.bnu.edu.cn

Brenda MacGibbon
Département de Mathématiques
Université du Québec à Montréal et GERAD
C.P. 8888, Succursale centre-ville
Montreal, PQ H4B1X8 CANADA
macgibbon.brenda@uqam.ca

Peter March
Department of Mathematics
100 Mathematics Bldg.
Ohio State University
231 West 18th Avenue
Columbus, Ohio 43210-1174 USA
march@math.ohio-state.edu
PMarch1784@aol.com

William A. Massey
Bell Laboratories
Murray Hill, NJ 07974 USA
will@research.bell-labs.com

Ely Merzbach
Department of Mathematics
Bar-Ilan University
52900 Ramat Gan ISRAEL
merzbach@bimacs.csbiu.ac.il

Stanislav A. Molchanov
Department of Mathematics
University of North Carolina
Charlotte, NC 28223-9998 USA
smolchan@uncc.edu

Carl E. Mueller
Department of Mathematics
University of Rochester
Rochester, NY 14627 USA
cmlr@troi.cc.rochester.edu

Rimas Norvaiša
Institute of Mathematics and Informatics
Akademijos 4
Vilnius 2600, LITHUANIA
norvaisa@ktl.mii.lt

Edwin Perkins
Department of Mathematics
University of British Columbia
Vancouver, BC V6T 1Y4 CANADA
perkins@math.ubc.ca

Bruno Remillard
Département de mathématiques
Université de Québec à Trois Rivières
CP 500 Trois Rivières
Québec G9A 5H7 CANADA
brunoremillard@UQTR.uquebec.ca

Donna Salopek
Department of Mathematics and Statistics
York University
4700 Keele Street
Toronto, ON M3J 1P3 CANADA
dsalopek@fields.utoronto.ca

David Sankoff
Centre de recherches mathématiques
Université de Montréal
CP 6128, Succ Centreville
Montréal, PQ H3C 3J7 CANADA
sankoff@ere.umontreal.ca

Byron Schmuland
Department of Mathematical Sciences
University of Alberta
Edmonton, AB T6G 2G1 CANADA
schmu@hal.stat.ualberta.ca

Rengarajin Srinivasan
Department of Mathematics and Statistics
University of Saskatchewan
Saskatoon, SK S7N 0W0 CANADA
raj@math.usask.ca

Barbara Szyszkowicz
Department of Mathematics and Statistics
Carleton University
1125 Colonel By Drive
Ottawa, ON K1S 5B6 CANADA
bszyszko@math.carleton.ca

Jean Vaillancourt, Vice-doyen
Faculté des Sciences
Université de Sherbrooke
Sherbrooke, PQ J1K 2R1 CANADA
vaillanc@dmi.usherb.ca

Vladimir Vinogradov
Department of Mathematics
321 Morton Hall
Ohio University
Athens, OH 45701 USA
vlavin@math.ohiou.edu

Anton Wakolbinger
Fachbereich Mathematik
J. W. Goethe - Universität
D-60054 Frankfurt am Main, GERMANY
wakolbin@math.uni-frankfurt.de

Hao Wang
Algorithmics Inc.
185 Spadina Ave.
Toronto, ON M5T 2C6 CANADA
hwang@algorithmics.com

Canadian Mathematical Society
Conference Proceedings
Volume **26**, 2000

# Stochastic Models of Evolving Information Systems

## Donald A. Dawson

ABSTRACT. This paper presents an overview of a class of infinite dimensional stochastic processes introduced to model the large space and time scale behaviour of a spatially distributed population of interacting individuals with evolving internal structure.

## 1. Introduction

The subject of stochastic modelling has its roots in models of the gambler's ruin problem, physical Brownian motion, uncertainty in financial markets, telephone traffic and genealogy. Over the past 100 years stochastic models have evolved to become basic tools in a wide variety of disciplines including statistical physics, chemical kinetics, mathematical finance [**32**],[**38**], operations research, computer science, molecular biology [**35**], genetics, evolutionary biology, ecology, epidemiology, immunology, neuroscience, signal processing and communications networks [**30**]. The many different areas of application have led to the development of different viewpoints and classes of models. In areas such as statistical physics and chemical kinetics the notions of energy and interaction are of primary importance. In areas such as signal processing and communications networks notions of error rate and waiting times are involved. In the case of genetics, immunology, neuroscience and randomized algorithms in computer science the notions of information and stochastic mechanisms to generate diversity are central.

Most of the stochastic models developed prior to the 1960's involved counting processes, real-valued processes or other finite-dimensional processes which in turn led to the development of the rich theories of Markov processes and stochastic analysis. Beginning with the seminal work of Dobrushin and Spitzer, processes with more complex state spaces describing spatial configurations or more complex temporal structures were introduced. This led to the study of infinite particle systems and infinite dimensional processes such as measure-valued processes and the study of the large scale behaviour of these systems. More complex stochastic systems such as these can potentially provide more realistic models of real world systems involving the interaction of a population of individuals. On the other hand

---

2000 *Mathematics Subject Classification.* Primary 60K35; Secondary 60J80.

Research supported by an NSERC Research Grant and Max Planck Award for International Collaboration.

in order to fit models to empirical data it is desirable to work with as simple a class of processes as possible [**28**]. To reconcile these competing scientific objectives renormalization group analysis (which arose from statistical physics [**40**] ) classifies models into universality classes so that the large scale behaviour of a complex system can be related to that of a simple member of the same class.

The focus of this paper is the potential role of infinite dimensional stochastic processes in developing mathematical models which can be used to explore the collective behaviour of a spatially distributed population of interacting individuals having an evolving internal structure over large space and time scales and the identification of universality classes of these processes. The fundamental mechanisms involved include the death and replication of individuals, competition among individuals for a set of limited resources and migration of individuals in space. In general such systems are not time reversible and the notions of Hamiltonians and Gibbs states which serve as a basic framework for the "energy-based" systems of statistical physics are not applicable for these "information-based" systems. A starting point for a "statistical theory of evolving information systems" are the classes of infinite dimensional processes which have their roots in the classical Galton-Watson branching process and Fisher-Wright gene frequency models as well as the replicator dynamics of Eigen, the spatial pattern formation model of Turing and the different classes of queueing network models (cf. [**30**]).

We now begin the description of the population model. The population consists of a collection of individuals which we refer to as *information units (IU)*. Each IU has an internal structure coded as an element of the set $E$. Furthermore each IU has a location in physical or virtual space, $S$, which can be, for example, a metric space, countable group or finite or countable graph describing a network topology. The state of a finite population at time $t$, $X(t)$, describes the spatial distribution of information units:

$$X(t) = \sum_{i \in \Lambda(t)} \delta_{x_i(t)}$$

where $x_i(t) \in E \times S$ describes the internal structure and location of the IU $i \in \Lambda(t)$, the set of IU alive at time $t$.

The system dynamics should incorporate replication of information, competition and interaction between information units and adaptive changes of their internal structure. Branching processes provide idealized models of the basic biological mechanism of reproduction. However they do not take account of the fact that biological individuals are not identical but have complex internal structure determined by their genetic material. Indeed, a biological population is not simply a collection of exchangeable individuals but has a rich structure of family relationships. The genetic structure of individuals determines their "fitness" and how they interact with other individuals. At the same time two related individuals share some genetic information. From the viewpoint of information theory the larger the population the larger are both the information (diversity) and redundancy (information multiplicity cf.[**3**]). In the next section we consider a class of models which abstract some aspects of these structures and may also have relevance to modelling in other areas such as the spread of technological innovation and information flows in communications networks.

## 2. Structured population models

### 2.1. Replication dynamics.

2.1.1. *Single type mechanism.* The simplest model of population death and replication is the classical branching particle system. The continuous state limits of branching particle systems ([6]) have advantages from the analytical viewpoint and will be considered here. The single type *continuous state branching* is given by the Feller's diffusion on $R^+$ with generator

$$Gf(x) = \frac{\gamma x}{2} \frac{\partial^2 f(x)}{\partial x^2}$$

and can also be represented by the Itô stochastic differential equation

$$dx(t) = \sqrt{\gamma x(t)} dw(t)$$

where $\{w(t) : t \geq 0\}$ is a standard Brownian motion.

More generally the "density-dependent" continuous state *generalized branching process* on $[0, \infty)$ (resp. $[0, K]$) is given by

$$Gf(x) = \frac{g(x)}{2} \frac{\partial^2 f(x)}{\partial x^2}$$

where $g$ is a Lipschitz continuous function satisfying $g(x) > 0$ if $x \in (0, K)$, $g(0) = 0$ and $g(K) = 0$ if $K < \infty$. There are three canonical examples

$$g(x) = x, \quad \text{Feller Branching, } K = \infty$$
$$g(x) = x(1-x), \quad \text{Fisher-Wright, } K = 1$$
$$g(x) = x^2, \text{ Geometric Brownian motion, } K = \infty$$

2.1.2. *Multitype mechanisms.* The internal structure of an information unit is coded by an element of $E$, the space of types. The *type* can code an internal state describing for example:

- a state transition program which defines its interaction with other IU
- a genome which codes for a complex developmental process
- a hierarchy of program modules
- a genetic algorithm (cf. [26]).

We begin with a finite or countable system of stochastic differential equations which model a finitely many type population with mutation and type-dependent fertility in an environment with finite carrying capacity, $i \in E = \{1, 2, \ldots, K\}$

$$dx^i(t) = \sqrt{\gamma x^i(t)} \, dw^i(t) \quad (\text{ continuous branching})$$

$$+ h_C \left( \sum_j x^j(t) \right) x^i(t) dt \quad (\text{ finite capacity})$$

$$+ \sum_j (\eta_{ji} \, x^j(t) - x^i(t)) dt \quad (\text{mutation from } i \text{ to } j)$$

$$+ (\theta^i - x^i(t)) dt \quad (\text{ immigration of type } i)$$

$$+ V_i x^i(t) dt \quad (\text{type dependent fertility})$$

where $0 \leq \kappa \leq \infty$, $\theta^i \geq 0$, $\sum_j \eta_{ij} = 1, \eta_{ij} = \eta_{ji}$, and $h_C$ is a continuous non-negative function on $R^+$ satisfying $h_C(0) = 1$, $h_C(C) = 0$. Here $C$ represents the *(carrying) capacity of the environment.*

2.1.3. *Finite dimensional Markov diffusion processes.* The special case ($\kappa = 0$) corresponds to a multitype branching process with immigration and mutation which can be characterized as the unique solution to the martingale problem with generator

$$Gf(x_1, \ldots, x_K) \quad = \quad \frac{1}{2} \sum_{i,j} \gamma_i x_i \delta_{ij} \frac{\partial^2 f}{\partial x_i \partial x_j} + \sum_i (\theta_i - x_i) \frac{\partial f}{\partial x_i}$$

$$+ \sum_{j,i} (\eta_{ij} x_j - x_i) \frac{\partial f}{\partial x_i} + \sum_i V_i x_i \frac{\partial f}{\partial x_i}$$

$$x_i \in R^+, \ i \in \{1, \ldots, K\}.$$

The special case ($\kappa = \infty$, $\gamma_i \equiv \gamma$) defines the finite type Fisher-Wright diffusion with mutation and selection which can be characterized as the unique solution of the martingale problem

$$Gf(p_1, \ldots, p_K) \quad = \quad \frac{\gamma}{2} \sum_{i,j} p_i (\delta_{ij} - p_j) \frac{\partial^2 f}{\partial p_i \partial p_j} + \sum_i (\theta_i - p_i) \frac{\partial f}{\partial p_i}$$

$$+ \sum_{j,i} (\eta_{ij} p_j - p_i) \frac{\partial f}{\partial p_i} + \sum_i (V_i - \sum_j V_j p_j) p_i \frac{\partial f}{\partial p_i}$$

$$p_i \geq 0, \ \sum_{i=1}^K p_i = 1.$$

This process arises as a continuous limit of the discrete Fisher-Wright finite population model where $\gamma$ is inversely proportional to effective population size and $\theta_i$ prescribes the mutation rate to type $i$.

2.1.4. *Infinitely many types: The measure-valued branching process with immigration.* In the study of complex population systems it is not natural to restrict the system to a finite number of potential types and it is useful to introduce infinite type models with for example, $E := [0, 1]$. In this case the branching process becomes a measure-valued process with state space $M(E)$, the space of finite Borel measures on $E$. The corresponding process is the unique solution of the $M(E)$-valued martingale problem with generator

$$GF(\mu) \quad = \quad \frac{1}{2} \iint \gamma(x) \, \mu(dx) \delta_x(dy) \frac{\partial^2 F(\mu)}{\partial \mu(x) \partial \mu(y)}$$

$$+ \sum (\Theta(dx) - \mu(dx)) \frac{\partial F(\mu)}{\partial \mu(x)}$$

$$+ \sum V(x) \mu(dx) \frac{\partial F(\mu)}{\partial \mu(x)}$$

$$\frac{\partial F(\mu)}{\partial \mu(x)} := \frac{d}{d\varepsilon} F(\mu + \varepsilon \delta_x)|_{\varepsilon=0}.$$

2.1.5. *Catalytic branching.* In the measure-valued branching process $\gamma$ is usually assumed to be constant and then represents the branching rate. In some situations branching only occurs in the presence of a *catalyst* and $\gamma$ can be replaced by a measure-valued process $\gamma_t(dx)$ representing the mass distribution of the catalyst at time $t$. Properties of catalytic branching and mutually catalytic branching have been studied in a series of papers [7],[8],[20],[27], [15].

2.1.6. *The Fleming-Viot infinitely many types Fisher-Wright process with mutation and selection.* A generalization of the infinitely many types Fisher-Wright process is the Fleming-Viot process with mutation, selection and weighted sampling. It has state space $M_1(E)$, the space of probability measures on $E$, and can be characterized as the solution to the martingale problem with generator

$$
\begin{aligned}
GF(\mu) \;=\; & \frac{1}{2} \iint \mu(dx) \left( \int \gamma(x,z)\delta_x(dy)\mu(dz) - \gamma(x,y)\mu(dy) \right) \frac{\partial^2 F(\mu)}{\partial\mu(x)\partial\mu(y)} \\
& + \theta \int \left( \int \Theta(dx,y)\mu(dy) - \mu(dx) \right) \frac{\partial F(\mu)}{\partial\mu(x)} \\
& + \int \left( V(x) - \int V(y)\mu(dy) \right) \mu(dx) \frac{\partial F(\mu)}{\partial\mu(x)}.
\end{aligned}
$$

Here the *weighted sampling term* $\gamma(x,y)\mu(dy)$ denotes the rate at which an individual of type $x$ is replaced by an individual of type $y$.

**2.2. Measure-valued processes – methodology.** As indicated above, the context of measure-valued processes is appropriate for modelling infinite type systems. The natural way to characterize a measure-valued process is as the unique solution of a martingale problem. In many cases the existence of solutions are obtained by tightness and weak convergence arguments leaving the question of uniqueness to be handled. In the case of measure-valued branching processes such as super-Brownian motion and Fleming-Viot processes the method of duality is used to establish uniqueness (see [6],[18] for an exposition of these techniques). In the case of super-processes, the log-Laplace equation associated to a deterministic dual is a powerful tool which has been used to obtain detailed information about the structure of the process (see [17] and [16] for recent surveys).

One additional method which can be used to establish uniqueness in the presence of selection is the Girsanov formula relating the law of the process to that of the process in the absence of selection. In the case of the Fleming-Viot process of Section 2.1.6 the expectation of the process with fitness function $V$ and $\gamma(.,.) \equiv 1$ can be represented via the Girsanov formula

$$
E^V_{FV}(g(X_t)) = E^0_{FV} \left[ \exp\left( \int_0^t V dM(s) - \frac{1}{2} \int_0^t [V^2 X - (VX)^2]ds \right) g(X_t) \right]
$$

where $\{M(t) : t \geq 0\}$ is a martingale measure determined by the process $X$ (cf. [6] for details).

Another important tool is the *historical process* which is an enriched version of a measure-valued process. The historical process codes information on the family structure of a population. This allows one to study the genealogy of the entire population and has been used to obtain results on both small scale and large scale spatial structures (see [14], [29]).

**2.3. Hierarchically Structured population.** A hierarchically structured process is one in which objects at given level in the hierarchy are comprised of a subpopulation of objects at the next lower level. For example, in anthropology tribes (level 3 objects) consist of a collection of families (level 2 objects) and each family consists of a collection of individuals (level 1 objects). Some other examples of higher level objects are demes in Wright's shifting balance theory, directories which contain a subpopulation of files in network file systems, and DNA microsatellites which consist of multiple copies of a DNA segment. One way of representing a two level population process in which individuals are given by particles $\delta_x$ with $x \in E$ is as a measure $\Xi^2 \in M(M(E))$ :

$$\Xi_t^2 = \sum_{j=1}^{N_2(t)} \delta_{\sum_{i=1}^{N^j(t)} \delta_{x_{i,j}(t)}}$$

where $N_2(t)$ denotes the number of level 2 objects and $N^j(t)$ denotes the number of level one objects in the $j$th level 2 object at time $t$. A *two level branching process* is one in which individuals can die or replicate and also entire families (level 2 objects) can die or replicate. A typical problem concerning the multiplicity of information is to determine the distribution of the number of copies of a given level 1 object if reproduction occurs at both levels [12].

**2.4. Spatial geometry, random fields and spatially structured population models.**

2.4.1. *Network geometry.* A discrete spatial structure is given by specifying the collection of *spatial sites*, $S$, which can be a finite or countable graph or group. Two basic examples are:

- the d-dimensional lattice $Z^d$
- the hierarchical group $\Omega_N$ :

$$\Omega_N = \{\xi = (\xi_1, \xi_2, \dots) : \xi_i \in \{0, \dots, N-1\}, \xi_i = 0 \ a.a. \ i\}$$

   e.g. if $N = 4$, then $(2,3,1,0,\dots)$ denotes the 3rd *level 1 object* in the 4th *level 2 object* in the 2nd *level 3 object* in the 1st *level 4 object*.

The hierarchical distance (ultrametric) on $\Omega_N$ is defined by:

$$d(\xi, \xi') = \max \{i : \xi_i \neq \xi_i'\}.$$

2.4.2. *Random fields.* A measure-valued random field with type space $E$ on $S$ is a collection of $M(E)$-valued random variables $\{x_\alpha : \alpha \in S\}$.

2.4.3. *Migration.*

Random Walks on $Z^d$: Let $\{S_t : t \geq 0\}$ denote the Markov semigroup on $C(S)$ defined by a random walk $\{W(\cdot)\}$ and for $k = 1, 2, \dots$ let

$$G^{(k)}\varphi := \frac{1}{(k-1)!} \int_0^\infty t^{(k-1)} S_t \varphi dt.$$

We say that $W$ is *transient* if $G^{(1)}\varphi$ is bounded for $\varphi \geq 0$ with finite support, $W$ is *strongly transient* if $G^{(2)}\varphi$ is bounded for $\varphi \geq 0$ with finite support and *level-2 strongly transient if* if $G^{(3)}\varphi$ is bounded for $\varphi \geq 0$ with finite support

The simple random walk on $Z^d$ is transient if and only if $d \geq 3$, strongly transient if and only if $d \geq 5$ and level-2 strongly transient if and only if $d \geq 7$.

The *hierarchical random walk* on $\Omega_N$ makes one step transitions $\xi \to \xi'$ at rate

$$q_{\xi,\xi'} = \frac{c_{k-1}}{N^{2k-2}(N-1)} \quad \text{if} \ d(\xi, \xi') = k.$$

Here $\frac{c_k}{N^{k-1}}$ gives the rate of jumping a distance $k$, that is, $k$ levels in the hierarchy. This random walk was introduced for $N = 2$ by Spitzer (see [37], p. 91) and for general $N$ by Sawyer and Felsenstein [36].

A hierarchical caricature of the random walk in $Z^d$ with one step distribution in the domain of attraction of the $\alpha$symmetric stable distribution, $0 < \alpha < 2$ or normal distribution ($\alpha = 2$) is given by: $c_k = N^{(1-\frac{\alpha}{d})k}$. This hierarchical random walk is transient if $\alpha < d$ and $\sum \frac{1}{c_k} < \infty$. It is strongly transient [9] if

$$c_k = (\kappa N^{1/2})^k, \quad k = 1, 2, \ldots \quad \text{and} \quad \kappa > 1.$$

Interacting Programmed Walks: In a *communication network* information units ("message packets") have prescribed destinations or routes and at each node there is a finite capacity so there can be a delay depending on the subpopulation at the node.

2.4.4. *Multisite population model with migration between sites.* We now consider a finite type spatially distributed population with replication, spatial migration, local interaction (e.g. competition, finite carrying capacity and transformation of type such as mutation, recombination, editing of a file). This is represented by the system of stochastic differential equations for $x_\xi^i(t)$, the mass of the population of type $i \in \{1, \ldots, K\}$ at site $\xi \in S$ at time $t \geq 0$:

$$dx_\xi^i(t) = \sum_{\xi'} q_{\xi,\xi'}(x_{\xi'}^i(t) - x_\xi^i(t))dt \quad \text{(Migration } \xi' \to \xi)$$

$$+\sqrt{2g(x_\xi^i(t))}\, dw_\xi^i(t) \quad \text{(continuous branching)}$$

$$+ h_C\left(\sum_j x_\xi^j(t)\right) x_\xi^i(t)dt \quad \text{(finite capacity } C)$$

$$+\sum_j (\eta_{ji}\, x_\xi^j(t) - x_\xi^i(t))dt \quad \text{(mutation } j \to i)$$

$$+ V_i x_\xi^i(t)dt \quad \text{(type dependent fertility)}$$

$$x_\xi^i(0) = \theta^i \quad \text{(initial population)}.$$

If we replace the random walks by interacting programmed walks we obtain a model of a communications network. In contrast to the standard model, the replication process here models the possibility that an information unit ("message packet") which is received can either be discarded or can trigger the transmission of a random number of copies (or responses) thus creating an interacting system. Here the mutation operation would correspond to introducing an error in the message.

2.4.5. *Information dynamics.* The system of Section 2.4.4 can be viewed as a simple model of a dynamical information system. Mutation is the source of new information and the replication mechanism can create multiple copies of a given information unit. Due to the finite capacity of the environment at each site information units must compete for resources and some information is lost through death (deletion). Different information units have different fitness levels and those having higher fitness are more likely to survive. Collectively this process can be viewed as search process for optimum fitness values. In the multisite system this search process is being carried out in parallel at the different sites. This provides the possibility for exploring different regions of the fitness landscape. However the migration between sites makes possible the spread of a superior type throughout the space. In Section 3 we will investigate the role of the migration mechanism in the neutral case and show that at least in this case the large time behavior of the migration random walk (i.e. transience or recurrence) determines the long time behavior of this information process.

2.4.6. *Multilevel branching: Replication of spatially distributed level 2 objects.* Consider a two level (single type) branching process in which level two objects (superparticles) consist of a spatially distributed family of related level one particles. The process has state space $M(M(S))$ and can be represented as follows:

$$\Xi_t^2 = \sum_{j=1}^{N_2(t)} \delta_{\sum_{i=1}^{N^j(t)} \delta_{x_{i,j}(t)}}$$

where $x_{i,j}(t)$ denotes the location of the $i$th particle in the $j$th family at time $t$. In multilevel branching (cf. 2.3) both particles and superparticles undergo death and replication. The large space-time scale structure of these processes have been studied in [41] and [22].

2.4.7. *Multiscale analysis.*

Renormalization group: Let $\{x_\xi : \xi \in S\}$ be a real (or measure-valued) random field on $S$ and consider a "window" centered at 0, $\mathcal{W}_k := \{\xi \in S : d(\xi, 0) \leq k\}$ and the *k-block average:*

$$x_{\xi,k} := \frac{1}{|\mathcal{W}_k|} \sum_{\xi \in \mathcal{W}_k} x_\xi.$$

The sequence $\{x_{\xi,k} : k = 1, 2, \dots\}$ describes the average of the system at different scales.

Asymptotic multiscale analysis: First consider the case of an exchangeable system indexed by $S = \{1, \dots, N\}$. The classical mean-field limit involves the study of the average, the empirical law and the law of the process at a tagged site as $N \to \infty$.

The mean-field limit is used as a simplified model in statistical physics (cf. [5]). An important refinement of this method is the so-called *hierarchical mean-field limit* in which we study the sequence of block averages $\{x_{\xi,k} : \xi \in \Omega_N\}$ in the case $S = \Omega_N$ (cf. Section 2.4.1) in the limit as $N \to \infty$ (see [11]).

## 2.5. Randomly evolving systems and evolutionary games.

2.5.1. *Infinite population replicator systems.* A large class of biological and economic population systems can be modelled by the *replicator system of ordinary differential equations:*

$$\frac{dp_i(t)}{dt} = p_i(t)[\sum_j V_{i,j} p_j(t) - \sum_{j,k} V_{j,k} p_j(t) p_k(t)]$$

where $p_i(t)$, $i = 1, \dots, N$ represents the frequency of type $i$ at time $t$ ([25], [39]).

In order to develop an evolutionary model of such systems random mechanisms such as mutation and recombination must be introduced. Space does not permit consideration of the important mechanism of recombination but we briefly sketch one way to introduce mutation in the next subsection. Interestingly, the ideas of introducing random mechanisms to search for stable adaptive behavior has arisen in different fields. For example, in working towards the next generation of telecommunications networks methods such as Boltzmann neural networks and genetic algorithms are being actively pursued (e.g. [31]).

2.5.2. *Replicator systems with randomly evolving strategies.* In order to introduce mutation it is useful to allow for an infinite number of types so that a mutation can always lead to a new type not seen before. To do this the space of types is assumed to be $E = [0, 1] \times C([0, 1])$. For $(x, V) \in E$, $x$ denotes the *quasispecies,* that is a subpopulation of related individuals and $V_x(y)$ denotes the effect of an individual of type $x$ on the fitness of an individual of type $y$. The relative fitness of type $(x, V_x)$ in the population $\mu \in M_1 ([0, 1] \times C([0, 1]))$ is

$$\int F(V_y(x), V_x(y))\mu(dy, dV_y).$$

The state of the infinitely many types replicator system at time $t$ is given by $X_t \in M_1 ([0, 1] \times C([0, 1]))$. The basic dynamics are:

1. Mutation within quasispecies $x$ : the "interaction strategy" $V_x(t) \in C([0, 1])$ evolves according to an infinite dimensional Ornstein-Uhlenbeck process:

$$dV_x(t, y) = (\Delta - c)V_x(t, y)dt + dW_x(t, y)$$
$$V_x(t, x) \equiv 0$$

where $\Delta$ is the one-dimensional Laplacian with periodic boundary conditions and $\{W_x(t) : t \geq 0\}$ is cylindrical Brownian motion.

2. Creation of new quasispecies: $(x, G) \longrightarrow (y, G)$ with mutation measure $dy \otimes Q \in M_1([0, 1] \times C([0, 1]))$.

3. Fleming-Viot weighted sampling with rates, $\gamma(x, y)$, a decreasing function of the distance between quasispecies $|x - y|$.

## 3. Structure of the population in large space and times scales

Consider a population system with type space $E = [0, 1]$ on the countable abelian group $S$, reproduction and migration according to a random walk on $S$ as described in Section 2.4. It is conjectured that the large space-time scale behaviour can be classified into universality classes whose qualitative behaviour depends on

- the transience or recurrence of the random walk
- whether the environment at each site has finite or infinite carrying capacity
- the nature of the generalized branching mechanism
- the nature of the mutation and selection which occurs throughout the system.

An essential element in the investigation of this phenomenon is the fact that these population systems can be decomposed into families, that is, a collection of information units containing shared information. One basic question is whether all information is eventually lost or if some information is *immortal.* In this section we summarize some recent results which provide some insight into these questions.

### 3.1. Ergodic behavior of Fleming-Viot processes.

3.1.1. *The single site mutation-selection.* Consider the infinite type Fleming-Viot system (see Section 2.1.6) with $E = [0, 1]$, constant sampling rate $\gamma(x, y) \equiv \gamma$, fitness function $V$ and mutation source $\Theta(dx, y) \equiv \eta(dx)$. Then [19] the equilibrium measure is

$$\Psi(d\mu) = exp(V(\mu))\Gamma_\eta(d\mu), \text{ where}$$

$$V(\mu) := \int V(x)\mu(dx)$$

and $\Gamma_\eta$ is the law of a random pure atomic measure with representation $\Gamma_\eta = \sum M_j \delta_{U_j}$ where the $\{U_j\}$ are i.i.d. $\eta$–distributed r.v.

$$M_j = B_j \prod_{i=1}^{j-1} (1 - B_i)$$

and the $\{B_j\}$ are i.i.d. Beta r.v. with density function

$$f(x) = c(1-x)^{\frac{1-\eta}{\eta}}, \qquad 0 \le x \le 1, \eta = \frac{\gamma}{\theta}.$$

3.1.2. *Mean-field limit of the model with selection.* Consider the exchangeable system of sites, $S = \{1\ldots, N\}$, undergoing dynamics as in Section 3.1.1 and with constant migration rate $q_{\xi,\xi'} \equiv \frac{c_0}{N} \; \forall \xi, \xi' \in S$, mutation rate $\theta$ and mutation source distribution $\eta$. In the *mean field limit,* $N \to \infty$ the system dynamics is described by a measure-valued McKean-Vlasov process [5], [11].

THEOREM 3.1. *([11]). In the mean field limit, $N \to \infty$, there is a one-to-one correspondence between the set of equilibrium laws of a tagged site given by*

$$\Psi_\nu(d\mu) = Z^{-1}e^{V(\mu)}\Gamma_{(\alpha\nu+(1-\alpha)\eta)}(d\mu)$$

*and the set of fixed points, $\nu$,*

$$\nu = G(\nu), \qquad \nu \in M(E)$$

$$G(\nu) = Z^{-1} \int \mu e^{V(\mu)}\Gamma_{(\alpha\nu+(1-\alpha)\eta)}(d\mu)$$

$$\alpha = \frac{c_0}{c_0 + \theta}.$$

*Moreover, for sufficiently large $\theta$, there is a unique fixed point.*

REMARK. In [11] partial results are obtained for the large time behavior of a system with type-dependent mutant distribution. This model is motivated by a system in which the type represents an internal structure which is hierarchically organized.

3.1.3. *Ergodic behavior of neutral FV stepping stone model without mutation.* In the case of spatially distributed systems the long time behavior involves a competition between processes such as genetic sampling which tend to decrease diversity and spatial migration which tends to increase local diversity by the flow of individuals of different types from other spatial regions. The following result illustrates this phenomenon and indicates the importance of the nature of the spatial migration process.

THEOREM 3.2. *Let $S$ be a countable abelian group and consider the neutral stepping stone model without mutation on $S$. Assume that the initial random field is spatially stationary and ergodic and has mean measure $\mu \in M_1[0,1]$.*
*(a) ([2]) If $\{q_i\}$ is a symmetric transient r.w. on $S$, then the stepping stone process converges in distribution to a nontrivial ergodic invariant random measure $\nu_\mu$ which*

*has single site mean measure $\mu$. If $\{q_i\}$ is recurrent, then the extremal invariant measures are $\delta_a$, $a \in [0,1]$, that is, there is local fixation.*
*(b) Let $S = Z^d$, $d \geq 3$ and $X_0 = \nu$, with $\nu$ nonatomic. Then*

$$\lim_{L \to \infty} \frac{1}{L^d} \sum_{|j| \leq L} < X_{eq}(j) \otimes X_{eq}(0), I_\Delta > = 0.$$

*where $\Delta$ denotes the diagonal in $[0,1] \times [0,1]$.*

REMARKS. (a) means that if the migration is recurrent the system loses diversity and eventually a given region is dominated by a single type. In the transient case diversity is maintained throughout the system. In this case each surviving information unit has an infinite number of copies ("immortal family" with a common ancestor) spread throughout the system. However (b) means that each of the immortal families has zero spatial density and together with (a) this implies that there are a countable number of infinite families in equilibrium.

**3.2. Critical multilevel branching random walk.** In the case of critical branching random walk, $\{\Xi_t^1\}$ on $Z^d$ with uniform initial measure, $\lambda$, the system converges as $t \to \infty$ to a nontrivial spatially homogeneous equilibrium measure if and only if the random walk is transient, that is, $d \geq 3$ [23], [14]. This analysis is based on the historical process description of the family structure of the population.
Similarly the two level branching system

$$\Xi_t^2 = \sum_i \delta_{\sum_j \delta_{x_{i,j}(t)}}$$

with initial condition

$$\Xi_0^2 = \sum_{j \in Z^d} \delta_{\delta_j}$$

converges to a nontrivial equilibrium measure if and only if $d \geq 5$ (cf. [41],[22]). The study of the large time scale behavior of multilevel branching Brownian motion with more general initial conditions is studied in [24].
In order to further explore the dimension dependence for single and multilevel branching we consider the *occupation time fluctuations*

$$U_1(t) := \int_0^t \Xi_s^1 ds$$

$$U_2(t) := \int_0^t \int_{M([0,1])} \mu \Xi_s^2(d\mu) ds.$$

Then (cf. [9]),

$$\frac{1}{t} U_i(t) \to \lambda, \text{ if } i = 1, d > 2 \text{ or } i = 2, d > 4$$

$$\frac{U_i(t) - t\lambda}{g_i(t)} \to \text{ Gaussian random field, } i = 1, 2$$

where

$$
\begin{aligned}
g_i(t) &= t^{\frac{3}{4}} & &\text{if } i = 1, d = 3 \text{ or } i = 2, d = 5 \\
g_i(t) &= (t \log t)^{1/2} & &\text{if } i = 1, d = 4 \text{ or } i = 2, d = 6 \\
g_i(t) &= t^{1/2} & &\text{if } i = 1, d \geq 5 \text{ or } i = 2, d \geq 7.
\end{aligned}
$$

These results reflect the fact that higher level branching creates longer range dependencies.

## 4. Open problems

The introduction of measure-valued and other infinite dimensional stochastic processes has opened up the possibility of building stochastic models of complex phenomena such as those described above. The use of mean-field, lattice and hierarchical mean-field measure-valued systems provide a tool for the study of large scale behaviour for idealized models of "large systems". In Section 3 we have given an introduction to some results which have been obtained for the class of models described in Section 2.4. However in order to realize the potential of this approach for more complex systems (such as those described in Section 2.5) many problems must be addressed. We do not give a detailed list but briefly sketch three directions of current research.

- *Process characterization.* The methods which have been developed to prove uniqueness for the solution of the martingale problems describing measure-valued branching, Fleming-Viot processes and some classes of branching with interaction include duality, Girsanov transformations of measure [6] and strong uniqueness [34],[33]. However there are many natural extensions of these models for which these methods are not applicable. See [13] for a class of examples for which only incomplete results have been obtained.

- *Universality.* There is reason to conjecture that the results on the large space-time scale behavior of branching random walks (see Section 3.2) and the infinite system of Fleming-Viot processes (see Section 3.1.3) remain valid for much larger classes of processes. In the case of one or two type systems, examples of this are found in [2] and for the measure-valued case in [10]. However a systematic classification of universality classes remains a challenging problem.

- *Randomly Evolving Dynamical Systems.* It is a question of considerable interest to determine the long time behavior of dynamical systems undergoing mutation such as those described in Section 2.5. For example, are there any limiting dynamical systems or is there a perpetual introduction of new species? See [25] for a discussion of some aspects of this problem. More generally, a central question is a conceptual understanding of the interplay between the randomly evolving internal dynamics of the individuals and the interaction among large groups of such individuals to produce complex multilevel systems (e.g. ecosystems, societies, knowledge systems). In recent years efforts have been made to come to grips with this problem at a conceptual level (e.g. [26], [1]). However a more complete understanding would require a mathematical theory of randomly evolving dynamical systems which would capture this process.

## References

[1] D.R. Brooks and E.O. Wiley, *Evolution as Entropy*, Univ. of Chicago Press, 1988.

[2] J.T. Cox, A. Greven and T. Shiga, *Finite and infinite systems of interacting diffusions*, Prob. Th. Rel. Fields **102** (1995), 165–197.

[3] D.A. Dawson, *Information flow in discrete Markov systems*, Journal of Applied Probability **10**(1973), 63–83.

[4] D.A. Dawson, *The Critical Measure Diffusion*, Z. Wahr. verw. Geb. **40** (1977), 125–145.

[5] D.A. Dawson, *Critical dynamics and fluctuations for a mean-field model of cooperative behavior*, Journal of Statistical Physics **31** (1983), 29–85

[6] D.A. Dawson, *Measure-Valued Markov Processes*. In: *Ecole d'Eté de Probabilités de St.Flour XXI - 1991*, Lecture Notes in Mathematics **1541** (1993), Springer-Verlag..

[7] D.A. Dawson, D.A., and K. Fleischmann, *A continuous super-Brownian motion in a super-Brownian medium*, J. Theor. Prob. **10** (1997), 213–276.

[8] D.A. Dawson, D.A., and K. Fleischmann, *Longtime behavior of a branching process controlled by branching catalysts*, Stoch. Proc. Appl. **71** (1997), 241–257.

[9] D.A. Dawson, L.G. Gorostiza and A. Wakolbinger, *Occupation time fluctuations in branching systems*, in preparation.

[10] D.A. Dawson and A. Greven, *State dependent mulitype spatial branching and their long-time behavior*, in preparation.

[11] D.A. Dawson and A. Greven (1999). *Hierarchically interacting Fleming-Viot processes with selection and mutation: Multiple space time scale analysis and quasi-equilibria*, Electronic J. Prob. **4** (1999), paper no. 4,1–81.

[12] D.A. Dawson and K.J. Hochberg,*A multilevel branching model*, Adv. Appl. Prob. **23** (1991), 701–715.

[13] D.A. Dawson and P. March, *Resolvent estimates for Fleming-Viot operators and uniqueness of solutions to related martingale problems*, J. Functional Anal. **132** (1995), 417-472.

[14] D.A. Dawson and E.A. Perkins, *Historical Processes*, Memoirs of the American Mathematical Society, no. **93** (1991), AMS, Providence.

[15] D.A. Dawson and E.A. Perkins, *Long-time behaviour and co-existence in a mutually catalytic branching model*, Ann. Probab. **26** (1998), 1088-1138.

[16] D. A. Dawson and E.A. Perkins, *Measure-valued processes and Renormalization of Branching Particle Systems*, in *Six Views of SPDE*, eds. R. Carmona and B. Rozovskii, American Mathematical Society, Providence, 1999.

[17] E.B. Dynkin, *An Introduction to branching measure-valued processes*, CRM Monograph **6**, Amer. Math. Soc. Providence, 1994.

[18] S.N. Ethier, and T.G. Kurtz, *Markov Processes: Characterization and Convergence*, Wiley, New York, 1986.

[19] S.N. Ethier, and T.G. Kurtz, *Fleming-Viot processes in population genetics*, SIAM. J. Control Optim. **31** (1993), 345–386.

[20] K. Fleischmann and C. Mueller, *Finite time extinction of catalytic branching processes*, this volume.

[21] H. Föllmer, *Macroscopic convergence of Markov chains on infinite product spaces*, Colloquia Math. Soc. J. Bolyai, 355-362, North Holland, 1979.

[22] L.G. Gorostiza, K.J. Hochberg and A. Wakolbinger, *Persistence of a critical super-2 process*, J. Appl. Prob. **32** (1995), 534-540.

[23] L.G. Gorostiza and A. Wakolbinger,*Long Time Behaviour of Critical Branching Particle Systems and Applications*. In D. Dawson (ed.) *Measure-Valued Processes, Stochastic Partial Differential Equations and Interacting Systems*, CRM Proc. Lecture Notes and Monographs **5**, pp. 119-137, American Mathematical Society, Providence, 1993.

[24] A. Greven and K.J. Hochberg, *New Behavioral Patterns for Two-Level Branching Systems*, this volume.

[25] J. Hofbauer and K. Sigmund, *Evolutionary games and population dynamics*, Cambridge Univ. Press, Cambridge, 1998.

[26] J.H. Holland, *Hidden Order*, Addison Wesley, 1995.

[27] A. Klenke, *A Review on Spatial Catalytic Branching*, this volume.

[28] R.J. Kulperger, *SDE estimation: effects of msspecified diffusion functions*, this volume.

[29] T.G. Kurtz, *Particle representations for measure-valued population processes with spatially varying birth rates*, this volume.

[30] W.A. Massey and R. Srinivasan, *Steady state analysis with heavy traffic limits for semi-open networks*, this volume.

[31] C. Mueller, P. Veitch, E.H. Magill and D.G. Smith, *Emerging AI Techniques for Network Management*, IEEE, 1995.

[32] R. Norvaiša and D.M. Salopek, *Estimating the Orey index of a Gaussian stochastic process with stationary increments: An application to financial data set*, this volume.

[33] E.A. Perkins, *Measure-valued branching diffusions with singular interactions*, Can. J. Math. **46** (1994),120–168.

[34] E.A. Perkins, *Measure-valued branching diffusions with spatial interactions* Probab. Theory Relat. Fields **94** (1992), 189–245.

[35] D. Sankoff and M. Blanchette, *Comparative genomics via phylogenetic invariants for Jukes-Cantor semigroups*, this volume.

[36] S. Sawyer and J. Felsenstein, *Isolation by distance in a hierarchically clustered population*, J. Appl. Prob. **20** (1983), 1-10.

[37] F. Spitzer, *Principles of Random Walk*, Van Nostrand, Princeton, N.J., 1964.

[38] H. Wang, *Valuation of a barrier European option on jump-diffusion underlying stock price*, this volume.

[39] J.W. Weibull, *Evolutionary Game Theory*, The MIT Press, 1995.

[40] K.G. Wilson and I. Kogut, *The renormalization group*, Phys. Rep. **12C** (1974), 75.

[41] Y. Wu, *Asymptotic behavior of the two level measure branching process*, Ann. Probab. **22** (1994), 854-874.

THE FIELDS INSTITUTE FOR RESEARCH IN MATHEMATICAL SCIENCES, 222 COLLEGE STREET, TORONTO, CANADA M5T 3J1

*E-mail address*: don@fields.utoronto.ca

Canadian Mathematical Society
Conference Proceedings
Volume **26**, 2000

# A Comparison of Free Branching and Stepping Stone Models

## Matthias Birkner and Anton Wakolbinger

*For Don Dawson*

ABSTRACT. It is well known that local fixation of a stepping stone system with a symmetric migration kernel $p$ implies local extinction of a critical branching system with the same migration kernel. This is part of a more general comparison result due to Cox, Fleischmann and Greven (1996), and also follows from characterisations via duality: Local fixation of the stepping stone system happens iff two independent $p$-chains spend a.s. an infinite amount of time together, whereas local extinction of the branching system holds iff the conditional expectation of this time is infinite, given one of the two chains. We show that the converse of this implication is false: there exists a symmetric recurrent migration dynamics which takes the critical branching system to local extinction but does not take the stepping stone system to local fixation.

## 1. Introduction

We focus in this paper on the long term behaviour of two systems of interacting diffusions

$$(1.1) \qquad d\zeta_i = \sum_{j \in S} p(i,j)(\zeta_j - \zeta_i)\, dt + \sqrt{\zeta_i(1 - \zeta_i)}\, dW_i, \quad i \in S$$

and

$$(1.2) \qquad d\eta_i = \sum_{j \in S} p(i,j)(\eta_j - \eta_i)\, dt + \sqrt{\eta_i}\, dW_i, \quad i \in S.$$

Here $S$ is a countable set and $p(i,j)$, $i,j \in S$ is an irreducible matrix with nonnegative entries such that $\sum_{j \neq i} p(i,j)$, $i \in S$, is bounded. The processes $\zeta = (\zeta_i)$ and $\eta = (\eta_i)$ take their values in $[0,1]^S$ and $[0,\infty)^S$, respectively, and $W_i$, $i \in S$, are independent Brownian motions.

Both (1.1) and (1.2) fit into a class of models which were recently investigated and compared by Cox, Fleischmann, and Greven ([**1**]):

$$(1.3) \qquad d\xi_i = \sum_{j \in S} p(i,j)(\xi_j - \xi_i)dt + \sqrt{g(\xi_i)}dW_i, \quad i \in S.$$

---

2000 *Mathematics Subject Classification.* Primary 60K35; Secondary 60J80, 60J70, 60J27.

Under mild assumptions on $g$ (which cover (1.1) and (1.2)) and on the initial condition $\xi(0)$, (1.3) has a unique strong solution (see [1], p. 514).

In intuitive terms, the comparison result (Theorem 1) of [1] states that a larger diffusion coefficient $g$ leads to a process whose components are more spread out in distribution. In particular, "the bigger $g$ is, the more the $\xi_i$ are driven towards the boundary". As a corollary (cf. [1], Thm. 2 and Remark 11) this shows that *local fixation of (1.1) implies local extinction of (1.2)*, i.e. if

$$(1.4) \qquad \mathcal{L}(\zeta(t)) \longrightarrow (1 - \Theta)\delta_{\underline{0}} + \Theta\delta_{\underline{1}} \quad \text{as } t \to \infty \qquad \text{for } \zeta_i(0) \equiv \Theta \in (0, 1)$$

then

$$(1.5) \qquad \mathcal{L}(\eta(t)) \longrightarrow \delta_{\underline{0}} \quad \text{as } t \to \infty \qquad \text{for } \eta_i(0) \equiv c \in [0, \infty).$$

Here $\underline{0}$ and $\underline{1}$ denote the constant configurations $(0, 0, \dots)$ and $(1, 1, \dots)$ respectively.

We will show that the converse of this implication is false:

THEOREM 1.1. *There exists an irreducible, symmetric, stochastic matrix $p$ on $S := \mathbb{N}_0$ such that the system $\eta$ given by (1.2) suffers local extinction, but the system $\zeta$ given by (1.1) does not fixate locally. That is $\eta$ obeys (1.5), but $\zeta$ does not obey (1.4).*

One part of the proof makes use of the fact that (1.1) is a system of interacting Fisher-Wright (or "stepping stone") diffusions and thus admits a well-known criterion for local fixation (see Section 2).

On the other hand, we note that for *symmetric* $p$, (1.2) can be written as

$$(1.6) \qquad d\eta_i = \left( \sum_{j \in S} p(j, i)\eta_j - \sum_{j \in S} p(i, j)\eta_i \right) dt + \sqrt{\eta_i}\, dW_i, \quad i \in S,$$

which describes a system of interacting Feller diffusions with migration kernel $p(i, j)$ — or, in other words, a Dawson-Watanabe process built over the Markov chain with jump rates $p(i, j)$. The other part of the proof consists in showing that, for our choice of $p$, this Dawson-Watanabe process suffers local extinction.

We will specify the migration kernel $p$ in our theorem in section 3; here we give an informal description. A particle sitting at site 0 jumps after an exponential waiting time to a randomly chosen site $i \neq 0$, from where it jumps back to the "home" state 0 after another exponential waiting time whose expectation is increasing in $i$.

Now consider a pair $(X, \widetilde{X})$ of independent $p$-chains. By choosing the distribution of jumps from 0 appropriately we will achieve the following:

1. the pair $(X, \widetilde{X})$ is transient, in particular the amount of time which $X$ and $\widetilde{X}$ spend at 0 together is finite a.s.,
2. the returns to 0 of $(X_t)$ will be sparse and at these instants $(\widetilde{X}_t)$ is likely to be far away, so that in fact $(X_t)$ jumps to the position occupied by $(\widetilde{X}_t)$ only finitely often.

By definition two $p$-particles can only meet either at 0, or whenever one of them jumps from 0 to $i \neq 0$ while the other happens to be at $i$, too. Therefore, as a consequence of 1. and 2., two $p$-chains will only spend a finite time together. This implies that (1.4) fails (see section 2).

On the other hand, our $p$-motion will turn out to be so lazy that a Dawson-Watanabe process starting from uniform mass locally dies out in probability. Indeed, at a late time $t$ all the families founded by individuals in $\{0, 1, \ldots, N(t)\}$ (where we choose a suitable $N(t) = o(t)$) will have died out with high probability, and since the $p$-motion is so lazy, the mass immigrating from outside $\{0, 1, \ldots, N(t)\}$ is too small to make up for this effect.

## 2. Duality criteria

Let us first recall the well-known criterion for local fixation of the stepping stone system in terms of duality.

LEMMA 2.1 ([9]). *Local fixation (1.4) holds for the solution $\zeta$ of (1.1) iff two independent copies of a Markov chain with jump rates $p$ almost surely spend an infinite time together.*

Next, we recall the well-known Kallenberg criterion for local extinction of the branching system (see [5], Prop. 2.5, for particle systems, and [6] for the connection between particle systems and Dawson-Watanabe-processes in terms of the Cox representation):

LEMMA 2.2. *Assume that $\lambda(i) \equiv 1$ defines an invariant measure for $p$, and let $\eta$ follow the dynamics (1.6). Then local extinction (1.5) holds for $\eta$ iff*

$$(2.1) \qquad \int_0^\infty p_t(X_t, i)dt = \infty \quad a.s. \text{ for all } i \in S.$$

*Here, $X$ is a Markov chain with jump rates $\tilde{p}(i,j) = p(j,i)$, that is, $p$ and $\tilde{p}$ are dual with respect to $\lambda \equiv 1$, and $p_t$ is the semigroup associated with $p$.*

We remark that by irreducibility of $p$ the divergence of the integral in (2.1) does not depend on the choice of $i \in S$.

If $p$ is a symmetric stochastic matrix, then $\lambda(i) \equiv 1$ *does* define an invariant measure for $p$, and condition (2.1) ties in nicely with the condition of Lemma 2.1. Indeed, in this case

$$\int_0^\infty p_t(X_t, i)dt = \int_0^\infty p_t(i, X_t)dt = \mathbb{E}\left( \left. \int_0^\infty 1_{\{X_t = \tilde{X}_t\}}dt \right| X \right),$$

where $(\tilde{X}_t)$ is another Markov chain with jump rates $p$, starting in $i$ and independent of $X$.

Thus in case of symmetric $p$, Lemma 2.2 can be rephrased in a way to parallel Lemma 2.1:

LEMMA 2.3. *Let $p$ be a symmetric stochastic matrix. Then local extinction (1.5) holds for the solution $\eta$ of (1.2) iff the conditional expectation of the total time spent together by two independent copies of a $p$-Markov chain, given one of them, is infinite a.s.*

In order to prove our theorem, because of Lemmata 2.1 and 2.3, it suffices to find an irreducible symmetric stochastic matrix $p$ on $\mathbb{N}_0$ such that for two independent copies $X$, $\tilde{X}$ of a Markov chain with jump matrix $p$ starting in an arbitrary $i \in \mathbb{N}_0$ there holds

$$(2.2) \qquad \int_0^\infty 1_{\{X_t = \tilde{X}_t\}}dt < \infty \quad \text{a.s.}$$

but

$$(2.3) \qquad \mathbb{E}\left(\int_0^\infty 1_{\{X_t = \tilde{X}_t\}} dt \,\Big|\, X\right) = \infty \quad \text{a.s.}$$

## 3. Proof of the Theorem

**3.1. Choice of the migration kernel $p$ and auxiliary results.** We choose a fixed $\gamma \in (1,2)$ and set, with $c_\gamma := (\sum_{i \geq 1} i^{-\gamma})^{-1}$

$$(3.1) \qquad p(0,i) = p(i,0) = c_\gamma i^{-\gamma} \text{ for } i > 0, \quad p(i,j) = 0 \quad \text{otherwise.}$$

Let $(X_t)$ be the Markov chain with jump rates $p(i,j)$, and write $p_t(i,j) := \mathrm{P}_i[X_t = j]$ for the transition probabilities of $X$. Let us define $\sigma_0 = \tau_0 := 0$ and

$$(3.2) \qquad \sigma_n := \inf\{t \geq \tau_{n-1} : X_t \neq 0\}, \quad \tau_n := \inf\{t \geq \sigma_n : X_t = 0\},$$

i.e. $\sigma_n$ and $\tau_n$ are the $n$-th exit time from 0 and the $n$-th return time to 0 respectively. We state now two propositions on the asymptotics of these return times and of the transition probabilities $p_t$. The proof of these auxiliary results will be deferred to the next section.

PROPOSITION 3.1.     1. *There is a constant $C_1$ such that*

$$(3.3) \qquad \mathrm{P}_0(\tau_1 > t) \sim C_1 t^{-(1-1/\gamma)}, \quad t \to \infty.$$

2. *For $0 < \beta < (1 - 1/\gamma)^{-1}$,*

$$(3.4) \qquad \liminf_{n \to \infty} \frac{\sigma_n}{n^\beta} = \infty.$$

PROPOSITION 3.2. *There exist constants $C_2, C_3$ such that*

$$(3.5) \qquad p_t(0,0) \sim C_2 t^{-1/\gamma}, \quad t \to \infty$$

*and*

$$(3.6) \qquad \sup_i p_t(0,i) \leq C_3 p_t(0,0) \quad \text{for all } t \geq 0.$$

**3.2. The stepping stone system $\zeta$ does not fixate locally.** Now let $(X_t)$ and $(\tilde{X}_t)$ be two independent $p$-chains, $X_0 = \tilde{X}_0 = 0$. Observe that

$$\mathrm{P}_{(0,0)}(X_0 = \tilde{X}_t = 0) = (p_t(0,0))^2 \sim C_2^2 t^{-2/\gamma}.$$

This gives

$$\int_0^\infty \mathrm{P}_{(0,0)}((X_t, \tilde{X}_t) = (0,0)) dt < \infty \quad \text{a.s.},$$

i.e. the pair $(X_t, \tilde{X}_t)$ is transient. In particular

$$\int 1_{\{(0,0)\}}(X_t, \tilde{X}_t) dt < \infty$$

holds almost surely.

Define $\sigma_n, \tau_n$, $n = 0, 1, \ldots$ in (3.2) with respect to the first component $(X_t)$ and let

$$A_n = \{X_{\sigma_n} = \tilde{X}_{\sigma_n}\}$$

be the event that the first component hits the second one immediately after its $n^{\text{th}}$ visit to 0. The jump rates defined in 3.1 admit only three possible ways how $(X_t, \tilde{X}_t)$ can hit the diagonal $\{(i,i) : i \geq 0\}$: Both components can meet at 0 or the first component might jump on the second component at $i \neq 0$ directly after a visit to 0 or vice versa. As the pair is transient only finitely many events of the first type will occur. Thus in order to prove the first part of the theorem it suffices to show that with probability 1 only finitely many $A_n$ occur. By symmetry, the same reasoning applies to $B_n = \{X_{\tilde{\sigma}_n} = \tilde{X}_{\tilde{\sigma}_n}\}$, where $\tilde{\sigma}_n$ is defined as in (3.2), but now with respect to the second component $(\tilde{X}_t)$.

Let $\mathcal{F} = \sigma(X_t, t \geq 0)$. By Proposition 3.2,

$$
\begin{aligned}
\mathrm{P}(X_{\sigma_n} = \tilde{X}_{\sigma_n} | \mathcal{F}) &= \sum_{i \geq 1} \mathrm{P}(X_{\sigma_n} = \tilde{X}_{\sigma_n} = i | \mathcal{F}) \\
&= \sum_{i \geq 1} p_{\sigma_n}(0, i) \mathrm{P}(X_{\sigma_n} = i | \mathcal{F}) \leq C_3 p_{\sigma_n}(0, 0) \sum_{i \geq 1} \mathrm{P}(X_{\sigma_n} = i | \mathcal{F}) \\
&= C_3 p_{\sigma_n}(0, 0) \sim C_3 C_2 (\sigma_n)^{-1/\gamma}.
\end{aligned}
$$

Choosing $\beta \in (\gamma, (1 - 1/\gamma)^{-1})$ (which is possible since $\gamma < 2$) we find using Proposition 3.1

$$
\tag{3.7} \mathrm{P}(A_n | \mathcal{F}) \leq \frac{C_3 C_2}{2} n^{-\beta/\gamma}
$$

for $n \geq n_0$ with some $n_0 < \infty$ depending on $(X_t)$. An application of the Borel-Cantelli lemma concludes the argument.

**3.3. The branching system $\eta$ goes to local extinction.** First we observe that the $p$-motion is "lazy" in the sense that for $\beta > 1/\gamma$

$$
\begin{aligned}
\mathrm{P}_0(X_t \geq t^\beta) &= \sum_{i \geq t^\beta} \mathrm{P}_0(X_t = i) \\
&= \sum_{i \geq t^\beta} \int_0^t p_s(0, 0) p(0, i)(1 - e^{-c_\gamma i^{-\gamma}(t-s)}) ds \\
&\leq C \sum_{i \geq t^\beta} i^{-\gamma} \int_0^t s^{-1/\gamma} ds \\
&\sim C'(t^\beta)^{-\gamma+1} t^{1-1/\gamma} = C' t^{(\beta-1/\gamma)(-\gamma+1)} \to 0,
\end{aligned}
$$

if $\beta > 1/\gamma$. Writing $\eta^k(t)$ for that part of the population $\eta(t)$ which originates from the mass initially at site $k$, we observe that

$$
\mathbb{E} \sum_{i \geq t^\beta} \eta_0^i(t) = \sum_{i \geq t^\beta} \mathrm{P}_i(X_t = 0) = \mathrm{P}_0(X_t \geq t^\beta) \to 0.
$$

In that sense, the contribution to $\eta_t(0)$ from outside $[0, t^\beta]$ is asymptotically negligible for large $t$. But on the other hand the population inside $[0, t^\beta]$ is too small to sustain itself until time $t$. Indeed, for every site $i$, $|\eta^i(t)| := \sum_{j \geq 0} \eta_j^i(t)$ is a critical Feller branching diffusion whose survival probability has the asymptotics $\mathrm{P}[\eta^i(t) \neq 0] \sim 1/t$, see e.g. [2], p. 70. We therefore have

$$
\mathrm{P}(\exists i \leq t^\beta : \eta_t^i(0) > 0) \leq \sum_{i \leq t^\beta} \mathrm{P}(\eta_t^i \neq 0) \sim C'' t^{\beta-1} \to 0.
$$

Thus $\eta_0(t)$ converges to 0 as $t \to \infty$, which is sufficient for local extinction of $\eta$.

**3.4. An alternative proof of (2.1).** The reasoning in subsection 3.3 together with Lemma 2.2 gives a "branching process proof" for the fact that $p$ defined in (3.1) obeys

$$(2.1) \qquad \int_0^\infty p_t(X_t, i)dt = \infty \quad \text{a.s. for any } i \in S.$$

In this subsection we give another proof of (2.1) which might be of some independent interest. We claim that (2.1) holds true for any symmetric irreducible recurrent $p_t$ satisfying (3.6) and

$$(3.8) \qquad p_t(i, i) \le K' p_{2t}(i, i) \quad \text{for some } i \text{ and all } t \ge 0.$$

To show this claim, observe first that, because of symmetry of $p$, the uniform measure on $S$ is $p$-reversible, and therefore $X_t$ appearing in (2.1) has again transition probability $p_t$. Let $G_t := \int_0^t p_s(X_s, i)ds$. We have $\mathbb{E}_i p_s(X_s, i) = \sum_x p_s(i, x)p_s(x, i) = p_{2s}(i, i)$, giving $\mathbb{E}_i G_t \to \infty$ by the assumed recurrence. Now assume that $\int_0^\infty p_t(X_t, i)\,dt < \infty$ holds almost surely and let $\widetilde{G}_t := G_t / \mathbb{E}_i G_t \,(\ge 0)$. Then $\widetilde{G}_t \to 0$ a.s. because the denominator $\mathbb{E}_i G_t$ diverges. But this implies that also $\widetilde{G}_t \to 0$ in $\mathcal{L}_1$ since by (3.6)

$$G_t \le C_3 \int_0^t p_s(i, i)ds \le C_3 K' \widetilde{G}_t$$

so that dominated convergence applies. This is a contradiction, since by construction $\mathbb{E}_i \widetilde{G}_t \equiv 1$. Because the event $\{G_t \to \infty\}$ is adapted to the tail field $\bigcap_t \sigma(X_s, s \ge t)$ (which is trivial because of Orey's 0-1 law (cf. [3], p. 320)), we infer (2.1), and thus have proved our claim.

Finally let us recall an open question (cf. [8]). We do not know whether there exists any irreducible recurrent migration dynamics which permits local survival of a critical branching system. The above argument shows that any symmetric migration dynamics having this property necessarily must violate (3.6) or (3.8).

## 4. Proofs of auxiliary results

**4.1. Proof of Proposition 3.1.** Write $\tau_1 = \sigma_1 + (\tau_1 - \sigma_1)$ (with $\sigma_i$ and $\tau_i$ as in the previous section). By the Markov property these two summands are independent. Obviously $\mathcal{L}(\sigma_1 | X_0 = 0) = \text{Exp}(1)$ and the distribution of $\tau_1 - \sigma_1$ is a mixture of exponential laws

$$\begin{aligned}
\mathrm{P}_0(\tau_1 - \sigma_1 > t) &= \sum_{k \ge 1} \mathrm{P}_0(\tau_1 - \sigma_1 > t \,|\, X_{\sigma_1} = k)\mathrm{P}_0(X_{\sigma_1} = k) \\
&= \sum_{k \ge 1} \mathrm{P}_k(\tau_1 > t)c_\gamma k^{-\gamma} = \sum_{k \ge 1} c_\gamma k^{-\gamma} e^{-c_\gamma k^{-\gamma} t} \\
&= \int_0^\infty c_\gamma x^{-\gamma} e^{-c_\gamma x^{-\gamma} t} dx + R_t = \frac{c_\gamma^{1/\gamma}}{\gamma} t^{-1+1/\gamma} \Gamma(1 - 1/\gamma) + R_t
\end{aligned}$$

where we used the substitution $c_\gamma t x^{-\gamma} = y$ in the last line and obtain an error term $R_t$. Set $g_t(x) = c_\gamma x^{-\gamma} e^{-c_\gamma x^{-\gamma} t}$. A straightforward calculation shows that $g_t'(\cdot)$ has two local extrema and $\sup_{x \ge 0} |g_t'(x)| = O(t^{-1-1/\gamma})$. Using the mean value theorem we can thus easily estimate

$$
\begin{aligned}
|R_t| &= \left| \sum_{k=1}^{\infty} g_t(k) - \int_0^{\infty} g_t(x)dx \right| \leq \int_0^{\infty} |g_t(\lceil x \rceil) - g_t(x)|dx \\
&\leq \int_0^{\infty} |g_t'(\Theta_x)|dx \leq 2 \sup_{x \geq 0} |g_t'(x)| + \int_0^{\infty} |g_t'(x)| + |g_t'(x+1)|dx \\
&\leq 2 \left( \sup_{x \geq 0} |g_t'(x)| + \int_0^{\infty} |g_t'(x)|dx \right) = O(1/t),
\end{aligned}
$$

where $x \leq \Theta_x \leq \lceil x \rceil$. This proves (3.3) because $R_t$ is of smaller magnitude than $t^{-1+1/\gamma}$ and $\sigma_1$ has exponential tails.

Furthermore $\mathcal{L}(\sigma_2 - \sigma_1 | X_0 = 0) = \mathcal{L}((\sigma_2 - \tau_1) + (\tau_1 - \sigma_1) | X_0 = 0) = \mathcal{L}(\tau_1 | X_0 = 0)$. Using (3.3) we can thus easily conclude (3.4) from the following lemma applied to $\sigma_n = (\sigma_n - \sigma_{n-1}) + \cdots + (\sigma_2 - \sigma_1) + \sigma_1$.

LEMMA 4.1. *Let $R_1, R_2, \ldots$ be an i.i.d. sequence of nonnegative random variables with $\mathrm{P}(R_1 > t) \sim K t^{-\alpha}$ where $0 < \alpha < 1$. Then*

$$
\lim_{n \to \infty} \frac{R_1 + \cdots R_n}{n^{\beta}} = \infty \qquad a.s.
$$

*for each $\beta \in (0, 1/\alpha)$.*

PROOF. Fix $0 < \beta < 1/\alpha$ and set $A_k = \{R_k \geq k^{\beta}\}$, $Y_k = 1_{A_1} + \cdots + 1_{A_n}$. We have $\mathbb{E} Y_n = \sum_{k=1}^{n} \mathrm{P}(R_k \geq k^{\beta}) \sim K' n^{1-\alpha\beta}$. There holds

$$
(4.1) \qquad \frac{Y_n}{\mathbb{E} Y_n} \longrightarrow 1 \qquad a.s.
$$

because the events $A_k$ are independent (cf. [**3**], Thm. 1.6.8). Denote by $L_n = \max\{k \leq n : R_k \geq k^{\beta}\}$ the last epoch $k$ before $n$ when the event $A_k$ occurred. From (4.1) we conclude that

$$
1 = \lim_{n \to \infty} \frac{Y_n}{K' n^{1-\alpha\beta}} = \lim_{n \to \infty} \frac{Y_{L_n}}{K' n^{1-\alpha\beta}} = \lim_{n \to \infty} \frac{Y_{L_n}}{K' L_n^{1-\alpha\beta}} \frac{L_n^{1-\alpha\beta}}{n^{1-\alpha\beta}} = \lim_{n \to \infty} \frac{L_n^{1-\alpha\beta}}{n^{1-\alpha\beta}},
$$

i.e. $L_n/n \to 1$ a.s. Now $R_1 + \cdots + R_n \geq R_{L_n} \geq (L_n)^{\beta} \sim n^{\beta}$, giving

$$
\liminf_{n \to \infty} \frac{R_1 + \cdots + R_n}{n^{\beta}} \geq 1.
$$

This concludes the proof because $\beta$ can be chosen arbitrarily close to $1/\alpha$. $\qquad \square$

**4.2. Proof of Proposition 3.2.** The asymptotics (3.5) relies on a result from renewal theory which is available only in a time discrete setting. So let us consider the time discrete analogue of a $p$-chain: Set $\overline{p}(i,j) = p(i,j)$ for $i \neq j$ and $\overline{p}(i,i) = 1 - \sum_{j \neq i} p(i,j)$ so that $\overline{p}$ is the transition matrix of a time discrete Markov chain $(\overline{X}_n)$. Since $p$ and $\overline{p}$ obey

$$
p_t(i,j) = \sum_{n=0}^{\infty} e^{-t} \frac{t^n}{n!} \overline{p}^n(i,j).
$$

it suffices to prove that

$$
(4.2) \qquad \overline{p}^n(0,0) \sim C n^{-1/\gamma}.
$$

To conclude (3.5) from (4.2) we write $p_t(0,0) = \mathbb{E}\,\bar{p}^{Z_t}(0,0)$ with $\mathcal{L}(Z_t) = \text{Poisson}(1)$. Then we find

$$\limsup_{t \to \infty} t^{1/\gamma} p_t(0,0) \leq \limsup_{t \to \infty} t^{1/\gamma} \left\{ \mathbb{E}\,\bar{p}^{Z_t}(0,0) 1_{\{Z_t \geq (1-\varepsilon)t\}} + \mathrm{P}(Z_t < (1-\varepsilon)t) \right\}$$
$$\leq C(1-\varepsilon)^{1/\gamma}$$

for each $\varepsilon > 0$ because by Chebyshev's inequality $\mathrm{P}(|Z_t - t| > \varepsilon t) \leq \text{Var}(Z_t)/(\varepsilon^2 t^2) = 1/(\varepsilon^2 t)$. Thus

$$\limsup_{t \to \infty} t^{1/\gamma} p_t(0,0) \leq C$$

and similarly

$$\liminf_{t \to \infty} t^{1/\gamma} p_t(0,0) \geq C.$$

The proof of (4.2), in turn, relies on a result from discrete renewal theory. Let $(\overline{X}_n)$ be a $\bar{p}$-chain, denote by $T_0 = 0$, $T_i = \min\{k > T_{i-1} : \overline{X}_k = 0\}$ the epochs of returns of $(\overline{X}_n)$ to 0, let $f_k = \mathrm{P}_0(\overline{X}_1, \ldots, \overline{X}_{k-1} \neq 0, \overline{X}_k = 0)$ be the probability that the first return to 0 occurs at epoch $k$. Obviously

$$f_k = \sum_i c_\gamma i^{-\gamma} (1 - c_\gamma i^{-\gamma})^{k-2} c_\gamma i^{-\gamma}$$

is decreasing (for $k \geq 2$). Arguing as in the proof of Proposition 3.1 we can show that

$$\mathrm{P}_0(T_1 > n) \sim C_1 n^{-(1-1/\gamma)}.$$

Now consider the renewal process $T_0, T_1, \ldots$ of successive returns of $(\overline{X}_n)$ to 0.

$$\bar{p}^n(0,0) = \mathrm{P}_0(\overline{X}_n = 0) = \mathrm{P}_0(\exists k : T_k = n),$$

so that we can conclude (3.5) from the following result of Williamson.

PROPOSITION 4.2 ([10], Thm. 3-A). *Let* $T_n = R_1 + \ldots + R_n$ *where* $R_i$ *are i.i.d.,* $\mathbb{N}$-*valued,* $f_k := \mathrm{P}(R_1 = k)$. *Suppose there holds*

1. $\gcd\{k : f_k > 0\} = 1$
2. $\sum_{k>n} f_k \sim L(n) n^{-\alpha}$ *with* $0 < \alpha < 1$ *and* $L$ *slowly varying, that is the distribution of* $R_1$ *lies in the domain of attraction of a stable law with exponent* $\alpha$.
3. $f_k \geq f_{k+1}$ *for* $k \geq k_0$ *with some* $k_0 < \infty$.

*Let* $u_n := \mathrm{P}(\exists k : S_k = n)$ *be the probability of a renewal at epoch* $n$. *Then there exists a constant* $K = K(\alpha)$ *such that*

$$u_n \sim K\,L(n)/n^{1-\alpha}, \quad n \to \infty.$$

REMARK. The assumption 2. of monotonicity for $f_k$ or some other additional property is necessary in the case $\alpha \leq 1/2$. In general, the assertion of proposition 4.2 holds in this case only with *lim inf* instead of *lim* (cf. [10] or [7], p. 230). We are not aware of a continuous time analogue of the above mentioned result; therefore we resorted to a "time discrete detour" in our proof.

In order to prove (3.6) we write for $i > 0$

$$\mathrm{P}_0(X_t = i) = \int_0^t p_s(0,0) c_\gamma i^{-\gamma} e^{-c_\gamma i^{-\gamma}(t-s)} ds.$$

In view of (3.5) it then suffices to observe that for any $q \geq 0$ and $\beta \in (0,1)$

$$
\begin{aligned}
\int_0^t s^{-\beta} q e^{-q(t-s)} ds &= \int_0^t (t-s)^{-\beta} q e^{-qs} ds \\
&\leq 2^\beta t^{-\beta} + \sum_{j=1}^\infty \left( \frac{t}{2^{j+1}} \right)^{-\beta} \int_{t(1-2^{-j})}^{t(1-2^{-j-1})} q e^{-qs} ds \\
&\leq t^{-\beta} \left( 2^\beta + \sum_{j=1}^\infty 2^{-\beta(j+1)} \frac{t}{2^{j+1}} q e^{-qt(1-2^{-j})} \right) \leq C_4 \, t^{-\beta},
\end{aligned}
$$

where we can put $C_4 = 2^\beta + (\sup_{x \geq 0} x e^{-x/2}) \sum_{j=1}^\infty 2^{-(\beta+1)(j+1)} < \infty$.

## 5. Thanks

Matthias Birkner thanks Donald Dawson for making it possible for him to come to Canada in spring 1998. Anton Wakolbinger says thanks to Don for friendship, encouragement and inspiration.

## References

[1] J.T. Cox, K. Fleischmann, A. Greven, A Comparison of Interacting diffusions and an application to their ergodic theory, Prob. Theory and Related Fields **105**, 513-528 (1996).

[2] D.A. Dawson, Measure-valued Markov Processes, in: Ecole d'Eté de Probabilités de Saint Flour XXI, P.L. Hennequin (ed), Lecture Notes in Mathematics 1541, Springer 1993.

[3] R. Durrett, *Probability: Theory and Examples*, 2nd ed., Duxbury Press, 1996.

[4] W. Feller, *An Introduction to Probability Theory and its Applications*, Vol. II, 2nd Ed., John Wiley & Sons, New York, 1966

[5] A. Greven, A. Klenke, A. Wakolbinger, The longtime behaviour of branching random walk in a catalytic medium, Preprint.

[6] L. Gorostiza, S. Roelly-Coppoletta, A. Wakolbinger, Sur la persistance du processus de Dawson-Watanabe stable. L'interversion de la limite en temps et de la renormalisation, Séminaire de Probabilités XXIV, J. Azéma et al (eds), Lecture Notes in Mathematics 1426, Springer 1990.

[7] A. Garsia, J. Lamperti, A discrete renewal theorem with infinite mean, Comment. Math. Helv. **37** 221-234 (1962/63)

[8] J.A. López-Mimbela, A. Wakolbinger, Which critically branching populations persist? In: Classical and Modern Branching Processes, K. B. Athreya and P. Jagers (eds), IMA Volumes in Math. and its Appl. 84, pp 203-216, Springer 1997.

[9] T. Shiga, An interacting system in population genetics, J. Math. Kyoto Univ. **20** 214-242 (1980)

[10] J.A. Williamson, Random walks and Riesz kernels, Pacific Journ. Math. **25**, 393-415 (1968)

JOHANN WOLFGANG GOETHE-UNIVERSITÄT, FACHBEREICH MATHEMATIK, 60054 FRANKFURT AM MAIN, GERMANY
*E-mail address:* `birkner@math.uni-frankfurt.de`

JOHANN WOLFGANG GOETHE-UNIVERSITÄT, FACHBEREICH MATHEMATIK, 60054 FRANKFURT AM MAIN, GERMANY
*E-mail address:* `wakolbinger@math.uni-frankfurt.de`

Canadian Mathematical Society
Conference Proceedings
Volume **26**, 2000

# Fourier Analysis Applied to Super Stable and Related Processes

## Douglas Blount and Amit Bose

*Dedicated to Professor Donald A. Dawson*

ABSTRACT. We use the Fourier transform to embed the $(\alpha, d, 1)$ superprocess in appropriate Sobolev spaces and obtain pathwise regularity results using maximal inequalities for the expected value of the supremum of "Ornstein-Uhlenbeck like" processes. Our techniques also give simple proofs of fluctuation theorems for the Brownian density process and a rescaling of the superprocess.

## 1. Hilbert space regularity results

Among the results in a fundamental paper, Dawson [1972], the following situation was considered.

$H$ is a Hilbert space with orthonormal basis $\{e_k\}_1^\infty$. $A$ is a linear operator satisfying $Ae_k = -\lambda_k e_k$ where, for all large $k$, there exist $a, b, \delta > 0$ with $ak^{1+\delta} \leq \lambda_k \leq bk^{1+\delta}$. $\{b_k(t)\}_1^\infty$ is a sequence of standard Brownian motions and

$$X(t) = \sum_{k=1}^\infty x_k(t)e_k$$

where

$$x_k(t) = \int_0^t e^{-\lambda_k(t-s)} db_k(s)$$

is an Ornstein-Uhlenbeck process. Since $E[x_k^2(t)] = (1 - e^{-2\lambda_k t})/(2\lambda_k)$, $P(X(t) \in H) = 1$ for each fixed $t \geq 0$. In fact, noting $(I - A)^\gamma e_k = (1 + \lambda_k)^\gamma e_k$, it's true that $P((I - A)^\gamma X(t) \in H) = 1$ if $\gamma < \delta/[2(1 + \delta)]$. Using a result of Newell [1962] which provides asymptotic estimates on the tail probabilities of 1-dimensional Ornstein-Uhlenbeck processes, Dawson showed that for any $\delta' < \delta$ and $T > 0$,

$$\sum_{k=1}^\infty P\left(\sup_{0 \leq t \leq T} x_k^2(t) \geq k^{-(1+\delta')}\right) < \infty;$$

2000 *Mathematics Subject Classification.* Primary 60H15, 60G17; Secondary 60G20, 60G57.

*Key words and phrases.* super processes; stable processes; Fourier transforms; Hilbert space regularity.

Research supported in part by NSERC, Canada.

with Borel-Cantelli this shows that

$$P((I - A)^\gamma X \in C([0, \infty) : H)) = 1$$

if $\gamma < \delta/[2(1 + \delta)]$.

In Blount and Bose [1997] the outline of Dawson's approach was applied to obtain Hilbert space regularity for the $(\alpha, d, 1)$ superprocess (see Dawson [1993]) by using Fourier transform techniques. The problem of having a continuous spectrum in contrast to the discrete case just considered was dealt with by breaking the parameter space into suitable blocks and obtaining a system of stochastic differential equations. These have a structure somewhat similar to the equations resulting from applying Ito's formula to $x_k^2(t)$, the square of an Ornstein-Uhlenbeck process. A maximal inequality was developed for the tail probabilities of the supremum over $t \in [0, T]$ of the processes; and a Borel-Cantelli argument yielded regularity results by embedding the Fourier transform of the superprocess in an appropriate Hilbert space. In this paper we reprove one of the results of Blount and Bose using a different argument (Theorem 1.5). We develop two maximal inequalities (Lemma 1.3) for the expected supremum of "Ornstein-Uhlenbeck like" processes and use these in place of estimating tail probabilities. We also prove a result showing on a pathwise basis that the superprocess cannot have point masses for $t > 0$. Although this has been proved by other means, it follows naturally from our results. In addition we rescale the superprocess to obtain a fluctuation theorem. We also prove a fluctuation theorem for the Brownian density process. Both of these are done very simply and naturally using Fourier transform techniques and Sobolev spaces as state spaces.

We now introduce some basic notation. Let $\mathcal{M}_\mathcal{F}$ be the space of finite positive measures on $R^d$ topologized using the Prohorov metric, $d$ (see Ethier and Kurtz [1986]). $C([0, \infty) : \mathcal{M}_\mathcal{F})$ is the set of continuous $\mathcal{M}_\mathcal{F}$-valued processes with topology defined by $\nu_n(\cdot) \to \nu(\cdot)$ if and only if $\sup_{0 \le t \le T} d(\nu_n(t), \nu(t)) \to 0$ for each $T > 0$. Our probability space, $(\Omega, P)$, consists of $\Omega = C([0, \infty) : \mathcal{M}_\mathcal{F})$ with $P$ being the distribution of the $(\alpha, d, 1)$ superprocess (which is subsequently defined).

For $x, \theta \in R^d$, let $e_{-\theta}(x) = e^{-i\theta \cdot x}$, and, for $\nu \in \mathcal{M}_\mathcal{F}$, $\hat{\nu}(\theta) = \nu(e_{-\theta}) = \int_{R^d} e_{-\theta}(x)\nu(dx)$. For $\gamma \in R$, let $H_\gamma = \{g : \int_{R^d} |g(\theta)|^2 (1 + |\theta|^2)^\gamma d\theta < \infty\}$; here $|\cdot|$ denotes the modulus of a complex number. $C([0, \infty) : H_\gamma)$ is the space of continuous $H_\gamma$-valued processes topologized by the seminorms $\sup_{0 \le t \le T} \|h(t)\|_\gamma$ for $T > 0$. Note that $H_0 = L_2(R^d)$, and that for $\gamma < -(d/2)$, $C([0, \infty) : \mathcal{M}_\mathcal{F})$ embeds continuously into $C([0, \infty) : H_\gamma)$ by identifying $\{\nu(t)\}_{t \ge 0}$ with $\{\hat{\nu}(t, \theta)\}_{t \ge 0, \theta \in R^d}$. We use the notation $\nu(f) = \int_{R^d} f(x)\nu(dx)$ for $\nu \in \mathcal{M}_\mathcal{F}$ and $f$ an appropriate function.

For $0 < \alpha \le 2$, let $A_\alpha$ be the fractional Laplacian satisfying $A_\alpha e_{-\theta} = -|\theta|^\alpha e_{-\theta}$ ($|\theta|$ is the Euclidean norm).

If $X(0) \in \mathcal{M}_F$ and $E[X(0, 1)] < \infty$, then for any bounded continuous function $f$ with two bounded continuous derivatives, the $(\alpha, d, 1)$ superprocess has sample paths in $C([0, \infty) : \mathcal{M}_F)$ and solves the martingale problem

$$(1.1) \qquad X(t, f) = X(0, f) + \int_0^t X(s, A_\alpha f)ds + M(t, f)$$

where $M(\cdot, f)$ is a continuous, square-integrable martingale with respect to $\sigma(X(s) : s \leq t)$ and has quadratic variation process

$$(1.2) \qquad [M(\cdot, f)](t) = \int_0^t X(s, f^2)ds.$$

Letting $\hat{X}(t, \theta) = X(t, e_{-\theta})$ and $\hat{M}(t, \theta) = M(t, e_{-\theta})$, we obtain

$$(1.3) \qquad \hat{X}(t, \theta) = \hat{X}(0, \theta) - |\theta|^\alpha \int_0^t \hat{X}(s, \theta)ds + \hat{M}(t, \theta),$$

where $\hat{M}(t, \theta)$ is a complex martingale satisfying

$$(1.4) \qquad [\text{Re } \hat{M}(\cdot, \theta)](t) = \int_0^t X(s, \cos^2[\theta \cdot (\cdot)])ds$$

and

$$(1.5) \qquad [\text{Im } \hat{M}(\cdot, \theta)](t) = \int_0^t X(s, \sin^2[\theta \cdot (\cdot)])ds.$$

Using variation of constants we rewrite (1.3) as

$$(1.6) \qquad \hat{X}(t, \theta) = e^{-|\theta|^\alpha t}\hat{X}(0, \theta) + \int_0^t e^{-|\theta|^\alpha(t-s)}d\hat{M}(s, \theta).$$

Letting $S_\alpha(t)$ be the Feller semigroup generated by $A_\alpha$ and noting $S_\alpha(t)e_{-\theta} = e^{-|\theta|^\alpha t}e_{-\theta}$, we also denote (1.6) by

$$(1.7) \qquad X(t) = S_\alpha(t)X(0) + \int_0^t S_\alpha(t-s)dM(s),$$

and we set

$$Y(t) = \int_0^t S_\alpha(t-s)dM(s).$$

We can interpret each term in (1.7) as a measure (or signed measure) which has a Fourier transform given by the corresponding term in (1.6). Also (1.7) can be interpreted as holding in $C([0, T] : H_\gamma)$ for any $\gamma < -(d/2)$ and $T > 0$. Likewise (1.3) can be interpreted as

$$(1.8) \qquad X(t) = X(0) + \int_0^t A_\alpha X(s)ds + M(t),$$

which holds in $C([0, T] : H_\gamma)$ for $\gamma < -(d/2)$ and $T > 0$; this is because the equation is an identity for the Fourier transforms, and all but the integral term are in this space (so all terms are). That $M$ is in this space follows from Doob's quadratic maximal inequality and continuity in $t$ (for fixed $\theta$) of $\hat{M}(t, \theta)$. The regularity of (1.7) can be considerably improved using subsequent Lemma 1.3.

Note that $\hat{X}(t, \theta)$ is continuous in each variable separately. The following lemma shows $\hat{X}$ is jointly continuous in $t$ and $\theta$, and by (1.3) and (1.6) this also holds for $\hat{M}$ and $\hat{Y}$.

LEMMA 1.1. *If $\nu \in C([t_1, t_2] : \mathcal{M}_\mathcal{F})$, then $\hat{\nu}(t, \theta)$ is jointly continuous in $t$ and $\theta$.*

PROOF. Consider

$$
\begin{aligned}
|\hat{\nu}(t,\theta) - \hat{\nu}(t_0,\theta_0)| &\leq |\hat{\nu}(t,\theta) - \hat{\nu}(t,\theta_0)| + |\hat{\nu}(t,\theta_0) - \hat{\nu}(t_0,\theta_0)| \\
&\leq \sup_{t_1 \leq t \leq t_2} \nu(t, |e_{\theta-\theta_0} - 1|) + |\hat{\nu}(t,\theta_0) - \hat{\nu}(t_0,\theta_0)|.
\end{aligned}
$$

If $(t,\theta) \to (t_0,\theta_0)$, the second term goes to 0 by assumption. Note $\{\nu(t)\}_{t_1 \leq t \leq t_2}$ is a compact set in $\mathcal{M}_{\mathcal{F}}$ and therefore is tight. Also,

$$
|e_{\theta-\theta_0}(x) - 1| \leq (|\theta - \theta_0||x|) \wedge 2.
$$

These facts imply the first term converges to 0 if $\theta \to \theta_0$.  □

LEMMA 1.2. *If $m(t)$ is a continuous, mean 0, martingale with quadratic variation process*

$$
[m](t) = \int_0^t h(s)ds \leq c(t),
$$

*where $c(t)$ is deterministic, then, for $a \geq 0$,*

$$
P(\sup_{0 \leq t \leq \mu} |m(t)| \geq a) \leq 2\exp\left(-\frac{a^2}{2c(\mu)}\right).
$$

PROOF. For $\theta \in R$,

$$
E\left[e^{\theta m(\mu) - (\theta^2 c(\mu)/2)}\right] \leq E\left[e^{\theta m(\mu) - (\theta^2/2)\int_0^\mu h(s)ds}\right] = 1,
$$

where the equality follows from Novikov's theorem. Thus $E[e^{\theta m(\mu)}] \leq e^{\theta^2 c(\mu)/2}$.
    If $\theta > 0$ and $a \geq 0$,

$$
\begin{aligned}
P\left(\sup_{0 \leq t \leq \mu} m(t) \geq a\right) &\leq E\left[e^{\theta m(\mu)}\right] e^{-\theta a} \\
&\leq e^{((c(\mu)\theta^2)/2) - \theta a} \\
&\leq \exp\left(\frac{-a^2}{2c(\mu)}\right),
\end{aligned}
$$

using Doob's submartingale inequality, Markov's inequality, and minimizing over $\theta$. The same argument with $-m(t)$ proves the lemma.  □

LEMMA 1.3. *Assume $m(t)$ is a continuous, mean 0, martingale with quadratic variation process*

$$
[m](t) = \int_0^t g(s)ds.
$$

(a) *If $\sup_{0 \leq s \leq T} g(s) \leq c$, a deterministic constant, then*

$$
y(t) = \int_0^t e^{-\beta(t-s)}dm(s)
$$

  *satisfies*

$$
E\left[\sup_{0 \leq t \leq T} y^2(t)\right] \leq \begin{cases} 16cT & \text{if } \beta \geq 0 \\ \dfrac{8ce^{4T}\log(4\beta)}{\beta} & \text{if } \beta \geq 1. \end{cases}
$$

(b) *If $p \geq 1$ and $\sup_{0 \leq s \leq T} E[g^p(s)] < \infty$, then*

$$
E\left[ \sup_{0 \leq t \leq T} y^{2p}(t) \right] \leq \begin{cases} c(p)T^p \sup_{0 \leq t \leq T} E[g^p(t)] & \text{if } \beta \geq 0 \\[2mm] \dfrac{c(p)e^{4pT} \sup_{0 \leq t \leq T} E[g^p(t)]}{\beta^{p-1}} & \text{if } \beta \geq 1; \end{cases}
$$

*and*

$$
E\left[ \sup_{0 \leq t \leq T} y^2(t) \right] \leq c(p,T) \sup_{0 \leq t \leq T} (E[g^p(t)])^{1/p}(1 \wedge \beta^{-(1-(1/p))}).
$$

PROOF. Consider $y(t)$ restricted to $a \leq t \leq b$ and note

$$
y(t) = e^{-\beta t}\left[ \int_0^a e^{\beta \mu} dm(\mu) + \int_a^t e^{\beta \mu} dm(\mu) \right].
$$

Thus

$$
(*) \quad \sup_{a \leq t \leq b} |y(t)| \leq \left| \int_0^a e^{-\beta(a-\mu)} dm(\mu) \right| + \sup_{a \leq t \leq b} \left| \int_a^t e^{\beta(\mu-a)} dm(\mu) \right|.
$$

First we prove (a).

Applying Lemma 1.2 to each term shows

$$
P\left( \sup_{a \leq t \leq b} |y(t)| \geq q \right) \leq 2\left[ \exp\left[ \frac{-(q/2)^2}{2c\left( \frac{1-e^{-2\beta a}}{2\beta} \right)} \right] + \exp\left[ \frac{-(q/2)^2}{2c\left( \frac{e^{2\beta(b-a)}-1}{2\beta} \right)} \right] \right]
$$

$$
\leq 4\exp\left[ \frac{-\beta q^2}{4ce^{2\beta(b-a)}} \right].
$$

Assume $\beta \geq 1$ and choose $n$ such that $n \leq \beta < n+1$. Set $t_k = (kT/n)$ for $0 \leq k \leq n$. Then our last result implies

$$
P\left( \sup_{t_k \leq t \leq t_{k+1}} |y(t)| \geq q \right) \leq 4\exp\left[ \frac{-\beta q^2}{4ce^{4T}} \right].
$$

Thus, since $n \leq \beta$,

$$
P\left( \sup_{0 \leq t \leq T} |y(t)| \geq q \right) \leq \sum_{k=0}^{n-1} P\left( \sup_{t_k \leq t \leq t_{k+1}} |y(t)| \geq q \right)
$$

$$
\leq 4\beta \exp\left[ \frac{-\beta q^2}{4ce^{4T}} \right].
$$

On the left hand side we can square $y$ and $q$ to obtain, for any $q$,

$$
P\left( \sup_{0 \leq t \leq T} y^2(t) \geq q^2 \right) \leq 4\beta \exp\left[ \frac{-\beta q^2}{4ce^{4T}} \right]
$$

Replacing $q^2$ by $a$ we obtain the tail estimate

$$
P\left( \sup_{0 \leq t \leq T} y^2(t) \geq a \right) \leq 1 \wedge \left( 4\beta \exp\left[ \frac{-\beta a}{4ce^{4T}} \right] \right) \quad \text{for } a \geq 0.
$$

We have, for any $q \geq 0$,

$$
\begin{aligned}
E\left[\sup_{0 \leq t \leq T} y^2(t)\right] &= \int_0^\infty P\left(\sup_{0 \leq t \leq T} y^2(t) \geq a\right) da \\
&\leq \int_0^q 1 \, da + \int_q^\infty 4\beta \exp\left[\frac{-\beta a}{4ce^{4T}}\right] da \\
&= q + 16ce^{4T} \exp\left[\frac{-\beta q}{4ce^{4T}}\right].
\end{aligned}
$$

Minimizing over $q$ (basic calculus) gives the bound of

$$
E\left[\sup_{0 \leq t \leq T} y^2(t)\right] \leq 4ce^{4T}\left(\frac{1 + \log(4\beta)}{\beta}\right) \leq \frac{8ce^{4T}\log(4\beta)}{\beta}.
$$

Now assume $\beta \geq 0$. Then, using integration by parts,

$$
y(t) = m(t) - \beta \int_0^t m(s)e^{-\beta(t-s)} ds.
$$

Thus $|y(t)| \leq 2\sup_{0 \leq \mu \leq t} |m(\mu)|$ and the bound follows from Doob's maximal inequality. This proves (a).

Now consider (b) and assume $\beta \geq 1$. Applying Burkholder's inequality to $(*)$ followed by Jensen's inequality shows

$$
\begin{aligned}
E\left[\sup_{a \leq t \leq b} |y(t)|^{2p}\right] &\leq c(p)E\left[\left(\int_0^a e^{-2\beta(a-\mu)}g(\mu)d\mu\right)^p + \left(\int_a^b e^{2\beta(\mu-a)}g(u)du\right)^p\right] \\
&\leq c(p)\left[\left(\frac{1 - e^{-2\beta a}}{2\beta}\right)^p \sup_{0 \leq \mu \leq a} E[g^p(u)]\right. \\
&\qquad + \left.\left(\frac{e^{2\beta(b-a)} - 1}{2\beta}\right)^p \sup_{a \leq \mu \leq b} E[g^p(u)]\right] \\
&\leq c(p) \sup_{0 \leq u \leq b} E[g^p(u)]e^{2p\beta(b-a)}\beta^{-p}.
\end{aligned}
$$

Choosing $n$ and $\{t_k\}_{k=0}^n$ as in the proof of (a), we obtain

$$
E\left[\sup_{t_k \leq t \leq t_{k+1}} y^{2p}(t)\right] \leq c(p) \sup_{0 \leq t \leq T} E[g^p(t)]e^{4pT}\beta^{-p}.
$$

Thus, recalling $n \leq \beta$,

$$
\begin{aligned}
E\left[\sup_{0 \leq t \leq T} y^{2p}(t)\right] &\leq \sum_{k=0}^{n-1} E\left[\sup_{t_k \leq t \leq t_{k+1}} y^{2p}(t)\right] \\
&\leq c(p)e^{4pT} \sup_{0 \leq t \leq T} E[g^p(t)]\beta^{-(p-1)}.
\end{aligned}
$$

Now assume $\beta \geq 0$. As observed in the proof of (a), $|y(t)| \leq 2\sup_{0 \leq \mu \leq t} |m(\mu)|$ and the bound in (b) for $\beta \geq 0$ follows from Burkholder's inequality.

The inequality for $E[\sup_{0 \leq t \leq T} y^2(t)]$ follows from the first result and Jensen's inequality. $\qquad \square$

LEMMA 1.4. *Assume* $E[X(0,1)] < \infty$.

(a) *If* $\tau_n = \inf\{t : X(t,1) \geq n\}$ *for* $n = 1, 2, \ldots$, *and* $Y_n(t) = \int_0^t S_\alpha(t-s)dM(s \wedge \tau_n)$, *then, for* $T > 0$ *and* $\gamma < (\alpha - d)/2$,

$$\int_{R^d} E\left[\sup_{0 \leq t \leq T} |\hat{Y}_n(t,\theta)|^2\right](1 + |\theta|^2)^\gamma d\theta < \infty.$$

(b)
$$P\left(\int_{R^d}\sup_{0 \leq t \leq T} |\hat{Y}(t,\theta)|^2(1 + |\theta|^2)^\gamma d\theta < \infty\right) = 1$$

*if* $\gamma < (\alpha - d)/2$.

PROOF. From (1.1), since $A_\alpha 1 \equiv 0$,

$$E[M^2(t,1)] = E\left[\int_0^t X(s,1)ds\right] = tE[X(0,1)]$$

and

$$\begin{aligned}
E\left[\sup_{0 \leq t \leq T} X(t,1)\right] &\leq E[X(0,1)] + \left(E\left[\sup_{0 \leq t \leq T} M^2(t)\right]\right)^{1/2} \\
&\leq E[X(0,1)] + 2\sqrt{TE[X(0,1)]},
\end{aligned}$$

using Doob's inequality.

Thus $\tau_n \uparrow \infty$ almost surely.

Consider $\hat{Y}_n(t,\theta) = \int_0^t e^{-|\theta|^\alpha (t-s)}d\hat{M}(s \wedge \tau_n, \theta)$ and observe

$$[\mathrm{Re}\hat{M}(\cdot \wedge \tau_n, \theta)](t) = \int_0^t I_{[0,\tau_n]}(s)X(s, \cos^2[\theta \cdot (\cdot)])ds.$$

Note the integrand is dominated by $n$. An analogous result holds for $[\mathrm{Im}\hat{M}(\cdot \wedge \tau_n, \theta)](t)$. Thus, by Lemma 1.3,

$$E\left[\sup_{0 \leq t \leq T} |\hat{Y}_n(t,\theta)|^2\right] \leq nC(T)\left(1 \wedge \frac{\log(4|\theta|^\alpha)}{|\theta|^\alpha}\right).$$

This shows the integral in (a) is bounded by

$$nc(T,d)\int_0^\infty\left(1 \wedge \frac{\log r}{r^\alpha}\right)(1 + r^2)^\gamma r^{d-1}dr,$$

which is finite if $\gamma < (\alpha - d)/2$. This proves (a).

Let $\gamma < (\alpha - d)/2$ and set

$$\begin{aligned}
A_n &= \int_{R^d}\sup_{0 \leq t \leq T} |\hat{Y}_n(t,\theta)|^2(1 + |\theta|^2)^\gamma d\theta; \\
A &= \int_{R^d}\sup_{0 \leq t \leq T} |\hat{Y}(t,\theta)|^2(1 + |\theta|^2)^\gamma d\theta.
\end{aligned}$$

By (a), $P(A_n < \infty) = 1$ for all $n$, and $\tau_n \uparrow \infty$ a.s. implies

$$P(A_n \neq A \text{ infinitely often}) = 0.$$

Thus $P(A < \infty) = 1$. $\qquad\square$

THEOREM 1.5. *Assume* $E[X(0,1)] < \infty$ *and* $T > 0$. *Then*

(a)
$$P(Y \in C([0,T] : H_\gamma)) = 1 \text{ if } \gamma < \frac{\alpha - d}{2}.$$

(b) $P(X \in C([0,T] : H_\beta) \cap C((0,T] : H_\gamma)) = 1$ if $\gamma < (\alpha - d)/2$ and $\beta < -(d/2)$.

PROOF. $X(t) = S_\alpha(t)X(0) + Y(t)$. Since $|e^{-|\theta|^\alpha t}\hat{X}(0,\theta)| \leq e^{-|\theta|^\alpha t}X(0,1)$ it follows easily that (b) holds with $S_\alpha(t)X(0)$ in place of $X$. It suffices now to prove (a).

Consider, for $0 \leq s, t \leq T$,

$$\|Y(t) - Y(s)\|_\gamma^2 = \int_{R^d} |\hat{Y}(t,\theta) - \hat{Y}(s,\theta)|^2 (1 + |\theta|^2)^\gamma d\theta.$$

Almost surely, $\hat{Y}(\mu, \theta)$ is continuous in $\mu$ for each fixed $\theta$. By (b) of Lemma 1.4 we can apply the dominated convergence theorem to the last integral and obtain

$$P\left(\lim_{s \to t} \|Y(t) - Y(s)\|_\gamma = 0, 0 \leq s, t \leq T\right) = 1.$$

$\square$

By conditioning on $X(0)$, the assumption $E[X(0,1)] < \infty$ in Theorem 1.5 can be weakened to require $X(0) \in \mathcal{M}_F$. We could also have used Lemma 1.3(b) in place of Lemma 1.3(a). In particular, if $E[X(0,1)^p] < \infty$ for $p \geq 1$, then applying Burkholder's inequality shows $\sup_{0 \leq t \leq T} E[X(t,1)^p] < \infty$, and by Lemma 1.3(b),

$$
\begin{aligned}
&E\left[\sup_{0 \leq t \leq T} \|Y(t)\|_\gamma^2\right] \\
&\leq \int_{R^d} E\left[\sup_{0 \leq t \leq T} |\hat{Y}(t,\theta)|^2 (1 + |\theta|^2)^\gamma\right] d\theta \\
&\leq c(T,p) \sup_{0 \leq t \leq T} [EX(t,1)^p]^{1/p} \int_{R^d} (1 \wedge |\theta|^{-\alpha(1-(1/p))})(1 + |\theta|^2)^\gamma d\theta \\
&< \infty \text{ if } \gamma < \frac{\alpha(1 - (1/p)) - d}{2}.
\end{aligned}
$$

By conditioning on $X(0)$, we may assume $X(0,1)$ has moments of all orders, and Theorem 1.5 would follow as before. $\square$

Note that Theorem 1.5 shows that for $d = 1$ and $\alpha > 1$, almost surely for all $t > 0$ $X$ has a density that is somewhat "smoother" than functions in $H_0 = L^2(R)$. In subsequent Theorem 1.7 we show that almost surely the $(\alpha, d, 1)$ superprocess cannot have any point masses for $t > 0$ and any value of $\alpha$ and $d$. While this result would follow from known results on the Hausdorff dimension on the support of $X$, it may be of interest that it follows from basic harmonic analysis and the fact that

$$\int_{R^d} \sup_{s \leq t \leq T} |\hat{X}(t,\theta)|^2 (1 + |\theta|^2)^{-d/2} d\theta < \infty$$

almost surely for any $0 < s < T$. Also, as just discussed, the assumption $E[X(0,1)] < \infty$ could be replaced by $X(0) \in \mathcal{M}_F$.

LEMMA 1.6. (a) If $\nu \in C([s,t] : \mathcal{M}_F)$, then

$$\sup_{s \leq \mu \leq t} \sum_{a \in R^d} \nu(\mu, \{a\})^2 = \sup_{s \leq \mu \leq t} \lim_{N \to \infty} (2N)^{-d} \int_{[-N,N]^d} |\hat{\nu}(\mu, \theta)|^2 d\theta.$$

(b) *If*

$$\int_{R^d} \sup_{s \leq \mu \leq t} |\hat{\nu}(\mu,\theta)|^2 (1+|\theta|^2)^{-d/2} d\theta < \infty,$$

*then* $\sup_{s \leq \mu \leq t} \sum_{a \in R^d} \nu(\mu,\{a\})^2 = 0$.

PROOF. Let $\nu \in \mathcal{M}_F$. Note $|\hat{\nu}(\theta)|^2 = \int_{R^d \times R^d} e_\theta(y-x)\nu(dx)\nu(dy)$. Then Fubini's theorem shows

$$(2N)^{-d} \int_{[-N,N]^d} |\hat{\nu}(\theta)|^2 d\theta = \int_{R^d \times R^d} I_N(x,y)\nu(dx)\nu(dy)$$

where, for $x = (x_1, \dots, x_d)$ and $y = (y_1, \dots, y_d)$,

$$I_N(x,y) = \prod_{k=1}^{d} \frac{\sin(N(x_k - y_k))}{N(x_k - y_k)}$$

converges boundedly and pointwise to

$$I(x,y) = \begin{cases} 1 & \text{if } x = y \\ 0 & \text{else.} \end{cases}$$

This proves (a). For any $k > 0$, the right hand side of the equality in (a) can be bounded by

$$\lim_{N \to \infty} (2N)^{-d} \int_{|\theta| \leq k} \sup_{s \leq \mu \leq t} |\hat{\nu}(\mu,\theta)|^2 d\theta + c(d) \int_{|\theta| > k} \sup_{s \leq \mu \leq t} |\hat{\nu}(\mu,\theta)|^2 (1+|\theta|^2)^{-d/2} d\theta.$$

The limit term equals 0 and, under the assumption in (b), by choosing $k$ large, the second term can be made arbitrarily small. This proves (b). □

THEOREM 1.7. *If $E[X(0,1)] < \infty$ and we replace $(\Omega, P)$ by its completion, then*

$$P\left(\sum_{a \in R^d} X(t,\{a\})^2 = 0 \text{ for } t > 0\right) = 1.$$

PROOF. For $0 < s \leq t \leq T$,

$$\sup_{s \leq t \leq T} |\hat{X}(t,\theta)|^2 \leq 2\left(e^{-2|\theta|^\alpha s} X(0,1)^2 + \sup_{s \leq t \leq T} |\hat{Y}(t,\theta)|^2\right).$$

By Lemma 1.4(b),

$$(*) \quad P\left(\int_{R^d} \sup_{s \leq t \leq T} |\hat{X}(t,\theta)|^2 (1+|\theta|^2)^{-d/2} d\theta < \infty\right) = 1.$$

Thus, by Lemma 1.6,

$$(**) \quad P\left(\sup_{s \leq t \leq T} \sum_{a \in R^d} X(t,\{a\})^2 = 0\right) = 1,$$

since the event in $(**)$ contains the event in $(*)$ which is measurable by Lemma 1.1. The theorem follows by letting $s = (1/n)$, $T = n$ and letting $n \to \infty$. □

## 2. Weak convergence results

In this section we prove two fluctuation theorems using the techniques of Section 1. We rescale the $(\alpha, d, 1)$ superprocess to obtain the first result, and then examine the Brownian density process first studied by Ito [1983].

Let $X$ be the $(\alpha, d, 1)$ superprocess, and for $\epsilon > 0$, let

$$X_\epsilon(t, f) = \epsilon^d X(\epsilon^{-\alpha} t, f(\epsilon \cdot)).$$

From (1.1), we obtain the analogue of (1.3),

$$(2.1) \qquad \hat{X}_\epsilon(t, \theta) = \hat{X}_\epsilon(0, \theta) - |\theta|^\alpha \int_0^t \hat{X}_\epsilon(s, \theta) ds + \epsilon^{(d-\alpha)/2} \hat{M}_\epsilon(t, \theta),$$

where $M_\epsilon$ is a continuous, square-integrable martingale with respect to $\sigma(X_\epsilon(s) : s \leq t)$ and

$$(2.2) \qquad [M_\epsilon(\cdot, f)](t) = \int_0^t X_\epsilon(s, f^2) ds.$$

We make the following assumptions in order to prove a fluctuation theorem and law of large numbers.

$$(2.3) \qquad\qquad d > \alpha, \quad 0 < \epsilon \leq 1 \text{ and } \epsilon \to 0.$$

REMARKS. If $d = \alpha$, then essentially the same process is again obtained. If $\epsilon \to 0$ and $d < \alpha$, then assuming conditions that force convergence of $X_\epsilon(0, 1)$, only convergence to the 0 process (for $t > 0$) is obtained. This is because

$$X_\epsilon(t, 1) = X_\epsilon(0, 1) + \epsilon^{(d-\alpha)/2} M_\epsilon(t, 1) \text{ and } [M_\epsilon(\cdot, 1)](t) = \int_0^t X_\epsilon(s, 1) ds.$$

Thus $X_\epsilon(t, 1)$ is a critical branching diffusion, and, if $d < \alpha$, the variance becomes infinite as $\epsilon \to 0$, forcing extinction for $t > 0$. This can be seen from computing the Laplace transform of $X_\epsilon(t, 1)$.

LEMMA 2.1. *Assume* $p \geq 1$ *and* $\sup_\epsilon E[X_\epsilon^p(0, 1)] \leq c$. *Then*

$$\sup_\epsilon E\left[ \sup_{0 \leq t \leq T} X_\epsilon^p(t, 1) \right] \leq c(T, p).$$

PROOF. Setting $\theta = 0$ in (2.1) and using (2.2) and Burkholder's inequality, we obtain

$$E\left[ \sup_{0 \leq s \leq t} X_\epsilon^p(s, 1) \right] \leq 2^{p-1} \left[ E[X_\epsilon^p(0, 1)] + c(p) E\left[ \left( \epsilon^{d-\alpha} \int_0^t X_\epsilon(s, 1) ds \right)^{p/2} \right] \right].$$

The result then follows from basic computations and Gronwall's inequality.  □

By analogy with (1.7), we can write

$$(2.4) \qquad\qquad X_\epsilon(t) = S_\alpha(t) X_\epsilon(0) + \epsilon^{(d-\alpha)/2} Y_\epsilon(t),$$

where

$$Y_\epsilon(t) = \int_0^t S_\alpha(t - s) dM_\epsilon(s).$$

LEMMA 2.2. *If* $\sup_\epsilon E[X_\epsilon(0,1)] < \infty$, *then*

(a) *the distributions of* $M_\epsilon$ *are relatively compact on* $C([0,\infty) : H_\gamma)$ *if* $\gamma < -d/2$.

(b) *The distributions of* $Y_\epsilon$ *are relatively compact on* $C([0,\infty) : H_\gamma)$ *if* $\gamma < (\alpha - d)/2$.

PROOF. Let $F_t^\epsilon = \sigma(X_\epsilon(s) : 0 \le s \le t)$ and consider, for $0 \le t \le t + \mu \le T$ and $\mu \le \delta$,

$$
\begin{aligned}
E[\|M_\epsilon(t+\mu) - M_\epsilon(t)\|_\gamma^2 | F_t^\epsilon] &= \int_{R^d} E[|\hat{M}_\epsilon(t+\mu,\theta) - \hat{M}_\epsilon(t,\theta)|^2 | F_t^\epsilon](1+|\theta|^2)^\gamma d\theta \\
&= \int_{R^d} \int_t^{t+\mu} X_\epsilon(s,1)ds(1+|\theta|^2)^\gamma d\theta \\
&\le E\left[\left(\sup_{0 \le s \le T} X_\epsilon(s,1)\right)c(\gamma)\delta | F_t^\epsilon\right] \\
&= E[A_\epsilon(\delta,T)|F_t^\epsilon],
\end{aligned}
$$

where $A_\epsilon(\delta,T) = \sup_{0 \le s \le T} X_\epsilon(s,1)c(\gamma)\delta$ satisfies, by Lemma 2.1,

$$
\lim_{\delta \to 0} \sup_\epsilon E[A_\epsilon(\delta,T)] = 0.
$$

(a) then follows from Kurtz's tightness condition, Theorem 3.8.6 of Ethier and Kurtz [1986].

Let $\tau_n(\epsilon) = \inf\{t : X_\epsilon(t,1) \ge n\}$. Using Lemma 2.1 and Markov's inequality,

$$
P(\tau_n(\epsilon) \le T) = P(\sup_{0 \le t \le T} X_\epsilon(t,1) \ge n) \le \frac{C(T)}{n}.
$$

Thus it suffices to prove (b) with

$$
Y_{n,\epsilon}(t) = \int_0^t S_\alpha(t-s)dM_\epsilon(\tau_n(\epsilon) \wedge s)
$$

in place of $Y_\epsilon$. Consider, for $0 \le t \le t + \mu \le T$ and $\mu \le \delta$,

$$
Y_{n,\epsilon}(t+\mu) - Y_{n,\epsilon}(t) = (S_\alpha(\mu) - I)Y_{n,\epsilon}(t) + \int_t^{t+\mu} S_\alpha(t+\mu-s)dM_\epsilon(\tau_n(\epsilon) \wedge s).
$$

Thus,

$$
\begin{aligned}
E[\|Y_{n,\epsilon}(t+\mu) \; - \;\; & Y_{n,\epsilon}(t)\|_\gamma^2 | F_t^\epsilon] \\
&\le 2E\left[\int_{R^d} (e^{-|\theta|^\alpha\mu} - 1)^2 |\hat{Y}_{n,\epsilon}(t,\theta)|^2 (1+|\theta|^2)^\gamma d\theta | F_t^\epsilon\right] \\
&\quad + 2E\left[\int_{R^d} \left(\int_t^{t+\mu} e^{-2|\theta|^\alpha(t+\mu-s)} I_{[0,\tau_n(\epsilon)]}(s)X_\epsilon(s,1)ds\right)\right. \\
&\qquad\qquad \left. \times \; (1+|\theta|^2)^\gamma d\theta | F_t^\epsilon\right] \\
&\le E[A_\epsilon(\delta,T)|F_t^\epsilon]
\end{aligned}
$$

where

$$
\begin{aligned}
A_\epsilon(\delta,T) &= 2\int_{R^d} (e^{-|\theta|^\alpha\delta} - 1)^2 \left(\sup_{0 \le t \le T} |\hat{Y}_{n,\epsilon}(t,\theta)|^2\right)(1+|\theta|^2)^\gamma d\theta \\
&\quad + 2n\int_{R^d} \left(\frac{1-e^{-2|\theta|^\alpha\delta}}{2|\theta|^\alpha}\right)(1+|\theta|^2)^\gamma d\theta.
\end{aligned}
$$

Using Lemma 1.3 exactly as in the proof of Lemma 1.4(a) shows

$$E\left[\sup_{0\le t\le T}|\hat{Y}_{n,\epsilon}(t,\theta)|^2\right] \le na(T,\theta),$$

where

$$\int_{R^d} a(T,\theta)(1+|\theta|^2)^\gamma d\theta \le C(T,\gamma).$$

Thus, $\lim_{\delta\to 0}\sup_\epsilon E[A_\epsilon(T,\delta)] = 0$ by the dominated convergence theorem and (b) also follows from Kurtz's tightness criterion. □

We can now prove a law of large numbers for $X_\epsilon$.

THEOREM 2.3. *Assume* $\sup_\epsilon E[X_\epsilon(0,1)] < \infty$. *Let* $\psi(0) \in \mathcal{M}_F$ *be deterministic and assume* $X_\epsilon(0) \xrightarrow{P} \psi(0)$ *as* $\mathcal{M}_F$ *valued random variables. Then for any* $T > 0$,

$$\sup_{0\le t\le T}\|X_\epsilon(t) - S_\alpha(t)\psi(0)\|_\gamma \xrightarrow{P} 0$$

*if* $\gamma < -(d/2)$, *and, for any* $0 < s \le T$,

$$\sup_{s\le t\le T}\|X_\epsilon(t) - S_\alpha(t)\psi(0)\|_\gamma \xrightarrow{P} 0$$

*if* $\gamma < (\alpha - d)/2$.

PROOF. $\|X_\epsilon(0) - \psi(0)\|_\gamma \xrightarrow{P} 0$ if $\gamma < -(d/2)$. The result then follows from Lemma 2.2(b) and (2.4). □

We now prove a fluctuation theorem for $X_\epsilon$. For a deterministic $\psi(0) \in \mathcal{M}_\mathcal{F}$, let

(2.5) $$\psi(t) = S_\alpha(t)\psi(0),$$

and

(2.6) $$V_\epsilon(t) = \epsilon^{(\alpha-d)/2}(X_\epsilon(t) - \psi(t)).$$

Then

(2.7) $$V_\epsilon(t) = V_\epsilon(0) + \int_0^t A_\alpha V_\epsilon(s)ds + M_\epsilon(t),$$

and, from (2.4),

(2.8) $$V_\epsilon(t) = S_\alpha(t)V_\epsilon(0) + Y_\epsilon(t).$$

For the moment (2.7) and (2.8) can be considered as identities for the Fourier transforms without regard to regularity in particular spaces. We can now state a fluctuation theorem for $X_\epsilon$.

THEOREM 2.4. *Assume* $\sup_\epsilon E[X_\epsilon(0,1)^{1+\delta}] < \infty$ *for some* $\delta > 0$, *and* $V_\epsilon(0) \xrightarrow{d} V'(0)$ *in* $H_{\gamma_0}$ *for some* $\gamma_0 < -(d/2)$. *Then* $(V_\epsilon, M_\epsilon) \xrightarrow{d} (V,M)$ *in* $C([0,\infty): H_{\gamma_0} \times H_\beta) \cap C((0,\infty): H_\gamma \times H_\beta)$ *if* $\beta < -(d/2)$ *and* $\gamma < (\alpha - d)/2$ *and* $(V,M)$ *has the following properties.*

*M is a Gaussian martingale and martingale measure with respect to the filtration* $\sigma(V(s) : s \le t)$. *M has independent increments and, almost surely, sample* *paths in $C([0, \infty) : H_\beta)$ if $\beta < -(d/2)$. If $f$ is a continuous and bounded function, then $M(t, f)$ has quadratic variation process*

$$(2.9) \qquad [M(\cdot, f)](t) = \int_0^t \psi(s, f^2) ds.$$

*The equation*

$$(2.10) \qquad V(t) = S_\alpha(t)V(0) + \int_0^t S_\alpha(t - s) dM(s)$$

*holds almost surely in $C([0, \infty) : H_{\gamma_0}) \cap C((0, \infty) : H_\gamma)$ if $\gamma < (\alpha - d)/2$.*
*The equation*

$$(2.11) \qquad V(t) = V(0) + \int_0^t A_\alpha V(s) ds + M(t)$$

*holds almost surely in $C([0, \infty) : H_{\gamma_0})$. $V(0) \stackrel{d}{=} V'(0)$ and $V(0)$ is independent of $M$.*

PROOF. By Lemma 2.2, (2.7), (2.8) and the continuous mapping theorem, (2.10) and (2.11) hold for any distributional limit of $(V_\epsilon, M_\epsilon)$.

Suppose $f$ is a $C^\infty$ function with compact support, and consider $(V_\epsilon, M_\epsilon(\cdot, f))$ and

$$\left(V_\epsilon, M_\epsilon^2(\cdot, f) - \int_0^{\cdot} X_\epsilon(s, f^2) ds\right) = \left(V_\epsilon, M_\epsilon^2(\cdot, f) - \int_0^{\cdot} [\psi(s, f^2) + \epsilon^{(d-\alpha)/2} V_\epsilon(s, f^2)] ds\right).$$

By Lemma 2.1 and Burkholder's inequality, the martingales are uniformly integrable. By problem 7 in Chapter 7 of Ethier and Kurtz [1986], any distributional limit $(V, M)$ will satisfy (2.9) with $f$; by standard arguments this extends to any continuous bounded function. Since $M$ has deterministic quadratic variation, it's Gaussian with independent increments (Lévy's theorem); and the other properties follow from the quoted problem of Ethier and Kurtz. $M$ is thus unique in distribution, but integration by parts shows

$$V(t) = S_\alpha(t)V(0) + M(t) + \int_0^t A_\alpha S_\alpha(t - s) M(s) ds.$$

Thus $V$ and $M$ are unique in distribution. $\qquad \square$

We now prove a law of large numbers and fluctuation limit for the Brownian density process.

Let $\{B_k(t)\}_{k=1}^n$ be independent standard Brownian motions in $R^d$ with some initial distribution, and let

$$(2.12) \qquad X_n(t) = \frac{1}{n} \sum_{k=1}^n \delta_{B_k(t)}$$

where $\delta_a$ is the probability measure with unit mass at $a$. Let $\Delta$ be the Laplacian and $f$ a continuous, bounded function with two bounded and continuous derivatives. By Ito's formula, $X_n$ satisfies

$$(2.13) \qquad X_n(t, f) = X_n(0, f) + \int_0^t X_n\left(s, \frac{\Delta}{2} f\right) ds + n^{-1/2} M_n(t, f)$$

where $M_n(\cdot, f)$ is a continuous martingale with respect to $\sigma(X_n(s) : s \leq t)$ and has quadratic variation

$$(2.14) \qquad [M_n(\cdot, f)](t) = \int_0^t X_n(s, |\nabla f|^2)ds;$$

here $\nabla f$ is the gradient of $f$.

Letting $f = e_{-\theta}$ we obtain

$$(2.15) \qquad \hat{X}_n(t, \theta) = \hat{X}_n(0, \theta) - \frac{|\theta|^2}{2} \int_0^t \hat{X}_n(s, \theta)ds + n^{-1/2}\hat{M}_n(t, \theta),$$

where $\hat{M}_n$ is a complex martingale satisfying

$$(2.16) \qquad [\operatorname{Re} \hat{M}_n(\cdot, \theta)](t) = |\theta|^2 \int_0^t X_n(s, \sin^2[\theta \cdot (\cdot)])ds$$

and

$$(2.17) \qquad [\operatorname{Im} \hat{M}_n(\cdot, \theta)](t) = |\theta|^2 \int_0^t X_n(s, \cos^2[\theta \cdot (\cdot)])ds.$$

We also have

$$(2.18) \qquad \hat{X}_n(t, \theta) = e^{-(|\theta|^2/2)t}\hat{X}_n(0, \theta) + \int_0^t e^{-(|\theta|^2/2)(t-s)}d\hat{M}_n(s, \theta).$$

Let $S(t)$ be the Feller semigroup generated by $\Delta/2$. Then as before we denote (2.15) and (2.18) by

$$(2.19) \qquad X_n(t) = X_n(0) + \int_0^t \frac{\Delta}{2}X_n(s)ds + n^{-1/2}M_n(t)$$

and

$$(2.20) \qquad X_n(t) = S(t)X_n(0) + n^{-1/2}Y_n(t)$$

where

$$Y_n(t) = \int_0^t S(t - s)dM_n(s).$$

Since $X_n(t, 1) \equiv 1$, we do not need any moment assumptions. The proofs are essentially identical to the previous results of this section but simpler without the need to compute moments. Thus we largely just describe the results.

LEMMA 2.5. (a) *The distributions of $M_n$ are relatively compact on $C([0, \infty); H_\gamma)$ if $\gamma < -(d/2) - 1$.*

(b) *The distributions of $Y_n$ are relatively compact on $C([0, \infty) : H_\gamma)$ if $\gamma < -d/2$.*

PROOF. Exactly as for Lemma 2.2 without the need for stopping times. □

THEOREM 2.6. *If $X_n(0) \xrightarrow{P} \psi(0)$ in $\mathcal{M}_F$, then for any $T > 0$,*

$$\sup_{0 \leq t \leq T} ||X_n(t) - S(t)\psi(0)||_\gamma \xrightarrow{P} 0$$

*if $\gamma < -(d/2)$.*

PROOF. This follows from (2.20) and Lemma 2.5(b). □

For deterministic $\psi(0) \in \mathcal{M}_F$, let

(2.21)
$$\psi(t) = S(t)\psi(0),$$

and

(2.22)
$$V_n(t) = \sqrt{n}(X_n(t) - \psi(t)).$$

Then

(2.23)
$$V_n(t) = V_n(0) + \int_0^t \frac{\Delta}{2} V_n(s)ds + M_n(t)$$

and, from (2.20)

(2.24)
$$V_n(t) = S(t)V_n(0) + Y_n(t).$$

We now state a fluctuation theorem for $V_n$.

THEOREM 2.7. *Assume* $V_n(0) \xrightarrow{d} V'(0)$ *in* $H_{\gamma_0}$ *for some* $\gamma_0 < -(d/2)$. *Then* $(V_n, M_n) \xrightarrow{d} (V, M)$ *in* $C([0,\infty) : H_{\gamma_0} \times H_\beta) \cap C((0,\infty) : H_\gamma \times H_\beta)$ *if* $\beta < -(d/2)-1$ *and* $\gamma < -(d/2)$, *and* $(V, M)$ *has the following properties.*

*M is a Gaussian martingale with respect to the filtration* $\sigma(V(s) : s \leq t)$. *M has independent increments and, almost surely, sample paths in* $C([0,\infty) : H_\beta)$ *if* $\beta < (-d/2) - 1$. *If* $f$ *is continuous and bounded with one bounded, continuous derivative, then* $M(\cdot, f)$ *has quadratic variation*

$$[M(\cdot, f)](t) = \int_0^t \psi(s, |\nabla f|^2)ds.$$

*The equation*

(2.25)
$$V(t) = S(t)V(0) + \int_0^t S(t-s)dM(s)$$

*holds almost surely in* $C([0,\infty) : H_{\gamma_0}) \cap C((0,\infty) : H_\gamma)$ *if* $\gamma < -(d/2)$. *The equation*

(2.26)
$$V(t) = V(0) + \int_0^t \frac{\Delta}{2} V(s)ds + M(t)$$

*holds almost surely in* $C([0,\infty) : H_\gamma)$ *if* $\gamma < -(d/2) - 1$ *and* $\gamma \leq \gamma_0$. $V(0) \xrightarrow{d} V'(0)$ *and* $V(0)$ *is independent of* $M$.

PROOF. The proof follows from Lemma 2.5, (2.23) and (2.24), exactly as the proof of Theorem 2.4 used Lemma 2.2, (2.7) and (2.8). $\square$

The scaling used for Theorems 2.3 and 2.4 is used for proving a fluctuation theorem and law of large numbers for a more complex particle model in Dawson, Fleischmann, and Gorostiza (1989). Their initial measure is infinite and the state space is the standard space of tempered distributions. One advantage of using the Sobolev-spaces $\{H_\gamma\}$ is that it avoids technical difficulties arising from the fact that $A_\alpha g$ is not rapidly decreasing if $g$ is but $\alpha < 2$. Of course if the initial measure (such as Lebesgue measure) does not have finite total mass, its distributional Fourier transform cannot necessarily be interpreted as a function of $\theta$. In Blount and Bose (1997), this problem was avoided by first applying a suitable weighting function of the form $\varphi_p(x) = (1 + |x|^2)^{-p}$, for $p > d/2$, to give the measure finite mass. The weighted measure satisfies an equation analogous to (1.3); but additional perturbation terms arise from applying "Leibnitz's" formula to $A_\alpha(\varphi_p e_{-\theta})$. Because of technical complexity we have not used this approach here, but there doesn't appear to be any reason why it wouldn't work for the results in Section 2 with suitable initial measures of infinite total mass.

The Brownian density process was studied in Ito [1983]. As a special case of a Mckean-Vaslov limit it is examined in Kallianpur and Xiong [1995], example 8.5.3. It is also examined (and named) in Walsh [1986]. By using the spaces $\{H_\gamma\}$ we are able to obtain very precise results on the regularity of the approximating and limiting processes, and the convergence results become technically very simple.

Finally, we note that versions of Lemma 1.2 and Lemma 1.6a are standard in the literature.

## References

[1] Blount, D. and Bose, A. (1997). Hilbert space regularity of the $(\alpha, d, 1)$ superprocess and its occupation time, Technical report, *Lab. for Res. in Stats. & Prob., Carleton U./U. of Ottawa.* **#308.**

[2] Dawson, D.A. (1972). Stochastic evolution equations. *Math. Biosciences* **15** 287-316.

[3] Dawson, D.A. (1993). Measure-valued Markov processes. *L.N. Math.* **1541**, École d'été de probabilitiés de Saint-Flour XXI, 1-260.

[4] Dawson, D.A., Fleischmann, K. and Gorostiza, L.G. (1989). Stable hydrodynamic limit fluctuations of a critical branching particle system. *Ann. Prob.* **17** 1083-1117.

[5] Ethier, S.N. and Kurtz, T.G. (1986). *Markov processes, characterization and convergence.* Wiley, New York.

[6] Itô, K. (1983). Distribution-valued processes arising from independent Brownian motions. *Math. Zeit.* **182** 17-33.

[7] Kallianpur, G. and Xiong, J. (1995). Stochastic differential equations in infinite dimensional spaces. *IMS Lecture Notes* Vol. **26**.

[8] Newell, G.F. (1962). Asymptotic extreme value distribution for one-dimensional diffusion processes. *J. Math. Mech.* **11** 481-496.

[9] Walsh, J.B. (1986). An introduction to stochastic partial differential equations. *L.N. Math.* **1180**. École d'été de probabilitiés de Saint-Flour XIV, 265–439.

DEPARTMENT OF MATHEMATICS, ARIZONA STATE UNIVERSITY, TEMPE, AZ, USA
*E-mail address*: blount@math.la.asu.edu

SCHOOL OF MATHEMATICS AND STATISTICS, CARLETON UNIVERSITY, OTTAWA, ON, CANADA
*E-mail address*: abose@math.carleton.ca

Canadian Mathematical Society
Conference Proceedings
Volume **26**, 2000

# Convergence to Equilibrium and
# Linear Systems Duality

## J. Theodore Cox\*, Achim Klenke[†], and Edwin A. Perkins[‡]

ABSTRACT. For a class of interacting particle systems on a countable set, including the so-called linear systems, self-duality has proved to be a strong tool to show longtime convergence to an invariant state $\nu_\theta$ from a constant initial state $\theta$. This convergence was extended to a class of translation invariant random initial states through the Liggett-Spitzer coupling.

Here we drop the assumption of translation invariance of the initial state. Instead, we assume only that there exists a global density $\theta$ in a certain $L^2$-sense. We use the duality to carry out a comparison argument and show convergence to the same $\nu_\theta$ as above. We treat some examples in more detail: the parabolic Anderson model, the mutually catalytic branching model, and the smoothing and potlatch processes.

## 1. Introduction

Our purpose here is to present some new weak convergence results for interacting particle systems that enjoy the "linear systems" type duality as described in Chapter IX of Liggett (1985). This particular form of duality was first exploited by Spitzer (1981) in his study of the smoothing, potlatch, and coupled random walk processes. For this class of processes, duality and a simple martingale argument prove weak convergence of the process from constant initial states. To obtain convergence from nonconstant initial states, a coupling method introduced in Liggett and Spitzer (1981) can be used. The coupling method is not duality based.

The Liggett–Spitzer coupling method generally requires that the law of the initial state be translation invariant. We show here, for interacting particle systems with the linear systems type duality, that duality itself can be used to prove weak convergence of the process from some nonconstant initial states. Translation invariance of the initial state is not needed, nor is it necessary to assume translation

---

2000 *Mathematics Subject Classification.* Primary 60K35, 60J80; Secondary 60J60, 60H10.

*Key words and phrases.* linear systems; duality; comparison; potlatch process; smoothing process; parabolic Anderson model; mutually catalytic branching; convergence to equilibrium.

\*Supported in part by NSF Grant DMS-96-26675 and by an NSERC Collaborative Grant. Part of the research was done while the author was on sabbatical from Syracuse University and in residence at The University of British Columbia.

[†]Supported in part by Don Dawson's Collaborative Research Prize from the Max Planck Gesellschaft.

[‡]Supported in part by an NSERC Research Grant and NSERC Collaborative Grant.

invariance of the basic mechanism defining the process. We will apply our method to three examples: the parabolic Anderson (or linear random potential) model considered in Shiga (1992), the mutually catalytic branching process introduced in Dawson and Perkins (1998), and the smoothing and potlatch processes.

This work originated in an attempt to extend Theorem 1.4 in Dawson and Perkins (1998) to nonconstant initial states. Since the Liggett–Spitzer coupling does not seem to apply to the mutually catalytic branching process, another method was needed. We found that the model's self–duality property could be used. This duality is similar to the linear systems duality, and we found that our method could also be applied in the linear systems setting.

The heart of our proofs is a convergence in probability statement for the dual process in a fairly general setting (Theorem 3.1). Since this proposition looks rather abstract without some motivation, we start in Section 2 by giving a direct proof of convergence for the parabolic Anderson model, which is the simplest of the models we treat. We then present the general convergence result in Section 3, and use it to obtain convergence results for the mutually catalytic branching process in Section 4 and the smoothing and potlatch processes in Section 5. We note that the method actually gives more than convergence to an equilibrium state from a broad class of initial distributions—it shows that if we pick the intial state from this initial distribution, the resulting random probability measure on the state space converges in probability to the equilibrium measure as time tends to infinity. This extension is used in Cox and Klenke (1998) to show that for a class of processes (including the mutually catalytic branching model in the recurrent setting) one obtains a.s. accumulation at all points in the support of the equilibrium measure.

As will become clear when we give the details of our first example, our method does not provide any new information on the specific nature of a given weak limit. Depending on the parameters involved, such a limit may or may not be "degenerate" (e.g., concentrated on the zero configuration). It is a fundamental problem to determine which is the case. What we show, roughly speaking, is that if two initial states have the same spatial density, measured in an appropriate way, and the process starting in one of these states has a weak limit, then the process starting in the other state has the same weak limit.

We introduce now some notation that will be common to our examples. Let $S$ be a countable set, and let $p(i, j)$, $i, j \in S$ be a (discrete time) irreducible Markov chain transition matrix. We assume $p$ is doubly stochastic so that $\tilde{p}(i, j) = p(j, i)$ is also a transition matrix. Define the continuous time kernel $p_t$, $t \geq 0$ by

$$(1.1) \qquad p_t(i, j) = e^{-t} \sum_{n=0}^{\infty} \frac{t^n}{n!}\, p^{(n)}(i, j),$$

where $p^{(n)}$ is the $n$th iterate of $p$. For $\phi \colon S \to [0, \infty)$, define $P_t\phi$ and $\phi P_t$ by $P_t\phi(i) = \sum_{j \in S} p_t(i, j)\phi(j)$ and $\phi P_t(j) = \sum_{j \in S} \phi(i)p_t(i, j)$. For $\phi, \psi \colon S \to [0, \infty)$, let $\langle \phi, \psi \rangle = \sum_{i \in S} \phi(i)\psi(i)$. For $\theta \in [0, \infty)$ let $\boldsymbol{\theta} \in [0, \infty)^S$ be defined by $\boldsymbol{\theta}(j) \equiv \theta$. Let $X_F$ be the set of $x \in [0, \infty)^S$ such that $x(j) = 0$ for all but finitely many $j \in S$, and let $X_f$ denote the set of those $x \in [0, \infty)^S$ such that $\langle x, \mathbf{1} \rangle < \infty$. Unless otherwise noted, all sums will be taken over $S$. Finally, $I$ denotes the identity matrix, $\mathcal{L}$ denotes law, $\overset{\mathrm{d}}{=}$ denotes equality in distribution, and $\Rightarrow$ denotes weak convergence of probability measures.

## 2. Parabolic Anderson model

We consider an interacting diffusion $x_t$, taking values in $[0, \infty)^S$, called the parabolic Anderson model in Shiga (1992). The evolution of $x_t$ is determined by the equation

$$(2.1) \qquad dx_t(i) = (p - I)x_t(i)\,dt + cx_t(i)\,dB_t(i), \qquad i \in S.$$

Here $c$ is a fixed positive constant and $\{B_t(i), i \in S\}$ is a collection of independent one–dimensional Brownian motions. To define the state space $X$, let $\gamma \in [0, \infty)^S$ be a strictly positive, summable reference measure satisfying, for some finite constant $\Gamma$,

$$(2.2) \qquad\qquad\qquad \gamma p \le \Gamma\gamma.$$

There is always such a reference measure (see Liggett and Spitzer (1981)). Let $X = X_\gamma = \{x \in [0, \infty)^S : \langle x, \gamma \rangle < \infty\}$. We endow $X$ with the topology generated by componentwise convergence. By results of Shiga and Shimizu (1980), for each starting point, $x_0 \in X$, there is a unique strong solution $x_t$ of (2.1) taking values in $X$. We let $\mathbf{P}^{x_0}$ denote its law on $C([0, \infty), X)$. Note that $(x_t)$ is linear in the following sense: If $(x_t^1)$ and $(x_t^2)$ are solutions of (2.1) with the same $(B_t)$, and $a$ and $b$ are constants, then

$$(2.3) \qquad\qquad\qquad x_t^3 = x_t^1 + x_t^2$$

is also a solution of (2.1).

As noted in (2.5) of Cox, Greven and Shiga (1995), for $x_0 \in X_F$, $E^{x_0}[x_t(i)] = P_t x_0(i)$, and

$$
\begin{aligned}
(2.4) \quad & \mathbf{E}^{x_0}[x_t(i)x_t(j)] \\
& = P_t x_0(i) P_t x_0(j) + c^2 \int_0^t \sum_k p_{t-r}(i, k)p_{t-r}(j, k)\mathbf{E}^{x_0}[x_r^2(k)]\,dr.
\end{aligned}
$$

It follows that

$$(2.5) \qquad \mathbf{E}^{x_0}[\langle x_t, \mathbf{1}\rangle^2] = \langle x_0, \mathbf{1}\rangle^2 + c^2 \int_0^t \mathbf{E}^{x_0}[\langle x_r^2, \mathbf{1}\rangle]\,dr.$$

By standard arguments and Gronwall's inequality, for all $x_0 \in X_f$ and $t \ge 0$,

$$(2.6) \qquad\qquad\qquad \mathbf{E}^{x_0}[\langle x_t, \mathbf{1}\rangle^2] < \infty.$$

It is easy to see from (2.1) and (2.6) that for $x_0 \in X_f$, $\langle x_t, \mathbf{1}\rangle$ is a continuous, square-integrable martingale.

Let $\tilde{x}_t$ denote the process determined by (2.1), but with $\tilde{p}$ instead of $p$. Then $x_t$ and $\tilde{x}_t$ are dual in the following sense. Given initial states $x_0 \in X$ and $\tilde{x}_0 \in X_F$,

$$(2.7) \qquad\qquad\qquad \langle x_t, \tilde{x}_0 \rangle \overset{\mathrm{d}}{=} \langle x_0, \tilde{x}_t \rangle.$$

A proof of this fact follows exactly as in Theorem IX.1.25 in Liggett (1985).

Suppose now that $\theta > 0$ is fixed, and $x_t^\theta$ is the parabolic Anderson model with initial state $x_0^\theta = \boldsymbol{\theta}$. Then, for any $\phi \in X_F$, (2.7) implies that

$$(2.8) \qquad\qquad \langle x_t^\theta, \phi \rangle \overset{\mathrm{d}}{=} \theta\langle \mathbf{1}, \tilde{x}_t \rangle, \qquad \tilde{x}_0 = \phi.$$

The right side of (2.8) is a nonnegative martingale, and hence converges a.s. as $t \to \infty$. Therefore, the left side of (2.8) must converge as $t \to \infty$. On account of this, there is a probability measure $\nu_\theta$ on $X$ such that

$$(2.9) \qquad \mathcal{L}[x_t^{\boldsymbol{\theta}}] \Rightarrow \nu_\theta \qquad \text{as } t \to \infty.$$

In particular,

$$(2.10) \qquad \begin{aligned} \mathbf{E}^\phi[e^{-\theta\langle \mathbf{1}, \tilde{x}_t\rangle}] &= \mathbf{E}[e^{-\langle x_t^\theta, \phi\rangle}] \\ &\to \int e^{-\langle x, \phi\rangle} \, \nu_\theta(dx) \qquad \text{as } t \to \infty. \end{aligned}$$

Note that by the linearity property (2.3), $\nu_\theta(A) = \nu_1(\theta^{-1}A)$.

In some cases, the limit $\nu_\theta$ may be concentrated on the configuration which is identically 0. In Remark 2.5 below we recall conditions which determine, in some cases, whether or not this happens. Our goal is to show that convergence to $\nu_\theta$ holds for a large class of initial states with an appropriately defined spatial density $\theta$.

For $\theta \in [0, \infty)$, define $\mathcal{M}_\theta$ to be the collection of probability measures $\nu$ on $X$ such that

$$(2.11) \qquad \sup_k \int x^2(k) d\nu(x) < \infty,$$

$$(2.12) \qquad \lim_{t \to \infty} \int (P_t x(k) - \theta)^2 d\nu(x) = 0, \quad k \in S.$$

Suppose, for example, $P_t$ is the semigroup of simple symmetric random walk on $\mathbb{Z}^d$. If $\{x(k) : k \in \mathbb{Z}^d\}$ are iid non-negative random variables with mean $\theta$ and finite variance, then their law $\nu$ is in $\mathcal{M}_\theta$. However, there are a number of non-translation invariant laws in $\mathcal{M}_\theta$. If $x(k) = 1(k_1 < 0)$ clearly $\delta_x \in \mathcal{M}_{1/2}$. More generally, if $x \in [0, \infty)^{\mathbb{Z}^d}$ is bounded and the average value of $x$ in a ball (Euclidean metric) approaches $\theta$ as the radius approaches $\infty$, then it is not hard to show that $\delta_x \in \mathcal{M}_\theta$.

Our main technical result is the following.

PROPOSITION 2.1. *Assume $p$ is doubly stochastic. Let $\theta \in [0, \infty)$ and let $\nu \in \mathcal{M}_\theta$. If $x_0$ has law $\nu$, and $\tilde{x}_0 \in X_F$, then*

$$(2.13) \qquad \langle x_0 - \boldsymbol{\theta}, \tilde{x}_t\rangle \to 0$$

*in $\nu \otimes \mathbf{P}^{\tilde{x}_0}$-probability as $t \to \infty$.*

COROLLARY 2.2. *If $S = \mathbb{Z}^d$, $p(i,j) = p(0, j - i)$, $\nu = \mathcal{L}[x_0]$ is translation invariant, shift ergodic, and satisfies $\int x(0) \, d\nu(x) = \theta$, then (2.13) holds.*

PROOF. For $N > 0$ define $x_0^N(i) = x_0(i) \wedge N$, and let $\theta_N = \int [x(0) \wedge N] \, d\nu(x)$. Then

$$\langle x_0 - \boldsymbol{\theta}, \tilde{x}_t\rangle = \langle x_0 - x_0^N, \tilde{x}_t\rangle + \langle x_0^N - \boldsymbol{\theta}_N, \tilde{x}_t\rangle + \langle \boldsymbol{\theta}_N - \boldsymbol{\theta}, \tilde{x}_t\rangle.$$

It is easy to see (Theorem 5.6 in Liggett 1973) that $\mathcal{L}[x_0^N] \in \mathcal{M}_{\theta_N}$, so that for each $N$, $\langle x_0^N - \boldsymbol{\theta}_N, \tilde{x}_t\rangle \to 0$ in $\nu \otimes \mathbf{P}^{\tilde{x}_0}$-probability as $t \to \infty$. On the other hand, letting $\mathbf{E}$ denote expectation with respect to $\nu \otimes \mathbf{P}^{\tilde{x}_0}$,

$$\mathbf{E}[\langle x_0 - x_0^N, \tilde{x}_t\rangle] \leq \int x(0) 1_{\{x(0) > N\}} \, d\nu(x) \, \langle \mathbf{1}, \tilde{x}_0\rangle \to 0$$

and

$$\mathbf{E}[\langle \boldsymbol{\theta} - \boldsymbol{\theta}_N, \tilde{x}_t \rangle] = (\theta - \theta_N) \langle \mathbf{1}, \tilde{x}_0 \rangle \to 0$$

as $N \to \infty$.                                                               □

Let us see now what duality and (2.13) imply. Let $\tilde{x}_0 \in X_F$, and let $x_0$ have law $\nu$ which satisfies (2.13). By (2.7),

(2.14)
$$\begin{aligned}
\mathbf{E}[e^{-\langle x_t, \tilde{x}_0 \rangle}] &= \mathbf{E}^{\nu} \otimes \mathbf{E}^{\tilde{x}_0}[e^{-\langle x_0, \tilde{x}_t \rangle}] \\
&= \mathbf{E}^{\nu} \otimes \mathbf{E}^{\tilde{x}_0}[e^{-\langle x_0 - \boldsymbol{\theta}, \tilde{x}_t \rangle} e^{-\theta \langle \mathbf{1}, \tilde{x}_t \rangle}].
\end{aligned}$$

By (2.13) $e^{-\langle x_0 - \boldsymbol{\theta}, \tilde{x}_t \rangle} \to 1$ in $\mathbf{P}^{\nu} \otimes \mathbf{P}^{\tilde{x}_0}$-probability as $t \to \infty$. Therefore, in view of (2.10),

(2.15)
$$\lim_{t \to \infty} \mathbf{E}^{\nu}[e^{-\langle x_t, \phi \rangle}] = \int e^{-\langle x, \phi \rangle} \, d\nu_\theta(x),$$

and we have established the following.

THEOREM 2.3. *Assume that $p$ is doubly stochastic, and either* (i) $\mathcal{L}[x_0] \in \mathcal{M}_\theta$, *or* (ii) $S = \mathbb{Z}^d$, $p(i,j) = p(0, j - i)$, *and $\mathcal{L}[x_0]$ is translation invariant, shift ergodic, and satisfies $\mathbf{E}[x_0(0)] = \theta$. Then $\mathcal{L}[x_t] \Rightarrow \nu_\theta$ as $t \to \infty$.*

REMARK 2.4. It is possible to formulate and prove a stronger type of convergence than given above, which is used in Cox and Klenke (1998). We do this for the mutually catalytic branching process in Theorem 4.3(a) below.

REMARK 2.5. Suppose that $S = \mathbb{Z}^d$ and $p(i,j) = p(0, j - i)$. If the symmetrization of $p$ is transient, and $c$ is sufficiently small, Theorem 1.1 of Shiga (1992) implies that $\mathcal{L}[x_t] \Rightarrow \nu_\theta$ for any initial law $\nu$ which is translation invariant, shift ergodic, and satisfies $\int x(0) \, d\nu(x) = \theta$. If the symmetrization of $p$ is recurrent, Theorem 2 of Cox, Fleischmann and Greven (1996) implies that $\nu_\theta = \delta_0$, and that $\mathcal{L}[x_t] \Rightarrow \delta_0$ for any initial law $\nu$ which is translation invariant and satisfies $\int x(0) \, d\nu(x) < \infty$. For information concerning the case when the symmetrization of $p$ is transient and $c$ is large, see Theorem 1.2 in Shiga (1992).

PROOF OF PROPOSITION 2.1. Recall that the total mass process $\langle \tilde{x}_t, \mathbf{1} \rangle$ is a continuous square-integrable martingale. By (2.1) and (2.6), $\langle \tilde{x}_t, \mathbf{1} \rangle = \langle \tilde{x}_0, \mathbf{1} \rangle + \sum_i M_t(i)$, where

$$M_t(i) = c \int_0^t \tilde{x}_s(i) dB_s(i)$$

and the sum converges in $L^2$. The $M_t(i)$ are continuous, square-integrable orthogonal martingales, with

$$\langle M(i) \rangle_t = c^2 \int_0^t \tilde{x}_s^2(i) \, ds.$$

The continuous martingale $\langle \tilde{x}_t, \mathbf{1} \rangle$ has integrable square function $A_t = \sum_i \langle M(i) \rangle_t$. Since $\langle \tilde{x}_t, \mathbf{1} \rangle$ converges a.s. $\mathbf{P}^{\tilde{x}_0}$ as $t \to \infty$,

(2.16)
$$\begin{aligned}
A_\infty &= \sum_i \langle M(i) \rangle_\infty \\
&= c^2 \int_0^\infty \sum_i \tilde{x}_s(i)^2 \, ds < \infty \qquad \text{a.s. } \mathbf{P}^{\tilde{x}_0}.
\end{aligned}$$

It is straightforward to check, by applying Itô's formula to $\tilde{P}_{t-s}\tilde{x}_s(i)$, that we have the representation

$$(2.17) \qquad \tilde{x}_t(i) = \tilde{P}_t\tilde{x}_0(i) + \int_0^t \sum_j \tilde{p}_{t-s}(i,j)\,dM_s(j).$$

Let $x_0$ have law $\nu$, let $\Delta = x_0 - \boldsymbol{\theta}$, and set $\mathbf{P} = \nu \otimes \mathbf{P}^{\tilde{x}_0}$. Since $\nu \in \mathcal{M}_\theta$, $C = \sup_i \int [\Delta(i)^2]d\nu$ is finite. It follows easily that

$$(2.18) \qquad \sup_{i,t} \int \tilde{P}_t\Delta(i)^2 d\nu \le C,$$

and also that ($\mathbf{E}$ denotes expectation with respect to $\mathbf{P}$)

$$(2.19) \qquad \mathbf{E}\Big[\sum_i \int_0^t \tilde{P}_{t-s}\Delta(i)^2\tilde{x}_s(i)^2\,ds\Big] \le C\int_0^t \mathbf{E}^{\tilde{x}_0}[\langle\tilde{x}_s,1\rangle^2]\,ds < \infty.$$

Consequently, letting

$$(2.20) \qquad N_s^t = \sum_i \int_0^s \Delta\tilde{P}_{t-r}(i)\,dM_r(i), \qquad s \le t,$$

we have

$$(2.21) \qquad \langle\Delta,\tilde{x}_t\rangle = \langle\Delta,\tilde{P}_t\tilde{x}_0\rangle + N_t^t.$$

Here, $\{N_s^t, s \le t\}$ is a continuous, square-integrable martingale under $\mathbf{P}$, with square variation function

$$\langle N^t\rangle_s = \int_0^s \sum_i \Delta\tilde{P}_{t-r}(i)^2\,d\langle M(i)\rangle_r < \infty \quad \text{a.s. } \mathbf{P}^{\tilde{x}_0}.$$

Here we have added $x_0$ to the underlying filtration at time 0.

We are now ready to prove (2.13). It is straightforward to check that

$$\int \langle\Delta,\tilde{P}_t\tilde{x}_0\rangle^2 d\nu = \int \langle P_t\Delta,\tilde{x}_0\rangle^2 d\nu$$

$$= \sum_{j,k} \tilde{x}_0(j)\tilde{x}_0(k)\int P_t\Delta(j)P_t\Delta(k)\,d\nu$$

$$\le \Big[\sum_j \tilde{x}_0(j)\Big(\int P_t\Delta(j)^2\,d\nu\Big)^{1/2}\Big]^2.$$

Using (2.12), (2.18), and the fact that $\tilde{x}_0 \in X_f$, it follows that

$$(2.22) \qquad \int \langle\Delta,\tilde{P}_t\tilde{x}_0\rangle^2 d\nu \to 0 \quad \text{as } t \to \infty.$$

The next step is to show that for all $\varepsilon' > 0$,

$$(2.23) \qquad \lim_{t\to\infty} \mathbf{P}[\langle N^t\rangle_t > \varepsilon'] = 0.$$

To do this, we consider the expectation

$$\int \langle N^t\rangle_t\,d\nu = \int_0^t \sum_i \int \tilde{P}_{t-r}\Delta(i)^2\,d\nu\,d\langle M(i)\rangle_r$$

$$= \sum_i \int_0^\infty \mathbf{1}_{\{r<t\}}\int \tilde{P}_{t-r}\Delta(i)^2\,d\nu\,d\langle M(i)\rangle_r.$$

We note that $\mathbf{1}_{\{r<t\}} \int \tilde{P}_{t-r}\Delta(i)^2 \, d\nu \to 0$ as $t \to \infty$, and is bounded by $C$ (by (2.18)). Hence, on account of (2.16) and the bounded convergence theorem,

$$(2.24) \qquad \mathbf{P}^{\tilde{x}_0}\left[\lim_{t\to\infty} \int \langle N^t \rangle_t \, d\nu = 0\right] = 1.$$

Therefore,

$$(2.25) \qquad \mathbf{P}[\langle N^t \rangle_t > \varepsilon'] = \mathbf{E}^{\tilde{x}_0}\left[\nu[\langle N^t \rangle_t > \varepsilon']\right] \to 0$$

as $t \to \infty$, which proves (2.23) .

By a standard stopping time argument for martingales, for any $\varepsilon > 0$,

$$(2.26) \qquad \mathbf{P}[|N_t^t| > \varepsilon, \langle N^t \rangle_t \leq \varepsilon^3] \leq \varepsilon.$$

The claim (2.13) now follows from (2.21)–(2.23) and (2.26) (set $\varepsilon' = \varepsilon^3$). $\qquad \square$

## 3. A convergence result for the dual process

The key step in the proof of the convergence to equilibrium for the parabolic Anderson model was the convergence in probability statement for the dual process given in Proposition 2.1. As the argument leading to that conclusion applies in a number of different settings exhibiting a similar duality, we now prove a general result. Recall that $P_t$ is the continuous time transition matrix associated with a doubly stochastic matrix $p(i,j)$ on a countable set $S$. Let $\{S_n\}$ be a sequence of finite sets which increase to $S$ as $n \to \infty$. We endow the set $[0,\infty)^S$ with the topology of pointwise convergence.

Let $\{x_t(i) : t \geq 0, i \in S\}$ be a right-continuous with left limits (RCLL) stochastic process taking values in $[0,\infty)^S$, defined on some probability space $(\Omega, \mathcal{F}, \mathbf{P}^x)$. Assume

$$(3.1) \qquad X_t \equiv \langle x_t, \mathbf{1} \rangle \text{ is a square-integrable RCLL martingale,}$$

and there is a family of RCLL $L^2$-martingales $\{M_t(i), t \geq 0, i \in S\}$ such that, a.s. for all $i \in S$ and $t \geq 0$,

$$(3.2) \quad x_t(i) = P_t x_0(i) + \sum_j \int_0^t P_{t-s}(i,j) \, dM_s(j) = P_t x_0(i) + \left[\int_0^t P_{t-s} \, dM_s\right](i).$$

Here it is understood that the series in (3.2) converges in $L^2$. For $\phi : S \to \mathbb{R}$ let

$$(3.3) \qquad Q_t(\phi) = \sum_{i,j} \phi(i)\phi(j)\langle M(i), M(j) \rangle_t,$$

$$(3.4) \qquad |Q|_t(\phi) = \sum_{i,j} \phi(i)\phi(j)|\langle M(i), M(j) \rangle|_t,$$

where $|\langle M(i), M(j) \rangle|_t$ is the total variation of $\langle M(i), M(j) \rangle$ up to time $t$. We will also use the notation

$$\int_0^t \phi \, dQ_t = \int_0^t \sum_{i,j} \phi(i)\phi(j) \, d\langle M(i), M(j) \rangle_t,$$

$$\int_0^t \phi \, d|Q|_t = \int_0^t \sum_{i,j} \phi(i)\phi(j) \, d|\langle M(i), M(j) \rangle|_t.$$

To ensure these expressions make sense, at least for bounded $\phi$, we assume

$$(3.5) \qquad \mathbf{E}[|Q|_t(\mathbf{1})] < \infty \qquad \text{for all } t \geq 0.$$

Here is our main convergence result.

THEOREM 3.1. *In addition to* (3.1)–(3.5), *assume that*

(3.6) $$|Q|_\infty(\mathbf{1}) < \infty \quad a.s.$$

*Let $\nu$ be a probability measure on $\mathbb{R}^S$ such that*

(3.7)
$$C(\nu) \equiv \sup_{k \in S} \int \phi(k)^2 \, d\nu(\phi) < \infty,$$
$$\lim_{t \to \infty} \int (\phi P_t(k))^2 \, d\nu(\phi) = 0 \quad \text{for all } k \in S.$$

*Then $\langle \phi, x_t \rangle \to 0$ in $\nu \otimes \mathbf{P}^x$-probability as $t \to \infty$.*

PROOF. Let $\mathbf{P} = \nu \otimes \mathbf{P}^x$, and let $(\phi, \omega)$ denote a generic point in $\mathbb{R}^S \times \Omega$. Let $N_s^t(i) = \left[\int_0^s P_{t-r} dM_r\right](i), s \le t$, so that we may write $x_t = P_t x_0 + N_t^t$. By (3.2),

$$\langle \phi \mathbf{1}_{S_n}, x_t \rangle = \langle \phi \mathbf{1}_{S_n}, P_t x_0 \rangle + \langle \phi \mathbf{1}_{S_n}, N_t^t \rangle.$$

As $n \to \infty$, the left side and first term on the right side above converge in $L^2(\mathbf{P})$ to $\langle \phi, x_t \rangle$ and $\langle \phi, P_t x_0 \rangle$, respectively. This is because, for $m < n$,

$$\mathbf{E}\left[\langle \phi(\mathbf{1}_{S_n} - \mathbf{1}_{S_m}), x_t \rangle^2 + \langle \phi(\mathbf{1}_{S_n} - \mathbf{1}_{S_m}), P_t x_0 \rangle^2\right]$$
$$\le C(\nu) \mathbf{E}^x\left[\langle \mathbf{1}_{S_n} - \mathbf{1}_{S_m}, x_t \rangle^2 + \langle \mathbf{1}_{S_n} - \mathbf{1}_{S_m}, P_t x_0 \rangle^2\right],$$

and by (3.1), the right side above tends to 0 as $m, n \to \infty$. Therefore,

$$\langle \phi, N_t^t \rangle = \lim_{n \to \infty} \langle \phi \mathbf{1}_{S_n}, N_t^t \rangle \quad \text{in } L^2$$

exists, and by Doob's strong $L^2$ inequality, $\mathbf{E}[\sup_{s \le t} |\langle \phi \mathbf{1}_{S_n}, N_s^t \rangle - \langle \phi, N_s^t \rangle|^2] \to 0$ as $n \to \infty$. Consequently, $\langle \phi, N_s^t \rangle$ is a RCLL $L^2$-martingale, and its predictable square function is the $L^2$-limit of $\sum_{j,j'} \int_0^s (\phi \mathbf{1}_{S_n} P_{t-r})(j)(\phi \mathbf{1}_{S_n} P_{t-r})(j') d\langle M(j), M(j') \rangle_r$, given by

(3.8) $$\sum_{j,j'} \int_0^s \phi P_{t-r}(j) \phi P_{t-r}(j') d\langle M(j), M(j') \rangle_r = \int_0^s (\phi P_{t-r}) \, dQ_r.$$

(The $L^2$ limit follows easily from (3.5), (3.7) and the dominated convergence theorem.) To sum up, we have established that

(3.9) $$\langle \phi, x_t \rangle = \langle \phi, P_t x_0 \rangle + \langle \phi, N_t^t \rangle,$$

where $\{\langle \phi, N_s^t \rangle, s \le t\}$ is a RCCL $L^2$-martingale, with predictable square function $\langle\langle \phi, N^t \rangle\rangle_s = \int_0^s (\phi P_{t-r}) \, dQ_r$.

We must show both terms on the right side of (3.9) tend to 0 in $\mathbf{P}$-probability as $t \to \infty$. The first term is easy. By (3.7),

(3.10) $$\sup_{k \in S} \int \phi P_t(k)^2 \, d\nu(\phi) \le C(\nu).$$

Therefore, we have

$$\int \langle \phi, P_t x_0 \rangle^2 \, d\nu(\phi) = \int \langle \phi P_t, x_0 \rangle^2 \, d\nu(\phi)$$

$$= \sum_{j,k} x_0(j) x_0(k) \int (\phi P_t)(j)(\phi P_t)(k) \, d\nu(\phi)$$

(3.11)

$$\leq \left[ \sum_j x_0(j) \left( \int (\phi P_t(j))^2 \, d\nu(\phi) \right)^{1/2} \right]^2$$

$$\to 0$$

as $t \to \infty$ by (3.1), (3.7) and dominated convergence.

To handle the martingale term in (3.9), we have

$$\int \langle\langle \phi, N^t \rangle\rangle_t \, d\nu(\phi) = \int \int_0^t (\phi P_{t-r}) dQ_r \, d\nu(\phi)$$

$$\leq \int_0^\infty \mathbf{1}_{\{r<t\}} \left( \int (\phi P_{t-r})^2 \, d\nu(\phi) \right)^{1/2} d|Q|_r.$$

Note that $\mathbf{1}_{\{r<t\}}[\int (\phi P_{t-r}(k))^2 \, d\nu(\phi)]^{1/2} \to 0$ pointwise as $t \to \infty$ by (3.7), and by (3.10) is bounded by $C(\nu)^{1/2}$. Therefore, by (3.6) and dominated convergence,

(3.12) $$\lim_{t \to \infty} \int \langle\langle \phi, N^t \rangle\rangle_t \, d\nu(\phi) = 0 \qquad \mathbf{P}^x\text{-a.s.}$$

Using this fact and a standard stopping time argument for martingales, we have

$$\mathbf{P}[|\langle \phi, N_t^t \rangle| > \varepsilon] \leq \mathbf{P}[\langle\langle \phi, N^t \rangle\rangle_t > \varepsilon^3] + \mathbf{P}[|\langle \phi, N_t^t \rangle| > \varepsilon, \langle\langle \phi, N^t \rangle\rangle_t \leq \varepsilon^3]$$

$$\leq \mathbf{E}^x \left[ \nu(\langle\langle \phi, N^t \rangle\rangle_t > \varepsilon^3) \right] + \varepsilon$$

$$< 2\varepsilon$$

for $t$ large. This completes our proof.  □

For the applications we have in mind, (3.1)–(3.5) will be easy to verify and (3.7) will be our working hypothesis on $\nu$. We now assume (3.1)–(3.5) and provide some alternative conditions which imply (3.6).

If we repeat the first part of the previous proof with $\phi = \mathbf{1}$, we see that

$$\langle x_t, \mathbf{1} \rangle = \langle x_0, \mathbf{1} \rangle + \sum_i \left[ \int_0^t P_{t-s} \, dM_s \right](i),$$

where the sum converges a.s. and in $L^2$. Let $P_r(S_n, j) = \sum_{i \in S_n} p_r(i,j)$. Then

$$\sum_{i \in S_n} \left[ \int_0^t P_{t-s} dM_s \right](i) = \sum_{j \in S_m} \int_0^t P_{t-s}(S_n, j) \, dM_s(j) + \sum_{j \in S_m^c} \int_0^t P_{t-s}(S_n, j) \, dM_s(j)$$

$$\equiv \Sigma_{n,m}^1 + \Sigma_{n,m}^2.$$

Dominated convergence and (3.5) imply that

$$\lim_{n \to \infty} \mathbf{E}^{x_0} \left[ |\Sigma_{n,m}^1 - \sum_{j \in S_m} M_t(j)|^2 \right] = 0, \qquad \lim_{m \to \infty} \sup_n \mathbf{E}^{x_0} \left[ |\Sigma_{m,n}^2|^2 \right] = 0.$$

This easily shows that $\sum_j M_t(j)$ converges in $L^2$ to $\sum_i [\int_0^t P_{t-s} dM_s(i)]$, and so (recall (3.1))

$$(3.13) \qquad X_t = X_0 + \sum_j M_t(j).$$

On account of this,

$$(3.14) \qquad \langle X \rangle_t = Q_t(\mathbf{1}) \qquad \text{for all } t \geq 0 \ a.s.$$

Now, by the martingale convergence theorem, $X_\infty = \lim_{t \to \infty} X_t$ exists and is finite a.s., and we would like to infer that $Q_\infty(\mathbf{1}) = \lim_{t \to \infty} Q_t(\mathbf{1})$ is finite a.s. too, but this requires an additional hypothesis if $X$ is not continuous.

LEMMA 3.2. *Assume $(N_t, t \geq 0)$ is a nonnegative, square-integrable, RCLL martingale. If $T_n = \inf\{t : N_t \geq n\}$ (with $\inf \emptyset = \infty$), and for each positive integer $n$, $\mathbf{E}[(N_{T_n} - N_{T_n-})^2 1_{\{T_n < \infty\}})] < \infty$, then $\lim_{t \to \infty} \langle N \rangle_t < \infty$ a.s.*

PROOF. The assumption on the jumps implies that, for each $n$, the martingale $M_t^{(n)} = N_{t \wedge T_n}^2 - \langle N \rangle_{t \wedge T_n}$ is $L^1$-bounded. Hence, as $t \to \infty$, $M_t^{(n)}$ converges a.s. to a finite limit $M_\infty^{(n)}$. Since $N_t$ is a nonnegative martingale, $N_t$ converges a.s. to a finite limit $N_\infty$. Therefore, on $\{T_n = \infty\}$, $\langle N \rangle_t \to \langle N \rangle_\infty = N_\infty^2 - M_\infty^{(n)}$ as $t \to \infty$. The fact that $N_\infty$ is finite a.s. implies that $\mathbf{P}[\cup_n \{T_n = \infty\}] = 1$, and hence $\mathbf{P}[\langle N \rangle_\infty < \infty] = 1$. $\qquad \square$

Here then are two conditions which together will imply (3.6). The first one is

$$(3.15) \qquad d\langle M(i), M(j) \rangle_t \geq 0 \qquad \text{for all } i, j \in S, t \geq 0 \ a.s.$$

The second, with $T_n = \inf\{t : \langle x_t, \mathbf{1} \rangle \geq n\}$, is

$$(3.16) \qquad \mathbf{E}[(\langle x_{T_n}, \mathbf{1} \rangle - \langle x_{T_n-}, \mathbf{1} \rangle)^2 1_{\{T_n < \infty\}}] < \infty \qquad \text{for all } n.$$

THEOREM 3.3. *Assume (3.1)–(3.5), (3.15) and (3.16). Then (3.6) holds, and so if $\nu$ is a probability measure on $\mathbb{R}^S$ satisfying (3.7), then $\langle \phi, x_t \rangle \to 0$ in $\nu \otimes \mathbf{P}^{x_0}$-probability as $t \to \infty$.*

PROOF. Lemma 3.2 implies $Q_\infty(\mathbf{1}) < \infty$ a.s., and (3.15) implies $|Q|_\infty(\mathbf{1}) = Q_\infty(\mathbf{1})$ a.s. Therefore, (3.6) is true, and Theorem 3.1 completes the proof. $\qquad \square$

## 4. Mutually catalytic branching

As in Dawson and Perkins (1998), let $(u_t, v_t)$ denote the mutually catalytic branching model defined by

$$(4.1) \qquad \begin{aligned} du_t &= (p - I)u_t \, dt + (cu_t v_t)^{1/2} dB_t, \\ dv_t &= (p - I)v_t \, dt + (cu_t v_t)^{1/2} dW_t. \end{aligned}$$

Here, $p(i, j)$ is an irreducible Markov chain transition matrix, $c$ is a fixed positive constant, and the $\{B_t(i)\}$ and $\{W_t(i)\}$ are independent families of independent one–dimensional standard Brownian motions. As in Dawson and Perkins (1998), we assume $S = \mathbb{Z}^d$, $p(i, j)$ is symmetric, and the exponential growth condition (H2) of that paper holds. This growth condition is satisfied, for example, when $p(i, j)$ is the transition function of a symmetric random walk such that $\sum_k \phi_{-\lambda}(k) p(0, k) < \infty$

for all $\lambda > 0$, where $\phi_\lambda(k) = c^{\lambda|k|}$, $|k| = \sum_{i=1}^d |k_i|$. To define an appropriate state space, we introduce

$$M_{\text{tem}} = \{u\colon \mathbb{Z}^d \to \mathbb{R}_+ \text{ such that } \langle u, \phi_\lambda \rangle < \infty \quad \forall \lambda < 0\}.$$

Let $|u|_\lambda = \sup\{|u(k)|\phi_\lambda(k) : k \in \mathbb{Z}^d\}$, and topologize $M_{\text{tem}}$ so that $u_n \to u$ in $M_{\text{tem}}$ iff $\lim_{n\to\infty} |u_n - u|_\lambda = 0$ for all $\lambda < 0$. For each pair of initial conditions $(u_0, v_0) \in M_{\text{tem}}^2$ there is a well defined Markov process $(u_t, v_t)$ determined by (4.1) taking values in $M_{\text{tem}}^2$ (see Theorems 1.1 and 2.4 in Dawson and Perkins (1998) for precise details), and its law $\mathbf{P}^{(u_0,v_0)}$ on $C([0,\infty), M_{\text{tem}}^2)$ is unique.

The mutually catalytic branching process $(u_t, v_t)$ is self–dual. Let $(u_t, v_t)$ be the mutually catalytic branching process with initial state $(u_0, v_0)$, and let $(\tilde{u}_t, \tilde{v}_t)$ be the mutually catalytic branching process with initial state $(\tilde{u}_0, \tilde{v}_0)$. Mytnik (1998) (see also Theorem 2.4 in Dawson and Perkins (1998)) showed that for $u_0, v_0 \in M_{\text{tem}}$ and $\tilde{u}_0, \tilde{v}_0 \in X_F$,

$$(4.2) \qquad \begin{aligned} \mathbf{E}^{(u_0,v_0)}&\big[e^{-\langle u_t + v_t, \tilde{u}_0 + \tilde{v}_0 \rangle + i\langle u_t - v_t, \tilde{u}_0 - \tilde{v}_0 \rangle}\big] \\ &= \mathbf{E}^{(\tilde{u}_0,\tilde{v}_0)}\big[e^{-\langle u_0 + v_0, \tilde{u}_t + \tilde{v}_t \rangle + i\langle u_0 - v_0, \tilde{u}_t - \tilde{v}_t \rangle}\big]. \end{aligned}$$

Suppose now that $a, b \geq 0$ are fixed. Let $(u_t^a, v_t^b)$ denote the mutually catalytic branching model with initial state $(u_0^a, v_0^b) = (\mathbf{a}, \mathbf{b})$. Then (4.2) implies that for $\tilde{u}_0$, $\tilde{v}_0$ in $X_F$,

$$(4.3) \qquad \begin{aligned} \mathbf{E}&\big[e^{-\langle u_t^a + v_t^b, \tilde{u}_0 + \tilde{v}_0 \rangle + i\langle u_t^a - v_t^b, \tilde{u}_0 - \tilde{v}_0 \rangle}\big] \\ &= \mathbf{E}^{(\tilde{u}_0,\tilde{v}_0)}\big[e^{-(a+b)\langle \mathbf{1}, \tilde{u}_t + \tilde{v}_t \rangle + i(a-b)\langle \mathbf{1}, \tilde{u}_t - \tilde{v}_t \rangle}\big]. \end{aligned}$$

By Theorem 2.2 in Dawson and Perkins (1998), $\langle \mathbf{1}, \tilde{u}_t \rangle$ and $\langle \mathbf{1}, \tilde{v}_t \rangle$ are nonnegative martingales, and hence converge a.s. $\mathbf{P}^{(\tilde{u}_0,\tilde{v}_0)}$ as $t \to \infty$. Therefore, the left side of (4.3) must converge in distribution as $t \to \infty$. It is not hard to see that this implies there is a stationary probability measure $\nu_{(a,b)}$ on $M_{\text{tem}}^2$ such that

$$\mathcal{L}[(u_t^a, v_t^b)] \Rightarrow \nu_{(a,b)} \qquad \text{as } t \to \infty$$

in the sense of weak convergence of probabilities on $M_{\text{tem}}^2$ (see Theorem 1.4 of Dawson and Perkins (1998)). In particular,

$$(4.4) \qquad \begin{aligned} \mathbf{E}^{(\phi,\psi)}\big[e^{-(a+b)\langle \mathbf{1}, \tilde{u}_t + \tilde{v}_t \rangle + i(a-b)\langle \mathbf{1}, \tilde{u}_t - \tilde{v}_t \rangle}\big] &= \mathbf{E}\big[e^{-\langle u_t^a + v_t^b, \phi + \psi \rangle + i\langle u_t^a - v_t^b, \phi - \psi \rangle}\big] \\ &\to \int e^{-\langle u' + v', \phi + \psi \rangle + i\langle u' - v', \phi - \psi \rangle} \, d\nu_{(a,b)}(u', v') \end{aligned}$$

as $t \to \infty$. If $T$ is the first exit time of planar Brownian motion $(B_t^1, B_t^2)$ from the first quadrant starting at $(a, b)$, then under appropriate recurrence hypotheses on $P_t$ (satisfied, for example, by simple symmetric random walk in 1 or 2 dimensions), $\nu_{(a,b)}(\cdot) = \mathbf{P}[(\mathbf{B}_T^1, \mathbf{B}_T^2) \in \cdot]$, where $\mathbf{B}_T^i(k) = B_T^i$ for all $k$. Hence $u \equiv 0$ or $v \equiv 0$ $\nu_{(a,b)}$-a.s. For transient $P_t$, $u(k)v(k) > 0$ for all $k$ $\nu_{(a,b)}$-a.s. See Theorems 1.5 and 1.6 of Dawson and Perkins (1998) for the precise results and further information about these limiting laws.

For $a, b \geq 0$, define $\mathcal{M}_{(a,b)}$ to be the collection of probability measures $\nu$ on $M_{\text{tem}} \times M_{\text{tem}}$ such that

$$(4.5) \qquad \sup_k \int (u^2(k) + v^2(k))\, d\nu(u, v) < \infty,$$

$$(4.6) \qquad \lim_{t \to \infty} \int [(P_t u(k) - a)^2 + (P_t v(k) - b)^2]\, d\nu(u, v) = 0, \quad k \in \mathbb{Z}^d.$$

As does $\mathcal{M}_\theta$ in Section 2, this class contains non-translation invariant laws. Assume, for example, that $P_t$ is the semigroup of simple symmetric random walk. If

$$(4.7) \qquad (u_0, v_0) = (1(k_1 < 0), 1(k_1 > 0)),$$

then $\delta_{(u_0, v_0)} \in \mathcal{M}_{(1/2, 1/2)}$. As for the parabolic Anderson model, if $u_0$, $v_0$ are bounded non-negative maps on $\mathbb{Z}^d$ whose averages over Euclidean balls approach $a$ and $b$, respectively, as the radius of the ball approaches $\infty$, then $\delta_{(u_0, v_0)} \in \mathcal{M}_{(a,b)}$.

Our main technical result for the mutually catalytic branching model is:

PROPOSITION 4.1. *Let* $a, b \geq 0$, *and* $\nu \in \mathcal{M}_{(a,b)}$. *If* $(u_0, v_0)$ *has law* $\nu$, *and* $(\tilde{u}_0, \tilde{v}_0) \in X_F \times X_F$, *then*

$$(4.8) \qquad |\langle u_0 - \mathbf{a}, \tilde{u}_t \rangle| + |\langle v_0 - \mathbf{b}, \tilde{u}_t \rangle| + |\langle u_0 - \mathbf{a}, \tilde{v}_t \rangle| + |\langle v_0 - \mathbf{b}, \tilde{v}_t \rangle| \to 0$$

*in* $\nu \otimes \mathbf{P}^{(\tilde{u}_0, \tilde{v}_0)}$-*probability as* $t \to \infty$.

As in Section 2, a truncation argument can be used to prove

COROLLARY 4.2. *If* $p(i, j) = p(0, j - i)$, *and* $\nu = \mathcal{L}[(u_0, v_0)]$ *is translation invariant, shift ergodic and satisfies* $\int u(0)\, d\nu((u, v)) = a$ *and* $\int v(0)\, d\nu((u, v)) = b$, *then* (4.8) *holds.*

PROOF OF PROPOSITION 4.1. Let $\mathbf{P} = \nu \otimes \mathbf{P}^{(\tilde{u}_0, \tilde{v}_0)}$. We will show that for all $\varepsilon > 0$,

$$(4.9) \qquad \lim_{t \to \infty} \mathbf{P}[|\langle u - \mathbf{a}, \tilde{u}_t \rangle| > \varepsilon] = 0,$$

and it will be clear that our proof will apply to the other terms in (4.8). By Theorem 2.2 of Dawson and Perkins (1998),

$$\tilde{u}_t(i) = P_t \tilde{u}_0(i) + \sum_j \int_0^t p_{t-s}(i, j) dM_s(j),$$

where the series converges in $L^2$, and the $\{(M_t(i))_{t \geq 0}, \ i \in S\}$ are orthogonal square-integrable continuous martingales with square variation functions

$$\langle M(i) \rangle_t = \int_0^t (\tilde{u}_s(i) \tilde{v}_s(i))\, ds.$$

The same result shows that $\langle \tilde{u}_t, \mathbf{1} \rangle$ is a continuous square-integrable martingale. Therefore, (3.1) and (3.2) hold. The orthogonality of the $\{M(i)\}$ and square integrability of the total mass martingale imply (3.5) and (3.15), and the continuity of $\langle \tilde{u}_t, \mathbf{1} \rangle$ implies that (3.16) holds trivially. Consequently, we may apply Theorem 3.3 with the measure $\nu$ of the result given by $\mathcal{L}[u_0 - \mathbf{a}]$. $\qquad \square$

As with the case of the parabolic Anderson model, we can use duality and Proposition 4.1 to prove a convergence result for $(u_t, v_t)$. Let $d$ be a complete metric inducing the topology of weak convergence on the space $M_1(M_{\text{tem}}^2)$ of probability measures on $M_{\text{tem}}^2$.

THEOREM 4.3. *Let* $\nu = \mathcal{L}[(u_0, v_0)]$. *Assume either that* (i) $\nu \in \mathcal{M}_{(a,b)}$, *or* (ii) $p(i, j) = p(0, j - i)$ *and* $\nu$ *is translation invariant, shift ergodic and satisfies* $\int u(0) \, d\nu((u, v)) = a$ *and* $\int v(0) \, d\nu((u, v)) = b$. *Then*

(a) $d\big(\mathbf{P}^{(u,v)}[(u_t, v_t) \in \cdot], \nu_{(a,b)}\big) \to 0$ *in* $d\nu(u, v)$-*probability as* $t \to \infty$.

(b) $\mathcal{L}^{\nu}[(u_t, v_t)] \Rightarrow \nu_{(a,b)}$ *as* $t \to \infty$.

PROOF. Clearly (b) follows from (a) by integrating out $(u, v)$ with respect to $\nu$.

For (a), choose $\tilde{u}_0, \tilde{v}_0 \in X_F$ and let $\mathbf{P} = \nu \otimes \mathbf{P}^{(\tilde{u}_0, \tilde{v}_0)}$. Note that under $\mathbf{P}$, $(u, v)$ is a random variable with distribution $\nu$. If we let

$$f(u, v, \phi, \psi) = e^{-\langle u+v-(\mathbf{a}+\mathbf{b}), \phi+\psi\rangle + i\langle u-v-(\mathbf{a}-\mathbf{b}), \phi-\psi\rangle},$$

then by (4.8), $f(u, v, \tilde{u}_t, \tilde{v}_t) \to 1$ in $\mathbf{P}$-probability as $t \to \infty$. By (4.2),

$$\mathbf{E}^{(u,v)}\big[c^{-\langle u_t+v_t, \tilde{u}_0+\tilde{v}_0\rangle + i\langle u_t-v_t, \tilde{u}_0-\tilde{v}_0\rangle}\big]$$

$$= \mathbf{E}^{(\tilde{u}_0, \tilde{v}_0)}\big[e^{-\langle (u+v), \tilde{u}_t+\tilde{v}_t\rangle + i\langle (u-v), \tilde{u}_t-\tilde{v}_t\rangle}\big]$$

$$= \mathbf{E}^{(\tilde{u}_0, \tilde{v}_0)}\big[f(u, v, \tilde{u}_t, \tilde{v}_t) e^{-(a+b)\langle \mathbf{1}, \tilde{u}_t+\tilde{v}_t\rangle + i(a-b)\langle \mathbf{1}, \tilde{u}_t-\tilde{v}_t\rangle}\big].$$

By (4.4) and the above,

$$\mathbf{E}^{(u,v)}\big[e^{-\langle u_t+v_t, \tilde{u}_0+\tilde{v}_0\rangle + i\langle u_t-v_t, \tilde{u}_0-\tilde{v}_0\rangle}\big]$$

(4.10)
$$\to \int e^{-\langle u'+v', \tilde{u}_0+\tilde{v}_0\rangle + i\langle u'-v', \tilde{u}_0-\tilde{v}_0\rangle} \, d\nu_{(a,b)}(u', v')$$

in $d\nu(u, v)$-probability as $t \to \infty$. If $\lambda > 0$, then Theorems 2.2(b)(iii) and 1.4 of Dawson and Perkins (1998) show that

$$\int \Big| \mathbf{E}^{(u,v)}[\langle u_t \pm v_t, \phi_{-\lambda}\rangle] - \int \langle u' \pm v', \phi_{-\lambda}\rangle \, d\nu_{(a,b)}(u', v') \Big| \, d\nu(u, v)$$

(4.11)
$$= \int \Big| \langle P_t(u \pm v), \phi_{-\lambda}\rangle - \langle a \pm b, \phi_{-\lambda}\rangle \Big| \, d\nu(u, v)$$

$$\leq \sum_k \phi_{-\lambda}(k) \int \big[ |P_t u(k) - a| + |P_t v_0(k) - b| \big] \, d\nu(u, v)$$

$$\to 0 \text{ as } t \to \infty,$$

where we have used (4.5), (4.6), and dominated convergence in the last line. Let $X_F^q$ be the set of $\tilde{u}_0$ in $X_F$ taking on rational values. Choose a sequence $t_n \to \infty$. By (4.10) and (4.11) we may choose a subsequence $t_{n_k}$ such that for $\nu$-a.a. $(u, v)$,

(4.12)
$$\lim_{k \to \infty} \mathbf{E}^{(u,v)}\big[e^{-\langle u_{t_{n_k}}+v_{t_{n_k}}, \tilde{u}_0+\tilde{v}_0\rangle + i\langle u_{t_{n_k}}-v_{t_{n_k}}, \tilde{u}_0-\tilde{v}_0\rangle}\big]$$

$$= \int e^{-\langle u'+v', \tilde{u}_0+\tilde{v}_0\rangle + i\langle u'-v', \tilde{u}_0-\tilde{v}_0\rangle} \, d\nu_{(a,b)}(u', v') \text{ for all } \tilde{u}_0, \tilde{v}_0 \in X_F^q,$$

and

$$\lim_{k \to \infty} \mathbf{E}^{(u,v)}[\langle u_{t_{n_k}} + v_{t_{n_k}}, \phi_{-\lambda}\rangle] = \int \langle u' + v', \phi_{-\lambda}\rangle \, d\nu_{(a,b)}(u', v') \text{ for all } \lambda \in \mathbb{Q}, \lambda > 0.$$

The latter implies that for $\nu$-a.a. $(u, v)$,

(4.13)
$$\sup_k \mathbf{E}^{(u,v)}[\langle u_{t_{n_k}} + v_{t_{n_k}}, \phi_{-\lambda}\rangle] < \infty \text{ for all } \lambda > 0.$$

Fix $(u, v)$ outside of a $\nu$-null set so that (4.12) and (4.13) both hold. A simple approximation argument using (4.13) to bound $\sup_k \mathbf{E}^{(u,v)}[\langle u_{t_{n_k}} + v_{t_{n_k}}, \mathbf{1}_F \rangle]$ for each finite set $F$, allows one to extend (4.12) to all $\tilde{u}_0, \tilde{v}_0$ in $X_F$. This extension, together with (4.13), allows us to apply Lemma 2.3(c) of Dawson and Perkins (1998) to conclude that

$$(4.14) \qquad d(\mathbf{P}^{(u,v)}((u_{t_{n_k}}, v_{t_{n_k}}) \in \cdot), \nu_{(a,b)}) \to 0 \text{ as } k \to \infty \; \nu - a.a. \; (u, v).$$

We have shown that every sequence $t_n \to \infty$ has a subsequence satisfying (4.14), and so (a) is now immediate. □

We include (a) because it is used in Cox and Klenke (1998) to show that, under the appropriate recurrence hyptheses mentioned above, as $t \to \infty$, the "predominant type" near 0 changes infinitely often. (Recall in this setting that there is extinction of one type in the equilibrium limit.) This result was the original motivation for this work.

These methods also apply to the continuous version of (4.1). Let $\dot{W}_1$ and $\dot{W}_2$ be independent space-time white noises on $\mathbb{R}_+ \times \mathbb{R}_+$ and consider the system of stochastic partial differential equations

$$(4.15) \qquad \begin{aligned} \frac{\partial u}{\partial t}(t, x) &= \frac{1}{2} \frac{\partial^2 u}{\partial x^2}(t, x) + (cu(t, x)v(t, x))^{1/2} \dot{W}_1(t, x), \\ \frac{\partial v}{\partial t}(t, x) &= \frac{1}{2} \frac{\partial^2 v}{\partial x^2}(t, x) + (cu(t, x)v(t, x))^{1/2} \dot{W}_2(t, x). \end{aligned}$$

See Dawson and Perkins (1998) for a precise interpretation of this pair of equations. Let $|f|_\lambda = \sup\{f(x)e^{\lambda|x|} : x \in \mathbb{R}\}$ and assume that $u_0, v_0 \in C_{\text{tem}}^+$, where

$$C_{\text{tem}}^+ = \{f : \mathbb{R} \to [0, \infty) : \; f \text{ is continuous, and } |f|_\lambda < \infty \text{ for all } \lambda < 0\}.$$

We topologize $C_{\text{tem}}^+$ so that $f_n \to f$ in $C_{\text{tem}}^+$ if and only if $|f_n - f|_\lambda \to 0$ for all $\lambda < 0$.

By Theorem 1.7 of Dawson and Perkins (1998), there is a unique (in law) solution $(u_t, v_t)$ of (4.15) satisfying $(u_., v_.) \in C([0, \infty), (C_{\text{tem}}^+)^2)$. Uniqueness was first shown by Mytnik (1998) through the continuous analogue of (4.2). Let $\mathbf{P}^{(u,v)}$ denote the law of the solution of (4.15) with $(u_0, v_0) = (u, v)$, and let $\mathbf{P}^\nu$ denote this law if $\mathcal{L}[u_0, v_0] = \nu$, a probability measure on $(C_{\text{tem}}^+)^2$.

As in Dawson and Perkins (1998), for the purpose of convergence to equilibrium, we weaken the topology on $C_{\text{tem}}^+$. Let $M_{\text{tem}}^c = C_{\text{tem}}^+$ but the weak topology given by: $f_n \to f$ in $M_{\text{tem}}^c$ if and only if $\lim \int f_n(x)\phi(x)\,dx = \int f(x)\phi(x)\,dx$ for all continuous $\phi$ satisfying $|\phi|_\lambda < \infty$ for some $\lambda > 0$. As before, let $\mathbf{P}^{a,b}$ denote the law of a planar Brownian motion started at $(a, b) \in \mathbb{R}_+^2$, and let $T$ be the Brownain motion's first exit time from the first quadrant. Let $\mathbf{a}, \mathbf{b}$ denote the constant functions on $\mathbb{R}$. Theorem 1.8 of Dawson and Perkins (1998) states that, for $a, b > 0$,

$$(4.16) \qquad \mathbf{P}^{\mathbf{a},\mathbf{b}}[(u_t, v_t) \in \cdot] \Rightarrow \nu_{(a,b)}^c = \mathbf{P}^{a,b}[(\mathbf{B}_T^1, \mathbf{B}_T^2) \in \cdot]$$

in $(M_{\text{tem}}^c)^2$ as $t \to \infty$.

For $a, b \geq 0$ let $\mathcal{M}_{(a,b)}$ be the set of probability measures $\nu$ on $(C_{\text{tem}}^+)^2$ such that

$$(4.17) \qquad \sup_x \int (u^2(x) + v^2(x)) \, d\nu(u, v) < \infty,$$

$$(4.18) \qquad \lim_{t \to \infty} \int [(P_t u(x) - a)^2 + (P_t v(x) - b)^2] \, d\nu(u, v) = 0, \quad \text{all } x \in \mathbb{R}.$$

where now $P_t$ is the semigroup of one-dimensional Brownian motion.

Although we may no longer invoke the "discrete" Theorem 3.3, it is easy to argue directly as in Section 2, using (4.16), the continuous analogue of (4.2), and Theorem 6.1 in Dawson and Perkins, to prove the analogue of Theorem 4.3 given below. Let $d$ be a metric on $M_1((M_{tem}^c)^2)$ inducing the topology of weak convergence on this space of probability laws.

THEOREM 4.4. *Assume $a, b \geq 0$, and $\nu = \mathcal{L}[(u_0, v_0)] \in \mathcal{M}_{a,b}^c$.*

(a) $d\big(\mathbf{P}^{(u,v)}[(u_t, v_t) \in \cdot], \nu_{(a,b)}^c\big) \to 0$ *in $d\nu(u, v)$-probability as $t \to \infty$.*

(b) $\mathcal{L}[(u_t, v_t)] \Rightarrow \nu_{(a,b)}^c$ *as $t \to \infty$ in the sense of weak convergence of probability measures on $(M_{tem}^c)^2$.*

## 5. Linear systems with values in $[0, \infty)^{\mathbb{Z}^{\mathbf{d}}}$

We consider a subclass of the linear systems treated in Chapter IX of Liggett (1985). We set $S = \mathbb{Z}^d$, and, following Liggett (1985), use $x, i, j, k, l$ to denote generic elements of $\mathbb{Z}^d$. Our process will be denoted $\eta_t$, and takes values in $[0, \infty)^S$. Let $A(x; i, j)$, $x, i, j \in \mathbb{Z}^d$, be nonnegative random variables. It is convenient to view $A(x) = A(x; i, j)$, $i, j \in \mathbb{Z}^d$, as a random matrix indexed by $\mathbb{Z}^d \times \mathbb{Z}^d$. Let $\mathfrak{M}$ denote the set of such infinite matrices. Given a configuration $\eta \in [0, \infty)^S$, let $A(x)\eta$ be the configuration defined by $A(x)\eta(i) = \sum_j A(x; i, j)\eta(j)$. The process $\eta_t$ is defined as follows. At each $x \in \mathbb{Z}^d$ there is a rate one exponential alarm clock. If the clock at site $x$ goes off at time $t$, the configuration $\eta_{t-}$is replaced by $A(x)\eta_{t-}$. At each such time and site, independent instances of the $A(x)$ are used and they are identically distributed in time for each site $x$. The smoothing and potlatch processes introduced in Spitzer (1981) are the main examples of this type of process we will consider.

Let $\mu$ be the infinite measure on $V = \mathbb{Z}^d \times \mathfrak{M}$ given by

$$(5.1) \qquad \mu(C \times D) = \sum_{x \in C} \mathbf{P}[A(x) \in D], \qquad C \subset \mathbb{Z}^d, \text{ Borel } D \subset \mathfrak{M}.$$

We assume there is a finite constant $M$ such that

$$(5.2) \quad \sup_i \int \left[ \sum_j |A(x; i, j) - \delta(i, j)| \right] + \left[ \sum_j |A(x; i, j) - \delta(i, j)| \right]^2 d\mu(x, A) \leq M.$$

This condition was introduced in Chapter IX of Liggett (1985) to ensure existence of the desired process, and finiteness of second moments of its coordinates (see (1.4), Lemma 1.6 and (3.2) of that reference). By Theorem IX.1.6 of Liggett (1985), there is a strictly positive, summable function $\alpha$ on $\mathbb{Z}^d$ such that

$$(5.3) \qquad \mathbf{E}\Big[ \sum_{x \in \mathbb{Z}^d} \sum_{i: \, i \neq j} \alpha(i) A(x; i, j) \Big] \leq M\alpha(j).$$

Let $X = X_\alpha = \{\eta \in [0, \infty)^{\mathbb{Z}^d} : \langle \eta, \alpha \rangle < \infty\}$, endowed with the topology of pointwise convergence. Under (5.2) there is a well defined Markov process $\eta_t$ taking values in $X$ which is specified by the description above (see Theorem IX.1.14 of Liggett (1985) for a precise statement).

Let $\tilde{A}(x)$ denote the transpose of $A(x)$. Then we may define another process $\tilde{\eta}_t$ using the $\tilde{A}(x)$ instead of the $A(x)$. In order that $\tilde{\eta}_t$ be well defined, we assume that (5.2) holds with $A$ replaced by $\tilde{A}$. As shown in Liggett (1985), $\eta_t$ and $\tilde{\eta}_t$ are dual processes, exactly as in the parabolic Anderson model. Given $\eta_0 \in X$ and $\tilde{\eta}_0 \in X_F$,

$$(5.4) \qquad\qquad \langle \eta_t, \tilde{\eta}_0 \rangle \overset{\mathrm{d}}{=} \langle \eta_0, \tilde{\eta}_t \rangle.$$

Now define $\tilde{\mu}$ as $\mu$ (5.1), but with $\tilde{A}$ in place of $A$, and also

$$(5.5) \qquad \gamma_x = \mathbf{E}[(A - I)(x)], \qquad \tilde{\gamma}_x = \mathbf{E}[(\tilde{A} - I)(x)],$$

and

$$(5.6) \quad \gamma = \sum_x \gamma_x = \int (A - I) d\mu(x, A), \qquad \tilde{\gamma} = \sum_x \tilde{\gamma}_x = \int (\tilde{A} - I) d\mu(x, A).$$

Since the $A(x)$ are nonnegative, $\gamma(i, j)$ and $\tilde{\gamma}(i, j)$ are nonnegative for $i \neq j$. Clearly (5.2) implies all these coefficients are finite and $\sup_i \sum_j |\gamma(i, j)| + |\tilde{\gamma}(i, j)| < \infty$. We also assume that

$$(5.7) \qquad\qquad \gamma \mathbf{1} = \mathbf{1}\gamma = 0.$$

Thus, $\gamma$ and $\tilde{\gamma}$ are $q$–matrices for rate-one continuous time Markov chains on $S$ whose transition semigroups are $\gamma_t$ and $\tilde{\gamma}_t$, respectively. (Note that, for simplicity, we have omitted the quantities $a(i, j)$ in Liggett (1985).)

We now construct $\eta_t$ as the unique solution of a stochastic differential equation driven by a Poisson point process. Although this is not essential for our arguments, it provides some additional methodology for the study of these linear systems, and does not quite follow from the general constructions in Kurtz and Protter (1996). We do this for $\eta_0 \in X_f = \{\eta \in [0, \infty)^{\mathbb{Z}^d} : \langle \eta, \mathbf{1} \rangle < \infty\}$, although our construction holds more generally. Give $X_f$ the topology of weak convergence of finite measures on $\mathbb{Z}^d$ and let $D(X_f)$ denote the Skorokhod space of right continuous $X_f$-valued paths on $[0, \infty)$ with left limits. Let $N$ be a Poisson point process on $[0, \infty) \times V$ with intensity $m \times \mu$, with respect to the filtration $\mathcal{F}_t$ on $(\Omega, \mathcal{F}, \mathbf{P})$ ($m$ is Lebesgue measure). Let $\hat{N}$ denote the orthogonal martingale measure

$$(5.8) \qquad \hat{N}(t, U) = N([0, t], U) - t\mu(U) \text{ for } \mu(U) < \infty.$$

Recall that $\int_0^t \int \int f(s, \omega, x, A) \hat{N}(ds, dx, dA)$ is defined and is a local martingale for a large class of integrands $f$ (see section II.3 of Ikeda and Watanabe (1981)). For example if $f \in \mathcal{L}^2$, the class of $\mathcal{F}_t$-predictable$\times$ Borel measurable functions such that $\mathbf{E}[\int_0^t \int_V f(s, \omega, x, A)^2 d\mu ds] < \infty$ for all $t > 0$, then $\int_0^t \int f(s, \omega, x, A) \hat{N}(ds, dx, dA)$ is a square-integrable martingale with square function $\int_0^t \int_V f(s, \omega, x, A)^2 d\mu \, ds$. For square summable $\eta \in [0, \infty)^S$, let

$$(5.9) \quad q(\eta; i, j) = \int_V \sum_{i'} (A(x; i, i') - \delta(i, i'))\eta(i') \sum_{j'} (A(x; j, j') - \delta(j, j'))\eta(j') \, d\mu.$$

An application of Hölder's inequality and (5.2) implies $|q(\eta; i, j)| \leq M\|\eta\|_2^2 < \infty$.

PROPOSITION 5.1. (a) *There is a unique process* $\{\eta_t : t \geq 0\}$ *with paths in* $D(X_f)$ *such that*

$$(5.10) \qquad \eta_t = \eta_0 + \int_0^t \int_V (A - I)\eta_{s-} N(ds, dx, dA) \quad \text{for all } t \geq 0, \quad a.s.$$

*Its law coincides with the law of the process constructed in Theorem IX.1.14 of Liggett (1985) (and described above).*

(b) *The total mass process*

$$(5.11) \qquad \langle \eta_t, \mathbf{1} \rangle = \langle \eta_0, \mathbf{1} \rangle + \int_0^t \int_V \langle (A - I)\eta_{s-}, \mathbf{1} \rangle \hat{N}(ds, dx, dA)$$

*is a non-negative square-integrable martingale with predictable square function*

$$(5.12) \qquad C_t = \int_0^t \int_V \langle (A - I)\eta_{s-}, \mathbf{1} \rangle^2 \, d\mu \, ds = \int_0^t \langle \mathbf{1}, q(\eta_{s-})\mathbf{1} \rangle ds \in L^1.$$

*Moreover if we let*

$$(5.13) \qquad \bar{C}_t = \int_0^t \int_V \langle |(A - I)|\eta_{s-}, \mathbf{1} \rangle^2 \, d\mu \, ds,$$

*then* $\bar{C}_t$ *is also integrable.*

(c) *We have*

$$(5.14) \qquad \eta_t = \eta_0 + \int_0^t \gamma \eta_s \, ds + M_t,$$

*where for each* $i \in \mathbb{Z}^d$,

$$(5.15) \qquad M_t(i) = \int_0^t \int_V (A - I)\eta_{s-}(i)\hat{N}(ds, dx, dA)$$

*is a square-integrable martingale satisfying*

$$(5.16) \qquad \langle M(i), M(j) \rangle_t = \int_0^t q(\eta_{s-}; i, j) \, ds.$$

PROOF. (a) Let $\{S_n\}$ be a sequence of finite sets increasing to $\mathbb{Z}^d$. As in Liggett (1985) we start with solutions $\eta_t^n$ of the truncated system

$$(5.17) \qquad \eta_t^n = \eta_0 + \int_0^t \int_V \mathbf{1}_{S_n}(x)(A - I)\eta_{s-}^n N(ds, dx, dA).$$

Such solutions exist and are unique because $N$ has finite intensity on compact time intervals if $x$ is restricted to $S_n$. With our finite initial conditions we may argue as in Theorem IX.1.14 of Liggett (1985) but with $\alpha \equiv 1$ to conclude that the limits

$$(5.18) \qquad \lim_{n \to \infty} \mathbf{E}[\eta_t^n] = m_t, \quad \lim_{n \to \infty} \mathbf{E}[\langle \eta_t^n, \mathbf{1} \rangle] = \langle m_t, \mathbf{1} \rangle,$$

exist, and

$$(5.19) \qquad \mathbf{E}[\langle \eta_t^n, \mathbf{1} \rangle] \leq \langle \eta_0, \mathbf{1} \rangle \, e^{3Mt} \text{ for all } n \in \mathbb{N}, t > 0.$$

Define the matrices $a_x = \mathbf{E}[\|(A - I)(x)\|]$ and

$$(5.20) \qquad a = \sum_x a_x = \int_V |(A - I)| \, d\mu.$$

For $m \leq n$,

$$\mathbf{E}\left[\sup_{s \leq t} \langle |\eta_s^n - \eta_s^m|, 1 \rangle\right]$$

$$\leq \mathbf{E}\left[\int_0^t \int_V \mathbf{1}_{S_m}(x)\langle|(A - I)||\eta_{s-}^n - \eta_{s-}^m|, 1\rangle \, N(ds, dx, dA)\right]$$

$$+ \mathbf{E}\left[\int_0^t \int_V \mathbf{1}_{S_n - S_m}(x)\langle|(A - I)|\eta_{s-}^n, 1\rangle \, N(ds, dx, dA)\right]$$

(5.21)
$$= \int_0^t \sum_{x \in S_m} \langle a_x \mathbf{E}[|\eta_s^n - \eta_s^m|], 1\rangle \, ds + \int_0^t \sum_{x \in S_n - S_m} \langle a_x \mathbf{E}[\eta_s^n], 1\rangle \, ds$$

$$\leq \left[\sup_j a\mathbf{1}(j)\right]\left[\int_0^t \mathbf{E}[\langle |\eta_s^n - \eta_s^m|, 1\rangle] \, ds + \int_0^t \langle |\mathbf{E}[\eta_s^n] - m_s|, 1\rangle \, ds\right]$$

$$+ \int_0^t \sum_{x \in S_n - S_m} \langle a_x m_s, 1 \rangle \, ds.$$

Denote the last term in the above by $\varepsilon_{m,n}(t)$. Note that (5.2) (applied to $\tilde{A}$) implies $\sup_j \sum_i a(i,j) \leq M$. This together with (5.19) implies that

$$(5.22) \qquad \int_0^t \langle am_s, 1 \rangle \, ds \leq M \int_0^t \langle \eta_0, 1 \rangle e^{3Ms} \, ds,$$

and so $\sup_{t \leq T} \varepsilon_{m,n}(t) \to 0$ as $m, n \to \infty$ by dominated convergence. Clearly (5.18) implies that for each $s$, $\lim_{n \to \infty} \langle |\mathbf{E}[\eta_s^n] - m_s|, 1 \rangle = 0$, and so (5.19) and dominated convergence show that the "middle term" on the right side of (5.21) approaches 0 as $m, n$ approach $\infty$. Substituting the above bounds into the right side of (5.21), we arrive at

$$(5.23) \qquad \mathbf{E}\left[\sup_{s \leq t} \langle |\eta_s^n - \eta_s^m|, 1\rangle\right] \leq M \int_0^t \mathbf{E}[\langle |\eta_s^n - \eta_s^m|, 1\rangle] \, ds + \delta_{m,n}(t),$$

where $\delta_{m,n}$ is an increasing function in $t$ which approaches 0 as $m, n \to \infty$. This shows that for each $T > 0$,

$$(5.24) \qquad \mathbf{E}\left[\sup_{t \leq T} \langle |\eta_t^n - \eta_t^m|, 1\rangle\right] \leq \delta_{m,n}(T)e^{MT} \to 0 \text{ as } m, n \to \infty.$$

By taking a subsequence we may assume that there is a process $\eta_t$ with sample paths in $D(X_f)$ such that $\lim_{n \to \infty} \sup_{t \leq T} \langle |\eta_t^n - \eta_t|, 1\rangle \to 0$ for all $T > 0$ a.s. The above argument and Fatou's lemma also show that

(5.25)

$$\mathbf{E}\left[\sup_{t \leq T}\left|\int_0^t \int \left(\mathbf{1}_{S_m}(x)\langle(A - I)(\eta_{s-}^m - \eta_{s-}), 1\rangle\right)N(ds, dx, dA)\right|\right] \to 0 \text{ as } m \to \infty.$$

Now let $n \to \infty$ in (5.17) to derive (5.10). The fact that the law of this solution coincides with that of the process constructed in Theorem IX.1.14 of Liggett (1985) is immediate from the construction of the latter as the weak limit of solutions to (5.17).

Turning to the uniqueness of solutions to (5.10), we let $\eta$ denote any solution of this equation. If $T_n = \inf\{t : \langle \eta_t, \mathbf{1} \rangle \geq n\}$, then

(5.26)
$$\mathbf{E}[\langle \eta_{t \wedge T_n}, \mathbf{1} \rangle] \leq \langle \eta_0, \mathbf{1} \rangle + \mathbf{E}\left[\int_0^{t \wedge T_n} \int_V \langle |(A - I)|\eta_{s-}, \mathbf{1} \rangle d\mu ds\right]$$
$$\leq \langle \eta_0, \mathbf{1} \rangle + M\mathbf{E}\left[\int_0^{t \wedge T_n} \langle \eta_{s-}, \mathbf{1} \rangle ds\right]$$

again by applying (5.2) to $\tilde{A}$. The right side of the above is clearly finite, and so $f_n(t) = \mathbf{E}[\langle \eta_{t \wedge T_n}, \mathbf{1} \rangle]$ is a finite function satisfying

(5.27)
$$f_n(t) \leq \langle \eta_0, \mathbf{1} \rangle + M \int_0^t f_n(s) \, ds.$$

From this and Fatou's lemma, we finally arrive at the bound

(5.28)
$$\mathbf{E}[\langle \eta_t, \mathbf{1} \rangle] \leq \langle \eta_0, \mathbf{1} \rangle e^{Mt}.$$

The uniqueness now follows as in Section 9.1 of Kurtz and Protter (1996). We apply their reasoning with $F_i(\eta_{s-}, A, x) = (A - I)\eta_{s-}(i)$, and note that (9.2) of that reference holds with their $a_{i,j}$ equal to our $a(i,j)$, and their (9.6) holds with $p = 1$, $q = \infty$, and $\alpha \equiv 1$ by (5.2). Of course, $F_i$ is not bounded as in that reference, but this was only used to derive (5.28) and so is not needed.

Parts (b) and (c) of Theorem IX.2.2 of Liggett (1985) shows $\langle \eta_t, \mathbf{1} \rangle$ is a martingale (this is also immediate from the representation and integrability conditions derived below). Note that

(5.29)
$$\mathbf{E}[\bar{C}_{t \wedge T_n}] = \mathbf{E}\left[\int_0^{t \wedge T_n} \int_V \langle |(A - I)|\eta_{s-}, \mathbf{1} \rangle^2 \, d\mu \, ds\right]$$
$$\leq \mathbf{E}\left[\int_0^{t \wedge T_n} \langle \eta_s, \mathbf{1} \rangle^2 2 \sup_i \int_V (\mathbf{1}|(A - I)|(i))^2 d\mu \, ds\right]$$
$$\leq \mathbf{E}\left[\int_0^{t \wedge T_n} \langle \eta_s, \mathbf{1} \rangle^2 2M \, ds\right] < \infty,$$

where we again use (5.2) for $\tilde{A}$. The square integrability of $\langle \eta_{t \wedge T_n}, \mathbf{1} \rangle$ is now clear from the above and (5.10). We also see from the above that

(5.30)
$$\mathbf{E}\left[\langle \eta_{t \wedge T_n}, \mathbf{1} \rangle^2\right] \leq 2\langle \eta_0, \mathbf{1} \rangle^2 + 2\mathbf{E}[\bar{C}_{t \wedge T_n}]$$
$$\leq 2\langle \eta_0, \mathbf{1} \rangle^2 + 4M\mathbf{E}\left[\int_0^t \langle \eta_{s \wedge T_n}, \mathbf{1} \rangle^2 \, ds\right].$$

It follows that

(5.31)
$$\mathbf{E}\left[\langle \eta_{t \wedge T_n}, \mathbf{1} \rangle^2\right] \leq 2\langle \eta_0, \mathbf{1} \rangle^2 e^{4Mt},$$

and Fatou's lemma shows that $\langle \eta_t, \mathbf{1} \rangle$ is square-integrable. Use this and let $n \to \infty$ in (5.29) to obtain the integrability of $\bar{C}(t)$ for each $t$.

The above integrability allows us to rewrite (5.10) as (see Section II.3 of Ikeda and Watanabe (1981))

(5.32)  $$\eta_t = \eta_0 + \int_0^t \int_V (A - I)\eta_{s-}d\mu \, ds + \int_0^t \int_V (A - I)\eta_{s-}\hat{N}(ds, dx, dA),$$

and this gives the expression in (c). The formulae for the predictable square functions are immediate from our earlier discussion on the $\hat{N}$ stochastic integrals. To
see the representation for $\langle \eta_t, \mathbf{1} \rangle$ in (b) simply sum (5.10) over the sites in $S$. The
stochastic integrals converge in $L^1$ by (5.2) and integrability of the total mass, and
(5.7) shows that the resulting integral with respect to $N$ equals the same integral
with respect to the martingale measure $\hat{N}$. The integrability of $\bar{C}(t)$, and hence of
$C(t)$, implies that $C(t)$ is the predictable square function of this stochastic integral
representation of the total mass process (for it shows the integrand is in $\mathcal{L}^2$).  $\square$

REMARK 5.2. Note that

(5.33)
$$\sum_{i,j} |\langle M(i), M(j) \rangle|_t$$
$$\leq \int \sum_{i,j,i',j'} \int_V |A(i,i') - \delta(i,i')||A(j,j') - \delta(j,j')| \, d\mu \, \eta_s(i')\eta_s(j') \, ds$$
$$= \bar{C}_t$$

and so the above integrability of $\bar{C}_t$ implies (3.5) in Section 3.

If $\tilde{\eta}_0 \in X_F$, the above shows that $\langle \tilde{\eta}_t, \mathbf{1} \rangle$ is a nonnegative martingale, and
hence converges almost surely. In view of the duality relation (5.4), this implies
weak convergence of $\eta_t$ starting from constant initial states. That is, if $\theta \in [0, \infty)$
and $\eta_t^\theta$ is the linear system with initial state $\eta_0^\theta = \boldsymbol{\theta}$, then there is a probability
measure $\nu_\theta$ on $X$ such that

$$\mathcal{L}[\eta_t^\theta] \Rightarrow \nu_\theta \qquad \text{as } t \to \infty$$

(as probability measures on $[0, \infty)^{\mathbb{Z}^d}$ with the product topology). Moreover, if
$\phi \in X_F$,

(5.34)
$$\mathbf{E}^\phi[e^{-\theta\langle \mathbf{1}, \tilde{\eta}_t \rangle}] = \mathbf{E}[e^{-\langle \eta_t^\theta, \phi \rangle}]$$
$$\to \int e^{-\langle \eta, \phi \rangle} \, d\nu_\theta(\eta) \qquad \text{as } t \to \infty.$$

For $\theta \in [0, \infty)$, define $\mathcal{M}_\theta$ to be the collection of probability measures $\nu$ on $X$
such that (recall that $\gamma_t$ is the semigroup of the $q$–matrix $\gamma$ defined in (5.6))

(5.35)
$$\sup_k \int \eta^2(k) \, d\nu(\eta) < \infty,$$

(5.36)
$$\lim_{t \to \infty} \int (\gamma_t \eta(k) - \theta)^2 \, d\nu(\eta) = 0, \quad k \in \mathbb{Z}^d.$$

Define $\tilde{C}$ and $\bar{\tilde{C}}$ as in Proposition 5.1, but with $(\tilde{\eta}, \tilde{A})$ in place of $(\eta, A)$. Hence
$\tilde{C}_t = \int_0^t \langle \mathbf{1}, \tilde{q}(\tilde{\eta}_s)\mathbf{1} \rangle ds$ is the predictable square function of the non-negative martingale $\langle \tilde{\eta}, \mathbf{1} \rangle$. For the following result, we recall that our standing hypotheses (5.2)
and (5.7) are in effect, and that $q$ is the covariation kernel defined in (5.9).

PROPOSITION 5.3. *Let $\theta \in [0, \infty)$ and $\nu = \mathcal{L}[\eta_0] \in \mathcal{M}_\theta$. Assume that $\tilde{\eta}_0 \in X_f$,
and*

(5.37)
$$\int_0^\infty \langle \mathbf{1}, |q(\tilde{\eta}_t)|\mathbf{1} \rangle \, dt < \infty \quad a.s. \ \mathbf{P}^{\tilde{\eta}_0}.$$

*Then*

(5.38)
$$\langle \eta_0 - \boldsymbol{\theta}, \tilde{\eta}_t \rangle \to 0$$

*in $\nu \otimes \mathbf{P}^{\tilde{\eta}_0}$-probability as $t \to \infty$.*

As in Section 2, the second moment condition can be weakened in the *random walk* setting

(5.39)
$$\gamma(i, j) = \gamma(0, j - i), \qquad i, j \in \mathbb{Z}^d.$$

A truncation argument can be used to prove

COROLLARY 5.4. *If (5.37) holds, $\gamma$ is the $q$-matrix of random walk, and $\nu = \mathcal{L}[\eta_0]$ is translation invariant, shift ergodic and satisfies $\int \eta(0) \, d\nu(\eta) = \theta$, then (5.38) holds.*

We may apply the duality argument used in the previous sections to obtain the following result.

THEOREM 5.5. *Assume (5.37) holds for all $\tilde{\eta}_0 \in X_f$, and either* (i) $\mathcal{L}[\eta_0] \in \mathcal{M}_\theta$, *or* (ii) $\gamma$ *is the $q$-matrix of a random walk and $\mathcal{L}[\eta_0]$ is translation invariant, shift ergodic and satisfies $\int \eta(0) \, d\nu(\eta) = \theta$. Then $\mathcal{L}[\eta_t] \Rightarrow \nu_\theta$ as $t \to \infty$.*

REMARK 5.6. It is possible to formulate and prove a stronger result, analogous to part (a) of Theorem 4.3.

REMARK 5.7. Theorems 3.17 and 3.29 in Chapter IX of Liggett (1985) give conditions under which $\mathcal{L}[\eta_t] \Rightarrow \nu_\theta$ if $\gamma$ is the $q$-matrix of a random walk, and $\mathcal{L}[\eta_0]$ is translation invariant and shift ergodic. Theorem 5.5 above solves, at least in part, Problem 5 of Chapter IX in Liggett (1985).

PROOF OF PROPOSITION 5.3. Apply Itô's lemma to $\tilde{\gamma}_{t-s}\tilde{\eta}_s$ in the decomposition for $\tilde{\eta}$ in Proposition 5.1(c) to see that

(5.40)
$$\tilde{\eta}_t = \tilde{\gamma}_t \tilde{\eta}_0 + \tilde{N}_t^t,$$

where

(5.41)
$$\tilde{N}_s^t(i) = \sum_j \int_0^s \tilde{\gamma}_{t-r}(i, j) d\tilde{M}_r(j) \equiv \int_0^s \tilde{\gamma}_{t-r} d\tilde{M}_r(i), \qquad 0 \le s \le t,$$

and $\tilde{M}$ is defined as in Proposition 5.1(b) with $\tilde{A}$, $\tilde{\eta}_s$, in place of $A$, $\eta_s$. Our assumption (5.2) easily gives us enough summability to verify that the bounded variation term in the above vanishes, and Remark 5.2 shows the above series of martingales is $L^2$-convergent. Therefore (3.2) holds, while (3.1) is immediate from Proposition 5.1, as is (3.5) (see Remark 5.2). Finally, (3.6) is precisely (5.37) (by (5.16)), and so we may apply Theorem 3.3 with $\nu$ equal to the law of $\eta_0 - \theta$.  $\square$

With a view to verifying (5.37) we would like to infer $\tilde{C}_\infty < \infty$ from the a.s. convergence of $\langle \tilde{\eta}_t, \mathbf{1} \rangle$. To apply Lemma 3.2 we will need

(5.42)
$$\exists \, K > 0 : \sup_j \sum_i [|(\tilde{A} - I)(x)(i, j)| + |(A - I)(x)(i, j)|] \le K \quad \mu\text{- a.a. } (x, A).$$

PROPOSITION 5.8. *Let $\tilde{\eta}_0 \in X_f$, and assume either condition* (5.42), *or*

$$(5.43) \qquad \sup_t \mathbf{E}^{\tilde{\eta}_0}[\langle \tilde{\eta}_t, \mathbf{1} \rangle^2] < \infty.$$

*Then*

$$(5.44) \qquad \tilde{C}_\infty = \int_0^\infty \langle \mathbf{1}, q(\tilde{\eta}_s)\mathbf{1} \rangle \, ds < \infty \ a.s.$$

PROOF. For $n \in \mathbb{N}$ we define $T_n = \inf\{t \geq 0 : \langle \tilde{\eta}_t, \mathbf{1} \rangle \geq n\}$. If (5.42) holds and $n \geq \langle \tilde{\eta}_0, \mathbf{1} \rangle$, then Proposition 5.1(b) shows that $|\langle \tilde{\eta}_{T_n} - \tilde{\eta}_{T_n -}, \mathbf{1} \rangle| \leq nK$ a.s. on $\{T_n < \infty\}$, where $K$ is as in (5.42) and so Lemma 3.2 implies the required conclusion. If the other hypothesis holds, then $\mathbf{E}[\tilde{C}_\infty] < \infty$ is obvious. $\qquad \square$

THEOREM 5.9. *Assume either that condition* (5.42) *holds, or for each $\tilde{\eta}_0 \in X_f$, $\sup_t \mathbf{E}^{\tilde{\eta}_0}[\langle \tilde{\eta}_t, \mathbf{1} \rangle^2] < \infty$. Assume also that $q(\eta; i, j) \geq 0$ for all $i, j \in \mathbb{Z}^d$ and square summable $\eta$. Then* (5.37) *holds, and therefore, if $\mathcal{L}[\eta_0] \in \mathcal{M}_\theta$ or $\mathcal{L}[\eta_0]$ and $\gamma$ are as in Corollary 5.4, then $\mathcal{L}[\eta_t] \Rightarrow \nu_\theta$ as $t \to \infty$.*

PROOF. As $q(\tilde{\eta}_t; i, j) \geq 0$, clearly the integral in (5.37) is just $\tilde{C}_\infty$. Proposition 5.8 therefore implies (5.37). The result now follows from Theorem 5.5. $\qquad \square$

**5.1. The smoothing and potlatch processes.** The two primary examples we consider are the *smoothing process* and the *potlatch process*, which are dual to one another. Let $p(i, j)$ be a doubly stochastic Markov chain matrix, and let $W$ be a bounded nonnegative random variable with mean 1. For the smoothing process, when the clock at $x$ goes off, $\eta(x)$ is replaced by $W_x p \eta(x)$, and all other coordinates are left fixed. Here, the $W_x, x \in \mathbb{Z}^d$ are iid with law $\mathcal{L}[W]$. For the potlatch process, when the clock at site $x$ goes off, the configuration $\eta$ is replaced by $A(x)\eta$, where $A(x)\eta(x) = p(x, x)W_x\eta(x)$, and for $i \neq x$, $A(x)\eta(i) = \eta(i) + W_x\eta(x)p(x, i)$. That is, the value $\eta(x)$ is removed, multiplied by $W_x$, and then redistributed according to the kernel $p(x, i)$. The two processes are respective duals.

For the smoothing process,

$$(5.45) \qquad A(x; i, j) = \begin{cases} 1, & i = j \neq x, \\ W_x p(i, j), & i = x, \\ 0, & \text{else,} \end{cases}$$

and hence

$$(A - I)(x; i, j) = ((W_x p) - I)(i, j)\delta(i, x).$$

It follows that

$$(5.46) \qquad (|(A - I)(x)| + |(\tilde{A} - I)(x)|)\mathbf{1}(i) \leq 2(W_x + 1)\delta(i, x).$$

The boundedness condition on $W$ shows that (5.42) holds and also gives (5.2) (although square integrability suffices for the latter). Note also that (5.7) is true because $W$ has mean 1. We also have

$$(5.47) \qquad \begin{aligned} q(\eta; i, j) &= \mathbf{E}[(((Wp) - I)\eta)(i)^2]\delta(i, j) \\ &= (\mathbf{E}[W^2 - 1]p\eta(i)^2 + (p - I)\eta(i)^2)\delta(i, j) \geq 0. \end{aligned}$$

Thus, Theorem 5.9 applies when $\tilde{\eta}_t$ is the smoothing process, and hence we obtain a convergence result for the dual of the smoothing process, which is the potlatch process.

THEOREM 5.10. *Let $\eta_t$ be the potlatch process, and assume that $\mathcal{L}[\eta_0] \in \mathcal{M}_\theta$, or that $p(i,j)$ is a random walk kernel and $\mathcal{L}[\eta_0]$ is translation invariant, shift ergodic and satisfies $E[\eta_0(0)] = \theta$. Then $\mathcal{L}[\eta_t] \Rightarrow \nu_\theta$ as $t \to \infty$.*

For the potlatch process,

$$(5.48) \qquad A(x; i, j) = \begin{cases} 1, & i = j \neq x, \\ W_x p(j, i), & j = x, \\ 0, & \text{else}, \end{cases}$$

and hence

$$(A - I)(x; i, j) = (W_x p(j, i) - \delta(j, i))\delta(j, x)$$

and we have already checked (5.2) and (5.42) (see (5.46)). As before, (5.7) also holds. Furthermore,

$$q(\eta; i, j) = \sum_k \mathbf{E}[(Wp(k, i) - \delta(k, i))(Wp(k, j) - \delta(k, j))]\eta(k)^2,$$

and

$$\langle \mathbf{1}, q(\eta)\mathbf{1} \rangle = (\mathbf{E}[W^2] - 1)\langle \eta^2, \mathbf{1} \rangle.$$

Now, in the case that $\mathbf{P}[W = 1] < 1$, $\mathbf{E}[W^2] > 1$, so it follows from Proposition 5.8 that

$$(5.49) \qquad \int_0^\infty \langle \eta_t^2, \mathbf{1} \rangle \, dt < \infty \quad \text{a.s. } \mathbf{P}^{\bar{\eta}_0}.$$

On the other hand, from the above expression for $q(\eta; i, j)$,

$$(5.50) \qquad \langle \mathbf{1}, |q(\eta)|\mathbf{1} \rangle \leq (\mathbf{E}[W^2] + 3)\langle \eta^2, \mathbf{1} \rangle.$$

Therefore, (5.37) holds by (5.49). Consequently, we obtain a convergence result for the smoothing process via Theorem 5.5 in the case that $\mathbf{P}[W = 1] < 1$.

In the case that $\mathbf{P}[W = 1] = 1$, then $\langle \eta_t, \mathbf{1} \rangle = \langle \eta_0, \mathbf{1} \rangle$ with probability one for all $t$, and there is no obvious way to obtain (5.37) in general. We now give a direct derivation of (5.49) that holds in the transient, random walk (i.e., (5.39) holds) case. Let $\hat{p}(i, j)$ be the symmetrized kernel, $\hat{p}(i, j) = (p(i, j) + p(j, i))/2$, and let $\hat{G}(i, j) = \int_0^\infty \hat{p}_s(i, j) \, ds$ (recall (1.1)).

PROPOSITION 5.11. *Let $\eta_t$ be the potlatch process. Assume (5.39), $\hat{G}(0, 0) < \infty$, and*

$$(5.51) \qquad \mathbf{E}[W^2] < \frac{\hat{G}(0, 0)}{(p\hat{G}\tilde{p})(0, 0)}.$$

*Then for any initial $\eta_0 \in X_f$,*

$$(5.52) \qquad \int_0^\infty \mathbf{E}[\langle \eta_t^2, \mathbf{1} \rangle] \, dt < \infty.$$

REMARK 5.12. The assumption $\hat{G}(0, 0) < \infty$ implies that the right side of (5.51) is strictly larger than 1. Thus, (5.49) must hold when $\hat{G}(0, 0) < \infty$ and $\mathbf{P}[W = 1] = 1$.

PROOF. Note that $\gamma_t = P_t$ since $\mathbf{E}[W] = 1$. By (5.40)

$$(5.53) \qquad \mathbf{E}[\eta_t^2(k)] = (P_t \eta_0(k))^2 + \mathbf{E}[(N_t^t(k))^2].$$

It is the second term on the right side above that requires the most effort. By (5.1) and (5.16)

$$\mathbf{E}[N_t^t(k)^2] =$$

$$\sum_{i,j} \int_0^t p_{t-s}(k,i) p_{t-s}(k,j) \sum_l \mathbf{E}[((Wp) - I)(l,i)((W\tilde{p}) - I)(l,j)] \mathbf{E}[\eta_s^2(l)] \, ds.$$

By summing on $k$, we obtain

$$(5.54) \quad \mathbf{E}[\langle (N_t^t)^2, \mathbf{1}\rangle] = \sum_i \int_0^t \mathbf{E}[((Wp) - I)\hat{p}_{2(t-s)}((W\tilde{p}) - I)](i,i) \mathbf{E}[\eta_s^2(i)] \, ds,$$

We define

$$\hat{p}(t) = \mathbf{E}[((Wp) - I)\hat{p}_t((W\tilde{p}) - I)](i,i), \qquad i \in \mathbb{Z}^d,$$

which by translation invariance of $p$ and $\hat{p}_t$ does not depend on $i$. Note that

$$(5.55) \qquad \int_0^\infty \hat{p}(t) dt = \mathbf{E}[((Wp) - I)\hat{G}((W\tilde{p}) - I)](0,0)$$

$$= \mathbf{E}[W^2](p\hat{G}\tilde{p})(0,0) - \hat{G}(0,0) + 2,$$

and that, by assumption, $\int_0^\infty \hat{p}(t) dt < 2$. Hence

$$(5.56) \qquad \int_0^T \mathbf{E}[\langle \eta_t^2, \mathbf{1}\rangle] \, dt = \int_0^T \langle (p_t \eta_0)^2, \mathbf{1}\rangle \, dt + \int_0^T dt \int_0^t ds \, \hat{p}(2(t-s)) \mathbf{E}[\langle \eta_s^2, \mathbf{1}\rangle]$$

$$\leq \frac{1}{2} \left( \langle \eta_0, \hat{G}\eta_0 \rangle + \int_0^\infty \hat{p}(t) dt \int_0^T \mathbf{E}[\langle \eta_t^2, \mathbf{1}\rangle] \, dt \right),$$

and we get

$$(5.57) \qquad \int_0^T \mathbf{E}[\langle \eta_t^2, \mathbf{1}\rangle] \, dt \leq \frac{\langle \eta_0, \hat{G}\eta_0 \rangle}{2 - \int_0^\infty \hat{p}(t) \, dt} < \infty.$$

Now let $T \to \infty$. In particular

$$(5.58) \qquad \int_0^\infty \mathbf{E}[\langle \eta_t^2, \mathbf{1}\rangle] \, dt \leq \frac{\hat{G}(0,0)}{\hat{G}(0,0) - \mathbf{E}[W^2](p\hat{G}\tilde{p})(0,0)} \, \langle \eta_0, \mathbf{1}\rangle^2 < \infty.$$

$\square$

To sum up, we now have the following.

THEOREM 5.13. *Let $\eta_t$ be the smoothing process. Assume either that $\mathbf{P}[W = 1] < 1$, or that $\mathbf{P}[W = 1] = 1$, (5.39) holds and $\hat{G}(0,0) < \infty$. If $\mathcal{L}[\eta_0] \in \mathcal{M}_\theta$, or $p(i,j)$ is a random walk kernel and $\mathcal{L}[\eta_0]$ is translation invariant, shift ergodic and satisfies $\mathbf{E}[\eta_0(0)] = \theta$, then $\mathcal{L}[\eta_t] \Rightarrow \nu_\theta$ as $t \to \infty$.*

Finally we note that our method fails for the smoothing process with $\mathbf{P}[W = 1] = 1$. First of all it is clear that we could not get (5.49) as we did above. In fact, we can give an example where (5.49) turns out to be wrong. Hence in this case we could not apply any version of Theorem 3.1. Here is the example.

Consider the *nearest neighbour* smoothing process $\eta_t$ and potlatch process $\tilde{\eta}_t$ on $\mathbb{Z}$ with $\mathbf{P}[W = 1] = 1$. Hence, if $A(i, j)$ is given by (5.48), $\int (A - I)(i, k)(A - I)(j, l)d\mu = \delta(k, l)(p - I)(k, i)(p - I)(l, j)$ assumes the values

$$
(5.59) \qquad
\begin{cases}
1, & i = j = k = l, \\
-\frac{1}{2}, & i \sim j = k = l, \\
-\frac{1}{2}, & j \sim i = k = l, \\
\frac{1}{4}, & j \sim k = l \sim i,
\end{cases}
$$

where $i \sim j$ means $|i - j| = 1$. In particular, $q(\mathbf{1}_{\{k\}}; i, i + 2) = \frac{1}{4}\mathbf{1}_{k=i+1}$. It follows that

$$
(5.60) \qquad \int_0^\infty \langle \mathbf{1}, |q(\tilde{\eta}_t)|\mathbf{1}\rangle \, dt \geq \frac{1}{4} \int_0^\infty \langle \tilde{\eta}_t^2, \mathbf{1}\rangle \, dt.
$$

Note that for the potlatch process, $\langle \tilde{\eta}_t, \mathbf{1}\rangle = \langle \tilde{\eta}_0, \mathbf{1}\rangle = 1$ $\mathbf{P}^{\delta_0}$-a.s., $t \geq 0$, and $\mathbf{E}^{\delta_0}[\tilde{\eta}_t(u)] = \gamma_t(0, u)$, where $\gamma_t = P_t$, the probability transition function of nearest-neighbour, rate-one, continuous random walk on $\mathbb{Z}$. Let $\rho_l = \mathbf{1}_{(-t^{3/4}, t^{3/4})}$. By Chebyshev's inequality, there exists a $C \in (0, \infty)$ such that

$$
\mathbf{E}^{\delta_0}[\langle \tilde{\eta}_t, \mathbf{1} - \rho_t\rangle] \leq C/t^{3/2}
$$

On account of this estimate,

$$
\int_0^\infty \mathbf{E}^{\delta_0}[\langle \tilde{\eta}_t, \mathbf{1} - \rho_t\rangle]dt < \infty.
$$

By the Borel-Cantelli lemma, it follows that

$$
(5.61) \qquad \lim_{n \to \infty} \int_n^{n+1} \langle \tilde{\eta}_t, \mathbf{1} - \rho_t\rangle = 0 \qquad \mathbf{P}^{\delta_0} - a.s.
$$

From this one can show, using the bounded transition rates of the potlatch process, that $\lim_{t \to \infty} \langle \tilde{\eta}_t, \mathbf{1} - \rho_t\rangle = 0$ $\mathbf{P}^{\delta_0}$-a.s., and thus

$$
(5.62) \qquad \lim_{t \to \infty} \langle \tilde{\eta}_t, \rho_t\rangle = 1 \qquad \mathbf{P}^{\delta_0}\text{-a.s.}
$$

Since $\langle \tilde{\eta}_t, \rho_t\rangle^2 \leq \langle \tilde{\eta}_t^2, \mathbf{1}\rangle \langle \mathbf{1}, \rho_t\rangle$, and $\langle \mathbf{1}, \rho_t\rangle \sim 2t^{3/4}$ as $t \to \infty$,

$$
\int_1^\infty \langle \tilde{\eta}_t^2, \mathbf{1}\rangle \, dt \geq \int_1^\infty \frac{\langle \tilde{\eta}_t, \rho_t\rangle^2}{\langle \mathbf{1}, \rho_t\rangle} \, dt = \infty \qquad \mathbf{P}^{\delta_0}\text{-a.s..}
$$

by (5.62). Thus, (5.49) does not hold.

## References

[1] Cox, J.T., Fleischmann, K. and Greven, A. (1996). Comparison of interacting diffusions and an application to their ergodic theory. *Prob. Th. Rel. Fields* **105** 515–528.

[2] Cox, J.T., Greven, A. and Shiga, T. (1995). Finite and infinite systems of interacting diffusions. *Prob. Th. Rel. Fields* **103** 165–197.

[3] Cox, J.T. and Klenke, A. (1999). Recurrence and ergodicity of interacting particle systems. *Prob. Th. Rel. Fields* (to appear).

[4] Dawson, D.A. and Perkins, E.A. (1998). Long-time behavior and co-existence in a mutually catalytic branching model. *Ann. Probab.* **26** 1088–1138.

[5] Ikeda, N. and Watanabe, S. (1981). *Stochastic differential equations and diffusion processes.* North Holland, Amsterdam.

[6] Kurtz, T. and Protter, P. (1996) Weak convergence of stochastic integrals and differential equations. II. Infinite-dimensional case. *Probabilistic models for nonlinear partial differential equations* (Montecatini Terme, 1995), 197-285, Lecture Notes in Math. 1627, Springer, New York.

[7] Liggett, T.M. (1973). A Characterization of the Invariant Measures for an Infinite Particle System with Interactions. *Trans. Amer. Math. Soc.* **179** 433–451.

[8] Liggett, T.M. (1985). *Interacting Particle Systems.* Springer Verlag, New York.

[9] Liggett, T.M. and Spitzer, F.L. (1981). Ergodic theorems for coupled random walks and other systems with locally interacting components. *Z. Wahrsch. verw. Gebiete* **56** 443–468.

[10] Mytnik, L. (1998). Uniqueness for a mutually catalytic branching model. *Prob. Th. Rel. Fields* **112** 245–253.

[11] Shiga, T. (1992). Ergodic theorems and exponential decay of sample paths for certain interacting diffusion systems. *Osaka J. Math.* **29** 789–807.

[12] Shiga, T. and Shimizu, A. (1980). Infinite-dimensional stochastic differential equations and their applications. *J. Math. Kyoto Univ.* **20** 395–416.

[13] Spitzer, F.L. (1981). Infinite systems with locally interacting components. *Ann. Probab.* **9** 349–364.

J. Theodore Cox, Mathematics Department, Syracuse University, Syracuse, NY 13244, USA
*E-mail address*: jtcox@gumby.syr.edu
*URL*: http://gumby.syr.edu/~jtcox

Achim Klenke, Universität Erlangen-Nürnberg, Mathematisches Institut, Bismarck-strasse $1\frac{1}{2}$, 91054 Erlangen, Germany
*E-mail address*: klenke@mi.uni-erlangen.de
*URL*: http://www.mi.uni-erlangen.de/~klenke

Edwin A. Perkins, Department of Mathematics, The University of British Columbia, 1984 Mathematics Road, Vancouver, B.C., Canada V6T 1Z2
*E-mail address*: perkins@math.ubc.ca
*URL*: http://www.math.ubc.ca/~perkins/perkins.html

Canadian Mathematical Society
Conference Proceedings
Volume **26**, 2000

# Weighted Quantile Processes and Their Applications to Change-Point Analysis

Miklós Csörgő and Barbara Szyszkowicz

*In honour of our super colleague, Don Dawson*

ABSTRACT. Given a sample of chronologically ordered independent observations, we study the problem of testing for them being identically distributed, versus having at most one change in their distribution at an unknown time. In Section 1 we summarize how this problem leads to having to describe it in weighted metrics via some two-time parameter Gaussian random fields (tied down Kiefer processes), when the problem in hand is parametrized in terms of empirical distributions. In Section 2 we pose the same problem in terms of the chronologically observed observations themselves. This approach combined with that of Section 1 in turn leads us to take an appropriately modified path to studying weighted quantile change-point processes in Section 3, where we show how the already utilized Gaussian random fields of Section 1 can be put to use again in this new context.

## 1. Introduction: weighted empirical change processes

Let $X_1, X_2, \ldots, X_n$ be independent real valued random variables with respective right continuously defined distribution functions $F^{(1)}, \ldots, F^{(n)}$, $n \geq 1$. We want to test

$$H_0 : \quad F^{(1)}(x) = \cdots = F^{(n)}(x) := F(x) \text{ for all } x \in I\!\!R$$

versus the change-point alternative

$$
\begin{aligned}
H_A \quad : \quad & \text{there exists an integer } k^*, \ 1 \leq k^* < n \text{ such that for all } x \in I\!\!R \\
& F^{(1)}(x) = \cdots = F^{(k^*)}(x), \quad F^{(k^*+1)}(x) = \cdots F^{(n)}(x) \\
& \text{and } F^{(k^*)}(x_0) \neq F^{(k^*+1)}(x_0) \text{ with some } x_0 \in I\!\!R.
\end{aligned}
$$

Divide the **chronologically ordered** random sample into two parts:

$$\{X_i, \ 1 \leq i \leq k\} \quad \text{and} \quad \{X_i, \ k < i \leq n\},$$

2000 *Mathematics Subject Classification.* Primary 60F17; Secondary 60G60, 60G10.

*Key words and phrases.* Weighted empirical and quantile processes; change-point analysis; tied down Kiefer processes (random fields); Brownian bridges.

Research supported by NSERC grants at Carleton University, Ottawa.

and define the **empirical distributions**

(1.1) $$\hat{F}_k(x) := \frac{1}{k} \sum_{1 \le i \le k} \mathbb{1}\{X_i \le x\},$$

(1.2) $$\check{F}_{n-k}(x) := \frac{1}{n-k} \sum_{k < i < n} \mathbb{1}\{X_i \le x\}, \ \check{F}_0(x) := 0, \ x \in \mathbb{R},$$

$k = 1, \ldots, n$. Since $k^*$ of $H_A$ is usually unknown, consider (cf., e.g., D.S. Darkhovsky (1984), D. Picard (1985), J. Deshayes and D. Picard (1981, 1986), B. Szyszkowicz (1991, 1994, 1998), M. Csörgő and B. Szyszkowicz (1994), M. Csörgő, L. Horváth and B. Szyszkowicz [CsHSz] (1997), and M. Csörgő and L. Horváth (1997)):

(1.3) $$n^{1/2} \max_{1 \le k < n} \sup_x |\hat{F}_k(x) - \check{F}_{n-k}(x)|$$

$$= \max_{1 \le k < n} \sup_x \frac{\left| \sum_{1 \le i \le k} \mathbb{1}\{X_i \le x\} - \frac{k}{n} \sum_{1 \le i \le n} \mathbb{1}\{X_i \le x\} \right|}{n^{1/2}\left(\frac{k}{n}\left(1 - \frac{k}{n}\right)\right)}.$$

Unfortunately, these weighted statistics $\xrightarrow{\text{D}} \infty$, as $n \to \infty$, even if $H_0$ were to be true! One shows (cf. CsHSz (1997)) however: If $H_0$ holds true, as $n \to \infty$, then

(1.4) $$\max_{1 \le k < n} \frac{n^{1/2}\left(\frac{k}{n}\left(1 - \frac{k}{n}\right)\right) \sup_x |\hat{F}_k(x) - \check{F}_{n-k}(x)|}{\left(\frac{k}{n}\left(1 - \frac{k}{n}\right)\right)^{1/2 - \nu}} = O_P(n^{-\nu})$$

for all $0 \le \nu < 1/2$.

Posing the problem of a possible change in distribution as in (1.3) and (1.4) leads in a most natural way to having to consider weight functions that relate to the *time* parameter set $1 \le k < n$ of a *two-parameter* real valued empirical chang e process on $\mathbb{R} \times [1 \le k < n)$. This is *unlike* the weighted sup–norm metric problems that are dealt with in D. Chibisov (1964), N. O'Reilly (1974), M. Csörgő, S. Csörgő, L. Horváth and D.M. Mason [CsCsHM] (1986), and in the related parts of the books G.R. Shorack and J.A. Wellner (1986), M. Csörgő and L. Horváth (1993), where the weight functions relate to the *single, space parameter* set of uniform empirical and quantile processes.

Clearly, the identification of the non–degenerate limit of the $\max_k \sup_x$ functionals (statistics) in (1.4) requires the construction of appropriate two–time parameter real valued stochastic processes (random fields). In this regard, we introduce $\{\Gamma_F(x,t); x \in \mathbb{R}, 0 \le t \le 1\}$, a mean zero Gaussian process, with covariance function

$$E\Gamma_F(x,s)\Gamma_F(y,t) = \{F(x \wedge y) - F(x)F(y)\}\{s \wedge t - st\}.$$

Thus,

(1.5) $$\{\Gamma_F(x,t); x \in \mathbb{R}, 0 \le t \le 1\} \overset{\text{D}}{=} \{K_F(x,t) - tK_F(x,1); x \in \mathbb{R}, 0 \le t \le 1\},$$

where $\{K_F(x,v); x \in \mathbb{R}, 0 \le v < \infty\}$ is a mean zero Gaussian process with covariance function

$$EK_F(x,u)K_F(y,v) = \{F(x \wedge y) - F(x)F(y)\}\{u \wedge v\},$$

i.e., $K_F(\cdot, \cdot)$ is a Kiefer process (cf. M. Csörgő and P. Révész (1981, Chapter 1)).

D.W. Müller (1970) was first to establish the convergence in law of the sample distribution process to $K_F(\cdot,\cdot)$. Via his construction of $K_F(\cdot,\cdot)$ J. Kiefer (1972) achieved the first strong approximation of the sample distribution process by such a random field.

Via J. Komlós, P. Major and G. Tusnády [KMT] (1975, 1976) (cf. (2.16)) instead of M. Csörgő and L. Horváth (1988), Lemma 3.1 of CsHSz (1997) on the real line reads as follows.

LEMMA 1.1. *If $H_0$ holds true, then we can define a sequence of Gaussian processes $\{\Gamma_n(x,t); \; x \in \mathbb{R}, \; 0 \leq t \leq 1\}$ such that, as $n \to \infty$, we have*

(i) $\displaystyle \sup_{\frac{1}{n} \leq t \leq \frac{n-1}{n}} \sup_x \left| \frac{[nt](n-[nt])}{n^{3/2}} \left( \hat{F}_{[nt]}(x) - \check{F}_{n-[nt]}(x) \right) - \Gamma_n(x,t) \right| = O_P(n^{-1/2} \log^2 n)$

*and*

(ii) $\displaystyle \sup_{\frac{1}{n} \leq t \leq \frac{n-1}{n}} \sup_x \frac{\left| \frac{[nt](n-[nt])}{n^{3/2}} (\hat{F}_{[nt]}(x) - \check{F}_{n-[nt]}(x)) - \Gamma_n(x,t) \right|}{(t(1-t))^{1/2-\nu}} = O_P(n^{-\nu})$

*for all $0 \leq \nu < 1/2$, where*

(1.6) $\qquad \{\Gamma_n(x,t); \; x \in \mathbb{R}, \; 0 \leq t \leq 1\}$

$\qquad\qquad \overset{\mathrm{D}}{=} \quad \{\Gamma_F(x,t); \; x \in \mathbb{R}, \; 0 \leq t \leq 1\}$ *for each $n \geq 1$.*

*Moreover, as $n \to \infty$, we have also*

(iii) $\displaystyle \sup_{\frac{1}{n} \leq t \leq \frac{n-1}{n}} \sup_x \frac{\left| \frac{[nt](n-[nt])}{n} \left( \hat{F}_{[nt]}(x) - \check{F}_{n-[nt]}(x) \right) - \Gamma_n(x,t) \right|}{q(t)} = O_P(1)$

*for all weight functions $q : (0,1) \to (0,\infty)$ satisfying*

(1.7) $\qquad\qquad\qquad \inf_{\delta \leq t \leq 1-\delta} q(t) > 0 \quad \text{for all} \; 0 < \delta < 1/2$

*and*

(1.8) $\qquad\qquad \lim_{t \downarrow 0} \frac{q(t)}{t^{1/2}} = \infty, \quad \text{as well as} \quad \lim_{t \uparrow 1} \frac{q(t)}{(1-t)^{1/2}} = \infty.$

In order to describe the weighted asymptotic Gaussian behaviour of the empirical change process on the whole unit interval in $t \in [0,1]$, we let

(1.9) $\qquad\qquad\qquad \Delta(t) := \sup_x |\Gamma_F(x,t)|, \quad 0 \leq t \leq 1,$

(1.10) $\qquad I_{0,1}(q,c) := \int_0^1 \frac{1}{t(1-t)} \exp\left( -c \frac{q^2(t)}{t(1-t)} \right) dt, \quad c > 0,$

and define $Q_{0,1}$ to be the *class of functions $q : (0,1) \to (0,\infty)$, that satisfy* (1.7) *and are nondecreasing near zero and nonincreasing near one.* We have

THEOREM 1.2. (Theorems 2.1 and 2.3 of CsHSz (1997)) *Assume $q \in Q_{0,1}$. Then*
(i) *$I_{0,1}(q,c) < \infty$ for some $c > 0$ if and only if*

(1.11) $\qquad \limsup_{t \downarrow 0} \Delta(t)/q(t) < \infty \quad \text{and} \quad \limsup_{t \uparrow 1} \Delta(t)/q(t) < \infty \; \text{a.s.}$

(ii)   $I_{0,1}(q,c) < \infty$ *for all* $c > 0$ *if and only if*

(1.12)        $\displaystyle\lim_{t\downarrow 0} \Delta(t)/q(t) = 0$   *and*   $\displaystyle\lim_{t\uparrow 1} \Delta(t)/q(t) = 0$   *a.s.*

REMARK 1.3. We note that for any fixed $x_0 \in I\!\!R$ the stochastic process

(1.13)   $\left\{ B_{x_0}(t);\ 0 \le t \le 1 \right\} := \left\{ \left( F(x_0)\left(1 - F(x_0)\right) \right)^{-1/2} \Gamma_F(x_0, t);\ 0 \le t \le 1 \right\}$

is a Brownian bridge, i.e., a mean zero Gaussian process with covariance function
$(s,t) \mapsto (s \wedge t) - st$. Moreover, by (1.11) and (1.12) respectively, we have also

(1.14)        $\displaystyle\limsup_{t\downarrow 0} \left| B_{x_0}(t) \right|/q(t) < \infty$   and   $\displaystyle\limsup_{t\uparrow 1} \left| B_{x_0}(t) \right|/q(t) < \infty$

and

(1.15)        $\displaystyle\lim_{t\downarrow 0} \left| B_{x_0}(t) \right|/q(t) = 0$   and   $\displaystyle\lim_{t\uparrow 1} \left| B_{x_0}(t) \right|/q(t) = 0.$

Hence, on using Theorem 3.3 of CsCsHM (1986) in combination with (1.14) we
obtain that $I_{0,1}(q,c) < \infty$ for some $c > 0$, while Theorem 3.4 of CsCsHM (1986)
in combination with (1.15) yields that $I_{0,1}(q,c) < \infty$ for all $c > 0$. The *difficult
part* of establishing Theorem 1.2 is to conclude (1.11) and (1.12), respectively,
from assuming the finiteness of $I_{0,1}(q,c)$ for some $c > 0$ and, respectively, that for
all $c > 0$ (for details we refer to Section 2 of CsHSz (1997), or to Section A.7 of
M. Csörgő and L. Horváth (1997)). We note also that the weight functions $q \in Q_{0,1}$
for which the integral $I_{0,1}(q,c)$ is finite for all $c > 0$ are the so–called Chibisov–
O'Reilly weight functions (cf. D. Chibisov (1964), N. O'Reilly (1974) and, for
further major steps along these lines, CsCsHM (1986)). For further developments
along these lines we refer to M. Csörgő, Q.-M. Shao and B. Szyszkowicz (1991),
and Keprta (1997a,b and 1998).

Based on Lemma 1.1 and Theorem 1.2, for the **empirical change process**
$w_n$ that, with $x \in I\!\!R$, is defined as

(1.16)  $w_n(x,t) := \begin{cases} 0, & 0 \le t < 1/n, \\ \frac{[nt](n-[nt])}{n^{3/2}}\left( \hat{F}_{[nt]}(x) - \check{F}_{n-[nt]}(x) \right), & \frac{1}{n} \le t \le \frac{n-1}{n}, \\ 0, & \frac{n-1}{n} < t \le 1, \end{cases}$

we have, as in CsHSz (1997, Theorem 3.1) (cf. also Theorem 2.6.1 in M. Csörgő
and L. Horváth (1997) and Theorem 8.2 in B. Szyszkowicz (1998))

THEOREM 1.4. *Under $H_0$ and $q \in Q_{0,1}$, with the sequence of Gaussian processes
$\Gamma_n(\cdot,\cdot)$ of Lemma 1.1, as $n \to \infty$, we have*

(i)        $\displaystyle\sup_{0<t<1} \sup_x \left| w_n(x,t) - \Gamma_n(x,t) \right|/q(t)$

$= \begin{cases} o_P(1) & \textit{if and only if } I_{0,1}(q,c) < \infty \textit{ for all } c > 0, \\ O_P(1) & \textit{if and only if } I_{0,1}(q,c) < \infty \textit{ for some } c > 0, \end{cases}$

(ii)        $\displaystyle\sup_{0<t<1} \sup_x \left| w_n(x,t) \right|/q(t) \xrightarrow{\mathrm{D}} \sup_{0<t<1} \sup_x \left| \Gamma_F(x,t) \right|/q(t)$

*if and only if $I_{0,1}(q,c) < \infty$ for some $c > 0$, where $\Gamma_F(\cdot,\cdot)$ is a tied down Kiefer
process as in (1.5).*

For example, (ii) is true with

$$q(t) := \left( t(1 - t) \log \log \frac{1}{t(1 - t)} \right)^{1/2}, \quad 0 < t < 1,$$

for which $I(q, c) < \infty$ only for $c > \sqrt{2}$. Thus, with this $q$, we can only conclude $O_P(1)$ of (i), which does not imply (ii).

On assuming that $F$ is continuous, the $o_P(1)$ statement of (i), and (ii) of Theorem 1.4 provide, in principle, a distribution free asymptotic test for testing $H_0$ versus $H_A$. Unfortunately, the distribution function of the limiting random variable in the latter limit theorem is not known, not even with $q = 1$. Hence it is of interest to study the tail behaviour of the tied down Kiefer process $\Gamma_F(\cdot, \cdot)$ (cf. (1.5)), possibly with weights $q \in Q_{0,1}$. In his Ph.D. thesis, Bin Chen (1998, Chapter 6) proved the following large deviation results that are of interest in this regard.

THEOREM 1.5. *Let $F$ be continuous. For $q \in Q_{0,1}$ assume that it is increasing on $(0, 1/2]$, and is symmetric with respect to $1/2$. Then, on assuming $I_{0,1}(q, c) < \infty$ for some $c > 0$, we have*

$$\lim_{u \to \infty} \frac{1}{u^2} \log P \left\{ \sup_{x \in \mathbb{R}} \sup_{0 < t < 1} \frac{|\Gamma_F(x, t)|}{q(t)} \geq u \right\} = - \inf_{0 < t < 1} \frac{2q^2(t)}{t(1 - t)},$$

*as well as*

$$\lim_{u \to \infty} \frac{1}{u^2} \log P \left\{ \sup_{0 < t < 1} \frac{|B(t)|}{q(t)} \geq u \right\} = - \inf_{0 < t < 1} \frac{q^2(t)}{2t(1 - t)},$$

*where $\{B(t);\ 0 \leq t \leq 1\}$ is a Brownian bridge.*

For studies of related distributional problems we refer to B.J. Eastwood and V.R. Eastwood (1998), and to M.D. Burke (1998).

REMARK 1.6. Interestingly enough, Theorem 1.5 tells us that even 'constants–wise', the weighted large deviation behaviour of a tied down Kiefer process is not very much different from that of a Brownian bridge. In view of Theorem 1.2 and Remark 1.3 it is not surprising of course that, after taking $\sup_{x \in \mathbb{R}}$ as in the statement of Theorem 1.5 that concerns $\Gamma_F(\cdot, \cdot)$, 'shape–wise' it should end up having the same type of weighted tail behaviour that is possessed by a Brownian bridge. It is also interesting to note in this regard that, while (i) of Theorem 1.4 completely characterizes the limiting Gaussian behaviour of the weighted empirical change processes $\{w_n(x, t)/q(t);\ x \in \mathbb{R},\ 0 < t < 1\}$ as $n \to \infty$, as well as that of its $\sup_{0 < t < 1} \sup_x$ behaviour as in its (ii), *the extreme value problem of* finding norming constants $a_n$ and $b_n$ such that, as $n \to \infty$,

$$a_n \sup_{0 < t < 1} \sup_x |w_n(x, t)|/(t(1 - t))^{1/2} - b_n \overset{D}{\to} \text{ a nondegenerate random variable}$$

remains open (cf. M. Csörgő and R. Norvaiša (1998) for some related results). On the other hand, with any fixed $x = x_0$ and appropriate norming constants $\tilde{a}_n$ and $\tilde{b}_n$, via D.A. Darling and P. Erdős (1956), we have

$$\lim_{n \to \infty} P \left\{ \tilde{a}_n \sup_{0 < t < 1} |w_n(x_0, t)|/(t(1 - t))^{1/2} - \tilde{b}_n \leq y \right\} = \exp \left( -2e^{-y} \right)$$

for $y \in \mathbb{R}$.

Based on M. Csörgő and L. Horváth (1988), in CsHSz (1997) Theorem 1.4 is proved for $I\!\!R^d$ valued random variables, $d \geq 1$. M. Csörgő and R. Norvaiša (1997) established versions of (i) of Theorem 1.4 that are based on Banach space valued random variables.

The aim of this exposition is to see how, and to what extent, can the weighted sup–norm theory of Lemma 1.1 and Theorem 1.4 be extended to comparing the asymptotic behaviour of the data "before" to that of data "after" the postulated change. This amounts to saying that we are set to study empirical quantiles in the latter "before" and "after" change–point context in sup–norm.

With $\hat{F}_k$ and $\check{F}_{n-k}$, $1 \leq k \leq n$, as in (1.1) and (1.2) respectively, let

$$\hat{F}_k^{-1}(y) := \inf\{x : \hat{F}_k(x) \geq y\}, 0 < y \leq 1, \ \hat{F}_k^{-1}(0) := \hat{F}_k^{-1}(0+)$$

and

$$\check{F}_{n-k}^{-1}(y) := \inf\{x : \check{F}_{n-k}(x) \geq y\}, 0 < y \leq 1, \ \check{F}_{n-k}^{-1}(0) := \check{F}_{n-k}^{-1}(0+)$$

be their respective **empirical quantile functions**. À la comparing empirical distributions "before" the change to those "after", the problem of change in terms of quantiles can be posed via studying the asymptotic behaviour of (cf. (1.16))

$$(1.17) \qquad \max_{1 \leq k < n} n^{1/2} \left( \frac{k}{n} \left( 1 - \frac{k}{n} \right) \right) \sup_{0 \leq y \leq 1} |\hat{F}_k^{-1}(y) - \check{F}_{n-k}^{-1}(y)|.$$

However, this sequence of "unweighted" statistics goes to infinity almost surely even under $H_0$, unless $F$ of $H_0$ has finite support. The latter degenerate limiting behaviour of this sequence of statistics is to be compared to what (i) of Lemma 1.1 tells us about the Gaussian behaviour of its empirical distributions counterpart under $H_0$.

Clearly then, as far as quantile change processes are concerned, we have to do more work in order to have results that are comparable to those of Lemma 1.1 and Theorem 1.4. Indeed, just like when we deal with quantile processes in general (cf., e.g., M. Csörgő (1983), and M. Csörgő and B. Szyszkowicz (1998)), in order to regulate the asymptotic behaviour of comparing quantiles "before" to quantiles "after" a postulated unknown time of change in distribution as in $H_0$ versus $H_A$, we will have to make assumptions about $F$ of $H_0$.

## 2. Preliminaries

From now on we assume that $F$ of $H_0$ is continuous. Then

$$U_1 := F(X_1), \ldots, U_n := F(X_n)$$

are independent identically distributed (i.i.d.) uniform–$(0,1)$ random variables (rv's). Chronologically ordered as they are, let

$$(2.1) \qquad \hat{E}_{[nt]}(y) := \begin{cases} \frac{1}{[nt]} \sum_{1 \leq i \leq [nt]} \mathbb{1}\{U_i \leq y\} & , \ 0 \leq y \leq 1, \ 1/n \leq t \leq 1, \\ 0 & , \ 0 \leq y \leq 1, \ 0 \leq t < 1/n, \end{cases}$$

and

$$(2.2) \quad \check{E}_{n-[nt]}(y) := \begin{cases} \frac{1}{n-[nt]} \sum_{[nt] < i \leq n-1} \mathbb{1}\{U_i \leq y\} & , \ 0 \leq y \leq 1, \ 0 \leq t \leq \frac{n-1}{n}, \\ 0 & , \ 0 \leq y \leq 1, \ \frac{n-1}{n} < t \leq 1, \end{cases}$$

be the **uniform empirical distributions** of these rv's. Let $F^{-1}$ be the quantile function of the distribution $F$, defined by

$$(2.3) \qquad F^{-1}(y) := \inf\{x : F(x) \geq y\}, \; 0 < y \leq 1, \; F^{-1}(0) = F^{-1}(0+).$$

Then (cf. (1.1) and (1.2))

$$(2.4) \qquad \hat{F}_{[nt]}(F^{-1}(y)) = \hat{E}_{[nt]}(y)$$

and

$$(2.5) \qquad \check{F}_{n-[nt]}(F^{-1}(y)) = \check{E}_{n-[nt]}(y),$$

while the corresponding **uniform empirical quantile functions** are

$$(2.6) \qquad \hat{E}_{[nt]}^{-1}(y) := \inf\{s : \hat{E}_{[nt]}(s) \geq y\}, \; 0 < y \leq 1, 0 \leq t \leq 1,$$

$$(2.7) \qquad \check{E}_{n-[nt]}^{-1}(y) := \inf\{s : \check{E}_{n-[nt]}(s) \geq y\}, \; 0 < y \leq 1, 0 \leq t \leq 1,$$

with $\hat{E}_{[nt]}^{-1}(0) := \hat{E}_{[nt]}^{-1}(0+)$ and $\check{E}_{n-[nt]}^{-1}(0) := \check{E}_{n-[nt]}^{-1}(0+)$, $0 \leq t \leq 1$.

The **general empirical quantile functions**, $\hat{F}_{[nt]}^{-1}$ and $\check{F}_{n-[nt]}^{-1}$, via (2.3) and (2.4) respectively, are defined in a similar way.

We put

$$(2.8) \qquad V_n(y,t) := \begin{cases} 0 & , \; 0 \leq t < 1/n, \\ \frac{[nt](n-[nt])}{n^{3/2}}\left(\hat{F}_{[nt]}^{-1}(y) - \check{F}_{n-[nt]}^{-1}(y)\right) & , \; \frac{1}{n} \leq t \leq \frac{n-1}{n}, \\ 0 & , \; \frac{n-1}{n} < t \leq 1, \end{cases}$$

and

$$(2.9) \qquad U_n(y,t) := \begin{cases} 0 & , \; 0 \leq t < 1/n, \\ \frac{[nt](n-[nt])}{n^{3/2}}\left(\hat{E}_{[nt]}^{-1}(y) - \check{E}_{n-[nt]}^{-1}(y)\right) & , \; \frac{1}{n} \leq t \leq \frac{n-1}{n}, \\ 0 & , \; \frac{n-1}{n} < t \leq 1. \end{cases}$$

Based on M. Csörgő and P. Révész (1978), and M. Csörgő, S. Csörgő, L. Horváth and P. Révész (1984), we have

PROPOSITION 2.1. *Let $F$ of $H_0$ be a continuous distribution function and assume*

(I) *$F$ is twice differentiable on $(a,b)$, where*

$$a = \sup\{x : F(x) = 0\},$$
$$b = \inf\{x : F(x) = 1\}, \quad -\infty \leq a < b \leq \infty,$$

(II) *$F'(x) = f(x) > 0$, $x \in (a,b)$,*

(III) *for some $\gamma > 0$, we have*

$$\sup_{0<y<1} y(1-y)\frac{|f'(F^{-1}(y))|}{f^2(F^{-1}(y))} \leq \gamma.$$

*Then*

$$\sup_{0 \leq t \leq 1} \sup_{\frac{1}{n-1} \leq y \leq \frac{n}{n+1}} |f(F^{-1}(y))V_n(y,t) - U_n(y,t)|$$

$$\stackrel{a.s.}{=} \begin{cases} O(n^{-1/2}(\log\log n)^{1+\gamma}), & \text{if } \gamma \leq 1, \\ O(n^{-1/2}(\log\log n)^{\gamma}(\log n)^{(1+\varepsilon)(\gamma-1)}), & \text{if } \gamma > 1, \end{cases}$$

*for all $\varepsilon > 0$.*

Thus this proposition concludes that, under conditions (I), (II) and (III), the **empirical quantile change process** as in (2.8) multiplied by the **density quantile function** $f(F^{-1})$ behaves like its uniform version of (2.9) does, as $n \to \infty$.

As to the asymptotic behaviour of the latter, for further use we first **restate** the first two statements of Lemma 1.1 for uniform–(0,1) rv's via the uniform empirical distributions of (2.1) and (2.2).

LEMMA 2.2. *If $H_0$ holds true, then we can define a sequence of Gaussian processes $\{\Gamma_n(y,t); \ 0 \le y \le 1, 0 \le t \le 1\}$ such that, as $n \to \infty$, we have*

(i)
$$\sup_{\frac{1}{n} \le t \le \frac{n-1}{n}} \ \sup_{0 \le y \le 1} \left| \frac{[nt](n-[nt])}{n^{3/2}} (\hat{E}_{[nt]}(y) - \check{E}_{n-[nt]}(y)) - \Gamma_n(y,t) \right|$$
$$= O_P(n^{-1/2} \log^2 n)$$

*and*

(ii)
$$\sup_{\frac{1}{n} \le t \le \frac{n-1}{n}} \ \sup_{0 \le y \le 1} \frac{\left| \frac{[nt](n-[nt])}{n^{3/2}} (\hat{E}_{[nt]}(y) - \check{E}_{n-[nt]}(y)) - \Gamma_n(y,t) \right|}{(t(1-t))^{1/2-\nu}}$$
$$= O_P(n^{-\nu})$$

*for all $0 \le \nu < 1/2$, where*

$$\{\Gamma_n(y,t); 0 \le y \le 1, 0 \le t \le 1\} \overset{\mathrm{D}}{=} \{\Gamma(y,t); 0 \le y \le 1, 0 \le t \le 1\} \quad \textit{for each } n \ge 1,$$

*and*

$$\{\Gamma(y,t); 0 \le y \le 1, 0 \le t \le 1\} \overset{\mathrm{D}}{=} \{K(y,t) - tK(y,1); 0 \le y \le 1, 0 \le t \le 1\},$$

*where $K(\cdot,\cdot)$ is a Kiefer process.*

Next we note that, with $w_n(\cdot)$ as in (1.16), we have (cf. (2.4) and (2.5))

$$(2.10) \quad w_n(F^{-1}(y), t) = \begin{cases} 0, & 0 \le t < 1/n, \\ \frac{[nt](n-[nt])}{n^{3/2}} (\hat{E}_{[nt]}(y) - \check{E}_{n-[nt]}(y)), & \frac{1}{n} \le t \le \frac{n-1}{n}, \\ 0, & \frac{n-1}{n} < t \le 1. \end{cases}$$

Also, on twice applying Kiefer's uniform Bahadur–Kiefer principle (cf. (2.15), J. Kiefer (1970)), we obtain

PROPOSITION 2.3. *If $H_0$ holds true, then, as $n \to \infty$,*

$$(2.11) \qquad \sup_{0 \le t \le 1} \ \sup_{0 \le y \le 1} |w_n(F^{-1}(y), t) + U_n(y,t)|$$
$$= O(n^{-1/4}(\log n)^{1/2}(\log \log n)^{1/4}) \ \textit{a.s.}$$

A combination of (i) of Lemma 2.2 and (2.11) yields

PROPOSITION 2.4. *If $H_0$ holds true, then, as $n \to \infty$, with the sequence of Gaussian processes $\{\Gamma_n(y,t); 0 \le y \le 1, 0 \le t \le 1\}$ of Lemma 2.2 we have*

$$(2.12) \qquad \sup_{0 \le t \le 1} \ \sup_{0 \le y \le 1} |U_n(y,t) + \Gamma_n(y,t)| = O_P(n^{-1/4}(\log n)^{1/2}(\log \log n)^{1/4}).$$

Thus, via (i) of Lemma 2.2 and (2.12), we see that $w_n(F^{-1}(\cdot), \cdot)$ and $U_n(\cdot,\cdot)$ have the same limit in distribution, namely both converge weakly (in distribution) to $\Gamma(\cdot,\cdot)$ of Lemma 2.2. Let

$$V_n^0(y,t) := V_n(y,t) \ \mathbb{1}_{[1/(n+1), n/(n+1)]}(y).$$

Now, as a consequence of (2.12) and Proposition 2.1, we conclude

PROPOSITION 2.5. *Let $F$ of $H_0$ be a continuous distribution function satisfying the conditions* (I), (II) *and* (III) *of* Proposition 2.1. *Then, as $n \to \infty$, with the sequence of Gaussian processes* $\{\Gamma_n(y,t); 0 \le y \le 1, 0 \le t \le 1\}$ *of* Lemma 2.1, *we have*

$$(2.13) \qquad \sup_{0 \le t \le 1} \sup_{1/(n+1) \le y \le n/(n+1)} \left| f(F^{-1}(y)V_n^0(y,t) + \Gamma_n(y,t) \right|$$
$$= O_P\left( n^{-1/4}(\log n)^{1/2}(\log\log n)^{1/4} \right).$$

Recalling the degenerate asymptotic behaviour of **the sup functional of the empirical quantile change process** $V_n(\cdot,\cdot)$ even under $H_0$ as in (1.17), we see that renormalized via being multiplied by the density quantile function $f(F^{-1})$, under the conditions of Proposition 2.1 we now have (as $n \to \infty$)

$$(2.14) \qquad \sup_{0 \le t \le 1} \sup_{1/(n+1) \le y \le n/(n+1)} |f(F^{-1}(y)V_n(y,t)| \overset{D}{\to} \sup_{0 \le t \le 1} \sup_{0 \le y \le 1} |\Gamma(y,t)|.$$

REMARK 2.6. With $t = 1$, let $\widehat{E}_n(\cdot)$ and $\widehat{E}_n^{-1}$ be as in (2.1) and (2.6) respectively and define $\alpha_n(y) := n^{1/2}(\widehat{E}_n(y) - y)$ and $\beta_n(y) := n^{1/2}\left( \widehat{E}_n^{-1}(y) - y \right)$, $y \in [0,1]$, the uniform empirical and quantile processes. Concerning Bahadur's representation of sample quantiles (R. Bahadur (1966)), J. Kiefer (1970) proved

$$(2.15) \quad \limsup_{n \to \infty} n^{1/4}(\log n)^{-1/2}(\log\log n)^{-1/4} \sup_{0 \le y \le 1} |\alpha_n(y) + \beta_n(y)| = 2^{-1/4} \text{ a.s.}$$

Consequently, *the rate of convergence in* (2.11) *is* also *optimal*.
    KMT (1975, 1976) constructed a Kiefer process $K(\cdot,\cdot)$ such that, as $n \to \infty$,

$$(2.16) \qquad \sup_{0 \le y \le 1} \left| \alpha_n(y) - n^{-1/2}K(y,n) \right| = O\left( n^{-1/2} \log^2 n \right) \text{ a.s.}$$

By combining (2.15) and (2.16), M. Csörgő and P. Révész (1975) observed that with the *same* Kiefer process as in (2.16), we have also

$$(2.17) \qquad \limsup_{n \to \infty} n^{1/4}(\log n)^{-1/2}(\log\log n)^{-1/4} \sup_{0 \le y \le 1} \left| \beta_n(y) + n^{-1/2}K(y,n) \right|$$
$$= 2^{-1/4} \text{ a.s.}$$

P. Deheuvels (1998 and its References) showed that *the rate of approximation as in* (2.17) *is best possible for any* Kiefer process replacing the one in (2.17) that was constructed by KMT (1975, 1976).
    We recall now that in Lemma 1.1, and hence also in Lemma 2.2, we used the KMT approximation of (2.16) in constructing our Gaussian processes $\Gamma_n(\cdot,\cdot)$. Consequently, in the light of Deheuvels' recent results concerning (2.17), by combining (i) of Lemma 2.2 with the just noted best possible rate of convergence in (2.11) of Proposition 2.3, we conclude that *the rate of convergence in* (2.12) *of* Proposition 2.4 *is also best possible*. Moreover, the latter statement also yields that the rate of convergence in (2.13) of Proposition 2.5 is also best possible.

    In order to deal with change in distribution via quantiles as suggested by (ii) of Lemmas 1.1, 2.2 and Theorem 1.4, that is to say in terms of weighted quantile changepoint processes, we make a fresh start in Section 3, the main section of this exposition, where the results of Section 2 are also utilized in the proofs.

## 3. Weighted quantile change processes

We continue to assume that $F$ of $H_0$ is continuous, introduce the **short hand notation** $\mathbb{1}_i(x) := \mathbb{1}\{X_i \leq x\}$, $x \in \mathbb{R}$, $1 \leq i \leq n$, and note that, due to its additivity in $[nt]$, $1/n \leq t \leq (n-1)/n$, $w_n(\cdot, t)$ of (1.16) can be written as

$$(3.1) \quad w_n(x, t) := \frac{[nt](n - [nt])}{n^{3/2}} \left( \hat{F}_{[nt]}(x) - \check{F}_{n - [nt]}(x) \right)$$

$$= \frac{1}{n^{1/2}} \left( \sum_{1 \leq i \leq [nt]} \mathbb{1}\{X_i \leq x\} - \frac{[nt]}{n} \sum_{1 \leq i \leq n} \mathbb{1}\{X_i \leq x\} \right)$$

$$= \begin{cases} \dfrac{1}{n^{\frac{1}{2}}} \Big\{ \displaystyle\sum_{i=1}^{[nt]} \big(\mathbb{1}_i(x) - F(x)\big) - \dfrac{[nt]}{n} \Big( \displaystyle\sum_{i=1}^{n/2} \big(\mathbb{1}_i(x) - F(x)\big) \\ \qquad\qquad + \displaystyle\sum_{i=\frac{n}{2}+1}^{n} \big(\mathbb{1}_i(x) - F(x)\big) \Big) \Big\}, & \frac{1}{n} \leq t \leq \frac{1}{2}, \\[20pt] \dfrac{1}{n^{\frac{1}{2}}} \Big\{ \dfrac{n - [nt]}{n} \Big( \displaystyle\sum_{i=1}^{n/2} \big(\mathbb{1}_i(x) - F(x)\big) + \displaystyle\sum_{i=\frac{n}{2}+1}^{n} \big(\mathbb{1}_i(x) - F(x)\big) \Big) \\ \qquad\qquad - \displaystyle\sum_{i=[nt]+1}^{n} \big(\mathbb{1}_i(x) - F(x)\big) \Big\}, & \frac{1}{2} \leq t \leq \frac{n-1}{n}, \\[20pt] 0 \qquad \text{for } t \in [0, 1/n) \cup ((n-1)/n, 1], \end{cases}$$

$$=: \begin{cases} w_n^{(1)}(x, t), & 0 \leq t \leq 1/2, \\ w_n^{(2)}(x, t), & 1/2 \leq t \leq 1. \end{cases}$$

Hence, correspondingly, we **define** the **general quantile change-point process** $\tilde{V}_n(\cdot, \cdot)$ by

$$(3.2) \quad \tilde{V}_n(y, t) :=$$

$$:= \begin{cases} \dfrac{1}{n^{\frac{1}{2}}} \Big\{ [nt]\big(\hat{F}_{[nt]}^{-1}(y) - F^{-1}(y)\big) - \dfrac{[nt]}{n} \Big( \dfrac{n}{2} \big(\hat{F}_{n/2}^{-1}(y) - F^{-1}(y)\big) \\ \qquad\qquad + \dfrac{n}{2} \big(\check{F}_{n/2}^{-1}(y) - F^{-1}(y)\big) \Big) \Big\}, & \frac{1}{n} \leq t \leq \frac{1}{2}, \\[20pt] \dfrac{1}{n^{\frac{1}{2}}} \Big\{ \dfrac{n - [nt]}{n} \Big( \dfrac{n}{2} \big(\hat{F}_{n/2}^{-1}(y) - F^{-1}(y)\big) + \dfrac{n}{2} \big(\check{F}_{n/2}^{-1}(y) - F^{-1}(y)\big) \Big) \\ \qquad\qquad - (n - [nt])\big(\check{F}_{[nt]}^{-1}(y) - F^{-1}(y)\big) \Big\}, & \frac{1}{2} \leq t \leq \frac{n-1}{n}, \\[20pt] 0 \qquad \text{for } t \in [0, 1/n) \cup ((n-1)/n, 1], \end{cases}$$

$$=: \begin{cases} \tilde{V}_n^{(1)}(y, t), & 0 \leq t \leq 1/2, \\ \tilde{V}_n^{(2)}(y, t), & 1/2 \leq t \leq 1. \end{cases}$$

Consequently, its uniform version, the **uniform quantile change-point process** $\tilde{U}_n(\cdot, \cdot)$ is necessarily defined by

$$(3.3) \qquad \tilde{U}_n(y, t) :=$$

$$
:= \begin{cases}
\frac{1}{n^{1/2}}\Big\{ [nt]\big(\hat{E}_{[nt]}^{-1}(y) - y\big) - \frac{[nt]}{n}\Big( \frac{n}{2}\big(\hat{E}_{n/2}^{-1}(y) - y\big)\Big) \\
\qquad + \frac{n}{2}\big(\check{E}_{n/2}^{-1}(y) - y\big) \Big)\Big\}, & \frac{1}{n} \le t \le \frac{1}{2}, \\[2mm]
\frac{1}{n^{1/2}}\Big\{ \frac{n-[nt]}{n}\Big( \frac{n}{2}\big(\hat{E}_{n/2}^{-1}(y) - y\big) + \frac{n}{2}\big(\check{E}_{n/2}^{-1}(y) - y\big)\Big) \\
\qquad - (n-[nt])\big(\check{E}_{n-[nt]}^{-1}(y) - y\big)\Big\}, & \frac{1}{2} \le t \le \frac{n-1}{n}, \\[2mm]
0 \qquad \text{for } t \in [0, 1/n] \cup ((n-1)/n, 1],
\end{cases}
$$

$$
=: \begin{cases}
\tilde{U}_n^{(1)}(y,t), & 0 \le t \le 1/2, \\
\tilde{U}_n^{(2)}(y,t), & 1/2 \le t \le 1.
\end{cases}
$$

We note that $\tilde{V}_n(y,t) \ne V_n(y,t)$ of (2.8), and $\tilde{U}_n(y,t) \ne U_n(y,t)$ of (2.9). The reason for that is due to the fact that $V_n(\cdot,t)$ and $U_n(\cdot,t)$ are not additive processes in $[nt]$, $1/n \le t \le (n-1)/n$. Nevertheless, imitating the additive nature of $w_n(\cdot,t)$ via (3.1), $\tilde{V}_n(\cdot,t)$ and $\tilde{U}_n(\cdot,t)$ as defined in (3.2) and (3.3) respectively continue to compare quantiles "before" to those "after" a postulated unknown time of change in distribution in the spirit of comparing empiricals in (3.1) broken up into two symmetrized pieces in $t \in [1/n, 1/2] \cup [1/2, (n-1)/n]$. Moreover, the unweighted asymptotic behaviour of $\tilde{V}_n$ and $\tilde{U}_n$ coincides with that of $V_n$ and $U_n$ (cf., e.g., Theorem 3.5 with $q \equiv 1$ being in agreement with Proposition 2.4.).

The first result is a symmetrized and also weighted version of Proposition 2.1.

THEOREM 3.1. *Assume $H_0$ and the conditions* (I), (II) *and* (III) *of Proposition 2.1. Then, as $n \to \infty$,*

$$
(3.4) \qquad \sup_{0 \le t \le 1} \sup_{\frac{1}{n+1} \le y \le \frac{n}{n+1}} |f(F^{-1}(y))\tilde{V}_n(y,t) - \tilde{U}_n(y,t)|
$$

$$
\overset{a.s.}{=} \begin{cases}
O(n^{-1/2}(\log\log n)^{1+\gamma}) & , \textit{if } \gamma \le 1 \\
O(n^{-1/2}(\log\log n)^{\gamma}(\log n)^{(1+\varepsilon)(\gamma-1)}) & , \textit{if } \gamma > 1,
\end{cases}
$$

*and*

$$
(3.5) \qquad \sup_{\frac{1}{n} \le t \le \frac{n-1}{n}} \sup_{\frac{1}{n+1} \le y \le \frac{n}{n+1}} \frac{|f(F^{-1}(y))\tilde{V}_n(y,t) - \tilde{U}_n(y,t)|}{(t(1-t))^{1/2-\nu}} \overset{a.s.}{=} O(1/n^{\nu}),
$$

*for all $0 \le \nu < 1/2$.*

PROOF. With $t \in (0,1]$ and $(nt) \to \infty$ as $n \to \infty$, on account of Proposition 2.1, for large $n$ we have

$$
(3.6) \qquad \sup_{1/(n+1) \le y \le n/(n+1)} n^{1/2}\Big| f\big(F^{-1}(y)\big)\tilde{V}_n^{(1)}(y,t) - \tilde{U}_n^{(1)}(y,t)\Big|
$$

$$
\le (C + o(1))(r_n(t,\gamma,\varepsilon) + tr_n(1,\gamma,\varepsilon)) \quad \text{a.s.}
$$

and, with $t \in [0,1)$ and $n(1-t) \to \infty$ as $n \to \infty$, for large $n$ we have as well

$$
(3.7) \qquad \sup_{1/(n+1) \le y \le n/(n+1)} n^{1/2}\Big| f\big(F^{-1}(y)\big)\tilde{V}_n^{(2)}(y,t) - \tilde{U}_n^{(2)}(y,t)\Big|
$$

$$
\le (C + o(1))(r_n(1-t,\gamma,\varepsilon) + (1-t)r_n(1,\gamma,\varepsilon)) \quad \text{a.s.},
$$

where $C$ is an absolute constant and, with $s \in (0,1]$ and $\varepsilon > 0$,

$$
(3.8) \qquad r_n(s,\gamma,\varepsilon) := \begin{cases}
(\log\log(ns))^{1+\gamma} & \text{if } \gamma \le 1, \\
((\log\log(ns))^{\gamma}(\log(ns))^{(1+\varepsilon)(\gamma-1)}) & \text{if } \gamma > 1.
\end{cases}
$$

Now, on taking $\sup_{0 \leq t \leq 1}$ on both sides of both (3.6) and (3.7), in view of (3.2), (3.3) and (3.8) we conclude (3.4). $\qquad\square$

By (3.6), as $n \to \infty$, we obtain

$$(3.9) \qquad \sup_{\frac{1}{n} \leq t \leq \frac{1}{2}} \sup_{\frac{1}{n+1} \leq y \leq \frac{n}{n+1}} \frac{n^{1/2} \left| f(F^{-1}(y)) \tilde{V}_n^{(1)}(y,t) - \tilde{U}_n^{(1)}(y,t) \right|}{(nt)^{1/2-\nu}} \stackrel{a.s.}{=} O(1)$$

and, by (3.7), we have as well

$$(3.10) \qquad \sup_{\frac{1}{2} \leq t \leq \frac{n-1}{n}} \sup_{\frac{1}{n+1} \leq y \leq \frac{n}{n+1}} \frac{n^{1/2} \left| f(F^{-1}(y)) \tilde{V}_n^{(2)}(y,t) - U_n^{(2)}(y,t) \right|}{(n-nt)^{1/2-\nu}} \stackrel{a.s.}{=} O(1)$$

for all $0 \leq \nu < 1/2$. Consequently, via (3.9) and (3.10), we arrive at (3.5) by writing

$$(3.11) \qquad \sup_{\frac{1}{n} \leq t \leq \frac{n-1}{n}} \sup_{\frac{1}{n+1} \leq y \leq \frac{n}{n+1}} \frac{n^\nu \left| f(F^{-1}(y)) \tilde{V}_n(y,t) - \tilde{U}_n(y,t) \right|}{(t(1-t))^{1/2-\nu}}$$

$$\leq \sup_{\frac{1}{n} \leq t \leq \frac{1}{2}} \sup_{\frac{1}{n+1} \leq y \leq \frac{n}{n+1}} \frac{n^\nu \left| f(F^{-1}(y)) \tilde{V}_n^{(1)}(y,t) - \tilde{U}_n^{(1)}(y,t) \right|}{t^{1/2-\nu}(1/2)^{1/2-\nu}}$$

$$+ \sup_{\frac{1}{2} \leq t \leq \frac{n-1}{n}} \sup_{\frac{1}{n+1} \leq y \leq \frac{n}{n+1}} \frac{n^\nu \left| f(F^{-1}(y)) \tilde{V}_n^{(2)}(y,t) - \tilde{U}_n^{(2)}(y,t) \right|}{(1-t)^{1/2-\nu}(1/2)^{1/2-\nu}}.$$

Our next step is the symmetrized version of Proposition 2.3 that now parallels Theorem 3.1.

THEOREM 3.2. *If $H_0$ holds true, then, as $n \to \infty$,*

$$(3.12) \quad \sup_{0 \leq t \leq 1} \sup_{0 \leq y \leq 1} \left| w_n \big( F^{-1}(y), t \big) + \tilde{U}_n(y,t) \right| \stackrel{a.s.}{=} O\big( n^{-1/4} (\log n)^{1/2} (\log \log n)^{1/4} \big)$$

*and*

$$(3.13) \qquad \sup_{\frac{1}{n} \leq t \leq \frac{n-1}{n}} \sup_{0 \leq y \leq 1} \frac{\left| w_n \big( F^{-1}(y), t \big) + \tilde{U}_n(y,t) \right|}{(t(1-t))^{1/2-\nu}} \stackrel{a.s.}{=} O(1/n^\nu)$$

*for all $0 \leq \nu < 1/4$.*

PROOF. With $t \in (0,1]$ and $(nt) \to \infty$ as $n \to \infty$, on account of Proposition 2.3, for large $n$ we have

$$(3.14) \quad \sup_{0 \leq y \leq 1} n^{1/2} \left| w_n^{(1)}(F^{-1}(y), t) + \tilde{U}_n^{(1)}(y,t) \right| \leq (C + o(1)(b_n(t) + tb_n(1)) \quad \text{a.s.}$$

and with $t \in [0,1)$ and $n(1-t) \to \infty$ as $n \to \infty$, we have as well

$$(3.15) \qquad \sup_{0 \leq y \leq 1} n^{1/2} \left| w_n^{(2)}(F^{-1}(y), t) + \tilde{U}_n^{(2)}(y,t) \right|$$

$$\leq (C + o(1)(b_n(t) + (1-t)b_n(1)) \quad \text{a.s.},$$

where $C$ is an absolute constant and, with $s \in (0,1]$

$$(3.16) \qquad b_n(s) := (ns)^{1/4} (\log(ns))^{1/2} (\log \log(ns))^{1/4}.$$

From here on, mutatis mutandis, the proof of this theorem is identical to that of Theorem 3.1. $\qquad\square$

A combination of Lemma 2.2 with Theorem 3.2 yields

COROLLARY 3.3. *If $H_0$ holds true, then, as $n \to \infty$, with the sequence of Gaussian processes $\{\Gamma_n(y,t);\ 0 \le y \le 1,\ 0 \le t \le 1\}$, of Lemma 2.2 we have*

$$(3.17) \qquad \sup_{0 \le t \le 1}\ \sup_{0 \le y \le 1} \left|\tilde{U}_n(y,t) + \Gamma_n(y,t)\right| = O_P\big(n^{-1/4}(\log n)^{1/2}(\log\log n)^{1/4}\big)$$

*and*

$$(3.18) \qquad \sup_{\frac{1}{n} \le t \le \frac{n-1}{n}}\ \sup_{0 \le y \le 1} \frac{\left|\tilde{U}_n(y,t) + \Gamma_n(y,t)\right|}{(t(1-t))^{1/2-\nu}} = O_P\big(1/n^\nu\big)$$

*for all $0 \le \nu < 1/4$.*

Put
$$\tilde{V}_n^0(y,t) := \tilde{V}_n(y,t)\mathbb{1}_{[1/(n+1),n/(n+1)]}(y).$$

Then, putting together Corollary 3.3 and Theorem 3.1, we conclude

PROPOSITION 3.4. *Assume $H_0$ and the conditions (I), (II) and (III) of Proposition 2.1. Then, as $n \to \infty$, with the sequence of Gaussian processes $\{\Gamma_n(y,t);\ 0 \le y \le 1,\ 0 \le t \le 1\}$ of Lemma 2.1, we have*

$$(3.19) \qquad \sup_{0 \le t \le 1}\ \sup_{0 \le y \le 1} \left|f(F^{-1}(y))\tilde{V}_n^0(y,t) + \Gamma_n(y,t)\right|$$
$$= O_P\big(n^{-1/4}(\log n)^{1/2}(\log\log n)^{1/4}\big)$$

*and*

$$(3.20) \qquad \sup_{\frac{1}{n} \le t \le \frac{n-1}{n}}\ \sup_{0 \le y \le 1} \frac{\left|f(F^{-1}(y))\tilde{V}_n^0(y,t) + \Gamma_n(y,t)\right|}{(t(1-t))^{1/2-\nu}} = O_P\big(1/n^\nu\big)$$

*for all $0 \le \nu < 1/4$.*

With the help of Proposition 3.4, an empirical quantile change process version of Theorem 1.4 now reads as follows.

THEOREM 3.5. *Assume $H_0$ and the conditions (I), (II) and (III) of Proposition 2.1, and let $q \in Q_{0,1}$. Then, as $n \to \infty$, with the sequence of Gaussian processes $\{\Gamma_n(y,t);\ 0 \le y \le 1,\ 0 \le t \le 1\}$ of Lemma 2.2, we have*

(i)
$$\sup_{0 < t < 1}\ \sup_{0 \le y \le 1} \left|f(F^{-1}(y))\tilde{V}_n^0(y,t) + \Gamma_n(y,t)\right|/q(t)$$

$$= \begin{cases} o_P(1) \text{ if and only if } I_{0,1}(q,c) < \infty \text{ for all } c > 0, \\ O_P(1) \text{ if and only if } I_{0,1}(q,c) < \infty \text{ for some } c > 0, \end{cases}$$

(ii) $\displaystyle \sup_{0 < t < 1}\ \sup_{0 \le y \le 1} \left|f(F^{-1}(y))\tilde{V}_n^0(y,t)\right|/q(t) \xrightarrow{\text{D}} \sup_{0 < t < 1}\ \sup_{0 \le y \le 1} |\Gamma(y,t)|/q(t),$

*if and only if $I_{0,1}(q,c) < \infty$ for some $c > 0$, where $\Gamma(\cdot,\cdot)$ is a tied down Kiefer process as in Lemma 2.2.*

PROOF. For any $0 < \delta < 1/2$, as $n \to \infty$, we have

$$(3.21) \qquad \sup_{\delta \le t \le 1-\delta}\ \sup_{0 \le y \le 1} \left|f(F^{-1}(y))\tilde{V}_n^0(y,t) + \Gamma_n(y,t)\right|/q(t) = o_P(1)$$

by (3.19). Consider now

$$(3.22) \quad \sup_{\frac{1}{n} \leq t \leq \delta} \sup_{0 \leq y \leq 1} \left| f(F^{-1}(y)) \tilde{V}_n^0(y,t) + \Gamma_n(y,t) \right| / q(t)$$

$$\leq \sup_{0 < t \leq \delta} \frac{t^{1/2}}{q(t)} \sup_{\frac{1}{n} \leq t \leq \frac{n-1}{n}} \sup_{0 \leq y \leq 1} \frac{\left| f(F^{-1}(y)) \tilde{V}_n^0(y,t) + \Gamma_n(y,t) \right|}{t^{1/2}}$$

and

$$(3.23) \quad \sup_{1-\delta \leq t \leq \frac{n-1}{n}} \sup_{0 \leq y \leq 1} \left| f(F^{-1}(y)) \tilde{V}_n^0(y,t) + \Gamma_n(y,t) \right| / q(t)$$

$$\leq \sup_{1-\delta \leq t < 1} \frac{(1-t)^{\frac{1}{2}}}{q(t)} \sup_{\frac{1}{n} \leq t \leq \frac{n-1}{n}} \sup_{0 \leq y \leq 1} \frac{\left| f(F^{-1}(y)) \tilde{V}_n^0(y,t) + \Gamma_n(y,t) \right|}{(1-t)^{\frac{1}{2}}}.$$

For proving (i), we *first assume* that $I_{0,1}(q,c) < \infty$ *for some* $c > 0$. Then (cf. Proposition 3.1 of CsCsHM (1986), or Lemma 4.1.3 in M. Csörgő and L. Horváth (1993))

$$(3.24) \quad \lim_{t \downarrow 0} t^{1/2}/q(t) = 0 \quad \text{and} \quad \lim_{t \uparrow 1} (1-t)^{1/2}/q(t) = 0.$$

Consequently, on taking $\delta$ arbitrarily small, and applying (3.20) with $\nu = 0$ together with (3.24), by (3.21), (3.22) and (3.23) we conclude

$$(3.25) \quad \sup_{\frac{1}{n} \leq t \leq \frac{n-1}{n}} \sup_{0 \leq y \leq 1} \left| f(F^{-1}(y)) \tilde{V}_n^0(y,t) + \Gamma_n(y,t) \right| / q(t) = o_P(1).$$

From the definition of $\tilde{V}_n(\cdot, \cdot)$ it follows that

$$(3.26) \quad \sup_{0 < t < 1/n} \sup_{0 \leq y \leq 1} \left| f(F^{-1}(y)) \tilde{V}_n^0(y,t) \right| / q(t) = 0$$

and

$$(3.27) \quad \sup_{\frac{(n-1)}{n} < t < 1} \sup_{0 \leq y \leq 1} \left| f(F^{-1}(y)) \tilde{V}_n^0(y,t) \right| / q(t) = 0.$$

We have (cf. Lemma 2.2)

$$(3.28) \quad \{\Gamma_n(y,t); 0 \leq y \leq 1, 0 \leq t \leq 1\}$$
$$\overset{\mathrm{D}}{=} \{\Gamma(y,t); 0 \leq y \leq 1, 0 \leq t \leq 1\} \quad \text{for each } n \geq 1.$$

Hence, by (i) of Theorem 1.2, as $n \to \infty$, we conclude

$$(3.29) \quad \sup_{0 < t < 1/n} \sup_{0 \leq y \leq 1} |\Gamma_n(y,t)| / q(t) = O_P(1)$$

and

$$(3.30) \quad \sup_{\frac{(n-1)}{n} < t < 1} \sup_{0 \leq y \leq 1} |\Gamma_n(y,t)| / q(t) = O_P(1).$$

Consequently, from (3.25)–(3.27) and (3.29) and (3.30), we conclude the if part of the $O_P(1)$ statement of (i).

On the other hand, *if $I_{0,1}(q,c) < \infty$ for all $c > 0$*, then, naturally, we continue to have (3.24)–(3.27), but instead of (3.29) and (3.30), by (3.28) and (ii) of Theorem 1.2 we now have

$$(3.31) \quad \sup_{0 < t < 1/n} \sup_{0 \leq y \leq 1} |\Gamma_n(y,t)| / q(t) = o_P(1)$$

and

$$(3.32) \qquad \sup_{(n-1)/n<t<1} \sup_{0\leq y\leq 1} |\Gamma_n(y,t)|/q(t) = o_P(1).$$

Thus, combining (3.25)–(3.27) with (3.31) and (3.32), we conclude the if part of the $o_P(1)$ statement of (i).

Due to (3.26) and (3.27) we have

$$(3.33) \qquad \sup_{0<t<1} \sup_{0\leq y\leq 1} \left| f(F^{-1}(y))\tilde{V}_n^0(y,t) + \Gamma_n(y,t) \right|/q(t)$$

$$\geq \sup_{0<t<1/n} \sup_{0\leq y\leq 1} |\Gamma_n(y,t)|/q(t) + \sup_{\frac{n-1}{n}<t<1} \sup_{0\leq y\leq 1} |\Gamma_n(y,t)|/q(t).$$

Hence, *conversely now*, if we assume that the $o_P(1)$ statement of (i) holds true, then by (3.33) we have (3.31) and (3.32), and the latter combined with (ii) of Theorem 1.2 via (2.28) results in concluding that $I_{0,1}(q,c) < \infty$ *for all* $c > 0$. Based on (3.33), (3.29) and (3.30), a similar argument yields that $I_{0,1}(q,c) < \infty$ *for some* $c > 0$, if we now assume that the $O_P(1)$ statement of (i) holds true. This also completes the proof of statement (i) of Theorem 3.5.

Now, for proving (ii), on *assuming* that $I_{0,1}(q,c) < \infty$ for some $c > 0$, we have already concluded (3.25). Combined with (3.26) and (3.27), (3.25) in turn implies that, as $n \to \infty$,

$$(3.34) \qquad \left| \sup_{0<t<1} \sup_{0\leq y\leq 1} \left| f(F^{-1}(y))\tilde{V}_n^0(y,t) \right|/q(t) \right.$$

$$\left. - \sup_{\frac{1}{n}\leq t\leq\frac{n-1}{n}} \sup_{0\leq y\leq 1} |\Gamma_n(y,t)|/q(t) \right| = o_P(1).$$

Moreover (cf. (3.28), (3.29) and (3.30)), we have also

$$(3.35) \qquad \sup_{\frac{1}{n}\leq t\leq\frac{n-1}{n}} \sup_{0\leq y\leq 1} |\Gamma_n(y,t)|/q(t) \xrightarrow{D} \sup_{0<t<1} \sup_{0\leq y\leq 1} |\Gamma(y,t)|/q(t),$$

and hence, via (3.34) we conclude the if part of statement (ii). On the other hand, *if* (3.35) *is assumed to hold true*, then it also means that the limiting random variable is almost surely finite. Hence, by (i) of Theorem 1.2 we also have that $I_{0,1}(q,c) < \infty$ for some $c > 0$. This also completes the proof of Theorem 3.5. □

The appearance of Theorem 3.5 coincides with that of Theorem 1.4. The latter, in principle, provides a distribution free asymptotic test for testing $H_0$ versus $H_A$. Unfortunately, on account of the unknown density quantile function $f(F^{-1}(\cdot))$ in front of $\tilde{V}_n^0(\cdot,\cdot)$, the same cannot be said about Theorem 3.5. The appearance of the unknown quantile function $F^{-1}$ in the definition of $\tilde{V}_n(\cdot,\cdot)$ (cf. (3.2)), just like that of $F$ in (3.1), is only for the sake of theorem proving. Indeed, for computational purposes we note that the general quantile change-point process $\tilde{V}_n(\cdot,\cdot)$ of (3.2) is actually, by definition, equal to

(3.36) $\tilde{V}_n(y,t) =$

$$= \begin{cases} \frac{1}{n^{1/2}}\left\{[nt]\hat{F}_{[nt]}^{-1}(y) - \frac{[nt]}{n}\left(\frac{n}{2}\left(\hat{F}_{n/2}^{-1}(y) + \check{F}_{n/2}^{-1}(y)\right)\right)\right\}, & \frac{1}{n} \le t \le \frac{1}{2}, \\ \frac{1}{n^{1/2}}\left\{\frac{n-[nt]}{n}\left(\frac{n}{2}\left(\hat{F}_{n/2}^{-1}(y) + \check{F}_{n/2}^{-1}(y)\right)\right) - (n-[nt])\check{F}_{[nt]}^{-1}(y)\right\}, & \frac{1}{2} \le t \le \frac{n-1}{n}, \\ 0 \quad \text{for } t \in [0,1/n) \cup ((n-1)/n, 1], \end{cases}$$

$$= \begin{cases} \tilde{V}_n^{(1)}(y,t), & 0 \le t \le 1/2, \\ \tilde{V}_n^{(2)}(y,t), & 1/2 \le t \le 1. \end{cases}$$

Thus, by definition, the process $\tilde{V}_n^0(\cdot, \cdot)$ in Proposition 3.4 and Theorem 3.5 is a computable quantile change process.

As to the problem of "getting rid of" the unknown density quantile function $f(F^{-1}(\cdot))$ in front of $\tilde{V}_n^0(\cdot, \cdot)$ in Theorem 3.5, the following conjecture would be of interest to settle.

CONJECTURE 3.6. *Let $F$ of $H_0$ satisfy the conditions* (I), (II) (III), *and let* $q \in Q_{0,1}$. *Let $h : (0,1) \to (0,1)$ be so that*

(IV)      $f(F^{-1})/h \in Q_{0,1}.$

*Then, as $n \to \infty$, with the sequence of Gaussian processes $\{\Gamma_n(y,t);\ 0 \le y \le 1,\ 0 \le t \le 1\}$ of Lemma 2.2, we have*

(i)     $\sup\limits_{0<t<1} \sup\limits_{0<y<1} \left| h(y)\tilde{V}_n^0(y,t) + \dfrac{h(y)}{f(F^{-1}(y))}\Gamma_n(y,t) \right| / q(t) = o_P(1)$

*if and only if $I_{0,1}(q,c) < \infty$ and $I_{0,1}(f(F^{-1})/h, c) < \infty$ for all $c > 0$,*

(ii)     $\sup\limits_{0<t<1} \sup\limits_{0<y<1} \left| h(y)\tilde{V}_n^0(y,t) \right| / q(t) \xrightarrow{D} \sup\limits_{0<t<1} \sup\limits_{0<y<1} \left| \dfrac{h(y)}{f(F^{-1}(y))}\Gamma(y,t) \right| / q(t)$

*if and only if $I_{(0,1)}(q,c) < \infty$ and $I_{0,1}(f(F^{-1})/h, c) < \infty$ for some $c > 0$, where $\Gamma(\cdot, \cdot)$ is a tied down Kiefer processes as in Lemma 2.2.*

## References

[1] Bahadur, R.R. (1996). A note on quantiles in large samples. *Ann. Math. Statist.* **37**, 577–580.

[2] Burke, M.D. (1998). A Gaussian bootstrap approach to estimation and tests. In: *Asymptotic Methods in Probability and Statistics*, B. Szyszkowicz (Editor) 697–706, Elsevier Science B.V.

[3] Chen, B. (1998). *Functional Limit Theorems*, Ph.D. thesis, Carleton University.

[4] Chibisov, D. (1964). Some theorems on the limiting behaviour of empirical distribution functions. *Selected Transl. Math. Statist. Probab.* **6**, 147–156.

[5] Csörgő, M. (1983). *Quantile Processes with Statistical Applications*. SIAM 42, Philadelphia, Pennsylvania.

[6] Csörgő, M., Csörgő, S., Horváth, L. and Mason, D.M. (1986). Weighted empirical and quantile processes. *Ann. Probab.* **14**, 31–85.

[7] Csörgő, M., Csörgő, S., Horváth, L. and Révész, P. (1985). On weak and strong approximations of the quantile process. In: *Proc. Seventh Conf. Probability Theory*, 81–95. Editura Academiei, Bucuresti and Nuscience Press, Utrecht.

[8] Csörgő, M. and Horváth, L. (1988). A note on strong approximations of multivariate empirical processes. *Stoch. Proc. Appl.* **28**, 101–109.

[9] Csörgő, M. and Horváth, L. (1993). *Weighted Approximations in Probability and Statistics*. Wiley, Chichester.

[10] Csörgő, M. and Horváth, L. (1997). *Limit Theorems in Change-Point Analysis*. Wiley, Chichester.

[11] Csörgő, M., Horváth, L. and Szyszkowicz, B. (1997). Integral tests for suprema of Kiefer processes with application. *Statistics & Decisions* **15**, 365–377.

[12] Csörgő, M., Norvaiša, R. (1997). Weighted invariance principle for Banach space valued random variables. In: *Tech. Rep. Ser. Lab. Res. Stat. Prob.* No. 297, Carleton U.-U. of Ottawa.

[13] Csörgő, M. and Norvaiša, R. (1998). Standardized sequential empirical processes. *Studia Sci. Math. Hung.* **34**, 51–69.

[14] Csörgő, M. and Révész, P. (1975). Some notes on the empirical distribution function and the quantile process. In: *Coll. Math. Soc. J. Bolyai* **11**, *Limit Theorems of Prob. Theory* (P. Révész, ed.) 59–71. North Holland, Amsterdam.

[15] Csörgő, M. and Révész, P. (1978). Strong approximations of the quantile process. *Ann. Statist.* **6**, 882–894.

[16] Csörgő, M. and Révész, P. (1981). *Strong Approximations in Probability and Statistics.* Academic Press, New York.

[17] Csörgő, M., Shao, Q.-M. and Szyszkowicz, B. (1991). A note on local and global functions of a Wiener process and some Rényi-type statistics. *Studia Sci. Math. Hungar.* **26**, 239–259.

[18] Csörgő, M. and Szyszkowicz, B. (1994). Applications of multi-time parameter processes to change-point analysis. In *Probability Theory and Mathematical Statistics – Proceedings of the Sixth Vilnius Conference*, pp. 159–222, B. Grigelionis *et al.* (Ed s.), VSP/TEV 1994.

[19] Csörgő, M. and Szyszkowicz, B. (1998). Sequential quantile and Bahadur–Kiefer processes. In: *Handbook of Statistics, Vol.* **16** (N. Balakrishnan and C.R. Rao, eds.), 631–688, Elsevier Science B.V.

[20] Darkhovsky, D.S. (1984). On two problems of estimating the moments of change in probabilistic characteristics of a random sequence (in Russian). *Teor. Veroyatn. Primen.* **29**, 464–473.

[21] Darling, D.A. and Erdős, P. (1956). A limit theorem for the maximum of normalized sums of independent random variables. *Duke Math. J.* **23**, 143–145.

[22] Deheuvels, P. (1998). On the local oscillations of empirical and quantile processes. In: *Asymptotic Methods in Probability and Statistics*, B. Szyszkowicz (Editor), 127–134, Elsevier Science B.V.

[23] Deshayes, J. and Picard, D. (1981). Convergence de processus à double indice: application aux tests de rupture dans un modèle. *C.R. Acad. Sc. Paris* **292**, 449–452.

[24] Deshayes, J. and Picard, D. Off-line statistical analysis of change-point models using non parametric and likelihood methods, in: M. Thoma and A. Wyner, eds., M. Basseville and A. Beneveniste, guest eds., *Detection of Abrupt Changes in Signals and Dynamical Systems* (Springer, Berlin, 1986) pp. 103–168.

[25] Eastwood, B.J. and Eastwood, V.R. (1998). Tabulating sup-norm functionals of Brownian bridges via Monte Carlo Simulation. In: *Asymptotic Methods in Probability and Statistics*, B. Szyszkowicz (Editor), 707–719, Elsevier Science B.V.

[26] Keprta, S. (1997a). *Integral Tests for Brownian Motions and some Related Processes*, Ph.D. thesis, Carleton University.

[27] Keprta, S. (1997b). A note on supremum of a Kiefer process. *Statist. Prob. Lett.* **35**, 59–63.

[28] Keprta, S. (1998). Integral tests for some processes related to Brownian motion. In: *Asymptotic Methods in Probability and Statistics*, B. Szyszkowicz (Editor), 253–279, Elsevier Science B.V.

[29] Kiefer, J. (1970). Deviations between the sample quantile process and the sample D.F. In: *Nonparametric Techniques in Stat. Inference* (M.L. Puri, ed.), 299–319. Cambridge University Press, Cambridge.

[30] Kiefer, J. (1972). Skorohod embedding of multivariate RV's and the sample D.F. *Z. Wahrsch. Verw. Gebiete* **24**, 1–35.

[31] Komlós, J., Major, P. and Tusnády, G. (1975). An approximation of partial sums of independent R.V.'s and the sample DF. I. *Z. Wahrsch. verw. Gebiete* **32**, 111–131.

[32] Komlós, J., Major, P. and Tusnády, G. (1976). An approximation of partial sums of independent R.V.'s and the sample DF. II. *Z. Wahrsch. verw. Gebiete* **34**, 33–58.

[33] Müller, D.W. (1970). On Glivenko–Cantelli convergence. *Z. Wahrscheinlichkeitstheoerie verw. Gebiete* **16**, 195–210.

[34] O'Reilly, N. (1974). On the weak convergence of empirical processes. *Ann. Prob.* **2**, 642–651.

[35] Picard, D. (1985). Testing and estimating change-points in time series. *Adv. Appl. Probab.* **17**, 841–867.

[36] Shorack, G.R. and Wellner, J.A. (1986). *Empirical Processes with Applications to Statistics.* Wiley, New York.

[37] Szyszkowicz, B. (1991). Changepoint problem and contiguous alternatives. *Statist. Prob. Lett.* **11**, 299–308.

[38] Szyszkowicz, B. (1994). Weak convergence of weighted empirical type processes under contiguous and changepoint alternatives. *Stoch. Proc. Appl.* **50**, 281–313.

[39] Szyszkowicz, B. (1998). Weighted sequential empirical type processes with applications to changepoint problems. In *Handbook of Statistics, Vol.* 16 (N. Balakrishnan and C.R. Rao, eds.), 573–630, Elsevier Science B.V.

SCHOOL OF MATHEMATICS AND STATISTICS, CARLETON UNIVERSITY, OTTAWA, ON CANADA K1S 5B6
*E-mail address*: `mcsorgo@math.carleton.ca`

SCHOOL OF MATHEMATICS AND STATISTICS, CARLETON UNIVERSITY, OTTAWA, ON CANADA K1S 5B6
*E-mail address*: `bszyszko@math.carleton.ca`

Canadian Mathematical Society
Conference Proceedings
Volume **26**, 2000

# Embedding Theorems, Scaling Structures, and Determinism in Time Series

Colleen D. Cutler

ABSTRACT. In this paper we discuss specific definitions of *deterministic* and *stochastic* for stationary time series. Our main purpose in doing so is to create a convenient rigorous framework in which to examine the interplay between state-space reconstruction (embedding theorems), scaling or fractal structures (the Grassberger-Procaccia algorithm), and the predictability properties of time series. Thus the definitions in and of themselves are not as important as the clarity and precision they provide within the above context. In spite of the various pitfalls and limitations involved, choosing and adhering to a specific appropriate definition of determinism provides a firm foundation for proving theorems and constructing examples in those areas of chaos theory and time series concerned with reconstruction of the underlying source (or generating mechanism) of a time series. In this paper we provide some examples where our approach enables us to show that such reconstruction cannot be done.

## 1. Introduction

The past few decades have seen considerable attention paid to the phenomenon of erratic, apparently stochastic, time series arising from the sampling of low-dimensional deterministic systems—the phenomenon popularly known as *chaos*. To some extent chaos is surprising (perhaps intuitively we do not expect randomness, or even pseudo randomness, to arise from a deterministic rule with a few degrees of freedom) and this intuition is backed up by the knowledge that the long-run behaviour of standard nonlinear systems cannot give rise to chaos. Fixed points and limit cycles, even quasiperiodic motions, cannot produce erratic random-looking time series. On the other hand, most of us are comfortable with the use of pseudo random number generators, the majority of which attempt to generate a sequence of pseudo independent random variables based on a few deterministic rules. In fact it is not too difficult to coax a deterministic map into producing a truly stochastic sequence by choosing the correct functional. Consider the binary shift map on the unit interval

$$(1.1) \qquad \eta(y) = \begin{cases} 2y & \text{for } 0 \leq y \leq 1/2 \\ 2(y - 1/2) & \text{for } 1/2 < y \leq 1 \end{cases}$$

2000 *Mathematics Subject Classification.* Primary 62M10, 58F13; Secondary 28A80.
This research was supported by a grant from NSERC.

and define the functional $h : [0,1] \to \mathbb{R}$ by $h(y) = 0$ if $0 \leq y \leq 1/2$ and $h(y) = 1$ if $1/2 < y \leq 1$. Noting that the uniform distribution (that is, Lebesgue measure restricted to the unit interval) is invariant and ergodic for $\eta$, choose an initial condition $Y$ randomly from the uniform distribution. Then the sampled time series $X_n = h(\eta^{n-1}(Y))$, $n = 1, 2, \ldots$ obtained by iterating $Y$ under $\eta$ will consist of an independent and identically-distributed ($i.i.d.$) sequence of 0's and 1's. (It is not possible to actually carry out this iteration on the computer because we cannot store the initial condition to infinite precision, and finite decimals are attracted to the fixed point at 0 under the action of $\eta$.)

Of course $\eta$ is a bounded nonlinear map. Moreover, it possesses the other necessary ingredient for chaos — it has a positive Lyapunov exponent and thus is sensitive to initial conditions (nearby points separate quickly under the action of $\eta$). So this is one part of the recipe for producing a stochastic time series. But the other part is the functional $h$ itself, which in this case destroys the structural link between the deterministic source $\eta$ and the resulting stochastic output $X_n$, $n = 1, 2, \ldots$. It is not possible to work backward from the $X_n$ series to $\eta$; the source of the $X_n$ output could just as easily be a sequence of random coin tosses. Hence we say that reconstruction of $\eta$ is not possible from the sampled time series.

At this point we need to back up and clarify our terminology. The sequence of 0's and 1's produced by $\eta$ and $h$ above is in fact $i.i.d.$ and hence stochastic by anyone's definition of the term. But this is not the case with the majority of time series resulting from sampling chaotic systems like $\eta$. These time series may appear erratic and may possess many stochastic properties (like decaying autocorrelations and a broadband spectrum) but often sufficient structural link exists between the deterministic source and the observed time series that the source *can* be reconstructed from the time series. Demonstrating the existence of this structural link (in the form of time-delay embeddings) is the purpose behind both the theorem of [**Ta**] and the theorems of [**SYC**]. We will argue later that a time series from which a finite-dimensional deterministic mechanism can be reconstructed (in a manner to be made precise later) should properly be viewed as deterministic, regardless of any apparent stochastic properties it might also possess.

Hence we believe it useful to have precise definitions of deterministic and stochastic in order to help us make distinctions between certain types of time series. It will also aid us in distinguishing, at least in some cases, between those sampled time series which permit system reconstruction and those which do not.

Our definitions of deterministic and stochastic (given here in Section 2 and earlier in [**Cu2**]) are motivated in part by the desire to equate a deterministic time series with a finite-dimensional dynamical system evolving on an appropriate state space (in other words, we take the concept of finite-dimensional dynamical system as a starting point for our notion of deterministic). They are further motivated by the common intuition that "deterministic" translates to "perfectly predictable". However, our definitions also fit seamlessly into the framework of time-delay embeddings and system reconstruction (discussed briefly above), and so are ideally suited for that environment.

Throughout the following we will be concerned with time series $X_n, n = 1, 2, \ldots$ taking values in a closed subset $K \subseteq \mathbb{R}^u$ for some $1 \leq u < \infty$. We will let $P$ denote the distribution of the time series on the Borel sets $\mathcal{B}^\infty$ of the infinite product space $K^\infty = \times_{m=1}^\infty K$ (equipped with the usual product topology) and let $P_n$ denote the distribution of $(X_1, \ldots, X_n)$ on the Borel sets $\mathcal{B}^n$ of $K^n = \times_{m=1}^n K$. The collection

$P_n$, $n = 1, 2, \ldots$ comprise the finite-dimensional or *joint distributions* of the time series. The time series is said to be *stationary* if, for each $n \geq 1$, the distribution of $(X_{k+1}, X_{k+2}, \ldots, X_{k+n})$ coincides with $P_n$ for each $k \geq 1$. Our attention will be confined to stationary time series.

The remainder of the paper is organized as follows: Section 2 presents definitions of deterministic and stochastic for stationary time series, provides added rationale for these definitions beyond those presented in [**Cu2**], and describes the link with finite-dimensional and infinite-dimensional dynamical systems. Section 3 makes the connection to time-delay embeddings and state-space reconstruction, and gives a small survey of the competing [**Ta**] and [**SYC**] methods. In Section 4 we examine the relationship between scaling properties (fractal dimensions), the Grassberger-Procaccia algorithm, and determinism in time series. Finally, Section 5 consists of some examples where we show that system reconstruction is not possible by proving that the sampled time series are stochastic. These examples also show that coordinate projections (the most natural of functionals) can be "bad" functionals in some cases.

Discussions on the relationships between chaos, determinism, and time series analysis can be found in the papers by [**PCFS**], [**Ca1, Ca2**] [**KG**], [**KBA**], [**ABST**], [**Cu2**], and in the collections of papers edited by [**ST**] and [**CK**].

## 2. Dynamical systems and determinism

As noted in the introduction, we take as our starting point for a fundamental notion of determinism the finite-dimensional dynamical system (f.d.d.s.) (defined below for discrete time). Part of the motivation for this comes from the physics literature, where such systems often appear to be informally equated with determinism. We are also motivated by the question of predictability — an f.d.d.s. can be said to be completely determined (and therefore predictable) after observation of only a finite number of coordinates (equivalently, after acquisition of only a finite amount of information). As such, a human observer can hope to acquire sufficient information within a finite length of time to completely determine and predict an f.d.d.s. Of course there is a distinction between theoretical and "real-world" observations; the latter are always contaminated by errors, and so actual predictability of the system will depend on continuity and smoothness properties of the f.d.d.s. We will also address this issue. We begin with general measure-preserving systems:

DEFINITION 2.1 (general dynamical system). Let $\mathcal{Y}$ be a metric space, and let $\varphi : \mathcal{Y} \to \mathcal{Y}$ be a Borel-measurable mapping. For each $y \in \mathcal{Y}$ the sequence of iterates $y, \varphi(y), \varphi^2(y), \ldots$ is called the *orbit with initial condition* $y$. The pair $(\mathcal{Y}, \varphi)$ constitutes the dynamical system. If $Q$ is a distribution on the Borel sets $\mathcal{B}$ of $\mathcal{Y}$ satisfying $Q = Q\varphi^{-1}$ (that is, $Q$ is an invariant measure for $\varphi$) then the quadruple $(\mathcal{Y}, \varphi, \mathcal{B}, Q)$ is called a *measure-preserving dynamical system* (m.p.d.s.).

The inclusion of a measure $Q$ introduces a kind of randomness into the dynamical system via the initial condition — the initial condition becomes a random variable $Y$ with distribution $Q$. However, this randomness is only observable under repeated sampling of the initial condition; it cannot be seen from a single orbit of the system. In the future we will always assume the presence of a measure (generally invariant) which governs the distribution of the initial condition.

Note also that if $(\mathcal{Y}, \varphi)$ is a dynamical system and the initial condition $Y$ is selected randomly according to some distribution $Q$ then iterating $Y$ under $\varphi$ produces a time series $Y_n = \varphi^{n-1}(Y)$, $n = 1, 2, \ldots$ with $\varphi^0(Y) = Y$, called *the time series corresponding to $\varphi$ and $Q$*. The joint distributions of this time series will be denoted by $Q_n$, $n = 1, 2, \ldots$ where $Q_n$ is the distribution of $(Y_1, \ldots, Y_n)$. It follows that $(\mathcal{Y}, \varphi, \mathcal{B}, Q)$ is a m.p.d.s. (that is, $Q$ is invariant) if and only if the corresponding time series $Y_n$, $n = 1, 2 \ldots$ is stationary with $Q_1 = Q$. Note also that the conditional distribution of $(Y_1, \ldots, Y_n)$ given $Y = y$ is a Dirac point mass with all probability concentrated at the point $(y, \varphi(y), \ldots, \varphi^{n-1}(y))$. See [**LM**] for a discussion of the properties of an iterated m.p.d.s.

There are two special cases of measure-preserving dynamical system in which we will be particularly interested, the one where $\mathcal{Y} = K$ and $K$ is a closed subset of $\mathbb{R}^u$ for $1 \leq u < \infty$, and the one where $\mathcal{Y} = K \subseteq \mathbb{R}^\infty$.

DEFINITION 2.2 (finite-dimensional dynamical system). We say that a m.p.d.s. $(K, \varphi, \mathcal{B}, Q)$ is a *finite-dimensional dynamical system* (f.d.d.s.) if $K$ is a closed subset of $\mathbb{R}^u$ where $1 \leq u < \infty$.

DEFINITION 2.3 (infinite-dimensional dynamical system). Here we restrict ourselves to the case where $K \subseteq \mathbb{R}^\infty$ and call this an *infinite-dimensional dynamical system* (i.d.d.s.). However, we note that it will sometimes be possible to meaningfully identify an i.d.d.s. with a system evolving on a finite-dimensional domain, in which case the process may be regarded as an f.d.d.s.

Any reasonable definition of *deterministic* for a stationary time series should, at the very least, always result in the conclusion that the stationary time series $Y_n = \varphi^{n-1}(Y)$, $n = 1, 2, \ldots$ corresponding to a measure-preserving f.d.d.s. is deterministic. The approach we take below does just that — the above $Y_n$ series becomes a special case of a deterministic time series. Some aspects of this approach and the definitions below were given earlier in [**Cu2**].

DEFINITION 2.4 (predictive dimension). Let $X_n$, $n = 1, 2, \ldots$ be a strictly stationary time series taking values in a closed subset $K \subseteq \mathbb{R}^u$, $1 \leq u < \infty$. We define the *predictive dimension* $p$ of the time series to be

$$p = \min\{n \geq 1 \,|\, \text{there exists} \, T : K^n \to K \, \text{such that} \, X_{n+1} = T(X_1, \ldots, X_n) \, \text{w.p.} \, 1\}.$$

If no such function $T$ exists for any $n \geq 1$, we set $p = \infty$. In the case that $p < \infty$ the function $T$ corresponding to $n = p$ is called the *predictor function* of the time series. If a predictor function $T$ exists then $X_{p+1} = T(X_1, \ldots, X_p)$ w.p. 1. and clearly $T$ is $P_p$-*a.s.* unique. Moreover, as a consequence of stationarity we also have

$$(2.1) \qquad X_{m+p+1} = T(X_{m+1}, X_{m+2}, \ldots, X_{m+p}) \; \text{w.p.} \, 1$$

for all $m \geq 0$.

The notion of predictive dimension leads to a precise definition of the terms *deterministic* and *stochastic* for stationary time series, as well as to a precise notion of *predictability*:

DEFINITION 2.5. We say that a stationary time series $X_n$, $n = 1, 2, \ldots$ taking values in $\mathbb{R}^u$, $1 \leq u < \infty$, is *deterministic* if $p < \infty$ and *stochastic* if $p = \infty$, where $p$ is the predictive dimension. Moreover, in the deterministic case, we say that the time series is also *predictable* if the predictor function $T$ is $P_p$-*a.s.* continuous.

Thus a *deterministic time series* is one which can predict itself perfectly based on a fixed finite number of past (error-free) observations. There is no assumption made or inferred about the nature of the underlying generating mechanism (or source) of the time series — the time series is evaluated only in reference to itself. It will be a point of later interest to consider whether a time series and its underlying source need share common deterministic/stochastic properties. (The example in the Introduction indicates that they need not.) Moreover, a *predictable time series* is a deterministic one in which the predictor function is continuous. This allows for some degree of prediction even in the case of observational error. Thus "predictable" might also be called "real-world deterministic". Note also that a *stochastic time series* is one which *cannot* predict itself based on a fixed finite number of past observations; this constitutes a very large class of processes. For some comments on the meaning of stochastic, see points 5., 6., 7., and 8. later in this section.

We note that, under Definition 2.5, the time series $Y_n = \varphi^{n-1}(Y)$, $n = 1, 2, \ldots$ corresponding to $\varphi$ is always deterministic with $p = 1$ and $T = \varphi$. Conversely, if $X_n$, $n = 1, 2, \ldots$, is a time series with $p = 1$ and predictor function $T$, then the time series must correspond to the f.d.d.s. $\varphi = T$. Thus it follows that a deterministic time series with $p \geq 2$ cannot correspond to a measure-preserving f.d.d.s. However, it is possible to identify such a time series with a coordinate projection of a measure-preserving f.d.d.s. evolving on the finite product space $K^p$; we call this system the *canonical f.d.d.s.* associated with the time series.

THEOREM 2.6 (canonical f.d.d.s.). *Let* $X_n$, $n = 1, 2, \ldots$ *be a stationary time series taking values in* $K \subseteq \mathbb{R}^u$, $1 \leq u < \infty$, *and having joint distributions* $P_n$, $n = 1, 2, \ldots$, *finite predictive dimension* $p$, *and predictor function* $T$. *Then the mapping* $\varphi_T : K^p \to K^p$ *defined by*

$$
(2.2) \qquad
\begin{aligned}
\varphi_T(x) &= T(x) & \text{if } p = 1 \\
\varphi_T(x_1, \ldots, x_p) &= (x_2, \ldots, x_p, T(x_1, \ldots, x_p)) & \text{if } p \geq 2
\end{aligned}
$$

*defines a measure-preserving f.d.d.s. on* $K^p$ *with invariant distribution* $Q = P_p$. *Taking the initial condition to be* $Y = (X_1, \ldots, X_p)$ *and defining the coordinate projection* $\pi : (\mathbb{R}^u)^p \to \mathbb{R}^u$ *by* $\pi(x_1, \ldots, x_p) = x_1$, *the original time series is then recovered as the projection* $X_n = \pi(\varphi_T^{n-1}(Y))$ *for* $n = 1, 2, \ldots$.

If $p = 1$, the canonical f.d.d.s. coincides with the measure-preserving f.d.d.s. generating the time series. If $p \geq 2$, Theorem 2.6 exhibits a dynamic correspondence between the points $(x_1, \ldots, x_p)$ of $K^p$ (and their evolution) and the realizations of the time series (and their evolution). In order to see the significance of this, we first note that *every* stationary time series can be represented as a coordinate projection of a dynamical system if we make the state space large enough — specifically, if we take the state space to be the infinite product space $K^\infty = \times_{m=1}^\infty K$. This construction leads to an i.d.d.s.

DEFINITION 2.7 (left-shift dynamical system). We let $L$ be the *left-shift operator* $L : K^\infty \to K^\infty$ given by

$$(2.3) \qquad L(x_1, x_2, x_3, \ldots) = (x_2, x_3, x_4, \ldots).$$

Stationarity of the time series is equivalent to $P$ being invariant under $L$, and so $(K^\infty, L, \mathcal{B}^\infty, P)$ is a measure-preserving dynamical system. The time series $X_1, X_2, \ldots$ can be represented as $X_n = \pi(L^{n-1}(Z))$ where $Z = (X_1, X_2, \ldots)$ is the initial condition (selected according to $P$) and $\pi$ is the projection onto the first coordinate; that is, $\pi(x_1, x_2, x_3, \dots) = x_1$. We note that the left-shift operator defines an infinite-dimensional dynamical system as discussed in Definition 2.3.

It follows from the above discussion and Theorem 2.6 that in the case of finite predictive dimension $p$ we obtain two representations of our time series, one as a coordinate projection of an f.d.d.s., and the other as a coordinate projection of an i.d.d.s. However, it is straightforward to establish that these two representations are dynamically equivalent in the sense below:

DEFINITION 2.8 (dynamical equivalence). $(\mathcal{Y}, \varphi, \mathcal{B}_1, Q)$ and $(\mathcal{X}, \psi, \mathcal{B}_2, V)$ are called *measure conjugate* or *dynamically equivalent* if there exists a mapping $\Psi : \mathcal{Y} \to \mathcal{X}$ satisfying the following:

1. $\Psi$ is a measurable mapping and $V = Q\Psi^{-1}$.
2. There exist sets $F_0 \in \mathcal{B}_1$ and $G_0 \in \mathcal{B}_2$ such that $Q(F_0) = V(G_0) = 1$ and $\Psi$ is 1–1 and onto between $F_0$ and $G_0$.
3. $\psi \circ \Psi = \Psi \circ \varphi$ (conjugacy property).

$\Psi$ is called a *measure conjugacy* between the two m.p.d.s.

[B] (p. 53) uses the term *isomorphism* for what we have called *measure conjugacy*. We chose a more explicit descriptive term because it is often necessary to impose further structure (e.g. topological and/or differential) on the dynamical systems and conjugacies in order to preserve desired properties. In the absence of such additional structure, measure conjugacy is a rather weak relation which may identify many (apparently dissimilar) systems.

In the case of our two representations for finite predictive dimension $p$, it is easy to establish a measure conjugacy between them:

$$(2.4) \qquad\qquad (K^\infty, L, \mathcal{B}^\infty, P) \overset{\Psi}{\longleftrightarrow} (K^p, \varphi_T, \mathcal{B}^p, P_p).$$

The mapping $\Psi : K^\infty \to K^p$ defined by $\Psi(x_1, x_2, \dots) = (x_1, \dots, x_p)$ has the property that $P_p = P\Psi^{-1}$ and also that $\Psi \circ L(x_1, x_2, \dots) = \varphi_T \circ \Psi(x_1, x_2, \dots)$ w. p. 1 using (2.1). It also follows from (2.1) that $\Psi$ is one-to-one on a set of $P$-measure one. Consequently (2.4) defines a measure conjugacy and the two dynamical systems may be regarded as dynamically equivalent. Additionally, we see that $\Psi$ is a bi-Lipschitz mapping provided $T$ itself is Lipschitz continuous. This preserves metric properties between the two spaces.

We are now in a position to discuss several points:

1. First note that the preceding paragraphs illustrate the fact that a time series may arise as a functional of more than one dynamical system (in fact there will generally be several possible sources for a given time series). The two different sources for a deterministic time series given in Theorem 2.6 and Definition 2.7 were shown to be dynamically equivalent, but we may ask whether this is always, or almost always, the case. Moreover, dynamical equivalence can be a very weak relation which does not necessarily imply equivalence of other properties of interest which may be determined by topological and metric structures on the spaces. Scaling quantities such as fractal dimensions, for example, may not be preserved between the systems unless the conjugacy is bi-Lipschitz.

2. Definition 2.7 shows us that every stationary time series can be identified with a measure-preserving dynamical system (an i.d.d.s.); hence such an identification in and of itself is not useful in arriving at a definition of determinism. The appropriate distinguishing criterion appears to be *finiteness*; can it be identified with an f.d.d.s.?

3. The preceding question brings us back to an important point noted prior to Definition 2.1 — the problem of observational error and continuity of the predictor function $T$. We have chosen to use the word "predictable" in the case where $T$ is *a.s.* continuous, since for any desired level of prediction accuracy we can find a level of observational accuracy which will provide that wished-for prediction accuracy. From the point of view of prediction in nonlinear chaotic systems, there seems to be no advantage in imposing a great deal more in the way of smoothness. The study of chaos has shown us that arbitrarily smooth nonlinear systems can produce orbits which diverge very rapidly, so that minor errors in the initial conditions are quickly magnified and predictability is lost (to the observer) within a few iterations. However we note that the embedding theorems of the next section, and the scaling results of the section following it, require some additional smoothness (at least Lipschitz continuous) on the predictor function $T$.

   We might also comment on the related problem of *estimating* $T$ from data in the more typical situation where $T$ is unknown. Such estimation methods generally rely on some smoothness properties of $T$ as well as on ergodicity of the time series.

4. We retain the notion of "deterministic" along with "predictable" because at this point it seems good, in principle, to have a definition which is separate from questions of smoothness and problems of observational or system error. (It remains to be seen if one can in fact separate these two notions.) Moreover, sufficiently interesting phenomena can be found even in the idealized case of exact error-free observations.

5. Note that our definition of deterministic was carefully restricted to time series taking values in $\mathbb{R}^u$ for finite $u$. Any dynamical system obviously has $p = 1$, and we wanted to exclude the case where the domain was $\mathbb{R}^\infty$, on the assumption that this required information on an infinite number of coordinates (and, in any case, would render *all* stationary time series deterministic by identification with the left-shift i.d.d.s.). However, some i.d.d.s. might properly be considered deterministic (that is, as f.d.d.s.) by an identification such as given in (2.4).

6. We also note that the introduction of observational or other types of noise (for example, system noise) creates added sources of randomness beyond that contained in the initial condition. The dynamics are no longer the same. In the case of system noise, the result may even be a time series whose qualitative behaviour is very different from that of the noise-free version. We consider these to be "higher-level" problems where the relevant question becomes one of quantifying the *degree* of noise or stochastic component present rather than a question of deterministic vs. stochastic — they are clearly stochastic by Definition 2.5. Another approach, deviating somewhat from Definition 2.5, would be to consider several different levels or grades of "deterministic" that allow for some kinds of noise. For results on

stochastic dynamical systems of the form

$$(2.5) \qquad X_n = T(X_{n-p}, \ldots, X_{n-1}) + \epsilon_n$$

(that is, deterministic systems plus system noise) see [**CT1, CT2**]. Also see the last three chapters of [**LM**] as well as [**SBDH**].

7. It has been pointed out that, in practice, a deterministic time series with a very large predictive dimension $p$ (in other words, a very high-dimensional system) behaves for all intents and purposes like a stochastic process. Of course this is due to limitations on observational accuracy and, especially, sample size. Stochastic models may be more useful for predicting such systems from limited data, but this does not affect the theoretical validity of our definitions.

8. Finally, we have been discussing Definition 2.5 almost exclusively from the point of view of what "deterministic" should mean. But it is also worth considering what we want "stochastic" to mean. Models such as (2.5) are generally considered stochastic by everyone — frequently the noise sequence $\epsilon_n$, $n = 1, 2, \ldots$ is *i.i.d.* or, at the very least, involves a great deal of independence. The sequence of 0's and 1's generated by the binary shift in (1.1) is *i.i.d.* and also certainly regarded as stochastic by all. However Definition 2.5 allows for stochastic time series which involve far less randomness than we might normally consider; some of these time series might bridge the gap between what we regard as truly deterministic and what we regard as truly stochastic. The examples in Section 5 should be viewed at least once from this perspective.

## 3. Time-delay embeddings and determinism

In the previous section we considered an arbitrary stationary time series $X_n$, $n = 1, 2, \ldots$. In this section we want to focus on those time series resulting from sampling an f.d.d.s. as discussed following (1.1). The basic situation is this: we are interested in the evolution of a measure-preserving f.d.d.s. $\varphi$ (as given in Definition 2.2) and, ideally, would like to record the corresponding time series $Y_n = \varphi^{n-1}(Y)$, $n = 1, 2, \ldots$. In practice, however, we are often unable to view the complete system evolving on its state space $K \subseteq \mathbb{R}^v$ but rather observe a *functional* or *projection* $h : K \to \mathbb{R}^u$ of the system (typically $u < v$, often $u = 1$). This produces an *observed* or *sampled* time series $X_n$, $n = 1, 2, \ldots$ taking values in $\mathbb{R}^u$, where $X_n = h(Y_n) = h(\varphi^{n-1}(Y))$ and $Y$ is the random initial condition of $\varphi$ chosen according to an invariant distribution $Q$. The assumption that $\varphi$ is measure-preserving ensures that $X_n$, $n = 1, 2, \ldots$ is stationary.

The sampled time series is generally less predictable (or, if you like, more stochastic) than the underlying f.d.d.s. because $h$ almost always condenses information by collapsing several pre-images to a common value. If we observe $Y_1 = y$ in the f.d.d.s. then $Y_2$ is uniquely determined by $Y_2 = \varphi(y)$. However, if we observe $X_1 = x$ in the sampled time series, we only obtain the information that $Y_1 \in \{y \mid h(y) = x\}$ and hence that $X_2 \in \{h(\varphi(y)) \mid h(y) = x\}$. Thus the conditional distribution of $X_2$ given $X_1 = x$ is not a Dirac point mass and the predictive dimension of the sampled time series exceeds 1. Another way of saying this is that $X_1$ does not uniquely determine the initial condition $Y_1$ and so the future evolution of the system is still uncertain to the observer. [**ABST**] point out that this feature often contributes to the erratic chaotic appearance of the sampled time series. The idea behind the method of *time-delay embeddings* is that while observation of a single value $X_1 = x$ will not pin down the underlying initial condition $Y_1$, observation of a sufficiently long string, say $X_1 = x_1, \ldots, X_d = x_d$, may do so. (This is essentially equivalent to obtaining a unique solution to a system of $d$ equations.) Moreover, the sequence of delay vectors $(X_k, X_{k+1}, \ldots, X_{k+d-1})$, $k = 1, 2, \ldots$ may then sketch out in $(\mathbb{R}^u)^d$ a "copy" of the evolution of $Y_1, Y_2, \ldots$ making it possible to study the hidden underlying system via the sampled time series. [**ABST**] refer to this process as *unfolding the attractor*; it is also known as *reconstructing the state space*. We describe this formally in the following:

DEFINITION 3.1. The *delay coordinate mapping* $\Phi_d : K \subseteq \mathbb{R}^v \to (\mathbb{R}^u)^d$ with embedding dimension $d$ is defined by

$$(3.1) \qquad \Phi_d(y) = (h(y), h(\varphi(y)), \ldots, h(\varphi^{d-1}(y)))$$

For $d = 1$ we have $\Phi_1(Y) = h(Y) = X$ which has distribution $P_1$. For $d \geq 2$ we have $\Phi_d(Y) = (X_1, \ldots, X_d)$ which has distribution $P_d$. Note also that $\Phi_d(\varphi^{k-1}(Y)) = \Phi_d(Y_k) = (X_k, X_{k+1}, \ldots, X_{k+d-1})$ so this is the method of generating the delay vectors discussed in the preceding paragraph.

We now make the link with the material in Section 2.

THEOREM 3.2. *Suppose for some $d \geq 1$ the delay coordinate mapping $\Phi_d$ is one-to-one on a set of $Q$-measure one. Then the sampled time series $X_n$, $n = 1, 2, \ldots$ is deterministic with predictive dimension $p \leq d$ and predictor function $T = \Phi_p \circ \varphi \circ \Phi_p^{-1}$. Furthermore, $\Phi_p$ is a measure conjugacy between the two measure-preserving f.d.d.s. $(K, \varphi, \mathcal{B}, Q)$ and $(K_0, \varphi_T, \mathcal{B}^p, P_p)$; that is, the canonical f.d.d.s. of the sampled time series is dynamically equivalent to the original underlying f.d.d.s.*

PROOF. Since $P_p$ is the distribution of $(X_1, \ldots, X_p)$ it follows from Definition 3.1 that $P_p = Q\Phi_p^{-1}$. Now, assuming that $\Phi_d$ is an embedding for some $d$, it follows that the predictive dimension $p$ of the sampled time series will be the smallest value of $d$ for which $\Phi_d$ is one-to-one w.p.1; hence $\Phi_p$ will be one-to-one on a set of $Q$-measure one. It is then easy to see that $T = \Phi_p \circ \varphi \circ \Phi_p^{-1}$ and $\Phi_p \circ \varphi = \varphi_T \circ \Phi_p$. Thus $\Phi_p$ is a measure conjugacy.       □

It is worth stating explicitly that state-space reconstruction via delay coordinates leads directly to creation of the canonical f.d.d.s. $(K_0, \varphi_T, \mathcal{B}^p, P_p)$; it is this dynamical system which is being sketched out by the delay vectors. The preceding theorem shows that this system is, under certain circumstances, dynamically equivalent to the underlying f.d.d.s. of interest, and so we can learn about the latter from the former. However, the delay coordinate space $K_0 \subseteq (\mathbb{R}^u)^d$ may not be the natural coordinate system for $\varphi$ and interpretation of the canonical f.d.d.s. in terms of a physical model can be difficult. Very often when the approach taken is to estimate $T$ directly from time series data we are engaging in "black box" modelling; the predictor $T$ is estimated as closely as possible by some statistical criterion and method but there is no physics involved. The alternative (one which makes physicists happier) is to use knowledge of the physical system, when possible, to develop a meaningful dynamical model, a set of difference equations or differential equations, back on the original state space $K$.

It is natural now to wonder when the delay coordinate mapping $\Phi_d$ will be one-to-one for some $d \geq 1$. In fact we often require additional structural properties on $\Phi_d$ beyond one-to-one; for example, we might require that $\Phi_d$ be bi-Lipschitz in order that metric properties, such as fractal dimensions, are preserved. This leads to the following definition:

DEFINITION 3.3 (embedding). A mapping $\Phi : K \subseteq \mathbb{R}^v \to K_0 \subseteq \mathbb{R}^w$ is called a (diffeomorphic) *embedding* if $\Phi$ is one-to-one and bi-Lipschitz, and both $\Phi$ and its inverse $\Phi^{-1}$ are diffeomorphisms.

The bi-Lipschitz property ensures that metric properties, such as scaling properties, are preserved between $K$ and $K_0$. The diffeomorphism property preserves local differential structure between the spaces; see [**SYC**] and [**Cu2**].

[**Ta**] and [**SYC**] provide theorems which basically state that the delay coordinate mapping will, in most cases, yield an embedding for sufficiently large $d$. That is, given a smooth f.d.d.s. $\varphi$ under certain mild technical conditions, "almost all" smooth functionals $h$ result in the delay coordinate mapping $\Phi_d$ being an embedding for sufficiently large $d$. Here "smooth" means twice-differentiable, sometimes only continuously differentiable. We will not supply other technical conditions of the theorems here. The most interesting difference between the [**Ta**] approach and the [**SYC**] approach lies in the interpretation of "almost all". [**Ta**] interprets this topologically, and the set of $h$ for which an embedding is obtained (the set of *good* $h$) is *generic*; that is, it is an open dense subset of the space of all smooth $h$. [**HSY**] and [**SYC**] point out that open dense sets can often be very small in terms of measure, and so they prefer a measure-theoretic approach. They show that the collection of good $h$ forms a *prevalent* subset of the set of all smooth $h$; that is, the complementary set consisting of all *bad* $h$ is a *shy* set. A shy set is the equivalent of a set of zero Lebesgue measure in an infinite-dimensional space (in this case the space of all smooth $h$). We close this section with the following points:

1. One difficulty with the theorems of [**Ta**] and [**SYC**] is that there is no criterion for determining whether a specific functional $h$ is good or bad. Moreover, being good or bad is not a property of the functional $h$ in isolation, it is a property of $h$ and $\varphi$ together. A functional may be good for one f.d.d.s. but bad for another. We should add that, genericity and prevalence properties notwithstanding, it is not difficult to construct pairs $(\varphi, h)$ where no embedding results.

2. A functional can be bad in more than one way. It may be that an embedding is achieved between the canonical f.d.d.s. and only one component (a subsystem) of the underlying f.d.d.s. In this case the sampled time series has finite predictive dimension but there is no measure conjugacy with the desired original f.d.d.s. Some examples of this are provided in Section 5. It may also happen that the functional $h$ produces a sampled time series which is actually stochastic and hence no embedding is possible with $\varphi$ or any other f.d.d.s. In fact if one can prove that a time series is stochastic according to Definition 2.5 then it always follows that no embedding can exist with any f.d.d.s. Examples of this are also given in Section 5.

3. We also note that [**SBDH**] develop embedding theorems, based on Takens' genericity approach, for certain stochastic time series arising from both

forced and noisy systems. This is a new and important direction which needs to be integrated with ours.

4. Finally we note again that a time series can arise mathematically as a functional of many different systems. If the time series has finite predictive dimension then an embedding will exist between the canonical f.d.d.s. and some of these systems but not others. It would be of interest to keep the sampled time series fixed and examine the collection of all theoretically possible or potential generating mechanisms. This is the "inverse problem" to the [**Ta**] and [**SYC**] embedding theorems.

## 4. Scaling structures and determinism

One of the first techniques for searching for determinism in observed time series was the *Grassberger-Procaccia (GP) algorithm*; see [**GP**] and the discussion of the algorithm and estimation methods in [**Sm**]. This technique was predicated on the assumption, or rule-of- thumb, that in a deterministic time series the fractal dimensions associated with the joint distributions $P_n$ would level off at fixed finite values once $n \geq d$ where $d$ is the correct embedding dimension, whereas in a stochastic time series the fractal dimensions would grow without bound as $n \to \infty$. Experimentalists would estimate the fractal dimension $D_n$ of the joint distribution $P_n$ then plot $\hat{D}_n$ vs. $n$. If the plotted values approached a finite asymptote as $n \to \infty$ this was taken as evidence of determinism or low-dimensional dynamics. If, in addition, the limiting asymptote was fractional in value, this was taken as evidence of chaos. We will not discuss this latter aspect of the algorithm here.

The definitions of deterministic and stochastic were not quite clear in the GP algorithm. Moreover, it was not certain what distinction, if any, was made between determinism in the sampled time series and determinism in the underlying generating mechanism, or how these related to the scaling properties of the joint distributions of the sampled time series. Using our terminology and the results obtained so far, it is possible to obtain a complete picture of the theoretical GP algorithm (by theoretical we mean that we use the true values $D_n$ rather than estimates).

There are many possible choices for the fractal dimension $D_n$ associated with $P_n$, and these different choices generally produce different values. The GP algorithm was initially developed in terms of the *correlation dimension $\nu_n$* where

$$(4.1) \qquad \nu_n = \lim_{r \to 0} \frac{\log C_n(r)}{\log r}$$

and

$$(4.2) \qquad C_n(r) = P_n \times P_n(\{(x,y) \mid \|x - y\| \leq r\}).$$

$C_n(r)$ is the (spatial) *correlation integral* of $P_n$. In fact $C_n(r)$ is the probability that two random independent points from $P_n$ are within distance $r$ of each other, and $\nu_n$ describes how this probability scales as a power of $r$ as $r \to 0$. The paper of [**OP**] purported to show a counterexample to the GP rule-of-thumb by exhibiting a stochastic process in which the sequence of correlation dimensions $\nu_n$ converged to a finite value as $n \to \infty$. [**Th**] and [**Cu1**], taking different approaches, showed that this counterexample was false. In particular, [**Cu1**] argued that this counterexample can be viewed as an artifact of the simulation method used by the authors. [**W**] gives some results on the behaviour of sample versions of the correlation integral in the time series setting.

Another popular choice for fractal dimension is *information dimension* $\sigma_n$. There are several ways of defining $\sigma_n$. The following is one approach if $P_n$ is supported on a compact subset $K_n \subseteq (\mathbb{R}^u)^n$. We first define the local scaling exponent or *pointwise dimension* at $x \in K_n$ by

$$(4.3) \qquad \sigma_n(x) = \lim_{r \to 0} \frac{\log P_n(B(x,r))}{\log r}$$

where $B(x,r)$ is the ball of radius $r$ centred at $x$, and then set

$$(4.4) \qquad \sigma_n = \int \sigma_n(x) \, P_n(dx) \, .$$

Thus information dimension is the average pointwise dimension of $P_n$. In many cases of interest the pointwise dimension will be constant almost surely, and the information dimension will coincide with this constant. We always have the inequality $\sigma_n \geq \nu_n$ and this inequality can be strict. We have argued elsewhere ([**Cu3**]) that information dimension is generally a more informative quantity than correlation dimension. However, it is often useful to look at both quantities, as well as yet other fractal dimensions, simultaneously.

The following is our basic result concerning the GP algorithm. While the rule-of-thumb is not strictly correct, in many ways it is very close to being correct.

THEOREM 4.1. *Let $P_n$, $n = 1, 2, \ldots$ be the joint distributions of a stationary time series $X_n$, $n = 1, 2, \ldots$ taking values in a compact set $K_0 \subseteq \mathbb{R}^u$. Let $D_n$ be the fractal dimension of $P_n$, where $D_n = \nu_n$ or $D_n = \sigma_n$. Then the following are true:*

1. *The sequence $D_n$ is increasing in $n$. That is, $D_n \leq D_{n+1}$ for every $n$.*
2. *If the time series arises as a functional of an f.d.d.s., where both the functional and the f.d.d.s. are at least Lipschitz continuous, then the sequence $D_n$ converges to a finite value $D_0$. (This convergence occurs whether the sampled time series is deterministic or stochastic. It may be stochastic.)*
3. *If $\lim_{n \to \infty} D_n = \infty$ and the only possible predictor functions are Lipschitz continuous, then the time series is stochastic. Moreover, the time series cannot arise as a Lipschitz functional of a Lipschitz f.d.d.s.*
4. *It is possible for a stochastic time series to yield $\lim_{n \to \infty} D_n = D_0 < \infty$. Some of these time series can be represented as a Lipschitz functional of a Lipschitz f.d.d.s., others cannot.*

**Note** Proofs of the first two items are given in [**Cu2**]. The third item is simply the contrapositive of the second. These first three items taken together comprise what is right about the rule-of-thumb in the GP algorithm. It is the fourth item which deviates from the rule-of-thumb and constitutes a most interesting grey area. It includes stochastic time series which are Lipschitz functionals of smooth f.d.d.s. and might be regarded as almost deterministic (both examples in Section 5 are of this type). But it also includes other time series which involve so much independence that they must be regarded as truly stochastic. Examples of this type, such as randomly iterated function systems, are given in [**Cu3**].

## 5. Examples

The following are two examples of real-valued time series arising as smooth functionals of chaotic f.d.d.s. evolving on the unit square. In each case the plotted time series appears random and erratic, and we will show that both are in fact stochastic by our definition. However, it follows from Theorem 4.1 that the GP algorithm would yield a finite fractal dimension $D_0$ for each of these time series. Also in each of these cases we find that any function of either coordinate projection $\pi_1(x, y) = x$ or $\pi_2(x, y) = y$ will fail to result in an embedding with the f.d.d.s. This does not contradict any of the embedding theorems — the set of all smooth functions of the coordinate mappings is a shy subset in the space of all smooth functionals $h : [0, 1]^2 \to \mathbb{R}$. However, coordinate projections are common choices for functionals, and we see that if we keep the functional fixed (choose a coordinate projection) then obtaining a good or bad result depends on the coordinate system we choose for our f.d.d.s.

EXAMPLE 5.1 Define $\tilde{\psi} : [0, 1]^2 \to [0, 1]^2$ by $\tilde{\psi}(x, y) = (\psi(x), \psi(y))$ where $\psi$ is the tent map

$$(5.1) \qquad \psi(x) = \begin{cases} 2x & \text{for } 0 \le x \le 1/2 \\ 2(1 - x) & \text{for } 1/2 \le x \le 1 \end{cases}$$

Lebesgue measure, restricted to the unit interval, is an ergodic invariant measure for $\psi$, and it follows that two-dimensional Lebesgue measure, confined to the unit square, is ergodic and invariant for $\tilde{\psi}$. Note that this latter f.d.d.s. is uncoupled as the $x$ and $y$ coordinates iterate independently of one another.

Suppose now we define $f : [0, 1] \to [0, 1]$ by

$$(5.2) \qquad f(x) = \begin{cases} 1 & \text{for } 0 \le x \le 1/2 \\ 2(1 - x) & \text{for } 1/2 \le x \le 1 \end{cases}$$

and then define the functional $h : [0, 1]^2 \to [0, 1]$ by $h(x, y) = f(x)\, y$. We will denote the deterministic time series resulting from iterating the f.d.d.s. $\tilde{\psi}$ (with a random initial condition) by $(X_1, Y_1), (X_2, Y_2), \ldots$. Applying $h$ produces the stationary time series $Z_1, Z_2, \ldots$ where $Z_n = h(X_n, Y_n)$. We note that the functional $h$ is Lipschitz continuous but not differentiable because of the behaviour of $f$ at the point $x = 1/2$. We could modify $f$, even make it infinitely differentiable, without affecting the relevant qualitative behaviour of the $Z$ series. However, to do so would complicate our analysis and we prefer to leave $f$ as it is.

A plot of 300 points of the functional $Z$ series $Z_n = h(X_n, Y_n)$ is given below in Figure 1.

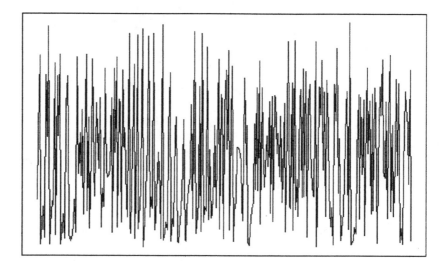

**Figure 1: 300 points of the $Z$ series**

It turns out that time-delay embeddings, applied to the $Z$ series, will fail to reconstruct the underlying system $\tilde{\psi}$. In other words, $h$ is a bad functional. We can prove this by showing that the $Z$ series is stochastic. Let $n \geq 2$ and suppose the initial condition $X_1$ of the $x$-coordinate satisfies $0 < X_1 \leq 2^{-n}$. (This event occurs with probability $2^{-n}$.) Then $X_1, X_2, \ldots, X_n$ all lie in the interval $(0, 1/2]$ and so $f(X_k) = 1$ for $k = 1, \ldots, n$. Thus $Z_k = Y_k$ for $k = 1, \ldots, n$ (and no additional information is obtained about the $x$ coordinate during this run of observed $Y$'s). We can detect the occurrence of such a run in the $Z$ series by noting that it corresponds to $Z_k = \psi(Z_{k-1})$ for $k = 2, \ldots, n$. Now if $f(X_k) = 1$ for $k = 1, \ldots, n$ (equivalently, if $Z_k = \psi(Z_{k-1})$ for $k = 2, \ldots, n$) then, using Lebesgue measure, there is a 50% probability that $0 < X_n \leq 1/4$ and a 50% probability that $1/4 < X_n \leq 1/2$. In the first case we get $f(X_{n+1}) = 1$ and $Z_{n+1} = \psi(Z_n)$ (that is, the run continues) and in the second case the outcome is $Z_{n+1} = 2(1 - 2X_n)\psi(Z_n)$. The variable $2(1 - 2X_n)$ is uniformly distributed over $[0, 1]$ and its value cannot be predicted by an observer based on the knowledge $Z_k = \psi(Z_{k-1})$ for $k = 2, \ldots, n$. This shows that, for each

$n \geq 2$, there are particular choices of $(z_1, \ldots, z_n)$ (which constitute a set of positive probability) for which the conditional distributions $P(Z_{n+1} \in \cdot / Z_n = z_n, \ldots, Z_1 = z_1)$ are not Dirac point masses, and hence an embedding of dimension $n$ cannot exist with the underlying f.d.d.s. $\tilde{\psi}$. Moreover, due to ergodicity of the system $\tilde{\psi}$, arbitrarily long runs of $f(X_{k+j}) = 1$, $j = 1, \ldots, m$ will occur infinitely often in the output sequence. Figure 2 shows a lag plot of $Z_{n+1}$ vs. $Z_n$; note that some of the underlying deterministic structure can be detected in the plot.

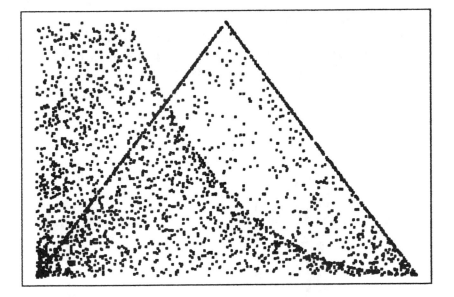

**Figure 2: Lag plot Z[n+1] vs. Z[n]**

Let us add that had we chosen either of the coordinate projections $\pi_1(x, y) = x$ or $\pi_2(x, y) = y$ as our functionals then we would also have been unable to reconstruct our f.d.d.s. but for a different reason. Each functional time series would have been deterministic but of predictive dimension $p = 1$, picking up only the 1-dimensional system $\psi$ evolving in the corresponding coordinate. This is a consequence of the coordinates of $\tilde{\psi}$ being uncoupled.

EXAMPLE 5.2 In this example we construct an f.d.d.s. where the coordinates are coupled but nonetheless one of the coordinate projections produces a stochastic time series and thus fails to reconstruct the f.d.d.s. Moreover, the f.d.d.s. is a type of *threshold model* commonly considered in nonlinear modelling; see [**To**]. We define $\varphi : [0,1]^2 \to [0,1]^2$ by $\varphi(x,y) = (S(x,y), L(y))$ where $L$ is the standard logistic map $L(y) = 4y(1-y)$ and $S$ is defined by

$$(5.3) \qquad S(x,y) = \begin{cases} L(x) & \text{if } y \leq 1/2 \\ (1-c(y))L(x) + c(y)x & \text{if } y > 1/2 \end{cases}$$

and $c(y) = 2(y - 1/2)$. The basic idea is to switch back and forth between two regimes: $L(x)$ for $y \leq 1/2$ and $x$ for $y > 1/2$. The factor $c(y)$ is introduced to create a smooth linear transition between the two regimes. It is not necessary to select the function $x$ for the second regime; it could be replaced by some other function $V(x)$. However, choosing $x$ produces some interesting behaviour; it shares the fixed point $x = 3/4$ with the logistic map but of course is not repelling—the result is that the $x$-coordinate series $X_n = \pi_1(\varphi^{n-1}(X_1, Y_1))$ tends to gravitate around $x = 3/4$ for stretches of time before being driven away by a change in regime and the repelling behaviour of the logistic map. Figure 3 shows a plot of 700 points of the $X$ series. The horizontal line corresponds to the fixed point $x = 3/4$.

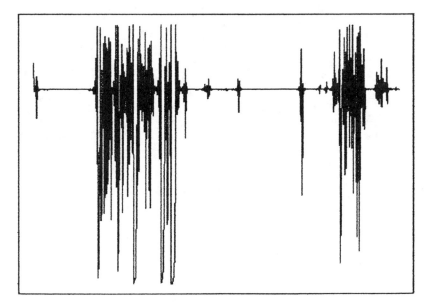

**Figure 3: 700 points of the X series in the threshold model**

Although we do not know the stationary ergodic behaviour of the overall system $\varphi$, it is well known that the logistic map has the arcsine (or beta$(\frac{1}{2}, \frac{1}{2})$) distribution as ergodic invariant measure. If the initial condition $Y_1$ of the $y$-coordinate is close to 0 (and hence several subsequent iterates of $L(Y_1)$ fall below 1/2), then the

1st coordinate of the system spends several iterations in the first regime; that is, $S(x,y) = L(x)$ or $X_n = L(X_{n-1})$. When the $Y$ coordinate finally exits to the second regime $Y > 1/2$, the value of $Y$ (and hence of $S(X,Y)$) cannot be predicted based on the observed $X$ series up to that point; this is very similar in spirit to Example 5. Consequently the $X$-coordinate series is stochastic, and an embedding with $\varphi$ cannot be constructed from it. Figure 4 shows a lag plot of $X_{n+1}$ vs. $X_n$; note that the "angel wings" illustrate the two boundaries $L(x) = 4x(1 - x)$ and $y = x$ as well as the stochastic region in between. The dark crossover point corresponds to the fixed point $x = 3/4$.

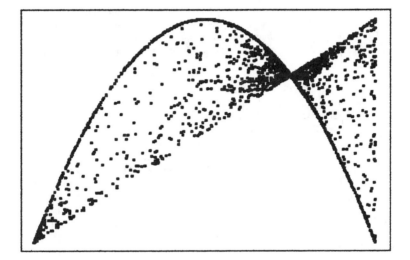

**Figure 4: Lag plot X[n+1] vs. X[n] in the threshold model**

We also note that $\varphi$ cannot be reconstructed from the $y$-coordinate projection, since lack of coupling in that direction ($x$ has no influence on $y$) results in predictive dimension $p = 1$ and reconstruction only of $L(y)$.

## Acknowledgements

I am indebted to David Brillinger for some gentle prodding on the concept of predictability, to the referee for forcing me to clarify some points, and to Luis Gorostiza and Gail Ivanoff for their infinite patience.

## Special acknowledgement

A version of this paper was presented at the International Conference on Stochastic Models in June 1998 in honour of Professor Donald Dawson. Don has been my teacher, my Ph.D. supervisor, and a co-author. He is responsible for introducing me to the notions of fractal dimension and scaling in measure theory and to many other areas of mathematics and probability. But in addition to his mathematical generosity, Don creates around him a unique atmosphere of integrity and enthusiasm for research, learning, and academic life which will not be duplicated. It has been a privilege to be his student and it is a pleasure to thank him.

## References

[ABST]  H.D.I. Abarbanel, R. Brown, J.J. Sidorowich, and L.S. Tsimring, *The analysis of observed chaotic data in physical systems,* Rev. Modern Phys. **65** (1993), 1331–1392.

[B]     P. Billingsley, *Ergodic Theory and Information,* Kriger, New York, (1978).

[Ca1]   M.C. Casdagli, *Nonlinear prediction of chaotic time series,* Physica D, **35** (1989), 335–356.

[Ca2]   M.C. Casdagli, *Chaos and deterministic versus stochastic and non-linear modelling,* J.R. Statist. Soc. B, **54** (1992), 303–328.

[CT1]   B. Cheng and H. Tong, *On consistent nonparametric order determination and chaos,* J. R. Statist. Soc. B, **54** (1992), 427–449.

[CT2]   B. Cheng, H. Tong, *Orthogonal projection, embedding dimension, and sample size in chaotic time series from a statistical perspective,* in Chaos and Forecasting (Ed.: H. Tong), World Scientific, Singapore, (1995) pp. 1–29.

[Cu1]   C.D. Cutler, *A theory of correlation dimension for stationary time series,* Philos. Trans. Roy. Soc. London A, **348** (1994), 343–355. *Expanded version with proofs* in Chaos and Forecasting (Ed.: H. Tong) World Scientific, Singapore, (1995) pp. 31–56.

[Cu2]   C.D. Cutler, *A general approach to predictive and fractal scaling dimensions in discrete-index time series,* in Nonlinear Dynamics and Time Series (Eds.: C.D. Cutler and D.T. Kaplan), Fields Institute Communications, **11**, American Mathematical Society, Providence RI, (1997), pp. 29–48.

[Cu3]   C.D. Cutler, *Computing pointwise fractal dimension by conditioning in multivariate distributions and time series,* to appear in Bernoulli.

[CK]    C.D. Cutler and D.T. Kaplan (Eds.) *Nonlinear Dynamics and Time Series,* Fields Institute Communications, **11**, American Mathematical Society, Providence, RI, (1997).

[GP]    P. Grassberger and I. Procaccia, *Characterization of strange attractors,* Phys. Rev. Lett. **50** (1983), 346–349.

[HSY]   B.R. Hunt, T. Sauer, and J.A. Yorke, *Prevalence: A translation-invariant "almost every" on infinite-dimensional spaces,* Bull. Amer. Math. Soc. **27** (1992), 217–238.

[KG]    D.T. Kaplan and L. Glass, *Direct test for determinism in a time series,* Phys. Rev. Lett. **68** (1992), 427–430.

[KBA]   M.B. Kennel, R. Brown, H.D.I. Abarbanel, *Determining embedding dimension for phase space reconstruction using a geometrical construction,* Phys. Rev. A **45** (1992), 3403–3411.

[LM]    A. Lasota and M.C. Mackey, *Chaos, Fractals, and Noise,* (2nd Ed.) Springer-Verlag, New York, (1994).

[OP]    A.R. Osborne and A. Provenzale, *Finite correlation dimension for stochastic systems with power-law spectra,* Physica D, **35** (1989), 357–381.

[PCFS]  N.H. Packard, J.P. Crutchfield, J.D. Farmer, and R.S. Shaw, *Geometry from a time series,* Phys. Rev. Lett. **45** (1980), 712–716.

[SYC]   T. Sauer, J.A. Yorke, and M. Casdagli, *Embedology,* J. Statist. Phys. **65** (1991), 579–616.

[Sm]    R.L. Smith . *Estimating dimension in noisy chaotic time series,* J. R. Statist. Soc. B, **54** (1992), 329–351.

[ST]    R.L. Smith and H. Tong (Eds.) *Proceedings of the Royal Statistical Society Meeting on Chaos,* in J. R. Statist. Soc. B, **54** (1992).

[SBDH]  J. Stark, D.S. Broomhead, M.E. Davies, and J. Huke (1996). *Takens embedding theorems for forced and stochastic systems,* submitted to the Proceedings of the 2nd World Congress of Nonlinear Analysis, Athens, Greece.

[Ta]    F. Takens, *Detecting strange attractors in turbulence,* in Lecture Notes in Mathematics, **898**, Springer-Verlag, New York, (1981), pp. 366–381.

[To]    H. Tong, *Non-linear Time Series: A Dynamical System Approach,* Oxford University Press, New York, (1990).

[Th]    J. Theiler, *Some comments on the correlation dimension of $1/f^\alpha$ noise,* Phys. Lett. A, **155** (1991), 480–493.

[W]     R.C.L. Wolff, *A note on the behaviour of the correlation integral in the presence of a time series,* Biometrika, **77** (1990), 689–697.

DEPARTMENT OF STATISTICS AND ACTUARIAL SCIENCE, UNIVERSITY OF WATERLOO, WATERLOO, ONTARIO, CANADA N2L 3G1

*E-mail address*: cdcutler@math.uwaterloo.ca

Canadian Mathematical Society
Conference Proceedings
Volume 26, 2000

# Kingman's Coalescent as a Random Metric Space

Steven N. Evans

*Dedicated with profound thanks to Professor Donald A. Dawson*

ABSTRACT. Kingman's coalescent is a Markov process with state–space the
collection of partitions of the positive integers. Its initial state is the trivial
partition of singletons and it evolves by successive pairwise mergers of blocks.
The coalescent induces a metric on the positive integers: the distance between
two integers is the time until they both belong to the same block. We investigate the completion of this (random) metric space. We show that almost surely
it is a compact metric space with Hausdorff and packing dimension both 1,
and it has positive capacities in precisely the same gauges as the unit interval.

## 1. Introduction

*Kingman's coalescent* was introduced in [**Kin82b, Kin82a**] as a model for
genealogies in the context of population genetics. This process has since been the
subject of a large amount of applied and theoretical work. We refer the reader to
[**Ald99**] for a recent survey and bibliography covering coalescent models in general,
and [**Tav84, Wat84**] for an indication of some of the applications of Kingman's
coalescent in genetics.

Here is a quick description of Kingman's coalescent (which we will hereafter
simply refer to as the coalescent). Recall that a *partition* of a set $S$ is a collection
$\{A_\lambda\}$ of subsets of $S$ (the $A_\lambda$ are called the *blocks* of the partition) such that
$A_\lambda \cap A_\mu = \emptyset$ for $\lambda \neq \mu$ and $\bigcup_\lambda A_\lambda = S$. Let $\Pi$ denote the collection of partitions
of $\mathbb{N} := \{1, 2, \ldots\}$. For $n \in \mathbb{N}$ let $\Pi_n$ denote the collection of partitions of $\mathbb{N}_n :=
\{1, 2, \ldots, n\}$. Each partition $\pi$ in $\Pi$ (resp. $\Pi_n$) corresponds to an *equivalence
relation* $\sim_\pi$ on $\mathbb{N}$ (resp. $\mathbb{N}_n$) by setting $i \sim_\pi j$ if $i$ and $j$ belong to the same block of
$\pi$. Write $\rho_n$ for the natural restriction map from $\Pi$ onto $\Pi_n$. Kingman [**Kin82b**]
showed that there was a (unique in law) $\Pi$–valued Markov process $\xi$ such that for
all $n \in \mathbb{N}$ the restricted process $\xi_n := \rho_n \circ \xi$ is a $\Pi_n$–valued, time–homogeneous
Markov chain with initial state $\xi_n(0)$ the trivial partition $\{\{1\}, \ldots, \{n\}\}$ and the
following transition rates: if $\xi_n$ is in a state with $k$ blocks, then

2000 *Mathematics Subject Classification.* Primary 60D05, 60J45; Secondary 60J25, 60K99.

*Key words and phrases.* coalesce; partition; exchangeable; capacity; Hausdorff dimension;
packing dimension; tree.

Research supported in part by NSF grant DMS-9703845.

- a jump occurs at rate $\binom{k}{2}$,
- the new state is one of the $\binom{k}{2}$ partitions that can be obtained by merging two blocks of the current state,
- and all such possibilities are equally likely.

There is a natural (random) metric $\delta$ on $\mathbb{N}$ defined by

$$\delta(i, j) := \inf\{t \geq 0 : i \sim_{\xi(t)} j\}.$$

In the original interpretation of the coalescent as the random genealogical tree of a countable collection of individuals (with time run backwards from the present), the distance $\delta(i, j)$ is just how long before the present the respective lines of descent of $i$ and $j$ diverged. Note that $\delta$ is actually an *ultrametric* on $\mathbb{N}$; that is,

$$\delta(i, j) \leq \delta(i, k) \vee \delta(k, j) \text{ for all } i, j, k \in \mathbb{N}.$$

Let $(\mathbb{S}, \delta)$ denote the *completion* of $(\mathbb{N}, \delta)$. Clearly, the extension of $\delta$ to $\mathbb{S}$ is also an ultrametric. Before presenting our main theorem giving some of the properties of $\mathbb{S}$, we will take some time to sketch a description of $\mathbb{S}$ that some readers might find helpful.

Recall that a *rooted tree* is a directed graph with the properties that, with the exception of a unique vertex (the *root*), every vertex has exactly one directed edge leading to it and the corresponding undirected graph is connected and acyclic. If two vertices $v$ and $w$ of a rooted tree are connected by a directed edge leading from $v$ to $w$, then $w$ is said to be a *child* of $v$.

As we recall in Section 2, almost surely the random partition $\xi(t)$ has finitely many blocks for all $t > 0$, $\xi$ evolves by blocks coalescing in pairs, and $\xi(t)$ consists of the single block $\mathbb{N}$ for all $t$ sufficiently large. Let $\mathcal{A}$ denote the collection of subsets $A$ of the integers such that $A$ is a block of $\xi(t)$ for some $t > 0$. We can think of the elements of $\mathcal{A}$ as the vertices of a tree rooted at $\mathbb{N}$: a block $A \in \mathcal{A}$ is the child of a block $C \in \mathcal{A}$ if $A \in \xi(s)$ and $C \in \xi(u)$ for some pair of times $s < u$ and there is another (unique) block $B \in \mathcal{A}$ such that $A$ coalesces with $B$ at some time $s < t \leq u$ to form $C$. Almost surely, each vertex in this rooted tree has two children.

In the usual terminology, an *end* of this rooted tree is an infinite directed path starting at the root, that is, an infinite sequence $\mathbb{N} = A_1, A_2, \ldots$ of blocks such that $A_{n+1}$ is a child of $A_n$ for all $n$. It is not hard to show that there is a one–to–one relationship between $\mathbb{S}$ and the set of ends.

The correspondence between coalescing partitions, tree structures and ultrametrics is a familiar idea, particularly in the physics literature (see, for example, [MPV87]). Some properties of the space $(\mathbb{N}, \delta)$ are considered explicitly in Section 4 of [Ald93].

It is our aim in this paper to investigate some of the *dimension* and *capacity* properties of $(\mathbb{S}, \delta)$. We remark that there is a large literature on such "fractal" properties of random trees constructed in various ways from Galton–Watson branching processes; for example, [Haw81] computed the Hausdorff dimension of the boundary of a Galton–Watson tree equipped with a natural metric (see also [Lyo90, LS98]). We refer the reader to [LP96] for an account, including an extensive bibliography, of this and other facets of probability on trees.

We remind the reader of the definitions and basic properties of Hausdorff dimension, packing dimension, energy and capacity. For more detail see [Mat95], [RT61] and [TT85]

Let $(T, \rho)$ be a metric space. Consider a non-decreasing, continuous function $g : \mathbb{R}_+ \to \mathbb{R}_+$ with $g(0) = 0$. The *Hausdorff outer measure* of a set $A \subseteq T$ with respect to the *measure function $g$* is defined by

$$m_g(A) := \lim_{\varepsilon \downarrow 0} m_g^\varepsilon(A),$$

where

$$m_g^\varepsilon(A) := \inf \left\{ \sum_i g(\mathrm{diam}(B_i)) \right\},$$

with the infimum taken over all countable collections of open balls $B_1, B_2, \ldots$ such that $A \subseteq \bigcup_i B_i$ (that is, $B_1, B_2, \ldots$ is a *cover* of $A$) and $\sup_i \mathrm{diam}(B_i) \leq \varepsilon$. The *Hausdorff dimension* of such a set $A$ is given by

$$\inf\{\alpha > 0 : m_{g_\alpha}(A) = 0\} = \sup\{\alpha > 0 : m_{g_\alpha}(A) = \infty\},$$

where $g_\alpha(s) := s^\alpha$, $s \geq 0$.

Let $(T, \rho)$, $g$ and $A$ be as above. The *packing premeasure* of $A$ with respect to the measure function $g$ is defined by

$$P_g(A) := \lim_{\varepsilon \downarrow 0} P_g^\varepsilon(A),$$

where

$$P_g^\varepsilon(A) := \sup \left\{ \sum_i g(\mathrm{diam}(B_i)) \right\},$$

with the supremum taken over all countable collections of pairwise disjoint open balls $B_1, B_2, \ldots$ with centres in $A$ (that is, $B_1, B_2, \ldots$ is a *packing* of $A$) such that $\sup_i \mathrm{diam}(B_i) \leq \varepsilon$. (We remark that it $(T, \rho)$ is an ultrametric space, then any point of a ball $B$ is a centre, and so in this case the requirement that the balls $B_1, B_2, \ldots$ are centred in $A$ is equivalent to the requirement that they intersect $A$.) The *packing outer measure* of $A$ with respect to the measure function $g$ is then defined to be

$$p_g(A) := \inf \left\{ \sum_i P_g(A_i) \right\},$$

where the infimum is over all countable collections of Borel sets $A_1, A_2, \ldots$ such that $A \subseteq \bigcup_i A_i$. The *packing dimension* of $A$ is given by

$$\inf\{\alpha > 0 : p_{g_\alpha}(A) = 0\} = \sup\{\alpha > 0 : p_{g_\alpha}(A) = \infty\},$$

with $g_\alpha$ as above.

By arguments similar to those in Lemma 5.11 of [**TT85**] it is possible to show that the inequality $m_g(A) \leq p_g(A)$ always holds and so, in particular, the Hausdorff dimension of a set is at most its packing dimension.

We now recall the definitions of energy and capacity. Again let $(T, \rho)$ be a metric space. Write $M_1(T)$ for the collection of (Borel) probability measures on $T$. A *gauge* is a function $f : [0, \infty[ \to [0, \infty]$, such that:

- $f$ is continuous and non-increasing,
- $f(0) = \infty$,
- $f(r) < \infty$ for $r > 0$,
- $\lim_{r \to \infty} f(r) = 0$.

Given $\mu \in M_1(T)$ and a gauge $f$, the *energy of $\mu$ in the gauge $f$* is the quantity

$$\mathcal{E}_f(\mu) := \int \mu(dx) \int \mu(dy) \, f(\rho(x,y)).$$

The *capacity of $A \subseteq T$ in the gauge $f$* is the quantity

$$\mathrm{Cap}_f(A) := (\inf\{\mathcal{E}_f(\mu)\})^{-1},$$

where the infimum is over probability measures $\mu \in M_1(T)$ with closed support contained in $A$ (note by our assumptions on $f$ that we need only consider diffuse $\mu \in M_1(T)$ in the infimum). The *capacity dimension* of a set $A \subseteq T$ is given by

$$\inf\{\alpha > 0 : \mathrm{Cap}_{f_\alpha}(A) = 0\} = \sup\{\alpha > 0 : \mathrm{Cap}_{f_\alpha}(A) = \infty\},$$

where $f_\alpha(s) := s^{-\alpha}$, $s > 0$.

The capacity dimension of a set equals its Hausdorff dimension (see Ch. 8 of [**Mat95**]), and hence the capacity dimension is also dominated by the packing dimension.

Our main result is the following, which, *inter alia*, asserts in the terminology of [**PP95**] (see, also, [**BP92, PPS96, Per96**]) that $\mathbb{S}$ is a.s. *capacity–equivalent* to the unit interval $[0,1]$.

THEOREM 1.1. *Almost surely, the metric space $(\mathbb{S}, \delta)$ is compact, and the Hausdorff and packing dimensions of $\mathbb{S}$ are both 1. There exist random variables $C^*, C^{**}$ such that almost surely $0 < C^* \leq C^{**} < \infty$ and for every gauge $f$*

$$(1.1) \qquad C^* \mathrm{Cap}_f([0,1]) \leq \mathrm{Cap}_f(\mathbb{S}) \leq C^{**} \mathrm{Cap}_f([0,1]).$$

Let us say a little about the interpretation of Theorem 1.1. Capacities, Hausdorff measures and packing measures are all ways of capturing how large a set is. By definition, knowing that $\mathbb{S}$ has positive capacity in some gauge indicates that $\mathbb{S}$ is large enough to allow mass to be spread "smoothly" on it. It is possible to establish analogues for $\mathbb{S}$ of density results for Euclidean space Hausdorff and packing measures (see Theorems 2.1 and 5.4 of [**TT85**] for statements of the Euclidean results). These results show that knowing $\mathbb{S}$ has positive Hausdorff or packing measure for some measure function is again equivalent to knowing that $\mathbb{S}$ supports a measure that is "smooth" in an appropriate sense.

The compactness claim of Theorem 1.1 and the fact that the Hausdorff and packing dimensions are at most 1 are established in Section 3. The capacity–equivalence (1.1) is proved in Section 6, and, by the general relationships between Hausdorff, packing and capacity dimensions, this also establishes the required lower bound on the Hausdorff and packing dimensions.

The results of this paper suggest a number of problems for future study. A process of coalescing partitions of $\mathbb{N}$ is constructed in [**DEF$^+$99**] using coalescing Brownian motions on the circle. The techniques of the present paper are used there to show that the corresponding metric space has Hausdorff and packing dimensions both equal to $\frac{1}{2}$ and that this space is capacity–equivalent to the middle–$\frac{1}{2}$ Cantor set (and hence, by the results of [**PPS96**], to the Brownian zero set). It is, of course, natural to investigate the existence of exact Hausdorff and packing measure functions for this Brownian model and the model considered here. In this regard, the random measure $\Gamma$ constructed in Section 5 and its analogue in the setting of [**DEF$^+$99**] are the natural candidates for applying the above mentioned analogues of the density theorems of [**RT61**] and [**TT85**], provided one can obtain the requisite upper and lower densities. Lastly, a number of other coalescing partition–valued processes arising from models in chemistry, cosmology and physics are considered in [**EP98, Pit99, BS98**] and it would be interesting to investigate the "fractal" properties of the corresponding metric spaces, which are typically not compact.

## 2. Some observations on the coalescent

We begin by recalling some results about the coalescent from [**Kin82b**]. Let $N(t)$ denote the number of blocks of the partition $\xi(t)$. Almost surely, $N(t) < \infty$ for all $t > 0$ and the process $N$ is a pure–death Markov chain that jumps from $k$ to $k - 1$ at rate $\binom{k}{2}$ for $k > 1$ (the state 1 is a trap). For $k \in \mathbb{N}$, put $\sigma_k := \inf\{t \geq 0 : N(t) = k\}$. The process $\xi$ is constant on each of the intervals $[\sigma_k, \sigma_{k-1}[$, $k > 1$. Write $I_1(t) < \cdots < I_{N(t)}(t)$ for an ordered listing of the least elements of the various blocks of $\xi(t)$. Almost surely, for all $t > 0$ the asymptotic block frequencies

$$F_i(t) := \lim_{n \to \infty} n^{-1} \left| \left\{ j \in \mathbb{N}_n : j \sim_{\xi(t)} I_i(t) \right\} \right|, \ 1 \leq i \leq N(t),$$

exist (where we use $|A|$ to denote the cardinality of a set $A$) and

$$F_1(t) + \cdots + F_{N(t)}(t) = 1.$$

It follows from the arguments that lead to Equation (35) in [**Ald99**] that

$$(2.1) \qquad\qquad \lim_{t \downarrow 0} t N(t) = 2, \ a.s.$$

Finally, we claim that

$$(2.2) \qquad\qquad \lim_{t \downarrow 0} t^{-1} \sum_{i=1}^{N(t)} F_i(t)^2 = 1, \ a.s.$$

To see this, set $X_{n,i} := F_i(\sigma_n)$ for $n \in \mathbb{N}$ and $1 \leq i \leq n$, and observe from (2.1) that it suffices to establish

$$(2.3) \qquad\qquad \lim_{n \to \infty} n \sum_{i=1}^{n} X_{n,i}^2 = 2, \ a.s.$$

By the "paintbox" construction in Section 5 of [**Kin82b**] (see also Section 4.2 of [**Ald99**] for an exposition with some helpful pictures) the random variable $\sum_{i=1}^{n} X_{n,i}^2$ has the same law as $U_{(1)}^2 + (U_{(2)} - U_{(1)})^2 + \cdots + (U_{(n-1)} - U_{(n-2)})^2 + (1 - U_{(n-1)})^2$, where $U_{(1)} \leq \ldots \leq U_{(n-1)}$ are the order statistics corresponding to i.i.d. random variables $U_1, \ldots, U_{n-1}$ that are uniformly distributed on $[0, 1]$. By a classical result on the spacings between order statistics of i.i.d. uniform random variables (see, for example, Section III.3.(e) of [**Fel71**]), the law of $\sum_{i=1}^{n} X_{n,i}^2$ is thus the same as that of $(\sum_{i=1}^{n} T_i^2)/(\sum_{i=1}^{n} T_i)^2$, where $T_1, \ldots, T_n$ are i.i.d. mean one exponential random variables.

Now for any $0 < \varepsilon < 1$ we have, recalling $\mathbb{P}[T_i^2] = 2$,

$$\mathbb{P}\left\{ \left(\sum_{i=1}^{n} T_i^2\right) \Big/ \left(\sum_{i=1}^{n} T_i\right)^2 > (1+\varepsilon)(1-\varepsilon)^{-2} 2n^{-1} \right\}$$

$$\leq \mathbb{P}\left\{ \sum_{i=1}^{n} \left(T_i^2 - \mathbb{P}[T_i^2]\right) > 2\varepsilon n \right\} + \mathbb{P}\left\{ \sum_{i=1}^{n} (T_i - \mathbb{P}[T_i]) < -\varepsilon n \right\}.$$

A fourth moment computation and Markov's inequality show that both terms on the right–hand side are bounded above by $c(\varepsilon)n^{-2}$ for a suitable constant $c(\varepsilon)$. A similar bound holds for

$$\mathbb{P}\left\{ \left(\sum_{i=1}^{n} T_i^2\right) \Big/ \left(\sum_{i=1}^{n} T_i\right)^2 < (1-\varepsilon)(1+\varepsilon)^{-2}2n^{-1} \right\}.$$

The claim (2.3) and hence (2.2) now follows by an application of the Borel–Cantelli Lemma. As the referee remarked, the tail estimates needed for the Borel–Cantelli Lemma can also be obtained from Markov's inequality and known moment formulae for Dirichlet distributions.

## 3. Compactness and upper bounds on dimensions

Given $B \subseteq \mathbb{S}$, write cl$B$ for the closure of $B$. Each of the sets

$$\begin{aligned}
U_i(t) &= \mathrm{cl}\{j \in \mathbb{N} : j \sim_{\xi(t)} I_i(t)\} \\
&= \mathrm{cl}\{j \in \mathbb{N} : \delta(j, I_i(t)) \leq t\} \\
&= \{y \in \mathbb{S} : \delta(y, I_i(t)) \leq t\}
\end{aligned}$$

is a closed ball with diameter at most $t$ (in an ultrametric space, the diameter and radius of a ball are equal). The closed balls of $\mathbb{S}$ are also the open balls and every ball is of the form $U_i(t)$ for some $t > 0$ (see, for example, Proposition 18.4 of [**Sch84**]) and, in fact, every ball is of the form $U_i(\sigma_k)$ for some $k \in \mathbb{N}$ and $1 \leq i \leq k$. In particular, the collection of balls is countable. Any ball of diameter at most $t$ is contained in a unique one of the $U_i(t)$, and any ball of diameter at least $t$ contains one or more of the $U_i(t)$ (see, for example, Proposition 18.5 of [**Sch84**]). Moreover, $k-1$ of the balls $U_i(\sigma_k)$, $1 \leq i \leq k$, are of the form $U_j(\sigma_{k+1})$ for some $1 \leq j \leq k+1$; and the remaining ball is of the form $U_h(\sigma_{k+1}) \cup U_\ell(\sigma_{k+1})$ for some pair $1 \leq h, \ell \leq k+1$.

Because $\mathbb{S}$ is complete by definition, in order to show that $\mathbb{S}$ is a.s. compact it suffices by Ascoli's theorem to show that $\mathbb{S}$ is a.s. totally bounded. However, for any $t > 0$ we have a.s. that $\mathbb{S}$ is covered by $N(t) < \infty$ closed balls of diameter at most $t$.

In order to establish that both the Hausdorff and packing dimensions of $\mathbb{S}$ are a.s. at most 1 it suffices to consider the packing dimension, because packing dimension always dominates Hausdorff dimension.

Recall that a packing of $\mathbb{S}$ is a pairwise disjoint collection of balls in $\mathbb{S}$. By definition of packing dimension, in order to establish that the packing dimension is at most 1 a.s. it suffices to show for each $\alpha > 1$ that there is a random variable $C$ such that $C < \infty$ a.s. and, for any packing $B_1, B_2, \ldots$ of $\mathbb{S}$ with balls of diameter at most 1, we have $\sum_k \mathrm{diam}(B_k)^\alpha \leq C$. As we observed above, if $2^{-p} \leq \mathrm{diam}(B_k) < 2^{-(p-1)}$ for some $p = 0, 1, 2, \ldots$, then $B_k$ contains one or more of the balls $U_i(2^{-p})$. Thus

$$|\{k \in \mathbb{N} : 2^{-p} \leq \mathrm{diam}(B_k) < 2^{-(p-1)}\}| \leq N(2^{-p})$$

and

$$\sum_k \mathrm{diam}(B_k)^\alpha \leq \sum_{p=0}^{\infty} N(2^{-p})2^{-(p-1)\alpha} < \infty,$$

by (2.1), as required.

## 4. An alternative expression for energy

We need to adapt to our setting the alternative expression for energy obtained by summation–by–parts in Section 2 of [**PP95**].

For $t > 0$ write $\mathcal{U}(t)$ for the collection of balls $\{U_1(t), \ldots, U_{N(t)}(t)\}$. Let $\mathcal{U}$ denote the union of these collections over all $t > 0$. As remarked in Section 3, the collection $\mathcal{U}$ is just the countable collection of all balls of $\mathbb{S}$. Given $U \in \mathcal{U}$ with $U \neq \mathbb{S}$, let $U^{\rightarrow}$ denote the unique element of $\mathcal{U}$ such that $U \subsetneq U^{\rightarrow}$ and if $V \in \mathcal{U}$ with $U \subseteq V \subseteq U^{\rightarrow}$ then either $V = U$ or $V = U^{\rightarrow}$. More concretely, such a ball $U$ is in $\mathcal{U}(\sigma_k)$ but not in $\mathcal{U}(\sigma_{k-1})$ for some unique $k > 1$, and $U^{\rightarrow}$ is the unique element of $\mathcal{U}(\sigma_{k-1})$ such that $U \subset U^{\rightarrow}$. Define $\mathbb{S}^{\rightarrow} := \dagger$, where $\dagger$ is an adjoined symbol. Put $\operatorname{diam}(\dagger) = \infty$.

Given a gauge $f$, write $\varphi_f$ for the diffuse measure on $[0, \infty[$ such that $\varphi_f([r, \infty[) = \varphi_f(]r, \infty[) = f(r)$, $r \geq 0$. For a diffuse probability measure $\mu \in M_1(\mathbb{S})$ we have, with the convention $f(\infty) = 0$,

$$
(4.1) \quad
\begin{aligned}
\mathcal{E}_f(\mu) &= \int \mu(dx) \int \mu(dy)\, f(\delta(x,y)) \\
&= \int \mu(dx) \int \mu(dy) \sum_{U \in \mathcal{U},\, \{x,y\} \subseteq U} f(\operatorname{diam}(U)) - f(\operatorname{diam}(U^{\rightarrow})) \\
&= \sum_{U \in \mathcal{U}} \left( f(\operatorname{diam}(U)) - f(\operatorname{diam}(U^{\rightarrow})) \right) \int \mu(dx) \int \mu(dy)\, \mathbf{1}\{\{x,y\} \subseteq U\} \\
&= \sum_{U \in \mathcal{U}} \left( f(\operatorname{diam}(U)) - f(\operatorname{diam}(U^{\rightarrow})) \right) \mu(U)^2 \\
&= \sum_{U \in \mathcal{U}} \int_{[0,\infty[} \varphi_f(dt)\, \mathbf{1}\{U \in \mathcal{U}(t)\} \mu(U)^2 \\
&= \int_{[0,\infty[} \varphi_f(dt) \sum_{U \in \mathcal{U}(t)} \mu(U)^2.
\end{aligned}
$$

## 5. Construction of a good measure on $\mathbb{S}$

In order to establish the left–hand side of the capacity–equivalence (1.1), it appears, a priori, that for each gauge $f$ we might need to find a random probability measure $\Gamma$ depending on $f$ such that $CCap_f([0,1]) \leq (\mathcal{E}_f(\Gamma))^{-1}$ for some a.s. non–zero random variable $C$ that does not depend on $f$. It turns out, however, that we can find a $\Gamma$ that works simultaneously for all gauges $f$. We construct $\Gamma$ as follows.

Let $\mathcal{B}$ denote the algebra of subsets of $\mathbb{S}$ generated by the collection of balls $\mathcal{U}$; so that $\mathcal{B}$ is just the countable collection of finite unions of balls. Of course, the $\sigma$–algebra generated by $\mathcal{B}$ is the Borel $\sigma$–algebra of $\mathbb{S}$. It is clear that, on an event $\Omega^*$ with $\mathbb{P}(\Omega^*) = 1$, the sets in $\mathcal{B}$ are compact, and, moreover, for all $k \in \mathbb{N}$ and indices $1 \leq i \leq k$ if $U_i(\sigma_k) = U_{i_1}(\sigma_{k+1}) \cup U_{i_2}(\sigma_{k+1})$ (that is, if $\{I_{i_1}(\sigma_{k+1}), I_{i_2}(\sigma_{k+1})\} = \{I_\ell(\sigma_{k+1}) : I_\ell(\sigma_{k+1}) \sim_{\xi(\sigma_k)} I_i(\sigma_k)\}$), then $F_i(\sigma_k) = F_{i_1}(\sigma_{k+1}) + F_{i_2}(\sigma_{k+1})$. It is therefore possible on the event $\Omega^*$ to define a finitely additive set function $\Gamma$ on $\mathcal{B}$ such that

$$(5.1) \qquad \Gamma(U_i(t)) = F_i(t),\ t > 0,\ 1 \leq i \leq N(t),$$

and

(5.2)                              $\Gamma(\mathbb{S}) = 1.$

Furthermore, if $A_1 \supseteq A_2 \supseteq \ldots$ is a decreasing sequence of sets in the algebra $\mathcal{B}$ such that $\bigcap_n A_n = \emptyset$, then, by compactness, $A_n = \emptyset$ for all $n$ sufficiently large and it is certainly the case that $\lim_{n\to\infty} \Gamma(A_n) = 0$. A standard extension theorem (see, for example, Theorems 3.1.1 and 3.1.4 of [**Dud89**]) gives that on the event $\Omega^*$ the set function $\Gamma$ extends to a probability measure (also denoted by $\Gamma$) on the Borel $\sigma$–algebra of $\mathbb{S}$. Define $\Gamma$ to be, say, the point mass $\delta_1$ off the event $\Omega^*$.

## 6. Capacities and lower bounds on dimensions

Establishing the capacity–equivalence (1.1) in the statement of Theorem 1.1 will certainly show that the capacity dimension of $\mathbb{S}$ is 1 a.s. The packing dimension of a set is at least its Hausdorff dimension, which is in turn equal to its capacity dimension. Therefore, (1.1) combined with the results of Section 3 will establish that the packing and Hausdorff dimensions of $\mathbb{S}$ are both 1 a.s.

¿From (4.1) and (2.2) we see that for some random variable $C'$ (not depending on $f$) with $0 < C' < \infty$ a.s. we have

$$
\mathrm{Cap}_f(\mathbb{S}) \geq (\mathcal{E}_f(\Gamma))^{-1} = \left( \int \varphi_f(dt) \sum_{U \in \mathcal{U}(t)} \Gamma(U)^2 \right)^{-1}
$$

(6.1)
$$
= \left( \int \varphi_f(dt) \sum_{i=1}^{N(t)} F_i(t)^2 \right)^{-1}
$$

$$
\geq C' \left( \int \varphi_f(dt)(t \wedge 1) \right)^{-1} = C' \left( \int_0^1 f(t)\, dt \right)^{-1}.
$$

Note from the Cauchy-Schwarz inequality that for any $\mu \in M_1(\mathbb{S})$

$$
1 = \left( \sum_{U \in \mathcal{U}(t)} \mu(U) \right)^2 \leq N(t) \sum_{U \in \mathcal{U}(t)} \mu(U)^2,
$$

and so, by (4.1),

$$
\mathrm{Cap}_f(\mathbb{S}) \leq \left( \int \varphi_f(dt) N(t)^{-1} \right)^{-1}.
$$

This sort of bound appears in Section IV.2 of [**Car67**]. Applying (2.1), we see that for some random variable $C''$ (again not depending on $f$) with $0 < C'' < \infty$ a.s. we have

(6.2)      $\mathrm{Cap}_f(\mathbb{S}) \leq C'' \left( \int \varphi_f(dt)(t \wedge 1) \right)^{-1} = C'' \left( \int_0^1 f(t)\, dt \right)^{-1}$

The capacity–equivalence (1.1) follows from (6.1) and (6.2) and the fact that there exist constants $0 < c^\# \leq c^{\#\#} < \infty$ such that

$$
c^\# \left( \int_0^1 f(t)\, dt \right)^{-1} \leq \mathrm{Cap}_f([0,1]) \leq c^{\#\#} \left( \int_0^1 f(t)\, dt \right)^{-1}
$$

(this is described as "classical" in [**PPS96**] and follows by arguments similar to those used around equation (9) in Section 2 of that paper to prove the analogous inequalities for $[0,1]^2$).

## Acknowledgement

Thanks are due to David Aldous and Jim Pitman for helpful discussions about Kingman's coalescent, and to an anonymous referee for several suggestions that improved the presentation.

## References

[Ald93]   D. Aldous. The continuum random tree III. *Ann. Probab.*, 21:248–289, 1993.

[Ald99]   D.J. Aldous. Deterministic and stochastic models for coalescence (aggregation, coagulation): a review of the mean–field theory for probabilists. *Bernoulli*, 5:3–48, 1999.

[BP92]    I. Benjamini and Y. Peres. Random walks on a tree and capacity in the interval. *Ann. Inst. H. Poincaré Probab. Statist.*, 28:557–592, 1992.

[BS98]    E. Bolthausen and A.-S. Sznitman. On Ruelle's probability cascades and an abstract cavity method. *Comm. Math. Phys.*, 197:247–276, 1998.

[Car67]   L. Carleson. *Selected Problems on Exceptional Sets*. van Nostrand, Princeton, 1967.

[DEF⁺99]  P. Donnelly, S.N. Evans, K. Fleischmann, T.G. Kurtz, and X. Zhou. Continuum–sites stepping–stone models, coalescing exchangeable partitions, and random trees. To appear *Ann. Probab.*, 1999.

[Dud89]   R.M. Dudley. *Real Analysis and Probability*. Wadsworth, Belmont CA, 1989.

[EP98]    S.N. Evans and J. Pitman. Construction of Markovian coalescents. *Ann. Inst. H. Poincaré Probab. Statist.*, 34:339–383, 1998.

[Fel71]   W. Feller. *An Introduction to Probability Theory and Its Applications*, volume II. Wiley, New York, 2nd edition, 1971.

[Haw81]   J. Hawkes. Trees generated by a simple branching process. *J. London Math. Soc. (2)*, 24:374–384, 1981.

[Kin82a]  J.F.C. Kingman. On the genealogy of large populations. In J. Gani and E.J. Hannan, editors, *Essays in Statistical Science*, pages 27–43. Applied Probability Trust, 1982. Special vol. 19A of *J. Appl. Probab.*

[Kin82b]  J.F.C. Kingman. The coalescent. *Stochastic Process. Appl.*, 13:235–248, 1982.

[LP96]    R.D. Lyons and Y. Peres. Probability on trees and networks. Book in preparation for Cambridge University Press, available via http://php.indiana.edu/~rdlyons/, 1996.

[LS98]    S.P. Lalley and T. Sellke. An extension of Hawke's theorem on the Hausdorff dimension of a Galton–Watson tree. Preprint, 1998.

[Lyo90]   R.D. Lyons. Random walks and percolation on trees. *Ann. Probab.*, 18:931–958, 1990.

[Mat95]   P. Mattila. *Geometry of Sets and Measures in Euclidean Spaces: Fractals and Rectifiability*, volume 44 of *Cambridge Studies in Advanced Mathematics*. Cambridge University Press, Cambridge – New York, 1995.

[MPV87]   M. Mezard, G. Parisi, and M.A. Virasoro. *Spin Glass Theory and Beyond*, volume 9 of *World Scientific Lecture Notes in Physics*. World Scientific, Singapore, 1987.

[Per96]   Y. Peres. Remarks on intersection–equivalence and capacity–equivalence. *Ann. Inst. H. Poincaré Phys. Théor.*, 64:339–347, 1996.

[Pit99]   J. Pitman. Coalescents with multiple collisions. To appear *Ann. Probab.*, 1999.

[PP95]    R. Pemantle and Y. Peres. Galton–Watson trees with the same mean have the same polar sets. *Ann. Probab.*, 23:1102–1124, 1995.

[PPS96]   R. Pemantle, Y. Peres, and J.W. Shapiro. The trace of spatial Brownian motion is capacity–equivalent to the unit square. *Probab. Theory Related Fields*, 106:379–399, 1996.

[RT61]    C.A. Rogers and S.J. Taylor. Functions continuous and singular with respect to a Hausdorff measure. *Mathematika*, 8:1–31, 1961.

[Sch84]   W. H. Schikhof. *Ultrametric Calculus: an Introduction to p-adic Analysis*, volume 4 of *Cambridge Studies in Advanced Mathematics*. Cambridge University Press, Cambridge – New York, 1984.

[Tav84]     S. Tavaré. Line–of–descent and genealogical processes, and their applications in popu-
            lation genetics. *Theoret. Population Biol.*, 26:119–164, 1984.
[TT85]      C. Tricot and S.J. Taylor. Packing measure, and its evaluation for a Brownian path.
            *Trans. Amer. Math. Soc.*, 288:679–699, 1985.
[Wat84]     G. A. Watterson. Lines of descent and the coalescent. *Theoret. Population Biol.*, 26:77–
            92, 1984.

DEPARTMENT OF STATISTICS #3860, UNIVERSITY OF CALIFORNIA AT BERKELEY, 367 EVANS
HALL, BERKELEY, CA 94720-3860, U.S.A.

*E-mail address*: evans@stat.berkeley.edu

Canadian Mathematical Society
Conference Proceedings
Volume **26**, 2000

# The Behaviour Near the Boundary of Some Degenerate Diffusions under Random Perturbation

Shui Feng

*Dedicated to Donald A. Dawson*

ABSTRACT. In this paper we study the behaviour near the boundary of some degenerate diffusions under small perturbations. In particular we are interested in the way that the processes approach or leave the boundary. Large deviation estimates are used in deriving the results.

## 1. Introduction

For any $\varepsilon > 0, T > 0$, consider the following random perturbation of the one dimensional stochastic differential equation (henceforth, SDE)

$$(1.1) \qquad d z_t^\varepsilon = b(z_t^\varepsilon)dt + \varepsilon\sigma(z_t^\varepsilon)d B_t, \;\; 0 \le t \le T, \;\; z_0^\varepsilon = x,$$

where $B_t$ is the standard Brownian motion. For uniform Lipschitz functions $b(x)$ and $\sigma(x)$ it is known ([**2**], [**4**]) that a large deviation principle holds for the law of $z_t^\varepsilon$ and the rate function obtained describes the behaviour of the process near the boundary when there is a boundary.

In the present article we are interested in the following models

$$(1.2) \qquad dy_t^\varepsilon = (cy_t^\varepsilon + d)dt + \varepsilon\sqrt{y_t^\varepsilon}d B_t,$$

where $d \ge 0$. Here the coefficient $\sigma(y)$ is degenerate, non-Lipschitz, and the equation has a unique strong solution.(e.g. [**5**].)

The process $y_t^\varepsilon$ appears in the diffusion approximation of critical branching process.(e.g. [**3**].) For $d > 0$ and small $\varepsilon$ the set $\{0\}$ is the entrance boundary of the process and the boundary is exit and absorbing when $d$ is 0. In this article we are interested in understanding the behaviour of the process $y_t^\varepsilon$ near the boundary as $\varepsilon$ goes to zero. In particular, we want to know how the process leaves the boundary when $d > 0$, and how the process approaches the boundary for $d = 0$. The main tools in the proof are techniques from large deviation theory.

2000 *Mathematics Subject Classification.* Primary 60F10; Secondary 92D10.
*Key words and phrases.* branching process; diffusion approximation; large deviations.
Research supported by the Natural Science and Engineering Research Council of Canada.

## 2. Exponential tightness

A family $\{P^\varepsilon\}_{\varepsilon>0}$ of probability measures on a Polish space $X$ is said to be exponentially tight with speed $a(\varepsilon) \downarrow 0$ if for any $R > 0$ there is a compact set $K_R$ in $X$ such that

$$(2.1) \qquad \limsup_{\varepsilon \to 0} a(\varepsilon) \log P^\varepsilon\{K_R^c\} < -R,$$

where $K_R^c$ is the complement of $K_R$ in $X$.

Let $C_y([0,T])$ be the space of all non-negative continuous functions on $[0,T]$ starting at $y$ equipped with the uniform convergence topology under which $C_y([0,T])$ becomes a Polish space. For any $y \geq 0$, $Q_y^\varepsilon$ is the law of the trajectory $y_t^\varepsilon$ starting at $y$. Then

THEOREM 2.1. *For any $y \geq 0$, the family $\{Q_y^\varepsilon\}_{\varepsilon>0}$ is exponentially tight on $C_y([0,T])$ with speed $a(\varepsilon) = \varepsilon^2$;*

PROOF. For any $M > 0$, let $h_M(y) = y \wedge M$, $b_M(y) = b(h_M(y))$, $\bar{b}_M(y) = d + |c|h_M(y)$, $\sigma_M(y) = \sqrt{h_M(y)}$, and $y_t^{\varepsilon,M}$ be the unique strong solution of the following SDE

$$(2.2) \qquad dy_t^{\varepsilon,M} = b_M(y_t^{\varepsilon,M})dt + \varepsilon\sigma_M(y_t^{\varepsilon,M})dB_t.$$

The law of the process $y_t^{\varepsilon,M}$ is denoted by $Q_y^{\varepsilon,M}$. Set

$$(2.3) \qquad \tau_M = \inf\{t \geq 0; y_t \geq M\},$$

then for any Borel subset $A$ of $C_y([0,T])$ the following holds

$$(2.4) \qquad Q_y^\varepsilon\{A \cap (\tau_M \geq T)\} = Q_y^{\varepsilon,M}\{A \cap (\tau_M \geq T)\}.$$

It is known from Lemma 4.12 of [7] that

$$(2.5) \qquad \lim_{M\to\infty} \limsup_{\varepsilon\to 0} \varepsilon^2 \log Q_y^\varepsilon\{\tau_M \leq T\} = -\infty.$$

Next we will first prove the exponential tightness of the family $\{Q_y^{\varepsilon,M}\}_{\varepsilon>0}$ for every $M > 0, y \geq 0$. Then by combining with (2.4) we finally derive (ii).

For any $\rho > 0, 0 < \gamma < T/2$, $\rho > \bar{b}_M(M)\gamma$, by the Markov property and Theorem 4.2.1 in [8], we get

$$(2.6) \quad Q_y^{\varepsilon,M}\{ \sup_{t,s\in[0,T],0\leq t-s\leq\gamma} |y_t - y_s| > \rho\}$$

$$\leq Q_y^{\varepsilon,M}\{ \sup_{t,s\in[0,T],0\leq t-s\leq\gamma} |y_t - y_s - \int_s^t \bar{b}_M(y_u)du| > \rho - \bar{b}_M(M)\gamma\}$$

$$\leq \sum_{k=0}^{[T/\gamma]-1} \sup_{z\geq 0} Q_z^{\varepsilon,M}\{ \sup_{t\in[k\gamma,(k+2)\gamma\wedge T)}$$

$$\times |y_t - y_{k\gamma} - \int_{k\gamma}^t \bar{b}_M(y_u)du| > (\rho - \bar{b}_M(M)\gamma)/2\}$$

$$\leq \frac{T}{\gamma} \sup_{z\geq 0} Q_z^{\varepsilon,M}\{ \sup_{0\leq t<2\gamma} |y_t - y_0 - \int_0^t \bar{b}_M(y_u)du| > (\rho - \bar{b}_M(M)\gamma)/2\}$$

$$\leq \frac{T}{\gamma} \exp(-\frac{(\rho - \bar{b}_M(M)\gamma)^2}{16\varepsilon^2 M\gamma}),$$

where $[T/\gamma]$ is the integer part of $T/\gamma$.

Let $n_0$ and $\varepsilon_0$ satisfy $Tn_0^2 > 1, R\varepsilon_0^{-2} > 1$. For any $R > 0, n \geq n_0, \varepsilon < \varepsilon_0$, set

$$\gamma_n = 1/n^2, \rho_n = \bar{b}_M(M)\gamma_n + \sqrt{16MR(1/n + 1/n^2 \log Tn^2)},$$

and define

$$K_R = \cap_{n \geq n_0}\{y. \in C([0,T]); \sup_{t,s \in [0,T], 0 \leq t-s < \gamma_n} |y_t - y_s| \leq \rho_n\}.$$

Then both $\gamma_n$ and $\rho_n$ converge to zero as $n$ goes to infinity, and the set $K_R$ is a compact subset of $C_y([0,T])$. By using (2.6) one gets

$$(2.7) \quad Q_y^{\varepsilon,M}\{y. \in K_R^c\} \leq \sum_{n=n_0}^{\infty} P\{\sup_{t,s \in [0,T], 0 \leq t-s < \gamma_n} |y_t^{\varepsilon,M} - y_s^{\varepsilon,M}| > \rho_n\}$$

$$\leq \sum_{n=n_0}^{\infty} \frac{T}{\gamma_n} \exp\left(-\frac{(\rho_n - b_M(M)\gamma_n)^2}{16M\gamma_n\varepsilon^2}\right)$$

$$\leq \frac{\exp(-\frac{R}{\varepsilon^2})}{1 - \exp(-\frac{R}{\varepsilon^2})},$$

which implies

$$(2.8) \qquad \limsup_{\varepsilon \to 0} \varepsilon^2 \log Q_y^{\varepsilon,M}\{K_R^c\} \leq -R,$$

and the family $\{Q_y^{\varepsilon,M}\}_{\varepsilon>0}$ is exponentially tight.

Finally for any $R > 0$, and $M$ large enough, we have from (2.5)

$$\limsup_{\varepsilon \to 0} \varepsilon^2 \log Q_y^{\varepsilon}\{\tau_M \leq T\} \leq -R.$$

This combined with (2.4) implies

$$\limsup_{\varepsilon \to 0} \varepsilon^2 \log Q_y^{\varepsilon}\{K_R^c\}$$

$$\leq \max\left[\limsup_{\varepsilon \to 0} \varepsilon^2 \log Q_y^{\varepsilon}\{K_R^c \cap (\tau_M \geq T)\}, \limsup_{\varepsilon \to 0} \varepsilon^2 \log Q_y^{\varepsilon}\{\tau_M \leq T\}\right]$$

$$\leq \max\{\limsup_{\varepsilon \to 0} \varepsilon^2 \log Q_y^{\varepsilon,M}\{K_R^c\}, \limsup_{\varepsilon \to 0} \varepsilon^2 \log Q_y^{\varepsilon}\{K_R^c\}\} \leq -R,$$

which proves the result. $\qquad\square$

REMARK. From the proof of the above theorem we can see that the exponential tightness holds for a much wider class of diffusion processes such as those processes satisfying assumptions of Lemma 4.12 in [7].

## 3. Large deviations

Large deviation principle has been established in [1] for the process $x_t^{\varepsilon}$ satisfying

$$dx_t^{\varepsilon} = (\mu_1(1 - x_t^{\varepsilon}) - \mu_2 x_t^{\varepsilon})dt + \varepsilon\sqrt{x_t^{\varepsilon}(1 - x_t^{\varepsilon})}dB_t,$$

where $\mu_1 > 0, \mu_2 > 0$, and the initial point $x_0^{\varepsilon}$ is not on the boundary $\{0, 1\}$. It is shown in [1] that the probability that the process approaches the boundary goes to zero superexponentially fast.

By using an argument similar to that in [1] we have that if $d > 0$ and $y_0^{\varepsilon} \neq 0$ then the process $y_t^{\varepsilon}$ satisfying (1.2) will not hit the boundary $\{0\}$ under large deviations. Therefore there are essentially two cases that still need to be studied, namely, the case of $d > 0, y_0^{\varepsilon} = 0$, and the case of $d = 0, y_0^{\varepsilon} > 0$.

We start with a lemma that describes the large deviation behaviour of $y_t^\varepsilon$ for every fixed time $t > 0$.

LEMMA 3.1. *For any $t > 0$, let $\mu_t^{\varepsilon,y}$ be the law of $y_t^\varepsilon$ with $y_0^\varepsilon = y$. Then the family $\{\mu_t^{\varepsilon,y}\}$ satisfies a large deviation principle with rate function*

$$S_t^y(x) = \sup_\theta \{\theta x - \log E^{\mu_t^{\varepsilon,y}}[e^{\frac{\theta x}{\varepsilon^2}}]\}$$

$$= ab + ax + 2d\log(d + \sqrt{d^2 + a^2 bx}) - 2d\log ax - 2\sqrt{d^2 + a^2 bx},$$

*where $b = ye^{ct}$, $a = \frac{2c}{e^{ct}-1}$ for $c \neq 0$, and $a = 2/t$ for $c = 0$. For simplicity we write $S_t(x)$ for $S_t^0(x)$.*

PROOF. For any $y \geq 0$, by Ito's formula we have that for $\theta < a$

$$M(t,y,\theta) = E^{\mu_t^{\varepsilon,y}}[e^{\frac{\theta x}{\varepsilon^2}}] = [1 - \frac{\theta}{a}]^{-\frac{2d}{\varepsilon^2}}\exp(\frac{1}{\varepsilon^2}\frac{b\theta}{1-\theta/a}),$$

and for $\theta \geq a$, $M(t,y,\theta) = \infty$. This combined with the Gärtner-Ellis theorem implies the result (c.f. [2]). $\qquad\square$

For any $y \geq 0$ , and $[\alpha,\beta]$ in $[0,T]$, set

$$H_y^{\alpha,\beta} = \{\varphi(\cdot); \varphi(\alpha) = \varphi(y) \text{ and } \varphi \text{ is absolutely continuous on } [\alpha,\beta]\},$$

and define

$$I_y^{\alpha,\beta}(\varphi) = \begin{cases} \frac{1}{2}\int_\alpha^\beta \frac{(\dot\varphi(t)-(c\varphi(t)+d))^2}{\varphi(t)}dt, & \varphi \in H_y^{\alpha,\beta} \\ \infty, & \varphi \notin H_y^{\alpha,\beta}. \end{cases}$$

For simplicity we will write $I_y(\varphi)$ for $I_y^{0,T}(\varphi)$.

THEOREM 3.2. *Assume that $d > 0$, and $y_0^\varepsilon = 0$. Then for any $\varphi(t)$ in $C_0([0,T])$, we have*

$$(3.1) \qquad \limsup_{\delta\to 0}\limsup_{\varepsilon\to 0} \varepsilon^2 \log Q_0^\varepsilon\{\sup_{t\in[0,T]}|y_t - \varphi(t)| \leq \delta\} \leq -I_0(\varphi)$$

$$(3.2) \qquad \liminf_{\delta\to 0}\liminf_{\varepsilon\to 0} \varepsilon^2 \log Q_0^\varepsilon\{\sup_{t\in[0,T]}|y_t - \varphi(t)| < \delta\} \geq -I_0(\varphi).$$

PROOF. We will prove (3.1) first. Three cases will be treated separately.

**Case I.** There is a $t_0 > 0$ such that $\varphi(t) = 0$ for all $t$ in $[0,t_0]$.
For any $\delta > 0$, we have

$$Q_0^\varepsilon\{\sup_{t\in[0,T]}|y_t - \varphi(t)| < \delta\} \leq Q_0^\varepsilon\{\sup_{t\in[0,t_0]}y_t < \delta\}$$

$$\leq Q_0^\varepsilon\{\sup_{t\in[0,t_0]}(-1)(y_t - \int_0^{t_0}(d + cy_s)ds) > c_\delta t_0 - \delta\},$$

where $c_\delta = \inf\{d+cy; 0 \leq y \leq \delta\}$. Noting that $d > 0$, we can choose $\delta$ small enough such that $c_\delta t_0 - \delta > dt_0/4 > 0$. Then by Doob's inequality for martingales, we get

$$Q_0^\varepsilon\{\sup_{t\in[0,T]}|y_t - \varphi(t)| < \delta\} \leq Q_0^\varepsilon\{\sup_{t\in[0,t_0]}[(-\rho)(y_t - \int_0^t(d + cy_s)ds)$$

$$-\frac{\varepsilon^2\rho^2}{2}\int_0^{t_0}y_s\,ds] > (dt_0/4)\,\rho - \frac{\varepsilon^2\rho^2\delta t_0}{2}\}$$

$$\leq e^{(-(dt_0/4)\rho+\frac{\varepsilon^2\rho^2\delta t_0}{2})}.$$

Taking $\rho = d/4\varepsilon^2\delta$, we get

$$(3.3) \qquad Q_0^\varepsilon\{ \sup_{t\in[0,T]} |y_t - \varphi(t)| < \delta\} \le e^{\left(-\frac{d^2 t_0}{32\varepsilon^2\delta}\right)},$$

which implies that

$$\limsup_{\delta\to 0}\limsup_{\varepsilon\to 0} \varepsilon^2 \log Q_0^\varepsilon\{ \sup_{t\in[0,T]} |y_t - \varphi(t)| \le \delta\} \le -\infty = -I_0(\varphi).$$

**Case II.** The function $\varphi(t)$ is strictly positive for all $t$ in $(0,T]$.
For any $\delta > 0, N \ge 1$, set

$$A(\varphi,\delta) = \{ \sup_{0\le t\le T} |y_t - \varphi(t)| \le \delta\}, A(\varphi,\delta,N) = \{ \sup_{1/N\le t\le T} |y_t - \varphi(t)| \le \delta\}.$$

For a countable dense subset $\{t_1, t_2, ....., t_k, ....\}$ of $[1/N, T]$ and $m \ge 1$, we define

$$A_m(\varphi,\delta,N) = \{ \max_{i=1,...,m} |y_{t_i} - \varphi(t_i)| \le \delta\}.$$

Then $A_m(\varphi,\delta,N)$ is decreasing in $m$ and

$$(3.4) \qquad \lim_{m\to\infty} A_m(\varphi,\delta,N) = \cap_{k=1}^\infty A_m(\varphi,\delta,N) = A(\varphi,\delta,N).$$

Noting that for any bounded continuous function $f$ on $R_+$ and $t$ in $[0,T]$ the integral $\int_0^1 f(y_t)dQ_y^\varepsilon$ is continuous in $y$, we conclude that, by the Markov property, $Q_y^\varepsilon\{A_m(\varphi,\delta,N)\}$ is upper semicontinuous in $y$ for all $m$. This combined with (3.4) implies that

$$(3.5) \qquad Q_y^\varepsilon\{A(\varphi,\delta,N)\} \text{ is upper semicontinuous in } y.$$

Similarly we have that

$$(3.6) \qquad Q_y^\varepsilon\{ \sup_{1/N\le t\le T} |y_t - \varphi(t)| < \delta\} \text{ is lower semicontinuous in } y.$$

For any $N \ge 1$, we choose $\delta$ small enough such that no functions in the set $A(\varphi, 2\delta, N)$ hit zero on the time interval $[1/N, T]$. Then for any $\varepsilon > 0$, one get

$$\varepsilon^2 \log Q_0^\varepsilon\{ \sup_{0\le t\le T} |y_t - \varphi(t)| < \delta\}$$

$$\le \quad \varepsilon^2 \log Q_0^\varepsilon\{A(\varphi,\delta)\}$$

$$\le \quad \varepsilon^2 \log Q_0^\varepsilon\{A(\varphi,\delta,N)\} = \varepsilon^2 \log \int_0^\infty Q_y^\varepsilon\{A(\varphi,\delta,N)\}\, \mu_{1/N}^{\varepsilon,0}(dy)$$

$$\le \quad \varepsilon^2 \log \sup_{|y-\varphi(1/N)|\le\delta} Q_y^\varepsilon\{A(\varphi,\delta,N)\} + \varepsilon^2 \log \mu_{1/N}^{\varepsilon,0}\{|y - \varphi(1/N)| \le \delta\}$$

$$\le \quad \sup_{|y-\varphi(1/N)|\le\delta} \varepsilon^2 \log Q_y^\varepsilon\{A(\varphi,\delta,N)\}$$

$$= \quad \varepsilon^2 \log Q_{y^\varepsilon}^\varepsilon\{A(\varphi,\delta,N)\} \text{ for some } |y^\varepsilon - \varphi(1/N)| \le \delta,$$

where in the last equality we used (3.5) and the property that the supremum of an upper semicontinuous function over a closed set can be reached at certain point inside the set. By the uniform large deviation principle for non-degenerate diffusions (cf. Theorem 4.14 in [**7**]), we get

(3.7)
$$\limsup_{\varepsilon \to 0} \varepsilon^2 \log Q_0^\varepsilon \{ \sup_{0 \le t \le T} |y_t - \varphi(t)| < \delta \}$$

$$\le - \inf_{|y-\varphi(1/N)| \le \delta} \inf_{\psi \in A(\varphi, \delta, N)} I_y^{1/N,T}(\psi).$$

Assume that $\inf_{|y-\varphi(1/N)| \le \delta} \inf_{\psi \in A(\varphi, \delta, N)} I_y^{1/N,T}(\psi)$ is finite for small $\delta$. Otherwise the upper bound is trivially true. For any $y$ satisfying $|y - \varphi(1/N)| \le \delta$, $\psi$ in $A(\varphi, \delta, N)$ satisfying $\psi(1/N) = y$, $I_y^{1/N,T}(\psi) < \infty$, we define for $t$ in $[1/N, T]$

$$\bar{\psi}(t) = \psi(t) + (\varphi(1/N) - y).$$

Then it is clear that $\bar{\psi}$ is in $A(\varphi, 2\delta, N)$ and thus does not hit zero. By direct calculation, we get that

$$I_{\varphi(1/N)}^{1/N,T}(\bar{\psi}) \le I_y^{1/N,T}(\psi) + \delta_N,$$

where $\delta_N$ goes to zero as $\delta$ goes to zero for any fixed $N$. This combined with (3.7) implies that

(3.8)
$$\limsup_{\delta \to 0} \limsup_{\varepsilon \to 0} \varepsilon^2 \log Q_0^\varepsilon \{ \sup_{0 \le t \le T} |y_t - \varphi(t)| < \delta \}$$

$$\le - \lim_{\delta \to 0} \inf_{\psi \in A(\varphi, 2\delta, N)} I_{\varphi(1/N)}^{1/N,T}(\psi) = -I_{\varphi(1/N)}^{1/N,T}(\varphi),$$

where the equality follows from the lower semicontinuity of $I_{\varphi(1/N)}^{1/N,T}(\cdot)$ at non-degenerate paths. Finally by letting $N$ go to infinity we end up with (3.1).

**Case III.** The function $\varphi(t)$ leaves the boundary $\{0\}$ immediately and returns to the boundary again at a later time $t$ in $(0, T]$.

In this case we have $I_0(\varphi) = \infty$ by an argument similar to the one used in the proof of Lemma 3.2 in [**1**]. Let $t_0 > 0$ be the first time that $\varphi$ hits the boundary after leaving zero. Then (3.1) follows by approaching $t_0$ from below and using result in **Case II**.

Next we turn to prove (3.2). The fact that the upper bounds in **Case I** and **Case III** are negative infinity implies the corresponding lower bound immediately. Now we focus on **Case II**.

Without loss of generality we further assume that $I_0(\varphi)$ is finite. The fact that $\{\sup_{t \le s \le T} |y_s - \varphi(s)| < \delta\}$ is a subset of $\{|y_t - \varphi(t)| < \delta\}$, combined with Lemma 3.1 implies that

$$S_t(\varphi(t)) = 2d(\log 2d - 1) + \left(\frac{2c}{e^{ct} - 1}\right)\varphi(t)$$

$$-2d \log[\left(\frac{2c}{e^{ct} - 1}\right)\varphi(t)] \le I_{\varphi(t)}^{t,T}(\varphi) < \infty.$$

Thus the sequence $\varphi(t)/t$ must be included in an interval $[h_1, h_2]$ for finite positive $h_1 < h_2$.

This combined with the variational expression of $I_0(\varphi)$, and Jensen's inequality implies that

(3.9)
$$\frac{(\varphi(t) - \int_0^t b(\varphi(s))ds)^2}{\int_0^t \varphi(s)\,ds} \le \int_0^t \frac{(\dot{\varphi}(s) - (c\varphi(s) + d))^2}{\varphi(s)}\,ds.$$

Hence

$$\limsup_{t \to 0} \left( \frac{\varphi(t) - \int_0^t b(\varphi(s))ds}{t} \right)^2 \le \limsup_{t \to 0} \frac{1}{t} \int_0^t \frac{\varphi(s)}{s} \, ds$$
$$\times \int_0^t \frac{(\dot{\varphi}(s) - (c\varphi(s) + d))^2}{\varphi(s)} \, ds = 0,$$

which implies that

(3.10)
$$\lim_{t \to 0} \frac{\varphi(t)}{t} = d, \quad \lim_{t \to 0} S_t(\varphi(t)) = 0.$$

By direct calculation we have

$$\log Q_0^\varepsilon \{ \sup_{1/N \le t \le T} |y_t - \varphi(t)| < \delta \}$$

$$= \log \int_{|y - \varphi(1/N)| < \delta} Q_y^\varepsilon \{ \sup_{1/N \le t \le T} |y_t - \varphi(t)| < \delta \} \mu_{1/N}^{\varepsilon,0}(dy)$$

$$\ge \inf_{|y - \varphi(1/N)| \le \delta/2} \log Q_y^\varepsilon \{ \sup_{1/N \le t \le T} |y_t - \varphi(t)| < \delta \}$$

$$+ \log \mu_{1/N}^{\varepsilon,0} \{ |y - \varphi(1/N)| \le \delta/2 \}$$

$$= \log Q_{y^\varepsilon}^\varepsilon \{ \sup_{1/N \le t \le T} |y_t - \varphi(t)| < \delta \} + \log \mu_{1/N}^{\varepsilon,0} \{ |y - \varphi(1/N)| \le \delta/2 \}$$

$$\text{for some } |y^\varepsilon - \varphi(1/N)| \le \delta/2,$$

where (3.6) is used in obtaining the last equality.

By using Theorem 4.14 of [7], (3.8), and Lemma 3.1, we get that

(3.11)
$$\liminf_{\delta \to 0} \liminf_{\varepsilon \to 0} \varepsilon^2 \log Q_0^\varepsilon \{ \sup_{1/N \le t \le T} |y_t - \varphi(t)| < \delta \}$$
$$\ge -I_0(\varphi) - S_{1/N}(\varphi(1/N)).$$

On the other hand, for any fixed $\delta_0 > 0$ and any $L > 0$, choose $N(\delta_0, L)$ large enough such that for all $N \ge N(\delta_0, L)$, $\sup_{0 \le t \le 1/N} \varphi(t) \le \delta_0$ and $(|c|L + d)/N \le \delta_0/2$. Then we have

(3.12)
$$Q_0^\varepsilon \{ \sup_{0 \le t \le 1/N} |y_t - \varphi(t)| \ge \delta_0 \} \le Q_0^\varepsilon \{ \sup_{0 \le t \le 1/N} y_t \ge \delta_0 \}$$

$$= Q_0^\varepsilon \{ \delta_0 \le \sup_{0 \le t \le 1/N} y_t \le L \} + Q_0^\varepsilon \{ \sup_{0 \le t \le 1/N} y_t > L \}$$

$$\le e^{-\frac{N}{\varepsilon^2} \frac{\delta_0^2}{8L}} + Q_0^\varepsilon \{ \sup_{0 \le t \le 1} y_t > L \},$$

which, by letting $\varepsilon$ go to zero, $N$ go to infinity, and $L$ go to infinity, implies that

(3.13)
$$\limsup_{N \to \infty} \limsup_{\varepsilon \to 0} \{ \sup_{0 \le t \le 1/N} |y_t - \varphi(t)| \ge \delta_0 \} = -\infty.$$

This combined with (3.10) and (3.11) implies that

$$\liminf_{\varepsilon \to 0} \varepsilon^2 \log Q_0 \{ \sup_{0 \le t \le T} |y_t - \varphi(t)| < \delta_0 \}$$

$$\ge \liminf_{N \to \infty} \liminf_{\delta \to 0} \liminf_{\varepsilon \to 0} \varepsilon^2 \log Q_0^\varepsilon \{ \sup_{1/N \le t \le T} |y_t - \varphi(t)| < \delta \}$$

$$\ge -I_0(\varphi) - \lim_{N \to \infty} S_{1/N}(\varphi(1/N)) = -I_0(\varphi).$$

Finally let $\delta_0$ go to zero we end up with (3.2). $\square$

REMARK. Theorem 3.2 combined with the exponential tightness obtained in section 2 implies that $Q_0^\varepsilon$ satisfies a large deviation principle with rate function $I_0(\cdot)$ (c.f. [**6**]).

Finally we consider the case of $d = 0$ and $y_0^\varepsilon = y > 0$.

THEOREM 3.3. *For any $\varphi(t) \in C_y([0, T], R_+)$ that hits $0$ at some time in $(0, T]$, we have*

$$(3.14) \quad \limsup_{\delta \to 0} \limsup_{\varepsilon \to 0} \varepsilon^2 \log Q_y^\varepsilon \{ \sup_{0 \le t \le t_0} |y_t - \varphi(t)| \le \delta \} \le -I_y^{0, t_0}(\varphi),$$

*where $t_0$ is the first time that $\varphi$ hits zero. If $I_0^{0, t_0}(\varphi)$ is finite, then the left derivative of $\varphi$ at $t_0$ is zero.*

PROOF. For any $0 \le t < t_0$, we have from the proof of **Case II**

$$\limsup_{\delta \to 0} \limsup_{\varepsilon \to 0} \varepsilon^2 \log Q_y^\varepsilon \{ \sup_{0 \le s \le t_0} |y_s - \varphi(s)| \le \delta \}$$

$$\le \limsup_{\delta \to 0} \limsup_{\varepsilon \to 0} \varepsilon^2 \log Q_y^\varepsilon \{ \sup_{0 \le s \le t} |y_s - \varphi(s)| \le \delta \} \le -I_y^{0, t}(\varphi).$$

Letting $t$ approach $t_0$, we end up with (3.14). Now assume that $I_y^{0, t_0}(\varphi)$ is finite. Then we have

$$\int_t^{t_0} \frac{\dot{\varphi}^2(s)}{\varphi(s)} ds < \infty.$$

Choose $t$ close to $t_0$ such that both $\sup_{s \in [t, t_0]} \varphi(s)$ and $\int_t^{t_0} \frac{\dot{\varphi}^2(s)}{\varphi(s)} ds$ are less than 1. Then

$$(3.15) \quad \varphi(t) = |\int_t^{t_0} \dot{\varphi}(s) ds| \le \sqrt{\int_t^{t_0} \frac{(\dot{\varphi}(s))^2}{\varphi(s)} ds} \sqrt{\int_t^{t_0} \varphi(s) ds}$$

$$\le \min \{ \sqrt{\int_t^{t_0} \varphi(s) ds}, \sqrt{t_0 - t} \}.$$

By iteration, we get

$$(3.16) \quad \varphi(t) \le (t_0 - t)^{\sum_{i=1}^\infty 1/2^i} = t_0 - t.$$

By using the variational expression of $I_{\varphi(t)}^{t, t_0}(\varphi)$ we get

$$\limsup_{t \to t_0} \left( \frac{\varphi(t) + \int_t^{t_0} b(\varphi(s)) ds}{t_0 - t} \right)^2 \le \limsup_{t \to t_0} \frac{1}{t_0 - t} \int_t^{t_0} \frac{\varphi(s)}{t_0 - s} ds$$

$$\times \int_t^{t_0} \frac{(\dot{\varphi}(s) - c\varphi(s))^2}{\varphi(s)} ds = 0,$$

which implies that $\lim_{t \to t_0} \frac{\varphi(t)}{t - t_0} = 0$. $\qquad \square$

REMARK. In this case a large deviation principle can also be established by an approximation argument.

## 4. Conclusion

From the results in section 3, we get the following picture of the boundary behaviour of the diffusion process $y_t^\varepsilon$.

If $d = 0$ and $y = 0$, the process will stay at 0; if $d = 0, y > 0$, under large deviations, the process can only approach zero in a way such that the left derivative at the boundary is 0; if $d > 0, y > 0$, then the probability that the process approaches the boundary is superexponentially small; if $d > 0, y = 0$, then the process will leave zero immediately following a path with the right hand derivative at time 0 equal to $d$.

**Acknowledgements.** The comments and suggestions of the referee are gratefully acknowledged.

## References

[1] Dawson, D.A. and Feng, S. (1998). Large deviations for the Fleming-Viot process with neutral mutation and selection. Stoch. Proc. Appl. 77, 207-232.

[2] Dembo, A. and Zeitouni, O. (1993). *Large Deviations and Applications.* Jones and Bartlett Publishers, Boston.

[3] Feller, W. (1951). Diffusion processes in genetics. *Proc. 2nd Berkeley Symp. math. Statist. Prob.*, 227–246, Berkeley, University of California Press.

[4] Freidlin, M.I. and Wentzell, A.D. (1984). *Random Perturbations of Dynamical Systems.* Springer-Verlag, New York.

[5] Ikeda, N. and Watanabe, S. (1981). *Stochastic Differential Equations and Diffusion Processes.* North-Holland, Amsterdam.

[6] Liptser, R.S. and Pukhalskii, A.A. (1992). Limit theorems on large deviations for semimartingales. Stochastics and Stoch. Rep. 38, 201-249.

[7] Stroock, D.W. (1984). *An Introduction to the Theory of Large Deviations.* Springer-Verlag, Berlin.

[8] Stroock, D.W. and Varadhan, S.R.S. (1979). *Multidimensional Diffusion Processes.* Springer-Verlag, Berlin.

DEPARTMENT OF MATHEMATICS AND STATISTICS, MCMASTER UNIVERSITY, 1280 MAIN STREET WEST, HAMILTON, ON L8S 4K1

*E-mail address*: shuifeng@mcmail.cis.mcmaster.ca

Canadian Mathematical Society
Conference Proceedings
Volume 26, 2000

# Finite Time Extinction of Catalytic Branching Processes

Klaus Fleischmann and Carl Mueller

*Dedicated to Don Dawson*

ABSTRACT. "Classical" critical branching processes die in finite time (as a rule). But if a catalytic medium governs the branching, the process might live forever. We survey finite time extinction for reactants moving in several classes of catalytic media, such as deterministic power laws, i.i.d. uniform, and stable catalysts. We report on a method of good and bad historical paths. Good paths have major collisions with the catalysts, and extinction follows by an individual time change depending on the collision local time and a comparison with Feller's branching diffusion. Bad paths do not interact much with the catalysts. However, they have a small expected mass, which is controlled by the hitting probability of the catalyst and the local time spent on it.

## 1. Introduction

**1.1. Classical branching processes.** A critical binary Galton-Watson process dies almost surely in finite time. Extinction occurs even though the mean is preserved in time (criticality). In fact, the state zero is reached in finite time, and this state is absorbing.

Actually, as a rule this property is shared by all, say, "classical" critical branching processes (see e.g. [**AN72**]) such as critical multi-type processes, or their "diffusion limits", as critical Feller's branching diffusion, which corresponds to the critical binary Galton-Watson case. Recall that Feller's branching diffusion $X$ is a process in $\mathsf{R}_+$ of the form

$$(1.1) \qquad \mathrm{d}X_t = \sqrt{X_t}\,\mathrm{d}W_t, \qquad t > 0, \quad X_{0+} \geq 0,$$

with $W$ a standard one-dimensional Brownian motion (Figure 1).

2000 *Mathematics Subject Classification.* Primary 60J80; Secondary 60J55, 60G57.

*Key words and phrases.* catalytic super-Brownian motion; historical superprocess; critical branching; finite time extinction; finite time survival; measure-valued branching; random medium; good and bad paths; stopped measures; collision local time; comparison; coupling; stopped historical superprocess; branching rate functional; super-random walk; interacting Feller's branching diffusion.

This is a detailed version of an invited paper held by the first author at the International Conference on Stochastic Models in Ottawa/Canada, June 1998.

Supported by an NSA grant.

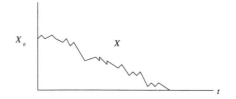

FIGURE 1. Feller's branching diffusion.

In all cases of classical critical (non-degenerate) branching processes $X$ we have for the survival probability

$$P(X_t \neq 0) \downarrow 0 \quad \text{as} \quad t \uparrow \infty,$$

implying *finite time extinction:*

(1.2) $$P\left(X_t = 0 \ \forall t \geq T, \text{ for some } T\right) = 1.$$

**1.2. Constant time-space medium.** If, in addition, the individuals in the population undergo an independent spatial migration, the situation does *not* change as long as the individuals evolve in the same way at all time-space points, and as long as we start with a finite initial mass. Indeed, in this case the total mass process $t \mapsto \|X_t\|$ is nothing other than a classical branching process.

However, the situation might change if the medium is not constant.

**1.3. A single point catalyst.** Consider, for instance, the case in which branching is allowed only in the presence of a single *point catalyst* (see [**DFL95**] for a quick introduction). Suppose that mass is spread out on the real line R by the heat flow, but additionally, at the origin, we have a kind of critical binary branching in the sense of the diffusion limit [recall (1.1)]:

(1.3) $$dX_t = \frac{1}{2} \Delta X_t \, dt + \sqrt{\delta_0 X_t} \, dW_t \quad \text{on } \mathsf{R}_{++} \times \mathsf{R}$$

[$\delta_0$ is Dirac's Delta function at $0 \in \mathsf{R}$ describing the point catalyst, and $\mathsf{R}_{++} := (0, \infty)$].

Written in this way, the model looks suspicious, since $\delta_0 X_t$ would be the density at zero of the population measure $X_t$ at time $t$. But such a density actually vanishes at fixed times in this model [**DF94**, Theorem 1.2.3]. However, in the corresponding quadratic variation process we should have something like

$$\int_0^T dt \ \delta_0 X_t$$

which can be interpreted as the occupation density at 0, or super-local time at 0, or collision local time between the measure-valued reactant path $X$ and the point catalyst $\delta_0$, all during the time period $[0, T]$. And this quantity actually makes sense and is non-trivial.

Thus, this single point-catalytic super-Brownian motion $X$ exists non-trivially, and the reactant population $X_t$ collides with the catalyst $\delta_0$ at random times $t$. This set of times has measure 0, and the collision intensity is a singular measure on the $t$ axis [**FL95**, Theorem 6]. Of course, it is essential that we are in dimension

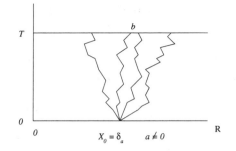

FIGURE 2. Browninian paths with absorption at 0.

one, where the Brownian local time at 0 of an intrinsic reactant particle exists non-trivially, and can therefore govern the branching.

Coming back to finite time extinction, it does *not* hold for this single point catalytic super-Brownian motion $X$ :

(1.4)
$$P\left\{X_T \neq 0,\ \forall T \mid X_0 \neq 0\right\} = 1$$

[**FL95**, Corollary5].

**1.4. Heuristic argument for the single point-catalytic model.** This can be seen by the following intuitive argument. Assume we start with a point mass $X_0 = \delta_a$ away from 0 (Figure 2). Then on the particle level, a huge number of particles with a small mass undergo independent Brownian motions from $a \neq 0$. Some of them do not hit the catalyst by time $T$, so they do not branch, hence cannot die. This group of particles cannot be neglected.

Or, speaking in terms of the diffusion limit $X$, part of the initial mass $\delta_a$ is smeared out by the heat flow with absorption at 0, and this deterministic part of $X_T$ survives at time $T$. More precisely, at time $T$ the total reactant mass $\|X_T\|$ is bounded below by this deterministic part:

(1.5)
$$\|X_T\| \geq \int_{\mathsf{R}} \mathrm{d}b\ p_T^0(a, b) > 0, \qquad T > 0, \quad a \neq 0,$$

where $p_T^0(a, b)$ is the Brownian transition density with killing at 0 (that is, the heat kernel with absorption at 0).

If we now start at the catalyst's position $a = 0$ instead, then not all particles are killed immediately. That is, some of them may escape from 0, and we are back in the previous case.

Altogether, the single point catalytic model does *not* have the finite time extinction property.

**1.5. Catalysts with gaps.** The reason finite time extinction fails in the single point catalyst model is essentially because this catalyst has a *positive gap* in space.

More precisely, assume in a $d$–dimensional catalytic super-Brownian motion $X$ we have a catalyst free open set $D \subseteq \mathsf{R}^d$ of positive area and with smooth boundary. Then the previous type of argument is in force:

(1.6)
$$\|X_T\| \geq \int_{\mathsf{R}^d} \mathrm{d}b\ p_T^{\partial D}(a, b) > 0, \qquad T > 0,$$

provided that $X_0 = \delta_a$ with $a \in D$. Here $p_T^{\partial D}(a, b)$ is the heat kernel with absorption at the boundary $\partial D$ of $D$.

Summarizing, here is our first observation: *Finite time extinction might be violated* for critical catalytic branching models. This is in particular the case if the catalyst has a "positive gap", as for instance in the single point-catalytic super-Brownian motion of Subsection 1.3.

Therefore, from now on we restrict our attention to catalysts which are at least everywhere dense in space.

## 2. Finite time extinction for three types of catalysts

Our purpose in this section is to introduce three different classes of catalysts for which finite time extinction can be verified.

**2.1. I.i.d. uniform catalyst on $Z^d$.** In our first case, the catalyst is given by a field $\varrho = \{\varrho(b) : b \in Z^d\}$ of independent random variables $\varrho(b)$ which are uniformly distributed in the interval $(0, 1)$. Note that in this case, the catalyst is present everywhere in the cubic lattice space $Z^d$ of dimension $d \geq 1$, but its strength fluctuates randomly. The "underlying" migration law is given by a continuous-time simple random walk in $Z^d$. In other words, "particles" jump at random to one of their nearest neighboring sites with rate one, i.e. after i.i.d. exponential waiting times of mean 1.

More precisely, given the catalyst $\varrho$, we consider the following linear system $X = X^\varrho = \{X_t(b) : t \geq 0, \ b \in Z^d\}$ of interacting diffusions in $R_+$ :

$$(2.1) \qquad dX_t(b) = \frac{1}{2} \Delta X_t(b) \, dt + \sqrt{\varrho(b) \, X_t(b)} \, dW_t(b),$$

$t > 0$, $b \in Z^d$, where $\Delta$ is the discrete Laplacian in $Z^d$, and $\{W(b) : b \in Z^d\}$ is a family of independent one-dimensional standard Brownian motions. In other words, we consider *simply interacting Feller's branching diffusions* $X^\varrho$ on $Z^d$ (also called *simple super-random walk*) in the catalytic medium $\varrho$.

We stress the fact, that in this paper in the case of a random catalyst we always use the *quenched* random medium approach: First the catalyst (such as our $\varrho$) is sampled, with law $\mathbb{P}$, say, and then, given the catalyst, the reactant process (such as $X^\varrho$) evolves, its (conditional) law denoted by $P$.

Note that given the catalyst $\varrho$, for each bounded initial state $X_0$ of the reactant, a unique strong solution $X^\varrho$ of (2.1) exists, that is (2.1) is a well-posed problem [**SS80**].

Once more, on a level of an approximating particle system, this model $X^\varrho$ is interpreted as follows: Many small particles migrate independently in $Z^d$ according to continuous-time simple random walks, but additionally, each particle at site $b \in Z^d$ dies or doubles with equal probability, and with an approximatedly high rate proportional to the amount $\varrho(b)$ of catalytic mass at $b$.

Although there are very large areas in $Z^d$ with a very small amount of catalytic mass, in which the reactant mainly moves and does only very little branching, this model $X^\varrho$ has the finite time extinction property. For simplicity, we start with a unit mass at the origin: $X_0 = \delta_0$.

THEOREM 2.1 (i.i.d. uniform catalysts). *For almost all catalysts $\varrho$,*

$$P\left(X_T^\varrho = 0 \ for \ some \ T\right) = 1.$$

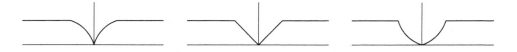

FIGURE 3. Variants of the power law catalyst ($\chi = 1/2$, 1, 2).

**2.2. Deterministic power law catalyst.** Our next example is a super-Brownian motion $X = X^\chi$ in R with a deterministic catalyst $\chi$ of the form

$$(2.2) \qquad\qquad \chi(b) = \chi_q(b) := |b|^q \wedge 1, \qquad b \in \mathsf{R},$$

where $q > 0$ is a given constant (Figure 3). Here $\chi(b)$ is the catalyst's density at $b \in \mathsf{R}$, which is positive except at the origin $b = 0$. The most interesting case is where the constant $q$ is very large, leading to a significant "depression" of branching rate close to the origin.

The model $X = X^\chi$ can loosely be described by the following stochastic pde:

$$(2.3) \qquad\qquad \mathrm{d}X_t(b) = \frac{1}{2}\Delta X_t(b)\,\mathrm{d}t + \sqrt{\chi_q(b)\,X_t(b)}\,\mathrm{d}W_t(b),$$

$(t, b) \in \mathsf{R}_{++} \times \mathsf{R}$. Here $\Delta$ is the one-dimensional Laplacian, $\mathrm{d}W_t(b)$ a standard white noise on $\mathsf{R}_{++} \times \mathsf{R}$, and $X_t(b)$ is the reactant's density at site $b \in \mathsf{R}$ at time $t > 0$. A rigorous construction of $X = X^\chi$ can be given in terms of a continuous measure-valued Markov process expressing its log-Laplace transition functionals

$$(2.4) \qquad\qquad -\log E\left\{\mathrm{e}^{-\langle X_t, \varphi\rangle} \,\middle|\, X_0\right\} = \langle X_0, v(t, \cdot)\rangle$$

in terms of solutions $v = v_\varphi$ of the log-Laplace equation. This is the following reaction-diffusion equation

$$(2.5) \qquad \frac{\partial}{\partial t}v(t, a) = \frac{1}{2}\Delta v(t, a) - \frac{1}{2}\chi_q(a)\,v^2(t, a), \qquad (t, a) \in \mathsf{R}_{++} \times \mathsf{R},$$

where $v\big|_{t=0+} = \varphi \geq 0$, with $\varphi$ running in a suitable set of test functions. (See, for instance, [**DF97**], Theorem 1(b), p.235, in connection with Corollary 2, p.257.)

Since one-dimensional Brownian motion is recurrent, the reactant "particles" may spent a lot of time in the "depression" of the catalyst, where they have only a very small chance of branching, hence dying. Nevertheless, finite time extinction holds:

THEOREM 2.2 (deterministic power law catalyst). *Assume* $\|X_0\| < \infty$. *Then, for each constant* $q \geq 0$,

$$P\left(X_T^\chi = 0 \text{ for some } T\right) = 1.$$

We mention that in the case of a catalytic density function with a "super-exponential depressions" at zero, *finite time survival* might become possible instead.

**2.3. Stable catalyst.** Our most interesting catalyst is the following random measures $\Gamma = \Gamma^\gamma$ on R with independent increments, which is defined by its log-Laplace transition functional

$$(2.6) \qquad\qquad -\log \mathbb{E}\,\mathrm{e}^{-\langle\Gamma, \varphi\rangle} = \int_\mathsf{R} \mathrm{d}b\,\varphi^\gamma(b), \qquad \varphi \geq 0,$$

where $\gamma \in (0, 1)$ is a fixed constant. Setting $\varphi = \mathbf{1}_B$ with $B$ a Borel subset of R, we see that the catalytic mass $\Gamma(B)$ in $B$ has a *stable* distribution with index $\gamma$. Note also that $\Gamma$ has a representation

$$(2.7) \qquad \Gamma = \sum_i \alpha_i \, \delta_{b_i}$$

with $\alpha_i > 0$ and where the $b_i$ are densely situated in R. Hence, the catalytic medium is now given by a dense set of random point catalysts with randomly varying action weights $\alpha_i$. Note that $\Gamma = \Gamma^\gamma$ can also be seen as the generalized derivative of a one-sided stable process (subordinator) on and in R with index $\gamma$.

An intuitive description of the required super-Brownian motion $X = X^\Gamma$ in the catalytic medium $\Gamma = \Gamma^\gamma$ can again be given by a stochastic equation:

$$(2.8) \qquad \mathrm{d}X_t = \frac{1}{2} \Delta X_t \, \mathrm{d}t + \sqrt{\sum_i \alpha_i \, \delta_{b_i} \, X_t} \, \mathrm{d}W_t, \qquad t > 0,$$

with $\Delta$ the one-dimensional Laplacian and $\mathrm{d}W$ a time-space white noise. So we imagine the reactant process $X = X^\Gamma$ as the evolution of some large population of small reactant masses which move chaotically in R. In addition, they branch with a high rate, in a critical binary fashion, and in fact only in case of collision with the point catalysts $\alpha_i \delta_{b_i}$. The rate of branching is indeed (locally) proportional to the action weights $\alpha_i$ of the related catalytic atoms $\delta_{b_i}$.

The first question is, of course, whether such a reactant model $X^\Gamma$ makes sense non-trivially. The answer is yes, for P–almost all $\Gamma$–realizations. This was shown in [**DF91**, **DF92**], where also the long-term clustering behavior was described in terms of a mass-time-space scaling limit theorem. A precise definition is again given using the log-Laplace transition functional as in (2.4), where the related log-Laplace equation can rigorously be written as an integral equation (mild approach):

$$(2.9) \qquad v(t, a) = \int_{\mathsf{R}} \mathrm{d}b \, p_t(b - a) \, \varphi(b) - \frac{1}{2} \int_0^t \mathrm{d}s \sum_i \alpha_i \, p_{t-s}(b_i - a) \, v^2(s, b_i),$$

$(t, a) \in \mathsf{R}_{++} \times \mathsf{R}$, where $p$ is the heat kernel:

$$p_t(b) = (2\pi t)^{-1/2} \, \mathrm{e}^{-b^2/2t}, \qquad (t, b) \in \mathsf{R}_{++} \times \mathsf{R}.$$

Via Kolmogorov's method of moments it can be shown that $X^\Gamma$ (given $\Gamma$) even has continuous sample paths [**DF97**, Theorem 1(b), p.235, and Corollary 2, p.257]. Moreover, at least at each fixed time $t > 0$, the reactant's samples $X_t^\Gamma$ are P–a.s. absolutely continuous measures (even for independently moving catalysts) [**DFR91**, Theorem 1.19].

**2.4. Boundary cases and the compact support property.** In order to get a feeling whether finite time extinction could hold for $X^\Gamma$, we will begin by discussing the boundary cases as $\gamma \uparrow 1$ and $\gamma \downarrow 0$.

First, we note that the stable random measures $\Gamma(\mathrm{d}b) = \Gamma^\gamma(\mathrm{d}b)$ approach in law the (deterministic) Lebesgue measure $\Gamma^1(\mathrm{d}b) = \mathrm{d}b$ as $\gamma \uparrow 1$. But $\mathrm{d}b$ is constant in space and time (or rather, its density function is constantly 1). So the related super-Brownian motion is the ordinary one, finite time extinction holds trivially (recall Subsection 1.2).

On the other hand, as $\gamma \downarrow 0$, the catalytic atoms are "rarefied" in a sense. Recall, that the stable measure $\Gamma = \Gamma^\gamma$ is a random measure with independent

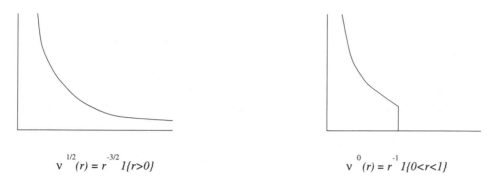

$$\nu^{1/2}(r) = r^{-3/2} 1\{r>0\} \qquad\qquad \nu^{0}(r) = r^{-1} 1\{0<r<1\}$$

FIGURE 4. Two variants of a Lévy measure density.

increments and Levy measure $\nu^\gamma(r)\,dr$ with density function

$$(2.10) \qquad\qquad \nu^\gamma(r) := \frac{c^\gamma}{r^{1+\gamma}}\, \mathbf{1}_{\{r>0\}}$$

with the constant $c^\gamma$ suitably chosen (Figure 4).

Consider now the following $\gamma = 0$ *boundary case* of a Levy measure $\nu^0(r)\,dr$ :

$$(2.11) \qquad\qquad \nu^0(r) := \frac{c^0}{r}\, \mathbf{1}_{\{0<r\le 1\}}\,.$$

The random measure with this Levy measure is no longer stable, but it still has a dense set of atoms. As [**DLM95**, Corollaries 1 and 2] showed, an essential change of behavior occurs for the related reactant process under such transition $\gamma \downarrow 0$ of the catalyst: The compact support property is lost.

Indeed, let $\mathrm{supp}X_t$ denote the closed support of the measure $X_t$, then the *global support* $\mathrm{gsupp}X$ of the measure-valued path $X$ is defined by the closure of the union

$$(2.12) \qquad\qquad \bigcup_{t\ge 0} \mathrm{supp}X_t\,.$$

By definition, $X$ has the *compact support property* if $\mathrm{gsupp}X$ is compact provided that the initial measure $X_0$ has compact support $\mathrm{supp}X_0$.

For the reactant process $X^\Gamma$ with catalyst $\Gamma = \Gamma^\gamma$ we now have the following situation: The compact support property holds for $\Gamma^1$ (ordinary super-Brownian motion, [**Isc88**, Corollary p.202]), and $\Gamma^\gamma$ with $0 < \gamma < 1$, but it fails for $\Gamma^0$ corresponding to (2.11) (as it fails for catalytic random measures with independent increments in the case of finite Levy measure $\nu$). For all these results on the compact support property, see [**DLM95**, Corollaries 1 and 2].

**2.5. Finite time extinction for a stable catalyst.** Coming back to our finite time extinction issue for $X^\Gamma$, we know it trivially holds in the ordinary case $\gamma = 1$. From the picture just presented, we may hope for its validity at least for $\gamma$ close to one, where the catalytic atoms are "very dense", or even for all $\gamma \in (0,1)$, where the compact support property holds. Actually, the latter is true:

THEOREM 2.3 (stable catalysts). *Let $0 < \gamma < 1$ and suppose that the closed support $\mathrm{supp}X_0$ of the finite initial measure $X_0$ is compact. Then for $\mathbb{P}$–almost* *all catalyst samples* $\Gamma = \Gamma^\gamma$,

$$P\left(X_T^\Gamma = 0 \text{ for some } T\right) = 1.$$

In suitable $\gamma = 0$ boundary cases however, *finite time survival* becomes possible instead, even for an everywhere dense catalyst.

Aimed to a proof of the theorem, at the first sight, one is perhaps seduced into calculating the extinction probability $P\left(X_t^\Gamma = 0\right)$ via the limit as $\theta \uparrow \infty$ of the log-Laplace transform

$$(2.13) \qquad\qquad -\log E \exp\left[-\theta \|X_t^\Gamma\|\right] = v_\theta\left(t, a \mid \Gamma\right),$$

where we assumed for simplicity that initially $X_0 = \delta_a$. Here $v_\theta = v_\theta\left(\cdot, \cdot \mid \Gamma\right)$ solves equation (2.9) with $\varphi(b) \equiv \theta$. Differentiating (2.9) formally, it can symbolically be written as

$$(2.14) \qquad\qquad \frac{\partial}{\partial t}v = \frac{1}{2}\Delta v - \frac{1}{2}\sum_i \alpha_i \delta_{b_i} v^2, \qquad v\big|_{t=0+} = \varphi.$$

However, we do not known how to attack this problem analytically, and we will follow a probabilistic route instead, which we feel is of some independent interest.

On the other hand, equations with such kind of singularities (or similar singularities) in the coefficient of reaction terms attracted some attention from an *applied* point of view, since they are relevant in many branches of technology as in microelectronics. See, for instance, [**OR72, PT88, GGH96**]. Via the log-Laplace connection as in (2.4) to the partial differential equation (2.14), our probabilistic approach can be viewed as a contribution to the study of certain asymptotics of solutions to reaction-diffusion equations in heterogeneous singular media.

(For relations between finite time extinction and the compactness of the range of super-Brownian motions in non-catalytic media but general branching mechanism, see [**She97**].)

## 3. The method of good and bad historical reactant paths

The Theorems 2.1–2.3 can be proved with a unified method which is called the *method of good and bad historical reactant paths*. It first occurred in [**FM97**, Section 5], and was inspired by modulus of continuity studies in [**DP91**] and the paper [**MP92**]. Our aim in this section is to explain this rather flexible probabilistic argument on a heuristic level, in the case of the stable catalyst. A rigorous development is given in the detailed version [**DFM98**] of the present survey article.

**3.1. Individual time change by collision local times.** First we mention that the assumption on the compact closed support of the initial measure $X_0$ is imposed in order to exploit the compact support property of $X^\Gamma$ (recall Subsection 2.4). At least in a heuristic sense, this way we are working on a *compact* phase space. (A rigorous reduction can be provided in terms of a coupling argument.) So from now on, let us replace the phase space R by the unit circle T.

For the moment, we fix attention on a single point catalyst $\alpha\delta_b$ and a single reactant path $W$. By recurrence, $W$ hits this point catalyst repeatedly and accumulates some rate of branching according to the local time

$$L_{[W,\alpha\delta_b]}\left(\mathrm{d}s\right) := \alpha\delta_b(W_s)\,\mathrm{d}s.$$

Or for the total catalyst $\Gamma$ (which is assumed to be fixed), a reactant particle's path $W$ accumulates the rate of branching according to the *collision local time*

$$(3.1) \qquad\qquad L_{[W,\Gamma]} (ds) := \sum_i \alpha_i \delta_{b_i}(W_s)\, ds.$$

Thinking in terms of the measure-valued path $t \mapsto \delta_{W_t}$ instead of $t \mapsto W_t$, this $L_{[W,\Gamma]}$ is a marginal measure with respect to time of a collision local time measure in the sense of [**BEP91**] which we recommend for a systematic study. Note that this collision local time $L_{[W,\Gamma]}$ formally occurs if one writes the integral version (2.9) of equation (2.14) probabilistically by representing the heat flow via an expectation using the Brownian path $W$. This collision local time $L_{[W,\Gamma]}$ is actually a tool for the construction of the reactant process $X^\Gamma$ (given $\Gamma$) by using $L_{[W,\Gamma]}$ as Dynkin's additive functional governing the branching; for details of the existence of $L_{[W,\Gamma]}$ and the construction of $X^\Gamma$, see [**DF97**], Corollary 2, p.257, and Theorem 1(b), p.235, respectively.

Next observe that $L_{[W,\Gamma]}[0,T] \uparrow \infty$ as $T \uparrow \infty$, for almost all $W$. If this collision local time $L_{[W,\Gamma]}$ grew linearly in time, we would have a constant rate of branching which is good for finite time extinction (recall Subsection 1.2).

This consideration suggests *changing the time* to get a linear growth on a new time scale. But note that we had been thinking in terms of a single reactant path $W$, so the time change has to be provided *individually* for each reactant path.

In fact, we write $\tau_s$ for the time the collision local time $L_{[W,\Gamma]}$ reaches the value $s$ (recall that $\Gamma$ is fixed):

$$(3.2) \qquad\qquad \tau_s := \inf\left\{ t:\ L_{[W,\Gamma]}\left([0,t]\right) = s \right\}, \qquad s \geq 0.$$

For each $s$, we get a Brownian stopping time $\tau_s$. Now consider all of the reactants paths at time $\tau_s$, and denote the arising measure by $X_{\tau_s}$. This is an example of a so-called *stopped (or exit) measure,* systematically introduced by [**Dyn91a**, Theorem 1.5].

Of course, this requires first of all that we work with a *historical* setting of the catalytic superprocess $X^\Gamma$. That is, in the description of a reactant particle, its whole historical path has to be included. We write $\widetilde{X}^\Gamma$ for this enriched process. Besides Dynkin, historical processes in terms of superprocesses were systematically studied in [**DP91**] and [**Per95**] and were actually based on earlier point process variants (for references see, for instance, p.201 of [**Daw93**]).

As a function in $s$, we get the so-called *stopped historical reactant process:*

$$(3.3) \qquad\qquad s \mapsto \widetilde{X}^\Gamma_{\tau_s}.$$

By construction, on this new time scale, the accumulated rate of branching grows almost linearly.

Consequently, one idea in our approach to finite time extinction is to use such *stopped* historical catalytic superprocesses governed by the collision local time $L_{[W,\Gamma]}$ of reactant paths $W$ and the catalyst $\Gamma$. In this way we want to come close to the constant medium case where finite time extinction is known to hold (Subsection 1.2).

**3.2. Simplification of the catalyst, time partitioning.** In a sense, we have now reduced the problem to a study of the collision local time $L_{[W,\Gamma]}$. Since this is still a rather "crowded" object, we need some further simplifications.

First of all, we can exploit the fact that the survival probability is monotone in the local branching rate: A smaller rate increases the probability of survival. Hence, for our aim of showing finite time extinction, we may reduce the catalytic mass $\Gamma(db)$.

Our first step is to remove all large atoms of $\Gamma$, say all atoms $\alpha_i \delta_{b_i}$ with action weights $\alpha_i \geq 1$. In fact, our idea is that the densely located small atoms cause the finite time extinction. So there should be no harm in removing the isolated large catalytic atoms.

Secondly, it would be nice to have only a *countable* number of different action weights. So let us reduce the weights $\alpha_i$ in the interval $[2^{-n}, 2^{-n+1})$ to the smallest value $2^{-n}$. Then all atoms with weights $2^{-n}$ are isolated, and moreover, their positions form a homogeneous *Poisson* point process $\pi_n$, with intensity of order $2^{\gamma n}$ as $n \uparrow \infty$. Recall that $\gamma \in (0,1)$ is the stable index of the catalyst.

Next, since we are working on the compact phase space $\mathsf{T}$, very large gaps in $\pi_n$ are unlikely, by large deviation probabilities. So we may assume without loss of generality that the gaps in $\pi_n$ are bounded by some $\Delta_n$. Write $\overline{\pi}_n$ for this "Poisson point process with bounded gaps".

But still, we have the whole bunch of weighted point processes $2^{-n} \overline{\pi}_n$, $n \geq 1$. The idea is now to look in successive time periods only at a *single* of these $2^{-n} \overline{\pi}_n$, that is to drop all the others.

More precisely, the aim is to construct Brownian stopping times $0 < T_1 < T_2 < \cdots \leq T_\infty < \infty$, where $T_\infty$ is a deterministic time. In the stage $[T_n, T_{n+1})$ we exploit only the catalyst $2^{-n} \overline{\pi}_n < \Gamma$. We nevertheless hope to show extinction by combining all these stages.

### 3.3. Good historical paths.
The central idea is now to distinguish between good and bad reactant paths $W$ during the stage $[T_n, T_{n+1})$.

*Good* paths $W$ are those which spent a significant amount of time at the $n^{\text{th}}$ weighted point process catalyst $2^{-n} \overline{\pi}_n$. Here, significant means that

$$(3.4) \qquad L_{[W, 2^{-n} \overline{\pi}_n]}\left([T_n, T_{n+1})\right) \geq \xi_n,$$

where $\xi_n$ is some reasonable level for this collision local time. By the individual time change as explained in Subsection 3.1, we will compare the conditional probability

$$(3.5) \qquad P\left\{ \widetilde{X}^\Gamma_{T_{n+1}}(W \text{ is good}) > 0 \,\middle|\, \left\| \widetilde{X}^\Gamma_{T_n} \right\| \leq 2^{-n} \right\}$$

that there are still good paths at the end of this period with the probability that Feller's branching diffusion, say $Y$, starting with mass $2^{-n}$ survives by time $\xi_n$:

$$(3.6) \qquad P\left\{ Y_{\xi_n} > 0 \,\middle|\, Y_0 = 2^{-n} \right\}.$$

(Recall that $\xi_n$ is the significant collision local time which the good path spends at $2^{-n} \overline{\pi}_n$, which is considered as the new time scale.) But the latter probability is of order $2^{-n}/\xi_n$ as $n \uparrow \infty$, and can be made small by the choice of $\xi_n$.

### 3.4. Bad historical reactant paths.
The remaining historical reactant paths $W$ [violating (3.4)] are called *bad*. For these paths, the aim is to control the conditional probability

$$(3.7) \qquad P\left\{ \widetilde{X}^\Gamma_{T_{n+1}}(W \text{ is bad}) \geq 2^{-n-1} \,\middle|\, \left\| \widetilde{X}^\Gamma_{T_n} \right\| \leq 2^{-n} \right\},$$

that the bad paths do not have a small mass. Now by Markov's inequality, we may estimate from above by

$$(3.8) \qquad \leq 2^{n+1} E\left\{\tilde{X}^{\Gamma}_{T_{n+1}}(W \text{ is bad}) \mid \|\tilde{X}^{\Gamma}_{T_n}\| \leq 2^{-n}\right\}.$$

Next we use the so-called *special Markov property* for stopped measures, which says that $\tilde{X}^{\Gamma}$ starts anew at the Brownian stopping time $T_n$, and the simple *expectation formula* for critical stopped historical superprocesses; see [**Dyn91a**], Theorem 1.6 and formula (1.50a), as well as [**Dyn91b**], Theorem 1.5. Thus, the latter line reduces to

$$(3.9) \qquad 2^{n+1} 2^{-n} \Pi\left(W \text{ is bad}\right) = 2\Pi\left(L_{[W, 2^{-n}\bar{\pi}_n]}\left([T_n, T_{n+1})\right) < \xi_n\right),$$

where $\Pi$ is the law of the Brownian path $W$ (which starting point at time $T_n$ we leave open at this stage of a rough argument). To estimate the latter probability, we look at a bad path $W$ in more detail and proceed as follows, see Figure 5.

W starts at time $T_n$ from $W_{T_n}$ (think of time running to the top of the picture). First we measure the *hitting time*, say $h_1$, of the catalyst $2^{-n}\bar{\pi}_n$ (the neighboring catalytic points are signed by bullets), and assume that $2^{-n}\delta_{a_1}$ is hit. Then, for a given deterministic time increment $s_n$, we count the *local time* $\ell_1$ which $W$ spends at that point catalyst $2^{-n}\delta_{a_1}$ within this period $[T_n + h_1, T_n + h_1 + s_n)$. After this, that is starting from time $T_n + h_1 + s_n$, we again stop at the hitting time of $2^{-n}\bar{\pi}_n$, denoted by $h_2$. Then we measure the local time $\ell_2$ on the point catalyst from $2^{-n}\bar{\pi}_n$ during a period of length $s_n$, etc.

Altogether, we consider $m_n$ (to be chosen) such pairs $(h_i, \ell_i)$, and we hope that they are realized within the current stage $[T_n, T_{n+1})$, that is,

$$(3.10) \qquad T_n + m_n s_n + \sum_{i=1}^{m_n} h_i \leq T_{n+1}.$$

Recall that a neighboring point catalyst in $2^{-n}\bar{\pi}_n$ is at most distance $\Delta_n$ away. Therefore the $h_i$ are essentially i.i.d. random variables, and by a large deviation argument, (3.10) is violated only with an exponentially small probability, provided all quantities are chosen in an appropriate way.

Now from (3.10) we conclude that

$$(3.11) \qquad \sum_{i=1}^{m_n} \ell_i \leq L_{[W, 2^{-n}\bar{\pi}_n]}\left([T_n, T_{n+1})\right) < \xi_n,$$

the latter since $W$ is bad by assumption. Therefore the probability in (3.9) can be estimated from above by

$$\leq \Pi\left(\sum_{i=1}^{m_n} \ell_i < \xi_n\right).$$

Since the local times $\ell_i$ are i.i.d. with all exponential moments finite, the latter probability can again be controlled by a standard large deviation argument.

Thus, the probability of the bad events is asymptotically small. One can finally conclude that with high probability, the process dies out by time $T_\infty$.

We want to give an impression how all the quantities are selected. First recall that $\gamma \in (0,1)$ is the stable index of the catalyst. Next, for the assumed largest gap between the atoms in $2^{-n}\bar{\pi}_n$ we set $\Delta_n := 2^{-\beta n}$ where $0 < \beta < \gamma$. In the

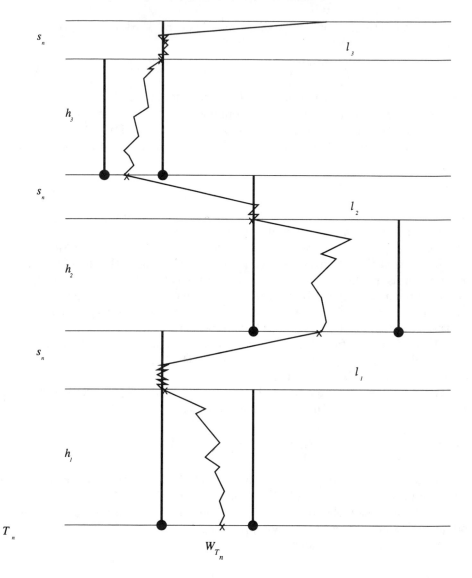

FIGURE 5. Our view to a bad path.

present model with a stable catalyst, the stage times $T_n$ can actually be chosen deterministically. On the other hand we let several quantities additionally depend on a parameter $\varepsilon > 0$ which we finally send to 0. So for the particular time increments $s_n^\varepsilon$ and the number $m_n^\varepsilon$ of substages within the stage $[T_n^\varepsilon, T_{n+1}^\varepsilon)$ we set

$$s_n^\varepsilon := \frac{e^{-\beta n}}{\varepsilon^2} \quad \text{and} \quad m_n^\varepsilon := \left[\frac{e^{\alpha n}}{\varepsilon}\right] \quad \text{with} \ \ \alpha \in \left(\frac{\beta}{2}, \beta\right).$$

Finally,

$$T_{n+1}^\varepsilon - T_n^\varepsilon := 2\, m_n^\varepsilon s_n^\varepsilon \quad \text{for all sufficiently large } n.$$

## 4. Final remarks

We admit that this heuristic argument is a bit vague. Our aim was to describe the *ideas* involved in proving finite time extinction for catalytic branching processes. Some efforts are needed to make all these ideas precise, in particular to find the relevant scales so that everything fits together. Slight variations are needed for different catalysts, but the basic ideas are the same.

We mention that in the case of a stable catalyst, the good reactant paths are those which spent enough local time at a neighboring catalytic point. For the power law catalyst the good paths spend most of their time away from the catalyst's depression. Finally, in the lattice case, the bad paths are those which spend a large amount of time at sites $b \in Z^d$ where the catalytic mass $\varrho(b)$ is small.

As already mentioned, details for the models, claims, and arguments of this survey paper are given in [**DFM98**]. There actually the method of good and bad paths is handled first on an *abstract level* in terms of a superprocess with a general branching functional $K$. Then the used abstract conditions are verified for the concrete catalytic models where $K$ is the collision local time of an intrinsic reactant path $W$ with the catalyst [as for instance $L_{[W,\Gamma]}$ from (3.1)].

In the lattice model (2.1) of Subsection 2.1, some additional work is needed to show that at the beginning of the $n^{\text{th}}$ stage the total reactant mass is already small with high probability.

Note in particular, that opposed to the stable catalyst case, for this discrete space model, the compact support property is not available (small reactant masses might be present also outside of bounded regions, even if we start at the origin, say).

## 5. Open problems

Many questions remain open in this area of research. We mention only a few.

- Open is the problem of formulating conditions on the catalyst, which are necessary *and* sufficient for finite time extinction. Our sufficient conditions seem to be too strong for this purpose (recall that we used the Markov inequality, for instance).

- Note also that we restricted our attention to critical *binary* branching mechanisms and to the *simplest* migration laws (Brownian motion respectively simple random walk). More general models can, for instance, be found in [**DFL98**].

- Finally, we excluded the case of *time-varying catalysts*. For models of this type, see [**DF91**], and [**FK99**] and references therein.

**Acknowledgment.** We are very grateful to Don Dawson for all the fruitful cooperation over the years, in particular for joining us in the current project (see our preprint [**DFM98**]). We also thank an anonymous referee for a careful reading of the manuscript and his suggestions for an improvement of the representation.

## References

[AN72]    K.B. Athreya and P.E. Ney, *Branching processes*, Springer-Verlag, Berlin, 1972.

[BEP91]   M.T. Barlow, S.N. Evans, and E.A. Perkins, *Collision local times and measure-valued processes*, Canad. J. Math. **43** (1991), no. 5, 897–938.

[Daw93]   D.A. Dawson, *Measure-valued Markov processes*, École d'été de probabilités de Saint Flour XXI–1991 (P.L. Hennequin, ed.), Lecture Notes in Mathematics, vol. 1541, Springer-Verlag, Berlin, 1993, pp. 1–260.

[DF91]    D.A. Dawson and K. Fleischmann, *Critical branching in a highly fluctuating random medium*, Probab. Theory Relat. Fields **90** (1991), 241–274.

[DF92]    D.A. Dawson and K. Fleischmann, *Diffusion and reaction caused by point catalysts*, SIAM J. Appl. Math. **52** (1992), 163–180.

[DF94]    D.A. Dawson and K. Fleischmann, *A super-Brownian motion with a single point catalyst*, Stoch. Proc. Appl. **49** (1994), 3–40.

[DF97]    D.A. Dawson and K. Fleischmann, *A continuous super-Brownian motion in a super-Brownian medium*, Journ. Theoret. Probab. **10** (1997), no. 1, 213–276.

[DFL95]   D.A. Dawson, K. Fleischmann, and J.-F. Le Gall, *Super-Brownian motions in catalytic media*, Branching processes: Proceedings of the First World Congress (C.C. Heyde, ed.), Lecture Notes in Statistics, vol. 99, Springer-Verlag, 1995, pp. 122–134.

[DFL98]   D.A. Dawson, K. Fleischmann, and G. Leduc, *Continuous dependence of a class of superprocesses on branching parameters and applications*, Ann. Probab. **26** (1998), no. 2, 262–601.

[DFM98]   D.A. Dawson, K. Fleischmann, and C. Mueller, *Finite time extinction of super-Brownian motions with catalysts*, WIAS Berlin, Preprint No. 431 (1998).

[DFR91]   D.A. Dawson, K. Fleischmann, and S. Roelly, *Absolute continuity for the measure states in a branching model with catalysts*, Stochastic Processes, Proc. Semin. Vancouver/CA 1990, Prog. Probab., vol. 24, 1991, pp. 117–160.

[DLM95]   D.A. Dawson, Y. Li, and C. Mueller, *The support of measure-valued branching processes in a random environment*, Ann. Probab. **23** (1995), no. 4, 1692–1718.

[DP91]    D.A. Dawson and E.A. Perkins, *Historical processes*, Mem. Amer. Math. Soc. **93** (1991), no. 454, iv+179.

[Dyn91a]  E.B. Dynkin, *Branching particle systems and superprocesses*, Ann. Probab. **19** (1991), 1157–1194.

[Dyn91b]  E.B. Dynkin, *Path processes and historical superprocesses*, Probab. Theory Relat. Fields **90** (1991), 1–36.

[FK99]    K. Fleischmann and A. Klenke, *Smooth density field of catalytic super-Brownian motion*, Ann. Appl. Probab. **9** (1999), no. 2, 298–318.

[FL95]    K. Fleischmann and J.-F. Le Gall, *A new approach to the single point catalytic super-Brownian motion*, Probab. Theory Related Fields **102** (1995), no. 1, 63–82.

[FM97]    K. Fleischmann and C. Mueller, *A super-Brownian motion with a locally infinite catalytic mass*, Probab. Theory Relat. Fields **107** (1997), 325–357.

[GGH96]   A. Glitzky, K. Gröger, and R. Hünlich, *Free energy and dissipation rate for reaction diffusion processes of electrically charged species*, Applicable Analysis **60** (1996), 201–217.

[Isc88]   I. Iscoe, *On the supports of measure-valued critical branching Brownian motion*, Ann. Probab. **16** (1988), 200–221.

[MP92]    C. Mueller and E.A. Perkins, *The compact support property for solutions to the heat equation with noise*, Probab. Theory Related Fields **93** (1992), no. 3, 325–358.

[OR72]    P. Ortoleva and J. Ross, *Local structures in chemical reactions with heterogeneous catalysis*, J. chemical physics **56** (1972), no. 9, 4397–4400.

[Per95]   E.A. Perkins, *On the martingale problem for interactive measure-valued branching diffusions*, Mem. Amer. Math. Soc. **549** (1995).

[PT88]    L. Pagliaro and D.L. Taylor, *Aldolase exists in both the fluid and solid phases of cytoplasm*, J. Cell Biology **107** (1988), 981–999.

[She97]   Y.-C. Sheu, *Lifetime and compactness of range for super-Brownian motion with a general branching mechanism*, Stochastic Process. Appl. **70** (1997), no. 1, 129–141.

[SS80]    T. Shiga and A. Shimizu, *Infinite-dimensional stochastic differential equations and their applications*, J. Mat. Kyoto Univ. **20** (1980), 395–416.

WEIERSTRASS INSTITUTE FOR APPLIED ANALYSIS AND STOCHASTICS (WIAS), MOHRENSTR. 39, D–10117 BERLIN, GERMANY
   *E-mail address*: fleischmann@wias-berlin.de
   *URL*: http://www.wias-berlin.de/~fleischm

DEPARTMENT OF MATHEMATICS, UNIVERSITY OF ROCHESTER, ROCHESTER, NY 14627 USA
   *E-mail address*: cmlr@troi.cc.rochester.edu

Canadian Mathematical Society
Conference Proceedings
Volume **26**, 2000

# Moment Asymptotics and Lifshitz Tails for the Parabolic Anderson Model

Jürgen Gärtner and Stanislav A. Molchanov

*Dedicated to Professor Donald A. Dawson*

ABSTRACT. We study the Cauchy problem for the heat equation $(\partial/\partial t)u = \mathcal{H}u$ on $\mathbb{R}_+ \times \mathbb{Z}^d$ associated with the Anderson Hamiltonian $\mathcal{H} = \kappa\Delta + \xi(\cdot)$. Here $\xi(\cdot)$ is a correlated shift-invariant random potential. Under mild regularity assumptions on $\xi(\cdot)$, we derive the second-order asymptotics of the statistical moments $\langle u(t,0)^p \rangle$, $p = 1, 2, \ldots$, as $t \to \infty$. As an application, we obtain improved asymptotic formulas for the Lifshitz tails of the spectral distribution function associated with $\mathcal{H}$. For i.i.d. potentials the moment asymptotics was treated in [8].

## 1. Moment asymptotics for correlated potentials

**1.1. Introduction and main result.** We consider the random Cauchy problem

$$(1.1) \qquad \begin{aligned} \frac{\partial u(t,x)}{\partial t} &= \kappa\Delta u(t,x) + \xi(x)u(t,x), \qquad (t,x) \in \mathbb{R}_+ \times \mathbb{Z}^d, \\ u(0,x) &= 1, \qquad x \in \mathbb{Z}^d, \end{aligned}$$

associated with the Anderson tight binding Hamiltonian

$$(1.2) \qquad \mathcal{H} := \kappa\Delta + \xi(\cdot).$$

Here $\kappa$ denotes a positive diffusion constant and $\Delta$ is the lattice Laplacian:

$$\Delta f(x) := \sum_{y:\, |y-x|=1} [f(y) - f(x)], \qquad x \in \mathbb{Z}^d.$$

The potential $\xi(\cdot)$ is a random field on a certain probability space. The underlying probability and expectation will be denoted by $\mathrm{Prob}(\cdot)$ and $\langle\cdot\rangle$, respectively.

In this paper we assume that $\xi(\cdot)$ is shift-invariant and that the cumulant generating function $H$ of its marginals is finite on the positive half-axis:

$$(1.3) \qquad H(t) := \log\left\langle e^{t\xi(0)} \right\rangle < \infty \qquad \text{for } t \geq 0.$$

2000 *Mathematics Subject Classification*. Primary 60H25, 82C44; Secondary 60F10, 60K40.

Our objective is to derive the second-order asymptotics of the moments $\langle u(t,0)^p \rangle$ as $t \to \infty$ for $p = 1, 2, \ldots$

For general correlated potentials it is unknown whether or not the nonnegative solution of the Cauchy problem (1.1) is unique a.s. (cf. e.g. the remark at the end of Section 2.2 in [7]). In the following we consider the smallest nonnegative solution which is given by the Feynman-Kac formula

$$(1.4) \qquad u(t,x) = \mathbb{E}_x \exp \left\{ \int_0^t \xi(x(s)) \, ds \right\},$$

where $(x(t), \mathbb{P}_x)$ denotes simple symmetric random walk on $\mathbb{Z}^d$ with generator $\kappa \Delta$ and $\mathbb{E}_x$ stands for expectation with respect to $\mathbb{P}_x$. For a detailed investigation of existence and uniqueness see [7], Section 2.

Let $\mathcal{P}_f(\mathbb{Z}^d)$ denote the space of probability measures on $\mathbb{Z}^d$ with finite support. Let further $\delta_z$ be the Dirac measure at $z \in \mathbb{Z}^d$. We consider shift-invariant functionals $G_t \colon \mathcal{P}_f(\mathbb{Z}^d) \to \mathbb{R}$, $t > 0$, defined by

$$(1.5) \qquad G_t(\mu) := \log \left\langle \exp \left\{ t \sum_{z \in \mathbb{Z}^d} \mu(z) \xi(z) \right\} \right\rangle.$$

As a consequence of Hölder's inequality, $G_t$ is convex. In particular,

$$G_t(\mu) \leq G_t(\delta_0) = H(t) < \infty.$$

We impose the following general regularity assumption on the tail behavior of $\xi(\cdot)$.

**Assumption (G).** For each $\mu \in \mathcal{P}_f(\mathbb{Z}^d)$, the (possibly infinite) limit

$$I(\mu) := \lim_{t \to \infty} \frac{G_t(\delta_0) - G_t(\mu)}{t}$$

exists.

To understand what this assumption roughly amounts to the reader should examine the examples in Section 2.

As a consequence of the mentioned properties of $G_t$, the functional $I$ is shift-invariant and concave. Moreover, $0 \leq I \leq \infty$ and $I(\delta_0) = 0$.

In addition to $I$, we introduce the Donsker-Varadhan functional $S_d$ defined by

$$S_d(\mu) := \sum_{\substack{\{x,y\} \subset \mathbb{Z}^d \\ |x-y|=1}} \left( \sqrt{\mu(x)} - \sqrt{\mu(y)} \right)^2, \qquad \mu \in \mathcal{P}_f(\mathbb{Z}^d).$$

This functional serves as rate function for large deviations of occupation time measures associated with random walks on $\mathbb{Z}^d$, see later. It will turn out that the quantity

$$(1.6) \qquad \chi_G(\kappa) := \inf_{\mathcal{P}_f(\mathbb{Z}^d)} \left[ \kappa S_d + I \right], \qquad \kappa > 0,$$

determines the second-order asymptotics of the moments $\langle u(t,0)^p \rangle$. Note that $\chi_G$ is concave and

$$0 \leq \chi_G(\kappa) \leq 2d\kappa, \qquad \kappa > 0.$$

The left equality is valid if $I \equiv 0$, and the right equality holds if $I(\mu) = \infty$ for all $\mu$ not a Dirac measure.

We are now ready to formulate our result about the moments of the solution $u$ to the Cauchy problem (1.1).

THEOREM 1.1. *Let Assumption (G) be satisfied. Then*

$$(1.7) \qquad \langle u(t,0)^p \rangle = \exp\{H(pt) - \chi_G(\kappa)pt + o(t)\}$$

*as $t \to \infty$ for $p = 1, 2, \ldots$*

REMARK 1.2. a) In (1.1), the initial condition $u(0,x) \equiv 1$ may be replaced by the weaker assumption that the initial datum $u_0(x)$ is a nonnegative shift-invariant random field which is independent of $\xi(\cdot)$ and satisfies

$$0 < \langle u_0(0)^p \rangle < \infty \qquad \text{for } p = 1, 2, \ldots$$

For details we refer to [8], Section 1.1.

b) It may be seen from our proof that Theorem 1.1 also works for the fundamental solution $q(t, x, y)$ of the Anderson Hamiltonian (1.2), i.e. the solution of (1.1) with initial condition $u(0, \cdot) = \delta_y$. In particular, under Assumption (G),

$$(1.8) \qquad \langle q(t,0,0) \rangle = \exp\{H(t) - \chi_G(\kappa)t + o(t)\} \qquad \text{as } t \to \infty.$$

This result will be used in Section 3 to study the behavior of the Lifshitz tails.

The particular case of i.i.d. potentials was treated in [8], see also Section 2.1 below. The moment asymptotics for the spatially continuous analogue of problem (1.1) with Gaussian and Poisson shot-noise potentials can be found in [6]. As $t \to \infty$, the solution $u(t, \cdot)$ is believed to develop a spatially intermittent structure consisting of 'islands' of high peaks which are located far from each other. These peaks are caused by high exceedances of the potential $\xi(\cdot)$. If the potential is ergodic, then the moments can be obtained by spatial averaging of the solution. Our proof below indicates that the peaks contributing to this averaging are rarely distributed in space but their density does not decay exponentially fast. Therefore the essential contribution to the second order asymptotics (1.7) comes from a singe typical peak. Roughly speaking, the first term, $e^{H(pt)}$, reflects the height of such a peak and the second, $e^{-\chi_G(\kappa)pt}$, is related to its deterministic shape. Namely, if the variational problem (1.6) admits a 'minimizer' $\mu_*$ (which must not be finitely supported but is unique modulo shifts), then the shape of a typical peak is believed to be a time-dependent multiple of $\sqrt{\mu_*}$. In the i.i.d. case, this picture is supported by the investigation of the correlation structure of the solution in [5]. See also [8] for a more detailed explanation of such intermittency effects. For a general discussion of intermittency and related topics we refer to Carmona and Molchanov [1] and the lectures by Molchanov [9] and also to the monograph by Sznitman [11].

**1.2. Preliminary remarks.** The rest of this section is devoted to the *proof of Theorem 1.1*. As a first step we will express the solution $u(t,x)$ of our Cauchy problem by means of local times of random walks on $\mathbb{Z}^d$. To this end, we exploit the Feynman-Kac representation (1.4). Given $p \in \mathbb{N}$, consider $p$ independent copies $x_1(t), \ldots, x_p(t)$ of the random walk $x(t)$, and denote by $\mathbb{P}_0^p$ and $\mathbb{E}_0^p$ probability and expectation given $x_1(0) = \cdots = x_p(0) = 0$, respectively. Let

$$l_{t,i}(z) := \int_0^t \mathbb{1}(x_i(s) = z) \, ds$$

be the local time of the $i$-th random walk spent at $z \in \mathbb{Z}^d$ during the time interval $[0, t]$, and introduce the total local time

$$l_t(z) := \sum_{i=1}^{p} l_{t,i}(z).$$

It then follows from (1.4) that

$$(1.9) \qquad u(t, 0)^p = \mathbb{E}_0^p \exp \left\{ \sum_{z \in \mathbb{Z}^d} l_t(z) \xi(z) \right\}.$$

We note that the *occupation time measures*

$$(1.10) \qquad L_t(\cdot) := \frac{l_t(\cdot)}{pt}$$

satisfy the *weak* large deviation principle as $t \to \infty$ with rate function being the Donsker-Varadhan functional $S_d$, cf. Donsker and Varadhan [3].

**1.3. Lower bound.** Let us start with a few remarks to prepare the proof of the lower bound in (1.7).

First, given $\mu \in \mathcal{P}_f(\mathbb{Z}^d)$ with $I(\mu) < \infty$, observe that $I(\nu) < \infty$ for all $\nu$ with $\operatorname{supp} \nu \subseteq \operatorname{supp} \mu$. Indeed, for such $\nu$ one finds $\gamma \in (0, 1)$ and $\eta \in \mathcal{P}_f(\mathbb{Z}^d)$ such that

$$\mu = \gamma \nu + (1 - \gamma) \eta.$$

Since $I$ is concave and nonnegative, this implies that

$$\infty > I(\mu) \geq \gamma I(\nu) + (1 - \gamma) I(\eta) \geq \gamma I(\nu).$$

Hence, $I(\nu)$ is finite. Now, if $I(\mu) < \infty$, then the restriction of the functional $I$ to $\{\nu \in \mathcal{P}_f(\mathbb{Z}^d): \operatorname{supp} \nu = \operatorname{supp} \mu\}$ is finite and continuous. This is obvious from the concavity of $I$ and the observation that the above set is convex and open in its affine hull. In particular, there exists an open neighborhood $U(\mu)$ of $\mu$ in $\{\nu \in \mathcal{P}_f(\mathbb{Z}^d): \operatorname{supp} \nu \subseteq \operatorname{supp} \mu\} = \mathcal{P}(\operatorname{supp} \mu)$ such that $I$ is bounded and continuous on $U(\mu)$.

Next, we remark that

$$(1.11) \qquad \inf_{\mathcal{P}_f(\mathbb{Z}^d)} [\kappa S_d + I] = \inf_{\mathcal{P}_{f,c}(\mathbb{Z}^d)} [\kappa S_d + I],$$

where $\mathcal{P}_{f,c}(\mathbb{Z}^d)$ is the subset of $\mathcal{P}_f(\mathbb{Z}^d)$ consisting of measures with connected support. To see this, fix $\mu \in \mathcal{P}_f(\mathbb{Z}^d) \setminus \mathcal{P}_{f,c}(\mathbb{Z}^d)$ arbitrarily and let $D_1, \ldots, D_m$ denote the connected components of $\operatorname{supp} \mu$. Then we find $\mu_1, \ldots, \mu_m \in \mathcal{P}_{f,c}(\mathbb{Z}^d)$ with $\operatorname{supp} \mu_i = D_i$ for $i = 1, \ldots, m$ and $\gamma_1, \ldots, \gamma_m \in (0, 1)$ with $\gamma_1 + \cdots + \gamma_m = 1$ such that

$$\mu = \sum_{i=1}^{m} \gamma_i \mu_i.$$

Since $I$ is concave, we have

$$I(\mu) \geq \sum_{i=1}^{m} \gamma_i I(\mu_i).$$

Moreover, since the supports of the measures $\mu_i$ are separated from each other by a distance larger than one, the functional $S_d$ splits into parts:

$$S_d(\mu) = \sum_{i=1}^{m} \gamma_i S_d(\mu_i).$$

Hence,

$$\kappa S_d(\mu) + I(\mu) \geq \sum_{i=1}^{m} \gamma_i \left[\kappa S_d(\mu_i) + I(\mu_i)\right]$$
$$\geq \inf_{\mathcal{P}_{f,c}(\mathbb{Z}^d)} \left[\kappa S_d + I\right].$$

This clearly proves (1.11).

Because of (1.11), the derivation of the *lower bound* reduces to the following lemma.

LEMMA 1.3. *Let Assumption (G) be satisfied. For every $\nu \in \mathcal{P}_f(\mathbb{Z}^d)$ such that $I(\nu) < \infty$ and supp $\nu$ is connected we have*

$$\langle u(t,0)^p \rangle \geq \exp\left\{H(pt) - pt\left[\kappa S_d(\nu) + I(\nu)\right] - o(t)\right\}$$

*as $t \to \infty$ for $p = 1, 2, \ldots$*

PROOF. Because of shift-invariance, we may assume that $0 \in$ supp $\nu$. Then, using (1.9), (1.10), Fubini's theorem, and the definition of the functionals $G_t$, we obtain

$$\langle u(t,0)^p \rangle = \left\langle \mathbb{E}_0^p \exp\left\{pt \sum_{z \in \mathbb{Z}^d} L_t(z)\xi(z)\right\}\right\rangle$$
$$= \mathbb{E}_0^p \exp\left\{G_{pt}(L_t(\cdot))\right\}$$
$$\geq e^{H(pt)} \mathbb{E}_0^p \exp\left\{-pt\frac{G_{pt}(\delta_0) - G_{pt}(L_t(\cdot))}{pt}\right\} \mathbb{1}(\text{supp } L_t \subseteq \text{supp } \nu).$$

We know from our preliminary remarks that there exists an open neighborhood $U(\nu)$ of $\nu$ in $\{\tilde{\nu} \in \mathcal{P}_f(\mathbb{Z}^d): \text{supp } \tilde{\nu} \subseteq \text{supp } \nu\}$ such that $I$ is bounded and continuous on $U(\nu)$. Moreover, $U(\nu)$ may be chosen so that the convergence

$$\frac{G_t(\delta_0) - G_t(\tilde{\nu})}{t} \to I(\tilde{\nu}) \qquad \text{as } t \to \infty$$

in Assumption (G) is uniform in $\tilde{\nu} \in U(\nu)$. This is again a consequence of concavity. Hence, we obtain

$$\langle u(t,0)^p \rangle \geq e^{H(pt)-o(t)} \mathbb{E}_0^p \exp\left\{-ptI(L_t(\cdot))\right\} \mathbb{1}\left(\text{supp } L_t \subseteq \text{supp } \nu, L_t(\cdot) \in U(\nu)\right).$$

Since supp $\nu$ is connected and finite, we may apply the weak large deviation principle for the occupation time measures $L_t$ on supp $\nu$ with killing on the complement of supp $\nu$ to see that the expectation on the right is bounded from below by

$$\exp\left\{-pt\left[\kappa S_d(\nu) + I(\nu)\right] - o(t)\right\},$$

and we are done. $\square$

**1.4. Upper bound.** The proof of the *upper bound* is more subtle, mainly because the method of 'periodization' used in [8] breaks down for correlated potentials. Our way out of this dilemma consists in deriving an upper bound for the moments by use of *Dirichlet* boundary conditions.

Given $R \in \mathbb{N}$, let $\mathbb{T}_R^d := \{-R, \dots, R\}^d$ denote the centered lattice cube of length $2R + 1$. Let further $u^{R,0}$ denote the solution to the initial boundary value problem for the equation

$$\frac{\partial u}{\partial t} = \kappa \Delta u + \xi(x)u \qquad \text{on } \mathbb{R}_+ \times \mathbb{T}_R^d$$

with initial datum identically one and Dirichlet boundary condition on the *exterior* boundary $\mathbb{T}_{R+1}^d \setminus \mathbb{T}_R^d$. Let further

$$\partial \mathbb{T}_R^d := \mathbb{T}_R^d \setminus \mathbb{T}_{R-1}^d$$

be the *interior* boundary of $\mathbb{T}_R^d$. The next lemma provides the key for the derivation of the upper bound.

LEMMA 1.4. *For each $R \in \mathbb{N}$ and $p = 1, 2, \dots$ we have*

$$\langle u(t,0)^p \rangle \le e^{d\kappa \varepsilon_R pt + o(t)} \sum_{x \in \mathbb{T}_R^d} \langle u^{R,0}(t,x)^p \rangle$$

*as $t \to \infty$, where*

$$\varepsilon_R := |\partial \mathbb{T}_R^d| / |\mathbb{T}_R^d| \to 0$$

*as $R \to \infty$.*

PROOF. $1^0$ Fix $R$ and $p$ arbitrarily. We consider the periodic frame

$$\Phi := \bigcup_{x \in (2R+1)\mathbb{Z}^d} \left( x + \partial \mathbb{T}_R^d \right).$$

Each lattice site is covered by exactly $|\partial \mathbb{T}_R^d|$ of the shifted frames

$$\Phi_y := y + \Phi, \qquad y \in \mathbb{T}_R^d.$$

From this we conclude that for each $t > 0$ there exists a (random) site $y \in \mathbb{T}_R^d$ such that

$$\sum_{z \in \Phi_y} L_t(z) \le |\partial \mathbb{T}_R^d| / |\mathbb{T}_R^d| = \varepsilon_R.$$

Because of this, using the probabilistic representation (1.9), the occupation time measures(1.10), and the shift-invariance of the potential, we get

$$\langle u(t,0)^p \rangle = \left\langle \mathbb{E}_0^p \exp \left\{ pt \sum_z L_t(z)\xi(z) \right\} \right\rangle$$

$$\le |\mathbb{T}_R^d| \left\langle \mathbb{E}_0^p \exp \left\{ pt \sum_z L_t(z)\xi(z) \right\} \mathbb{1} \left( \sum_{z \in \Phi} L_t(z) \le \varepsilon_R \right) \right\rangle$$

(1.12) $$\le |\mathbb{T}_R^d| e^{d\kappa \varepsilon_R pt} \left\langle \mathbb{E}_0^p \exp \left\{ pt \sum_z L_t(z)\xi_R(z) \right\} \right\rangle,$$

where the potential $\xi_R(\cdot)$ has been obtained by lowering $\xi(\cdot)$ on the frame $\Phi$ by an amount of $d\kappa$:

$$\xi_R(z) := \begin{cases} \xi(z) - d\kappa, & \text{if } z \in \Phi, \\ \xi(z), & \text{otherwise.} \end{cases}$$

$2^0$ We next introduce the centered cubes

$$V_r := \bigcup_{\substack{z \in (2R+1)\mathbb{Z}^d \\ |z^1|,\dots,|z^d| \le r}} \left(z + \mathbb{T}_R^d\right), \qquad r > 0,$$

where $z^1, \dots, z^d$ are the components of $z$, and denote by $\tau^p(r)$ the first time when one of the random walks $x_1(t), \dots, x_p(t)$ leaves $V_r$. We set

$$r(t) := t \log t$$

and show that the paths of our random walks may be restricted to the cube $V_{r(t)}$. More precisely, we want to prove that

$$(1.13) \qquad \left\langle \mathbb{E}_0^p \exp\left\{ pt \sum_z L_t(z)\xi_R(z) \right\} \right\rangle$$

$$\le \left\langle \mathbb{E}_0^p \exp\left\{ pt \sum_z L_t(z)\xi_R(z) \right\} \mathbb{1}\left(\tau^p(r(t)) > t\right) \right\rangle + e^{H(pt) - \gamma t}$$

for arbitrary $\gamma > 0$ and all sufficiently large $t$. Indeed, with

$$R_n(t) := nr(t), \qquad n \in \mathbb{N},$$

we obtain

$$\left\langle \mathbb{E}_0^p \exp\left\{ pt \sum_z L_t(z)\xi_R(z) \right\} \right\rangle$$

$$= \left\langle \mathbb{E}_0^p \exp\left\{ pt \sum_z L_t(z)\xi_R(z) \right\} \mathbb{1}\left(\tau^p(r(t)) > t\right) \right\rangle$$

$$+ \sum_{n=1}^{\infty} \left\langle \mathbb{E}_0^p \exp\left\{ pt \sum_z L_t(z)\xi_R(z) \right\} \mathbb{1}\left(\tau^p(R_n(t)) \le t < \tau^p(R_{n+1}(t))\right) \right\rangle.$$

The first term on the right coincides with the one in (1.13). To estimate the second term we proceed as follows. The expression under the sum may be estimated from above by

$$\left\langle \exp\left\{ pt \max_{z \in V_{R_{n+1}(t)}} \xi(z) \right\} \right\rangle \mathbb{P}_0^p\left(\tau^p(R_n(t)) \le t\right)$$

$$\le |V_{R_{n+1}(t)}| e^{H(pt)} \mathbb{P}_0^p\left(\tau^p(R_n(t)) \le t\right).$$

Combining this with the bound

$$\mathbb{P}_0^p(\tau^p(r) \le t) \le p\, 2^{d+1} \exp\left\{ -r \log \frac{r}{d\kappa t} + r \right\}, \qquad t > 0, r > 0,$$

(see e.g. [7], Lemma 4.3), one readily checks that the above sum does not exceed

$$e^{H(pt) - \gamma t}$$

for arbitrary $\gamma > 0$ and all sufficiently large $t$. This proves (1.13).

$3^0$ Next observe that the first term on the right of (1.13) coincides with the $p$-th moment of $u_{r(t)}^0(t, 0)$, where $u_{r(t)}^0$ is the solution of the initial boundary value problem for our parabolic equation on $V_{r(t)}$ with lowered potential $\xi_R(\cdot)$, Dirichlet boundary condition, and initial datum identically one. Therefore, combining (1.12) with (1.13), we find that

$$(1.14) \qquad \langle u(t,0)^p \rangle \leq e^{d\kappa\varepsilon_R pt + o(t)} \left[ \left\langle u^0_{r(t)}(t,0)^p \right\rangle + e^{H(pt) - \gamma t} \right]$$

for arbitrary $\gamma > 0$. To get rid of the last exponential term we may take $\gamma > (2 + \varepsilon_R) d\kappa p$ and use the trivial bound

$$e^{H(pt) - 2d\kappa pt} \leq \langle u(t,0)^p \rangle$$

which is obtained from the probabilistic representation of the $p$-th moment by forcing all random walks to stay at 0 until time $t$. Then, after moving the bound obtained for the last term in (1.14) to the left hand side, we arrive at

$$(1.15) \qquad \langle u(t,0)^p \rangle \leq e^{d\kappa\varepsilon_R pt + o(t)} \left\langle u^0_{r(t)}(t,0)^p \right\rangle.$$

$4^0$ Now a Fourier expansion of $u^0_{r(t)}(t,\cdot)$ with respect to the eigenvalues and eigenfunctions of the operator

$$\mathcal{H}^{R,t} := \kappa\Delta + \xi_R(\cdot) \qquad \text{in } l^2(V_{r(t)})$$

with Dirichlet boundary condition yields

$$u^0_{r(t)}(t,0) \leq |V_{r(t)}| \, e^{t\lambda^0_{r(t)}(\xi_R(\cdot))},$$

where $\lambda^0_{r(t)}(\xi_R(\cdot))$ denotes the principal (i.e. largest) eigenvalue. Hence,

$$(1.16) \qquad \left\langle u^0_{r(t)}(t,0)^p \right\rangle \leq e^{o(t)} \left\langle e^{pt\lambda^0_{r(t)}(\xi_R(\cdot))} \right\rangle.$$

For each $z \in (2R+1)\mathbb{Z}^d$, let $\lambda^{R,0}(\xi(z + \cdot))$ denote the principal eigenvalue of

$$\mathcal{H}^R_z := \kappa\Delta + \xi(\cdot) \qquad \text{in } l^2(z + \mathbb{T}^d_R)$$

with Dirichlet boundary condition. A straightforward computation shows that, because of the special form of the potential $\xi_R(\cdot)$,

$$\mathcal{H}^{R,t} \leq \bigoplus_{\substack{z \in (2R+1)\mathbb{Z}^d \\ |z^1|,\ldots,|z^d| \leq r(t)}} \mathcal{H}^R_z$$

in the sense of positive definiteness in $l^2(V_{r(t)})$. (I.e., the difference between the operators on the right and the left is nonnegative definite.) Consequently,

$$\lambda^0_{r(t)}(\xi_R(\cdot)) \leq \max_{\substack{z \in (2R+1)\mathbb{Z}^d \\ |z^1|,\ldots,|z^d| \leq r(t)}} \lambda^{R,0}(\xi(z + \cdot)).$$

Therefore

$$(1.17) \qquad \left\langle e^{pt\lambda^0_{r(t)}(\xi_R(\cdot))} \right\rangle \leq |V_{r(t)}| \left\langle e^{pt\lambda^{R,0}(\xi(\cdot))} \right\rangle.$$

Using the Fourier expansion of $u^{R,0}(t,\cdot)$, one finds that

$$(1.18) \qquad e^{t\lambda^{R,0}(\xi(\cdot))} \leq \sum_{x \in \mathbb{T}^d_R} u^{R,0}(t,x).$$

One also has to take into account that the eigenfunction corresponding to the eigenvalue $\lambda^{R,0}(\xi(\cdot))$ is positive. Combining (1.15)–(1.18), we finally arrive at the assertion of our lemma. $\qquad \square$

Because of the last lemma, it only remains to derive appropriate upper bounds for the moments of $u^{R,0}$. This will be done now.

LEMMA 1.5. *Let Assumption (G) be satisfied. Then, for each $R \in \mathbb{N}$, $p = 1, 2, \ldots$, and $x \in \mathbb{T}_R^d$, we have*

$$\langle u^{R,0}(t,x)^p \rangle \leq \exp \left\{ H(pt) - pt \min_{\substack{\mu \in \mathcal{P}_f(\mathbb{Z}^d) \\ \text{supp } \mu \subseteq \mathbb{T}_R^d}} [\kappa S_d(\mu) + I(\mu)] + o(t) \right\}$$

*as $t \to \infty$.*

PROOF. Let us assume for simplicity that $x = 0$. Again using the probabilistic representation of the $p$-th moment, we find that

$$\langle u^{R,0}(t,0)^p \rangle = e^{H(pt)} \, \mathbb{E}_0^p \exp \left\{ -pt \frac{G_{pt}(\delta_0) - G_{pt}(L_t(\cdot))}{pt} \right\} \mathbb{1} \left( \text{supp } L_t \subseteq \mathbb{T}_R^d \right).$$

Together with Assumption (G) this indicates that our assertion follows from the upper bound in the Laplace-Varadhan method applied for the weak large deviation principle of the occupation time measures $L_t$. The expression

$$\frac{G_t(\delta_0) - G_t(\mu)}{t}$$

is nonnegative and converges to $I(\mu)$ as $t \to \infty$. In general, this convergence is not uniform and the limiting functional $I$ is not continuous on $\mathcal{P}(\mathbb{T}_R^d)$. Nevertheless, an analysis of the standard proof of the upper bound in the Laplace-Varadhan method (see e.g. Deuschel and Stroock [2], Lemma 2.1.8) shows that it will work if the following assertion is valid. For each $\mu \in \mathcal{P}(\mathbb{T}_R^d)$ and $\delta \in (0, 1)$ there exists a neighborhood $U_\delta(\mu)$ of $\mu$ in $\mathcal{P}(\mathbb{T}_R^d)$ such that

$$(1.19) \qquad \liminf_{t \to \infty} \inf_{\tilde{\mu} \in U_\delta(\mu)} \frac{G_t(\delta_0) - G_t(\tilde{\mu})}{t} \geq (1 - \delta) I(\mu).$$

The idea is to restrict $L_t$ to a finite covering of a big compact level set $\{\kappa S_d \leq s\}$ by such neighborhoods $U_\delta(\mu)$. To prove (1.19), note that

$$U_\delta(\mu) := \left\{ (1 - \gamma)\mu + \gamma\eta \colon 0 \leq \gamma < \delta, \, \eta \in \mathcal{P}(\mathbb{T}_R^d) \right\}$$

is a neighborhood of $\mu$ in $\mathcal{P}(\mathbb{T}_R^d)$ for any $\delta \in (0, 1)$. Since the functionals $G_t$ are convex, we obtain

$$G_t(\delta_0) - G_t((1 - \gamma)\mu + \gamma\eta) \geq (1 - \gamma) \left( G_t(\delta_0) - G_t(\mu) \right) + \gamma \left( G_t(\delta_0) - G_t(\eta) \right)$$
$$\geq (1 - \gamma) \left( G_t(\delta_0) - G_t(\mu) \right).$$

Hence,

$$\inf_{\tilde{\mu} \in U_\delta(\mu)} \frac{G_t(\delta_0) - G_t(\tilde{\mu})}{t} \geq (1 - \delta) \frac{G_t(\delta_0) - G_t(\mu)}{t} \longrightarrow (1 - \delta) I(\mu)$$

as $t \to \infty$ for all $\delta \in (0, 1)$. This clearly implies (1.19). $\qquad \square$

The proof of Theorem 1.1 is now complete.

## 2. Examples

In this section we are going to illustrate Theorem 1.1 by several examples. In particular, we will compute the functional $I$ of Assumption (G) which is crucial for determining the important quantity $\chi_G(\kappa)$ via (1.6). Recall that the functional $G_t$ is defined by (1.5).

**2.1. Uncorrelated potentials.** Here we consider the particular case when the potential consists of i.i.d. random variables $\xi(x)$, $x \in \mathbb{Z}^d$. Besides of (1.3), we impose the following regularity condition on the cumulant generating function $H$.

**Assumption (H).** There exists $\varrho$, $0 \le \varrho \le \infty$, such that

$$(2.1) \qquad \lim_{t \to \infty} \frac{H(ct) - cH(t)}{t} = \varrho\, c \log c$$

for all $c \in (0, 1)$.

Then

$$G_t(\mu) = \sum_z H(t\mu(z)),$$

and Assumption (H) implies Assumption (G) with

$$I = \varrho I_d,$$

where $I_d$ is the $d$-dimensional entropy functional:

$$(2.2) \qquad I_d(\mu) = - \sum_{x \in \mathbb{Z}^d} \mu(x) \log \mu(x), \qquad \mu \in \mathcal{P}_f(\mathbb{Z}^d).$$

This shows that Theorem 1.2 in [8] is a particular case of Theorem 1.1. Note that Assumption (H) with $0 < \varrho < \infty$ is fulfilled if $\xi(0)$ is double exponentially distributed:

$$(2.3) \qquad \mathrm{Prob}(\xi(0) > r) = \exp\{-e^{r/\varrho}\}.$$

For the role of double exponential tails in this context, see [8] and [5]. Both papers give a detailed analysis of the quantity $\chi_G(\kappa)$ for this situation and discuss the implications for the peak structure of the solution $u$.

**2.2. Random clouds.** Let $\eta(\cdot)$ be an i.i.d. random field such that

$$H_\eta(t) := \log \left\langle e^{t\eta(0)} \right\rangle < \infty \qquad \text{for all } t \in \mathbb{R}.$$

Given a function $\varphi \colon \mathbb{Z}^d \to \mathbb{R}$ with finite support, consider the random potential

$$\xi(x) := \sum_{y \in \mathbb{Z}^d} \varphi(y - x)\eta(y), \qquad x \in \mathbb{Z}^d.$$

Then the cumulant generating function $H$ of $\xi(0)$ has the form

$$H(t) = \sum_y H_\eta(t\varphi(y)).$$

Let us first consider the case when $\varphi$ is nonnegative. Suppose that the cumulant generating function $H_\eta$ satisfies Assumption (H) of Section 2.1 for some $\varrho$ with $0 \le \varrho \le \infty$. Then the function $H$ also fulfills this assumption but with parameter $\varrho \sum_y \varphi(y)$. We obtain

$$G_t(\mu) = \sum_y H_\eta(t(\varphi * \mu)(y)), \qquad \mu \in \mathcal{P}_f(\mathbb{Z}^d),$$

where $\varphi * \mu$ denotes convolution of $\varphi$ with $\mu$. Hence, if $\varrho$ is finite, then the potential $\xi(\cdot)$ satisfies Assumption (G) with

$$I(\mu) = \varrho\left[I_d(\varphi * \mu) - I_d(\varphi)\right],$$

where $I_d$ is the entropy functional given by (2.2). Assumption (G) is also fulfilled for $\varrho = \infty$. In this case $I(\mu) = \infty$ if $\mu$ is not a Dirac measure. This may be seen from the formula

$$(2.4) \qquad \frac{G_t(\delta_0) - G_t(\mu)}{t} = \sum_y \left[ \frac{\sum_z \mu(z) H_\eta(t\varphi(y-z)) - H_\eta(t(\varphi * \mu)(y))}{t} \right].$$

By convexity, the expression in the square brackets is always nonnegative. Now suppose that $|\operatorname{supp} \mu| \geq 2$. Then one finds $y \in \mathbb{Z}^d$ such that $\operatorname{supp} \mu$ intersects $y - \operatorname{supp} \varphi$ but is not entirely contained in that set. It will be enough to show that, for this $y$, the term in the square brackets in (2.4) converges to $+\infty$ as $t \to \infty$. Note that

$$0 < \gamma := \sum_{y-z \in \operatorname{supp} \varphi} \mu(z) < 1.$$

Hence, again by convexity,

$$\sum_z \mu(z) H_\eta(t\varphi(y-z)) = \gamma \sum_{y-z \in \operatorname{supp} \varphi} \frac{\mu(z)}{\gamma} H_\eta(t\varphi(y-z))$$

$$\geq \gamma H_\eta \left( t \sum_{y-z \in \operatorname{supp} \varphi} \frac{\mu(z)}{\gamma} \varphi(y-z) \right)$$

$$= \gamma H_\eta \left( t \frac{1}{\gamma} (\varphi * \mu)(y) \right).$$

Thus, for our particular lattice site $y$, the term in the square brackets on the right of (2.4) may be estimated from below by

$$\gamma \frac{H_\eta \left( t\gamma^{-1}(\varphi * \mu)(y) \right) - \gamma^{-1} H_\eta \left( t(\varphi * \mu)(y) \right)}{t}.$$

But, since $\varrho = \infty$ and $\gamma^{-1} > 1$, this expression converges to infinity as $t \to \infty$.

If $\varphi$ attains both positive and negative values, then the situation will be more complex. Let us consider in detail the case when the random field $\eta(\cdot)$ is both unbounded from above and below and the cumulant generating functions of $\eta(0)$ and $-\eta(0)$ satisfy Assumption (H) with finite nonnegative parameters $\varrho_+$ and $\varrho_-$, respectively. Then we may use the decomposition

$$\frac{G_t(\delta_0) - G_t(\mu)}{t} = \sum_y \frac{H_\eta(t\varphi(y)) - H_\eta(t(\varphi * \mu)(y))}{t}$$

$$= \sum_y \frac{H_\eta(t\varphi(y)) - |\varphi(y)| H_\eta(t \operatorname{sign} \varphi(y))}{t}$$

$$- \sum_y \frac{H_\eta(t(\varphi * \mu)(y)) - |(\varphi * \mu)| H_\eta(t \operatorname{sign}(\varphi * \mu)(y))}{t}$$

$$+ \left[ \|\varphi^-\|_1 - \|(\varphi * \mu)^-\|_1 \right] \frac{H_\eta(-t)}{t}$$

$$+ \left[ \|\varphi^+\|_1 - \|(\varphi * \mu)^+\|_1 \right] \frac{H_\eta(t)}{t},$$

where $\| \cdot \|_1$ denotes the $l^1$-norm. As $t \to \infty$, the first sum on the right converges to $-\varrho_- I_d(\varphi^-) - \varrho_+ I_d(\varphi^+)$. The second sum converges to the same expression with $\varphi$ replaced by $\varphi * \mu$. The rest is either zero or converges to infinity, since the terms in the square brackets are nonnegative and both $H_\eta(-t)/t$ and $H_\eta(t)/t$ tend to infinity. But the sum of the two terms in the square brackets equals

$$\|\varphi\|_1 - \|\varphi * \mu\|_1$$

which is zero if and only if, for each $y \in \mathbb{Z}^d$, supp $\mu$ intersects at most one of the two sets $y - \operatorname{supp} \varphi^-$ and $y - \operatorname{supp} \varphi^+$. Let $\mathcal{D}_\varphi$ denote the set of all probabilities $\mu \in \mathcal{P}_f(\mathbb{Z}^d)$ with this property. Then the above considerations show that the potential $\xi(\cdot)$ satisfies Assumption (G) with

$$\begin{aligned} I(\mu) = \varrho_- &\left[ I_d((\varphi * \mu)^-) - I_d(\varphi^-) \right] \\ &+ \varrho_+ \left[ I_d((\varphi * \mu)^+) - I_d(\varphi^+) \right] \end{aligned}$$

if $\mu \in \mathcal{D}_\varphi$ and $I(\mu) = \infty$ otherwise.

Let us finally cast a glance at the particular situation when $\operatorname{supp} \varphi^-$ and $\operatorname{supp} \varphi^+$ are neighboring in each direction (i.e., for each $e \in \mathbb{Z}^d$ with $|e| = 1$, there exist $x, y \in \mathbb{Z}^d$ such that $x - y = e$ and $\varphi(x)\varphi(y) < 0$). Then $\mu \in \mathcal{D}_\varphi$ implies that all connected components of supp $\mu$ consist of single lattice sites and, therefore, $S_d(\mu) = 2d$. As a consequence, we obtain $\chi_G(\kappa) = 2d\kappa$.

**2.3. Random plateaus.** Let again $\eta(\cdot)$ be a field of i.i.d. random variables with finite cumulant generating function $H_\eta$. Given a finite subset $V$ of $\mathbb{Z}^d$, consider the potential

$$\xi(x) := \max_{y \in V} \eta(x - y), \qquad x \in \mathbb{Z}^d.$$

Suppose that $H_\eta$ satisfies Assumption (H) with $\varrho = \infty$. This is true in particular for i.i.d. Gaussian fields $\eta(\cdot)$. We obtain

$$\max_{m:\ \operatorname{supp} \mu \to V} \left\langle \exp\left\{ t \sum_z \mu(z)\eta(z - m(z)) \right\} \right\rangle \le e^{G_t(\mu)}$$

$$\le \sum_{m:\ \operatorname{supp} \mu \to V} \left\langle \exp\left\{ t \sum_z \mu(z)\eta(z - m(z)) \right\} \right\rangle.$$

Note that the upper and lower bounds differ at most by the factor $|V|^{|\operatorname{supp} \mu|}$. Moreover, note that the maximum on the left may as well be taken over all $m \colon \mathbb{Z}^d \to V$. For such $m$,

$$\sum_z \mu(z)\eta(z - m(z)) = \sum_x \mu(D_m(x))\eta(x),$$

where $D_m(x) := \{z \colon z - m(z) = x\}$ and $\mu(D_m(x)) = \sum_{z \in D_m(x)} \mu(z)$. Taking this into account, we find that

$$G_t(\mu) = \max_{m \colon \mathbb{Z}^d \to V} \sum_x H_\eta(t\mu(D_m(x))) + O(1)$$

as $t \to \infty$. In particular, $G_t(\delta_0) = H_\eta(t) + O(1)$. Recall that $\sum_x \mu(D_m(x)) = 1$ for all $m$. Hence,

$$\frac{G_t(\delta_0) - G_t(\mu)}{t} = \min_{m \colon \mathbb{Z}^d \to V} \sum_x \frac{\mu(D_m(x))H_\eta(t) - H_\eta(t\mu(D_m(x)))}{t} + o(1).$$

Since $\varrho = \infty$, the minimum on the right is either zero or tends to infinity as $t \to \infty$. It is zero if and only if there exist $m \colon \mathbb{Z}^d \to V$ and $x \in \mathbb{Z}^d$ such that $\operatorname{supp} \mu \subseteq D_m(x)$. But this happens if and only if $\operatorname{supp} \mu \subseteq x + V$ for some $x \in \mathbb{Z}^d$.

We have therefore shown that Assumption (G) is fulfilled with $I(\mu) = 0$ if $\operatorname{supp} \mu \subseteq x + V$ for some $x$ and $I(\mu) = \infty$ otherwise.

Since the functional $S_d$ is shift-invariant, we conclude from this that

$$\chi_G(\kappa) = \kappa \min_{\operatorname{supp} \mu \subseteq V} S_d(\mu) = \kappa \lambda^0(V),$$

where $\lambda^0(V)$ denotes the smallest eigenvalue of $-\Delta$ in $l^2(V)$ with Dirichlet boundary condition. If $|V| \geq 2$, then $0 < \lambda^0(V) < 2d$.

## 3. Lifshitz tails of the integrated density of states

**3.1. Introduction and main result.** In this section we will study the logarithmic tail behavior of the spectral distribution function ('integrated density of states') associated with the Anderson Hamiltonian

$$\mathcal{H} = \kappa \Delta + \xi(\cdot).$$

For general background information on Lifshitz tails we refer to Pastur and Figotin [10].

We will assume throughout that the potential $\xi(\cdot)$ is shift-invariant and ergodic and satisfies (1.3). The distribution function $F(r) := \operatorname{Prob}(\xi(0) \leq r)$ is supposed to be continuous and such that $F(r) < 1$ for all $r$. Later on we will also impose Assumption (G) of Section 1 as well as an additional condition on the tail behavior of $F$. Recall that $\mathbb{T}_R^d = \{-R, \ldots, R\}^d$ is the centered lattice box of length $2R + 1$.

Given $R \geq 1$, let $\lambda_1^{(R)} \geq \lambda_2^{(R)} \geq \cdots \geq \lambda_{|\mathbb{T}_R^d|}^{(R)}$ denote the eigenvalues of $\mathcal{H}$ in $l^2(\mathbb{T}_R^d)$ with Dirichlet boundary condition. The associated spectral distribution function $N^{(R)}(h)$ is defined as the relative number of eigenvalues not exceeding $h$:

$$N^{(R)}(h) := \frac{1}{|\mathbb{T}_R^d|} \sum_{i=1}^{|\mathbb{T}_R^d|} \mathbb{1}\left(\lambda_i^{(R)} \leq h\right), \qquad h \in \mathbb{R}.$$

The *spectral distribution function* $N$ of $\mathcal{H}$ in $l^2(\mathbb{Z}^d)$ may then be defined by

$$N(h) := \lim_{R \to \infty} N^{(R)}(h), \qquad h \in \mathbb{R}.$$

It is well-known that this limit exists a.s. and is nonrandom as a consequence of the ergodicity of $\xi(\cdot)$. Moreover, its Laplace transform coincides with the 'trace' of the fundamental solution $q(t, x, y)$ of $\mathcal{H}$:

$$(3.1) \qquad \int_{-\infty}^{\infty} e^{th} N(dh) = \langle q(t, 0, 0) \rangle \qquad \text{for } t \geq 0,$$

provided that the expression on the right is finite for all $t \geq 0$ which is certainly true under our assumptions. Let $\bar{N}^{(R)}(h) := 1 - N^{(R)}(h)$ and $\bar{N}(h) := 1 - N(h)$ denote the upper tails of the spectral distribution functions $N^{(R)}$ and $N$, respectively. Let further $\bar{F}(h) := 1 - F(h)$ be the upper tail of the distribution of $\xi(0)$.

Note that

$$\xi(\cdot) - 4d\kappa \leq \mathcal{H} \leq \xi(\cdot)$$

in the sense of positive definiteness of operators on $l^2(\mathbb{Z}^d)$. (I.e., the operators $\xi(\cdot) - \mathcal{H}$ and $\mathcal{H} - \xi(\cdot) + 4d\kappa$ are nonnegative definite.) This implies that

$$\frac{1}{|\mathbb{T}_R^d|} \sum_{x \in \mathbb{T}_R^d} \mathbb{1}(\xi(x) - 4d\kappa > h) \leq \bar{N}^{(R)}(h) \leq \frac{1}{|\mathbb{T}_R^d|} \sum_{x \in \mathbb{T}_R^d} \mathbb{1}(\xi(x) > h),$$

and the strong law of large numbers yields the trivial bounds

$$(3.2) \qquad \bar{F}(h + 4d\kappa) \leq \bar{N}(h) \leq \bar{F}(h), \qquad h \in \mathbb{R}.$$

We remark that these bounds are universal in the sense that they are valid for any shift-invariant ergodic potential.

Let us next introduce the non-decreasing function

$$\varphi(r) := \log \frac{1}{1 - F(r)}, \qquad r \in \mathbb{R},$$

and its left-continuous inverse

$$\psi(s) := \min\{r \colon \varphi(r) \geq s\}, \qquad s > 0.$$

Note that $\psi$ is strictly increasing and $\varphi(\psi(s)) = s$ for all $s > 0$.

We now formulate our restriction on the tail behavior of the distribution function $F$.

**Assumption (F).** There exists $\varrho$, $0 \leq \varrho < \infty$, such that

$$(3.3) \qquad \lim_{s \to \infty} [\psi(cs) - \psi(s)] = \varrho \log c$$

for all $c \in (0, 1)$.

Roughly speaking, if $0 < \varrho < \infty$, then assumption (3.3) requires that the upper tail of $F$ behaves like that of a double exponential distribution (2.3) with parameter $\varrho$. The case $\varrho = 0$ is that of an 'almost bounded' potential, i.e. with a tail that is below a double exponential.

It has been shown in [8], Lemma 2.3, that Assumption (F) not only implies (1.3) but also Assumption (H) of Section 2.1 and the relation

$$(3.4) \qquad \psi(t) = \frac{H(t)}{t} - \varrho \log \varrho + \varrho + o(1) \qquad \text{as } t \to \infty.$$

Recall that the function $\chi_G$ is defined by the variational expression (1.6). The following theorem provides a more precise description of the asymptotics of $\bar{N}(h)$ than (3.2) because it takes into account the structure of the high peaks of the potential $\xi(\cdot)$.

THEOREM 3.1. *Let Assumption (F) be satisfied for some $\varrho$ with $0 \leq \varrho < \infty$. Let also Assumption (G) be fulfilled.*

*a) If $0 < \varrho < \infty$, then*

$$(3.5) \qquad \log \bar{N}(h) \sim \log \bar{F}(h + \chi_G(\kappa)) \qquad \text{as } h \to \infty.$$

*b) If $\varrho = 0$, then*

$$(3.6) \qquad \bar{F}(h + \chi_G(\kappa) + \delta) \leq \bar{N}(h) \leq \bar{F}(h + \chi_G(\kappa) - \delta)$$

*for arbitrary $\delta > 0$ and all sufficiently large $h$.*

It can easily be seen that, for $0 \leq \varrho < \infty$ and $c \neq 0$, $\log \bar{F}(h)$ and $\log \bar{F}(h + c)$ are not asymptotically equivalent. Hence, Theorem 3.1 is indeed an improvement upon the universal bounds (3.2). On the other hand, for '$\varrho = \infty$' this is not true. The latter case corresponds to tails of $F$ which decay faster than that of a double exponential distribution. In such situations, with some more effort, one may expect to obtain more accurate asymptotic formulas for the integrated density of states. This will be the subject of a separate paper.

**3.2. Proof of Theorem 3.1.** In order to prove Theorem 3.1 we introduce the shifted spectral distribution functions

$$(3.7) \qquad N_t(h) := N\left(\frac{H(t)}{t} - \chi_G(\kappa) + h\right), \qquad t > 0, \, h \in \mathbb{R},$$

where, as before, $H$ is the cumulant generating function of $\xi(0)$ and $\chi_G(\kappa)$ is given by (1.6). We will consider $N_t$, $t > 0$, as probability measures on the left-compactified real line $[-\infty, \infty)$. We further abbreviate $\bar{N}_t(h) := 1 - N_t(h)$. The following lemma provides the key for our proof of Theorem 3.1.

LEMMA 3.2. *Let Assumption (F) be satisfied for some $\varrho$ with $0 \leq \varrho < \infty$. Let also Assumption (G) be fulfilled.*

*a) If $0 < \varrho < \infty$, then the probability measures $N_t$ on $[-\infty, \infty)$ satisfy the full large deviation principle as $t \to \infty$ with scale $t$ and rate function $J \colon [-\infty, \infty) \to \mathbb{R}_+$ given by $J(-\infty) := 0$ and*

$$(3.8) \qquad J(h) = \varrho \exp\left\{\frac{h}{\varrho} - 1\right\} \qquad for \; h \in \mathbb{R}.$$

*b) If $\varrho = 0$, then*

$$(3.9) \qquad \lim_{t \to \infty} \frac{1}{t} \log \bar{N}_t(h) = \begin{cases} 0 & for \; h < 0, \\ -\infty & for \; h > 0. \end{cases}$$

PROOF. We know from Remark 1.2 b) that the fundamental solution $q(t, x, y)$ of $\mathcal{H}$ satisfies (1.8). Because of (3.1) and (3.7), this implies that

$$\frac{1}{t} \log \int e^{\beta t h} \, N_t(dh) = \frac{H(\beta t) - \beta H(t)}{t} + o(1)$$

for each $\beta > 0$. Together with Assumption (H), this yields

$$(3.10) \qquad \lim_{t \to \infty} \frac{1}{t} \log \int e^{\beta t h} \, N_t(dh) = \varrho \, \beta \log \beta =: \Gamma(\beta)$$

for $\beta > 0$.

a) Suppose now that $0 < \varrho < \infty$. Since the limiting function $\Gamma$ is continuously differentiable on $(0, \infty)$ and $\Gamma'(\beta) \to -\infty$ for $\beta \to 0$, we conclude from (3.10) that $N_t$ satisfies the full large deviation principle with scale $t$ and rate function $J$ being the Legendre transform of $\Gamma$:

$$J(h) = \sup_{\beta > 0} [h\beta - \Gamma(\beta)], \qquad h \in \mathbb{R},$$

which coincides with (3.8). See e.g. Freidlin and Wentzell [4], Chap. 5, for standard large deviation arguments of such type.

b) We now turn to the case $\varrho = 0$. Fix $\beta > 0$ and $\delta > 0$ arbitrarily. We conclude from (3.10) for $\varrho = 0$ that

$$\left(\int_{-\infty}^{-\delta} + \int_{-\delta}^{\delta} + \int_{\delta}^{\infty}\right) e^{\beta h t} \, N_t(dh) = e^{o(t)}.$$

The first and the third integral decay exponentially fast as $t \to \infty$. For,

$$\int_{-\infty}^{-\delta} e^{\beta t h} \, N_t(dh) \leq e^{-\beta \delta t},$$

while by (3.10) for $\varrho = 0$,

$$(3.11) \qquad \int_\delta^\infty e^{\beta th} N_t(dh) \leq e^{-\gamma \delta t} \int_{-\infty}^\infty e^{(\beta + \gamma) th} N_t(dh) = e^{-\gamma \delta t + o(t)}$$

for $\gamma > 0$. Thus,

$$e^{\beta \delta t} \bar{N}_t(-\delta) \geq \int_{-\delta}^\delta e^{\beta th} N_t(dh) = e^{o(t)}.$$

Since $\beta > 0$ may be chosen arbitrarily small, this shows that

$$\lim_{t \to \infty} \frac{1}{t} \log \bar{N}_t(-\delta) = 0.$$

This is the first part of (3.9). The second part is immediate from

$$e^{\beta \delta t} \bar{N}_t(\delta) \leq \int_\delta^\infty e^{\beta th} N_t(dh)$$

and (3.11) by choosing $\gamma$ arbitrarily large. $\qquad \square$

PROOF. *[Theorem 3.1]* It is not hard to show that, under Assumption (F), $\psi(\varphi(h)) = h + o(1)$ as $h \to \infty$. Hence, we conclude from (3.4) that

$$(3.12) \qquad h = \frac{H(\varphi(h))}{\varphi(h)} - \varrho \log \varrho + \varrho + o(1) \qquad \text{as } h \to \infty.$$

a) Recall that $\bar{N}_t(h) = 1 - N_t(h)$ and $N_t(h)$ is defined by (3.7). If $0 < \varrho < \infty$, then we may combine (3.12) with assertion a) of Lemma 3.2 to find that

$$\log \bar{N}(h - \chi_G(\kappa)) = \log \bar{N}_{\varphi(h)} \left( -\varrho \log \varrho + \varrho + o(1) \right)$$
$$\sim -\varphi(h) J(-\varrho \log \varrho + \varrho + o(1))$$
$$\sim -\varphi(h) = \log \bar{F}(h)$$

as $h \to \infty$. Here we have used that the rate function $J$ is non-decreasing and continuous and

$$J \left( -\varrho \log \varrho + \varrho \right) = 1.$$

This clearly proves (3.5).

b) Suppose now that $\varrho = 0$. In this case (3.12) tells us that

$$h = \frac{H(\varphi(h))}{\varphi(h)} + o(1) \qquad \text{as } h \to \infty.$$

From this and assertion b) of Lemma 3.2 we conclude that

$$\log \bar{N}(h - \chi_G(\kappa) - \delta) = \log \bar{N}_{\varphi(h)}(-\delta + o(1)) \geq -\varphi(h) = \log \bar{F}(h)$$

for every $\delta > 0$ and all sufficiently large $h$. This crude estimate yields the lower bound in (3.6). The proof of the upper bound is similar. $\qquad \square$

## References

[1] R. A. Carmona and S. A. Molchanov, *Parabolic Anderson problem and intermittency*, AMS Memoir 518, Amer. Math. Soc., Providence RI, 1994.

[2] J.-D. Deuschel and D. W. Stroock, *Large deviations*, Academic Press, Boston, 1989.

[3] M. D. Donsker and S. R. S. Varadhan, *Asymptotic evaluation of certain Markov process expectations for large time, III*, Commun. Pure Appl. Math. **29** (1976), 389–461.

[4] M. I. Freidlin and A. D. Wentzell, *Random perturbations of dynamical systems*, Springer, New York, 1984.

[5] J. Gärtner and F. den Hollander, *Correlation structure of intermittency in the parabolic Anderson model*, Probab. Theory Relat. Fields **114** (1999), 1–54.

[6] J. Gärtner and W. König, *Moment asymptotics for the continuous parabolic Anderson model*, To appear in Ann. Appl. Prob., 1999.

[7] J. Gärtner and S. A. Molchanov, *Parabolic problems for the Anderson model. I. Intermittency and related topics*, Commun. Math. Phys. **132** (1990), 613–655.

[8] ———, *Parabolic problems for the Anderson model. II. Second-order asymptotics and structure of high peaks*, Probab. Theory Relat. Fields **111** (1998), 17–55.

[9] S. A. Molchanov, *Lectures on random media*, D. Bakry, R.D. Gill, and S.A. Molchanov, Lectures on Probability Theory, Ecole d'Eté de Probabilités de Saint-Flour XXII-1992, Lect. Notes in Math. **1581**, Springer, Berlin, 1994, pp. 242–411.

[10] L. Pastur and A. Figotin, *Spectra of random and almost-periodic operators*, Springer, Berlin, 1992.

[11] A.-S. Sznitman, *Brownian motion, obstacles and random media*, Springer, Berlin, 1998.

TECHNISCHE UNIVERSITÄT, FB MATHEMATIK, STR. DES 17. JUNI 136, D-10623 BERLIN, GERMANY
*E-mail address*: jg@math.tu-berlin.de

DEPARTMENT OF MATHEMATICS, UNIVERSITY OF NORTH CAROLINA, CHARLOTTE, NC 28223
*E-mail address*: smolchan@uncc.edu

Canadian Mathematical Society
Conference Proceedings
Volume **26**, 2000

# High-Density Fluctuations of Immigration Branching Particle Systems

Luis G. Gorostiza and Zeng-Hu Li

*Dedicated to Professor Donald A. Dawson*

ABSTRACT. We obtain a class of generalized Ornstein-Uhlenbeck processes as high–density fluctuation limits of branching particle systems with immigration. We consider in particular the stationary case.

## 1. Introduction

Fluctuation limits of particle systems (with or without branching) have been studied extensively. Usually they lead to generalized Ornstein-Uhlenbeck processes; see e.g. Bojdecki and Gorostiza [**1**, **2**], Dawson et al [**5**], Dawson and Gorostiza [**6**], Gorostiza [**10**], Holley and Stroock [**12**], and Itô [**14**]. Since the branching particle systems can be unstable, non-stationary scalings are employed in studying their fluctuation limits, which yield time-inhomogeneous Ornstein-Uhlenbeck processes. Fluctuation limits of measure-valued branching processes with immigration were studied in Gorostiza and Li [**11**], and Li [**16**], which gave time-homogeneous Ornstein-Uhlenbeck processes. The measure-valued processes considered in [**11**, **16**] are superprocess-type limits of a class of branching particle systems with immigration. In this paper we consider high-density fluctuation limits for the branching particle systems with immigration. The Ornstein-Uhlenbeck processes obtained here are different from the ones in [**11**, **16**], and we shall see that the superprocess-type limit and the high-density fluctuation limit are interchangeable. We will consider in particular the stationary case.

Let us recall some basic facts on the branching particle systems. We refer the reader to Dawson [**4**] for the necessary background. Let $\mathbb{R}^d$ be the $d$-dimensional Euclidean space. We denote by $C(\mathbb{R}^d)$ the set of bounded continuous functions on $\mathbb{R}^d$, and $C_0(\mathbb{R}^d)$ the set of functions in $C(\mathbb{R}^d)$ vanishing at infinity. Suppose that $\xi = (\Omega, \mathcal{F}, \mathcal{F}_t, \xi_t, \mathbf{P}_x)$ is a diffusion process with semigroup $(P_t)_{t\geq 0}$ generated by a differential operator $A$. Throughout the paper we fix a strictly positive reference

---

2000 *Mathematics Subject Classification*. Primary 60J80; Secondary 60G20, 60G10.

*Key words and phrases*. immigration particle system; entrance law; Ornstein-Uhlenbeck process; Langevin equation; fluctuation limit; stationary distribution.

function $\rho \in \mathcal{D}(A)$ with $A\rho \in C_\rho(\mathbb{R}^d)$, where $C_\rho(\mathbb{R}^d)$ denotes the set of functions $f \in C(\mathbb{R}^d)$ satisfying $|f| \leq \text{const} \cdot \rho$. We assume further that $\rho^{-1}g$ is bounded for every rapidly decreasing function $g \in C(\mathbb{R}^d)$. The subsets of non-negative elements of the above function spaces are indicated by the superscript '+', e.g. $C_\rho(\mathbb{R}^d)^+$. Let $M_\rho(\mathbb{R}^d)$ be the space of $\sigma$-finite Borel measures $\mu$ on $(\mathbb{R}^d, \mathcal{B}(\mathbb{R}^d))$ such that $\mu(f) := \int f d\mu < \infty$ for all $f \in C_\rho(\mathbb{R}^d)^+$. We equip $M_\rho(\mathbb{R}^d)$ with the topology defined by $\mu_k \to \mu$ if and only if $\mu_k(f) \to \mu(f)$ for all $f \in C_\rho(\mathbb{R}^d)$. Let $N_\rho(\mathbb{R}^d)$ be the subspace of $M_\rho(\mathbb{R}^d)$ consisting of integer-valued measures.

Let $\beta \in C(\mathbb{R}^d)^+$ and let $g(x, z)$ be a continuous function of $(x, z) \in \mathbb{R}^d \times [0, 1]$. Suppose that for each fixed $x \in \mathbb{R}^d$, $g(x, \cdot)$ coincides on $[0, 1]$ with the probability generating function of a critical branching law. Throughout this paper we assume that $c(x) := \beta(x)g''(x, 1^-)$ is a bounded continuous function on $\mathbb{R}^d$. For any $f \in C_\rho(\mathbb{R}^d)$ the evolution equation

$$
\exp\{-u_t(x)\} = \mathbf{P}_x \exp\{-f(\xi_t)\} - \int_0^t \mathbf{P}_x \left[\beta(\xi_{t-s}) \exp\{-u_s(\xi_{t-s})\}\right] ds
$$
$$
+ \int_0^t \mathbf{P}_x \left[\beta(\xi_{t-s})g(\xi_{t-s}, \exp\{-u_s(\xi_{t-s})\})\right] ds, \quad t \geq 0,
$$

has a unique positive solution $u_t = U_t f \in C_\rho(\mathbb{R}^d)$. In the sequel we shall simply write the above equation as

$$
(1.1) \qquad \mathrm{e}^{-u_t} = P_t \mathrm{e}^{-f} - \int_0^t P_{t-s}\left[\beta\left(\mathrm{e}^{-u_s} - g(\mathrm{e}^{-u_s})\right)\right] ds, \quad t \geq 0.
$$

It is well-known that the formula

$$
(1.2) \qquad \int_{N_\rho(\mathbb{R}^d)} \mathrm{e}^{-\nu(f)} Q_t(\sigma, d\nu) = \exp\{-\sigma(U_t f)\}, \quad f \in C_\rho(\mathbb{R}^d),
$$

defines a transition semigroup $(Q_t)_{t\geq 0}$ on $N_\rho(\mathbb{R}^d)$. A Markov process $X$ is called a *branching particle system* with parameters $(\xi, \beta, g)$ if its transition probabilities are determined by (1.1) and (1.2).

For $f \in C_\rho(\mathbb{R}^d)$ let

$$
(1.3) \qquad J_t f(x) = 1 - \exp\{-U_t f(x)\}, \quad t \geq 0, x \in \mathbb{R}^d.
$$

Then from (1) we have

$$
(1.4) \quad J_t f(x) + \int_0^t ds \int_{\mathbb{R}^d} \varphi(x, J_s f(y)) P_{t-s}(x, dy) = P_t \left(1 - \mathrm{e}^{-f}\right)(x),
$$

where

$$
(1.5) \qquad \varphi(x, z) = \beta(x)[g(x, 1-z) - (1-z)], \quad x \in \mathbb{R}^d, 0 \leq z \leq 1.
$$

Note that $\varphi''(x, 0^+) = c(x)$ by our assumption.

## 2. Immigration particle systems

Suppose that $X$ is a branching particle system with transition semigroup $(Q_t)_{t\geq 0}$ as described in the last section. Let $(N_t)_{t\geq 0}$ be a family of probability measures on $N_\rho(\mathbb{R}^d)$. We call $(N_t)_{t\geq 0}$ a *skew convolution semigroup* associated with $X$ or $(Q_t)_{t\geq 0}$ provided that

$$(2.1) \qquad N_{r+t} = (N_r Q_t) * N_t, \quad r, t \geq 0,$$

where '$*$' denotes the operation of convolution. The relation (2.1) is necessary and sufficient to ensure that

$$(2.2) \qquad Q_t^N(\sigma, \cdot) := Q_t(\sigma, \cdot) * N_t, \quad t \geq 0, \ \sigma \in N_\rho(\mathbb{R}^d),$$

defines a transition semigroup $(Q_t^N)_{t\geq 0}$ on $N_\rho(\mathbb{R}^d)$. A Markov process $Y$ is naturally called an *immigration particle system* associated with $X$ if it has $(Q_t^N)_{t\geq 0}$ as transition semigroup. The intuitive meaning of the immigration particle system is clear from (2.2), that is, $Q_t(\sigma, \cdot)$ is the distribution of descendants of the people distributed at time zero as $\sigma \in N_\rho(\mathbb{R}^d)$, and $N_t$ is the distribution of descendants of the people immigrating to $\mathbb{R}^d$ during the time interval $(0, t]$. We refer to Li [15] for the basic facts and characterizations about skew convolution semigroups and immigration particle systems. (In [15] only particle systems taking finite measure values were considered, but all the results there remain valid in the present context under obvious modifications.)

Now we consider a particular form of the immigration particle system. Let $\gamma \in M_\rho(\mathbb{R}^d)$ be a purely excessive measure for $\xi$. Then there is an entrance law $(\kappa_t)_{t>0}$ for $\xi$ such that $\gamma = \int_0^\infty \kappa_t dt$; see Dynkin [8]. For $t > 0$ and $f \in C_\rho(\mathbb{R}^d)$ we let

$$(2.3) \qquad R_t(\kappa, f) = \kappa_t \left(1 - e^{-f}\right) - \int_0^t \kappa_{t-s} \left(\varphi(J_s f)\right) ds.$$

Let $N_\rho(\mathbb{R}^d)^\circ = N_\rho(\mathbb{R}^d) \backslash \{0\}$, where $0$ denotes the null measure, and let $(Q_t^\circ)_{t\geq 0}$ be the restriction of $(Q_t)_{t\geq 0}$ to $N_\rho(\mathbb{R}^d)^\circ$. By Theorem 3.3 of [15] we see that

$$\int_{N_\rho(\mathbb{R}^d)^\circ} e^{-\nu(f)} K_t(d\nu) = \exp\{-R_t(\kappa, f)\}, \quad f \in C_\rho(\mathbb{R}^d),$$

defines an infinitely divisible probability entrance law $(K_t)_{t>0}$ for $(Q_t^\circ)_{t\geq 0}$. It follows from Theorem 3.2 of [15] that there is a finite entrance law $(H_t)_{t>0}$ such that

$$\int_{N_\rho(\mathbb{R}^d)^\circ} (1 - e^{\nu(f)}) H_t(d\nu) = R_t(\kappa, f), \ f \in C_\rho(\mathbb{R}^d).$$

Then using Theorem 3.1 of [15] we see that the formula

$$(2.4) \qquad \int_{N_\rho(\mathbb{R}^d)} e^{-\nu(f)} Q_t^{(\kappa)}(\sigma, d\nu) = \exp\left\{-\sigma(U_t f) - \int_0^t R_s(\kappa, f) ds\right\}, \ f \in C_\rho(\mathbb{R}^d),$$

defines a transition semigroup $(Q_t^{(\kappa)})_{t\geq 0}$ on $N_\rho(\mathbb{R}^d)$, which is a special case of the one defined by (2.2). In the sequel, a Markov process $Y$ will be called an *immigration particle system* with parameters $(\xi, \beta, g, \kappa)$ if it has transition semigroup determined by (2.4).

A special excessive measure for $\xi$ is given by $\gamma = \int_0^\infty \nu P_t dt$ for some $\nu \in M_\rho(\mathbb{R}^d)$. In this case we have $R_t(\kappa, f) = \nu(J_t f)$ by (1.4) and (2.3), and the immigration is governed by a time-space Poisson random measure with intensity $ds\nu(dx)$; see e.g. Dawson and Ivanoff [7].

We will need the following lemmas.

LEMMA 2.1. *For any $t \geq 0$, $\sigma \in N_\rho(\mathbb{R}^d)$ and $f \in C_\rho(\mathbb{R}^d)$ we have*

$$(2.5) \qquad \int_{N_\rho(\mathbb{R}^d)} \nu(f) Q_t^{(\kappa)}(\sigma, d\nu) = \sigma(P_t f) + \int_0^t \kappa_s(f) ds.$$

PROOF. From (1.1) and (2.3) we get

$$\frac{\partial}{\partial \theta} U_t(\theta f)\Big|_{\theta=0} = P_t f, \quad \frac{\partial}{\partial \theta} R_t(\kappa, \theta f)\Big|_{\theta=0} = \kappa_t(f).$$

Using these results and (2.4) we obtain (2.5). $\square$

Let $\mathbf{Q}_{(\mu)}^{(\kappa)}$ denote the law of the immigration particle system $\{Y_t : t \geq 0\}$ given that $Y_0$ is a Poisson random measure with intensity $\mu \in M_\rho(\mathbb{R}^d)$. Then we have from (2.4)

$$(2.6) \quad \mathbf{Q}_{(\mu)}^{(\kappa)} \exp\{-Y_t(f)\} = \exp\left\{ -\mu(J_t f) - \int_0^t R_s(\kappa, f) ds \right\}, \quad f \in C_\rho(\mathbb{R}^d).$$

LEMMA 2.2. *For $t \geq 0$ and $f \in C_\rho(\mathbb{R}^d)$ we have $\mathbf{Q}_{(\gamma)}^{(\kappa)} Y_t(f) = \gamma(f)$ and*

$$(2.7) \qquad \mathbf{Q}_{(\gamma)}^{(\kappa)}\{Y_t(f)^2\} = \gamma(f)^2 + \gamma(f^2) + \int_0^t \gamma(c(P_s f)^2) ds.$$

PROOF. This is similar to the proof of Lemma 2.1. From (1.4) and (2.3) it follows that

$$\frac{\partial}{\partial \theta} J_t(\theta f)\Big|_{\theta=0} = P_t f, \quad \frac{\partial}{\partial \theta} R_t(\kappa, \theta f)\Big|_{\theta=0} = \kappa_t(f),$$

$$-\frac{\partial^2}{\partial \theta^2} J_t(\theta f)\Big|_{\theta=0} = P_t(f^2) + \int_0^t P_{t-u}(c(P_u f)^2) du,$$

$$-\frac{\partial^2}{\partial \theta^2} R_t(\kappa, \theta f)\Big|_{\theta=0} = \kappa_t(f^2) + \int_0^t \kappa_{t-u}(c(P_u f)^2) du.$$

Replacing $f$ by $\theta f$ in (2.6), differentiating with respect to $\theta$ at zero and using (2.3) we get

$$\mathbf{Q}_{(\mu)}^{(\kappa)} Y_t(f) = \mu(P_t f) + \int_0^t \kappa_s(f) ds,$$

and

$$\mathbf{Q}_{(\mu)}^{(\kappa)}\{Y_t(f)^2\} = \left[\mu(P_t f) + \int_0^t \kappa_s(f) ds\right]^2 + \mu(P_t(f^2)) + \int_0^t \mu P_{t-s}(c(P_s f)^2) ds$$

$$+ \int_0^t \left[\kappa_s(f^2) + \int_0^s \kappa_{s-u}(c(P_u f)^2) du\right] ds.$$

Setting $\mu = \gamma$, the results are clear by the entrance law property $\kappa_s P_t = \kappa_{s+t}$ and the integral representation for $\gamma$. $\square$

Let $\mathcal{S}(\mathbb{R}^d)$ be the space of infinitely differentiable, rapidly decreasing functions all of whose derivatives are also rapidly decreasing.

LEMMA 2.3. *Suppose that $\{Y_t, \mathcal{F}_t : t \geq 0\}$ is a realization of the immigration particle system with parameters $(\xi, \beta, g, \kappa)$. If $f \in \mathcal{S}(\mathbb{R}^d)$, then the limit $\kappa_{0+}(f) := \lim_{r \downarrow 0} \kappa_r(f)$ exists and*

$$(2.8) \qquad M_t(f) := Y_t(f) - \int_0^t Y_s(Af)\mathrm{d}s - t\kappa_{0+}(f), \quad t \geq 0,$$

*is a martingale. In particular $\{Y_t(f), t \geq 0\}$ has a right–continuous modification for any $f \in \mathcal{S}(\mathbb{R}^d)$.*

PROOF. Under the hypothesis we have $Af \in \mathcal{S}(\mathbb{R}^d)$ and $f = P_t f - \int_0^t P_s Af \mathrm{d}s$ for any $t \geq 0$. It follows immediately from the entrance law property that

$$(2.9) \qquad \kappa_{0+}(f) = \kappa_t(f) - \int_0^t \kappa_s(Af)\mathrm{d}s, \quad t \geq 0.$$

If $t \geq r \geq 0$, using (2.5) we get

$$\mathbf{E}\left\{Y_t(f) - \int_0^t Y_s(Af)\mathrm{d}s - t\kappa_{0+}(f)\middle|\mathcal{F}_r\right\}$$

$$= Y_r(P_{t-r}f) + \int_0^{t-r} \kappa_s(f)\mathrm{d}s - \int_0^r Y_s(Af)\mathrm{d}s - \int_r^t \mathbf{E}\{Y_s(Af)|\mathcal{F}_r\}\mathrm{d}s - t\kappa_{0+}(f).$$

By (2.5) and (2.9) it follows that

$$\int_r^t \mathbf{E}\{Y_s(Af)|\mathcal{F}_r\}\mathrm{d}s = \int_r^t \left[Y_r(P_{s-r}Af) + \int_0^{s-r} \kappa_u(Af)\mathrm{d}u\right]\mathrm{d}s$$

$$= Y_r(P_{t-r}f - f) + \int_r^t [\kappa_{s-r}(f) - \kappa_{0+}(f)]\mathrm{d}s$$

$$= Y_r(P_{t-r}f - f) + \int_0^{t-r} \kappa_s(f)\mathrm{d}s - (t-r)\kappa_{0+}(f).$$

Now it is clear that

$$\mathbf{E}\left\{Y_t(f) - \int_0^t Y_s(Af)\mathrm{d}s - t\kappa_{0+}(f)\middle|\mathcal{F}_r\right\} = Y_r(f) - \int_0^r Y_s(Af)\mathrm{d}s - r\kappa_{0+}(f).$$

That is, (2.8) is a martingale. $\square$

We conclude this section by observing that the measure-valued immigration processes considered in [16] arise as superprocess-type limits of the immigration particle systems. Let $\{Y_t(k) : t \geq 0\}$, $k = 1, 2, \ldots$ be a sequence of immigration particle systems with parameters $(\xi, k\beta, g, k\kappa)$. Suppose that $Y_0(k)$ is a Poisson random measure with intensity $k\gamma \in M_\rho(\mathbb{R}^d)$. Since for any $l \geq 0$ we have

$$k^2 \varphi(x, z/k) = k^2 \beta(x)[g(x, 1 - z/k) - (1 - z/k)] \to c(x)z^2/2$$

uniformly on the set $\mathbb{R}^d \times [0, l]$ as $k \to \infty$, by a theorem in [15] the sequence $\{k^{-1}Y_t(k) : t \geq 0\}$ converges as $k \to \infty$ to a Markov process $\{Y_t^{(0)} : t \geq 0\}$ with $Y_0^{(0)} = \gamma$ and with transition semigroup $(Q_t^\kappa)_{t \geq 0}$ determined by

$$(2.10) \quad \int_{M_\rho(\mathbb{R}^d)} \mathrm{e}^{-\nu(f)} Q_t^\kappa(\mu, \mathrm{d}\nu) = \exp\left\{-\mu(V_t f) - \int_0^t S_u(\kappa, f)\mathrm{d}u\right\}, f \in C_\rho(\mathbb{R}^d)^+,$$

where $V_t f$ is the solution to

$$V_t f(x) + \frac{1}{2} \int_0^t \mathrm{d}s \int_{\mathbb{R}^d} c(y) V_s f(y)^2 P_{t-s}(x, \mathrm{d}y) = P_t f(x), \quad t \geq 0, x \in \mathbb{R}^d,$$

and $S_u(\kappa, f)$ is defined by

$$S_u(\kappa, f) = \kappa_u(f) - \frac{1}{2} \int_0^u \kappa_{u-s}(c(V_s f)^2) \mathrm{d}s, \quad u > 0, f \in C_\rho(\mathbb{R}^d)^+.$$

The process $\{Y_t^{(0)} : t \geq 0\}$ has a diffusion realization, which is called a *measure-valued immigration diffusion process*; see [16].

## 3. Fluctuation limits

In this section we consider the high density fluctuation limits of the immigration particle systems. Let $\gamma$ and $\kappa$ be given as in the last section and let $\{Y_t^{(k)} : t \geq 0\}, k = 1, 2, \ldots$ be a sequence of immigration particle systems with corresponding parameters $(\xi, \beta, g, k\kappa)$. Suppose that $Y_0^{(k)}$ is a Poisson random measure with intensity $k\gamma$. We define the fluctuation process $\{Z_t^{(k)} : t \geq 0\}$ by

$$(3.1) \qquad\qquad Z_t^{(k)} = \frac{1}{\sqrt{k}} [Y_t^{(k)} - k\gamma], \quad t \geq 0.$$

Then $\{Z_t^{(k)} : t \geq 0\}$ is a Markov process taking signed-measure values from the space $N_k(\mathbb{R}^d) := \{\mu/\sqrt{k} - \sqrt{k}\gamma : \mu \in N_\rho(\mathbb{R}^d)\}$.

LEMMA 3.1. *The Markov process* $\{Z_t^{(k)} : t \geq 0\}$ *has transition semigroup* $(R_t^{(k)})_{t \geq 0}$ *which is determined by*

$$(3.2) \quad \int_{N_k(\mathbb{R}^d)} \mathrm{e}^{-\nu(f)} R_t^{(k)}(\mu, \mathrm{d}\nu) =$$

$$= \exp\left\{ - \mu(U_t^{(k)} f) + A_t^{(k)}(f) + \int_0^t k\gamma(\varphi(J_s(f/\sqrt{k}))) \mathrm{d}s \right\},$$

*where* $U_t^{(k)} f = \sqrt{k} U_t(f/\sqrt{k})$ *and*

$$(3.3) \quad A_t^{(k)}(f) = k\gamma(f/\sqrt{k} - 1 + \mathrm{e}^{-f/\sqrt{k}}) - k\gamma(U_t(f/\sqrt{k}) - J_t(f/\sqrt{k})).$$

PROOF. Let us compute the conditional Laplace functional of the process $\{Z_t^{(k)} : t \geq 0\}$. Take $t \geq 0$ and $r \geq 0$. Using the Markov property of $\{Y_t^{(k)} : t \geq 0\}$ and (9) we have

$$\mathbf{E}\left[ \exp\{-Z_{r+t}^{(k)}(f)\} | Z_s^{(k)} : s \leq r \right]$$

$$= \exp\left\{ \sqrt{k}\gamma(f) \right\} \mathbf{E}\left[ \exp\{-Y_{r+t}^{(k)}(f/\sqrt{k})\} | Y_s^{(k)} : s \leq r \right]$$

$$= \exp\left\{ \sqrt{k}\gamma(f) \right\} \exp\left\{ - Y_r^{(k)}(U_t(f/\sqrt{k})) - \int_0^t R_s(k\kappa, f/\sqrt{k}) \mathrm{d}s \right\}$$

$$= \exp\left\{ \sqrt{k}\gamma(f) - Z_r^{(k)}(\sqrt{k} U_t(f/\sqrt{k})) - \gamma(k U_t(f/\sqrt{k})) - \int_0^t R_s(k\kappa, f/\sqrt{k}) \mathrm{d}s \right\}.$$

That is, $\{Z_t^{(k)} : t \geq 0\}$ is a Markov process with transition semigroup $(R_t^{(k)})_{t\geq 0}$ given by

$$\int_{N_k(\mathbb{R}^d)} e^{-\nu(f)} R_t^{(k)}(\mu, d\nu) = \exp\Big\{ -\mu(U_t^{(k)} f) + \gamma(\sqrt{k}f) - \gamma(kU_t(f/\sqrt{k}))$$

$$(3.4) \hspace{6cm} -\int_0^t R_s(k\kappa, f/\sqrt{k}) ds \Big\}.$$

In addition to $A_t^{(k)}(f)$ given by (3.3), let

$$B_t^{(k)}(f) = k\gamma(1 - e^{-f/\sqrt{k}}) - k\gamma(J_t(f/\sqrt{k})) - \int_0^t R_s(k\kappa, f/\sqrt{k}) ds,$$

where $J_t$ is defined by (1.3). We may rewrite (3.4) as

$$(3.5) \quad \int_{N_k(\mathbb{R}^d)} e^{-\nu(f)} R_t^{(k)}(\mu, d\nu) = \exp\Big\{ -\mu(U_t^{(k)} f) + A_t^{(k)}(f) + B_t^{(k)}(f) \Big\}.$$

Using the equation (1.4) we have

$$\int_t^\infty k\kappa_r(1 - e^{-f/\sqrt{k}}) dr - k\gamma(J_t(f/\sqrt{k}))$$

$$= \int_0^\infty k\kappa_r(P_t(1 - e^{-f/\sqrt{k}})) dr - \int_0^\infty k\kappa_r(J_t(f/\sqrt{k})) dr$$

$$= \int_0^\infty dr \int_0^t \kappa_{r+t-s} k(\varphi(J_s(f/\sqrt{k}))) ds$$

$$= \int_0^t ds \int_{t-s}^\infty k\kappa_u(\varphi(J_s(f/\sqrt{k}))) du.$$

On the other hand, by (2.3) it follows that

$$\int_0^t k\kappa_r(1 - e^{-f/\sqrt{k}}) dr - \int_0^t R_r(k\kappa, f/\sqrt{k}) dr = \int_0^t dr \int_0^r k\kappa_{r-s}(\varphi(J_s(f/\sqrt{k}))) ds$$

$$= \int_0^t ds \int_0^{t-s} k\kappa_u(\varphi(J_s(f/\sqrt{k}))) du.$$

Summing the two last equations we get

$$(3.6) \hspace{3cm} B_t^{(k)}(f) = \int_0^t k\gamma(\varphi(J_s(f/\sqrt{k}))) ds.$$

Then (3.2) follows from (3.6) and (3.5). $\square$

LEMMA 3.2. *The one-dimensional distributions of the process* $\{Z_t^{(k)} : t \geq 0\}$ *are determined by*

$$(3.7) \quad \mathbf{E}\exp\{-Z_t^{(k)}(f)\} =$$

$$= \exp\Big\{ k\gamma(f/\sqrt{k} - 1 + e^{-f/\sqrt{k}}) + \int_0^t k\gamma(\varphi(J_s(f/\sqrt{k}))) ds \Big\}.$$

PROOF. By (2.6) and the present assumption we have

$$\mathbf{E}\exp\{-Y_t^{(k)}(f)\} = \exp\Big\{ -k\gamma(J_t f) + \int_0^t R_s(k\kappa, f) ds \Big\}.$$

Then for (3.1) we get

$$\mathbf{E}\exp\{-Z_t^{(k)}(f)\} = \exp\left\{-k\gamma(J_t(f/\sqrt{k}) - f/\sqrt{k}) - \int_0^t R_s(k\kappa, f/\sqrt{k})ds\right\}.$$

Using the notation in the proof of the last lemma we have

$$\mathbf{E}\exp\{-Z_t^{(k)}(f)\} = \exp\left\{k\gamma(f/\sqrt{k} - 1 + e^{-f/\sqrt{k}}) + B_t^{(k)}(f)\right\}.$$

Then (3.7) follows by (3.6). $\square$

Let $\mathcal{S}'(\mathbb{R}^d)$ be the dual space of $\mathcal{S}(\mathbb{R}^d)$ and write $\langle\,,\,\rangle$ for the duality on $(\mathcal{S}'(\mathbb{R}^d), \mathcal{S}(\mathbb{R}^d))$. We may also regard $\{Z_t^{(k)} : t \geq 0\}$ as a process in $\mathcal{S}'(\mathbb{R}^d)$.

THEOREM 3.3. *The finite dimensional distributions of* $\{Z_t^{(k)} : t \geq 0\}$ *converge as* $k \to \infty$ *to those of a Markov process* $\{Z_t : t \geq 0\}$ *with state space* $\mathcal{S}'(\mathbb{R}^d)$. *The transition semigroup* $(R_t^{(\kappa)})_{t\geq 0}$ *of* $\{Z_t : t \geq 0\}$ *is given by*

$$(3.8) \quad \int_{\mathcal{S}'(\mathbb{R}^d)} e^{-\langle\nu,f\rangle} R_t^{(\kappa)}(\mu, d\nu)$$

$$= \exp\left\{-\langle\mu, P_t f\rangle + \frac{1}{2}\gamma(f^2 - (P_t f)^2) + \frac{1}{2}\int_0^t \gamma(c(P_s f)^2)ds\right\},$$

*and its one dimensional distribution is determined by*

$$(3.9) \quad \mathbf{E}\exp\{-\langle Z_t, f\rangle\} = \exp\left\{\frac{1}{2}\gamma(f^2) + \frac{1}{2}\int_0^t \gamma(c(P_s f)^2)ds\right\}.$$

PROOF. Take any bounded sequence $\{f_k\} \in \mathcal{S}(\mathbb{R}^d)$ such that $f_k \to f \in \mathcal{S}(\mathbb{R}^d)$. Using the equations (1.1) and (1.4), and criticality of $g$ one can check that

$$(3.10) \quad \lim_{k\to\infty} \sqrt{k}U_t(f_k/\sqrt{k}) = \lim_{k\to\infty} \sqrt{k}J_t(f_k/\sqrt{k}) = P_t f.$$

By Taylor's expansion,

$$\lim_{k\to\infty} k[U_t(f_k/\sqrt{k}) - J_t(f_k/\sqrt{k})] = \lim_{k\to\infty} k[U_t(f_k/\sqrt{k}) - 1 + \exp\{-U_t(f_k/\sqrt{k})\}]$$

$$= \frac{1}{2}(P_t f)^2.$$

Then by (3.3) it follows that

$$(3.11) \quad \lim_{k\to\infty} A_t^{(k)}(f_k) = A_t(f) := \frac{1}{2}\gamma(f^2) - \frac{1}{2}\gamma((P_t f)^2).$$

Since $\varphi''(x, 0^+) = c(x)$ by the assumption, using (3.11) and Taylor's expansion we get

$$(3.12) \quad \lim_{k\to\infty} B_t^{(k)}(f_k) = \lim_{k\to\infty}\int_0^t k\gamma(\varphi(J_s(f_k/\sqrt{k})))ds = \frac{1}{2}\int_0^t \gamma(c(P_s f)^2)ds.$$

By (3.7) the one-dimensional distributions of $\{Z_t^{(k)} : t \geq 0\}$ converge to those of $\{Z_t : t \geq 0\}$.

For $0 = t_0 \leq t_1 < \cdots < t_n$ and $f_1, \cdots, f_n \in \mathcal{S}(\mathbb{R}^d)$ let

$$h_j^{(k)} = f_j + U_{t_{j+1}-t_j}^{(k)}(f_{j+1} + \cdots + U_{t_n-t_{n-1}}^{(k)}f_n).$$

Using (3.2) inductively we get

$$\mathbf{E}\exp\left\{-\sum_{j=1}^{n}\langle Z_{t_j}^{(k)}, f_j\rangle\right\} = \exp\left\{k\gamma(h_1^{(k)}/\sqrt{k} - 1 + e^{-h_1^{(k)}/\sqrt{k}})\right.$$

$$(3.13) \qquad \left. + \sum_{j=1}^{n} A_{t_j-t_{j-1}}^{(k)}(h_j^{(k)}) + \sum_{j=1}^{n}\int_0^{t_j-t_{j-1}} k\gamma(\varphi(J_s(h_j^{(k)}/\sqrt{k})))\mathrm{d}s\right\}.$$

By (3.10) it is clear that

$$(3.14) \qquad h_j^{(k)} \to h_j := f_j + P_{t_{j+1}-t_j}(f_{j+1} + \cdots + P_{t_n-t_{n-1}}f_n)$$

boundedly as $k \to \infty$. Applying (3.11), (3.12) and (3.14) to (3.13) we have

$$\lim_{k\to\infty}\mathbf{E}\exp\left\{-\sum_{j=1}^{n} Z_{t_j}^{(k)}(f_j)\right\}$$

$$= \exp\left\{\frac{1}{2}\gamma(f_1^2) + \sum_{j=1}^{n} A_{t_j-t_{j-1}}(h_j) + \frac{1}{2}\sum_{j=2}^{n}\int_0^{t_j-t_{j-1}} \gamma(c(P_s h_j)^2)\mathrm{d}s\right\}.$$

As in Iscoe [13], we see that the finite-dimensional distributions of $\{Z_t^{(k)} : t \geq 0\}$ converge to those of the Markov process $\{Z_t : t \geq 0\}$.  □

Observe that if $\{Z_t(k) : t \geq 0\}$ is given by the last theorem with the parameters $(\xi, c, \gamma)$ replaced by $(\xi, kc, k\gamma)$, then $\{k^{-1}Z_t(k) : t \geq 0\}$ converges to a Markov process $\{Z_t^{(0)} : t \geq 0\}$ with $Z_0^{(0)} = 0$ and with semigroup $(R_t^\kappa)_{t\geq 0}$ given by

$$\int_{\mathcal{S}'(\mathbb{R}^d)} e^{-\langle \nu, f\rangle} R_t^\kappa(\mu, \mathrm{d}\nu) = \exp\left\{-\langle\mu, P_t f\rangle + \frac{1}{2}\int_0^t \gamma(c(P_s f)^2)\mathrm{d}s\right\}.$$

This together with the result in [16] shows that the superprocess–type limit and the fluctuation limit are interchangeable.

For any $f \in \mathcal{S}(\mathbb{R}^d)$ define $Qf \in \mathcal{S}'(\mathbb{R}^d)$ by

$$(3.15) \qquad \langle Qf, g\rangle = \gamma(cfg) - \gamma(fAg + gAf), \quad g \in \mathcal{S}(\mathbb{R}^d).$$

Then we have

THEOREM 3.4. *The fluctuation limit process $\{Z_t : t \geq 0\}$ obtained in Theorem 3.3 has a continuous realization which solves the Langevin equation*

$$(3.16) \qquad \begin{aligned} \mathrm{d}Z_t &= A^* Z_t \mathrm{d}t + \mathrm{d}W_t, \quad t \geq 0, \\ Z_0 &= \textit{white noise based on } \gamma, \end{aligned}$$

*where $A^*$ denotes the adjoint operator of $A$ and $\{W_t : t \geq 0\}$ is an $\mathcal{S}'(\mathbb{R}^d)$-valued Wiener process with covariance functional*

$$(3.17) \quad \mathbf{E}\{\langle W_r, f\rangle\langle W_t, g\rangle\} = (r \wedge t)\langle Qf, g\rangle, \quad r, t \geq 0, \ f, g \in \mathcal{S}(\mathbb{R}^d).$$

PROOF. Observe that $\{Z_t : t \geq 0\}$ is an $\mathcal{S}'(\mathbb{R}^d)$-valued mean zero Gaussian process. Set $K(r, f; t, g) = \mathbf{E}\{\langle Z_r, f\rangle\langle Z_t, g\rangle\}$. By a standard argument one may check from (3.8) that

$$\int_{\mathcal{S}'(\mathbb{R}^d)} \langle \nu, f\rangle R_t^{(\kappa)}(\mu, \mathrm{d}\nu) = \mu(P_t f),$$

and

$$K(t, f; t, g) = \gamma(fg) + \int_0^t \gamma(cP_s f P_s g)\mathrm{d}s.$$

It follows from these results and the Markov property that

$$(3.18) \quad K(r, f; t, g) = \gamma(f P_{t-r} g) + \int_0^r \gamma(c P_s f P_{t-r+s} g)\mathrm{d}s, \quad t \geq r \geq 0.$$

By (3.18) and the fact $\|f - P_{t-r} f\| \leq \|Af\|(t - r)$ (where $\| \quad \|$ denotes sup norm) one easily sees that

$$
\begin{aligned}
\mathbf{E}\{|\langle Z_t, f \rangle - \langle Z_r, f \rangle|^2\} &= 2\gamma(f[f - P_{t-r} f]) + \int_r^t \gamma(c(P_s f)^2)\mathrm{d}s \\
&\quad + 2\int_0^r \gamma(c P_s f[P_s f - P_{t-r+s} f])\mathrm{d}s \\
&\leq 2\|Af\|\gamma(|f|)(t - r) + \mathrm{const} \cdot \|cf\|\gamma(\rho)(t - r) \\
&\quad + 2\|cAf\|(t - r)\int_0^r \gamma(P_s f)\mathrm{d}s
\end{aligned}
$$

for $t \geq r \geq 0$. Then $\{Z_t : t \geq 0\}$ has a continuous realization; see e.g. Walsh [17, p. 274]. Observe that

$$(3.19) \quad \int_0^t \gamma(c[P_s f P_s Ag + P_s g P_s Af])\mathrm{d}s = \gamma(c P_t f P_t g) - \gamma(c f g).$$

By (3.18) one checks that

$$(3.20) \quad \frac{\partial}{\partial t} K(t, f; t, g) = \gamma(c P_t f P_t g).$$

Using (3.18), (3.19) and (3.20) we get

$$\frac{\partial}{\partial t} K(t, f; t, g) - K(t, Af; t, g) - K(t, f; t, Ag) = \gamma(cfg) - \gamma(fAg + gAf).$$

By the results of [1, p. 234] (see also [2]) we conclude that $\{Z_t : t \geq 0\}$ satisfies the generalized Langevin equation (3.16) with $\{W_t : t \geq 0\}$ given by (3.17). $\square$

Since it may happen that $c = 0$, (3.15) and (3.17) indicate that $\gamma(fAf) \leq 0$. To see that this is true observe that

$$\gamma(fAf) = \frac{1}{2}\frac{\mathrm{d}}{\mathrm{d}t}\gamma((P_t f)^2)\Big|_{t=0}$$

and

$$\gamma((P_t f)^2) \leq \gamma(P_t(f^2)) \leq \gamma(f^2),$$

where the second inequality holds because $\gamma$ is an excessive measure for $(P_t)_{t \geq 0}$.

Note that the Ornstein-Uhlenbeck process $Z$ is different from the ones obtained in [11, 16] as small branching fluctuation limits, where the distribution of the driving process $\{W_t : t \geq 0\}$ does not involve the generator $A$.

## 4. Weak convergence

We already have the convergence of the finite-dimensional distributions. Since the limit process is continuous, the tightness and consequently the weak convergence of the sequence $\{Z_t^{(k)} : t \geq 0\}$ in the cadlag space $D([0,\infty), \mathcal{S}'(\mathbb{R}^d))$ can be obtained easily as follows.

THEOREM 4.1. *The sequence $\{Z_t^{(k)} : t \geq 0\}$ converges weakly to the process $\{Z_t : t \geq 0\}$ in the space $D([0,\infty), \mathcal{S}'(\mathbb{R}^d))$.*

PROOF. By a theorem in [**9**], due to the continuity of the limit and the martingale structure in Lemma 2.3 it suffices to show that

$$(4.1) \qquad \sup_{k \geq 1} \mathbf{E} \sup_{0 \leq s \leq t} \{\langle Z_s^{(k)}, f \rangle^2\} < \infty$$

for all $t > 0$ and $f \in \mathcal{S}(\mathbb{R}^d)$. Let

$$(4.2) \qquad N_t^{(k)}(f) := \langle Z_t^{(k)}, f \rangle - \int_0^t \langle Z_s^{(k)}, Af \rangle \mathrm{d}s, \quad t \geq 0.$$

Since $\gamma(f) = \int_0^\infty \kappa_s(f)\mathrm{d}s < \infty$, we have $\lim_{t\to\infty} \kappa_t(f) = 0$. Letting $t \to \infty$ in (2.9) gives that $\gamma(Af) = -\kappa_{0+}(f)$. Then (3.1) and (4.2) yield that

$$
\begin{aligned}
\sqrt{k} N_t^{(k)}(f) &= Y_t^{(k)}(f) - k\gamma(f) - \int_0^t Y_s^{(k)}(Af)\mathrm{d}s + tk\gamma(Af), \\
&= Y_t^{(k)}(f) - k\gamma(f) - \int_0^t Y_s^{(k)}(Af)\mathrm{d}s - tk\kappa_{0+}(f).
\end{aligned}
$$

By Lemma 2.3 we see that $\{N_t^{(k)}(f) : t \geq 0\}$ is a martingale. On the other hand, by Lemma 2.2 we get

$$(4.3) \quad \mathbf{E}\{\langle Z_t^{(k)}, f \rangle^2\} = \gamma(f^2) + \int_0^t \gamma(c(P_s f)^2)\mathrm{d}s, \quad t \geq 0, f \in C_\rho(\mathbb{R}^d).$$

By (4.2) and Doob's inequality we see that

$$
\begin{aligned}
\mathbf{E} \sup_{0 \leq s \leq t} \{\langle Z_s^{(k)}, f \rangle^2\} &\leq 2\mathbf{E} \sup_{0 \leq s \leq t} \{N_s^{(k)}(f)^2\} + 2\mathbf{E}\left\{ \sup_{0 \leq s \leq t} \left[ \int_0^s \langle Z_u^{(k)}, Af \rangle \mathrm{d}u \right]^2 \right\} \\
&\leq 8\mathbf{E}\{N_t^{(k)}(f)^2\} + 2\mathbf{E}\left\{ \sup_{0 \leq s \leq t} s \int_0^s \langle Z_u^{(k)}, Af \rangle^2 \mathrm{d}u \right\} \\
&\leq 16\mathbf{E}\{\langle Z_t^{(k)}, f \rangle^2\} + 16t\mathbf{E}\left\{ \int_0^t \langle Z_u^{(k)}, Af \rangle^2 \mathrm{d}u \right\} + 2t\mathbf{E}\left\{ \int_0^t \langle Z_u^{(k)}, Af \rangle^2 \mathrm{d}u \right\} \\
&\leq \text{const} \cdot \mathbf{E}\{\langle Z_t^{(k)}, f \rangle^2\} + \text{const} \cdot t\mathbf{E}\left\{ \int_0^t \langle Z_u^{(k)}, Af \rangle^2 \mathrm{d}u \right\}.
\end{aligned}
$$

Then (4.1) follows from (4.3). $\square$

EXAMPLE. A typical example is where $A = \Delta - b$ is the generator of a killed Brownian motion with $b \in C(\mathbb{R}^d)^+$ bounded away from zero. In this case, we may let $\rho(x) = 1/(1+|x|^p)$ for any $p > d$ and let $\gamma \in M_\rho(\mathbb{R}^d)$ be the Lebesgue measure.

As in [11] and [16] one may take a sequence $b_k \downarrow 0$ and replace $A$ by $A - b_k$ in taking the fluctuation limit. By doing so one includes the situation where $\gamma \in M_\rho(\mathbb{R}^d)$ is a general excessive (not necessarily purely excessive) measure. In particular, one may include the case $A = \Delta$ and $\gamma =$ Lebesgue measure in the above example.

## 5. Stationary processes

We now give a brief discussion of the fluctuation limit for stationary particle systems. Let $(Q_t^{(\kappa)})_{t\geq 0}$ be the semigroup determined by (2.4). By the definition (2.3) it is easy to check that

$$\int_0^\infty R_t(\kappa, f)\mathrm{d}t = \gamma(1 - \mathrm{e}^{-f}) - \int_0^\infty \gamma(\varphi(J_s f))\mathrm{d}s.$$

It follows from (2.4) and the fact $U_t\rho \leq P_t\rho$ that if $\sigma(P_t\rho) \to 0$ as $t \to \infty$, then $Q_t^{(\kappa)}(\sigma, \cdot) \to Q_\infty^{(\kappa)}$ as $t \to \infty$, where $Q_\infty^{(\kappa)}$ is the stationary distribution of $(Q_t^{(\kappa)})_{t\geq 0}$ given by

$$\int_{M_\rho(\mathbb{R}^d)} \mathrm{e}^{-\nu(f)} Q_\infty^{(\kappa)}(\mathrm{d}\nu) = \exp\left\{-\gamma(1-\mathrm{e}^{-f}) + \int_0^\infty \gamma(\varphi(U_s f))\mathrm{d}s\right\}, \quad f \in C_\rho(\mathbb{R}^d)^+.$$

On the other hand, if $\{Z_t : t \geq 0\}$ is the process obtained in Theorem 3.3, then from (3.8) the distribution of $Z_t$ converges as $t \to \infty$ to $R_\infty^{(\kappa)}$ given by

$$\int_{\mathcal{S}'(\mathbb{R}^d)} \mathrm{e}^{-\langle\nu,f\rangle} R_\infty^{(\kappa)}(\mathrm{d}\nu) = \exp\left\{\frac{1}{2}\gamma(f^2) + \frac{1}{2}\int_0^\infty \gamma(c(P_s f)^2)\mathrm{d}s\right\}, \quad f \in \mathcal{S}(\mathbb{R}^d).$$

It follows that $R_\infty^{(\kappa)}$ is a stationary distribution of the semigroup $(R_t^{(\kappa)})_{t\geq 0}$ given by (3.8). Moreover, if $\langle\mu, P_t\rho\rangle \to 0$ as $t \to \infty$, then $R_t^{(\kappa)}(\mu, \cdot) \to R_\infty^{(\kappa)}$ as $t \to \infty$.

If we consider a sequence of stationary immigration processes $\{Y_t^{(k)} : t \geq 0\}$ with semigroup $(Q_t^{(k\kappa)})_{t\geq 0}$ and one-dimensional distribution $Q_\infty^{(k\kappa)}$, and if we take the fluctuation limit as in section 3, then we get a stationary $\mathcal{S}'(\mathbb{R}^d)$-valued Markov process with semigroup $(R_t^{(\kappa)})_{t\geq 0}$ and one-dimensional distribution $R_\infty^{(\kappa)}$. That is, the fluctuation limit and the long–time limit are interchangeable. We refer the reader to Bojdecki and Jakubowski [3] for discussions on invariant measures of generalized Ornstein-Uhlenbeck processes in conuclear spaces.

**Acknowledgments.** The first named author thanks CIMAT (Guanajuato, Mexico) for its hospitality, and acknowledges support of the Consejo Nacional de Ciencia y Tecnología (Mexico, Grant No. 27932-E). The second named author acknowledges supports of the National Natural Science Foundation (Grant No. 19361060) and the Mathematical Center of the State Education Commission of China.

## References

[1] Bojdecki, T. and Gorostiza, L.G., *Langevin equation for $\mathcal{S}'$-valued Gaussian processes and fluctuation limits of infinite particle systems*, Probab. Theory Related Fields **73** (1986), 227-244.

[2] Bojdecki, T. and Gorostiza, L.G., *Gaussian and non-Gaussian distribution-valued Ornstein-Uhlenbeck processes*, Canadian J. Math. **43** (1991), 1136-1149.

[3] Bojdecki, T. and Jakubowski, J., *Invariant measures for generalized Langevin equations in conuclear space*, Stoch. Proc. Appl. (to appear).

[4] Dawson, D.A., *Measure-valued Markov Processes*, Lecture Notes Math. **1541** (1993), Springer-Verlag.

[5] Dawson, D.A., Fleischmann, K. and Gorostiza, L.G, *Stable hydrodynamic limit fluctuations of a critical branching particle system in a random medium*, Ann. Probab. **17** (1989), 1083-1117.

[6] Dawson, D.A. and Gorostiza, L.G, *Generalized solutions of a class of nuclear space-valued stochastic evolution equations*, Appl. Math. Optim. **22** (1990), 241-263.

[7] Dawson, D.A. and Ivanoff, G, *Branching diffusions and random measures*, In: Advances in Probability and Related Topics **5** (1978), 61-103, Joffe, A. and Ney, P. eds., Marcel Dekker, New York.

[8] Dynkin, E.B., *Minimal excessive measures and functions*, Trans. Amer. Math. Soc. **258** (1980), 217-244.

[9] Fernández, B. and Gorostiza, L.G., *Convergence of generalized semimartingales to a continuous process*, In: Stochastic Partial Differential Equations and Applications, G. Da Prato and L. Tubaro eds., PRNMS 268, Longman Scientific and Technical, Harlow (1992), 158-164.

[10] Gorostiza, L.G., *Fluctuation theorem for a superprocess with small branching rate*, IV Simposio de Probabilidad y Procesos Estocásticos, Notas de Investigación **12**, Aportaciones Matemáticas, Sociedad Matemática Mexicana (1996), 119-127.

[11] Gorostiza, L.G. and Li, Z.-H., *Fluctuation limits of measure-valued immigration processes with small branching rate*, "Modelos Estocásticos", V Simposio de Probabilidad y Procesos Estocásticos (1998), Notas de Investigación **14**. Aportaciones Matemáticas, Sociedad Matemática Mexicana (1998), 261–268

[12] Holley, R. and Stroock, D.W., *Generalized Ornstein-Uhlenbeck processes and infinite particle branching Brownian motions*, Public. RIMS Kyoto Univ. **14** (1978), 741-788.

[13] Iscoe, I., *A weighted occupation time for a class of measure-valued branching processes*, Probab. Th. Rel. Fields **71** (1986), 85-116.

[14] Itô, K., *Distribution valued processes arising from independent Brownian motions*, Math. Zeit. **182** (1983), 17-33.

[15] Li, Z.-H., *Immigration processes associated with branching particle systems*, Adv. Appl. Probab., **30** (1998), 657-675.

[16] Li, Z.-H., *Measure-valued immigration diffusions and generalized Ornstein-Uhlenbeck diffusions*, Acta Mathematicae Applicatae Sinica **15** (1999), 310-320.

[17] Walsh, J.B., *An Introduction to Stochastic Partial Differential Equations*, Lect. Notes Math. **1180** (1986), Springer-Verlag.

DEPARTAMENTO DE MATEMÁTICAS, CENTRO DE INVESTIGACIÓN Y DE ESTUDIOS AVANZADOS, A.P. 14-740, 07000 MÉXICO D. F., MÉXICO
*E-mail address*: **gortega@servidor.unam.mx**

DEPARTMENT OF MATHEMATICS, BEIJING NORMAL UNIVERSITY, BEIJING 100875, P. R. CHINA
*E-mail address*: **lizh@email.bnu.edu.cn**

Canadian Mathematical Society
Conference Proceedings
Volume **26**, 2000

# On Phase-Transitions in Spatial Branching Systems with Interaction

## A. Greven

*Dedicated to Professor Donald A. Dawson*

ABSTRACT. We consider a spatial stochastic process with values in $(\mathbb{N})^S$, where $S$ is a countable Abelian group, for example $\mathbb{Z}^d$. The process evolves like a binary branching random walk on $S$, which is supercritical and where in addition at each site after exponential waiting times all particles are killed. In a branching random walk on $S$ particles migrate on $S$ independently according to random walks and split or die after exponential rates. Our process is a population growth model with interaction between population and environment. The different families (descendents of one original particle) do not evolve independently anymore due to the site killing. The resulting process is again critical (mean preserving) for the appropriate choice of parameters and on this case we shall focus here. We call this process the coupled branching process.

The longtime behavior of the coupled branching process exhibits a number of new phenomena and correspondingly the proofs require some new techniques. We show that in the case of a recurrent symmetrized motion the system becomes locally extinct. If the symmetrized migration is transient the system exhibits a phase transition. Let $b$ be the splitting rate of a particle into two particles, $(1-p)b$ the death rate of individual particles and $pb$ the rate of death of a whole colony (site killing), where $p \in [0,1]$. For small $p < p^{(2)}$ we see equilibria with finite second moments for the number of particles per site, while for $p > p^{(*)} \geq p^{(2)}$ the system becomes locally extinct. Between $p^{(2)}$ and $p^{(*)}$ we find non extinction and divergence of the second moments. For suitable parameters $1 > p^{(*)} > p^{(2)} > 0$ holds. The regime $p < p^{(2)}$ has been investigated in detail in Greven (1991). We refine the picture here and also discuss the case $p > p^*$ in more detail.

Due to the dependence in the evolution of different families and due to the infinite moments in the large time limit for large enough $p$, both branching random walk techniques and particle system techniques break down and hence some new methods are needed. They are based on a combination of historical processes, comparison and exchangeability techniques. These techniques allow one to get estimates of the Palm distribution (size-biasing) of $(\eta_t)_{t \geq 0}$ and replace Laplace-transform, moment calculations and coupling arguments used in the analysis of branching random walks.

2000 *Mathematics Subject Classification.* Primary 60K35, 60H15.

*Key words and phrases.* spatial branching processes; branching processes with interaction;0 phase transition; historical process; Palm measures; exponential functionals.

# 1. Introduction and results

We begin describing in 1(a) the background of the problem treated in this paper, then we describe in 1(b) rigorously the model and subsequently formulate the results in two separate subsections 1(c), 1(d). The proofs are in sections 2 - 4.

**1.1. (a) Motivation and background.** In this paper we consider a supercritical branching random walk, where in addition all particles at a colony are killed at a certain exponential rate such that the process becomes critical (mean preserving). The focus is on the interesting features of the longtime behavior and we begin by pointing out the background of this problem.

In the longtime behavior of spatial systems with countably many components interacting via migration (hence linearly) and being mean preserving one observes for many models the following dichotomy in the longtime behavior of the system: in one case convergence to laws concentrated on traps occurs and in the other case convergence to non trivial equilibria takes place, depending on whether the symmetrized migration is recurrent or transient. Examples are the voter model [see Liggett (1985)], branching random walks [Kallenberg (1977), Durrett (1979)], interacting Fisher-Wright diffusions, Feller branching diffusion [Cox and Greven (1994), Shiga (1992)], critical Ornstein-Uhlenbeck diffusions, or the Dawson-Watanabe measure valued process [Dawson (1977)].

More precisely, in the case of transient symmetrized migration the system converges to equilibria with the same intensity as the initial state and the extremal equilibria are nontrivial (positive variance) and are parameterized by the intensity. In the case of recurrent symmetrized motion we see local convergence to the traps of the system and the initial intensity is not preserved as $t \to \infty$. Traps are points in the configuration space which are invariant under the evolution.

The background of this phenomenon is a *competition* between the migration mechanism which homogenizes the system, pushing it towards a Poisson system (which are the equilibria of a system of independent random walks) and the random (diffusive) mechanism taking place in each of the different colonies, which moves the state of each component towards the traps of this random mechanism. In the transient case migration wins, in the recurrent case it looses.

In the scenario of competition described above, one is also interested in classes of systems, which are in a *different universality class*, as the above mentioned examples. Instead of either migration or random mechanism within the colony winning, now the migration and the random mechanism in each colony are qualitatively of *equal strength*, so that the outcome of the competition of the two depends on the real number, which regulates the relative intensity between these two mechanisms. In other words we obtain a system with a *phase transition* and the parameter is of the nature of a temperature, in the sense that the noise in the components of the system is proportional to that parameter.

The longtime behavior of such systems is very delicate and hence difficult to analyze and not very much is known. Essentially four systems have been considered, where partial results are known and where only the first three are particle systems:

- generalized potlatch and smoothing processes [Holley and Liggett (1981)],
- binary path contact process [Griffeath (1983)],
- coupled branching process [Greven (1991)],

- interacting diffusions process with local diffusion function $g(x) = cx^2$, the parabolic Anderson model [Shiga (1992); Cox, Greven and den Hollander (1999)].

Many interesting phenomena were exhibited in these papers, however no complete *phase diagram* has yet been rigorously established. The purpose of this paper is to complete the picture a bit more for the example of a branching random walk with interaction between different families, called coupled branching process in [**11**]. Finally we discuss the longtime behavior in the various different regimes in a more detailed fashion and refine the expected qualitative picture with some quantitative information. This leads to many quite interesting *open questions*.

The coupled branching process is also interesting as a model for branching with *interaction between families*, which generally also exhibits new longtime behavior and furthermore as population growth model, where some *interaction between the population and the environment* takes place.

In order to carry out the analysis we need to develop some new tools beyond the techniques of coupling and duality, used in the earlier mentioned systems (voter model, branching etc.). Modifications of these new techniques should also be helpful for other systems in this universality class. Examples are generalized smoothing or interacting diffusions with $g(x) = cx^2$ as diffusion function for a colony in state $x$.

The method of analysis carried out in this paper consists of embedding the system into a probability space consisting of two independent components, a *Poisson space-time trap field* and the *historical process* associated with a supercritical branching random walk, for which we can explicitly represent both the *Palm distribution* and asymptotically certain *conditional distributions*. This will allow us to get lower and upper bounds on a measure, which asymptotically approximates the Palm distribution of the coupled branching process. Palm distribution here means size-biasing with respect to the number of particles at a fixed site at a given time. The key then is to decompose the Palm distribution of the coupled branching process into components according to the fate of a typical particle of the supercritical branching random walk. With the help of *exchangeability* and *comparison* arguments, it is then possible to derive an extinction-survival criterion. A further element of the analysis is the asymptotic evaluation of exponential functionals of random walks, most of which is however deferred to [**2**].

**1.2. (b) The model.** We shall construct a Markov process with state space $\mathcal{E} \subseteq (\mathbb{N})^{\mathbb{Z}^d}$. We denote the process $(\eta_t)_{t\geq 0}$ and we call it the *coupled branching process*. We interpret a configuration $\eta \in (\mathbb{N})^{\mathbb{Z}^d}$, $\eta = (\eta(x))_{x\in\mathbb{Z}^d}$ as particle numbers, i.e. $\eta(x)$ is the number of particles located at $x \in \mathbb{Z}^d$. Passing to a subset $\mathcal{E}$ is needed, to exclude influence from infinity in this model with unbounded and countably many components. The route we follow here to choose a state space, has first been used by Liggett and Spitzer (1981) in the study of potlatch and smoothing processes (see [**19**]).

As parameter we have positive real numbers $b, m$, a number $p \in [0, 1]$ and a transition kernel $a(\cdot, \cdot)$ on $\mathbb{Z}^d \times \mathbb{Z}^d$, which satisfies the following homogeneity and irreducibility condition:

(1.1) $$a(x, y) = a(0, y - x) \qquad \forall x, y \in \mathbb{Z}^d,$$

(1.2)           $$\sum_{n=0}^{\infty} a^{(n)}(x,y) + a^{(n)}(y,x) > 0 \qquad \forall x, y \in \mathbb{Z}^d.$$

The dynamics of the process involves the following transitions, which occur independently of each other:

- Every particle performs independent of all other particles a continuous time random walk on $\mathbb{Z}^d$ with jump rate $m$ and transition kernel $a(x,y)$ for the jumps.
- Every particle splits at exponential rate $b$ into two particles. All particles act independently.
- Every particle dies at exponential rate $(1-p)b$. All particles act independently.
- All particles at a site $x$ die at exponential rate $pb$, all sites act independently.

As initial configuration for the coupled branching process we often choose random configurations $\eta_0$, whose law $\mu$ satisfies:

- $\mu$ is translation invariant,
- $E_\mu[\eta(x)] = \theta < \infty$,
- $\sum_y a_t(x,y)\eta(y) \xrightarrow[t\to\infty]{} \theta$, in $L_1(\mu)$.

The collection of all these laws on $(\mathbb{N})^{\mathbb{Z}^d}$ we denote:

(1.3)                                    $\mathcal{T}_\theta.$

In the rest of this subsection we give a rigorous, formal description of the process.

According to the heuristic description of the evolution given above, we write down the pregenerator acting on functions $f$ on $(\mathbb{N})^{\mathbb{Z}^d}$ which are bounded and depend only on finitely many components:

$$(Gf)(\eta) = \sum_x \left\{ m \left[ \sum_y a(x,y))\eta(x)[f(\eta + \delta_y - \delta_x) - f(\eta)] \right] \right.$$

(1.4)           $$+ b\eta(x)[f(\eta + \delta_x) - f(\eta)] + (1-p)b\eta(x)[f(\eta - \delta_x) - f(\eta)]$$

$$\left. + pb[f(\eta - \eta(x)\delta_x) - f(\eta)] \right\}.$$

Next we specify the state space $\mathcal{E}$ of the Markov process as subset of $(\mathbb{N})^{\mathbb{Z}^d}$. Define for $M > 1$

(1.5)           $$\gamma(x) = \sum_{n=0}^{\infty} M^{-n} a^{(n)}(x,y)\beta(y), \quad ||\eta|| = \sum_x \gamma(x)\eta(x),$$

where $\beta(\cdot)$ is a given strictly positive and summable function on $\mathbb{Z}^d$. Then define (we suppress the dependence on $\beta$ in the notation):

(1.6)           $$\mathcal{E} = \left\{ \eta \in (\mathbb{N})^{\mathbb{Z}^d} \,\Big|\, \sum_x \eta(x)\gamma(x) < \infty \right\}.$$

Note that the (measure theoretic) support of a law $\mu \in \mathcal{T}_\theta$ is contained in $\mathcal{E}$ for *every* $\beta$, which we are allowed to choose.

In [11] it has been shown, that the pregenerator $G$ defines a unique Markov process on $\mathcal{E}$, such that the corresponding semigroup $(S_t)_{t\geq 0}$ on positive measurable functions on $\mathcal{E}$ has the following property. Restrict the semigroup to the measure determining class $\mathcal{L}$ of all Lipschitz functions $f \geq 0$ on $\mathcal{E}$, i.e. functions satisfying

$$(1.7) \qquad |f(\eta) - f(\eta')| \leq L(f)\|\eta - \eta'\|, \quad 0 \leq L(f) < \infty.$$

The operator $G$ can be extended to $\mathcal{L}$ by the expression (1.4), (i.e. $|Gf| < \infty$) and $S_t$ satisfies for all $t > 0$ the relations

$$(1.8) \qquad S_t(\mathcal{L}) \subseteq \mathcal{L} \quad \text{and} \quad S_t f = f + \int_0^t GS_u(f)\,du, \qquad \forall f \in \mathcal{L}.$$

We call the Markov process on $\mathcal{E}$ corresponding to $(S_t)_{t\geq 0}$ acting on $\mathcal{L}$ the *coupled branching process*.

REMARK 1.1. Above and in the sequel, $\mathbb{Z}^d$ can be replaced by any countable abelian group.

**1.3. (c) Results on the longtime behavior.** The next point to discuss is the longtime behavior of the process $(\eta_t)_{t\geq 0}$. We proceed in four steps namely we present the background, define some relevant quantities and formulate the main results in two groups, Theorems 1-3 and Proposition 1.1.

*(i)    Background*    Important for the longtime behavior of the coupled branching process is that it is *critical* in the sense, that the mean of a single component is a preserved quantity.

LEMMA 1.2. *[Greven (1991)]    If $\mu \in \mathcal{T}_\theta$ then for $\mathcal{L}(\eta_0) = \mu$,*

$$(1.9) \quad \mathcal{L}[\eta_t] \in \mathcal{T}_\theta, \quad \forall t > 0, \quad \text{in particular } E[\eta_t(x)] = \theta \qquad \forall t \geq 0,\ x \in \mathbb{Z}^d. \quad \square$$

This means now in particular that if we consider this system on a finite group, for example the torus $[-N, N]^d$, then $(\sum_x \eta_t(x))_{t\geq 0}$ is a nonnegative martingale and hence converges. Since $\underline{0}$ is the only trap, the finite system becomes extinct as $t \to \infty$. For this reason the system forms with small probabilities very large configurations and with probability close to one the empty configuration.

In addition if we put $m = 0$ we find that even the infinite system on $\mathbb{Z}^d$ becomes extinct, while if we put $b = 0$ then the infinite system will approach a Poisson system with intensity $\theta$, the equilibria of pure random walk systems. Hence we see that the branching and site-killing part of the mechanism pushes the system towards local extinction, while the migration pushes it to a decent equilibrium state, which preserves the initial intensity. As a consequence the longtime behavior depends on the outcome of the competition of these two groups of mechanism.

For the parameter value $p = 0$ we obtain a classical branching random walk and then it is well known that the outcome of the competition between migration and branching depends only on whether the symmetrized migration kernel is recurrent (branching wins) or transient (migration wins).

However if $p > 0$ then *new phenomena occur*, namely by replacing the individual deaths in a branching random walk by the joint killing of all individuals at a site, the fluctuations in the system get much larger. Hence the following tendency of the system is strengthened: The system forms (at random locations) high peaks, namely at sites where no site killing occurred for some time and on the other hand $(E[\eta_t(x)] \equiv \theta!)$ large regions are formed without any particles. The latter are regions with enough site killing taking place. [This forming of peaks resembles the intermittency effects studied by Gärtner and den Hollander (1999)]. The height of the peaks and their frequency in space depend on the size of the parameter $p$. Hence we expect to see, in the case where $\widehat{a}$ is transient, different behavior depending on the actual value of $p$. To clarify this point is the content of the remaining three paragraphs of subsection 1(c), in particular to exhibit a *phase-diagram in the parameter* $p$ for the longtime behavior of the system.

*(ii)    Preparation*    We introduce here the quantity, which will turn out to determine the longtime behavior of the coupled branching process. Note at this point, that the influence of one component onto another at a later time is transmitted by random walks. If several random walks visit the same site at the same time, they are jointly exposed to the risk of site-killing. Therefore the fluctuations in the state of single components is related to occupation time functionals of random walks.

Let $(X_s^{(1)})_{s \geq 0}$ and $(X_s^{(2)})_{s \geq 0}$ be two independent copies of a random walk with transition rate $\bar{a}(x,y) = a(y,x)$ and starting in 0. We define for $p \in (0,1]$ the random variables (living on the probability space generated by the second random walk):

$$(1.10) \qquad V_t^p = E\left[ \exp\left( pb \int_0^t \mathrm{1\!I}(X_s^{(1)} = X_s^{(2)}) ds \right) \Big| (X_s^{(2)})_{s \geq 0} \right],$$

$$(1.11) \qquad V_\infty^p = \uparrow - \lim_{t \to \infty} V_t^p.$$

The first point is to observe (see section 3) the following 0-1 law:

LEMMA 1.3.

$$(1.12) \qquad P(V_\infty^p < \infty) = 0 \quad or\, 1. \quad \square$$

More properties of the $p$-dependence of $V_t^p$ will be given in paragraph (iv) of this subsection.

With these objects we are able to formulate below in three theorems the main results for the following three possible regimes:

$$(1.13) \qquad \begin{aligned} &E[V_\infty^p] < \infty, \\ &E[V_\infty^p] = \infty \quad \text{and} \quad V_\infty^p < \infty \quad \text{a.s.}, \\ &V_\infty^p = \infty \quad \text{a.s.} \end{aligned}$$

There are two reasons why $V_\infty^p$ can be $+\infty$. Let $\widehat{V}_\infty^p$ denote the random variable

$$(1.14) \qquad \exp\left( pb \int_0^\infty \mathrm{1\!I}\left( X_s^{(1)} = X_s^{(2)} \right) ds \right).$$

Then the regime $V_\infty^p = \infty$ splits into the two cases: $\widehat{V}_\infty^p = +\infty$ *and* $V_\infty^p = +\infty$ on the one hand and the case $\widehat{V}_\infty^p < \infty$ *and* $V_\infty^p = +\infty$ on the other hand.

Next note that the case $\widehat{V}_\infty^p = +\infty$ a.s. can be characterized easily in terms of the symmetrized migration kernel, namely:

$$(1.15) \qquad \widehat{V}_\infty^p = +\infty \text{ a.s.} \iff \widehat{a}(\cdot,\cdot) \text{ is recurrent.}$$

This means that below when we treat the three cases of (1.13) we are dealing in Theorem 1.4 and 1.5 with the case $\widehat{a}(\cdot,\cdot)$ transient only, which exhibits a more complex behavior for our process, while Theorem 1.7 applies to both cases.

*(iii)   Identification of three regimes in the longtime behavior*   Consider the behavior of the coupled branching process as $t \to \infty$. We use the terminology, that a process is called *persistent*, if the intensity (i.e. $E[\eta(x)]$) is preserved in the limit $t \to \infty$ and we say *local extinction* occurs, if all weak limit points as $t \to \infty$ have intensity 0.

The three cases of (1.13) correspond to two persistent regimes and a third regime of local extinction, each of which is presented next in a separate theorem. The persistent regimes are: equilibria with finite second moments and the regime of equilibria with infinite second moments. The significance of the latter is explained in 1(d).

THEOREM 1.4. *[Greven (1991)]*
*Assume that $E[V_\infty^p] < \infty$. Then for $\mathcal{L}[\eta_0] \in \mathcal{T}_\theta$ the process is persistent:*

$$(1.16) \qquad \mathcal{L}[\eta_t] \underset{t\to\infty}{\Longrightarrow} \nu_\theta,$$

*and $\nu_\theta$ is translation invariant and satisfies:*

$$(1.17) \qquad \nu_\theta \text{ is an extremal invariant measure of the process}$$

$$(1.18) \qquad E_{\nu_\theta}[\eta(x)] = \theta, \quad E_{\nu_\theta}[\eta^2(x)] < \infty,$$

$$(1.19) \qquad E_{\nu_\theta}[\eta(x)\eta(y)] \underset{|x-y|\to\infty}{\longrightarrow} \theta^2 \quad \text{and } \nu_\theta \text{ is spatially mixing.}$$

*All extremal invariant measures of the process are given by $\{\nu_\theta, \theta \geq 0\}$.*   □

Note that the properties of $\nu_\theta$ above imply in particular that $\nu_\theta \in \mathcal{T}_\theta$.

THEOREM 1.5. *Assume $E[V_\infty^p] = +\infty$, but $V_\infty^{\bar{p}} < \infty$ a.s. for some $\bar{p} > p$. Choose $\mathcal{L}[\eta_0] = \mathcal{H}_\theta$, where $\mathcal{H}_\theta$ is the law of a Poisson system with intensity $\theta$ (i.i.d. - Poisson($\theta$)-marginals). Then the process is persistent, more precisely:*

$$(1.20) \qquad \mathcal{L}[\eta_t] \underset{t\to\infty}{\Longrightarrow} \nu_\theta,$$

*the law $\nu_\theta$ is translation invariant and has in addition the following properties:*

$$(1.21) \qquad \nu_\theta \text{ is an extremal invariant measure of the process,}$$

$$(1.22) \qquad E_{\nu_\theta}[\eta(x)] = \theta, \quad \forall x \in \mathbb{Z}^d,$$

$$(1.23) \qquad E_{\nu_\theta}[\eta^2(x)] = +\infty, \quad \forall x \in \mathbb{Z}^d,$$

$$(1.24) \qquad \nu_\theta \text{ is spatially mixing.}$$

Again we know from above properties that in particular $\nu_\theta \in \mathcal{T}_\theta$.

REMARK 1.6. One would conjecture that in fact the $\nu_\theta$ are exactly *all* the extremal invariant measures of the process. However due to the infinite second moment, the coupling techniques used for Theorem 1.4 do not work.

THEOREM 1.7. *If $V_\infty^p = +\infty$ a.s. and $\mathcal{L}[\eta_0] \in \mathcal{T}_\theta$ with $\theta \geq 0$ then the process goes to local extinction:*

$$(1.25) \qquad\qquad \mathcal{L}[\eta_t] \underset{t \to \infty}{\Longrightarrow} \delta_{\underline{0}}. \quad \square$$

Note that by (1.15) this implies in particular, that for $\hat{a}$ recurrent, the system goes to local extinction.

*(iv)    A phase diagram for the transient case*   As a consequence of the above three theorems we will see that for the case $\hat{a}$ transient the system undergoes a phase transition in the parameter $p$ with three distinct phases. We need some preparation to get sufficient information on the functionals $V_\infty^p$ and $E[V_\infty^p]$.

LEMMA 1.8. *Denote by $\prec\!\!\!\prec$ the stochastic order. Then*

$$(1.26) \qquad\qquad V_\infty^p \prec\!\!\!\prec V_\infty^{p'} \quad if\, p < p'.$$

*As a consequence there exist critical values $p^*, p^{(2)}$ namely (recall (1.12)):*

$$(1.27) \qquad\qquad p^{(*)} = \inf(p|P(V_\infty^p = \infty) = 1)$$

$$(1.28) \qquad\qquad p^{(2)} = \inf(p|E[V_\infty^p] = +\infty).$$

*Furthermore:*

$$(1.29) \qquad\qquad p^{(*)} \geq p^{(2)}.$$

REMARK 1.9. In [**11**] it was shown that

$$(1.30) \qquad\qquad p^{(2)} = m\left((b\widehat{G}(0,0))\right)^{-1},$$

where $\widehat{G}$ is the Greens function of the kernel $\hat{a}$, with $\hat{a}(x,y) = \frac{1}{2}(a(x,y) + a(y,x))$.
    The main point is to show that:

PROPOSITION 1.10. *For suitable $m, b, a(\cdot, \cdot)$ one has the relation:*

$$(1.31) \qquad\qquad 1 > p^{(*)} > p^{(2)} > 0. \quad \square$$

This implies that we find for suitable values of $m, b, p$ and kernels $\hat{a}$ *three distinct regimes* (phases) for the longtime behavior of the system in the case of transient symmetrized motion instead of only one, as in all the examples like voter model, branching random walk etc. mentioned in the beginning of subsection 1(a). In particular we obtain between the two regimes, local extinction and persistence with stabilization in an equilibrium with finite second moments and spatially decaying correlations, a new third and intermediate phase, namely an equilibrium with infinite second moments. In subsection (d) we discuss the question why infinite *second* moments play a special role and what happens with higher moments.

It is an open problem whether always $V_\infty^{p*} = +\infty$ if $p^* \in (0, 1)$, in which case Theorems 1.4 - 1.7 give a complete picture of the longtime behavior. This is treated in [**2**].

**1.4. (d) Outlook on results concerning a finer analysis of the longtime behavior.** In this subsection we discuss in a more detailed fashion properties of the three different phases described in the subsection 1(c).

(i)    *A more detailed picture of the stable phase $p < p^{(2)}$.*    In the paper [11] the case $p < p^{(2)}$ was analyzed in more detail. One finds a sequence

$$(1.32) \qquad\qquad p^{(n)} \in (0, p^{(2)}), \quad n \geq 3,$$

such that for $p \geq p^{(n)}$ the equilibrium measure $\nu_\theta$ has infinite $n$-th moment, while for $p < p^{(n)}$ it has finite $n$-th moment. The sequence $p^{(n)}$ has the property that for $\hat{a}$ transient

$$(1.33) \qquad\qquad 0 < p^{(n)} \leq p^{(3)} < p^{(2)} \quad \text{and } p^{(n)} \downarrow 0 \quad \text{as } n \to \infty.$$

This means in particular that for $p > 0$ (since $p^{(n)} \to 0$) the extremal equilibrium of the process never has finite moments of arbitrary order, as is the case in the pure branching system, which is obtained from the coupled branching process by putting $p = 0$.

(ii)    *The intermediate phase $p \in [p^{(2)}, p^*)$*    To highlight the significance of the regime, where $p < p^*$ but $p \geq p^{(2)}$ focus on $S = \mathbb{Z}^d$ and note that for $p < p^{(2)}$ one expects that under the equilibrium measure and under the assumption below, which guarantees (observation by G. Lawler proved in [23]), that the Greens function $\widehat{G}(x, y)$ of $\hat{a}$ behaves asymptotically as $|x - y| \to \infty$ as in the case of Brownian motion, ($|x| = \max_{i=1,\dots,d}(|x_i|)$ for $x \in \mathbb{Z}^d$):

$$(1.34) \qquad \begin{array}{ll} \sum_{x \in \mathbb{Z}^d} a(0, x)|x|^2 < \infty & \text{for} \quad d = 3 \\ \sum_{x \in \mathbb{Z}^d} a(0, x)|x|^2 \log|x| < \infty, & \text{for} \quad d = 4 \\ \sum_{x \in \mathbb{Z}^d} a(0, x)|x|^{2+\delta} < \infty, & \text{with} \quad \delta = d - 2, \text{ for } d > 4 \end{array}$$

one has the following CLT:

$$(1.35) \qquad\qquad \frac{1}{f[(2N+1)^{\frac{d}{2}}]} \sum_{|x| \leq N} (\eta(x) - \theta) \underset{N \to \infty}{\Longrightarrow} \mathcal{N}(0, \sigma^2).$$

The function $f$ increases more rapidly than linear and depends on the Greens function of $\hat{a}$. The function $f$ can be given explicitly as:

$$(1.36) \qquad\qquad f(r) = r^\alpha, \quad \alpha = \frac{d+2}{2}, \quad \text{for } d \geq 3.$$

Similar behavior was found in the voter model by Bramson and Griffeath (1979) for $d = 3$ and with different methods in Major (1980) and by Zähle (1999) also for $d \geq 4$. A related picture has been verified for branching systems recently by Dawson, Gorostiza and Wakolbinger (1999).

On the other hand for $p > p^{(2)}$ we enter the realm of non Gaussian limits for the expression in (1.35) and the rescaling function $f$ changes and depends in a subtle way on $p$ and finer properties of $a(\cdot, \cdot)$, since the fluctuations of the equilibrium state are of a quite different nature. More information is in subsection 4(b).

(iii)    *The regime of local extinction*    In this regime the process forms in rare spots very high peaks. In the case of classical branching models (that is $p = 0$) in the regime of $\hat{a}$ recurrent this cluster formation has been studied extensively [see Fleischman (1978), Dawson and Greven (1996), Klenke (1997), Winter (1999)]. Roughly speaking the peaks grow at the order of magnitude given by

$$(1.37) \qquad\qquad G_t = \int\limits_0^t \widehat{a}_s(0,0)ds$$

and the regions covered by peaks of that size, grow at rates which are highly depending on the properties of $G_t$ and the random walk generated by $\widehat{a}(\cdot,\cdot)$. In particular we can conclude that the peaks grow typically in dimension $d = 2$ (for simple random walk for example) like $\sqrt{t}$ and in dimension $d = 2$ like $\log t$. Also the extension of a cluster in space is well understood, namely in $d = 1$, we have clusters with diameter of order $\sqrt{t}$, where as in $d = 2$ cluster diameters are of order $t^{\frac{\alpha}{2}}$, where $\alpha \in [0, 1]$ is a random variable, a phenomenon called diffusive clustering.

In the case where $p > 0$ we will see clusters growing at much more rapid rates of growth and they will be of much smaller extension in space in the case where $\widehat{a}$ is transient. In the regime $\widehat{a}$ transient and $p > p^*$ the cluster height will be related to $V_t^p$ and its growth and the cluster size of the related peaks will be finite (compare [**12**]. In section 4 we present some bounds and heuristic ideas.

## 2. Extinction and survival criteria

In this chapter we prove Theorems 1.5 and 1.7. Despite the independence properties missing in $(\eta_t)_{t \geq 0}$ but present in classical branching random walk and exploited in [**17**] to prove results on the longtime behavior, it is possible in the regime $p < p^{(2)}$ to use moment calculations combined with coupling techniques to prove Theorem 1.4, which was carried out in [**11**]. For $p \geq p^{(2)}$ however the second moments $E[\eta_t^2(x)]$ diverge as $t \to \infty$ and we cannot work with moments and correlations of the limiting states anymore. One consequence of this property is, that the coupling techniques of [**11**] break down, since they require the establishment of certain mixing properties of the limit points of $\mathcal{L}(\eta_t)$ as $t \to \infty$, usually obtained by covariance estimates. For that reason we have to proceed differently for $p \in [p^{(2)}, 1]$.

In order to control the longtime behavior of the process $(\eta_t)_{t \geq 0}$ for $p \geq p^{(2)}$, we calculate lower and upper bounds to asymptotic approximations of the Palm-distribution of $(\eta_t)_{t \geq 0}$. To this end we give in subsection 2(a) an explicit construction of this process as functional of two independent components, the *historical process* of a *supercritical branching random walk* and a *Poisson point process* on $\mathbb{R}^+ \times \mathbb{Z}^d$.

This construction will allow us to write down criteria for extinction and survival (subsection 2(b)), using the concept of Palm measures (size-biasing) applied successively twice. Then this criteria can be rewritten, using another object simpler to calculate with. Next the criteria can be expressed, via an explicit representation of the Palm measure for classical branching random walks, approximately for $t \to \infty$ in terms of exponential functionals of random walks (subsection 2(c)). With these results and comparison with simpler systems, we obtain finally in subsection 2(d) manageable extinction and survival criteria in terms of random walks and which are used in subsection 2(e) to prove Theorem 1.5 and 1.7. The main results of 2(b) - 2(d) are each formulated as Proposition.

**2.1. (a) An embedding of $(\eta_t)_{t \geq 0}$ in a historical super critical branching random walk.** The key point on deciding whether the process becomes extinct (locally) or converges to an equilibrium with positive intensity is, to embed the process in a richer structure, which contains information on the history and family relations of the individuals. In order to make this precise, we construct first in (i) a particular version of the underlying supercritical branching random walk containing this information and then based on this we construct in (ii) a particular version of $(\eta_t)_{t \geq 0}$ by thinning, using a space-time Poisson field.

*(i)    Historical process*

Consider the dynamics of a *supercritical branching random walk* given by the transitions below, which occur independently:

- Every particle migrates according to a continuous time random walk with exponential jump rate $m$ and transition kernel $a(\cdot, \cdot)$, all particles act independently
- At exponential rate $b$ a particle splits into 2 particles, all particles act independently
- At exponential rate $(1 - p)b$ a particle dies, all particles act independently.

We call the resulting stochastic process $(\xi_t)_{t \geq 0}$ and below we construct a specific version. A construction of this process with the pregenerator can be given along the lines of subsection 1(b).

For this branching random walk we can explicitly construct below the *historical process*, namely with every particle alive at a *fixed* time $t > 0$, we associate the path leading back to the ancestor at time 0. Beyond $t$ we continue the path as a constant. We associate with the whole collection of random paths starting from the different points $x \in \mathbb{Z}^d$ backwards the empirical measure, assigning mass 1 to the path of every surviving (time $t$) individual. In other words by now varying $t$, we introduce a process $(\tilde{\xi}_t)_{t \geq 0}$ containing more information on the *history* of individuals. The state space is the space of locally finite integer-valued measures on $D([0, \infty), \mathbb{Z}^d)$, the set of right continuous maps $[0, \infty) \to \mathbb{Z}^d$ with left limits.

We can construct this *historical process* $(\tilde{\xi}_t)_{t \geq 0}$ explicitly as follows: Construct for every potential initial particle a continuous time supercritical binary branching process $(N_t^{x,k})_{t \geq 0}$, $x \in \mathbb{Z}^d$, $k \in \mathbb{N}$ with state space $\mathcal{G}$ of binary trees, by just leaving out the migration in the mechanism described above. Then we take a collection of independent random walks $((X_t^{y,k})_{t \geq 0})_{y \in \mathbb{Z}^d, k \in \mathbb{N}}$ (jump rate $m$, transition probabilities $a(x, y)$ and starting point $y$), which is independent of the previous collection of branching processes. With both ingredients we construct below $(\tilde{\xi}_t)_{t \geq 0}$ on the probability space generated by:

$$(2.1) \qquad \mathcal{X} = \left\{ \left(N_t^{x,k}\right)_{t \geq 0}, \left(X_t^{y,j}\right)_{t \geq 0}; \quad x, y \in \mathbb{Z}^d; \quad k, j \in \mathbb{N} \right\}.$$

Namely we start the system in a random initial configuration $\eta_0 \in \mathcal{E}$ independent of (2.1) and associate with every initial particle a branching process from the above collection (choosing $k = 1, \ldots, \eta_0(x)$ etc.) and a random walk, which the particle will follow until it dies. If a birth occurs the new particle uses the next not yet used random walk at its location, which it follows till it dies. Now associate with every path of an individual alive at the fixed time $t$ the $\delta$ measure on that path and then sum over all these paths. This results in an integer-valued and locally finite measure on $D([0,t], \mathbb{Z}^d)$. By continuing every path beyond $t$ as the constant path, we can use $D([0,\infty), \mathbb{Z}^d)$ as state space of this new process. The resulting random process is called $(\tilde{\xi}_t)_{t \geq 0}$.

REMARK 2.1. It is a standard fact, that we can always represent the random measure $\tilde{\xi}$ by collections of random path associated with the sites, which are for each site *exchangeable* (representing the surviving individuals). We label the "individuals" obtained this way by $(i, x)$; $i \in \mathbb{N}$, $x \in \mathbb{Z}^d$. This is very important later on.

We can obtain a version of the process $(\xi_t)_{t \geq 0}$ by considering the functionals of $\tilde{\xi}_t$ which are defined by taking the number of particles alive at time $t$ at site $x$, which means define

$$(2.2) \qquad \xi_t(x) = \tilde{\xi}_t \left( \{ Y \in D([0,\infty), \mathbb{Z}^d) | Y(t) = x \} \right).$$

*(ii)    Embedding*
The coupled branching process will be embedded into the process $(\tilde{\xi}_t)_{t \geq 0}$ constructed on the probability space (2.1). We consider for this purpose an enrichment of this probability space, which allows to define a *thinning* of $(\tilde{\xi}_t)_{t \geq 0}$ at each time $t$ as follows.

Construct a *Poisson point process* on $\mathbb{Z}^d \times \mathbb{R}^+$ with intensity measure $(pb)(n \otimes \lambda)(\cdot)$, with $n$ the counting measure on $\mathbb{Z}^d$ and $\lambda$ the Lebesgue measure on $\mathbb{R}^+$. The point process defines (by putting $\delta$-masses on the points) an atomic random measure, called the site-killing field, which is denoted:

$$(2.3) \qquad \Pi \text{ on } \mathbb{Z}^d \times \mathbb{R}^+.$$

With these ingredients we define the basic probability space $(\Omega, \mathcal{A}, P)$, by combining $\mathcal{X}$ of (2.1) and $\Pi$ of (2.3), where the basic law $P$ is $\mathcal{L}(\mathcal{X}) \otimes \mathcal{L}(\Pi)$.

On this probability space we define now a *version of the coupled branching process*, by considering the following functional. For every particle $(i, x)$ alive at time $t$ and sitting in the site $x$, we consider the following random variable. Let $(Y_s^{i,x})_{s \in [0,t]}$ be the path which leads from the particle $i$ at site $x$ and alive at time $t$ back to its ancestor at time 0. Define the $\{0, 1\}$-random variables:

$$(2.4) \qquad Z_t^{i,x} = \Pi \left[ \Pi(\{Y_s^i, s), \ s \in [0,t]\}) > 0 \right].$$

If we put (recall (2.2)):

$$(2.5) \qquad \tilde{\eta}_t(x) = \sum_{i=1}^{\xi_t(x)} Z_t^{i,x},$$

then $(\tilde{\eta}_t)_{t \geq 0}$ is a version of the coupled branching process:

LEMMA 2.2. *If the starting configuration for the process $(\tilde{\xi}_t)_{t \geq 0}$ is concentrated on constant path and satisfies $\{\xi_0(\{Y \in D([0,\infty), \mathbb{Z}^d) | Y(0) = x\})\}_{x \in \mathbb{Z}^d} = \eta_0$, then*

$$(2.6) \qquad \mathcal{L}[(\tilde{\eta}_t)_{t \geq 0}] = \mathcal{L}[(\eta_t)_{t \geq 0}]. \quad \square$$

PROOF. We consider $f \in \mathcal{L}$ and verify that the expectations $E[f(\tilde{\eta}_t)]$ satisfy the backward equation (1.8). We omit the details.                               $\square$

**2.2. (b) The law of the process seen from a typical particle and persistence-extinction criteria.** The key to extinction criteria is to consider auxiliary configurations $\widehat{\eta}_t^*$, which is obtained, by conditioning in a supercritical branching process under the Palm distribution on a particular individual to survive site-killing and another auxiliary configuration $\widehat{\eta}_t^{**}$ describing asymptotically the process seen from a *typical surviving particle* of $\eta_t$, which is obtained via the Palm distribution. In (i) we construct the new objects. Later in (ii) we shall relate these objects to prove the key criterion and the main result of this subsection 2(b) namely Proposition 2.3 below.

### (i)    Construction

The object $\widehat{\eta}_t^*$ we introduce first. For the explicitly constructed process $(\widetilde{\xi}_t)_{t\geq 0}$ we can consider the following functional. We pick from the total population alive at time $t$ a particle at random called the *typical particle* of the supercritical branching random walk and consider the spatial configuration generated as seen from this particle as the new origin (a rigorous description follows in (2.12) below). This procedure results in a random configuration

$$(2.7) \qquad\qquad \xi_t^* \in (\mathbb{N})^{\mathbb{Z}^d}.$$

This object induces then by the embedding and the thinning mechanism of the previous subsection (see (2.4) and (2.5)) a random configuration

$$(2.8) \qquad\qquad \eta_t^* \in (\mathbb{N})^{\mathbb{Z}^d}.$$

If we take the law of this object conditioned on the typical surviving particle at time $t$ at site 0 of the process $\widetilde{\xi}_t$, denoted $(i^*, 0)$, to satisfy (see (2.12)):

$$(2.9) \qquad\qquad Z^{i^*,0} = 1,$$

then we obtain the law of a random configuration

$$(2.10) \qquad\qquad \widehat{\eta}_t^* \in (\mathbb{N})^{\mathbb{Z}^d} \quad \text{with} \quad \widehat{\eta}_t^*(0) \geq 1 \quad \text{a.s.}$$

The notion of a randomly chosen particle used above is made rigorous by the notion of the Palm distribution. Let $P_T$ be the law of the historical process $(\widetilde{\xi}_t)_{t\in[0,T]}$. Then we define the law of $\widetilde{\xi}_t^*$ as

$$(2.11) \qquad\qquad (P_T)_x,$$

by passing to the *size-biased* random variable as follows (here $\mathcal{M}$ denotes locally finite measures):

$$(2.12)$$
$$(P_T)_x[A] = \theta^{-1}e^{-pbt}\int \xi_t(x)(w)\,\mathbb{1}_A(w)P_T(dw), \quad A \in \mathcal{B}(\mathcal{M}(D([0,\infty), \mathbb{Z}^d)))),$$

where of course $\theta e^{pbt} = E[\xi_t(x)]$. Recall that $P_T$ is translation invariant and hence $(P_T)_x = (P_T)_0$. Then we pick $x \in \mathbb{Z}^d$ for example $x = 0$ and we consider $(P_T)_0$ instead of $P_T$ as the basic distribution on our probability space and consequently use $\mathcal{L}((P_T)_0) \otimes \mathcal{L}(\Pi)$ in (2.4). We obtain this way the laws of the functionals $\xi_t^*, \eta_t^*, \widehat{\eta}_t^*$ for the typical particle.

In this new probability space based on $(P_T)_0$ we can consider the Palm distribution with respect to $\eta_t^*$, that means we size bias again (recall (2.12)), but now with respect to $\eta_t^*(0)$. Then we call the configuration resulting from this new law by

(2.13)                           $\widehat{\eta}_t^{**} \in (\mathbb{N})^{\mathbb{Z}^d}$   with $\widehat{\eta}_t^{**}(0) \geq 1$.

   *(ii)    Criterion*

In order to obtain extinction and survival criteria for $(\eta_t)_{t\geq 0}$ we use the following idea. Recall that the process $(\eta_t)_{t\geq 0}$ preserves the mean particle number per site. However $(\eta_t(0))_{t\geq 0}$ is in general *not uniformly integrable*, in fact this distinguishes persistence from local extinction. Furthermore it is not possible to decide whether or not $(\eta_t(0))_{t\geq 0}$ is uniformly integrable by analyzing the second moments, this gives only for $p = 0$ (classical branching random walk) the correct answer. In addition we cannot calculate $1 + \varepsilon$ moments of $\eta_t(0)$ with $\varepsilon \in (0, 1)$. In other words we need something more adapted to the special structure of $(\eta_t)_{t\geq 0}$.

   Such a tool is the Palm measure of $(\eta_t)_{t\geq 0}$. This we cannot calculate explicitly, but we can get approximate lower and upper bounds. Namely since $(\xi_t)_{t\geq 0}$ satisfies for $p > 0$ a law of large numbers in $L_1$, we can use in this case as approximations the random variable $\widehat{\eta}_t^{**}(0)$, which stays stochastically bounded for persistence of $(\eta_t)_{t\geq 0}$ and diverges in case of local extinction of $(\eta_t)_{t\geq 0}$.

   Therefore we will use stochastic bounds for $\widehat{\eta}_t^{**}(0)$ by $\mathbb{N}$-valued random variables $X(\omega, \omega')$ where the variable conditioned on $\omega$ has a finite mean. Then typically of course we use this when $X(\omega, \omega')$ itself does *not* have a mean. The key idea is to show for the $\widehat{\phantom{x}}$-quantities, that either $\widehat{\eta}_t^{**}(0)$ stays stochastically bounded or $\widehat{\eta}_t^{**}(0)$ diverges. We do this by trying to decompose $\{\mathcal{L}(\widehat{\eta}_t^{**}(0))_{t\geq 0}\}$ into components, which are either all uniformly integrable or if that fails, show that total loss of the mean occurs in the limit. The trick then is to relate the behavior of these variables $\widehat{\eta}_t^{**}(0)$ to the one of $\widehat{\eta}_t^*(0)$, which can be calculated. Here we use the special structure of the random variables $\eta_t(0)$, $\eta_t^*(0)$, $\widehat{\eta}_t^*(0)$, $\widehat{\eta}_t^{**}(0)$.

   We can prove with this technique the main result of this subsection:

   PROPOSITION 2.3. *Assume that the initial law $\mu$ of $(\xi_t)_{t\geq 0}$ satisfies $\mu \in \mathcal{T}_\theta$ and $E_\mu[\xi_0^2(0)] < \infty$. Then:*

(2.14)                     $\mathcal{L}[\widehat{\eta}_t^*(0)] \underset{t\to\infty}{\Longrightarrow} \delta_\infty$   *implies*   $\mathcal{L}[\eta_t] \underset{t\to\infty}{\Longrightarrow} \delta_0.$

*The property*

(2.15) $\mathcal{L}[\widehat{\eta}_t^*(0)]$   *is stochastically bounded by a random variable $X(\omega, \omega')$*

                    *with finite mean over $\omega'$ for almost surely all $\omega$,*

*where $\omega$ abbreviates $(Y_s^{i^*,0})_{s\geq 0}$ and $\omega'$ abbreviates $\Pi$, implies that*

(2.16)                          $\underset{t\to\infty}{\liminf}(P[\eta_t(0) > 0]) > 0$

*and furthermore all weak limit points of $\mathcal{L}(\eta_t)$ as $t \to \infty$ have mean $\theta$.*                    □

   PROOF OF PROPOSITION 2.3. We begin with establishing in step 1 a law of large numbers and in (the key-)step 2 we rewrite the assertion before we complete the proof in step 3.                                                         □

   STEP 1    A key property of $(\xi_t)_{t\geq 0}$ needed in step 2 and for concluding from $\widehat{\eta}_t^{**}$ something about the Palm measure of $\eta_t$, is the following law of large numbers:

LEMMA 2.4. *Assume that* $\mathcal{L}(\xi_0) \in \mathcal{T}_\theta$ *with* $\theta > 0$ *and* $E[\xi_0^2(0)] < \infty$. *Then for* $p > 0$:

(2.17)          $$\mathcal{L}[e^{-pbt}\xi_t] \underset{t\to\infty}{\longrightarrow} \delta_{\underline{\theta}} \quad and \quad e^{-pbt}\xi_t(x) \underset{t\to\infty}{\longrightarrow} \theta \ in \ L_1. \quad \square$$

PROOF OF LEMMA 2.4. We can calculate the system of differential equations for the moments $\{E[\xi_t(x)], \ x \in \mathbb{Z}^d\}$, $\{E[\xi_t(x)(\xi_t(y) - \delta_{x,y})] | \ x, y \in \mathbb{Z}^d\}$ and show based on an explicit solution of these equations, that for every $x \in \mathbb{Z}^d$:

(2.18)          $$\mathrm{Var}\left[e^{-pbt}\xi_t(x)\right] \underset{t\to\infty}{\longrightarrow} 0, \quad E\left[e^{-pbt}\xi_t(x)\right] = \theta \qquad \forall t \geq 0,$$

which proves the claim (2.17).          $\square$

The system of differential equations for the moments is given by ($a(x,x) = 0$ is assumed here for convenience and recall $\bar{a}(x,y) := a(y,x)$):

(2.19)          $$\frac{d}{dt}E\left[\xi_t(x)\right] = \sum_y \bar{a}(x,y)E\left[(\xi_t(y) - \xi_t(x))\right] + pbE[\xi_t(x)], \quad x \in \mathbb{Z}^d$$

(2.20)
$$\begin{aligned}
\frac{d}{dt}&E\left[\xi_t(x)(\xi_t(y) - \delta_{x,y})\right] \\
= \ &\sum_z \bar{a}(x,z)E\left[\xi_t(z)(\xi_t(y) - \delta_{z,y})\right] - E[\xi_t(x)(\xi_t(y) - \delta_{x,y})] \\
&+ \sum_z \bar{a}(y,z)E\left[\xi_t(x)(\xi_t(z) - \delta_{z,y})\right] - E[\xi_t(x)(\xi_t(y) - \delta_{x,y})] \\
&+ 2pbE[\xi_t(x)(\xi_t(y) - \delta_{x,y})] \\
&+ bE[\xi_t(x)\delta_{x,y}], \quad x, y \in \mathbb{Z}^d.
\end{aligned}$$

The relation (2.19) gives with the Feynman-Kac formula and $E\xi_0(\cdot) \equiv \theta$:

(2.21)          $$E[\xi_t(x)] = \theta e^{bpt}, \quad \text{for all } x \in \mathbb{Z}^d.$$

In the second system consider the rescaling $e^{-2bpt}E[\xi_t(x)(\xi_t(y) - \delta_{x,y})]$. The resulting new system of differential equation looks like (2.20) without the third term and the fourth term replaced by $e^{-2pbt}bE[\xi_t(x)\delta_{x,y}] = e^{-pbt}\theta\delta_{x,y}$.

For $t = 0$ the second moments define a bounded function on $\mathbb{Z}^d \times \mathbb{Z}^d$ and the last term in (2.20) does not require knowing second moments. Then use the formula of partial integration for two transition kernels $V_t, U_t$ on $L_\infty(\mathbb{Z}^d \times \mathbb{Z}^d)$ with generators $L_V^s$ (at time $s$), $L_U$ and with $(L_V^s - L_U)(f) = \theta e^{-pbs}\delta_{x,y}$ for $f \in L_\infty(\mathbb{Z}^d \times \mathbb{Z}^d)$. This gives:

(2.22)          $$V_t = U_t + \int\limits_0^t U_{t-s}(L_V^s - L_U)V_s ds,$$

where here in our context $(U_t)_{t\geq 0}$ is the semigroup associated with two independent random walks with transition kernel $\bar{a}$ and jump rate $m$ and $(V_t)_{t\geq 0}$ the transition operators arising from solving the rescaled second moment equations.

Using (2.22) and the fact that $e^{-pt}$ is for $p > 0$ integrable on $[0, \infty)$ and $\hat{a}_t(0,0)$ converges to 0 ($\mathbb{Z}^d$ is a *countable* and not a finite group) we see that the integral converges to 0 as $t \to \infty$ and hence the convergence of the second moment of $\{e^{-pbt}\xi_t(x), \ x \in \mathbb{Z}^d\}$ to $\theta^2$ follows from the fact, that $U_t = \bar{a}_t \otimes \bar{a}_t$ and $\mu \in \mathcal{T}_\theta$, $E_\mu[\xi_0(x)]^2 < \infty$, the latter implying

(2.23)          $$\sum_{z,z'} \bar{a}_t(x,z)\bar{a}_t(y,z')\xi_0(z)\xi_0(z') \to \theta^2 \quad \text{in } L_2(\mu), \quad \text{as } t \to \infty,$$

(see [11] for more detail of such moment calculations). This convergence together with (2.21) gives the relation (2.18), completing the proof of Lemma 2.4.

STEP 2    Next we shall prove that due to the special structure of the random variable $\eta_t(0)$ the following relations hold, which will allow in the sequel for explicit calculations:

LEMMA 2.5.    *Under the assumption of the Proposition 2.3:*

$$(2.24) \qquad \mathcal{L}[\widehat{\eta}_t^*(0)] \underset{t\to\infty}{\Longrightarrow} \delta_\infty \quad \text{implies } \mathcal{L}[\widehat{\eta}_t^{**}(0)] \underset{t\to\infty}{\Longrightarrow} \delta_\infty$$

*and furthermore the relation (2.15) is implied by:*

$$(2.25) \qquad \mathcal{L}[\widehat{\eta}_t^*(0)] \text{ is stochastically bounded by } \mathcal{L}[Y(\omega,\omega')],$$

*where $E_{\omega'}|Y(\omega,\omega')| < \infty$ for almost all $\omega$ and here again $\omega = (Y_s^{i^*,0})_{s\geq 0}$ and $\omega' = \Pi$.*

PROOF OF LEMMA 2.5.    We need first some preparation, introducing the tools, which use exchangeability and Palm measures for mixed Poisson distributions.    □

An explicit calculation gives the following facts on mixed Binomial distributions and mixed Poisson distributions. Suppose that $X$ is mixed Poisson distributed with mixing law $F$ and $Y$ is mixed binomial distributed with parameter $n$ and with mixing law $G$. Suppose the mean of both distributions is $\theta$. Denote by $\widehat{\ }$-the corresponding object after size-biasing, which means passing from a law $P(dx)$ to the law $xP(dx)/\int xP(dx)$. Then:

$$(2.26) \qquad \widehat{X}\widehat{=}1 + Z \quad \mathcal{L}(Z) \text{ is mixed Poisson with mixing measure } \widehat{F},$$

$$(2.27) \quad \widehat{Y}\widehat{=} \quad 1 + Z \quad \mathcal{L}(Z) \text{ is mixed binomial with mixing measure } \widehat{G}_n$$

$$\text{with } \widehat{G}_n(dp) = np\theta^{-1}G(dp).$$

Recall furthermore de Finetti's result: An exchangeable array of $n$-random variables with values in $\{0,1\}$, which can be continued indefinitely, has the property that the sum has a mixed binomial distribution for some mixing distribution $F$ on $[0,1]$. In the case that the array cannot be continued, we have to work with the hyper geometric distributions instead of the binomial distribution. However in the limit of the length of the array tending to infinity we have the approximation result of the hyper geometric by the binomial distribution, so that in the limit $t \to \infty$ we are able to ignore this problem and work in both cases with mixed Poisson distributions.

But in our situation here we have the additional structure, that the laws $F, G$ can be obtained as image measures, which are induced by distributions living on the point process $\Pi$ defining the thinning, which allows for useful calculations.

We next rephrase the problem in terms of the notion introduced above. Define for $\Pi$ as in subsection 2(a) and $Y = (Y^{i^*,0})_{s\geq 0}$ the distributions:

$$(2.28) \quad \Lambda = \mathcal{L}(\Pi), \Lambda_Y = \mathcal{L}(\Pi|\Pi(\{(s,x)|Y_s^{i^*,0} = x\}) = 0), \quad m = \mathcal{L}((Y_s^{i^*,0})_{s\geq 0}).$$

Furthermore abbreviate the space of all path by $\mathcal{Y}$ and the space of all site-killing configurations by $\mathcal{K}$.

Recall the law of large numbers for $\xi_t(0)$. Consider the events $\{\xi_t(0) = n\}$ respectively $\{\xi_t^*(0) = n\}$ and condition on one of them. Then conditioned on

$\Pi$, $\eta_t(0)$ and $\eta_t^*(0)$ are according to subsection 2(a),(b) given by exchangeable arrays. Then by taking weak limit points $\eta_\infty, \eta_\infty^*$ and using the law of large numbers for $\xi_t(0)$ respectively $\xi_t^*(0)$ we see that there exist laws $\varrho_\Pi$ $\varrho_{Y,\Pi}$ on $\mathbb{R}^+$ such that:

$$(2.29) \qquad \mathcal{L}[\widehat{\eta}_\infty^*(0) - 1] = \int_{\mathcal{Y}} \int_{\mathcal{K}} \int_{\mathbb{R}^+} \text{Poiss}\,(\theta)\varrho_{Y,\Pi}(d\theta)\Lambda_Y(d\Pi)m(dY),$$

$$(2.30) \qquad \mathcal{L}[\eta_\infty(0)] = \int_{\mathcal{K}} \int_{\mathbb{R}^+} \text{Poiss}\,(\theta)\tilde{\varrho}_\Pi(d\theta)\Lambda(d\Pi).$$

Recall now the characterization of $\mathcal{L}(\widehat{\eta}_\infty^{**}(0))$ in terms of the mixing laws of $\eta_\infty(0)$ (compare (2.26), (2.27)). For the proof of (2.24) and (2.25) we have to show, that (denote by $\mathcal{P}_1(\cdot)$ probability measures with finite first moment)

$$(2.31) \qquad \left( \int_{\mathcal{K}} \varrho_{Y,\Pi}(\cdot)\Lambda_Y(d\Pi) \right) \left\{ \begin{array}{ll} \in \mathcal{P}_1(\mathbb{N}), & m(\cdot) - \text{a.s.} \\ = \delta_\infty & m(\cdot) - \text{a.s.} \end{array} \right.$$

corresponds to ($\widehat{\phantom{x}}$-denotes size-biasing)

$$(2.32) \qquad \left( \int_{\mathcal{K}} \tilde{\varrho}_\Pi(\cdot)\Lambda(d\Pi) \right)^{\widehat{\phantom{x}}} \left\{ \begin{array}{l} \in \mathcal{P}(\mathbb{N}), \\ \delta_\infty. \end{array} \right.$$

For this purpose we have to relate these two objects on the l.h.s. of (2.31) and (2.32). This is possible, since the quantities arise in a very special way.

The following identity holds which immediately establishes the correspondence described in (2.31) and (2.32) above. Namely

$$(2.33) \qquad \left( \int_{\mathcal{K}} \tilde{\varrho}_\Pi(\cdot)\Lambda(d\Pi) \right)^{\widehat{\phantom{x}}} = \int_{\mathcal{Y}} \int_{\mathcal{K}} \varrho_{Y,\Pi}(\cdot)\Lambda_Y(d\Pi)m(dY).$$

Then by inspection the r.h.s. of (2.33) is just a decomposition of the l.h.s. according to the path of the distinguished particle. This completes the proof of Lemma 2.5.

The reason for the relation (2.33) is the following

LEMMA 2.6.

$$(2.34) \qquad \mathcal{L}\left[\widehat{\eta}_t^{**}(0)\right] = \mathcal{L}\left[\widehat{\eta}_t^*(0)\right], \quad \text{for all } t \geq 0. \quad \square$$

PROOF. We calculate the probabilities for the appearance of n given particles in $\eta_t(0)$ for given $\Pi$ in two ways. First using the representation as mixed hypergeometric-geometric (mixed Poisson in case of the limit points as $t \to \infty$) and second using the explicit construction. This identifies the moments of the Palm of the mixing distribution by occupation time functionals of paths. The same procedure for $\widehat{\eta}_t^*(0)$ gives the same expression for the moments of the mixing distribution. $\square$

Return to Lemma 2.3. By inspection the r.h.s. of (2.33) is just a decomposition of the l.h.s. according to the path of the distinguished particle. This completes the proof of Lemma 2.5.

Step 3   Start with (2.14). First apply (2.24). If $\mathcal{L}(\widehat{\eta}_t^{**}(0))$ converges to $\delta_\infty$, then in the notation of (2.26), we have $\widehat{F} = \delta_\infty$ and hence since $\int xF(dx) \leq \theta$ we know that $F = \delta_0$. Therefore $\mathcal{L}(\eta_t(0))$ converges to $\delta_0$ as $t \to \infty$. Since this argument carries over to $\mathcal{L}(\eta_t(x))$ for every $x$, we know that $\mathcal{L}(\eta_t) \underset{t\to\infty}{\Longrightarrow} \delta_{\underline{0}}$, which completes the proof.

It remains now to prove (2.15) and (2.16). First apply (2.25). If $\mathcal{L}(\widehat{\eta}_t^{**})$ is a sequence of laws which are stochastically smaller than $\mathcal{L}(X(\omega, \omega'))$ with $\int X(\omega, \omega') P(dw') < \infty$, then all weak limit points of $\mathcal{L}(\widehat{\eta}_t^{**})$ must be stochastically smaller than the mixture of laws $\mathcal{L}(X(\omega, \cdot))$, the mixing being induced by the law of $\omega$, and which can be decomposed in laws $\nu(\omega)$ such that every component must have a finite mean $\tilde{\theta}(\omega)$. If $P(\eta_t(0) > 0)$ would tend to 0 along a subsequence $t_n \to \infty$, the Palm measures $\mathcal{L}(\widehat{\eta}_t^{**}(0))$ must diverge along the subsequence $(t_n)_{n\in\mathbb{N}}$. Namely the limit as $n \to \infty$ has mean 0, so that no size-biased (recall definition of a Palm measure, compare (2.12)) mass can remain in finite intervals, a contradiction to the assumption. Therefore the limit points of the laws of $\mathcal{L}(\eta_t)$ as $t \to \infty$ cannot be concentrated on 0 and in fact by a similar argument we see that they even preserve the intensity.

### 2.3. (c) Backbone construction, bounds by random walk quantities.

In order to be able to utilize the criterion obtained in the Proposition 2.3 above, we want to bring the random walk quantity $V_\infty^p$ into play. To do this we give an explicit representation for the random variables $\xi_t^*, \eta_t^*$ under the Palm distribution $(P_T)_x$. Here we will exploit the representation of the Palm distribution of branching processes, which start in Poisson configurations and which was given by Kallenberg (1977) in the discrete time setting and which was extended to the continuous time setting in Gorostiza and Wakolbinger (1991). The Palm measure is represented by means of the family tree of a distinguished particle plus an independent copy of the process.

These results allow, as we shall see below, the following construction. We realize the Palm measure of $(\xi_s^*)_{s\geq 0}$ at time $t$ with respect to the site 0. We obtain the path of the distinguished (immortal) particle and independent of it a collection of other paths. We label the path, remaining in the population at time $t$ and ending at site 0 at time $t$, from $1, 2, \ldots, \xi_t^*(0) - 1$ in a random manner. The immortal particle is denoted $i^*$. This way we obtain variables $Z_t^{i^*,0}$, $Z_t^{1,0}, \ldots, Z_t^{n,0}$ with $n = \xi_t^*(0) - 1$. The array $Z_t^{1,0}, \ldots Z_t^{n,0}$ is exchangeable on all the events $\xi_t^*(0) = n + 1$, $n \in \mathbb{N}$.

The goal of this subsection is to prove based on the construction of 2(a), 2(b) the following:

Proposition 2.7. *Assume that* $\mu = \mathcal{H}_\theta$ *with* $\theta \in (0, \infty)$, *where* $\mathcal{H}_\theta$ *is the law of a Poisson point process on* $\mathbb{Z}^d$ *with intensity* $\theta$.

*Denote with* $\tilde{E}$ *expectations with respect to the random process* $(Y_s^x)_{s\in[0,t]}$, *where* $\{(Y_s^x)_{s\in[0,t]}, x \in \mathbb{Z}^d\}$ *are independent random walks starting in* $x$, *with jump rates* $m$ *and transition kernels* $\bar{a}(x, y) = a(y, x)$, *while* $(Y_s^{i,x})_{s\in[0,t]}$ *denotes the path of descent of the* $(i, x)$-*particle for fixed time horizon* $t \in (0, \infty)$.

(a)    Then (denoting with $(i^*, 0)$ the distinguished particle (based on $(P_t)_0))$):

(2.35)        $Prob\left[ Z_t^{j,x} = 1 | Z_t^{i^*,0} = 1,\ (Y_s^{i^*,0})_{s\in[0,t]} \right]$

$$\geq\ e^{-pdt}\tilde{E}\left[ \exp\left( pd\int_0^t \mathrm{I\!I}(Y_s^x = Y_s^{i^*,0})ds \right)\right]\ a.s.$$

(b)    If $\hat{a}$ is transient, then for $p < p^*$, there exists for every path $(Y_s^{i^*,0})_{s\geq0}$ a $C = C((Y_s^{i^*,0})_{s\geq0}) \in [1, \infty)$ such that:

(2.36)
$$Prob\left[ Z_t^{j,x} = 1 | Z_t^{i^*,0} = 1,\ (Y_s^{i^*,0})_{s\in[0,t]} \right]$$
$$\leq\ Ce^{-pbt}\tilde{E}\left[ \exp\left( pd\int_0^t \mathrm{I\!I}(Y_s^x = Y_s^{i^*,0}) ds \right)\right]\ \ a.s.$$

(c)    The law of the distinguished ancestral path is identified as follows:

(2.37)        $$\mathcal{L}[(Y_s^{i^*,0})_{s\geq0}] = \mathcal{L}[(Y_s^0)_{s\geq0}].$$

Recall that by the Proposition 2.7(c) and the definition (2.37) of $V_t^p$ in section 1(c):

COROLLARY 2.8.

(2.38)        $$V_t^p \hat{=} \tilde{E}\left[ \exp\left( pd\int_0^t \mathrm{I\!I}(Y_s^0 = Y_s^{i^*,0})ds \right)\right].\quad \square$$

PROOF OF PROPOSITION 2.7. The proof of (a) and (b) each consists of two steps. We work first with the paths of descent $(Y_s^{i,x})_{s\in[0,t]}$ and $(Y_s^{i^*,0})_{s\in[0,t]}$, which we then in a second step compare with the random walk path mentioned in the Proposition. The latter paths do not take into account possible common ancestors and they are associated with larger expectations for the $Z$-variables than the free random walk ones so that this direction is fairly straightforward. More subtle is to prove part (b) and to argue the other way round, namely that the expectations in question in the limit $t \to \infty$ are not substantially larger than the free random walk ones. The basic idea is that in a supercritical population most surviving particles at a site at a large time $t$ are *not related* and given they are related, then the last common ancestor occurred at an early state. Here the fact, that the underlying symmetrized walk is transient works for us. To prove this a crucial role is played by the law of large numbers giving that $\mathcal{L}(e^{-pbt}\xi_t(x))$ converges to $\delta_\theta$ as $t \to \infty$.    $\square$

PART (A)

We simply calculate the probability in question separately on the two events that the individuals $(j, x)$ and $(i^*, 0)$ are related or are not related. To calculate these probabilities we use the backbone construction.

STEP 1    The distribution of the process $\xi(t)$ resp. $\tilde{\xi}(t)$ is infinitely divisible, since we start in a Poisson system and evolve according to a branching dynamic. Then we can calculate the Palm distribution by superposing independently the $\mathcal{L}(\tilde{\xi}_t)$ and the *Palm canonical measure*. (See [**6**], chapter 3). Next we investigate these two components.

We can identify a version of the law of the Palm canonical measure associated with the supercritical branching random walk. This is a result obtained by Chauvin, Rouault and Wakolbinger (1991) in the form needed here and the original idea for discrete time models goes back to Kallenberg (1977). The path of the picked individual is a random walk with transition kernel $\bar{a}(x, y)$ starting from 0 and the relatives are obtained by letting at rate

$$(2.39) \qquad\qquad b(2-p)\frac{2}{2-p} = 2b$$

break of independent versions of the supercritical branching random walk with random walk transition kernels $a(x, y)$.

Next we need the information, that the law of the path (see (2.3) - (2.5)) in the process $(\tilde{\xi}_t)_{t \geq 0}$, but under the *Palm measure* $(P_t)_0$, is a random walk starting in $x$ with jump rate $m$ and transition kernel $\bar{a}(x, y)$, which can be written formally as follows. Consider the particles sitting at $x$ at time $t$ and label them in an exchangeable way from $1, \ldots, \xi_t^*(x) - 1$ and denote the path of descent of the $j$-th particle by $(\tilde{Y}_s^{j,x})_{s \leq t}$. Then the claim is that:

$$(2.40) \qquad\qquad \mathcal{L}\left(\left(\tilde{Y}_s^{j,x}\right)_{s \in [0,t]}\right) = \mathcal{L}\left((Y_s^x)_{s \in [0,t]}\right).$$

If we have a pure random walk system, i.e. $b = 0$ then the assertion is true, since $\mathcal{H}_\theta$ are reversible equilibrium states (extremal ones in fact) of the process. Instead of proving (2.40) we show an extension of this statement below.

Let $T$ be the time where the particle $(j, x)$ splits of from $(i^*, 0)$, where we put $T = +\infty$ if they are not related. Denote by $Y_s^{(t,x,y)}$ the random walk bridge from $x$ to $y$ over the time span $t$. Then:

$$(2.41) \quad \mathcal{L}\left(\left(\tilde{Y}_s^{(j,x)}\right)_{s \in [0,t]} \Bigg| \left(Y_s^{i^*,0}\right)_{s \in [0,t]}, T\right) = \mathcal{L}\left(\left(\bar{Y}_s^{(j,x)}\right)_{s \in [0,t]}\right)$$

$$
\begin{aligned}
\bar{Y}_s^{(j,x)} &= Y_s^x, & s \leq t, && \text{if } T > t \\
\bar{Y}_s^{(j,x)} &= Y_s^{(t-T,x,z)}, & s \leq t - T && \text{and } Y_T^{i^*,0} = z, && \text{if } T \leq t \\
\bar{Y}_s^{(j,x)} &= Y_s^{i^*,0}, & s > t - T, && \text{if } T \leq t.
\end{aligned}
$$

The description on the event $T \leq t$ is immediate, due to the characterization of the Palm measure and the Palm canonical measure. It hence remains to verify (2.40) but now under the law $\mathcal{L}(\tilde{\xi}_t)$ rather than its Palm.

Consider first random walk systems. Denote by $(U_t)_{t \geq 0}, ((\bar{U}_t)_{t \geq 0})$ the semigroup of the following two Markov processes with values in the measures on collections of paths: $\mathcal{M}_{\mathbb{N}}(D(\mathbb{R}, \mathbb{Z}^d))$, where $\mathcal{M}_{\mathbb{N}}$ denotes integer-valued locally finite measures. The evolution of measures is induced by the following dynamics on path. The systems start at time $t = 0$ in collections of constant path, such that the numbers of path passing through the points $z \in \mathbb{Z}^d$, defines a configuration in $\mathcal{E}$ (see subsection 1(b)). The evolution is as follows. A path is denoted $Z = (Z_s)_{s \in \mathbb{R}}$. Define $\eta_s(x)(\bar{\eta}_s(x))$ to be the number of paths located at $x$ at time $s$. At rate $ma(x, y)\eta_t(x)$ $(m\bar{a}(x, y)\bar{\eta}_t(x))$ the path $Z$ is replaced by $\tilde{Z}$, where

$$(2.42) \qquad\qquad \tilde{Z}_s = Z_s \quad \forall s < t, \quad \tilde{Z}_s = y \quad \forall s \geq t.$$

Consider on the paths $(Z_s)_{s \in \mathbb{R}}$ the map $\Pi_t$ with $\Pi_t((Z_s)_{s \in \mathbb{R}}) = (Z_{t-s})_{s \in \mathbb{R}}$. With $\widehat{\Pi}_t$ we denote the corresponding map on $\mathcal{M}((D(\mathbb{R}, \mathbb{Z}^d)))$. Let $\nu$ be the law of a random initial state, which arises by associating with every particle of a Poisson system a constant path. Then the key is the following:

LEMMA 2.9. *Define* $\nu_t = (\nu)U_t$. *Then*

$$(2.43) \qquad\qquad \widehat{\Pi}_t(\nu_t) = (\nu)\bar{U}_t. \quad \square$$

PROOF. Poisson systems define *reversible equilibria* of systems of independent random walks (put $b = 0$ in the coupled branching system) for every transition rate $m$ and transition kernel $a(\cdot, \cdot)$. This proves Lemma 2.9 using the definition of the path-valued process generated by a random walk and the corresponding historical process of a system of independent random walks as described above. Another way to see this is to use the representation of the Palm measure for $b = 0$. $\qquad \square$

Next we sketch the proof of (2.40). If we now want to return to our supercritical branching random walk and its historical process as introduced in subsection 2(a) to prove (2.40), we note first that since the deaths occur independent of the motion it suffices to consider the case where the death rate of particles is 0 and later prune the system of path at rate $d$. For every path we define its order $n \in \mathbb{N}$. Namely paths present at time $t = 0$ have order 0, their direct descendents order 1 etc. Now the assertion (2.41) holds for order 0 paths by Lemma 2.9, which imply that the corresponding path-processes have the desired relation. (For the calculus of historical processes and path processes see [6]). According to these ancestral lines the population splits into independent family clusters. Note that at each birth the parent particle and its child are statistically indistinguishable. Furthermore we can label particles of each order according to their order of appearance.

Now we proceed by induction in the order. Namely first condition on the paths of lower order. Then the path of the descendents have after their birth time the structure of a random walk, before the birth time the path agrees with a lower order one, for which the result holds by induction hypotheses. By the exchangeability property of particles the claim follows.

STEP 2    On the event that the two particles $(j, x)$ and $(i^*, 0)$ are *not related* the path of descent are just independent random walks by (2.41) and by the identification of the Palm canonical measure. Hence we need just to observe that the occupation functional can only increase on the event that $(j, x)$ and $(i^*, 0)$ are related, since the two paths agree backwards from the splitting time on, but their joint law before merging remains to be the random walk law.

PART (B)
Let $(X_s^{(1)})_{s \geq 0}$ and $(X_s^{(2)})_{s \geq 0}$ be independent $(m, \bar{a})$-random walks starting in $x$ and 0. Then we shall show that for $\bar{p} > p$:

$$(2.44) \qquad \begin{aligned} e^{pbt} E\left[ Z_t^{j,x} | Z_t^{i^*,0} = 1, \ \left( Y_s^{i^*,0} \right)_{s \in [0,t]} \right] &\leq C \tilde{E}\left[ \exp(\bar{p}bX) \right] < \infty, \\ \text{where } X = \int_0^\infty 1 \left( X_s^{(1)} = X_s^{(2)} \right) ds, \end{aligned}$$

which proves immediately the claim if we choose a value $\bar{p} \in (p, p^*)$.

We observe first that on the event where $(i^*, 0)$ and the particle from $x$ are not related, the path $Y_s^x$ behaves like a random walk, while if the particle at $x$ and $(i^*, 0)$ are relatives, then $Y_s^x$ will be equal to $Y_s^{i^*,0}$ for $s \geq R_t$, where $R_t$ is a random time describing the time point at which the two paths of descent splitted based on the law $(P_t)_0$, that is based on the size-biased sampling from the time $t$ population. We estimate the joint occupation time between the path of descent of the $(j, x)$ and the path of descent of the $(i^*, 0)$ particle on the event $R$ that they are related by

$$(2.45) \qquad\qquad\qquad (t - R_t) + \bar{X},$$

where $\bar{X}$ is the joint occupation time of $(Y_t^{i^*,0})_{t \geq 0}$ and the $(j, x)$ path before the time $R_t$.

Define:

$$(2.46) \qquad\qquad Q_t(u) = \text{Prob}\left(R_t \in u + du | (Y_s^{i^*,0})_{s \geq 0}\right).$$

In order to prove the first inequality in (2.44) it suffices to show that for $g(t) = g(t, (Y_s^{i^*,0})_{s \geq 0}) = a_t(Y_t^{i^*,0}, x)$ the following holds:

$$(2.47) \qquad\qquad Q_t(u) \leq C\left((Y_s^{i^*,0})_{s \geq 0}\right) g(u) e^{-pb(t-u)},$$

$$(2.48) \qquad f(t) = E\left[e^{pbX} | Y_t^{i^*,0} = Y_t^x\right] \leq C < \infty, \quad \forall t > 0$$

since then

$$(2.49) \qquad l.h.s.(2.44) \leq E\left[e^{pbX}\right] + \int_0^t g(u) f(u) du < \infty.$$

By the representation of the canonical Palm distribution via the Kallenberg backward tree, we find that

$$(2.50) \qquad Q_t(u) = 2bE\left(\xi_t^{y(u)}(0)/(\xi_t(0) + \bar{\xi}_t(0))\right), \quad y(u) = Y_u^{i^*,0},$$

where $(\xi_s^x)_{s \geq 0}$ is a version of the supercritical branching random walk starting with one particle at site $x$ at time $0$ and which is *independent* of $(\xi_t)_{t \geq 0}$ and where $\bar{\xi}_t$ denotes the backward tree population, of which $\xi_t^{y(u)}$ is a piece. We continue by using the law of large numbers for $\xi_t(0)$ and the fact that (we omit the tedious details to get the uniform integrability via the Laplace transform and its anti-derivative)

$$(2.51) \qquad\qquad E\left(\frac{e^{bpt}}{\xi_t(0) + 1}\right) \xrightarrow[t \to \infty]{} \theta^{-1},$$

to verify the estimate (2.47) above.

To get (2.48) we need the finiteness of $E[\exp(\bar{p}bX)]$ for some $\bar{p} > p$. The result follows immediately if we assume that the value $p^*$ can be characterized as the point, where the exponential growth rate changes from $0$ to a positive value. Namely Hölder's inequality shows that the conditional expectation in question cannot grow faster than $(g(u))^{-(p/\bar{p})}$ and hence the l.h.s. (2.44) cannot grow faster than polynomial. This question is treated in a forthcoming paper [2]. An alternative route is to proceed by contradiction using that the bridge structure decomposes into two independent pieces each behaving as $E[\exp(pbX)|\text{path}]$, where the path is given by the two halves of the bridge. We do not have th space for all the details.

PART (C)

We only need to observe that the process $\Pi$ of site-killing points is independent of the historical process $(\tilde{\xi}_t)_{t \geq 0}$. For the process $(\tilde{\xi}_t)_{t \geq 0}$ we can apply the characterization of the Palm distribution of branching random walks and the path $(Y_s^{j,x})_{s \in [0,t]}$ as given in the proof of part (a).

**2.4. (d) An extinction-persistence criterion based on random walks.**
The next task is to prove with Proposition 2.7 the following criterion for divergence versus tightness of $\mathcal{L}(\widehat{\eta}_t^*)$ as $t \to \infty$:

PROPOSITION 2.10. *Assume that* $\mu = \mathcal{H}_\theta$. *Then in the notation of Proposition 2.7:*

$$(2.52) \qquad \mathcal{L}[\widehat{\eta}_t^*(0)] \underset{t \to \infty}{\Longrightarrow} \delta_\infty, \quad if \quad V_\infty^p = +\infty \quad \mathcal{L}[(Y_t^0)_{t \geq 0}] - a.s.$$

*If* $V_\infty^{\bar{p}} < \infty$, $\mathcal{L}[(Y_t^0)_{t \geq 0}]$-*a.s. for some* $\bar{p} > p$, *then:*

$$(2.53) \qquad \mathcal{L}[\widehat{\eta}_t^*(0)|(Y_s^{i^*,0})_{s \geq 0})] \text{ is stochastically bounded } \mathcal{L}[(Y_s^{(0)})_{s \geq 0}] - a.s. \qquad \square$$

PROOF OF PROPOSITION 2.10. PART (2.52)    The proof of the relation (2.52) proceeds in two steps, namely by comparing the system with a caricature $(\tilde{\eta}_t^*)_{t \geq 0}$ and the caricature with stochastically smaller systems $(\tilde{\eta}_t^{*,T})_{t \geq 0}$ for which we show first for every fixed $T$ convergence to mixed Poisson distributions for the random variable $\tilde{\eta}_t^{*,T}(0) - 1$ as $t \to \infty$. Next we establish that mixing law tends to $\delta_\infty$ as $T \to \infty$.                                                        $\square$

*Step 1 (caricature)*    The caricature of $\widehat{\eta}_t^*$ arises as follows. Define $\varrho_t = e^{pbt}$. Consider instead of the ancestral paths of a supercritical branching random walk population in the point 0 at time $t$, a $\mathcal{H}_{\varrho_t}$ distributed number of the path of reversed random walks evolving independently. This does (asymptotically) not change the number of random variables but changes the random variable on that part of the paths where ancestral path have merged, since they correspond to related individuals. We can define and characterize Palm distributions as before, since we start in Poisson systems, which are invariant for random walks. Therefore we know that under the Palm measures we have a distinguished reversed random walk path and an independent realization of a Poisson distributed number of reversed random walk path.

The system smaller than the caricature arises by allowing additional site killing after time $T$ on *the path of the typical particle* of the supercritical branching random walk, on whose survival we condition in $\widehat{\eta}_t^*$. This means modifying the thinning between time $[T \wedge t, t]$ on the typical path. This way certain individuals get now an extra chance to be hit by site-killing and hence the modified population is stochastically smaller. We now carry out this construction formally.

Turn now to the new system where with each particle in the supercritical branching random walk under the Palm distribution alive at time $t$ at site 0, we associate *independently* a random walk path with kernel $\bar{a}(x,y)$. We denote the typical surviving path $(Y_s^{i^*,0})$ in the caricature system with $(X_s^{*,0})_{s \geq 0}$. The law of this path is a $m, \bar{a}$ random walk starting in 0. This system is asymptotically equivalent (by Lemma 2.4) to the one started with path, whose numbers in the various starting points are chosen independently of the random walk path and they are $\mathcal{H}_{\varrho_t}$-distributed with $\varrho_t = \theta e^{pbt}$. We apply to this latter collection of path the thinning procedure described in (2.4), but with $\Pi$ replaced by

$$(2.54) \qquad\qquad\qquad \Pi_T^* = \Pi 1_{I_T},$$

with

$$(2.55) \qquad\qquad I_T = \left\{ (s,x) \in \mathbb{R}^+ \times \mathbb{Z}^d \big| x \neq X_s^{(*,0)} \quad \text{for } s \leq T \right\}.$$

Then if we condition on the number of particles at 0 at time $t$ we obtain for the modified variables $Z_t^{j,x}$ an exchangeable array, which has the structure that it can be continued indefinitely. We denote the modified random variables replacing $Z_t^{j,x}$, with $\tilde{Z}_{t,T}^{j,x}$. Define:

$$(2.56) \qquad\qquad \sum_{\substack{1 \\ j \neq i^*}}^{N_t(0)} \tilde{Z}_{t,T}^{j,0} = S_{t,T},$$

$$\text{where} \quad \mathcal{L}(N_t) = H_{\varrho_t}, \text{ and } N_t \text{ is independent of } \left\{ \tilde{Z}_{t,T}^{j,0} \right\}_{j \in \mathbb{N}}.$$

Note that if we condition on the *path* of the picked particle and if we omit the corresponding variable in the total sum then we obtain still on the events $\{N_t(0) = n\}$, $n \in \mathbb{N}$ an exchangeable array, which can be continued indefinitely.

Then we obtain for the law of $S_{t,T}$ under the Palm distribution $(P_t)_0$, conditioned on the path of $(Y_s^{i^*,0})_{s \leq T}$ and based on the site-killing field $\Pi_T^*$ as weak limit points as $t \to \infty$, a mixed Poisson distribution with $T$-dependent mixing distribution $\Lambda_T$ on $[0, \infty]$.

We have to show that all the weight of the mixing distribution tends to $+\infty$, as $T \to \infty$. For technical reasons it is easier to proceed indirectly and show instead, that the dichotomy holds that either the caricature goes to local extinction anyway or $\Lambda_T \to \delta_\infty$ as $T \to \infty$.

Observe first, that if we condition on the *site killing points* in the complement of the path of $(i^*, 0)$ up to time $T$, that is on $\Pi_T^*$, then the random variables $(\tilde{Z}_{t,T}^{j,x})_{j \in \mathbb{N}}$ become independent identically distributed random variables. Let $q_t(\Pi_T^*)$ be the probability that a path up to time $t$ does not hit a site-killing point for given $\Pi_T^*$. We shall see later that the quantity $e^{+pbt} q_t(\Pi_T^*)$ converges as $t \to \infty$ to $q_\infty(\Pi_T^*)$. (See (2.69) - (2.71)). Next we use that by the (space-time) independence property of the site-killing field $\Pi$:

$$(2.57) \qquad \begin{aligned} &\text{Prob} \left( \tilde{Z}_{t,T}^{j,x} = 1 \big| j \neq i^*, \Pi_T^*, \, \tilde{Z}_t^{i^*,0} = 1 \right) \\ &\underset{t \to \infty}{\widetilde{\phantom{mm}}} \quad \tilde{E} \left[ \exp \left( pb \int_0^T \mathbb{1} \left( X_s^{j,x} = Y_s^{i^*,0} \right) ds \right) \right] q_t(\Pi_T^*), \end{aligned}$$

and that we know by construction that the number of summands in (2.56) grows like $\theta e^{pbt}$. Then, by the law of large numbers, the mixing law $\Lambda_T$ is given by ($\mathcal{L}$ with respect to $(Y_s^{i^*,0})_{s \geq 0}$):

$$(2.58) \qquad \mathcal{L} \left( \tilde{E} \left[ \exp \left( pb \int_0^T \mathbb{1} \left( X_s^{j,x} = Y_s^{i^*,0} \right) ds \right) \right] q_\infty(\Pi_T^*) \right).$$

Now there are two possibilities. If $q_\infty(\Pi_T^*) \equiv 0$, then also $q_\infty(\Pi) \equiv 0$ and the caricature becomes locally extinct. Or otherwise $q_\infty(\Pi_T^*) > 0$ with positive probability.

Note that the event $\{q_\infty(\Pi_T^*) = 0\}$ satisfies the 0-1 law due to the independence properties of $\Pi$. Since by assumption $q_\infty(\Pi_T^*) > 0$ with positive probability, we conclude $q_\infty(\Pi_T^*) > 0$ a.s. Letting $T \to \infty$ we see that the assumption $V_\infty^p \equiv +\infty$ forces $\Lambda_T \Rightarrow \delta_\infty$, and hence as $t \to \infty$ the caricature becomes locally extinct q.e.d.

STEP 2 (THE REAL SYSTEM)    The point now is that due to the possibility of two particles in $\xi_t(0)$ to descent from a common ancestor, we loose the independence property, once we condition on site-killing fields, which was used to exploit the fact that $V_\infty^p = +\infty$ a.s. In other words in (2.57) we have no longer in the scaled limit for given path $Y^{i^*,0}$ the parameter of a Poisson distribution, but a priori only the lower bound for the mean of a *mixed* Poisson distribution. But the result on the caricature tells us, that with $\tilde{q}(\Pi)$ denoting the survival of a single path for given site-killing field $\Pi$:

$$(2.59) \qquad\qquad q_t(\Pi) = o(e^{-pbt}).$$

Since the number of summands is $O(e^{pbt})$, this proves the assertion.

PART (2.53)    We observe that all weak limit points of $\mathcal{L}(\widehat{\eta}_t^*(0) - 1)$ as $t \to \infty$ are mixed Poisson distributions with a mixing measure which is a probability measure on $\mathbb{R}^+$, since it can be decomposed such that the mean is bounded by a multiple of the r.h.s. of (2.36), which is by assumption finite.

## 2.5. (e) Proof of Theorem 1.5 and Theorem 1.7.

PROOF OF THEOREM 1.5.    Here we combine (2.53) with (2.15)-(2.16), to obtain that all weak limit points of $\{\mathcal{L}(\eta_t), t \geq 0\}$ as $t \to \infty$ have positive weight on $\{\eta(x) > 0\}$ and even have still intensity $\theta$, and hence we have in this regime *persistence*.                                                                           $\square$

We have now to work to verify the more detailed statements given in the theorem. The basic tool is the representation of the process in terms of $\tilde{\xi}_t$ and the field $\Pi$ of site killing points given in subsection 3(a) and the fact that in the supercritical branching process two particles are typically not related and if they are then the last common ancestor is far back. We proceed in two steps.

STEP 1    We begin the analysis by studying a *caricature* $(\zeta_t)_{t \geq 0}$ of the system $(\eta_t)_{t \geq 0}$. Namely we consider the configuration $\xi_t$ and associate with the particles independent random walk path with jump rate $m$ and transition kernel $\bar{a}(x, y)$. In other words we leave out the effects due to *common ancestors*, which we will incorporate later. Based on these paths we construct a random configuration as in (2.7). We call this caricature of $(\eta_t)_{t \geq 0}$ here $(\zeta_t)_{t \geq 0}$ and we prove the following:

PROPOSITION 2.11.    *Let* $\mu = \mathcal{H}_\theta$ *and* $p \in [p^{(2)}, p^*)$. *Then*

$$(2.60) \qquad\qquad \mathcal{L}[\zeta_t] \underset{t \to \infty}{\Longrightarrow} \lambda_\theta,$$

*with*

$$(2.61) \qquad E_{\lambda_\theta}[\zeta(x)] = \theta, \quad E_{\lambda_\theta}[\zeta^2(x)] = +\infty \qquad \forall x \in \mathbb{Z}^d,$$

$$(2.62) \qquad\qquad \lambda_\theta \quad \textit{is spatially mixing},$$

$$(2.63) \qquad \lambda_\theta[\zeta(x) = n] = \int\limits_0^\infty e^{-\varrho} \frac{\varrho^n}{n!} \Lambda(d\varrho), \quad \Lambda \in \mathcal{P}(\mathbb{R}^+). \quad \square$$

PROOF OF PROPOSITION 2.11. Recall that $\zeta_t(x)$ is defined by considering whether the random walk paths 1, $\ldots, \xi_t(x)$ hit till time $t$ a site killing point. In order to verify relation (2.60) we observe that if we condition on the site killing field $\Pi$ then the random variables $\left(\Pi_t = \Pi \, \mathrm{I}_{\mathbb{Z}^d \times [0,t]}\right)$:

$$(2.64) \qquad \bar{Z}_t^{j,x} = \begin{cases} 1 & \text{if the } j\text{-th path starting in } x \text{ does not hit } \mathrm{supp}(\Pi_t) \\ 0 & \text{if the } j\text{-th path starting in } x \text{ does hit } \mathrm{supp}(\Pi_t) \end{cases}$$

are for fixed $x$ independent and identically distributed. Define $\Pi$-random variables $q_t = q_t(\Pi)$ by

$$(2.65) \qquad q_t = P\left(\bar{Z}_t^{j,x} = 1 \big| \Pi\right), \quad x \in \mathbb{Z}^d, \, j \in \mathbb{N}.$$

$$\square$$

We want to show, that the sum over the array (2.64) has a nontrivial limit as $t \to \infty$. Then by the law of large numbers for $\xi_t(x)$ as $t \to \infty$ (recall (2.17)) we have to show, that there exists a $\Pi$-random variable $q_\infty(\Pi)$ with values in $[0, \infty)$ and not identically 0, such that:

$$(2.66) \qquad e^{pbt} q_t \underset{t \to \infty}{\Longrightarrow} q_\infty^* \qquad \mathcal{L}(\Pi)\text{-a.s.},$$

since then we can conclude that (with nontrivial r.h.s.)

$$(2.67) \qquad \mathcal{L}(\zeta_t(x) | \Pi) \underset{t \to \infty}{\Longrightarrow} \mathrm{Poiss}(q_\infty^*(\Pi)).$$

From here we conclude, since $\zeta_t(x)$ is $\mathbb{N}$-valued and $\theta \to (n!)^{-1} e^{-\theta} \theta^n$ is bounded continuous for every $n$, that

$$(2.68) \qquad \mathcal{L}(\zeta_t(x)) \underset{t \to \infty}{\Longrightarrow} \int \mathrm{Poiss}(q_\infty^*(\Pi)) Q(d\Pi), \quad Q = \mathcal{L}(\Pi).$$

The convergence of $e^{-pbt} q_t$ as $t \to \infty$ will be obtained with the Martingale convergence theorem as follows. Consider the filtration given by $\Pi_t$. Observe that by independence property of the Poisson point process on $\mathbb{R}^+ \times \mathbb{Z}^d$:

$$(2.69) \qquad \left(q_t(\Pi) e^{+pbt}\right)_{t \geq 0} \quad \text{is a nonnegative Martingale with expectation 1.}$$

Therefore by the Martingale convergence theorem the relation (2.66) indeed holds.

Now we apply the previous reasoning to $n$-tuples of space points $x_1, x_2, \ldots, x_n$. We consider the sum of the values $\zeta_t(x_1) + \cdots + \zeta_t(x_n)$. The probability of success for a member of this array at a sampled site, is of the form $e^{-pbt} \frac{1}{n}(q_t(\Pi_{x_1}) + \cdots + q_t(\Pi_{x_n}))$ where $\Pi_x = \Pi(x + \cdot)$ and hence converges. This holds for very $n \in \mathbb{N}$ and $n$-tuple in $\mathbb{Z}^d$ of points. Now with standard arguments we conclude that

$$(2.70) \qquad \mathcal{L}(\zeta_t) \underset{t \to \infty}{\Longrightarrow} \lambda_\theta \in \mathcal{P}(\mathcal{E}),$$

(recall $E(\zeta_\infty(x)) \leq \theta$ by Fatou), which proves (2.60) and (2.63).

The next step is to show that not only $E(\zeta_t) = \theta$ and $E(\zeta_\infty) \leq \theta$, but that there is no loss of mean in the limit $t \to \infty$, i.e.

$$(2.71) \qquad E[\zeta_\infty] = \theta.$$

This follows however immediately from (2.53) if we replace $\widehat{\eta}_t^*(x)$, $\widehat{\eta}_t^{**}(x)$ by the caricature $\widehat{\zeta}_t^*(x)$, $\widehat{\zeta}_t^{**}(x)$ and use Propositions 2.3 - 2.10 for the caricature, which we can trivially do.

The fact that $E(\zeta_\infty^2(x)) = +\infty$ can be derived with the help of the potential theory of random walks, as was shown in [11] in the proof of the corresponding property for $\eta_\infty^2(x)$. We refer the reader to that paper.

Finally we investigate the mixing properties of $\lambda_\theta$. Define the probability that the random walks $(Y_s^x)_{s\geq 0}$ and $(Y_s^y)_{s\geq 0}$ ever meet as $H(x,y)$. Since $\widehat{a}$ is transient we know that $H(x,y) \to 0$ as $|x-y| \to \infty$. Note that $\bar{Z}_t^{x,j}$ and $\bar{Z}_t^{y,k}$ are independent on the event, where $(Y_s^x)_{s\geq 0}$ and $(Y_s^y)_{s\geq 0}$ never meet. Furthermore for $p < p^*$:

$$(2.72) \quad \begin{aligned} & P\left[\bar{Z}_t^{j,x} = 1 \big| \bar{Z}_t^{k,0} = 1, \ (Y_s^{k,0})_{s\geq 0}\right] \\ & \leq \left\{1 + H\left(x, (Y_s^{k,0})_{s\geq 0}\right) \text{ Const } \left((Y_s^{k,0})_{s\geq 0}\right)\right\} e^{-pbt} \end{aligned}$$

with

$$(2.73) \qquad H\left(x, (Y_s^0)_{s\geq 0}\right) = P(Y_s^0 \text{ and } Y_s^x \text{ meet for some } s | (Y_s^0)_{s\geq 0}).$$

Of course

$$(2.74) \qquad\qquad H(x,y) = E[H(x, (Y_s^0)_{s\geq 0})].$$

Consequently

$$(2.75) \qquad \limsup_{t\to\infty} \left\{e^{pbt} P\left[\bar{Z}_t^{j,x} = 1 \big| \bar{Z}_t^{k,0} = 1, \ (Y_s^{k,0})_{s\geq 0}\right]\right\} \xrightarrow[|x|\to\infty]{} \theta \quad \text{ in } L_1.$$

Recall that $e^{pbt} P(\bar{Z}_t^{j,x} = 1) = \theta$. With the techniques of chapter 2 (using the exchangeability and special structure of $\zeta_\infty$) it can be shown, that this implies that $\zeta_\infty(x)$ and $\zeta_\infty(0)$ are asymptotically independent as $|x| \to \infty$. Namely for the $n$-dimensional distribution of the field $\zeta_\infty$ we have a mixture of a multinomial distribution, which converges to a $n$-dimensional mixture of Poisson distributions, i.e. a law of the form $\int_{\mathbb{R}^+} \cdots \int_{\mathbb{R}^+} \bigotimes_1^n \text{Poiss } (\theta_i)Q_x(d\theta)$, $x = (x_1,\ldots,x_n)$. It suffices to verify, that the mixing $Q$ laws corresponding to the $n$-tuple, factorizes asymptotically as $\min|x_i - x_j| \to \infty$. Focus on $n = 2$, which is then verified immediately with (2.75) and (2.58) for the case $T = \infty$. But the probability measure $Q_x$ on $(\mathbb{R}^+)^n$ corresponds for every $x$ to $n$ associated $\mathbb{R}^+$-valued random variables. In that case it is well-known, that if suffices to consider the case $n = 2$.

To see the associateness property proceed as follows. Define $0_t^{(n)}$ the joint occupation time (with multiplicity) till time $t$ of the $n$ random walks starting in $x_1,\ldots,x_n$. Then we can see along the lines of reasoning in (2.64) - (2.67) that the probability for the event $Z_t^{1,x_i} = 1$, $i = 1,\ldots,n$ is equal to $(e^{-pbt})^n$ $(\Pi_{i=1}^n q_t(\Pi_{x_i}))$ $E[\exp(0_t^{(n)})] \geq (e^{-pbt})^n \Pi_{i=1}^n q_t(\Pi_{x_i})$. Let now $t$ tend to infinity. Hence $Q_x$ describes $n$-associated random variables. This completes the proof of Proposition 2.11.

STEP 2  The next task is now to pass from the caricature to the *real system*, that means we have to modify the previous analysis, such that we can also handle the fact that two paths of descent in $\tilde{\xi}_t$ might belong to two related individuals. The effect of this is now that, even if we condition on the site killing field $\Pi$, the random variables which are 0 or 1, depending on whether the path does or does not hit a site killing point, are now no longer independent but only exchangeable and the array can not be continued indefinitely. As a consequence given $\Pi$ we will already see in the limit $t \to \infty$ possibly a *mixed* Poisson distribution rather than a pure Poisson distribution. This requires however only to reprove (2.60) and to replace (2.63), we obtain now for the law for $\eta_t(0)$ the following (here $m_\Pi$ are element of $\mathcal{P}(\mathbb{R}^+)$)

$$(2.76) \qquad \int_{\mathcal{K}} \int_0^\infty \text{Poiss } (\theta) m_\Pi(d\theta) P(d\Pi), \quad P = \mathcal{L}(\Pi).$$

In other words we only loose the identification of the mixing measure. The argument for the convergence uses that the occupation times of paths of related particles ending in the same space window observed over the last length $T$ piece converge for every $T$ together with the bound of Proposition 2.2. This allows to obtain the convergence of the mixing measures corresponding to the $T$-caricatures (i.e. site-killing in the piece $[t - T, t]$ and otherwise critical branching) by the methods of moments. Then let $T \to \infty$ using the bound of Proposition 2.2. We skip further details.

The arguments for the mean and infinite second moment carry over immediately. It remains to modify the argument about the mixing property to adapt it from $\lambda_\theta$ to $\nu_\theta$ we need a generalized version of the statement (2.40), we need it for pairs of path, which was given in (2.41).

The invariance property of $\nu_\theta$ follows immediately from the fact that the semigroup $S(t)$ of the process $(\eta_t)_{t \geq 0}$ has the following property. Suppose $\mu_n \in \mathcal{T}_\theta$, $\mu_n \Rightarrow \mu$ and $\eta(0)$ is uniformly integrable under $(\mu_n)_{n \in \mathbb{N}}$. Then for every fixed $t$ (Compare for the needed truncation techniques [11]):

$$(2.77) \qquad \mu_n S(t) \Rightarrow \mu S(t).$$

PROOF OF THEOREM 1.7. We assume first that $\mu = \mathcal{H}_\theta$ for $\theta \geq 0$. Then we can combine (2.14) with (2.52) to obtain the assertion for the special initial laws $\mathcal{H}_\theta$. □

Next we have to extend this to initial laws $\mu \in \mathcal{T}_\theta$. Note that with a simple first moment calculation and the explicit representation of subsection 2(a) we find that the processes $(\eta_t^i)_{t \geq 0}$, $i = 1, 2$ arising from the two initial configurations $\mathcal{H}_\theta$ and $\delta_{\underline{n}}$ $(n \in \mathbb{N})$ can be constructed on one probability space, such that for every fixed $n$:

$$(2.78) \qquad E\left[\eta_t^1(0) - \eta_t^2(0)\right]^- \leq \varepsilon_\theta, \quad \varepsilon_\theta \to 0 \quad \text{as } \theta \to \infty.$$

Therefore we obtain the assertion also for the initial states $\delta_{\underline{n}}$ by monotonicity.

This immediately implies that for every $\mu \in \mathcal{T}_\theta$ and with $\mu_n$ denoting the law of $\eta \wedge n$ for $\eta$ distributed according to $\mu$, the assertion holds for $\mu_n$ by monotonicity (recall the explicit representation of subsection 2(a)). Finally repeating the estimate of (2.78) now for the laws $\mu_n$ and $\mu$, we see by now letting $n \to \infty$ that the assertion holds for $\mu$. This completes the proof of Theorem 3.

## 3. Properties of the occupation functional

The main purpose of this subsection is to prove the statements about the exponential functional given in (1.10). These are Lemma 1.3, 1.8 and Proposition 1.10 relation (1.31).

For the proof of Lemma 1.3 note, that the result does not depend on any piece $(X_s^{(2)})_{s \leq T}$ of the random walk on which we condition. Hence this event is a terminal event for the jump sizes and jump times of the random walk. By the 0-1 law for i.i.d. random sequences the result follows.

For the proof of Lemma 1.8 note that we can define a version of the conditional expectation by taking expectation over the first random walk. Since for fixed $(X_s^{(2)})_{s \geq 0}$ the function

$$(3.1) \qquad p \to \exp \left( pb \int_0^\infty 1\!\!1 \left( X_s^{(1)} = X_s^{(2)} \right) ds \right)$$

is monotone increasing for every realization of $X_s^{(1)}$, the result is immediate.

In order to prove Proposition 1.10, we proceed as follows: Obviously we can have $1 > p^{(2)} > 0$, as we see by inspection, it suffices that (recall (1.30)):

$$(3.2) \qquad \widehat{G}(0,0) < \infty, \quad m < b\widehat{G}(0,0).$$

We have next to verify the two relations (3.3) and (3.4):

$$(3.3) \qquad p^* < 1 \quad \text{for suitable } m, b, a(\cdot, \cdot),$$

$$(3.4) \qquad p^{(2)} < p^{(*)} \quad \text{for suitable } m, b, a(\cdot, \cdot).$$

Here if $\widehat{a}(\cdot, \cdot)$ is transient, for the relation (3.3) we have to show, that with $\tilde{E}$ denoting expectation over $(X_s^{(1)})_{s \geq 0}$ only, the following holds for $b \leq b_0$, $b_0 > 0$:

$$(3.5) \qquad \tilde{E} \left[ \exp \left( b \int_0^\infty 1\!\!1 \left( X_s^{(1)} = X_s^{(2)} \right) ds \right) \right] < \infty, \quad \mathcal{L}((X_s^{(2)})_{s \geq 0}) - \text{a.s.}$$

while $p^{(2)} < p^*$ simply says that the random variable $V_\infty^P$ can be finite, without the mean being finite.

These are assertions about exponential functionals and will be treated in [2] using large deviation theory and the technique of variational problems.

## 4. Finer Analysis of the stable regime and of local extinction

In this chapter we want to give a very rough sketch of the approach, which can be taken up using the analysis of section 2, to study the equilibria in the regime $p < p^{(2)}$ respectively the clusters in the regime $p > p^*$. We do not have the space to carry out all the details of the analysis in the present paper.

**4.1. (a) A central limit theorem for $\nu_\theta$.** Here we want to explain some of the elements needed to prove (1.35). Note first that we can carry out the construction of the explicit representation of $\eta_\infty$, with $\mathcal{L}(\eta_\infty) = \nu_\theta$ as a limiting object from the historical process of a supercritical branching random walk. We have to start the process at time $-t$, observe the process at 0 and let $t \to \infty$. Recall subsection 2(a).

The correct rate of renormalization is found by a second moment calculation, which results for $p < p^{(2)}$ into the same order of magnitude found for the case $p = 0$ of pure branching, since the covariances $E_{\nu_\theta}[\eta(x)\eta(y)]$ again decay as $H(0, y - x)$

where $H(x, y)$ is the hitting probability of $y$ of a random walk starting in $x$ with transition kernel $\widehat{a}(\cdot, \cdot)$.

Under the Palm measure $(P_\infty)_0$ for $(\tilde{\xi}_t)_{t \geq 0}$ the correlation between the variable $Z_t^{x,j}$ and $Z_t^{0,i^*}$ conditioned on both $Z_\infty^{0,i^*} = 1$ and the individuals $(0, i^*)$ and $(x, j)$ not being related, is of the same order as ($Y^z$ is a $m, \bar{a}(\cdot, \cdot)$ random walk starting in $z$):

$$(4.1) \qquad E\left[\exp\left(pb \int_0^t 1\!\mathrm{I}(Y_s^0 = Y_s^x)ds\right)\right] e^{-pbt},$$

which can be rewritten as ($H_t(0, x)$ is the hitting probability of $0$ starting in $x$ by time $t$)

$$(4.2) \qquad H_t(0, x) E\left[\exp\left(pb \int_0^t 1\!\mathrm{I}(Y_s^{(0)} = \tilde{Y}_s^{(0)})ds\right)\right] e^{-pbt},$$

where $(Y_s^0)_{s \geq 0}$ and $(\tilde{Y}_s^0)_{s \geq 0}$ is an independent copy of $(Y_s^0)_{s \geq 0}$. Therefore the asymptotics of this quantity is given by

$$(4.3) \qquad H(0, x) E\left[\exp\left(pb \int_0^\infty 1\!\mathrm{I}\left(Y_s^{(0)} = \tilde{Y}_s^{(0)}\right) ds\right)\right] e^{-pbt}.$$

Note that this implies in particular, that the expression in (4.1) will be decaying for $|x| \to \infty$ the same way as $H(0, x)$ and therefore bounds the correlations. Note that this is also the decay of correlations found in the voter model or classical branching and hence we expect qualitatively the same behavior for coupled branching equilibria in case $p < p^{(2)}$, as in the voter model or branching.

However as a strategy of proof the method of moments does not work any more, since we do not have finite higher order moments $E_{\nu_\theta}[\eta(x)]^n$, at least if $p > p^{(n)}$ and has therefore to be modified quite a bit.

Namely we can still decompose the equilibrium state denoted $\eta_\infty$ as follows

$$(4.4) \qquad \eta_\infty = \sum_i \eta_\infty^i,$$

where $(\eta_\infty^i)_{i \in \mathbb{N}}$ are *exchangeable* elements and consist of all particles which descended in an infinitely old system from the same ancestor. The fact that a larger renormalization is needed in the central limit theorem than for i.i.d. random fields is due to two things. First, as already in classical branching ($p = 0$), due to the correlations between components caused by parts of the population belonging to *one clan* $\eta_\infty^i$. However, since we are now using a supercritical, rather than critical, branching random walk to generate the family structure, we obtain a positive intensity of clans surviving the branching, while in the critical branching random walk the intensity is $0$. Hence this effect is now not changing orders of magnitudes of the scale. However we get a second source of correlations, since particles of different families may pass through the same points of site-killing, so that different families do not evolve independent. This is exactly the point, where we have to see how much the presence of site-killing changes the behavior, compared with the case of independent path. The key point is here that the spatial decay of correlations is still of the same order as in pure branching and given in both cases by random walk hitting probabilities.

We have to make this precise. If we return to the explicit construction of $(\eta_t)$ in terms of $(\tilde{\xi}_t)$, then conditioned in addition on the event that the ancestral path are disjoints, these variables $Z^{x,j}$ and $Z^{0,i^*}$ become independent. Therefore our strategy has to be to divide the population of the block $[-N, N]^d$ into independent families interacting only via the site-killing and show that blocks in great distance share few site-killing events. Then use a caricature where the site killing takes only place in the last time interval $[-T, 0]$ for the process started in $-\infty$. This allows to use (for fixed $T$) the *method of moments* for the caricature. Then let $T \to \infty$.

**4.2. (b) Cluster formation in the regime of local extinction.** We are here primarily interested in the regime $\hat{a}$ transient and $p > p^*$, the case $\hat{a}$ recurrent is of a different nature. We want to analyze the height of the peaks forming as $t \to \infty$.

Recall the historical process $\tilde{\xi}_t$ of subsection 2(a) and the approximating systems of subsection 2(d). Suppose we condition on the historical supercritical branching random walk $\tilde{\xi}(t)$ and average only over the site killing events. If the typical particle survives, then the number of other particles surviving at this site is approximately

$$(4.5) \qquad (\tilde{\xi}_t(0)e^{-pbt}) \exp\left( pb \int_0^t \mathrm{I\!I}(Y_s^{(0)} = \tilde{Y}_s^{(0)})ds \right),$$

where $(Y_s^{(0)})_{s \geq 0}, (\tilde{Y}_s^{(0)})_{s \geq 0}$ are realizations of independent copies of a random walk with jump rate $m$ and transition kernel $\bar{a}(\cdot, \cdot)$ starting in 0. For a given time-space Poisson point process giving the site-killing events on the path of the typical walk, the exponential functional in (4.5) multiplied by $e^{-pbt}$, gives the magnifying factor of the probability that realizations of the two random walks survive the site killing, given that the typical one does. Therefore using the relation of this object and the Palm distribution, the cluster height of a typical particle of the coupled branching process is described, discarding the effect of relatives in the branching process for the moment, by

$$(4.6) \qquad \mathcal{L}\left( \exp\left( pb \int_0^t \mathrm{I\!I}(Y_s^{(0)} = \tilde{Y}_s^{(0)})ds \right) \bigg| (\tilde{Y}_s^{(0)})_{s \geq 0} \right).$$

The question is now to find a scaling function $h(t)$, such that

$$(4.7) \qquad \mathcal{L}\left( h^{-1}(t) \exp\left( pb \int_0^t \mathrm{I\!I}(Y_s^{(0)} = \tilde{Y}_s^{(0)})ds \right) \bigg| (\tilde{Y}_s^{(0)})_{s \geq 0} \right)$$

converges to a law which is not concentrated on 0. This raises the questions: Is this possible? Is $h(t)$ is deterministic?

The expression in (4.6) gives a stochastic rough approximation only for the cluster height. For the exact expression, we would have to show that the effect of related particles does not change the order of magnitude of the cluster heights at least in the case $\hat{a}$ transient.

# References

[1] M. Bramson, D. Griffeath, *Renormalizing the 3-Dimensional Voter Model*, Ann. Prob., Vol. **7**, No. 3, 418-432 (1979).

[2] J.T. Cox, A. Greven, F. den Hollander, *The long-time behavior of a class of interacting diffusions*, manuscript, (1999).

[3] J.T. Cox, A. Greven, *Ergodic theorems for systems of locally interacting diffusions*, Ann. of Prob. **22**(2), 833-853 (1994).

[4] B. Chauvin, A. Rouault, A. Wakolbinger, *Growing conditioned trees*, Stochastic Processes and their Applications **39**, 117-130 (1991).

[5] D.A. Dawson, *The critical measure diffusion*, Z. Wahr. verw. Geb. **40**, 125-145 (1977).

[6] D.A. Dawson, *Measure valued Markov processes*, École d'Été de Probabilités de Saint Flour 1991, Lecture Notes in Mathematics **1541**, 1-260, Springer-Verlag (1993).

[7] D.A. Dawson, A. Greven, *Multiple space-time analysis for interacting branching models*, Electronic Journal of Probability, Vol. **1**, paper 14, (1996).

[8] D.A. Dawson, K.G. Gorostiza, A. Wakolbinger, *Occupation time fluctuations in branching systems*, manuscript, (1999).

[9] R. Durrett, *An infinite particle system with additive interactions*, Adv. Appl. Prob. **11**, 355-383 (1979).

[10] J. Fleischman, *Limiting distribution for branching random fields*, Trans. Amer. Math. Soc. **239** (1978).

[11] A. Greven, *Phase transition for the Coupled Branching Process, Part I: The ergodic theory in the range of finite second moments*, Prob. Theory and Rel. Fields **87**, 417-458 (1991).

[12] J. Gärtner, F. den Hollander: *Correlation structure of intermittency in the parabolic Anderson model*, To appear Prob. Theory and Rel. Fields (1999).

[13] D. Griffeath, *The binary contact path process*, Ann. Prob. **11**, 692-705 (1983).

[14] L.G. Gorostiza, S. Roelly, A. Wakolbinger, *Persistence of critical multi type particle and measure branching processes*. Prob. Theory. Relat. Fields **92**, 313-335 (1992).

[15] L.G. Gorostiza, A. Wakolbinger, *Persistence criteria for a class of critical branching particle systems in continuous time*, Ann. Prob. **19**, 266-288 (1991).

[16] R. Holley, T. Liggett, *Generalized potlatch and smoothing processes*, Z. Wahrsch. verw. Gebiete **55**, 165-196 (1981).

[17] O. Kallenberg, *Stability of critical cluster fields*, Math. Nachr. **77**, 7-45 (1977).

[18] A. Klenke, *Multiple Scale Analysis of clusters in spatial branching models*, Ann. Prob. **25**, No. 4, 1670-1711 (1997).

[19] T. Liggett, *Infinite particle systems*, Berlin-Heidelberg-New York, Springer (1985).

[20] P. Mayor, *Renormalizing the Voter Model. Space and Space-Time Renormalization*, Studia Scientiarum Mathematicum Hungarica **15**, 321-341 (1980).

[21] T. Shiga, *Ergodic theorems and exponential decay of sample paths for certain interacting diffusion systems*, Osaka J. Math., **29**, 789-807 (1992).

[22] A. Winter, *Multiple scale analysis of branching processes under the Palm distribution*, PHD-Theses (Erlangen), 1999.

[23] I. Zähle, *Renormalization of the voter model in equilibrium*, manuscript 1999.

UNIVERSITÄT ERLANGEN-NÜRNBERG, MATHEMATISCHES INSTITUT, BISMARCKSTRASSE $1\frac{1}{2}$, 91054 ERLANGEN, GERMANY

*E-mail address*: greven@mi.uni-erlangen.de

*URL*: http://www.mi.uni-erlangen.de/~greven

Canadian Mathematical Society
Conference Proceedings
Volume **26**, 2000

# New Behavioral Patterns for Two-Level Branching Systems

Andreas Greven and Kenneth J. Hochberg

*Dedicated to Don Dawson, who introduced us to measure-valued processes and hierarchically structured branching systems*

ABSTRACT. The two-level superprocess is the diffusion limit of a two-level branching Brownian motion, where particles are grouped into superparticles which themselves duplicate or vanish according to a branching dynamic, in addition to the motion and branching of the individual particles themselves.

We define three classes of initial states for two-level superprocesses and describe the corresponding patterns of longtime behavior, including two different types of equilibria. Superparticles that are initially small and frequent lead to classical behavioral patterns and yield a persistence/extinction dichotomy for dimensions $d \geq 4$ versus $d < 4$, in which the initial intensity is preserved in high dimensions. In contrast, superparticles that are initially very highly clustered can lead to long-term extinction even in high dimensions. An intermediate case exists as well, in which the superparticles initially have a critical frequency and size, which can result in partial, but not total, loss of the initial intensity. These latter two classes of initial states lead to longtime behavioral patterns in high dimensions that do not exist for ordinary, single-level branching systems or superprocesses.

## 1. Introduction

This paper describes new patterns of long-term behavior for two-level superprocesses and two-level branching systems that do not exist for single-level superprocesses or other interacting systems, such as interacting diffusions and infinite particle systems.

Multilevel branching systems were first introduced by Dawson, Hochberg and Wu ([DHW,90], [DH,91]), and many aspects of their behavior have been investigated, including path properties (Dawson, Hochberg and Vinogradov ([DHV1,94], [DHV2,95], [DHV3,96])) and certain aspects of the longtime behavior (Dawson and Hochberg ([DH,91]), Wu ([Wu3,94]), Gorostiza, Hochberg and Wakolbinger ([GHW,95]), and Hochberg and Wakolbinger ([HW,95])).

The study of multilevel branching systems has been motivated by potential application in the modeling of hierarchically structured interacting systems in biology and computer science; see [DH,91] and [DHV2,95] for more detailed references.

1991 *Mathematics Subject Classification.* Primary 90K35, 60G57.

It is particularly worth noting that such multilevel branching systems provide a rather natural example of the stochastic dynamics of systems having components that possess a rich internal structure.

The two-level superprocess is the diffusion limit of an infinite collection of particles in $\mathbf{R}^d$, each of which moves in space according to a standard Brownian motion and, at random exponentially distributed times with mean $\gamma_1^{-1}$, undergoes critical binary ("level-1") branching. In addition, the particles are grouped in some natural way into *superparticles* (or *families* or *clans*) which, independently of everything else, undergo critical binary ("level-2") branching at exponential times with mean $\gamma_2^{-1}$.

As is by now well known, the high-density, small-mass, short-lifetime diffusion limit of the infinite system of individual branching diffusing particles is the single-level *Dawson-Watanabe superprocess* $(X_t)_{t\geq0}$, which is a Markov process with state space $\mathcal{M}(\mathbf{R}^d)$, where $\mathcal{M}$ denotes the set of nonnegative Radon measures.

Similarly, the corresponding high-density, small-mass, short-lifetime diffusion limit for the two-level system described earlier is the *two-level superprocess* (or *super-2 process*) $(Y_t)_{t\geq0}$, which is a Markov process on $\mathcal{M}^2(\mathbf{R}^d) := \mathcal{M}(\mathcal{M}(\mathbf{R}^d))$ and can be defined via a martingale problem (see, e.g., [Wu1,92], [Wu2,93], [H,95] and [DHV4,96] for complete details). It should be emphasized that as a result of the simultaneous branching of all particles in a superparticle at the time of a level-2 branching, the individual particles no longer behave independently, as they do in ordinary single-level branching systems and superprocesses.

In the following sections, we describe several aspects of the longtime behavior of multilevel branching systems that are *both new and distinctive*, in the sense that such behavioral patters are not exhibited by ordinary single-level branching systems and superprocesses. This qualitatively different behavior is a result of the global interaction between different space regions, induced by the branching on the second level. This paper collects some of the basic results in this direction; a much more complete, detailed analysis can be found in Greven and Hochberg (1999).

As is the case for single-level systems, the longtime behavior depends on the relative strength of the two competing forces, migration and branching—one flattening out the mass distribution and the other causing local accumulation of mass. In low dimensions, the branching dominates, while in higher dimensions, the migration tends to dominate. For ordinary single-level branching, such behavioral changes occur at critical dimension two, while for two-level branching, this critical dimension is now four.

In dimensions $d \leq 4$, the limit of $\mathcal{L}(Y(t))$ as $t \to \infty$ will be $\delta_{\delta_{\underline{0}}}$ or $\delta_{\delta_{\underline{\infty}}}$ where $\underline{0}$ denotes the zero measure and $\underline{\infty}$ denotes the measure that satisfies $\underline{\infty}(A) = \infty$ for every Borel set with positive Lebesgue measure. Because of the influence of the level-1 branching, each of the dimensions 1,2,3,4 has a distinct way of approaching $\delta_{\delta_{\underline{0}}}$ or $\delta_{\delta_{\underline{\infty}}}$, respectively.

In dimensions $d > 4$, we find behavior that is different from the behavior in dimensions $d > 2$ in the single-level case. In the multilevel situation in high dimensions, non-trivial non-degenerate equilibria with finite intensity are also possible. The long-term behavior of $\mathcal{L}(Y(t))$ is now determined by *both* of the following properties:

    a) the initial intensity of the particles;

    b) the spatial distribution and extension of the superparticles.

Even extinction is now possible in high dimensions, in the event that the individual particles are initially grouped into superparticles that are "too large" in a certain well-defined sense. For that reason, it is not any more true (as is the case for ordinary single-level systems) that either the initial intensity is preserved in the limit as $t \to \infty$ or it is zero; now partial loss of intensity is possible. Furthermore, we find extremal equilibrium states of different types for the same value of the aggregated intensity (i.e., mean total mass).

We therefore introduce next, in Sections 2 and 3, three different classes of initial states, which will correspond to three distinct patterns of longtime behavior for $\mathcal{L}(Y(t))$ in dimensions $d > 4$. In Section 2, we cover the case of initial states that lead to classical behavior, while in Section 3, the initial states lead to new features in the longtime behavior. In Section 4, we discuss in more detail the behavior of the total-mass process, which already exhibits most of the principle phenomena.

## 2. Classical initial states

We begin by formally defining a class of initial states for which the longtime behavior can be discussed in a reasonable framework and which is used quite generally for interacting systems. Within that class, we shall then single out three subclasses exhibiting different features for the longtime behavior. We start by introducing some of the necessary ingredients.

Let $T_a : \mathbf{R}^d \to \mathbf{R}^d$ be the map given by $T_a(x) = x + a$ with $x, a \in \mathbf{R}^d$, and let $\widehat{T}_a$ and $\widehat{\widehat{T}}_a$ be the maps induced on $\mathcal{M}(\mathbf{R}^d)$ and $\mathcal{M}(\mathcal{M}(\mathbf{R}^d))$ by $T_a$ and $\widehat{T}_a$, respectively. By $\mathcal{B}(E)$, we denote the $\sigma$-algebra of Borel sets on a topological space $E$.

DEFINITION 2.1. (General class of initial states.) Let $\mathcal{T}_\theta$ denote the set of probability measures on $(\mathcal{M}(\mathcal{M}(\mathbf{R}^d)), \mathcal{B}(\mathcal{M}(\mathcal{M}(\mathbf{R}^d))))$ that satisfy the following three conditions:

(i) $P \in \mathcal{T}_\theta$ is **translation invariant**, i.e.,

$$(2.1) \qquad P(\widehat{\widehat{T}}_a A) = P(A) \quad \forall A \in \mathcal{B}(\mathcal{M}(\mathcal{M}(\mathbf{R}^d))), \ a \in \mathbf{R}^d.$$

(ii) $P \in \mathcal{T}_\theta$ is **shift ergodic**, i.e.,

$$(2.2) \qquad \widehat{\widehat{T}}_a A = A \quad \forall a \in \mathbf{R}^d, \ A \in \mathcal{B}(\mathcal{M}(\mathcal{M}(\mathbf{R}^d))), \ \text{implies } P(A) = 0 \text{ or } 1.$$

(iii) $P$ has **locally finite aggregated intensity measure**, i.e.,

$$E_P \left( \int \mu\nu(d\mu) \right), \quad \mu \in \mathcal{M}(\mathbf{R}^d), \ \nu \in \mathcal{M}(\mathcal{M}(\mathbf{R}^d))$$

is a Radon measure on $\mathbf{R}^d$.

Since $E_P(\int \mu\nu(d\mu))$ is $\widehat{T}_a$-invariant for all $a \in \mathbf{R}^d$, it follows that $E_P(\int \mu\nu(d\mu))$ is a multiple of Lebesgue measure. Thus, the parameter $\theta \in \mathbf{R}^+$ is defined as the real number that satisfies

$$(2.3) \qquad E_P \left( \int \mu\nu(d\mu) \right) = \theta \cdot dx, \quad \theta \in \mathbf{R}^+,$$

and we refer to $\theta$ as the **aggregated intensity**.

Our first class of initial states—states which lead to classical longtime behavioral patterns—corresponds to those situations in which the superparticles are initially *small and frequent* (see (2.5) below):

DEFINITION 2.2. (Type-1 initial states.) Let $\mathcal{T}_\theta^*$ denote the set of all probability measures on $(\mathcal{M}(\mathcal{M}(\mathbf{R}^d)),\ \mathcal{B}(\mathcal{M}(\mathcal{M}(\mathbf{R}^d))))$ that satisfy conditions (i)-(iii) in Definition 2.1 above and also the following condition:

(iv) For $P \in \mathcal{T}_\theta^*$,

$$(2.4) \qquad \int\limits_{\mathcal{M}(\mathcal{M}(\mathbf{R}^d))} \int\limits_{\mathcal{M}(\mathbf{R}^d)} \prod_{i=1}^n \mu(A_i)\nu(d\mu)P(d\nu) < \infty, \qquad |A_i| < \infty,$$

and we have a convergence relation for the following measures on $(\mathbf{R}^d)^n$ for every $n \geq 2$, $q \geq 1$:

$$(2.5) \quad t^{n-1} \int\limits_{\mathcal{M}(\mathbf{R}^d)} \left[ \prod_{i=1}^n \int_{\mathbf{R}^d} P_t(x_i, dy_i)\mu(dx_i) \right] \nu(d\mu) \underset{t\to\infty}{\Longrightarrow} 0 \quad \text{vaguely in } L^q(P),$$

and hence, in particular, putting $q = 1$:

$$(2.6) \quad t^{n-1} \int\limits_{\mathcal{M}(\mathcal{M}(\mathbf{R}^d))} \int\limits_{\mathcal{M}(\mathbf{R}^d)} \left[ \prod_{i=1}^n \int_{\mathbf{R}^d} P_t(x_i, dy_i)\mu(dx_i) \right] \nu(d\mu)P(d\nu) \underset{t\to\infty}{\Longrightarrow} 0,$$

where $x_i, y_i \in \mathbf{R}^d$, $i = 1, 2, ...n$; $P_t(x, dy) = (2\pi t)^{-\frac{d}{2}} \exp(-|x - y|^2/2t)dy$; and $\Longrightarrow$ denotes vague convergence of measures on $(\mathbf{R}^d)^n$.

Note that condition (iv) assures that the initial intensity of mass alone determines the long-term behavior. Note also that (2.5) implies that all moments of the total-mass process remain finite.

Recall that $\gamma_k$ $(k = 1, 2)$ denotes the rate of binary branching at the $k$-th level. To help understand the role played by (2.6), we can consider the case $\gamma_1 = 0$, which has little effect qualitatively in $d > 4$. Then the superparticles create spatially coupled fluctuations, in the sense that all those points where individual particles belong to the same superparticle are coupled, and hence, due to the preservation of the mean total mass of critical branching, *peaks of mass* tend to be created. Namely, we see that the "number" of superparticles descending from one superparticle at time zero will be positive with probability $t^{-1}$ as $t \to \infty$, and, conditioned on being positive, will be of order $t$. This explains the presence of the factor $t^{n-1}$ in connection with the moment measure appearing in (2.5). On the other hand, the mass created is always spread out by the heat kernel $P_t(x, dy)$, so that mass fluctuations might not become visible unless the superparticle is large. This explains the kernel $P_t(x, dy)$ appearing in (2.5).

Hence, the question is whether, in a window of observation, this coupling between masses in different regions induced by the level-2 branching (i) is not visible in the longtime limit (convergence to zero in (2.6)), (ii) is barely visible (finite non-zero limit in (2.6) for at least $n \leq 3$), or (iii) determines the local evolution over all times (divergence in (2.6) for all $n$). Which of the three regimes occurs will depend on the "size" and the "number" of the superparticles—that is, on their *spatial extension* and their *frequencies* in large volumes—as we shall explain below.

We see from conditions (iii) and (iv) in Definitions 2.1 and 2.2 that in the case $P \in \mathcal{T}_\theta$, the mass-to-volume ratio of a superparticle decreases as the volume increases to $\mathbf{R}^d$ at a rate that depends on the dimension:

PROPOSITION 2.3. *Assume that $P \in \mathcal{T}_\theta$. Let $B_r$ denote the ball of radius $r$ centered at $0$ in $\mathbf{R}^d$.*

(a) *If*

$$(2.7) \qquad \nu\left(\left\{\mu \in \mathcal{M}(\mathbf{R}^d) : \overline{\lim}_{r\to\infty}\left(\frac{\mu(B_r)}{r^{d-2}}\right) > 0\right\}\right) = 0 \quad P\text{-}a.s.,$$

*then $P \in \mathcal{T}_\theta^*$.*

(b) *If $P \in \mathcal{T}_\theta^*$, then for some $\varepsilon > 0$,*

$$(2.8) \qquad \underline{\lim}_{r\to\infty}\left\{\nu(\{\mu : \mu(B_r) \geq \varepsilon\})/r^2\right\} > 0 \quad P\text{-}a.s.$$

Corresponding to the classical type-1 initial states, we have the following class of equilibrium states:

DEFINITION 2.4. (Type-1 equilibrium states.)

a) Let $(X_t)_{t\geq 0}$ be the Dawson-Watanabe single-level superprocess. Then for $d > 2$, define

$$(2.9) \qquad Q_\theta := w - \lim_{t\to\infty}\mathcal{L}(X_t|X_0 = \theta \cdot dx) \text{ with } \theta \in \mathbf{R}^+.$$

$Q_\theta$ is an infinitely divisible measure and an extremal equilibrium state for the single-level superprocess (see Dawson [D,93]).

The canonical measure of $Q_\theta$ is denoted by $R_\theta^\infty \in \mathcal{M}(\mathcal{M}(\mathbf{R}^d))$, i.e., for $\mu \in \mathcal{M}(\mathbf{R}^d)$ and $\Phi \in C_0(\mathbf{R}^d)$,

$$(2.10) \quad \int \left(1 - \exp\left(-\langle\mu, \Phi\rangle\right)\right)R_\theta^\infty(d\mu) = \log\left[E_{Q_\theta}\exp\left(-\langle\mu, \Phi\rangle\right)\right].$$

b) Let $(Y_t)_{t\geq 0}$ be the two-level super-2 process. Then, for $d > 4$, define the measure

$$(2.11) \qquad P_\theta^\infty := w - \lim_{t\to\infty}\mathcal{L}(Y_t|Y_0 = R_\theta^\infty).$$

The limit exists and is an equilibrium state of the two-level superprocess; cf. Gorostiza, Hochberg and Wakolbinger [GHW,95].

We can now state the following *persistence/extinction dichotomy* for two-level branching systems with type-I initial states:

THEOREM 2.5. *Assume that $\gamma_1 > 0$ and $\gamma_2 > 0$.*

a) *Suppose that $P \in \mathcal{T}_\theta$ and that $\mathcal{L}(Y_0) = P$. Then the following dichotomy holds:*

(i) *If $d > 4$, then for $P \in \mathcal{T}_\theta^*(\subseteq \mathcal{T}_\theta)$,*

$$(2.12) \qquad\qquad \mathcal{L}(Y_t) \underset{t\to\infty}{\Longrightarrow} P_\theta^\infty \in \mathcal{T}_\theta^*.$$

(ii) *If $d \leq 4$, then*

$$(2.13) \qquad\qquad \mathcal{L}(Y_t) \underset{t\to\infty}{\Longrightarrow} \delta_{\delta_{\underline{0}}},$$

*with $\underline{0}$ being the zero measure on $\mathbf{R}^d$.*

In other words, for classical-type initial states, the process tends to local extinction in dimensions $d \leq 4$, whereas in dimensions $d > 4$, the process preserves its initial intensity in the long term.

**Remark** Here it is crucial to assume that $\gamma_1$ and $\gamma_2$ are *both* strictly positive; otherwise, this dichotomy does not hold.

## 3. Two types of non-classical initial states

In order to obtain non-classical initial states, we have to relax condition (2.5), under which the coupled fluctuations due to level-2 branching are not visible, to other conditions that allow for these fluctuations to be either barely or strongly visible. Replace (2.5) by the condition

$$(3.1) \qquad t^{n-1} \int\limits_{\mathcal{M}(\mathcal{M}(\mathbf{R}^d))} \int\limits_{\mathcal{M}(\mathbf{R}^d)} \left[ \prod_{i=1}^{n} \int\limits_{\mathbf{R}^d} P_t(x_i, dy_i)\mu(dx_i) \right] \nu(d\mu)$$

$$\underset{t\to\infty}{\Longrightarrow} \Lambda^{(n)}(dy_1, dy_2, \ldots, dy_n)$$

$$\text{in } L^q(P) \text{ for all } q \geq 1, \ n \in \mathbf{N},$$

where either $\Lambda^{(n)}$ is a Radon measure on $(\mathbf{R}^d)^n$ and $\Longrightarrow$ denotes vague convergence of measures, or, alternatively, all or some of the $\Lambda^{(n)}$ are the infinite measure, in which case we mean that

$$(3.2) \qquad t^{n-1} \int\limits_{\mathcal{M}(\mathbf{R}^d)} \left[ \prod_{i=1}^{n} \int\limits_{\mathbf{R}^d} P_t(x_i, dy)\mu(dx_i) \right] \nu(d\mu) \underset{t\to\infty}{\Longrightarrow} \infty$$

(i.e., the left-hand side diverges in $P$-probability on every set with positive $nd$-dimensional Lebesgue measure). Then, we can introduce two new classes of non-classical initial states that will yield new, non-classical patterns of longtime behavior.

One such class of initial states involves superparticles that initially have a *critical frequency and size*, so that the coupled fluctuations due to level-2-branching are barely visible, resulting in intermediate behavioral patterns:

DEFINITION 3.1. (Type-2 initial states.)
(a) Let $\Lambda = (\Lambda^{(2)}, \Lambda^{(3)}, \ldots)$ be a sequence of Radon measures on $(\mathbf{R}^d)^2, (\mathbf{R}^d)^3, \ldots$ (in particular, all $\Lambda^{(n)}$ are $\sigma$-finite), which in addition satisfies

$$(3.3) \qquad \Lambda^{(n)}((\mathbf{R}^d)^n) > 0, \quad \text{for } n \geq 2.$$

Define

$(3.4) \ \ \mathcal{T}^{**}_{\theta,\Lambda} = \{P \in \mathcal{P}(\mathcal{M}(\mathcal{M}(\mathbf{R}^d))) : (2.1) - (2.4) \text{ and } (3.1) \text{ hold for every } n\}.$

(b) Let $\Lambda_n = (\Lambda^{(2)}, \ldots, \Lambda^{(n)})$ be a sequence of nontrivial Radon measures on $(\mathbf{R}^d)^2, \ldots, (\mathbf{R}^d)^n$, and set

$(3.5) \quad \mathcal{T}^{**}_{\theta,\Lambda_n} = \{P \in \mathcal{P}(\mathcal{M}(\mathcal{M}(\mathbf{R}^d))) : (2.1) - (2.4) \text{ and } (3.1) \text{ hold,}$

$$\text{and } \Lambda^{(m)} \text{is infinite iff } m > n\}.$$

We can now construct *non-classical equilibrium states*, as follows. If $P = \mathcal{L}(Y_0) \in T^{**}_{\theta,\Lambda}$, we can construct a system in which the superparticles are $(d-2)$-dimensional objects that are situated such that there are $R^2$ of them in large balls of radius $R$. Let $\lambda^m$ denote the $m$-dimensional Lebesgue measure. We shall construct an element $\nu \in \mathcal{M}(\mathcal{M}(\mathbf{R}^d))$ with the desired properties, as follows. Set

$$(3.6) \qquad H_{x_1,x_2} = \{(x_1, x_2, y_3, \ldots, y_d) \mid y_i \in \mathbf{R}, \ i = 3, \ldots, d\}$$

and define a "superparticle" $\mu_{x_1,x_2}$ for every $x_1, x_2$ by putting

$$(3.7) \qquad \mu_{x_1,x_2}(A) = \theta\lambda^{d-2}(H_{x_1,x_2} \cap A), \quad A \in \mathcal{B}(\mathbf{R}^d).$$

We obtain a translation-invariant configuration if we set

$$(3.8) \qquad \nu = \int_{\mathbf{R}^2} \lambda^2(dx_1, dx_2)\mu_{x_1,x_2}.$$

This state has aggregated intensity $\theta\lambda$.

To find an invariant measure (recall that the intensity is constant in time), we pick a subsequence along which the Cesaro limit of $\mathcal{L}(Y_t)$ converges; i.e., with $t$ running through a suitable sequence, we define

$$(3.9) \qquad P_\theta^{\infty,*} := w - \lim_{t \to \infty} \left( \frac{1}{t} \int_0^t \mathcal{L}(Y_t|Y_0 = \nu)dt \right).$$

The measure just constructed is in fact—aside from the possibility of rotating configurations—typical for initial states leading to non-classical equilibria.

THEOREM 3.2.    (a)    *The measure $P_\theta^{\infty,*}$ has the following properties:*
   (i) *$P_\theta^{\infty,*}$ is a translation-invariant ergodic measure.*
   (ii) *$P_\theta^{\infty,*}$ is an invariant measure of the process.*
   (iii) *The intensity measure of $P_\theta^{\infty,*}$ is $\theta \cdot \lambda$; all $n$-th moment measures are locally finite; and the correlations of the total masses of all orders are positive.*
 (b)    *$P_\theta^{\infty,*}$ is different from $P_\theta^\infty$.*

Our final class of initial states involves highly clustered situations, in which there are initially *very few superparticles*, so that the coupled fluctuations due to level-2 branching become massively visible:

DEFINITION 3.3. (Type-3 initial states.)

$$(3.10) \quad \mathcal{T}_{\theta,\infty}^{**} = \{P \in \mathcal{P}(\mathcal{M}(\mathcal{M}(\mathbf{R}^d))) : (2.1)\text{–}(2.4) \text{ and } (3.2) \text{ hold } \forall n \geq 2\}.$$

For example, one special class of type-3 initial states is given by the following:

DEFINITION 3.4. (Special class of type-3 initial states.) For a law $P \in \mathcal{T}_\theta$, we say that

$$(3.11) \qquad\qquad P \in \mathcal{T}_\theta^\emptyset$$

if and only if for all $\varepsilon > 0$,

$$(3.12) \qquad \overline{\lim}_{r \to \infty} \nu(\{\mu : \mu(B_r) \geq \varepsilon\})/r^2 = 0 \quad P\text{-a.s.}$$

For this class, we obtain a *high-dimensional regime of local extinction:*

THEOREM 3.5.    (a)    *For $P \in \mathcal{T}_\theta^\emptyset$ and $d > 4$, we have*

$$(3.13) \qquad\qquad \mathcal{L}(Y_t) \underset{t \to \infty}{\Longrightarrow} \delta_{\underline{\emptyset}}.$$

 (b)    *For $P \in \mathcal{T}_{\theta,\infty}^{**}$, we have*

$$(3.14) \qquad\qquad \mathcal{L}(Y_t) \underset{t \to \infty}{\Longrightarrow} \delta_{\underline{\emptyset}}.$$

As a consequence of Theorems 2.5 and 3.5, we can easily superimpose, independently, two configurations, one satisfying the conditions of Theorem 3.5 and the other satisfying those of Theorem 3.2, and obtain an *intermediate regime of partial but not total loss of intensity*:

THEOREM 3.6. *For every* $\theta^* \in (0, \theta)$, *there exists an initial law* $P \in \mathcal{T}_\theta$ *such that (recall (2.11))*

$$(3.15) \qquad \mathcal{L}(Y_t) \underset{t \to \infty}{\Longrightarrow} P_{\theta^*}^\infty.$$

The above phenomenon is accompanied by new convergence patterns. For example, consider the case where

$$Y_0 = Y_0^1 + Y_0^2 + Y_0^3, \quad Y_0^1, Y_0^2, Y_0^3 \quad \text{independent},$$

such that

$$(3.16) \qquad \mathcal{L}(Y_0^1) \in \mathcal{T}_{\theta_1}^*, \qquad \mathcal{L}(Y_0^2) \in \mathcal{T}_{\theta_2, \Lambda}^{**}, \qquad \mathcal{L}(Y_0^3) \in \mathcal{T}_{\theta_3}^\emptyset.$$

Then, since different superparticles evolve independently, the process $Y_t$ will satisfy

$$(3.17) \qquad \mathcal{L}(Y_t) \underset{t \to \infty}{\Longrightarrow} \mathcal{L}(Y_\infty),$$

with the two properties:

$$(3.18) \qquad Y_\infty = Y_\infty^1 + Y_\infty^2, \quad Y_\infty^i, \ i = 1, 2 \quad \text{independent},$$

$$(3.19) \qquad \mathcal{L}(Y_\infty^1) = P_{\theta_1}^\infty, \quad \mathcal{L}(Y_\infty^2) \in \mathcal{T}_{\theta_2, \Lambda}^{**}.$$

In this case, the configuration $Y_t$ looks for large $t$ like an independent superposition of $Y_\infty^i, \ i = 1, 2$ with aggregated intensity $\theta_1 + \theta_2$, plus some random measure concentrated on a thin subset in space which charges these thin subsets with increasing amounts of mass as $t \to \infty$. Hence, even though the two processes $\tilde{Y}_t$, $\tilde{\tilde{Y}}_t$ defined by choosing

$$(3.20) \qquad \tilde{Y}_0 = Y_0^1 + Y_0^2 \quad \text{and} \quad \tilde{\tilde{Y}}_0 = Y_0^1 + Y_0^2 + Y_0^3$$

converge weakly to the same limit law when observed *locally*, we do see a different behavior if we look at the system from a *global* point of view, where we witness the formation of rare but high peaks. Hence, we should study the *cluster formation* in *the regime of large superparticles*.

## 4. Total-mass process

We now consider the *total-mass process*

$$(4.1) \qquad \bar{Y}_t := \int_{\mathcal{M}(\mathbf{R}^d)} \mu Y_t(d\mu) \in \mathcal{M}(\mathbf{R}^d),$$

that describes the mass aggregated over all the superparticles. This process $\bar{Y}_t$ alone already allows us to determine whether $Y_t$ becomes locally extinct or remains nontrivial as $t \to \infty$.

For $\Phi \in C_0(\mathbf{R}^d)$, define the *n-th moment measure* of $\bar{Y}_t$ by $E\big(\langle \bar{Y}_t, \Phi \rangle^n\big)$. The second moment measure of $\bar{Y}_t$ then displays the following long-term behavior:

PROPOSITION 4.1.      (a) *If $P \in \mathcal{T}_\theta^*$, then the second moment measure of the total-mass process converges to a quantity which depends on $P$ only through the aggregated intensity $\theta$:*

(4.2)      $$E\big(\langle \bar{Y}_t, \Phi \rangle^2\big) \underset{t \to \infty}{\Longrightarrow} \theta^2 \langle \lambda, \Phi \rangle^2 + A^{(2)}_{\Phi,\theta}, \quad \Phi \in C_0(\mathbf{R}^d),$$

$$A^{(2)}_{\Phi,\theta} = 4\gamma_1 \gamma_2 \theta \int\limits_0^\infty \int\limits_0^\infty \int\limits_{\mathbf{R}^d} P_{u+s}(\Phi)(x) \lambda(dx)\,du\,ds,$$

*where $A^{(2)}_{\Phi,\theta}$ is finite for $d > 4$ and infinite for $d \le 4$, if $\gamma_1 > 0, \gamma_2 > 0, \theta > 0, \Phi \ge 0, \Phi \not\equiv 0$.*
*If $\gamma_1 = 0$, then $A^{(2)}_{\Phi,\theta} = 0$.*

   (b) *If $P \in \mathcal{T}^{**}_{\theta,\Lambda}$, then the second moment measure of the total-mass process converges for $d > 4$ to a quantity which depends both on $\theta$ and on $\Lambda$ (more specifically, on $\Lambda^{(2)}$), i.e., on the superparticle structure:*

(4.3)      $$E\big(\langle \bar{Y}_t, \Phi \rangle^2\big) \underset{t \to \infty}{\Longrightarrow} \theta^2 \langle \lambda, \Phi \rangle^2 + A^{(2)}_{\Phi,\theta} + B^{(2)}_{\Phi,P}, \quad \Phi \in C_0(\mathbf{R}^d),$$

*and diverges if $d \le 4$, $\theta > 0$, $\Phi \ge 0, \Phi \not\equiv 0$ and $\gamma_1 > 0$, $\gamma_2 > 0$.*
*The quantity $B^{(2)}_{\Phi,P}$ is given by*

(4.4)      $$B^{(2)}_{\Phi,P} = 2\gamma_2 \langle \Lambda^{(2)}, \Phi \otimes \Phi \rangle.$$

   (c) *If $P \in \mathcal{T}^{**}_{\theta,\infty}$, then the second moment measure of the total-mass process diverges for all dimensions $d$ if $\gamma_2 > 0$:*

(4.5)      $$E\big(\langle \bar{Y}_t, \Phi \rangle^2\big) \underset{t \to \infty}{\Longrightarrow} \infty, \quad \Phi \in C_0(\mathbf{R}^d), \; \Phi \ge 0, \; \Phi \not\equiv 0.$$

Similarly, the third moment measure behaves as follows:

PROPOSITION 4.2. *Assume that $d > 4$ and $P = \mathcal{L}(Y_0) \in \mathcal{T}_\theta$.*
   (a) *For $P \in \mathcal{T}_{\theta,}^*$ we have*

(4.6)      $$E\big(\langle \bar{Y}_t, \Phi \rangle^3\big) \underset{t \to \infty}{\Longrightarrow} \theta^3 \langle \lambda, \Phi \rangle^3 + A^{(3)}_{\Phi,\theta} < \infty \quad \text{for } \Phi \in C_0(\mathbf{R}^d),$$

*i.e., the right-hand side depends only on $\theta$, i.e., on $\mathcal{L}(\bar{Y}_0)$, and not on $\mathcal{L}(Y_0)$.*
*If $\gamma_1 = 0$, then $A^{(3)}_{\Phi,\theta} = 0$.*
   (b) *For $P \in \mathcal{T}^{**}_{\theta,\Lambda_3}$ with $\Lambda_3$ a nontrivial Radon measure or $P \in \mathcal{T}^{**}_{\theta,\Lambda}$, we have*

(4.7)      $$E\big(\langle \bar{Y}_t, \Phi \rangle^3\big) \to \theta^3 \langle \lambda, \Phi \rangle^3 + A^{(3)}_{\Phi,\theta} + B^{(3)}_{\Phi,P},$$

*where now $B_\Phi$ does depend on $\mathcal{L}(Y_0)$ and not only on $\mathcal{L}(\bar{Y}_0)$. Specifically,*

(4.8)      $$B^{(3)}_{\Phi,P} = 4\gamma_2 \langle \Lambda^{(3)}, \Phi \otimes \Phi \otimes \Phi \rangle + 4\gamma_2 \theta \langle \lambda, \Phi \rangle \langle \Lambda^{(2)}, \Phi \otimes \Phi \rangle.$$

   (c) *For $P \in \mathcal{T}^{**}_{\theta,\infty}$, in the case $\gamma_2 > 0$, we have*

(4.9)      $$E\big(\langle \bar{Y}_t, \Phi \rangle^3\big) \underset{t \to \infty}{\Longrightarrow} \infty \quad \text{for } \Phi \in C_0(\mathbf{R}^d), \; \Phi \ge 0, \; \Phi \not\equiv 0.$$

Representation of the moment measures of $\bar{Y}_t$ via a recursive scheme then yields the following long-term behavior for the $n$-th moment measure of $\bar{Y}_t$ :

PROPOSITION 4.3. *Assume that $d > 4$ and $P \in \mathcal{T}_\theta$. Then, for every $n \ge 2$:*

(a) *For $P \in \mathcal{T}_\theta^*$, we have*

(4.10)    $E\big(\langle \bar{Y}_t, \Phi \rangle^n\big) \underset{t \to \infty}{\Longrightarrow} \theta^n \langle \lambda, \Phi \rangle^n + A_{\Phi,\theta}^{(n)} < \infty \quad for \, \Phi \in C_0^+(\mathbf{R}^d),$

where $A_{\Phi,\theta}^{(n)}$ depends only on $\mathcal{L}(\bar{Y}_0)$ through $\theta$ and not on $\mathcal{L}(Y_0)$.
If $\gamma_1 = 0$, then $A_{\Phi,\theta}^{(n)} = 0$.
(b) *For $P \in \mathcal{T}_{\theta,\Lambda}^{**}$, we have*

(4.11)  $E\big(\langle \bar{Y}_t, \Phi \rangle^n\big) \underset{t \to \infty}{\Longrightarrow} \theta^n \langle \lambda, \Phi \rangle^n + A_{\Phi,\theta}^{(n)} + B_{\Phi,P}^{(n)} < \infty \quad for \, \Phi \in C_0^+(\mathbf{R}^d),$

where $B_{\phi,P}$ does depend on $\mathcal{L}(Y_0)$ and not only on $\mathcal{L}(\bar{Y}_0)$.
(c) *For $P \in \mathcal{T}_{\theta,\infty}^{**}$, we have for $\gamma_2 > 0$:*

(4.12)      $E\big(\langle \bar{Y}_t, \Phi \rangle^n\big) \underset{t \to \infty}{\Longrightarrow} \infty \quad for \, \Phi \in C_0(\mathbf{R}^d), \, \Phi \geq 0, \, \Phi \not\equiv 0.$

# References

[D]    D.A. Dawson, *Measure-Valued Markov Processes*, Ecole d'Eté de Probabilités de Saint-Flour XXI-1991, P.L. Hennequin, Lecture Notes in Mathematics, **vol. 1451**, Springer, Berlin [1993], pages 191-260

[DH]   D.A. Dawson and K.J. Hochberg, *A multilevel branching model*, Adv. Appl. Prob. **vol. 23**, [1991], pages 701-715

[DHV1] D.A. Dawson, K.J. Hochberg and V. Vinogradov, *On path properties of super-2 processes I*, D.A. Dawson, Measure-valued Processes, Stochastic Partial Differential Equations and Interacting Systems, CRM Proceedings and Lecture Notes, **vol. 5**, AMS, Providence [1994], pages 69-82

[DHV2] D.A. Dawson, K.J. Hochberg and V. Vinogradov, *On path properties of super-2 processes II*, M.G. Cranston and M.A. Pinsky, Proceedings of Symposia in Pure Mathematics Series, **vol. 57**, AMS, Providence [1995], pages 385-403

[DHV3] D.A. Dawson, K.J. Hochberg and V. Vinogradov, *High-density limits of hierarchically structured branching-diffusing populations*, Stoch. Proc. Appl., **vol. 62** [1996], pages 191-222

[DHV4] D.A. Dawson, K.J. Hochberg and V. Vinogradov, *On weak convergence of branching particle systems undergoing spatial motion*, Stochastic Analysis: Random Fields and Measure-Valued Processes, Israel Mathematical Conference Proceedings, **vol. 10**, AMS, Providence [1996], pages 65-79

[DHW]  D.A. Dawson, K.J. Hochberg, and Y. Wu, *Multilevel branching systems*, T. Hida, H.H. Kuo, J. Potthoff and L. Streit, White Noise Analysis: Mathematics and Applications, World Scientific Publ. [1990], pages 93-107

[GHW]  L.G. Gorostiza, K.J. Hochberg and A. Wakolbinger, *Persistence of a critical super-2 process*, J. Appl. Probab., **vol 32** [1995], pages 534-540

[GH]   A. Greven and K.J. Hochberg, *The longtime behavior of multilevel branching systems*, manuscript [1999]

[H]    K.J. Hochberg, *Hierarchically structured branching populations with spatial motion*, Rocky Mountain J. Math., **vol 25** [1995], pages 269-283

[HW]   K.J. Hochberg and A. Wakolbinger, *Non-persistence of two-level branching particle systems in low dimensions*, A. Etheridge, Stochastic Partial Differential Equations, London Mathematical Society Lecture Note Series, **vol. 216**, Cambridge U. Press [1995], pages 126-140

[Wu1]  Y. Wu, *Dynamic Particle Systems and Multilevel Measure Branching Processes*, Carleton U. thesis [1992]

[Wu2]  Y. Wu, *Multilevel birth and death particle system and its continuous diffusion*, Adv. Appl. Probab., **vol 25** [1993], pages 549-569

[Wu3]  Y. Wu, *Asymptotic behavior of the two level measure branching process*, Ann. Probab., **vol. 22** [1994], pages 854-874

MATHEMATISCHES INSTITUT, UNIVERSITÄT ERLANGEN-NÜRNBERG, BISMARCKSTRASSE 1 1/2, D-91054 ERLANGEN, GERMANY

*E-mail address*: greven@mi.uni-erlangen.de

DEPARTMENT OF MATHEMATICS AND COMPUTER SCIENCE, BAR-ILAN UNIVERSITY, 52900 RAMAT-GAN, ISRAEL & COLLEGE OF JUDEA AND SAMARIA, 44837 ARIEL, ISRAEL

*E-mail address*: hochberg@macs.biu.ac.il

Canadian Mathematical Society
Conference Proceedings
Volume **26**, 2000

# Set-Indexed Markov Processes

## B. Gail Ivanoff and Ely Merzbach

*Dedicated to Professor Donald A. Dawson*

ABSTRACT. Three types of Markov properties (Markov, sharp Markov and
strong Markov) are defined for set-indexed stochastic processes, and their relationships are studied. It is proved that a wide class of set-indexed processes
with independent increments are Markov and sharp Markov. Under general
conditions, it is shown that sharp Markov point processes are also strong
Markov.

## 1. Introduction

It is well known that any $\mathbf{R}_+$-indexed process with independent increments
is a Markov process. When defining a Markov process for multiparameter or set-
indexed processes, it is natural to hope that the same property holds. However, this
is not the case in the literature to date. The main goal of this work is to present
some satisfactory definitions for Markov properties of a set-indexed process, and to
show that set-indexed processes with independent increments are Markov.

There are several legitimate generalizations of the Markov property for a process
indexed by a partially ordered set, and the problem of finding a good definition
for a multiparameter stochastic process to be a Markov process has produced an
impressive bibliography. In fact, the story began in 1948 when P. Lévy defined the
Markov property in the following way: we say that the process $\{X_z, \ z \in \mathbf{R}_+^2\}$ has
the sharp Markov property with respect to $A$ if the $\sigma$-algebras $\sigma\{X_z, \ z \in A\}$ and
$\sigma\{X_z, \ z \in A^c\}$ are conditionally independent given $\sigma\{X_z, \ z \in \partial A\}$. Do there exist
nondegenerate two-parameter processes with independent increments which satisfy
the sharp Markov property with respect to a sufficiently large class of sets [**17**]?

2000 *Mathematics Subject Classification.* Primary 60J99; Secondary 60G51.

*Key words and phrases.* set-indexed Markov process; sharp Markov; strong Markov; independent increments.

The first named author was supported by a grant from the Natural Sciences and Engineering
Research Council of Canada.

Research done while the second named author was visiting the University of Ottawa. With
this 13th joint paper, he especially thanks the first author for her warm hospitality and the quiet
environment perfectly suited for research.

We shall briefly discuss this question in the context of three examples, the first of which is the Brownian sheet. In 1976, J.B. Walsh showed that the Brownian sheet does not satisfy the sharp Markov property in the sense of Lévy even with respect to triangles ([25], see also [7] for some generalizations).

To obtain a broader class of processes with a Markov property, several other definitions were proposed. A weaker definition called the 'germ-field Markov property' was introduced by McKean [19] and Pitt [22] and was studied in the Gaussian case by several authors [3], [15], [18], [4]. In this definition, the $\sigma$-algebra $\sigma\{X_z, \ z \in \partial A\}$ is replaced by the larger germ $\sigma$-algebra of the boundary of $A$, that is, by $\cap_{\varepsilon>0}\sigma\{X_z, \ d(z, A) < \varepsilon\}$. Another type of definition, involving the partial order induced by the Cartesian coordinates, was introduced by Nualart and Sanz [21], Guyon and Prum [12], and extended by Korezlioglu, Lefort and Mazziotto [16]. Yet another definition was proposed by Wong and Zakai [26] and concerns processes parametrized by smooth curves in $\mathbf{R}_+^2$.

Most of the papers on Markov random fields deal with the Gaussian case; the literature on the Markov property for point processes, which constitutes our second example, is quite limited.

Carnal and Walsh [2] proved that the Poisson sheet satisfies the sharp Markov property relative to all bounded open relatively convex sets, and this property fails to be true for general unbounded open relatively convex sets. Russo [24] proved the following extension: every two-parameter point process with independent increments satisfies the sharp Markov property relative to any finite union of rectangles whose sides are parallel to the axes. This result was extended by Merzbach and Nualart [20] for point processes to a class of compact sets using a detailed analysis of the germ $\sigma$-algebras. In particular, it is shown that for a set $A$ of this class and for an appropriate point process $\{M_z, \ z \in \mathbf{R}_+^2\}$, the $\sigma$-algebra $\sigma\{M_z, \ z \in A\}$ is equal to the germ $\sigma$-algebra $\cap_{\varepsilon>0}\sigma\{M_z, d(z, A) < \varepsilon\}$. This class of sets contains the triangles and the lower layers. ($A$ is a lower layer if $z \in A \Rightarrow [0, z] \subseteq A$). Further extensions are given by Dalang and Walsh [6].

It is tempting to hope that the preceding results would hold not only for point processes but for any pure jump process with independent increments, but this is not the case as our third example illustrates.

This example is a pure jump process which was studied by Dalang and Walsh [6]: let $X = \{X_z, \ z \in \mathbf{R}_+^2\}$ be the two-parameter signed Poisson process (each jump point is independently assigned a mass of +1 or -1 with probability 1/2). Consider the triangle $\Delta$ with vertices (0,0), (0,1) and (1,0). Imagine the following two scenarios. The process may have two jump points in the triangle at the points $\left(\frac{1}{5}, \frac{3}{5}\right)$ and $\left(\frac{3}{5}, \frac{1}{5}\right)$ and masses +1 at each of these two points. With this configuration, $X(\Delta) = 2$. Now suppose, for example, that the two points are $\left(\frac{1}{5}, \frac{1}{5}\right)$ with mass +1 and $\left(\frac{2}{5}, \frac{2}{5}\right)$ with mass $-1$. With this configuration $X(\Delta) = 0$. However, on the diagonal $\partial\Delta = \{(s,t) : s + t = 1\}$, the values of the process are exactly the same for these two different configurations. ($X_z$ is equal to 1 for points $z = (s,t)$ on the set $\left\{(s, 1 - s) : \frac{1}{5} \leq s < \frac{2}{5} \text{ or } \frac{3}{5} \leq s < \frac{4}{5}\right\}$ and zero elsewhere).

This process, with independent increments, has the sharp Markov property on rectangles, but since $X(\Delta)$ is not $\mathcal{F}_{\partial\Delta}$-measurable, it is not sharp Markov on triangles or lower sets.

It is clear from the discussion of these three examples that Lévy's definition of the sharp Markov property is not completely satisfactory. Fundamental problems arise from the fact that the process is indexed by points whereas the Markov property is defined with respect to lines or other kinds of sets.

Here our approach is very simple. Let $T$ be a metric space, $\mathcal{A}$ a collection of closed subsets of $T$, and $X = \{X_A, \ A \in \mathcal{A}\}$ a stochastic process indexed by the parameter set $\mathcal{A}$. In this work – as in the classical case on $\mathbf{R}_+$ in which processes are time indexed and the Markov property is defined by a "time point" – both our processes and the Markov properties are defined by the same class of objects which is a collection of closed subsets of a metric space. We believe that this framework is more natural and more flexible in that we use only one class of objects in our definitions.

We shall use the partial order of set-inclusion to define a "Markov" set-indexed process, analogous to that defined in [16] in the case of $\mathbf{R}_+^2$. We then define a "sharp Markov" property in analogy to Lévy's definition, and show that in fact these two definitions are equivalent in certain circumstances. As well, we will observe that in our framework, all processes with independent increments are both Markov and sharp Markov. As will be shown later, this fact does not contradict the previous discussion.

The notion of a Markov property on random sets leads to the notion of the strong Markov property. It was introduced first by Evstigneev [10] for random fields, but with a different kind of $\sigma$-algebra. In [26], Wong and Zakai conclude their work by defining the strong Markov property for a multi-parameter process. Merzbach and Nualart [20] proved that every strictly simple Markov point process whose intensity is absolutely continuous with respect to the Lebesgue measure possesses the strong Markov property relative to all the bounded stopping lines.

Here, too, our definitions are particularly well-adapted to a study of the strong Markov property. Using the theory of stopping sets developed in [13], we show that a large class of processes, including point processes, are strong Markov.

The paper is organized as follows.

In the first section, we present the general framework for the study of set-indexed processes (see, for example, [13] or [14]).

In Section two, we give a new definition of the Markov proprty for a set-indexed stochastic process. This definition is inspired by that of Korezlioglu, Lefort and Mazziotto [16] given for the two-parameter case. As will be shown later, this definition is both intuitive and well-adapted to our framework; it is a generalization of the $\mathbf{R}^n$-case and leads to interesting results. We shall prove that any process with independent increments satisfies the Markov property and, conversely, that the Markov property implies a conditional independence property (called conditional orthogonality) for the filtration generated by the process.

Section three is devoted to the sharp Markov property. We prove there that a process is Markov if and only if it is sharp Markov and the conditional orthogonality property holds for the filtration generated by the process.

Finally, the last section deals with the strong Markov property. We define random sets and stopping sets, and prove that the sharp Markov property is equivalent to the strong Markov property relative to discrete adapted random sets. If the $\sigma$-algebras are also outer-continuous, this property is extended to stopping sets.

## 2. The set-indexed framework

The framework used in this paper is the same as that presented in other articles by the authors in the theory of set-indexed martingales; see, for example, [**14**]. We use this structure as it has a well-developed theory of stopping sets, which will lead to a natural definition of the strong Markov property (cf. Section 5).

We assume that $T$ is a $\sigma$-compact complete separable metric space. $\mathcal{A}$ is a semilattice of compact connected subsets of $T$, closed under arbitrary intersections satisfying:

- $\emptyset \in \mathcal{A}$.
- $A, B \in \mathcal{A}, A, B \neq \emptyset \Rightarrow A \cap B \neq \emptyset$
- Separability from above: There exists an increasing sequence of finite sub-semilattices $\mathcal{A}_n \subseteq \mathcal{A}$ closed under intersections such that $\emptyset \in \mathcal{A}_n$ and there exists a sequence of functions $g_n : \mathcal{A} \to \mathcal{A}_n(u) \cup \{T\}$ ($\mathcal{A}_n(u)$ is the class of unions of sets in $\mathcal{A}_n$) preserving countable intersections and finite unions satisfying $A \subseteq (g_n(A))^\circ$, $g_n(A) \subseteq g_m(A)$ if $n \geq m$, and $A = \cap_n g_n(A)$ $\forall A \in \mathcal{A}$. Also:
    1. If $A \in \mathcal{A}$, $B \in \mathcal{A}(u)$ ($\mathcal{A}(u)$ is the class of finite unions of sets in $\mathcal{A}$) and $A \subseteq B^\circ$, then there exists $N < \infty$ such that $g_n(A) \subseteq B^\circ$ $\forall n \geq N$.
    2. If $A, A_1, ..., A_k \in \mathcal{A}$ are such that $A \not\subseteq \cup_1^k A_j$, then for $I \subseteq \{1, ..., k\}$, if $A \cap \cap_I A_i \not\subseteq A_j^\circ$, $j = 1, ..., k$, then $g_n(A \cap \cap_I A_i) \cap \cap_j A_j^c \neq \emptyset$.

COMMENTS 2.1. 1. We note that since nonempty sets in $\mathcal{A}$ have nonempty intersection, sets in $\mathcal{A}(u)$ are connected. Also, if $A, B \in \mathcal{A}$ and $A \not\subseteq B$, then since $A$ is connected, $A \cap B \not\subseteq B^\circ$.

2. Condition 1. above is satisfied whenever $T$ is compact, or if there exists an increasing sequence $(B_n)$ of sets in $\mathcal{A}$ such that $\cup_n B_n = T$ and for any set $A \in \mathcal{A}$, there exists $m$ such that $A \subseteq B_m^\circ$.

3. Condition 2. above is a technical condition which is required only for Theorem 3.3, Lemma 3.4 and Corollary 4.3. It is easily seen to be satisfied in all the examples given below.

There are many examples of classes of sets satisfying the above properties. We give a few below; for details and other examples, see [**14**].

EXAMPLES 2.2. 1. $T = [0, 1]^d$ or $\mathbf{R}_+^d$, $\mathcal{A} = \{[0, t] : t \in T\} \cup \{\emptyset\}$.

2. $T = [a, b]^d$, $a < 0 < b$, or $T = \mathbf{R}^d$, $\mathcal{A} = \{[0, t] : t \in T\} \cup \{\emptyset\}$. This will give a kind of $2^d$-directional structure.

3. $T = [0, 1]^d$ or $\mathbf{R}_+^d$, $\mathcal{A} = \{A : A$ a compact lower layer$\} \cup \{\emptyset\}$. ($A$ is a lower layer if $t \in A \Rightarrow [0, t] \subseteq A$, $\forall t$.)

4. $T = [0, 1]^d$ or $\mathbf{R}_+^d$, $\mathcal{A} = \{A : A$ a compact, convex lower layer$\} \cup \{\emptyset\}$.

We introduce another class of sets generated by $\mathcal{A}$. Recall that $\mathcal{A}(u)$ is the class of finite unions of sets in $\mathcal{A}$. $\mathcal{C}$ is the class of all sets of the form $C = A \setminus B$, $A \in \mathcal{A}$ and $B \in \mathcal{A}(u)$. *We note here that all of our Markov properties will be defined with respect to the class $\mathcal{A}(u)$.* This is the natural choice for two main reasons. First, increments of set-indexed functions are defined over sets in $\mathcal{C}$, and since it is natural to expect that processes with independent increments be Markov, we need to be able to consider the 'lower boundary' of a set in $\mathcal{C}$, which is defined by a set in $\mathcal{A}(u)$. Second, in analogy with the classical case, we will be defining a strong Markov property with respect to a 'stopping set'. Stopping sets were defined and studied in detail in [**13**] and, in particular, must take their values in $\mathcal{A}(u)$.

Before continuing, we observe that the functions $g_n$ may be extended to the class $\mathcal{A}(u)$: for $B = \cup_1^k B_i$, $B_1, ..., B_k \in \mathcal{A}$, $g_n(B) = \cup_1^k g_n(B_i)$.

Now, given a probability space $(\Omega, \mathcal{F}, P)$, let $X = \{X_A : A \in \mathcal{A}\}$ be an $\mathcal{A}$-indexed stochastic process. We shall always assume that $X_\emptyset = 0$ and that $X$ has a finitely additive extension to $\mathcal{C}$. We say that $X$ has *independent increments* if $X_{C_1}, ...., X_{C_k}$ are independent r.v.'s whenever $C_1, ..., C_k$ are disjoint sets in $\mathcal{C}$. We say that $X$ is *outer-continuous* if for any decreasing sequence $(A_n) \subseteq \mathcal{A}$, $X_{\cap_1^\infty A_n} = \lim_{n \to \infty} X_{A_n}$. Note that by additivity, $X$ is outer-continuous as well for a decreasing sequence $(B_n) \subseteq \mathcal{A}(u)$, provided that $\cap_n B_n \in \mathcal{A}(u)$. (Note: $\mathcal{A}(u)$ is not necessarily closed under countable intersections.)

Next, we define the $\sigma$-fields generated by the process $X$ which are associated with various classes of subsets of $T$. In general, for $B \subseteq T$ such that $B = D$, $D^\circ$, $\partial D$, $D^c$ or $\overline{D^c}$ for some $D \in \mathcal{A}(u)$, define

$$\mathcal{F}_B = \sigma\{X_A : A \in \mathcal{A}(B)\},$$

where for $D \in \mathcal{A}(u)$,

$$
\begin{aligned}
\mathcal{A}(D) &= \{A \in \mathcal{A} : A \subseteq D\} \\
\mathcal{A}(D^\circ) &= \{A \in \mathcal{A} : A \subseteq D^\circ\} \\
\mathcal{A}(\partial D) &= \mathcal{A}(D) \setminus \mathcal{A}(D^\circ) \\
\mathcal{A}(D^c) &= (\mathcal{A}(D))^c = \{A \in \mathcal{A} : A \not\subseteq D\} \\
\mathcal{A}(\overline{(D^c)}) &= (\mathcal{A}(D^\circ))^c = \mathcal{A}(D^c) \cup \mathcal{A}(\partial D)
\end{aligned}
$$

We remark here that for $D_1, D_2 \in \mathcal{A}(u)$, $\mathcal{F}_{D_1 \cup D_2} = \mathcal{F}_{D_1} \vee \mathcal{F}_{D_2}$. This is an easy consequence of the fact that $X$ is additive.

Next we define 'conditional orthogonality':

DEFINITION 2.3. **(CO)**: Condition (CO) is said to be satisfied if for every $B_1, B_2 \in \mathcal{A}(u)$,

$$E(\cdot \mid \mathcal{F}_{B_1} \mid \mathcal{F}_{B_2}) = E(\cdot \mid \mathcal{F}_{B_2} \mid \mathcal{F}_{B_1}) = E(\cdot \mid \mathcal{F}_{B_1} \cap \mathcal{F}_{B_2}).$$

(Equivalently, $\mathcal{F}_{B_1}$ and $\mathcal{F}_{B_2}$ are conditionally independent given $\mathcal{F}_{B_1} \cap \mathcal{F}_{B_2}$. We denote this by

$$\mathcal{F}_{B_1} \perp \mathcal{F}_{B_2} \mid \mathcal{F}_{B_1} \cap \mathcal{F}_{B_2}.)$$

We remark that the above condition is close to condition (CI) as defined in [**13**] and [**14**].

## 3. The Markov property

We begin with the 'Markov' property, which exploits the partial order of set-intersection. Our definition is motivated by [**16**], where Markov properties are defined for $\mathbf{R}_+^2$-indexed processes. These definitions are very close when restricted to Example 2.2.1.

DEFINITION 3.1. $X$ is **Markov** if for every $B \in \mathcal{A}(u)$ and $A_1, ..., A_k \in \mathcal{A}$ such that $A_i \not\subseteq B$, $i = 1, ..., k$, and for every bounded measurable function $f : \mathbf{R}^k \to \mathbf{R}$,

$$E[f(X_{A_1}, ..., X_{A_k}) \mid \mathcal{F}_B] = E[f(X_{A_1}, ..., X_{A_k}) \mid \mathcal{F}_{(\partial B)} \cap \mathcal{F}_{(\cup_1^k A_i)}].$$

We have the following:

THEOREM 3.2. *If $X$ is Markov, then (CO) holds.*

PROOF. Let $D_1, D_2 \in \mathcal{A}(u)$. We have that $\mathcal{F}_{D_1} \perp \mathcal{F}_{D_2} \mid \mathcal{F}_{D_1} \cap \mathcal{F}_{D_2}$ if for every integrable $\mathcal{F}_{D_2}$-measurable random variable $Y$, $E(Y \mid \mathcal{F}_{D_1}) = E(Y \mid \mathcal{F}_{D_1} \cap \mathcal{F}_{D_2})$. By a monotone class argument, it suffices to show for $A_1, ..., A_k \in \mathcal{A}(D_2)$, $A_i \not\subseteq D_1$, $i = 1, ..., k$, that

$$E[f(X_{A_1}, ..., X_{A_k}) \mid \mathcal{F}_{D_1}] = E[f(X_{A_1}, ..., X_{A_k}) \mid \mathcal{F}_{D_1} \cap \mathcal{F}_{D_2}]$$

for any bounded measurable function $f : \mathbf{R}^k \to \mathbf{R}$. However, since $X$ is Markov,

$$E[f(X_{A_1}, ..., X_{A_k}) \mid \mathcal{F}_{D_1}] = E[f(X_{A_1}, ..., X_{A_k}) \mid \mathcal{F}_{\partial D_1} \cap \mathcal{F}_{\cup_1^k A_i}].$$

Because $\cup_1^k A_i \subseteq D_2$, it follows that $\mathcal{F}_{\partial D_1} \cap \mathcal{F}_{\cup_1^k A_i} \subseteq \mathcal{F}_{D_1} \cap \mathcal{F}_{D_2}$, and the proof is complete. $\qquad \square$

In analogy with Markov processes on $\mathbf{R}_+$, we have:

THEOREM 3.3. *If $X$ has independent increments, then $X$ is Markov.*

PROOF. Assume that $B \in \mathcal{A}(u)$ and that $A_1, ..., A_k \in \mathcal{A}$ are such that $A_i \not\subseteq B$, $i = 1..., k$. Now, if $f : \mathbf{R}^k \to \mathbf{R}$ is a bounded measurable function, then

$$E[f(X_{A_1}, ..., X_{A_k}) \mid \mathcal{F}_B] =$$

$$(3.1) \qquad E[f(X_{A_1 \setminus B} + X_{A_1 \cap B}, ..., X_{A_k \setminus B} + X_{A_k \cap B}) \mid \mathcal{F}_B].$$

Since $X$ has independent increments, $X_{A_i \setminus B}$ is independent of $\mathcal{F}_B$, for $i = 1, ..., k$. Thus, it will be sufficient to show that $X_{A_i \cap B}$ is $\mathcal{F}_{\partial B} \cap \mathcal{F}_{\cup_1^k A_i}$-measurable, for $i = 1, ..., k$. This is a result of the following lemma. $\qquad \square$

LEMMA 3.4. *For any $A \in \mathcal{A}$ and $B \in \mathcal{A}(u)$ such that $A \not\subseteq B$, $X_{A \cap B}$ can be expressed as a linear combination of terms of the form $X_D$ where $D \in \mathcal{A}(\partial B) \cap \mathcal{A}(A)$ (i.e. $X_{A \cap B}$ is $\mathcal{F}_{\partial B} \cap \mathcal{F}_A$ measurable).*

PROOF. Suppose that $B = \cup_{i=1}^k B_i$, $B_1, ..., B_k \in \mathcal{A}$. Since $X$ is additive, we use the usual inclusion-exclusion formula for the union of sets to obtain

$$(3.2) \qquad X_{A \cap B} = \sum_1^k X_{A \cap B_i} - \sum_{i \neq j} X_{A \cap B_i \cap B_j} + ... + (-1)^{k+1} X_{A \cap B_1 \cap ... \cap B_k}.$$

Let $I \subseteq \{1, ..., k\}$. We claim that if $A \cap \cap_I B_i \subseteq B_j^\circ$ for some $j \in \{1, ..., k\}$, then in fact $X_{A \cap \cap_I B_i}$ is cancelled by another term in the right hand side of (3.2), and does not contribute to the final value of the sum. Thus, all remaining terms will be of the form $X_{A \cap \cap_I B_i}$ such that $A \cap \cap_I B_i \not\subseteq B_j^\circ$ for any $j \in \{1, ..., k\}$. By 2. of the separability assumption, it follows that for every $n$, $g_n(A \cap \cap_I B_i) \cap B^c \neq \emptyset$. By 1. of the separability assumption, $A \cap \cap_I B_i \not\subseteq B^\circ$, and so $A \cap \cap_I B_i \in \mathcal{A}(\partial B) \cap \mathcal{A}(A)$.

It remains to prove the claim above. We proceed according to the following algorithm:

Step 1: Compare $A \cap B_1$ with $B_i$, $i = 2, ..., k$. If $A \cap B_1 \subseteq B_j$, say, then all terms $X_{A \cap \cap_I B_i}$ with $1 \in I$ may be removed from the sum, since if $j \notin I$, $X_{A \cap \cap_I B_i}$ is cancelled by $X_{A \cap \cap_{I \cup \{j\}} B_i}$. Continue in this way, and compare each term of the form $A \cap B_i$ with each $B_j, j \neq i$. If $A \cap B_i \subseteq B_j$, and if terms $X_{A \cap \cap_I B_i}$ with $j \in I$ have not already been removed from the sum, remove all terms with $i \in I$. If the terms with $j \in I$ have been removed, then there exists some $n$ such that $X_{A \cap B_n}$ remains, and $A \cap B_i \subseteq A \cap B_j \subseteq B_n$. If $n \neq i$, then delete all terms with $i \in I$. If $n = i$, retain $X_{A \cap B_i}$ for the moment, and continue the comparisons.

Now suppose that $X_{A \cap B_i}$ has not been deleted. Either $A \cap B_i \not\subseteq B_j$ or $A \cap B_i = A \cap B_j$, $\forall j$. By Comment 2.1.1, it follows that $A \cap B_i \not\subseteq B_j^\circ$.

We proceed inductively:

Step $h$: We now consider any terms of the form $X_{A \cap \cap_I B_i}$ where $I = \{i_1, ..., i_h\} \subseteq 1, ..., k$ which were not removed from the sum in the first $h - 1$ steps. We compare $A \cap \cap_I B_i$ with each $B_j, j \notin I$. If $A \cap \cap_I B_i \not\subseteq B_j$, then retain $X_{A \cap \cap_I B_i}$ for the moment. If $A \cap \cap_I B_i \subseteq B_j$, there are two possibilities. If $X_{A \cap \cap_I B_i \cap B_j}$ is still in the sum, then remove all remaining terms of the form $X_{A \cap \cap_{I'} B_i}$ with $I \subseteq I'$. On the other hand, if $X_{A \cap \cap_I B_i \cap B_j}$ has already been removed from the sum, then there exists some subset $I'' \subset I$ and $\ell \neq j$, $\ell \notin I''$, such that $A \cap \cap_{I''} B_i \cap B_j \subseteq B_\ell$ and $A \cap \cap_I B_i \cap B_\ell$ has not yet been removed from the sum. If $\ell \notin I$, then $A \cap \cap_I B_i = A \cap \cap_I B_i \cap B_j \subseteq B_\ell$, and we remove all remaining terms of the form $X_{A \cap \cap_{I'} B_i}$ with $I \subseteq I'$. If $\ell \in I \setminus I''$, then

(3.3) $$A \cap \cap_I B_i = A \cap \cap_I B_i \cap B_j,$$

and

(3.4) $$A \cap \cap_{I \setminus \{\ell\}} B_i \cap B_j = A \cap \cap_I B_i \cap B_j.$$

Combining equations (3.3) and (3.4), we obtain

(3.5) $$A \cap \cap_I B_i = A \cap \cap_{I \setminus \{\ell\}} B_i \cap B_j,$$

and in this case we retain the term $X_{A \cap \cap_I B_i}$ for the moment. We observe that $A \cap \cap_{I \setminus \{\ell\}} B_i \not\subseteq B_j$. This follows since if $A \cap \cap_{I \setminus \{\ell\}} B_i \subseteq B_j$, then $A \cap \cap_{I \setminus \{\ell\}} B_i = A \cap \cap_{I \setminus \{\ell\}} B_i \cap B_j \subseteq B_\ell$. However, according to our algorithm, in this case $A \cap \cap_{I \setminus \{\ell\}} B_i$ would have been removed at the previous step.

After completing the $k$ steps in this procedure, we consider the remaining terms. Let $X_{A \cap \cap_I B_i}$ be one such term. If $j \notin I$, either $A \cap \cap_I B_i \not\subseteq B_j$ or $A \cap \cap_I B_i \subseteq B_j$ but there exists $\ell \in I$ such that $A \cap \cap_I B_i = A \cap \cap_{I \setminus \{\ell\}} B_i \cap B_j$ and $A \cap \cap_{I \setminus \{\ell\}} B_i \not\subseteq B_j$. By Comment 2.1.1 and (3.5), in this case $A \cap \cap_I B_i = A \cap \cap_{I \setminus \{\ell\}} B_i \cap B_j \not\subseteq B_j^\circ$.

Now, if $j \in I$, it is clear that $A \cap \cap_{I \setminus \{j\}} B_i \not\subseteq B_j$ (otherwise $A \cap \cap_I B_i$ would have been removed), and again by Comment 2.1.1, $A \cap \cap_I B_i \not\subseteq B_j^\circ$.

Thus, for every $j \in 1, ..., k$, $A \cap \cap_I B_i \not\subseteq B_j^\circ$, proving the claim and completing the proof of the theorem. $\square$

## 4. The sharp Markov propery

In this section, we consider the sharp Markov property. It is defined in a manner analogous to the definition found in [20], but note that our $\sigma$-algebras are defined using processes indexed by sets, not points.

DEFINITION 4.1. $X$ is **sharp Markov** if for every $B \in \mathcal{A}(u)$,

$$\mathcal{F}_B \perp \mathcal{F}_{B^c} \mid \mathcal{F}_{\partial B}.$$

THEOREM 4.2. $X$ *is Markov if and only if (CO) holds and $X$ is sharp Markov.*

PROOF. First we assume that $X$ is Markov. That $X$ is sharp Markov follows immediately by a monotone class argument, and it was proven in Theorem 3.2 that (CO) holds.

Conversely, let $B \in \mathcal{A}(u)$ and suppose that $A_1, ..., A_k \in \mathcal{A}$ are such that $A_i \not\subseteq B$, $i = 1, ..., k$. Let $f : \mathbf{R}^k \to \mathbf{R}$ be bounded and measurable. If (CO) holds,

(4.1) $$E[f(X_{A_1}, ..., X_{A_k}) \mid \mathcal{F}_B] = E[f(X_{A_1}, ..., X_{A_k}) \mid \mathcal{F}_B \cap \mathcal{F}_{(\cup_1^k A_i)}].$$

If $X$ is sharp Markov,

$$(4.2) \qquad E[f(X_{A_1}, ..., X_{A_k}) \mid \mathcal{F}_B] = E[f(X_{A_1}, ..., X_{A_k}) \mid \mathcal{F}_{\partial B}].$$

Therefore, combining (4.1) and (4.2), we obtain

$$E[f(X_{A_1}, ..., X_{A_k}) \mid \mathcal{F}_B] = E[f(X_{A_1}, ..., X_{A_k}) \mid \mathcal{F}_{\partial B} \cap \mathcal{F}_{(\cup_1^k A_i)}],$$

and $X$ is Markov. $\qquad\qquad\qquad\qquad\qquad\qquad\qquad\qquad\qquad\qquad\qquad\square$

**Comment:** We note that (CO) cannot be removed from the statement of the preceding theorem, as there exist processes which are sharp Markov but not Markov. For example, let $T = \mathbf{R}_+^2$ and $\mathcal{A} = \{[0, t] : t \in T\}$. Let $S$ be a random variable uniformly distributed on $[0, 1]$, and place a single point $U$ at random on the line $L_S = \{(x, y) \in T : x + y = S\}$. Then define $X_A = I_{\{U \in A\}}$. Let $A \in \mathcal{A}$, $B \in \mathcal{A}(u)$, $A, B \subseteq [0, 1]^2$, $A \not\subseteq B$. For a single-jump process on $\mathbf{R}_+^2$, it is well-known that $\mathcal{F}_B = \mathcal{F}_{\partial B}$, so trivially $X$ is sharp Markov. Now, on the set $F = \{L_S \subseteq B\} \cap \{X_{A \cap B} = 0\}$, $E(X_A \mid \mathcal{F}_{\partial B}) = 0$. However, although $X_{A \cap B}$ is $\mathcal{F}_{\partial B} \cap \mathcal{F}_A$-measurable, $E(X_A \mid \{X_{A \cap B} = 0\}) > 0$. Thus, $X$ is not Markov.

The following is a corollary of Theorems 3.3 and 4.2

COROLLARY 4.3. *If $X$ has independent increments, then $X$ is sharp Markov.*

All of the examples in the Introduction have independent increments and so are both Markov and sharp Markov, so at first glance, this corollary seems to contradict the first and third examples. In fact, when $T \subseteq \mathbf{R}_+^2$, we may identify a $T$-indexed process with an $\mathcal{A}$-indexed process where $\mathcal{A}$ is the class of rectangles $\{[0, t] : t \in T\}$. Thus, $\Delta \notin \mathcal{A}(u)$, and so there is no contradiction. In the set-indexed framework, if $\Delta \in \mathcal{A}(u)$ (eg. if $\mathcal{A}$ is the class of (convex) lower layers), then by our definition $X_\Delta$ is $\mathcal{F}_{\partial \Delta}$-measurable. In fact, if $X$ is additive, then
1. If $\mathcal{A} = $ the lower layers, then $\mathcal{F}_{\partial \Delta} = \mathcal{F}_\Delta$
2. If $\mathcal{A} = $ the convex lower layers, then $\mathcal{F}_{\partial \Delta} \subset \mathcal{F}_\Delta$

## 5. The strong Markov property

In this section, we introduce random sets and stopping sets (as defined in [13] and [14]) and prove that the sharp Markov property is equivalent to a Markov property with respect to both classes of sets. This property is called the strong Markov property. It was introduced first by Dynkin for processes on the line and later by Evstigneev [10] for random fields, but using a different kind of $\sigma$-algebra. In [20], Merzbach and Nualart prove strong Markov properties for two-parameter strictly simple point processes with independent increments.

Here, we shall generalize this result and show that any set-indexed point process with the sharp Markov property also satisfies a strong Markov property.

We begin this section by defining and proving a strong Markov property with respect to a class of discrete random sets, without assuming outer-continuity of the $\sigma$-algebras generated by the process. Next, a strong Markov property with respect to stopping sets is obtained if the $\sigma$-algebras are outer-continuous. These results are then shown to apply to point processes.

Suppose that an integrable outer-continuous process $X = \{X_A : A \in \mathcal{A}\}$ is given. Let $\theta : \Omega \to \mathcal{A}(u)$ be a random set. Then for a random set $\eta$ of the form $\eta = \theta$, $\theta^\circ$, $\partial\theta$, $\theta^c$ or $\overline{\theta^c}$, define the following natural $\sigma$-algebra:

$$\mathcal{F}_\eta^X = \sigma\{X_A I_{\mathcal{A}(\eta)}(A), I_{\mathcal{A}(\eta)}(A),\ A \in \mathcal{A}\}.$$

We have the following:

LEMMA 5.1. *1. If $\theta \equiv B \in \mathcal{A}(u)$, then for $\eta =$ (respectively) $\theta$, $\theta^\circ$, $\partial\theta$, $\theta^c$, $\overline{\theta^c}$,
$\mathcal{F}_\eta^X =$ (respectively) $\mathcal{F}_B$, $\mathcal{F}_{B^\circ}$, $\mathcal{F}_{\partial B}$, $\mathcal{F}_{B^c}$, $\mathcal{F}_{\overline{B^c}}$. (cf. Section 2).*
*2. $\mathcal{F}_{\partial\theta}^X \subseteq \mathcal{F}_\theta^X$.*
*3. $\mathcal{F}_{\overline{\theta^c}}^X \subseteq \mathcal{F}_{\theta^c}^X \vee \mathcal{F}_{\partial\theta}^X$.*

PROOF. 1. Obvious.
2. First we show that $I_{\mathcal{A}(\partial\theta)}(A)$ is $\mathcal{F}_\theta^X$-measurable. This follows since

$$I_{\mathcal{A}(\partial\theta)}(A) = I_{\mathcal{A}(\theta)}(A) \prod_{n=1}^{\infty}(1 - I_{\mathcal{A}(\theta)}(g_n(A))) \in \mathcal{F}_\theta^X.$$

Finally,

$$X_A I_{\mathcal{A}(\partial\theta)}(A) = X_A I_{\mathcal{A}(\theta)}(A) I_{\mathcal{A}(\partial\theta)}(A) \in \mathcal{F}_\theta^X.$$

3. This follows from the observation that

$$I_{\mathcal{A}(\overline{\theta^c})}(A) = I_{\mathcal{A}(\partial\theta)}(A) + I_{\mathcal{A}(\theta^c)}(A).$$

$\square$

We now define the *strong Markov property*:

DEFINITION 5.2. Let $\Theta$ be a class of random sets taking their values in $\mathcal{A}(u)$. X is **strong Markov** with respect to $\Theta$ if for every $\theta \in \Theta$,

$$\mathcal{F}_\theta^X \perp \mathcal{F}_{\theta^c}^X \mid \mathcal{F}_{\partial\theta}^X.$$

REMARK 5.3. Note that $\mathcal{F}_{\theta^c}^X$ may be replaced by $\mathcal{F}_{\overline{\theta^c}}^X$ in the preceding definition. This follows from Lemma 5.1.3 since

$$\mathcal{F}_\theta^X \perp (\mathcal{F}_{\theta^c}^X \vee \mathcal{F}_{\partial\theta}^X) \mid \mathcal{F}_{\partial\theta}^X.$$

Next we define the class of *adapted* random sets:

DEFINITION 5.4. A function $\alpha : \Omega \to \mathcal{A}(u)$ is called an adapted random set if it takes a countable number of configurations, and for any $B \in \mathcal{A}(u)$, we have $\{\alpha = B\} \in \mathcal{F}_B$.

COMMENT 5.5. It is clear that if $\alpha$ is a random set taking countably many configurations in $\mathcal{A}(u)$, then $\alpha$ is an adapted random set if and only if for any $B \in \mathcal{A}(u)$, $\{\alpha \subseteq B\} \in \mathcal{F}_B$.

Throughout the rest of this section, it will be assumed that the class $\mathcal{A}$ satisfies the following property: *If $B \in \mathcal{A}(u)$ and $B = \cup_{i=1}^k A_i$ is a representation of $B$ such that $A_i \not\subseteq \cup_{j \neq i} A_j \ \forall i$, then $A_i \not\subseteq B^\circ \ \forall i$.* All our examples satisfy this assumption.
We will frequently use the following simple result.

LEMMA 5.6. *Let $\alpha$ be an adapted random set and let $B \in \mathcal{A}(u)$. Then $\{\alpha = B\} \in \mathcal{F}_B \cap \mathcal{F}_{\partial\alpha}^X$. Moreover, on the set $\{\alpha = B\}$, we have:*

$$\mathcal{F}_B = \mathcal{F}_\alpha^X \quad and \quad \mathcal{F}_{\partial B} = \mathcal{F}_{\partial\alpha}^X.$$

PROOF. If $A \in \mathcal{A}$, then $\{A \in \mathcal{A}(\partial\alpha)\} \in \mathcal{F}_{\partial\alpha}^X$. Now if $B = \cup_{i=1}^{k} A_i$, $A_i \in \mathcal{A}$ is a representation of $B$ such that $A_i \not\subseteq \cup_{j\neq i} A_j \ \forall i$, it follows that

$$\{\alpha = B\} = \cap_i \{A_i \in \mathcal{A}(\partial\alpha)\} \cap \cap_{A \in \mathcal{A}', \ A \not\subseteq B} \{A \in \mathcal{A}(\partial\alpha)\}^c \in \mathcal{F}_{\partial\alpha}^X,$$

where $\mathcal{A}'$ is a countable subclass of $\mathcal{A}$ containing all sets $A$ which are part of a possible configuration of $\alpha$.

The remaining statements are easily proven.        $\square$

We are now in position to state the main result of this section about the strong Markov property. Notice that the following result does not require any outer-continuity for the process $X$ nor for the $\sigma$-algebras. Clearly, with the additional assumption of outer-continuity, we shall obtain more precise results.

THEOREM 5.7. *Let $X$ be an integrable process. $X$ satisfies the sharp Markov property if and only if $X$ satisfies the strong Markov property with respect to the class of adapted random sets.*

REMARK 5.8. By Lemma 5.1.3, instead of $\mathcal{F}_{\alpha^c}^X$, we can take $\mathcal{F}_{\overline{\alpha^c}}^X$. We note that this result is completely analogous to Markov processes on $\mathbf{R}_+$ and discrete stopping times.

PROOF. It is trivial that the strong Markov property implies the sharp Markov property.

Conversely, we need to prove that for any $\mathcal{F}_{\alpha^c}^X$-measurable random variable $Y$, we have:

$$E(Y|\mathcal{F}_\alpha^X) = E(Y|\mathcal{F}_{\partial\alpha}^X).$$

By a monotone class argument, it suffices to obtain this equality for $Y$ of the form:

$$Y = f\left(X_{A_1} I_{\{A_1 \subseteq \alpha\}^c}, I_{\{A_1 \subseteq \alpha\}^c}, \ldots, X_{A_k} I_{\{A_k \subseteq \alpha\}^c}, I_{\{A_k \subseteq \alpha\}^c}\right)$$

where $f : \mathbf{R}^{2k} \to \mathbf{R}$ is a continuous and bounded function, and for any $A_1, \ldots A_k \in \mathcal{A}$.

Since $\alpha$ is discrete, then for all $i$ :

$$I_{\{A_i \subseteq \alpha\}^c} = \sum_{B \in \mathcal{A}_\alpha(u), \ A_i \not\subseteq B} I_{\{\alpha = B\}},$$

and

$$Y = \sum_{B \in \mathcal{A}_\alpha(u)} I_{\{\alpha = B\}} Y,$$

where $\mathcal{A}_\alpha(u)$ is the set of possible configurations of $\alpha$.

It is easily seen that on $\{\alpha = B\}$, $I_{\{\alpha = B\}} Y = I_{\{\alpha = B\}} Y_B$, where $Y_B$ is $\mathcal{F}_{B^c}$-measurable, and so

$$Y = \sum_{B \in \mathcal{A}_\alpha(u)} I_{\{\alpha = B\}} Y_B.$$

Next, for any sub-$\sigma$-field $\mathcal{F}'$ of $\mathcal{F}$, let $\mathcal{F}' \cap \{\alpha = B\}$ denote the trace of $\mathcal{F}'$ on $\{\alpha = B\}$. Now, for $\mathcal{F}' = \mathcal{F}_\alpha^X$, $\mathcal{F}_{\partial\alpha}^X$, or $\mathcal{F}_B$, $I_{\{\alpha = B\}}$ is $\mathcal{F}'$-measurable. Thus, on $\{\alpha = B\}$ we have that

(5.1)    $I_{\{\alpha = B\}} E(Y_B \mid \mathcal{F}') = E(I_{\{\alpha = B\}} Y_B \mid \mathcal{F}') = E(I_{\{\alpha = B\}} Y_B \mid \mathcal{F}' \cap \{\alpha = B\}).$

By Lemma 5.6, $\mathcal{F}_\alpha^X \cap \{\alpha = B\} = \mathcal{F}_B \cap \{\alpha = B\}$, and so on $\{\alpha = B\}$ we have

(5.2)        $E(I_{\{\alpha = B\}} Y_B \mid \mathcal{F}_\alpha^X) = E(I_{\{\alpha = B\}} Y_B \mid \mathcal{F}_B).$

By the sharp Markov property,

$$
\begin{aligned}
E(I_{\{\alpha=B\}}Y_B \mid \mathcal{F}_B) &= I_{\{\alpha=B\}}E(Y_B \mid \mathcal{F}_B) \\
&= I_{\{\alpha=B\}}E(Y_B \mid \mathcal{F}_{\partial B}),
\end{aligned}
$$

(5.3)

and so on $\{\alpha = B\}$, by (5.2), (5.3) and Lemma 5.6 it follows that $E(I_{\{\alpha=B\}}Y_B \mid \mathcal{F}_\alpha^X)$ is $\mathcal{F}_{\partial B} \cap \{\alpha = B\} = \mathcal{F}_{\partial\alpha}^X \cap \{\alpha = B\}$-measurable. Finally, we may apply (5.1) to conclude that this implies that on $\{\alpha = B\}$,

$$
E(I_{\{\alpha=B\}}Y_B \mid \mathcal{F}_\alpha^X) = E(I_{\{\alpha=B\}}Y_B \mid \mathcal{F}_{\partial\alpha}^X).
$$

By summing, we get the result. $\qquad\square$

The theory of stopping for set-indexed processes was developed in [**13**], and as will be shown, there is a close relationship between stopping sets and adapted random sets.

DEFINITION 5.9. A function $\xi : \Omega \to \mathcal{A}(u)$ is called a stopping set if it is of the form $\xi(\omega) = \cup_{i=1}^k \xi_i(\omega)$, $\xi_i : \Omega \to \mathcal{A}$, $i = 1, \ldots, k < \infty$, for any $A \in \mathcal{A}$, $\{\omega : A \subseteq \xi(\omega)\} \in \mathcal{F}_A$ and $\{\omega : \phi = \xi(\omega)\} \in \mathcal{F}_\phi$. If $\xi$ takes a countable number of configurations, $\xi$ is said to be discrete.

DEFINITION 5.10. The family $\{\mathcal{F}_B, \ B \in \mathcal{A}(u)\}$ is outer-continuous if $\forall B \in \mathcal{A}(u)$, we have $\mathcal{F}_B = \cap_n \mathcal{F}_{B_n}$ for any decreasing sequence $\{B_n\}$ in $\mathcal{A}(u)$ such that $B = \cap_n B_n$.

It was proved in [**13**] that if the family $\{\mathcal{F}_B, \ B \in \mathcal{A}(u)\}$ is outer-continuous, then for any stopping set $\xi$, we have

$$
\{\xi \subseteq B\} \in \mathcal{F}_B \ \forall B \in \mathcal{A}(u).
$$

Therefore, we have the following relationship between stopping sets and adapted random sets:

LEMMA 5.11. *If $\xi$ is a discrete stopping set and the family of $\sigma$-algebras $\{\mathcal{F}_B, \ B \in \mathcal{A}(u)\}$ is outer-continuous, then $\xi$ is an adapted random set.*

PROOF. $\{\xi = B\} = \{\xi \subseteq B\} \cap \{B \subseteq \xi\} \in \mathcal{F}_B$. $\qquad\square$

We immediately have the following corollary to Theorem 5.7:

COROLLARY 5.12. *Let $X$ be an integrable process and assume that the family $\{\mathcal{F}_B, \ B \in \mathcal{A}(u)\}$ is outer-continuous. $X$ satisfies the sharp Markov property if and only if $X$ satisfies the strong Markov property with respect to the class of discrete stopping sets.*

We now turn to the class of general stopping sets, and we shall see that the strong Markov property becomes a more delicate question. In fact, we shall prove something slightly weaker.

First, we note that for stopping sets, the $\sigma$-fields $\mathcal{F}_.^X$ are monotonic in the following sense:

PROPOSITION 5.13. *If $\eta$, $\xi$ are stopping sets such that $\eta \subseteq \xi$, then $\mathcal{F}_\eta^X \subseteq \mathcal{F}_\xi^X$.*

PROOF. We must show that for any $A \in \mathcal{A}$, $I_{\{A \subseteq \eta\}}$ and $X_A I_{\{A \subseteq \eta\}}$ are $\mathcal{F}_\xi^X$-measurable. Since $\{A \subseteq \eta\} \in \mathcal{F}_A$, there exist a countable number of sets $A_1, A_2, \ldots \subseteq A$ such that $I_{\{A \subseteq \eta\}} = f(X_{A_1}, X_{A_2}, \ldots)$ for some measurable function $f$. Now,

$$
\begin{aligned}
I_{\{A \subseteq \eta\}} &= f(X_{A_1}, X_{A_2}, \ldots) \prod_i I_{\{A_i \subseteq \xi\}} \\
&= f(X_{A_1} I_{\{A_1 \subseteq \xi\}}, X_{A_2} I_{\{A_2 \subseteq \xi\}}, \ldots) \prod_i I_{\{A_i \subseteq \xi\}},
\end{aligned}
$$

which is $\mathcal{F}_\xi^X$-measurable. Likewise,

$$
\begin{aligned}
X_A I_{\{A \subseteq \eta\}} &= X_A I_{\{A \subseteq \xi\}} f(X_{A_1}, X_{A_2}, \ldots) \prod_i I_{\{A_i \subseteq \xi\}} \\
&= X_A I_{\{A \subseteq \xi\}} f(X_{A_1} I_{\{A_1 \subseteq \xi\}}, X_{A_2} I_{\{A_2 \subseteq \xi\}}, \ldots) \prod_i I_{\{A_i \subseteq \xi\}}
\end{aligned}
$$

is $\mathcal{F}_\xi^X$-measurable. $\qquad\square$

Now, under very general conditions, it can be shown that every stopping set can be approximated from above by a decreasing sequence of discrete stopping sets. In particular, it will be seen that the following assumption suffices:

ASSUMPTION 5.14. *Let $(B_n)$ be a decreasing sequence of sets in $\mathcal{A}(u)$. Then there exists a sequence $(A_n)$ of sets in $\mathcal{A}$ such that*

$$
\cap_n B_n = \overline{\cup_n A_n}.
$$

Sufficient conditions for Assumption 5.14 to hold are discussed in [**14**], and it is easy to see that all of our examples satisfy this property.

*For the remainder of this section, it will be assumed that:*

- *$X$ is outer-continuous.*
- *The family $\{\mathcal{F}_B : B \in \mathcal{A}(u)\}$ is outer-continuous.*
- *Assumption 5.14 holds.*

The following $\sigma$-algebra may be now defined as in [**13**]:

DEFINITION 5.15. *If $\xi$ is a stopping set, then*

$$
\mathcal{F}_\xi = \{F \in \mathcal{F} : F \cap \{\xi \subseteq B\} \in \mathcal{F}_B, \ \forall B \in \mathcal{A}(u)\}.
$$

REMARK 5.16. *If $\xi$ is a discrete stopping set, then $\mathcal{F}_\xi = \{F \in \mathcal{F} : F \cap \{\xi = B\} \in \mathcal{F}_B, \ \forall B \in \mathcal{A}(u)\}$.*

We have the following proposition (proved in [**13**]):

PROPOSITION 5.17. *Let $\xi$ be a stopping set. Then $\{g_n(\xi)\}$ is a decreasing sequence of stopping sets such that $\xi = \cap_n g_n(\xi)$. Moreover, if the $\sigma$-algebras are outer-continuous, then*

$$
\mathcal{F}_\xi = \cap_n \mathcal{F}_{g_n(\xi)}.
$$

PROPOSITION 5.18. *Let $\xi$ be a discrete stopping set (and hence an adapted random set). Then $\mathcal{F}_\xi = \mathcal{F}_\xi^X$.*

PROOF. Let $F \in \mathcal{F}_\xi$. Then $\forall B \in \mathcal{A}(u)$, the random variable $I_{F \cap \{\xi = B\}}$ is $\mathcal{F}_B$-measurable (above remark). We may apply Lemma 5.6 to prove that $I_{F \cap \{\xi = B\}}$ is $\mathcal{F}_\xi^X$-measurable. Since $I_F = \sum_B I_{F \cap \{\xi = B\}}$, then $F \in \mathcal{F}_\xi^X$.

Conversely, it is enough to show that for any $A \in \mathcal{A}$ and $R$ a Borel subset of $\mathbf{R}$, $\{A \subseteq \xi\}$ and $\{X_A \in R\} \cap \{A \subseteq \xi\}$ belong to $\mathcal{F}_\xi$. Since for any $B \in \mathcal{A}(u)$,

$$\{A \subseteq \xi\} \cap \{\xi \subseteq B\} = \{A \subseteq \xi\} \cap \{\xi \subseteq B\} \cap \{A \subseteq B\} \in \mathcal{F}_B,$$

then $\{A \subseteq \xi\} \in \mathcal{F}_\xi$. Now $\{X_A \in R\} \cap \{A \subseteq \xi\} \cap \{\xi \subseteq B\} = \{X_A \in R\} \cap \{A \subseteq \xi\} \cap \{A \subseteq B\} \cap \{\xi \subseteq B\} \in \mathcal{F}_B$.

This completes the proof of the equality. $\qquad\square$

Although the sigma-algebras $\{\mathcal{F}_\xi\}$ satisfy a property of outer-continuity, at this point, it is not clear that the same outer-continuity property holds for the $\sigma$-algebras $\{\mathcal{F}_\xi^X\}$. We shall need the outer-continuous version of these $\sigma$-algebras and therefore denote:

$$\overline{\mathcal{F}}_\xi^X = \cap_{n=1}^\infty \mathcal{F}_{g_n(\xi)}^X.$$

It follows immediately from Propositions 5.17 and 5.18 that

$$\overline{\mathcal{F}}_\xi^X = \cap_{n=1}^\infty \mathcal{F}_{g_n(\xi)}^X = \cap_{n=1}^\infty \mathcal{F}_{g_n(\xi)} = \mathcal{F}_\xi.$$

Trivially, we have the following outer-continuity property:

$$\overline{\mathcal{F}}_\xi^X = \cap_{n=1}^\infty \overline{\mathcal{F}}_{g_n(\xi)}^X.$$

**Warning:**
This does not imply outer-continuity for the family $\{\mathcal{F}_\xi^X\}$. Indeed, for a non-discrete stopping set, this $\sigma$-algebra may be strictly smaller than $\mathcal{F}_\xi$. As well, while it is straightforward to show that $X_\xi$ is both $\mathcal{F}_\xi^X$- and $\mathcal{F}_\xi$-measurable when $\xi$ is an adapted random set (i.e. $\xi$ is discrete), in the general case outer-continuity of both $X$ and $\mathcal{F}_\xi$ is used to show that $X_\xi$ is $\mathcal{F}_\xi$-measurable (cf. [**14**]).

As in [**13**], for any two stopping sets $\xi$ and $\xi'$, the stochastic interval $[\xi, \xi']$ is defined to be the random collection $\mathcal{A}(\xi') \setminus \mathcal{A}(\xi^\circ)$. Therefore, it is natural to define:

$$\mathcal{F}_{[\xi,\xi']}^X = \sigma\{X_A I_{\mathcal{A}(\xi') \setminus \mathcal{A}(\xi^\circ)}(A), I_{\mathcal{A}(\xi') \setminus \mathcal{A}(\xi^\circ)}(A), \ A \in \mathcal{A}\}.$$

Notice that $\mathcal{F}_{[\xi,g_n(\xi)]}^X \subseteq \mathcal{F}_{g_n(\xi)}^X$, and if $g_n : \mathcal{A} \to \mathcal{A}_n \cup \{T\}$, $\forall n$, then $\mathcal{F}_{\partial g_n(\xi)}^X \subseteq \mathcal{F}_{[\xi,g_n(\xi)]}^X$. However, although Proposition 5.13 implies that the $\sigma$-fields $\mathcal{F}_{g_n(\xi)}^X$ are monotone decreasing, it is not clear that the same is true of the $\sigma$-fields $\mathcal{F}_{[\xi,g_n(\xi)]}$. Therefore, we define

$$\tilde{\mathcal{F}}_{[\xi,g_n(\xi)]}^X = \vee_{m \geq n} \mathcal{F}_{[\xi,g_m(\xi)]}^X,$$

and $\overline{\mathcal{F}}_{\partial \xi}^X = \cap_{n=1}^\infty \tilde{\mathcal{F}}_{[\xi,g_n(\xi)]}^X.$

THEOREM 5.19. *Assume that $g_n : \mathcal{A} \to \mathcal{A}_n \cup \{T\}$, $\forall n$. If $X$ has the sharp Markov property, then for any stopping set $\xi$, we have:*

$$\overline{\mathcal{F}}_\xi^X \perp \mathcal{F}_{\xi^c}^X \mid \overline{\mathcal{F}}_{\partial \xi}^X.$$

PROOF. The main idea is the same as in Theorem 5.7. Let $Y$ be of the form:

$$f\big(X_{A_1} I_{\{A_1 \subseteq \xi\}^c}, I_{\{A_1 \subseteq \xi\}^c}, \dots, X_{A_k} I_{\{A_k \subseteq \xi\}^c}, I_{\{A_k \subseteq \xi\}^c}\big),$$

where $f : \mathbf{R}^{2k} \to \mathbf{R}$ is bounded and continuous, and $A_i \in \mathcal{A}$.

Then $Y = \lim_{n\to\infty} Y_n$, where $Y_n$ is the bounded random variable:

$$Y_n = f\big(X_{A_1} I_{\{A_1 \subseteq g_n(\xi)\}^c}, I_{\{A_1 \subseteq g_n(\xi)^c\}}, \ldots, X_{A_k} I_{\{A_k \subseteq g_n(\xi)\}^c}, I_{\{A_k \subseteq g_n(\xi)\}^c}\big).$$

Therefore

$$
\begin{aligned}
E(Y|\overline{\mathcal{F}}^X_\xi) &= \lim_n E(Y_n|\overline{\mathcal{F}}^X_{g_n(\xi)}) &&\text{(cf. [9], pg. 228, Exercise 6.2)}\\
&= \lim_n E(Y_n|\mathcal{F}^X_{g_n(\xi)}) &&\text{(Proposition 5.18)}\\
&= \lim_n E(Y_n|\mathcal{F}^X_{\partial g_n(\xi)}) &&\text{(Theorem 5.7)}
\end{aligned}
$$

Since $\mathcal{F}^X_{\partial g_n(\xi)} \subseteq \mathcal{F}^X_{[\xi,g_n(\xi)]} \subseteq \tilde{\mathcal{F}}^X_{[\xi,g_n(\xi)]} \subseteq \mathcal{F}^X_{g_n(\xi)}$, then this conditional expectation is equal to

$$
\begin{aligned}
E(Y|\overline{\mathcal{F}}^X_\xi) &= \lim_n E\big(Y_n|\tilde{\mathcal{F}}^X_{[\xi,g_n(\xi)]}\big)\\
&= E(Y|\overline{\mathcal{F}}^X_{\partial\xi}).
\end{aligned}
$$

$\square$

**Comment:** The requirement that $g_n : \mathcal{A} \to \mathcal{A}_n \cup \{T\}$, $\forall n$ is satisfied by examples 1, 3 and 4 of 2.2. This condition is imposed only to ensure that

(5.4) $$\mathcal{F}^X_{\partial g_n(\xi)} \subseteq \mathcal{F}^X_{[\xi,g_n(\xi)]}, \ \forall n$$

which may not be true in general if $g_n$ takes its values in $\mathcal{A}_n(u) \cup \{T\}$. However, it is not difficult to show that (5.4) is satisfied in the case of example 2 of 2.2, so that in fact Theorem 5.19 holds for all our examples.

As an application, we shall consider a set-indexed point process. We shall assume that the Borel sets of $T$ are generated by $\mathcal{A}$.

DEFINITION 5.20. [14] An integrable process $X = \{X_A, \ A \in \mathcal{A}\}$ is called a point process if the following are satisfied.

- $\forall C \in \mathcal{C}$, $0 \le X_C < \infty$ a.s.
- $X$ is outer-continuous.
- $\forall A \in \mathcal{A}, \quad X_A \in \mathbf{N}$.
- $\forall t \in T, \quad X_{\{t\}} \le 1$ a.s.

The preceding results will be applicable if we can show that the family $\{\mathcal{F}_B : B \in \mathcal{A}(u)\}$ is outer-continuous.

PROPOSITION 5.21. *If $X$ is a point process, then the family $\{\mathcal{F}_B : B \in \mathcal{A}(u)\}$ is outer-continuous.*

PROOF. The proof follows closely the analogous result for point processes on $\mathbf{R}_+$. Let $B \in \mathcal{A}(u)$, and let $(B_n)$ be a decreasing sequence in $\mathcal{A}(u)$ such that $B = \cap_n B_n$. We must show that $\cap_n \mathcal{F}_{B_n}$ is included in $\mathcal{F}_B$.

Let $\delta(\omega)$ be the minimum distance between $B$ and the support of the process outside $B$ (i.e. $\delta(\omega) = \inf\{d(t,B) : t \in B^c, X_{\{t\}} = 1\}$). Then $\delta(\omega)$ is a random variable and $\delta(\omega) > 0$ a.s. since $X$ is finite on sets in $\mathcal{C}$. We may choose a subsequence $(B_{n_k})_k$ such that $d_H(B, B_{n_k}) < \frac{1}{k}$ ($d_H$ denotes the Hausdorff distance). Denote also

$$H_k = \{\omega : d_H(B, B_{n_k}) < \frac{1}{k} < \delta(\omega)\}.$$

These sets increase to $\Omega$ when $k \to \infty$. If it can be shown that $H_k \cap \mathcal{F}_B = H_k \cap \mathcal{F}_{B_{n_k}}$ $\forall k$ ($H_k \cap \mathcal{F}_B := \{F \cap H_k : F \in \mathcal{F}_B\}$), then $\cap_n \mathcal{F}_{B_n} = \cap_k \mathcal{F}_{B_{n_k}} = \mathcal{F}_B$. Now, if $A \subseteq B_{n_k}$, then $X_A = X_{A \cap B}$ on $H_k$ . Therefore, $X_A I_{H_k} = X_{A \cap B} I_{H_k}$ $\forall A \subseteq B_{n_k}$, and so $H_k \cap \mathcal{F}_B = H_k \cap \mathcal{F}_{B_{n_k}}$ $\forall k$, as required. $\qquad\square$

## References

[1] A. Alabert and D. Nualart (1992). Some remarks on the conditional independence and the Markov property. *Stoch. Anal. Rel. Topics* (H. Korezlioglu, A.S. Ustunel, eds.) Probability **31**, Birkhauser, Boston, pp. 343-364.

[2] E. Carnal and J.B. Walsh (1991). Markov properties for certain random fields. *Stochastic Analysis* (E. Meyer-Wolf, E. Merzbach, A. Schwartz, eds.) Academic Press, New York, pp. 91-110.

[3] F. Constantinescu and W. Thalheimer (1976). Remarks on generalized Markov processes. *J. Funct. Analysis* **23**, 33-38.

[4] R.C. Dalang and Q. Hou (1997). On Markov properties of Lévy waves in two dimensions. *Stoch. Proc. Appl.* **72**, 265-287.

[5] R.C. Dalang and F. Russo (1988). A prediction problem for the Brownian sheet. *J. Multiv. Anal.* **26**, 16-47.

[6] R.C. Dalang and J.B. Walsh (1992). The sharp Markov property for Lévy sheets. *Ann. Probab.* **20**, 591-626.

[7] R.C. Dalang and J.B. Walsh (1992). The sharp Markov property of the Brownian sheet and related processes. *Acta Math.* **168**, 153-218.

[8] M. Dozzi (1983). Propriétés Markoviennes des processus sur $\mathbf{R}^2$. *Ann. Inst. Henri Poincaré* **19**, 209-221.

[9] R. Durrett (1991). *Probability: Theory and Examples*, Wordsworth & Brooks/Cole.

[10] I.V. Evstigneev (1977). Markov times for random fields. *Th. Probab. Applic.* **22**, 563-569.

[11] P. Greenwood and I.V. Evstigneev (1990). A Markov evolving random field and splitting random elements. *Th. Probab. Applic.* **37**, 40-42.

[12] X. Guyon and B. Prum (1979). Propriétés markoviennes de certains processus indicés par $\mathbf{R}_+^2$. *Rap. tec. 79*, Univ. Paris-Sud.

[13] B.G. Ivanoff and E. Merzbach (1995). Stopping and set-indexed local martingales. *Stoch. Proc. Appl.* **57**, 83-98.

[14] B.G. Ivanoff and E. Merzbach (1999). *Set-Indexed Martingales*, Chapman & Hall/CRC, London.

[15] G. Kallianpur and V. Mandrekar (1974). The Markov property for generalized Gaussian random fields. *Ann. Institut Fourier* **24**, 143-164.

[16] H. Korezlioglu, P. Lefort and G. Mazziotto (1981). Une propriété markovienne et diffusions associées. *Lect. Notes in Math.*. **863**, Springer-Verlag, 245-276.

[17] P. Lévy (1948). Exemples de processus doubles de Markoff. *C.R.A.S.* **226**, 307-308.

[18] V. Mandrekar (1983). Markov properties for random fields. *Probabilistic Analysis and Related Topics 3* (A.J. Bharucha-Reid, ed.), 161-193.

[19] J.H.P. McKean (1963). Brownian motion with a several-dimensional time. *Th. Probab. Applic.* **8**, 335-354.

[20] E. Merzbach and D. Nualart (1990). Markov properties for point processes on the plane. *Ann. Probab.* **18**, 342-358.

[21] D. Nualart and M. Sanz (1979). A Markov property for two-parameter Gaussian processes. *Stochastica 3*, 1-16.

[22] L.D. Pitt (1971). A Markov property for Gaussian processes with multidimensional parameter. *Arch. Rat. Mech. Anal.* **43**, 367-397.

[23] Y.A. Rozanov (1982). *Markov Random Fields,* Springer-Verlag.

[24] F. Russo (1984). Etude de la propriété de Markov étroite en relation avec les processus planaires à accroissements indépendants. *Lect. Notes in Math.* **1059**, Springer-Verlag, 353-378.

[25] J.B. Walsh (1986). Martingales with a multidimensional parameter and stochastic integrals in the plane. *Lecture Notes in Math.* **1215**, Springer-Verlag, 229-491.

[26] E. Wong and M. Zakai (1985). Markov processes in the plane. *Stochastics* **15**, 311-333.

DEPARTMENT OF MATHEMATICS AND STATISTICS, UNIVERSITY OF OTTAWA, P.O. BOX 450, STN. A, OTTAWA, ONTARIO K1N 6N5, CANADA
*E-mail address*: `givanoff@science.uottawa.ca`

DEPARTMENT OF MATHEMATICS AND COMPUTER SCIENCE, BAR ILAN UNIVERSITY, 52 900 RAMAT GAN, ISRAEL
*E-mail address*: `merzbach@macs.biu.ac.il`

Canadian Mathematical Society
Conference Proceedings
Volume **26**, 2000

# Condensation Transition for Zero Range Invariant Measures

## Intae Jeon and Peter March

ABSTRACT. Let $(Z_n^1, Z_n^2, \cdots, Z_n^n)$ be distributed according to the $n$-particle, zero range invariant measure with rate function $g(k)$, and let $Z_n^*$ be the maximum of the $Z_n^i$'s. We say that condensation occurs if $P\{Z_n^* = n\} \to 1$ as $n \to \infty$. We prove that condensation occurs if and only if both $(n - Z_n^*, \ n \geq 1)$ is tight and $ng(n) \to 0$. By itself, the condition $ng(n) \to 0$ is not sufficient for tightness. However, we give some conditions on the rate function $g$ which do imply tightness. In case $(n - Z_n^*, \ n \geq 1)$ is tight, but condensation does not occur, we investigate the form of the possible limiting distributions.

## 1. Introduction

Imagine $n$ indistinguishable particles distributed randomly over $m$ distinguishable sites according to the following recipe. Let $Z_n^k$ denote the number of particles at site $k$, thus;

$$Z_n^1 + Z_n^2 + \cdots + Z_n^m = n.$$

Let $g(k), k \geq 0$ be a sequence of positive real numbers, and set

$$P\{Z_n^k = \eta_k, \ 1 \leq k \leq m\} = C \prod_{k=1}^{m} g!(\eta_k)^{-1},$$

where $g!(0) = 1$, $g!(k) = g(k)g(k-1)\cdots g(1)$ and $C$ is the normalizing constant.

These probabilities arise as invariant measures of finite, irreducible, zero range processes [**8**, **3**]. Roughly speaking, this process consists of $n$ Markovian particles making transitions between $m$ sites subject to the following interaction. If there are $k$ particles at a given site then the waiting time for the first of these particles to move is exponential of rate $g(k)$, independent of the particles at all other sites. Clearly, if $g(k) = kg(1)$ then particles are mutually independent and this linear case will serve as a convenient benchmark against which others cases can be measured. Thus, if $g(k)$ is superlinear then particles tend to leave occupied sites faster than independent particles would. Conversely, if $g(k)$ is sublinear then particles tend to linger at occupied sites more so than independent particles. In the former case, one expects the occupancy numbers $Z_n^k$ to be rather evenly distributed whereas in

---

2000 *Mathematics Subject Classification.* Primary 60K35; Secondary 60C05.

We thank Boris Pittel for several stimulating conversations on the subject of this paper and for advice on using the conditioning device.

the latter case, one expects them to be unevenly distributed, with just a few of the largest occupancy numbers accounting for most of the particles.

To simplify matters a bit, let's consider only the case $m = n$, so that the density of particles is one.

A measure of the distribution of particles over sites is the size of the largest occupancy number,

$$Z_n^* = \max_{1 \leq k \leq n} Z_n^k.$$

In a previous paper [5] with Boris Pittel, we showed that for $g(k)$ decreasing rapidly to zero almost all particles become concentrated on a single site. For example, let

$$g(k) = k^{-\alpha}, \ 0 < \alpha < \infty.$$

If $\alpha > 1$ then

$$\lim_{n \to \infty} P\{Z_n^* = n\} = 1.$$

If $\alpha = 1$ then $n - Z_n^*$ converges in distribution to a Poisson random variable of parameter 1. Finally, if $\alpha < 1$ then for any $l > 0$,

$$\lim_{n \to \infty} P\{Z_n^* \geq n - l\} = 0.$$

This change in behavior of the maximum occupancy number, which we call the condensation transition, is the focus of the present paper.

We were motivated to examine the behavior of $Z_n^*$ by analogy with problems of coagulation and gelation in colloid physics. (See, for example, the survey paper of Aldous [1] for a lengthier discussion than the one given here and for references to the literature.) One is interested in the sol-gel phase transition displayed by certain well stirred suspensions of sticky particles. As particles collide they tend to stick together forming clusters of various sizes and, further, clusters collide and combine to form even larger clusters. If the tendency for large clusters to fragment is small compared to the tendency for them to coagulate then it may happen that a giant cluster forms whose size is comparable to the size of system. A standard model in the theory is the Smolouchowski equations for the density per unit volume of clusters of size $n = 1, 2, 3, \ldots$ (See [4] for a proof of the existence of the phase transition under natural hypotheses on the rates of coagulation.) This phenomenon is similar to the emergence of the giant component in the theory of random graphs. As we showed in [5], and elaborate upon here, there is analogous behavior for the maximum occupancy number under zero range invariant measures.

Fix a sequence $g(k)$. We say condensation occurs, or the system condenses, if

$$\lim_{n \to \infty} P\{Z_n^* = n\} = 1.$$

Our first result characterizes condensation.

THEOREM 1.1. *Condensation occurs if and only if the sequence* $(n - Z_n^*, n \geq 1)$ *is tight and* $ng(n) \to 0$.

PROOF. One direction is easy, for clearly condensation implies tightness. Now, if $P\{Z_n^* = n\} \to 1$ as $n \to \infty$, then it is easy to calculate from the definition that

(1.1)
$$\frac{P\{Z_n^* = n - 1\}}{P\{Z_n^* = n\}} = \frac{n(n-1)g!(n)}{ng!(n-1)g!(1)}$$

(1.2)
$$= (n-1)g(n)/g(1) \to 0.$$

To go in the other direction, we first introduce some notation. Let $B_k = \{Z_n^* = k\}$ and let

$$C_{k,l} = \{Z_n^* = k, \ |\{i \ : \ Z_n^i \geq 1\}| = l\}.$$

By decomposing the event $\{Z_n^* = k\}$, $k < n$, according to the number of occupied sites, we find

$$B_k = \bigcup_{l=2}^{n-k+1} C_{k,l}.$$

Note that for $k > \frac{1}{2}n$ there can be at most one site containing $k$ particles. Thus, by elementary reasoning

$$|C_{k,l}| = \binom{n}{l} l \binom{n-k-1}{l-2}.$$

$\square$

Since it will be useful to have a notation for the un-normalized zero range invariant measure, let

$$\mu_n(\eta) = \prod_{k=1}^{n} g!(\eta_k)^{-1}.$$

LEMMA 1.2. *For any fixed $k \geq 0$, as $n \to \infty$,*

$$\mu_n(B_{n-k}) = \frac{k+1}{g!(n-k)g(1)^k} \binom{n}{k+1} (1 + o(1)).$$

PROOF. A lower bound is easy to derive. For sufficiently large $n$ we have $n - k > n/2$, hence

$$\mu_n(B_{n-k}) \geq \mu(C_{n-k,k+1}) = \frac{(k+1)}{g!(n-k)g(1)^k} \binom{n}{k+1}.$$

The inequality above is, in fact, an equality when $k = 0, 1$, which settles the matter in these cases. So we may assume $k \geq 2$.

To get an upper bound, first note that any configuration in $B_{n-k}$ is a permutation of some configuration of the form

$$\eta = (n - k, \eta_n, \cdots, \eta_n), \ \sum_{i=2}^{n} \eta_i = k.$$

Let

$$c_k = \max\{\prod_{i=2}^{n} g!(\eta_i)^{-1} \ : \ \eta(1) = n - k\}.$$

Now $c_k$ is independent of $n$ since by convention $g!(0) = 1$ and there are at most $k$ non-zero terms among the $\eta_i$'s. And clearly $\mu_n(\eta) \leq c_k/g!(n-k)$ for any $\eta \in B_{n-k}$. Therefore, for sufficiently large $n$,

$$(1.3) \qquad \mu_n(B_{n-k}) = \mu_n(C_{n-k,k+1}) + \sum_{l=2}^{k} \mu_n(C_{n-k,l})$$

$$(1.4) \qquad \leq \mu_n(C_{n-k,k+1}) + \sum_{l=2}^{k} l \binom{n}{l} \binom{k-1}{l-2} \frac{c_k}{g!(n-k)}.$$

Let
$$a_l = l\binom{n}{l}\binom{k-1}{l-2}\frac{c_k}{g!(n-k)}, \quad 2 \le l \le k.$$

Then
$$\frac{a_{l-1}}{a_l} = \frac{(l-1)(l-2)}{(n-l+1)(k-l+2)}$$

which is easily seen to be increasing in $l$. Since

$$\frac{a_{k-1}}{a_k} \le \frac{C}{n}$$

for some constant $C$, we have

$$\sum_{l=2}^{k} a_l = a_k(1 + O(\frac{1}{n})).$$

Therefore,

$$(1.5) \qquad \mu_n(B_{n-k}) \le \mu_n(C_{n-k,k+1}) + a_k(1 + O(\frac{1}{n}))$$

$$(1.6) \qquad = \frac{k+1}{g!(n-k)g(1)^k}\binom{n}{k+1} + k\binom{n}{k}\frac{(k-1)c_k}{g!(n-k)}(1 + O(\frac{1}{n}))$$

$$(1.7) \qquad = \frac{(k+1)}{g!(n-k)g(1)^k}\binom{n}{k+1}(1 + o(1)).$$

Thus, the lemma is proved.                                                    □

Now suppose $(n - Z_n^*, n \ge 1)$ is tight and $ng(n) \to 0$. By Lemma 1.1,

$$(1.8) \qquad \frac{P\{Z_n = n - k\}}{P\{Z_n = n\}} = \frac{\mu_n(B_{n-k})}{\mu_n(B_n)}$$

$$(1.9) \qquad = \frac{(k+1)g!(n)}{ng!(n-k)g(1)^k}\binom{n}{k+1}(1 + o(1))$$

$$(1.10) \qquad \le C\prod_{i=0}^{k-1}(n-i)g(n-i),$$

which tends to zero.

Now, tightness implies that for any $\epsilon > 0$ there exists $N$ such that

$$P\{Z_n^* \ge n - N\} \ge 1 - \epsilon.$$

But,

$$(1.11) \qquad P\{Z_n^* \ge n - N\} = \sum_{k=0}^{N} P\{Z_n^* = n - k\}$$

$$(1.12) \qquad = P\{Z_n^* = n\}(1 + \sum_{k=1}^{N}\frac{P\{Z_n = n - k\}}{P\{Z_n = n\}})$$

$$(1.13) \qquad = P\{Z_n^* = n\}(1 + o(1)).$$

Hence,

$$P\{Z_n^* = n\} \ge \frac{1 - \epsilon}{1 + o(1)},$$

which shows the system condenses.

A natural question is whether tightness is implied by the condition that $ng(n)$ tends to zero. The following example resolves the question in the negative.

EXAMPLE 1. Define a rate function $g$ by $g(0) = 1$ and

$$g(n) = \begin{cases} n^{-n^2} & \text{if } n = 10^k, \quad k \geq 0 \\ n^{-2} & \text{otherwise.} \end{cases}$$

If the sequence $(n - Z_n^*, \; n \geq 1)$ is tight then any subsequence has a further subsequence which converges weakly to a probability distribution. Consider the subsequence $n_m = 10^m - 1$ and let $k_m = 10^{m-1}$. By Lemma 1.1, for any $l \geq 0$, and sufficiently large $n$,

$$(1.14) \qquad P\{Z_{n_m}^* = n_m - l\} \leq \frac{P\{Z_{n_m}^* = n_m - l\}}{P\{Z_{n_m}^* = n_m - k_m\}}$$

$$(1.15) \qquad \qquad \leq \frac{\mu_{n_m}(B_{n_m-l})}{\mu_{n_m}(C_{n_m-k_m,2})}$$

$$(1.16) \qquad \qquad \leq \frac{C(l+1)n_m^{l+1}g!(n_m - k_m)g!(k_m)}{n_m(n_m-1)g!(n_m-l)}$$

$$(1.17) \qquad \qquad \leq \frac{n_m^l g!(k_m)}{g(n_m-l)\cdots g(n_m-k_m+1)}$$

$$(1.18) \qquad \qquad \leq \frac{10^{ml}10^{2m10^{m-1}}}{10^{(m-1)}10^{2(m-1)}}$$

$$(1.19) \qquad \qquad \leq 10^{-C'm},$$

for a positive constant $C'$. Therefore, tightness fails since these probabilities all tend to zero.

## 2. Some consequences of tightness

Let's suppose we have tightness and ask what kind of limit distributions can arise. Evidently, a point mass at zero is one possibility but there are others, as the next result shows.

PROPOSITION 2.1. *Suppose $(n - Z_n, n \geq 1)$ is tight and suppose $P\{Z_{n_l}^* = n_l - k\} \to p_k$, as $l \to \infty$, where $\sum_{k=0}^{\infty} p_k = 1$. Let $k_0 = \min\{k : p_k > 0\}$. Then for all $k \geq k_0$,*

$$\lambda_k = \lim_{l \to \infty} \prod_{j=0}^{k-k_0-1} (n_l - k_0 - j)g(n_l - k_0 - j)$$

*exists and*

$$p_k = \frac{\lambda_k g(1)^{k_0-k}}{k(k-1)\cdots(k_0+1)}p_{k_0}.$$

PROOF. By Lemma 1.1,

$$(2.1) \qquad \frac{p_k}{p_{k_0}} = \lim_{l \to \infty} \frac{P\{Z_{n_l}^* = n_l - k\}}{P\{Z_{n_l}^* = n_l - k_0\}}$$

$$(2.2) \qquad = \lim_{l \to \infty} \frac{(k+1)\binom{n_l}{k+1}g!(n_l - k_0)g(1)^{k_0}}{(k_0+1)\binom{n_l}{k_0+1}g!(n_l - k)g(1)^k}(1 + o(1))$$

$$(2.3) \qquad = \lim_{l \to \infty} \frac{(n_l - k_0 - 1)\cdots(n_l - k)g(n_l - k_0)\cdots g(n_l - k + 1)g(1)^{k_0 - k}}{k(k-1)\cdots(k_0 + 1)}$$

$$(2.4) \qquad = \frac{\lambda_k g(1)^{k_0 - k}}{k(k-1)\cdots(k_0 + 1)}.$$

$\square$

This result implies that the class of limit distributions is somewhat broad due, in fact, to the oscillatory behavior of $ng(n)$. Indeed, if $ng(n)$ is not bounded and not bounded away from zero then the possibilities for $\lambda_k$ seem limitless. The next proposition shows what can happen if $ng(n)$ is bounded.

PROPOSITION 2.2. *In addition to the hypotheses of Proposition 2.1, assume* $ng(n) \le C$ *for some positive constant $C$. Then $p_0 > 0$,*

$$\rho_k = \lim_{n \to \infty} n_l g(n_l - k)$$

*exists and*

$$p_k = p_0 \frac{\prod_{i=1}^k \rho_i}{k! g(1)^k}.$$

PROOF. First of all, let us show that if $p_{k-1} = 0$ then $p_k = 0$. Indeed, by Lemma 1.1 again, we have

$$(2.5) \qquad \frac{P\{Z_{n_l}^* = n_l - k\}}{P\{Z_{n_l}^* = n_l - (k-1)\}} = \frac{(n_l - k)g(n_l - k + 1)(1 + o(1))}{g(1)k}$$

$$(2.6) \qquad \le \frac{C}{g(1)k},$$

hence the numerator must tend to zero if the denominator does. This remark show that $p_0 > 0$ since otherwise all the $p_k$'s would vanish.

Now, for any fixed $k$, we have

$$(2.7) \qquad \frac{p_k}{p_0} = \lim_{l \to \infty} \frac{P\{Z_{n_l}^* = n_l - k\}}{P\{Z_{n_l}^* = n_l\}}$$

$$(2.8) \qquad = \lim_{l \to \infty} \frac{(n_l - 1)\cdots(n_l - k)g(n_l)\cdots g(n_l - k + 1)}{k! g(1)^k}.$$

By induction on $k$, we see that

$$\rho_k = \lim_{n \to \infty} n_l g(n_l - k)$$

exists. Therefore,

$$p_k = p_0 \frac{\prod_{i=1}^k \rho_i}{k! g(1)^k}.$$

$\square$

## 3. Sufficient conditions for tightness

As we have seen in Example 1, neither the assumption $ng(n) < C$ nor even $ng(n) \to 0$ implies tightness. In this section we will provide two sufficient conditions on $g$ which do imply tightness. The first, in Theorem 3.1, relies on a strong control over the oscillations of $ng(n)$. The second, in Theorem 3.2 , relies on monotonicity of $ng(n)$. These regularity conditions are sufficiently strong that, using Theorem 1.1 or Proposition 2.2, they actually imply either condensation or weak convergence to a Poisson distribution.

The proof of our first result uses a classic conditioning device pioneered by Khintchine, Shepp and Lloyd, and Kolchin. The works of Arratia and Tavaré [**2**], Kolchin [**6**], and Pittel [**7**] provide several recent examples of its use. The argument given below is a specialization of some more general arguments presented in [**5**]. The advantage of the peculiar choice of the parameter $x_n$ appearing in the proof is that one can use it to calculate quickly and easily. The disadvantage is that this choice hides what is actually essential for an argument like this to work. We refer to [**5**] for a lengthier discussion.

THEOREM 3.1. *Suppose $\lambda > 0$ and $ng(n) = \lambda + \phi(n)$ where $\sum |\phi(n)| < \infty$. Then the sequence $(n - Z_n^*, n \geq 1)$ is tight and converges weakly to a Poisson random variable of parameter $\lambda/g(1)$.*

PROOF. Choose
$$x_n = \lambda(1 - \frac{\frac{3}{2}\log n}{n})(\frac{e}{n}),$$
and introduce the random variable $X$ such that
$$P\{X = k\} = \frac{q_k}{M}, \quad q_k = \frac{x_n^k}{g!(k)},$$
and $M$ is the normalizing factor,
$$M = \sum_{k \geq 0} q_k.$$

Let $X_1, \ldots, X_n$ be independent copies of $X$. It is a simple but important observation that
$$\{Z_n^1, \ldots, Z_n^n\} \overset{\mathcal{D}}{\equiv} \{X_1, \ldots, X_n | X_1 + \cdots + X_n = n\}.$$

$\square$

To advance the proof we need certain estimates of the weights $q_k$.

LEMMA 3.2. *For large $n$,*
*(1) $q_1 = C_1(1 + o(1))/n$, $q_2 = C_2(1 + o(1))/n^2$ , and $q_n = C_3(1 + o(1))/n$.*
*(2) If $r_k = q_k/q_{k+1}$ then*
$$r_k = \frac{n}{ek}(1 + o(1)).$$
*(3) For small $c$ and $2 \leq k \leq \epsilon n$, we have $q_k \leq q_2$.*
*(4) If $\epsilon n \leq k \leq n - \sqrt{n}$, then $q_k \leq n^{-2}$.*
*(5) $P\{X_1 + X_2 + \cdots + X_n = n\} \geq C > 0$, for some constant $C$.*

PROOF. First,

$$q_1 = \frac{x_n}{g!(1)} = (1 - \frac{\frac{3}{2}\log n}{n})(\frac{C_1}{n}) = \frac{C_1}{n}(1 + o(1)),$$

$$q_2 = \frac{x_n^2}{g!(2)} = (1 - \frac{\frac{3}{2}\log n}{n})^2(\frac{C_2}{n^2}) = \frac{C_2}{n^2}(1 + o(1)),$$

and by Stirling's formula,

$$(3.1) \qquad q_n = \frac{x_n^n}{g!(n)} = (1 - \frac{\frac{3}{2}\log n}{n})^n(\frac{\lambda e}{n})^n \frac{1}{g!(n)}$$

$$(3.2) \qquad = n^{-\frac{3}{2}}(\frac{\lambda e}{n})^n \frac{n!}{\lambda^n \prod_{k=1}^{n}(1 + \frac{\phi(k)}{\lambda})}(1 + o(1))$$

$$(3.3) \qquad = \frac{C_3}{n}(1 + o(1)).$$

For item (2) we have

$$(3.4) \qquad r_k = \frac{q_k}{q_{k+1}}$$

$$(3.5) \qquad = \frac{g(k+1)}{x_n}$$

$$(3.6) \qquad = \frac{n}{\lambda e}g(k+1)(1 + o(1))$$

$$(3.7) \qquad = \frac{n}{ek}(1 + o(1)).$$

As for item (3) note that if $k \le \epsilon n$, where $\epsilon << 1$, then by (2)

$$r_k \ge \frac{C}{\epsilon} >> 1.$$

Therefore $q_k$ is a decreasing sequence hence $q_k \le q_2$.

Concerning item (4), if $\epsilon n \le k \le n - \sqrt{n}$, then for large $n$,

$$(3.8) \qquad q_k = \frac{x_n^k}{g!(k)}$$

$$(3.9) \qquad \le (\frac{\lambda e}{n})^k \frac{1}{g!(k)}$$

$$(3.10) \qquad = (\frac{\lambda e}{n})^k \frac{k!}{\lambda^k \prod_{j=1}^{k}(1 + \frac{\phi(j)}{\lambda})}$$

$$(3.11) \qquad \le Cn(\frac{k}{n})^k$$

$$(3.12) \qquad \le Cn(1 - \frac{1}{\sqrt{n}})^{\epsilon n}$$

$$(3.13) \qquad \le Cne^{-\epsilon\sqrt{n}}$$

$$(3.14) \qquad \le n^{-2}.$$

Finally, for item (5) observe that if $n - \sqrt{n} \le k < n$, then

$$r_k = \frac{n}{ek}(1 + o(1)) \le \frac{1 + o(1)}{e} < 1.$$

Therefore

$$\sum_{k=[n-\sqrt{n}]}^{n} q_k = O(q_n) = O(\frac{1}{n}).$$

Making similar estimation of the series defining $M$ we get, using (1), (2), (3) and (4), that $M = 1 + O(n^{-1})$. Therefore, $M^n$ is bounded away from zero, hence

(3.15) $\qquad P\{X_1 + X_2 + \cdots + X_n = n\} \geq nP\{X_1 = n, X_2 = 0, \cdots, X_n = 0\}$

(3.16) $\qquad\qquad\qquad = \dfrac{n}{M^n} q_n$

(3.17) $\qquad\qquad\qquad \geq C > 0.$

$\square$

Returning to the proof of the theorem, let $X_*$ be the maximum of the $X_j$'s and recall the fact that

$$\{Z_n^1, \ldots, Z_n^n\} \stackrel{\mathcal{D}}{\equiv} \{X_1, \ldots, X_n \mid X_1 + \cdots + X_n = n\}.$$

For any fixed $N$,

(3.18) $\qquad P\{Z_n^* \leq n - N\} = \dfrac{P\{X_* \leq n - N, \sum X_i = n\}}{P\{\sum X_i = n\}}$

(3.19) $\qquad\qquad\qquad \leq CP\{X_* \leq n - N\}$

(3.20) $\qquad\qquad\qquad = CP\{X_* = 1\} + CP\{2 \leq X_* < n - \sqrt{n}\}$

(3.21) $\qquad\qquad\qquad\quad + CP\{n - \sqrt{n} \leq X_* \leq n - N\}).$

Evidently $P\{X_* = 1\} \leq Cq_1^n$ which tends to zero exponentially fast. Next,

$$P\{2 \leq X_* < n - \sqrt{n}\} \leq n \max\{n^{-2}, q_2\}$$

and according to Lemma 3.1, this tends to zero as well. Finally, by the proof of item 5 in Lemma 3.1, the ratios $r_k \leq 1/2$ in the range $n - \sqrt{n} \leq k \leq n - N$. Thus, the weights $j \to q_{n-N-j}$ are geometrically decreasing and so

$$\sum_{k=n-\sqrt{n}}^{n-N} q_j \leq Cq_{n-N}.$$

Now,

(3.22) $\qquad\qquad q_{n-N} = q_n \dfrac{g(n) \cdots g(n - N + 1)}{x_n^N}$

(3.23) $\qquad\qquad\qquad \leq q_n \lambda^N \dfrac{\prod_{j=1}^{N}(1 + \phi(n - j + 1)/\lambda)}{n(n-1)\ldots(n-N+1)x_n^N}$

(3.24) $\qquad\qquad\qquad \leq Cq_n e^{-N}.$

It follows that

(3.25) $\qquad P\{n - \sqrt{n} \leq X_* \leq n - N\} \leq nP\{n - \sqrt{n} \leq X_1 \leq n - N\}$

(3.26) $\qquad\qquad\qquad \leq Cnq_{n-N}$

(3.27) $\qquad\qquad\qquad \leq C'e^{-N}.$

Together these estimates prove tightness of $(n - Z_n^*, n \geq 1)$. Convergence to a Poisson random variable of parameter $\lambda$ follows from Proposition 2.2.

To begin the proof of the other sufficient condition for tightness we introduce a partial order on particle configurations

$$\eta = (\eta_1, \eta_2, \cdots, \eta_n), \ \sum_{i=1}^{n} \eta_i = n.$$

Roughly speaking, a configuration obtained from $\eta$ by moving a particle from a less occupied site to a more occupied site, followed by any relabellng of the sites, is an immediate successor in this partial order. Equally, we say that $\eta$ precedes this new configuration. The following definition makes this precise.

DEFINITION 3.3. (1) We say $\eta$ precedes $\zeta$, if there exist $i, j$ with $1 \leq \eta(i) \leq \eta(j)$ and a permutation $\sigma$ such that $\zeta = \sigma(\eta - e_i + e_j)$.
(2) We say $\eta \prec \zeta$ if there exist $\theta_k, k = 1, \cdots, m$ such that $\eta = \theta_1, \zeta = \theta_m$ and $\theta_k$ precedes $\theta_{k+1}$ for all $1 \leq k \leq m - 1$.

We want to compare probabilities of configurations under finite zero range measures corresponding to different rate functions. Let us use the notation $\mu_n$ for the un-normalized measure corresponding to $g(k)$ and $\nu_n$ for the un-normalized measure corresponding to $h(k)$.

LEMMA 3.4. *Let $g$ and $h$ be rate functions and suppose $g = h\phi$, where $\phi$ is decreasing. Then for any configurations $\eta \prec \zeta$,*

$$\frac{\mu_n(\eta)}{\mu_n(\zeta)} \leq \frac{\nu_n(\eta)}{\nu_n(\zeta)}.$$

PROOF. It suffices to consider the case that $\eta$ precedes $\zeta$. Because of their product structure the un-normalized measures assign equal weight to configurations which are permutations of one another. So, we can assume without loss of generality that the entries of $\eta$ are ordered in decreasing order,

$$\eta_1 \geq \eta_2 \geq \cdots \geq \eta_n.$$

Then there exist $i > j$ such that $\zeta = \eta - e_i + e_j$. By a simple computation,

$$\frac{\nu_n(\eta)}{\nu_n(\zeta)} = \frac{\mu_n(\eta)}{\mu_n(\zeta)} \cdot \frac{\phi(\eta_j + 1)}{\phi(\eta_i)}.$$

Since $\eta_i \leq \eta_j + 1$ and $\phi$ is decreasing, the latter ratio is less than or equal one, which proves the lemma. $\square$

THEOREM 3.5. *Suppose $ng(n)$ is monotone decreasing. Then the sequence $(n - Z_n^*, n \geq 1)$ is tight. Let*

$$\lambda = \lim_{n \to \infty} ng(n).$$

*If $\lambda = 0$ then the system condenses. If $\lambda > 0$ then $n - Z_n^*$ converges weakly to a Poisson random variable of parameter $\lambda/g(1)$.*

PROOF. Let $h(n) = 1/n$ and $\nu_n$ be corresponding un-normalized measure. Since $\phi(n) = ng(n)$ is decreasing, we have $g = h\phi$ and the comparison lemma applies to this situation.

In [5], we proved that with rate function $h$ the sequence $(n - Z_n^*, n \geq 1)$ is tight and, in fact, converges to a Poisson distribution of rate 1. This is also a consequence of Theorem 3.1.

It suffices to show that $P_g\{Z_n^* \le n - l\}$ can be made arbitrarily small by taking $l$ sufficiently large. Note that if $\zeta = (n, 0, \cdots, 0)$ then $\eta \prec \zeta$, for all configurations $\eta$. Using this fact we find,

$$(3.28) \qquad P_g\{Z_n^* \le n - l\} \le \frac{P_g\{Z_n^* \le n - l\}}{P_g\{Z_n^* = n\}}$$

$$(3.29) \qquad = \frac{\mu_n(\bigcup_{k=l}^n B_{n-k})}{n\mu_n(\zeta)}$$

$$(3.30) \qquad \le \frac{\nu_n(\bigcup_{k=l}^n B_{n-k})}{n\nu_n(\zeta)}$$

$$(3.31) \qquad \le \frac{P_h\{Z_n^* \le n - l\}}{P_h\{Z_n^* = n\}}.$$

The last term vanishes as $l \to \infty$ which proves tightness of $n - Z_n^*$ under $P_g$. The remaining assertions follow from Theorem 1.1 and Propostion 2.2. $\qquad\square$

It is an interesting question whether the boundedness of $ng(n)$ from below and above implies tightness. Surprisingly, the following example shows that there is $g(n)$ such that $1/8 \le ng(n) \le 1$, whose corresponding sequence $(n - Z_n^*, n \ge 1)$ is not tight.

EXAMPLE 2. Let

$$g(n) = \begin{cases} M/n & \text{if } 2^{2k} < n \le 2^{2k+1} \\ 1/n & \text{if } 2^{2k+1} < n \le 2^{2k+2}. \end{cases}$$

Assume $(n - Z_n^*, n \ge 1)$ is tight and consider the subsequence $n_m = 2^{2m}$. Since $M \le ng(n) \le 1$, we know by Proposition 2.2 that

$$p_0 = \lim_{m \to \infty} P\{Z_{n_m}^* = n_m\} > 0.$$

By the assumed tightness we also have

$$\lim_{m \to \infty} P\{Z_{n_m}^* = n_m/2\} = 0.$$

In order to arrive at a contradiction, let us compute the ratio of these probabilities and try to bound it from above. By paying attention to the fact that $g(n)$ changes over dyadic blocks and by keeping track of those blocks we find, using Stirling's formula,

$$(3.32) \qquad \frac{P\{Z_{n_m}^* = n_m\}}{P\{Z_{n_m}^* = n_m/2\}} \le \frac{n_m P\{Z_n^1 = n_m, Z_n^k = 0, \ k \ge 2\}}{\binom{n_m}{2} P\{Z_n^1 = Z_n^2 = n_m/2, \ Z_n^k = 0, \ k \ge 3\}}$$

$$(3.33) \qquad \le \frac{n_m}{\binom{n_m}{2}} \frac{g!(n_m/2)^2}{g!(n_m)}$$

$$(3.34) \qquad = \frac{M^k n_m!}{[(n_m/2)!]^2}, \ k = \frac{n_m - 1}{3}$$

$$(3.35) \qquad \le C M^k 2^{n_m}$$

$$(3.36) \qquad = C'(2M^{1/3})^{n_m}.$$

Therefore, if $M = 1/8$ the last term is bounded and we get a contradiction.

# References

[1] Aldous, D. J., *Deterministic and stochastic models for coalescence (aggregation, coagulation): A review of the mean field theory for probabilists*, Bernoulli, **5** (1999), 3-48.

[2] Arratia, R., Tavaré, S., *Independent process approximations for random combibatorial structures* Adv. Math., **104** (1994) 90-154.

[3] Holley, R., *A class of interactions in an infinite particle system*, Adv. Math., **5** (1970), 291-309.

[4] Jeon, I., *Existence of gelling solutions for coagulation fragmentation equations*, Commun. Math. Phys **194** 1998, 541-567.

[5] Jeon, I., March, P., Pittel, B., *Asymptotic cluster size at the most occupied site under zero range invariant measures*, preprint.

[6] Kolchin, V. F., *Random mappings*, Optimization Software, 1986.

[7] Pittel, B., *On a likely shape of the random Ferrers diagram* Adv. Appl. Math., **18** (1997) 432-488.

[8] Spitzer, F., *Interaction of Markov processes*, Adv. Math., **5** (1970), 246–290.

DEPARTMENT OF MATHEMATICS, OHIO STATE UNIVERSITY, COLUMBUS, OH 43210
*E-mail address*: jeon@math.ohio-state.edu

DEPARTMENT OF MATHEMATICS, OHIO STATE UNIVERSITY, COLUMBUS, OH 43210
*E-mail address*: march@math.ohio-state.edu

Canadian Mathematical Society
Conference Proceedings
Volume **26**, 2000

# A Review on Spatial Catalytic Branching

## Achim Klenke

ABSTRACT. We review some results on spatial branching processes $X^\varrho$ in random media $\varrho$. The local branching rate or law depends on the medium $\varrho$ that may vary in space and time and can be random.

The main emphasis lies on catalytic super-Brownian motion where $\varrho$ governs the local branching rate and is considered as a catalytic medium. We display the construction of $X^\varrho$ and give results on absolute continuity of the states, longtime behaviour and so on.

## Introduction

In the last 15 years there has been a lot of interest in spatial branching models – branching random walk, branching Brownian motion, super-Brownian motion and so on – where the branching mechanism may vary in space and/or time in a deterministic or random way. The protagonists in this field of research are Don Dawson and Klaus Fleischmann.

This survey focuses mainly on the best-studied subclass of models where the branching mechanism is critical with finite variance (super processes) or binary (particle systems). It is only the local *rate* at which the branching occurs that varies. The local branching rate is interpreted as the concentration of catalytic matter that enables the branching. This catalyst is a function (or distribution) in space and/or time and it may be random or deterministic. We even consider a case where two branching processes catalyse each other in a symmetric way.

The ambition of this article is to serve as a quick guide to the subject and to give a survey of the main results. It is by no means comprehensive and the author wishes to apologize to everyone whose work is not considered here. We do not aim at rigour in the exposition but appeal to the intuition of the reader. Some notions are explained loosely to give non-specialists a vague idea and enable them to go on reading.

2000 *Mathematics Subject Classification.* Primary 60J80; Secondary 60G57, 60K35.

*Key words and phrases.* super Brownian motion; branching functionals; random media; longtime behaviour; absolute continuity.

Research supported by an NSERC grant at Carleton University, Ottawa.

# 1. Varying branching law

Galton–Watson processes in random environments have been studied for over 30 years. Since we shall focus on spatial models we refer only briefly to the books of Athreya and Ney (1972) and Jagers (1975), and for some more recent sources to d'Souza (1994), Fleischmann and Vatutin (1999).

We tried to keep the notation consistent throughout this article. The reader should be warned that this entails incongruencies with the notation in the original articles.

## 1.1. The starting point.

Spatial catalytic branching processes in a random medium were considered first by Dawson and Fleischmann (1983) and (1985). They study discrete time branching random walk on $\mathbb{Z}^d$ where the branching law is critical but depends on the location of the particle. More precisely, let $a(i,j)$ be the transition kernel of a random walk on $\mathbb{Z}^d$, denote by $a^{(n)}$ the $n$-step transition probabilities, let $F = (f_i, \ i \in \mathbb{Z}^d)$ be a family of probability generating function with $f_i'(1-) = 1$, $i \in \mathbb{Z}^d$. Denote by $(p_i(k))_{k \in \mathbb{N}_0}$ the corresponding probability distribution on $\mathbb{N}_0$. The branching random walk (BRW) $(X_n)_{n \in \mathbb{N}_0}$ in the environment $F$ is the particle system where in each time step at each site $i$ the particles branch according to the (critical) distribution $p_i$. The resulting progeny moves according to the kernel $a$. The number of particles at site $i$ at time $n$ is denoted by $X_n(i)$.

We assume that the environment $F$ is sampled from some probability law $\mathbb{P} = \mathcal{L}(F)$ but is fixed for all times. For fixed $F$ denote by $\mathbf{P}^F$ the law of the branching random walk $X$ (quenched law). $\mathbf{P}^F$ allows a random initial state $X_0$ whose distribution is assumed to be independent of $F$. Finally denote by $\mathcal{P} = \mathbb{E}\,\mathbf{P}^F = \int \mathbb{P}(dF)\mathbf{P}^F$ the annealed distribution.

The aspiration of Dawson and Fleischmann (1985) is to obtain criteria for the persistence of $X$. Recall that (for fixed $F$) $X$ is called persistent if, roughly speaking, it maintains its spatial intensity of particles in the longtime limit. In Theorem 3.1 they obtain a Kallenberg criterion for persistence: Let $g_i(s) = f'(1-s)$, $s \in [0,1]$, and let $(Z_n)$ be a random walk with kernel $a$. Then $X$ is persistent iff for all $i,j \in \mathbb{Z}^d$:

$$(1.1) \qquad \mathbf{P}^F \left[ \sum_{n=1}^{\infty} g_{-Z_n}(a^{(n)}(-Z_n, j)) < \infty \,\middle|\, Z_0 = i \right] = 1.$$

The philosophy leading to (1.1) is simple: $X$ is persistent if each $\{X_n(i), n \in \mathbb{N}_0\}$ is uniformly integrable. This is equivalent to the stochastic boundedness of the size biased random variables $\{\widehat{X_n(i)}, n \in \mathbb{N}_0\}$. (Recall that these are defined by $\mathbf{P}^F[\widehat{X_n(i)} = k] = k\mathbf{P}^F[X_n(i) = k]/\mathbf{E}^F[X_n(i)]$, $k \in \mathbb{N}_0$.) Now there is a nice representation of $\widehat{X_n(i)}$ going back to Olav Kallenberg. Trace back the ancestral line of a particle located at $i$ (this is the random walk $(-Z_m)_{m=0,\ldots,n}$). At each time $m$ generate a random number $Y_m$ of particles distributed according to the size biased offspring distribution $(kp_{-Z_n}(k))_{k \in \mathbb{N}_0}$ which has probability generating function $f'_{-Z_n}$. Now let $Y_m - 1$ offspring particles perform BRW and evaluate at time $n - m$. Adding over $m$ yields the distribution of $\widehat{X_n(i)}$. This explains the quantities arising in (1.1). Some uniform integrability arguments are needed to establish the exact form of the criterion. In fact, this makes the proof a little involved.

If we sample $F$ from a spatially homogeneous ergodic law $\mathbb{P}$ it is easily derived that (see Proposition 4.1)

$$(1.2) \qquad\qquad \mathbb{P}[X \text{ is persistent}] \in \{0, 1\}.$$

The main result (Theorem 5.2) of Dawson and Fleischmann (1985) is the existence of a critical dimension for persistence. Assume that $\{f_i,\ i \in \mathbb{Z}^d\}$ is i.i.d. and that there exists a $\beta \in (0, 1]$ such that $\lim_{s \to 0} s^{-\beta} \mathbb{E}[g_i(s)] \in (0, \infty)$ exists. Further assume that $a$ is in the normal domain of attraction of a genuinely $d$-dimensional symmetric $\alpha$–stable law for some $\alpha \in (0, 2]$. Then

$$(1.3) \qquad\qquad X \text{ is a.s. persistent} \iff d > \frac{\alpha}{\beta}.$$

Under somewhat weaker assumptions ($\mathbb{E}[g_i(s)]$ is only regularly varying, $\mathbb{P}$ is only ergodic, $a$ is only in the domain of attraction) one gets (1.3) but excluding the case $d = \alpha/\beta$ (Theorem 5.1). Note that the result is qualitatively identical to the classical non-random homogeneous situation. Considering the fact that the medium is ergodic and the mean is constant a.s. this is understandable: the particles experience a mixture of the medium and this mixture converges to a deterministic limit. Since the branching is critical everywhere no sites are favoured and this limit coincides with mean with respect to $\mathbb{P}$.

**1.2. Large deviations for non-critical branching.** The situation changes drastically if we allow the mean of the offspring distribution to vary also. This case has examined studied in Greven and den Hollander (1991), (1992) and Baillon, Clément, Greven and den Hollander (1993). We sketch the model and the result of the latter paper. The model is the same as the one introduced above but lives only on $\mathbb{Z}$. For the random walk kernel $a$ we make the special choice $a(i, i) = 1 - h$, $a(i, i + 1) = h$, $i \in \mathbb{Z}$, for some parameter $h \in (0, 1)$. We allow the mean $m_i = \sum_{k \in \mathbb{N}_0} k p_i(k)$ to be random. We assume that $\{p_i,\ i \in \mathbb{Z}\}$ is i.i.d. and that $M := \text{ess sup}_{\mathbf{P}}\ m_i < \infty$.

It is clear that the particles do not experience a $\mathbb{P}$–average of the medium. Rather they try to stay at those fertile sites $i$ where $M - m_i$ is small. (Note that it is the distance $M - m_i$ to the optimal value $M$ rather than the absolute value of $m_i$ that determines the (relative) "fertility" of a site.) On the other hand, for a particle not to move costs a (entropy) price depending on $h$. We are the observers of a thrilling interplay of two opposed tendencies. It is not too hard to guess that we are peeking into the world of large deviations and that the best strategy for a particle can be characterised in terms of a variational problem.

However tempting it is to give a panorama, our focus lies on critical branching. We mention only briefly that the techniques developed in the three papers mentioned above have had a considerable spin-off to the theory of one-dimensional random polymers (see, e.g., Greven and den Hollander (1993), Baillon, Clément, Greven and den Hollander (1994), König (1996), van der Hofstad and den Hollander (1997), and van der Hofstad and Klenke (1998)). As an appetizer we present part of the result of Baillon et al. (1993) in a nutshell (cf. discussion on page 313). The interesting quantities are the Malthusian (global) growth rate $\rho(h)$ and the drift of the particles $\theta(h)$.

Assume that the law $\mathcal{L}(m_i)$ is non-trivial and that it has an atom at $M$. Here the flesh pots are abundant and travelling is too strenuous. On the other hand if

$\mathcal{L}(m_i)$ has a "thin tail at $M$" the flesh pots are scarce and crossing the desert might be worthwhile. More precisely, we distinguish two cases.

**Case 1:** $\mathbb{E}[(M - m_i)^{-1}] = \infty$. Here we have

(1.4)
$$\begin{aligned} \rho(h) &= \log(M(1 - h)), & h \in (0, 1), \\ \theta(h) &\equiv 0. \end{aligned}$$

**Case 2:** $\mathbb{E}[(M - m_i)^{-1}] < \infty$. There exists an $h_c \in (0, 1)$ at which a phase transition occurs: For $h \in (0, h_c)$ the values of $\rho(h)$ and $\theta(h)$ are as in (1.4). However for $h \in (h_c, 1)$ these values are strictly exceeded. The functions $\rho$ and $\theta$ are analytic on $(h_c, 1)$ and $\rho$ is continuous at $h_c$ while $\theta$ is continuous at $h_c$ iff $\mathbb{E}[(M - m_i)^{-2}] = \infty$.

**1.3. Hydrodynamic fluctuations.** As a last example where the branching law rather than the branching rate is affected by the random medium we mention a paper by Dawson, Fleischmann and Gorostiza (1989). They investigate a branching process in continuous time where particles move according to a symmetric $\alpha$–stable process (with generator $\Delta_\alpha = -(-\Delta)^{\alpha/2}$) in $\mathbb{R}^d$. The particles branch at rate one according to a critical local offspring distribution $p_x$ given by the probability generating function $f_x(s) = s + h(x) \cdot (1 - s)^{1+\beta}$, $s \in [0, 1]$, $x \in \mathbb{R}^d$. Here $\beta \in (0, 1]$ is a fixed parameter and the stationary and ergodic random function $h : \mathbb{R}^d \to [0, (1 + \beta)^{-1}]$ is the (time homogeneous) environment.

Now we perform the hydrodynamic limit procedure. Attach to each particle a mass $\varepsilon^d$, rescale time by $\varepsilon^{-\alpha}$ and space by $\varepsilon$. The corresponding process $X_t^\varepsilon$ converges as $\varepsilon \to 0$ to the process $\Lambda_t$ of deterministic mass flow governed by the $\alpha$–stable semigroup.

Dawson, Fleischmann and Gorostiza (1989) scrutinize the fluctuations $Y_t^\varepsilon := \varepsilon^k(X_t^\varepsilon - \Lambda_t)$, where $k := (d\beta - \alpha)/(1 + \beta)$. They show (Theorem 4.9) that (if the initial configurations converge) the process $(Y_t^\varepsilon)$ converges as $\varepsilon \to 0$ to a generalised Ornstein-Uhlenbeck process $Y_t$ which is the solution of a generalised Langevin equation

(1.5)
$$dY_t = \Delta_\alpha Y_t \, dt + dZ_t.$$

Here $Z_t$ is a (distribution valued) process with independent increments. It is Gaussian if $\beta = 1$ and asymmetric $\beta$–stable otherwise. The method of proof is a detailed study of the (random) cumulant equation of the Laplace functionals.

## 2. Catalytic super-Brownian motion

We come to the main object of the discussion. It is a continuous time branching model where the offspring distribution is fixed and only the local infinitesimal rate $\varrho$ at which branching occurs varies. We could formulate the model in terms of (continuous time) branching random walk (BRW) on $\mathbb{Z}^d$ or any countable Abelian group. This has been done in some generality in Greven, Klenke and Wakolbinger (1999). However we follow a semi-chronological route and first present the setting where the underlying motion process $(W_t)$ is Brownian motion in $\mathbb{R}^d$.

The Dawson-Watanabe process (or super-Brownian motion (SBM)) in $\mathbb{R}^d$ is the diffusion limit of (critical binary) branching Brownian motion (BBM). In order to model the varying branching rate $\varrho$ assume that each particle has a clock $A(t)$ which is an additive functional of $(W_t)$. If $\{\varrho_t(x), \ t \geq 0, \ x \in \mathbb{R}^d\}$ is a (nonnegative) function we can set $A(t) = \int_0^t \varrho_s(W_s)ds$. If $\varrho$ is more generally a measure (on

$[0, \infty) \times \mathbb{R}^d)$ we have to be more careful with the definition. Under some regularity assumptions $A$ is the collision local time $L_{[W,\varrho]}(0,t)$ of $W$ with $\varrho$, that is, $\varrho$ is the Revuz measure of the time inhomogeneous additive functional $A$. In fact, it is a non-trivial piece of work to check in the examples that $L_{[W,\varrho]}$ can be well-defined. For example, this needs that the support of $\varrho$ is not polar for Brownian motion and this restricts us in some cases to $d = 1$ or $d \leq 3$.

Denote by $(X_t^{\varrho,n})_{t \geq 0}$ the catalytic BBM (CBBM) where we assign to each particle a mass $n^{-1}$ and where the clock is $nA(t)$ rather than $A(t)$. Let $\mu \in \mathcal{M}_F(\mathbb{R}^d)$ (=space of finite measures with the vague topology) and assume that $X_0^{\varrho,n}$ is a Poisson point process with intensity $n\mu$. We define catalytic SBM (CSBM) $(X_t^\varrho)$ as the limit of $(X_t^{\varrho,n})$ as $n \to \infty$ and denote its law by $\mathbf{P}_\mu^\varrho$. Of course, it has to be justified that the limit exists and defines a Markov process with nice path properties. This intuitively appealing approach has been made by Delmas (1996) for stationary catalyst $\varrho_t \equiv \sigma$ with some additional energy assumption on $\sigma$ and by Dynkin (1991) for more general additive functionals $A$ but with a very restrictive exponential moment assumption. These assumptions have been relaxed in Dynkin (1994). Most of the recent papers rely on Dynkin's result and try to circumvent the moment assumptions by some approximation scheme (e.g., Dawson and Fleischmann (1997a), Fleischmann and Mueller (1995)). Due to the independence structure ("$\mathbf{P}_{\mu+\nu}^\varrho = \mathbf{P}_\mu^\varrho * \mathbf{P}_\nu^\varrho$") Laplace functionals are an important tool for the investigation of CSBM. For $\varphi \in C_c^+(\mathbb{R}^d)$ (=space of nonnegative continuous functions with compact support) we define the function $v_\varphi^\varrho(t; x)$ by

$$(2.1) \qquad v_\varphi^\varrho(t; x) = -\log \mathbf{E}_{\delta_x}^\varrho[\exp(-\langle X_t^\varrho, \varphi \rangle)].$$

Note that $-\log \mathbf{E}_\mu^\varrho[\exp(-\langle X_t^\varrho, \varphi \rangle)] = \langle \mu, v_\varphi^\varrho(t; \bullet) \rangle$. The analytical means by which we scrutinize $v_\varphi^\varrho$ is the *cumulant equation* ($p_t$ is the heat kernel)

$$(2.2) \qquad v_\varphi^\varrho(s, t; x) = (p_{t-s}\varphi)(x) - \int_s^t du \int_{\mathbb{R}} \varrho_u(dy)(v_\varphi^\varrho(u, t; y))^2 p_{u-s}(x, y)$$

or formally

$$(2.3) \qquad \begin{aligned} -\frac{d}{ds} v_\varphi^\varrho(s, t; x) &= \frac{1}{2}\Delta v_\varphi^\varrho(s, t; x) - \frac{\varrho_s(dx)}{dx}(v_\varphi^\varrho(s, t; x))^2, \\ v_\varphi^\varrho(0, t) &= \varphi. \end{aligned}$$

As a rule we set $v_\varphi^\varrho(t; x) = v_\varphi^\varrho(0, t; x)$. (2.3) is the Kolmogorov backward equation of the Laplace functional. Since CSBM is in general time-inhomogeneous we work with this formulation rather than with the forward equation. From (2.2) it is not hard to derive a recursion scheme for the moments of $\langle X^\varrho, \varphi \rangle$. We only mention that the expectation and variance are given by

$$(2.4) \qquad \begin{aligned} \mathbf{E}_\mu^\varrho[\langle X_t^\varrho, \varphi \rangle] &= \langle p_t\mu, \varphi \rangle, \\ \mathbf{Var}_\mu^\varrho[\langle X_t^\varrho, \varphi \rangle] &= \int_0^t ds \int_{\mathbb{R}} \varrho_s(dx)(p_s\mu)(x)(p_{t-s}\varphi)^2(x). \end{aligned}$$

Again it is not a priori clear that there exists a (unique) solution to (2.2) or (2.3). Establishing this by analytical methods for a certain catalyst $\varrho$ was the starting point of Dawson and Fleischmann (1991). They use a smoothing procedure for the catalyst replacing $\varrho_t$ by $p_\varepsilon \varrho_t$ (recall that $p_\varepsilon$ heat kernel). Letting $\varepsilon \to 0$ they show that the cumulant equation could be uniquely solved. It is not too hard to deduce from this the existence of a unique Markov process $(X_t^\varrho)$ connected to $v_\varphi^\varrho$ by (2.1). However, establishing path properties such as existence of a càdlàg or continuous version requires more work.

**2.1. Single point catalyst in d = 1.** The simplest catalyst which is not a function is a unit point mass $\varrho_t \equiv \delta_c$ at a point $c \in \mathbb{R}^d$. For $d \geq 2$ single points are polar for Brownian motion, so we have to assume $d = 1$. This model was studied first by Dawson and Fleischmann (1994). A remarkable insight via a nice representation in terms of the super process with respect to a $\frac{1}{2}$-stable sub-ordinator $(U_t)_{t\geq 0}$ is due to Fleischmann and Le Gall (1995) and we follow their exposition.

The unit mass $\delta_c$ is the Revuz measure of (Brownian) local time $L(\bullet, c)$ at $c$ and this local time is a perfectly well understood object. On a heuristic level the "infinitesimal particles" of $(X_t^{\delta_c})$ branch at a high rate while they are at $c$ and perform excursions from $c$ otherwise. The length of the excursions can be described in terms of the jumps of the inverse local time $\widetilde{L}(t,c) = \inf\{s > 0 : L(s,c) \geq t\}$ which is a $\frac{1}{2}$-stable sub-ordinator $(R_t)_{t\geq 0}$. Associated with $(R_t)$ is a super process $(U_t)_{t\geq 0}$ (derived as above from branching particles moving independently on $\mathbb{R}^+$ like $(R_t)$). Note that the former time-variable is now a space-variable. Define $V = \int_0^\infty U_t dt$. The reader might by now be willing to believe that the occupation density

$$(2.5) \qquad \lambda_x^{\delta_c}(t) := \frac{1}{dx} \int_0^t X_s^{\delta_c}(dx) ds$$

exists at $x = c$ and that it should equal in distribution

$$(2.6) \qquad \{\lambda_c^{\delta_c}(t),\ t \geq 0\} \overset{\mathcal{D}}{=} \{V([0,t]),\ t \geq 0\}.$$

This is in fact true (Theorem 1 of Fleischmann and Le Gall (1995)) if we define

$$(2.7) \qquad U_0(dt) = -d\|Q_t^c \mu\|.$$

Here $(Q_t^c)_{t\geq 0}$ is the heat flow killed at $c$ (note that $U_0(\{0\}) = \mu(\{c\})$) and $X_0^{\delta_c} = \mu$ a.s. Let $(q_t^c(x) = -d\|Q_t^c \delta_x\|/dt,\ t > 0,\ x \neq c)$ be the density of the first hitting time of Brownian motion at $c$:

$$q_t^c(x) = \frac{|x - c|}{(2\pi t^3)^{1/2}} \exp\left(-\frac{(x - c)^2}{2t}\right), \qquad t > 0.$$

Recalling the idea of infinitesimal particles performing excursions off $c$ and considering the duality of excursions and Brownian motion killed at $c$ we arrive at the following representation formula ($\ell$ is the Lebesgue measure):

$$(2.8) \qquad X_t^{\delta_c} := \left(\int_0^t V(ds) q_{t-s}^c\right) \ell + Q_t^c \mu, \qquad t \geq 0,$$

is a version of CSBM in the medium $\varrho \equiv \delta_c$ (Theorem 1b). From this fancy formula one can derive a bunch of nice properties. E.g., for $x \neq c$ the density $\xi_t^{\delta_c}(x) = X_t^{\delta_c}(dx)/dx$ exists, is $C^\infty$ and solves the heat equation (since $q_t^c$ shares this property). Further $X_t^{\delta_c}$ contains all information about the past: $X_s^{\delta_c}$ can be reconstructed from $X_t^{\delta_c}$ if $0 \leq s \leq t$. Finally, if $\mu = \delta_c$ then $\lambda_c^{\delta_c}(\infty)$ is a $\frac{1}{2}$-stable random variable and $\lambda_x^{\delta_c}(\infty) = \lambda_c^{\delta_c}(\infty)$ a.s. for all $x \in \mathbb{R}^d$.

It is intriguing to know more about the behaviour of $X_t^{\delta_c}$ near $c$ and at $c$. From infinite divisibility and a Palm formula Fleischmann and Le Gall (1995) derive that the support of $V$ (and hence $\lambda_c^{\delta_c}$) has Hausdorff dimension 1. It is singular with respect to Lebesgue measure (Theorem 6) and diffusive: $V(\{t\}) = 0$ for all $t > 0$. The singularity of $\lambda_c^{\delta_c}$ had been shown earlier with a considerable technical effort by Dawson, Fleischmann, Li and Mueller (1995). They do not use (2.6) but construct historical CSBM $(\tilde{X}_t^{\delta_c})$ and derive a Kallenberg type representation for the Palm canonical measure $Q_{s,w}^{r,\omega}$ of the historical occupation density $\tilde{\lambda}_c^{\delta_c}$. From this they derive that $\varepsilon^{-1}\lambda_c^{\delta_c}((t-\varepsilon,t]) \to \infty$, $\varepsilon \to 0$, $Q_{s,w}^{r,\omega}$–a.s. (Theorem 4.2.2). Using standard arguments this gives the claim.

**2.2. Extension to d ≥ 1.** As mentioned above a single point catalyst makes sense only if $d = 1$. For $d \geq 1$ it is however possible to construct CSBM even for a singular catalyst $\varrho$ if (roughly speaking) its carrying dimension is larger than $d - 2$. More precisely, Delmas (1996) (Theorem 3.2 and 4.7) was able to construct (a continuous version of) $(X_t^\varrho)$ for $\varrho_t \equiv \sigma$, where $\sigma$ fulfills his "hypothesis (H)"

$$(2.9) \qquad \exists \beta \in (0,1): \quad \sup_{x\in\mathbb{R}^d} \int_{\|y-x\|\leq 1} \frac{\sigma(dy)}{|x-y|^{d-2+2\beta}} < \infty.$$

For example, in $d = 1$ every finite measure $\sigma \in \mathcal{M}_f(\mathbb{R})$ fulfills (H) (take $\beta = 1/2$). In any dimension the Lebesgue measure $\ell$ fulfills (H). Note that $\sigma$ does not charge polar sets and that the Hausdorff dimension of its support is at least $d - 2 + 2\beta$.

Delmas can show that for $\varphi : \mathbb{R}^d \to \mathbb{R}$ bounded and measurable the evaluation process $(\langle X_t^\varrho, \varphi \rangle)_{t\geq 0}$ is a.s. continuous (Theorem 4.9).

He also shows an extension of the result that in one dimension the occupation measure densities exist (Proposition 5.1): For a measure $\eta$ fulfilling a slightly stronger assumption than (2.9) (his "hypothesis (H')") there exists the weighted occupation measure $\Gamma_\eta(dt, dx)$. Formally $\Gamma_\eta$ is defined by

$$(2.10) \qquad \Gamma_\eta(dt, dx) = \xi_t^\varrho(x)\eta(dx)dt.$$

For $d = 1$ and $\nu = \delta_x$ we get back $\lambda_x^\varrho$. The main result (Theorem 7.1) is a representation formula analogous to (2.8). Let $D = \mathrm{supp}(\sigma)$, and let $\nu$ be the Revuz measure of Brownian local time in $D$. For $x \in D$ let $H^x$ be the excursion measure on paths $\omega$ starting at $\omega(0) = x$ (see Maisonneuve (1975)). Denote by $L(\omega)$ the length of the excursion and define $H_t^x \in \mathcal{M}_f(\mathbb{R}^d)$ by $H_t^x(A) = H^x(\{\omega : L(\omega) > t, \omega(t) \in A\})$. Finally let $(Q_t^D)_{t\geq 0}$ denote the heat flow killed at $D$. Then Delmas' formula is

$$(2.11) \qquad X_t^\varrho \mathbf{1}_{D^c} = \int_{[0,t]\times D} \Gamma_\nu(ds, dx)H_{t-s}^x + Q_t^D\mu.$$

Using a refinement of the argument given above for the one-point catalyst Delmas deduces (Theorem 8.1) that on $D^c$ the reactant $X_t^\varrho$ has a density $\xi_t^\varrho(x)$ which is $\mathcal{C}^\infty$ and solves the heat equation

$$\left(\frac{\partial}{\partial t} - \frac{1}{2}\Delta\right)\xi_t^\varrho = 0.$$

Finally we wish to mention that Delmas gives a characterisation (Proposition 9.1) of $X^\varrho$ in terms of a martingale problem where the increasing process is $\Gamma_\sigma(dt, dx)$.

**2.3. Multiple point catalyst.** A natural extension of the single point catalyst is the multiple point catalyst. If the points are discrete, then it fits into the framework of Delmas (1996). If the points accumulate, the catalyst is locally not integrable. This causes a special behaviour and we delay the discussion of an example of a locally non-integrable catalyst.

The example we wish to examine here is that of a catalyst concentrated on points that are everywhere dense in $\mathbb{R}$. However these points $c$ do not carry the unit mass $\delta_c$ but rather multiples that ensure local integrability. To be concrete, let $\sigma$ be sampled from a random measure $\Gamma$ on $\mathbb{R}$ with independent increments and with Lévy measure $\nu$ and define $\varrho_t \equiv \sigma$. We can define $\Gamma$ in terms of Laplace transforms by

$$(2.12) \qquad -\log \mathbb{E}[e^{-\langle \Gamma, \varphi \rangle}] = \int_0^\infty \nu(dx)\langle \ell, 1 - e^{-x\varphi} \rangle, \qquad \varphi \in C_c^+(\mathbb{R}).$$

We assume that $\int \nu(dx)(1 \wedge x) < \infty$ so that $\Gamma$ is locally finite and is carried by a countable set. One example is the stable point process with index $\gamma \in (0, 1)$, that is $\nu(dx) = c \cdot x^{-(1+\gamma)}dx$, $x > 0$, for some $c > 0$. If $c$ is chosen appropriately this yields

$$(2.13) \qquad -\log \mathbb{E}[e^{-\langle \Gamma, \varphi \rangle}] = \langle \ell, \varphi^\gamma \rangle, \qquad \varphi \in C_c^+(\mathbb{R}).$$

This model has been considered in Dawson, Li and Mueller (1995). They address the question if $X^\varrho$ has a compact global support

$$(2.14) \qquad \mathcal{G} = \text{closure} \left( \bigcup_{t \geq 0} suppX_t^\varrho \right)$$

$\mathbf{P}_{\delta_x}^\varrho$–a.s. Recall that $\mathcal{G}$ is in fact compact a.s. for classical SBM (see Iscoe (1988)). Dawson, Li and Mueller give complicated sufficient conditions for compactness of $\mathcal{G}$ (Theorem 1) and for non-compactness (Theorem 2) in terms of $\nu$. One example for compactness (Corollary 1) is the $\gamma$–stable point process (recall that $\mathcal{P}_{\delta_x} = \mathbb{E}\,\mathbf{P}_{\delta_x}^\varrho$ is the annealed law)

$$(2.15) \qquad \mathcal{P}_{\delta_x}[\mathcal{G} \text{ is compact}] = 1.$$

On the other hand (Corollary 2) if $\nu$ is finite or if, for instance, $\nu(dx) = x^{-1}\mathbf{1}_{(0,1]}(x)dx$ then

$$(2.16) \qquad \mathcal{P}_{\delta_x}[\mathcal{G} \text{ is compact}] = 0.$$

The method employed is quite similar to the original approach of Iscoe. Consider a function $\psi \in C_c^+(\mathbb{R})$ and define

$$(2.17) \qquad u_\psi(t; x) = -\log \mathbf{E}_{\delta_x}^\varrho[\exp(-\int_0^t \langle X_s^\varrho, \psi \rangle ds)].$$

Hence $u_\psi$ solves the integral equation (recall that $p_t$ is the heat kernel)

$$(2.18) \qquad u_\psi(t; x) = \int_0^t (p_{t-s}\psi)(x)ds - \int_0^t ds \int_{\mathbb{R}} \Gamma(dy)p_{t-s}(x, y)u_\psi^2(s, y).$$

Now let $t \to \infty$, and let $\psi \uparrow \mathbf{1}_{[\alpha_1, \alpha_2]^c}$ for some reals $\alpha_1 < \alpha_2$. Then (by Theorem 0) $u_\psi$ approaches the solution $u(x) = u_{\alpha_1, \alpha_2}(x)$ of the formal boundary value problem

$$(2.19) \qquad \frac{1}{2}\frac{d^2}{dx^2}u(x) = u^2(x)\Gamma(dx), \quad x \in (\alpha_1, \alpha_2),$$

$$u(\alpha_1) = u(\alpha_2) = \infty.$$

Since $\mathbf{P}_{\delta_x}^\varrho[\mathcal{G} \subset [\alpha_1, \alpha_2]] = 1 - \exp(-u_{\alpha_1, \alpha_2}(x))$, it is clear that (2.15) holds if $u_{-\alpha, \alpha}(x) \to 0$, $\alpha \to \infty$. On the other hand, $u_{-\alpha, \alpha} \equiv \infty$ for all $\alpha$ implies (2.16). It is the content of Theorem 1 and 2 to give sufficient conditions on $\nu$ for either case to hold.

In a recent work Dawson, Fleischmann and Mueller (1998) study the question of finite time extinction for this model: Is it true that for $\mu \in \mathcal{M}_f(\mathbb{R})$

$$(2.20) \qquad \mathbb{E}\, \mathbf{P}_\mu^\varrho[\|X_t^\varrho\| = 0 \text{ for } t \text{ large enough}] = 1?$$

Obviously, this is not the case if $\varrho$ is supported by a non-dense set and $\mu((\operatorname{supp} \varrho)^c) > 0$. In this case $X_t^\varrho \geq Q_t^{\operatorname{supp} \varrho}\mu$ (recall that $Q_t^D$ is the semigroup of the heat flow killed at $D$). Thus $\|X_t^\varrho\| > 0$ a.s. On the other hand, the $\gamma$–stable catalyst $\varrho$ has a dense support. The dense support alone does not guarantee finite time extinction. However in this particular example Dawson, Fleischmann and Mueller can show that (2.20) holds.

**2.4. Moving multiple point catalyst.** We come to our first example where the catalyst is not time homogeneous. Assume that $\Gamma$ is the $\gamma$–stable point process, $\gamma \in (0,1)$, introduced in (2.13). Consider the representation

$$(2.21) \qquad \Gamma = \sum_{i=1}^{\infty} g^i \delta_{x^i},$$

where $\{g^i\}$ are the "action weights" of the points $\{x^i\}$. Now allow the points to perform independent Brownian motion and carry their weights with them. More precisely, let $\{(x_t^i)_{t\geq 0}\}$ be independent Brownian motions, $x_0^i = x^i$, and define

$$(2.22) \qquad \varrho_t = \sum_{i=1}^{\infty} g^i \delta_{x_t^i}.$$

This model has been studied by Dawson, Fleischmann and Roelly (1991) and Dawson and Fleischmann (1991). In fact, these papers cover a moderately more general situation, namely with the Brownian motion replaced by the fashionable symmetric $\alpha$–stable process. The finite variance branching is replaced by a certain offspring law in the normal domain of attraction of a $\beta$-stable law, $\beta \in (0,1]$, (in the cumulant equation replace $u^2$ by $u^{1+\beta}$). Here we no not draw a bead on these details, the reader may think of $\alpha = 2$, $\beta = 1$.

We need the following definitions. Let $\mathcal{M}(\mathbb{R}^d) = \{\text{Radon measures on } \mathbb{R}^d\}$ be equipped with the vague topology. This is a Polish space (see Kallenberg (1983)). For $p > d$ define the space $\mathcal{M}_p(\mathbb{R}^d)$ of $p$–tempered measures by

$$(2.23) \qquad \mathcal{M}_p(\mathbb{R}^d) := \left\{ \mu \in \mathcal{M}(\mathbb{R}^d) : \langle \mu, \phi_p \rangle < \infty \right\},$$

where $\phi_p(x) = (1 + \|x\|^2)^{-p/2}$. Note that $\ell \in \mathcal{M}_p(\mathbb{R}^d)$.

Dawson and Fleischmann construct $(X_t^\varrho)$ as a Markov process with values in $\mathcal{M}_p(\mathbb{R}^d)$ (Theorem 1.8.2). However they do not make a statement on whether a càdlàg version exists. The reason for this flaw originates in the construction. Rather than following the currently appreciated approach via "branching functionals" they show that the cumulant equation (2.2) could be uniquely solved. They replace $\varrho_t$ by the function $p_\varepsilon \varrho_t$, $\varepsilon > 0$, and then let $\varepsilon \to 0$. This yields a solution, uniqueness follows by standard arguments, and also existence of $(X_t^\varrho)$. Dawson, Fleischmann and Roelly (Theorem 1.19) show absolute continuity of the states of $X_t^\varrho$ with respect to Lebesgue measure if $\alpha = 2$ or if $(\beta\gamma)(1 + \alpha) > 1$. The "main result" of Dawson and Fleischmann (1991) is a scaling limit (Theorem 1.9.4). Fix $\eta > 0$ and define for $K > 0$

$$(2.24) \qquad {}^K X_t^\varrho = K^{-\eta} X_{tK}^\varrho(K^\eta \bullet).$$

If $\eta = \eta_c := ((\gamma(\alpha - 1) + 1)/\alpha\beta\gamma$ then for $t > 0$  $\mathcal{L}_\ell^\varrho(^K X_t^\varrho) \Longrightarrow \mathcal{L}_\ell(^\infty X_t^\varrho)$, $K \to \infty$ in $\mathbb{P}$–probability (note that $\Longrightarrow$ denotes weak convergence). Here for $t > 0$, $^\infty X_t$ is a homogeneous independent point process on $\mathbb{R}^d$ with Lévy measure $\nu_t^\infty$ characterised as follows. For fixed $\varrho$ let $v_t^\varrho$ be the Lévy measure of $\mathbf{P}_{\delta_0}^\varrho[\|X_t^\varrho\| \in \bullet]$. Then $\nu_t^\infty = \mathbb{E}[\nu_t^\varrho]$. In terms of the Laplace functionals this reads

$$(2.25) \qquad -\log \mathbf{E}[\exp(-\langle X_t^\infty, f \rangle)] = \int dx \, \mathbb{E}[v_{f(z)\mathbf{1}}^\varrho(t; 0)].$$

If $\eta > \eta_c$ then there holds a law of large numbers: $^\infty X_t = \delta_\ell$ a.s. (Theorem 1.10.1).

## 2.5. Hyperplanes.

Dawson and Fleischmann (1995) also have an example for a time homogeneous catalyst $\varrho$ in $\mathbb{R}^d$, $d > 1$, whose support may be everywhere dense. However they practically assume that $\varrho$ factors into a function of $d - 1$ coordinates and a measure in one dimension. So you can keep in mind the example where $\varrho$ consists of "hyperplanes": $\varrho = \sigma \otimes \ell_{d-1}$, where $\sigma \in \mathcal{M}(\mathbb{R})$ and $\ell_{d-1}$ is the $(d - 1)$–dimensional Lebesgue measure. Under some conditions on $\varrho$ they show the existence of CSBM (Lemma 2.3.4). Further under some very restrictive assumptions on the branching rate functional A (e.g., all moments exist and fulfill a growth condition, see Definition 2.4.7 and 2.6.1) they show absolute continuity of $X_t^\varrho$ with respect to $\ell_d$ (Theorem 2.6.2). This is proved by showing that there exists a proper solution of the cumulant equation with terminal condition $\delta_x$, $x \in \mathbb{R}^d$. Compare this result with SBM in $d \geq 2$ which is singular with respect to the Lebesgue measure.

## 2.6. Locally infinite catalyst.

As far as the author knows there is only one paper studying a model where the catalyst is locally not integrable. Fleischmann and Mueller (1997) construct one-dimensional CSBM (Theorem 1) with time homogeneous catalyst $\varrho$ whose density is given by $\varrho_t(dx)/dx = \theta|x - c|^{-\sigma}$, where $c \in \mathbb{R}$, $\theta \in (0, \infty)$ and $\sigma \in [1, 2]$. In particular, $\langle \varrho_t, \mathbf{1}_{[c-1,c+1]} \rangle = \infty$. They show that for $\langle X_0^\varrho, \mathbf{1} \rangle < \infty$ the total mass process $\langle X_t^\varrho, \mathbf{1} \rangle$ is a super martingale but not a martingale. It has finite variance iff $\sigma < 2$ (Theorem 3).

For Brownian motion $W$ let $\tau = \{t > 0 : W_t = c\}$ and recall that $A(t) = \theta \int_0^t |W_s - c|^{-\sigma} ds$. Note that for $x \neq c$

$$\mathbf{P}_x[A(\tau) = \infty] = \begin{cases} 1, & \sigma = 2, \\ 0, & \sigma < 2. \end{cases}$$

Since the critical Galton–Watson process dies out eventually this implies that if $\sigma = 2$ then "no infinitesimal particle of $X_t^\varrho$ ever reaches $c$". More precisely, there exists an increasing sequence of stopping times $(\tau_n)$ of Brownian motion $W$ such that any $\tau_n$ is strictly smaller than $\tau$ and such that $\lim_{n \to \infty} \mathbf{P}_{\delta_x}^\varrho[X_{\tau_n}^\varrho = 0] = 1$ (Theorem 4). Here $X_{\tau_n}^\varrho$ is understood in the sense of Dynkin's stopped measure (see Dynkin (1991)).

## 3. CSBM in a SBM medium

In an intriguing model we consider a random time-inhomogeneous catalyst $\varrho$: The catalyst is itself a sample path of SBM on $\mathbb{R}^d$. Let $\mathbb{P}_\mu$ denote its law with initial condition $\mu \in \mathcal{M}_p(\mathbb{R}^d)$ (recall (2.23)). Recall that SBM is a $(d \wedge 2)$–dimensional object. Hence the support is polar for Brownian motion if $d \geq 4$. We thus have to restrict ourselves to $d \leq 3$.

This CSBM can serve as a model for a biological one-way interaction between two species (green and red, say). The green species (the catalyst) does not even notice the red species and simply performs its migration and resampling scheme, however influences thereby the reproduction rate of the red species. This model has been studied by Dawson and Fleischmann (1997a), (1997b), Etheridge and Fleischmann (1998), and Fleischmann and Klenke (1999). On a BRW level it is treated in Greven, Klenke and Wakolbinger (1999).

A model with a symmetric interaction between two species is due to Dawson and Perkins (1998). We discuss it briefly in Section 4.

**3.1. The construction.** Dawson and Fleischmann (1997a) construct $X^\varrho$ in a somewhat more general framework. Instead of aiming at the concrete model directly they first construct Hölder continuous versions of CSBM for a certain class of Hölder continuous branching functionals $A$.

Define **K** to be the set of continuous additive functionals of $W$ such that for all $t_0 \geq 0$ (recall (2.23))

$$(3.1) \qquad \sup_{x \in \mathbb{R}^d} \mathbf{E}_{s,x} \left[ \int_s^t A(dr)\phi_p(W_r) \right] \to 0, \qquad s, t \to t_0.$$

Further define for $\xi \in (0,1]$ the subclass $\mathbf{K}^\xi$ by imposing the additional requirement that for $T > 0$ there exists $c_T > 0$ such that

$$(3.2) \qquad \mathbf{E}_{s,x} \left[ \int_s^t A(dr)(\phi_p(W_r))^2 \right] \leq c_T |t - s|^\xi \phi_p(x), \qquad 0 \leq s \leq t \leq T.$$

They show (Proposition 1) that for $A \in \mathbf{K}$ the cumulant equation

$$(3.3) \quad v_\varphi(s,t;x) = (p_{t-s}\varphi)(x) - \mathbf{E}_{s,x} \left[ \int_s^t A(dr)v_\varphi(r,t;W_r)^2 \right], \qquad \varphi \in C_c^+(\mathbb{R}^d),$$

has a unique solution. The proof relies on an approximation of $A$ by functionals that fit into Dynkin's (1994) framework. As an application (Proposition 2 (sic!), page 230) one gets the existence of a time-inhomogeneous multiplicative $\mathcal{M}_p(\mathbb{R}^d)$–valued Markov process $(X^A, \mathbf{P}_{t,\mu}^A, t \geq 0, \mu \in \mathcal{M}_p(\mathbb{R}^d))$ with log-Laplace function

$$(3.4) \qquad v_\varphi(s,t;x) = -\log \mathbf{E}_{s,\delta_x}^A [\exp(-\langle X_t^A, \varphi \rangle)].$$

The moments of $\langle X_t^A, \varphi \rangle$ can be expressed in terms of the derivatives of $v_{\theta\varphi}$ at $\theta = 0$. In particular the first and second moments are

$$\mathbf{E}_{s,\mu}^A[X_t] = p_{t-s}\mu,$$

$$(3.5) \quad \mathbf{Cov}_{s,\mu}^A[\langle X_{t_1}^A, \varphi_1 \rangle, \langle X_{t_2}^A, \varphi_2 \rangle] =$$

$$\int \mu(dx) \mathbf{E}_{s,x}^A \left[ \int_s^{t_1 \wedge t_2} A(dr)(p_{t_1-s}\varphi_1)(W_r)(p_{t_2-s}\varphi_2)(W_r) \right].$$

Dawson and Fleischmann develop a recursion formula for the $n$-th derivatives of $v_{\theta\varphi}$ at $\theta = 0$ and for the $n$-th centred moments $\mathbf{E}_{s,\mu}^A[|\langle Z_t^A, \varphi \rangle|^n]$, where $Z_t^A = X_t^A - \mathbf{E}_{s,\mu}^A[X_t^A]$. These moments are finite if, for instance, $A \in \mathbf{K}^\xi$ for some

$\xi \in (0,1]$ (Lemma 5). In this case the following estimate holds (Lemma 6)

$$\mathbf{E}_{s,\mu}^A[|\langle Z_{t+h}^A - Z_t, \varphi\rangle|^{2n}]$$

(3.6)
$$\leq \mathrm{const}\left(\left\|\frac{p_h\varphi - \varphi}{\phi_p}\right\|_\infty^{2n} + h^{\xi n}\left\|\frac{\varphi}{\phi_p}\right\|_\infty^{2n}\right)(\langle\mu,\phi_p\rangle + 1)^{2n}.$$

Using a measure-valued version of Kolmogorov's method of moments one obtains for every $\varepsilon \in (0,\xi/2)$ a Hölder-$\varepsilon$-continuous version of the centred process $(Z_t^A)$ (Theorem 1). Here we assumed a certain underlying metric that generates the vague topology on $\mathcal{M}_p(\mathbb{R}^d)$ but which we do not specify here. In particular, $(X_t^A)$ has a continuous version since $t \mapsto p_{t-s}\mu$ is continuous.

There is a simple criterion for absolute continuity of $X_t^A$ with respect to Lebesgue measure $\ell$. Formally we could define the density $\xi_t^A(x) = \langle X_t^A, \delta_x\rangle$, $x \in \mathbb{R}^d$. This expression makes sense as the limit of $\langle X_t^A, p_\varepsilon\delta_x\rangle$ as $\varepsilon \to 0$. If $t > s$ then $\mathbf{E}_{s,\mu}^A[X_t] = p_{t-s}\mu$ is absolutely continuous and it suffices to check that the variances converge along some sequence $\varepsilon_n \downarrow 0$

(3.7)
$$\lim_{n\to\infty} \int \mu(dx)\mathbf{E}_{s,x}^A\left[\int_s^t A(dr)p_{t-r+\varepsilon_n}(W_r,z)^2\right]$$
$$= \int \mu(dx)\mathbf{E}_{s,x}^A\left[\int_s^t A(dr)p_{t-r}(W_r,z)^2\right] < \infty.$$

If (3.7) holds then $\xi_t^A(z)$ exists as the $L^2$–limit of $\langle X_t^A, p_{\varepsilon_n}\delta_z\rangle$, $n \to \infty$, and is the density of $X_t^A$ (Proposition 4). It has first and second moment

$$\mathbf{E}_{s,\mu}^A[\xi_t^A] = p_{t-s}\mu,$$

(3.8)    $$\mathbf{Cov}_{s,\mu}^A[\xi_{t_1}^A(z_1),\xi_{t_2}^A(z_2)] =$$
$$\int \mu(dx)\mathbf{E}_{s,x}^A\left[\int_s^{t_1\wedge t_2} A(dr)p_{t_1-s}(W_r,z_1)p_{t_2-s}(W_r,z_2)\right].$$

A criterion similar to the one above is derived for the absolute continuity of the occupation time measure $\int_s^t X_r^A dr$ (Proposition 5).

We come back to the situation where $\varrho$ is a sample path of SBM and $A = L_{[W,\varrho]}$ is the *collision local time* of $W$ with $\varrho$. As above with the density of $X_t^A$ we would like to define $L_{[W,\varrho]}(s,t) = \int_s^t dr\,(\varrho_r(dx)/dx)\big|_{x=W_r}$. If $d = 1$ this makes perfect sense since $\varrho_r$ is absolutely continuous. However for $d \geq 2$ it is not. Hence we define for $\varepsilon > 0$

(3.9)
$$L_{[W,\varrho]}^\varepsilon(s,t) = \int_s^t dr(p_\varepsilon\varrho_r)(W_r).$$

We hope that it makes sense to define

(3.10)
$$L_{[W,\varrho]}(s,t) = \lim_{\varepsilon\to 0} L_{[W,\varrho]}^\varepsilon(s,t).$$

The reader might guess that this is non-trivial. However, Evans and Perkins (1994, Theorem 4.1) show that if $d \leq 3$ and if $\nu \in \mathcal{M}_f(\mathbb{R}^d)$ is absolutely continuous, then for $\mathbb{P}_\nu$–a.a. $\varrho$ the limit on the r.h.s. of (3.10) makes sense in $L^2$ and defines a continuous additive functional. A simple approximation argument extends this to absolutely continuous $\nu \in \mathcal{M}_p(\mathbb{R}^d)$. However we typically want singular initial conditions. For instance, if $d = 3$ and $\nu$ is a sample from the equilibrium of SBM, then $\nu$ is singular with respect to $\ell$. We outline the steps that yield in fact that (3.10) makes sense in $L^2$ for the examples we have in mind and that we even have $A = L_{[W,\varrho]} \in \mathbf{K}^\xi$ for some $\xi \in (0, 1/4)$.

Dawson and Fleischmann improve a result of Sugitani (1989) which states that the occupation measure $Y$ on $[0, \infty) \times \mathbb{R}^d$ defined by

$$Y([s, t] \times B) = \int_s^t \varrho_r(B) dr$$

is absolutely continuous and has a (jointly) continuous density $(t, x) \mapsto y(t, x)$. In fact, using moment estimates and Kolmogorov's method they get that for the centred density $\overline{y} = y - \mathbb{E}_\nu[y]$ on any set $[0, T] \times \mathbb{R}^d$, $T > 0$, the function $\overline{y}(t, x)\phi_p(x)$ is Hölder-$\xi$-continuous, $\xi \in (0, 1/4)$ (Theorem 2). Hence $y(t, x)\phi_p(x)$ is Hölder-$\xi$-continuous iff

(3.11)     $(t, x) \mapsto \displaystyle\int_0^t (p_r\nu)(x)\phi_p(x) dr$ is Hölder-$\xi$-continuous.

Denote by $\mathcal{M}_p^\xi(\mathbb{R}^d)$ the space of measures $\nu \in \mathcal{M}_p(\mathbb{R}^d)$ for which (3.11) holds. We do not give a characterisation of $\mathcal{M}_p^\xi(\mathbb{R}^d)$ but only mention some examples established by Fleischmann and Klenke (1999): In $d = 1$ we have $\mathcal{M}_p^\xi(\mathbb{R}) = \mathcal{M}_p(\mathbb{R})$. In any dimension $\mathcal{M}_p^\xi(\mathbb{R}^d)$ contains any $\nu \ll \ell_d$ with bounded density. If $\nu \in \mathcal{M}_p(\mathbb{R}^d)$ and $\delta > 0$, then $\mathbb{P}_\nu[\varrho_0 \in \mathcal{M}_p^\xi(\mathbb{R}^d)] = 1$. Note that $\delta_x \notin \mathcal{M}_p^\xi(\mathbb{R}^d)$ if $d \geq 2$.

Dawson and Fleischmann use the Hölder continuity to imitate the existence proof for the collision local time of Evans and Perkins. They can show that for $\nu \in \mathcal{M}_p^\xi(\mathbb{R}^d)$ the limit (3.10) makes sense in $L^2$ and that $A = L_{[W,\varrho]} \in \mathbf{K}^\xi$. In particular, for $\mathbb{P}_\nu$–a.a. $\varrho$ there exists a continuous version of $(X_t^\varrho)$.

**3.2. Longtime behaviour.** In this subsection we study the longtime behaviour of CSBM in a SBM medium.

It is well known (see Dawson (1977)) that SBM is persistent iff $d > 2$. More precisely, $\mathbb{P}_\ell[\varrho_t(B) > 0] \to 0$, $t \to \infty$, for any compact set $B$ if $d = 1, 2$. However, if $d \geq 3$ then there exist equilibria $\nu_i \in \mathcal{M}_1(\mathcal{M}_p(\mathbb{R}^d))$ with intensity $\int m\nu(dm) = i\ell$, $i \in [0, \infty)$ and such that $\mathbb{P}_{i\ell}[\varrho_t \in \bullet] \Rightarrow \nu_i$, $t \to \infty$. The situation is quite different for our CSBM $X^\varrho$.

First we consider $d = 3$. This is the maximal dimension for which CSBM exists. The catalyst is persistent and we assume that $(\varrho_t)_{t \in \mathbb{R}}$ is the stationary process with intensity $i_c > 0$. We start $X^\varrho$ in $i_r\ell$ for some $i_r > 0$. Instead of starting at time 0 and evaluating at time $t$ it is more convenient to start at time $-t$ and evaluate at time 0. The advantage is that we can fix $\varrho$ and exploit monotonicity of the cumulant equation (3.3). Recall that this is a backward equation and it is immediate that $\langle i_r\ell, v_\varphi^\varrho(-t, 0) \rangle$ is monotone decreasing in $t$. Hence its limit as $t \to \infty$ exists and so does the limit of $\mathbf{P}_{-t, i_r\ell}^\varrho[X_0^\varrho \in \bullet]$. Note that the variances are monotone in $t$ and converge to a finite limit since Brownian motion in $\mathbb{R}^3$ is transient. Hence for $\varphi \in C_c^+(\mathbb{R}^3)$ the random variable $\langle X_0^\varrho, \varphi \rangle$ is uniformly integrable under the sequence $\mathbf{P}_{-t, i_r\ell}^\varrho$ and thus $\mathbf{E}_{-\infty, i_r\ell}^\varrho[X_0^\varrho] = i_r\ell$. In other words, three-dimensional CSBM is persistent (see Theorem 1 of Dawson and Fleischmann (1997b)).

In dimension $d = 1$ the catalyst does not only die out locally in distribution but even a.s. for any compact set $B$,

$$\mathbb{P}_{i_c\ell}[\varrho_t(B) = 0 \text{ for } t \text{ large enough}] = 1.$$

Dawson and Fleischmann (1997a) show that (see Proposition 7) there exists a random time $\tau$ such that

$$\mathbb{E}_{i_c\ell}[\mathbf{P}_0[L_{[W,\varrho]}(\tau,\infty) = 0]] = 1.$$

Furthermore (Proposition 8) for all $x \in \mathbb{R}$,

$$(3.12) \qquad\qquad \mathbb{P}_{i_c\ell}[\mathbf{E}_x[L_{[W,\varrho]}(0,\infty)^2] < \infty] = 1.$$

Note that for finite initial mass $\mu \in \mathcal{M}_f(\mathbb{R})$ and fixed $\varrho$ the total mass process $\langle X_t^\varrho, \mathbf{1}\rangle$ is a nonnegative $L^2$–martingale with variance

$$(3.13) \qquad\qquad \mathbf{Var}_\mu^\varrho[\langle X^\varrho, \mathbf{1}\rangle] = 2\int \mu(dx)\mathbf{E}_x[L_{[W,\varrho]}(0,t)].$$

By (3.12) this martingale is bounded in $L^2$ and by the martingale convergence theorem it converges to a limit with finite variance and expectation $\|\mu\|$ (Theorem 5). In other words, we have persistence even of a finite initial mass. Having in mind a law of large numbers it is clear that for $\mathbb{P}_{i_c\ell}$–a.a. $\varrho$ we have (see Theorem 6)

$$(3.14) \qquad\qquad \mathbf{P}_{i_r\ell}^\varrho[X_t^\varrho \in \bullet] \Longrightarrow i_r\ell, \quad t \to \infty.$$

For $d = 2$ the situation is a little more involved. Neither do we have a non-trivial equilibrium nor a.s. extinction of the catalyst. In fact, any non-trivial open set is visited at arbitrarily large times. The key to the long-time behaviour lies in the self similarity (see Dawson and Fleischmann (1997b, Proposition 13))

$$(3.15) \quad \mathbb{P}_{i_c\ell}[\mathbf{P}_{i_r\ell}^\varrho[K^{-1}X_{Kt}^\varrho(K^{1/2}\bullet) \in \bullet] \in \bullet] = \mathbb{P}_{i_c\ell}[\mathbf{P}_{i_r\ell}^\varrho[X_t^\varrho(\bullet) \in \bullet] \in \bullet], \quad K, t > 0,$$

and a study of the detailed behaviour for fixed time.

Fleischmann and Klenke (1999) show in their Theorem 1 that CSBM in $d = 1, 2, 3$ is absolutely continuous on the complement $Z(\varrho)$ of the space-time support of the catalyst and that the density $\xi_t^\varrho(z)$ is $\mathcal{C}^\infty$ on $Z(\varrho)$ and solves the heat equation. This is established by similar means as in Delmas (1996). In $d = 2, 3$ the catalyst is singular with respect to Lebesgue measure $\ell_d$ and hence $X_t^\varrho$ is absolutely continuous everywhere. Plugging this result into (3.15) they derive (Corollary 2) that for $d = 2$, CSBM is in fact persistent and that

$$(3.16) \qquad \mathbb{P}_{i_c\ell}[\mathbf{P}_{i_r\ell}^\varrho[X_t^\varrho \in \bullet] \in \bullet] \Longrightarrow \mathbb{P}_{i_c\ell}[\mathbf{P}_{i_r\ell}^\varrho[\xi_1^\varrho(0)\ell \in \bullet] \in \bullet].$$

Hence the limit of $X_t^\varrho$ is a random multiple of the Lebesgue measure with full expectation. The randomness reflects the catalyst as experienced by a "reactant particle".

Earlier Dawson and Fleischmann (1997b, Theorem 18) established that if $d = 2$ then the normalised occupation measures $t^{-1}\int_0^t X_s^\varrho ds$ converge to $(\int_0^1 \xi_s^\varrho(0)ds) \cdot \ell$ (in the sense of (3.16)). Note however that Dawson and Fleischmann do not show absolute continuity of $X_t^\varrho$ but only of $\int_0^t X_s^\varrho ds$.

### 3.3. Catalytic branching random walk.

Recall that we defined CSBM as the diffusion limit of CBBM. Here we consider a model where neither the particle nor the spatial diffusion limit has been taken: catalytic branching random walk (CBRW).

Greven, Klenke and Wakolbinger (1999) study CBRW in some detail. They construct by elementary means the process in the following setting. The process lives on a countable Abelian group $G$ as site space. For the moment the catalyst can be any measurable function $\varrho : [0,\infty) \times G \to [0,\infty)$, $(t,g) \mapsto \varrho_t(g)$. The reactant $X^\varrho$ performs a continuous rate 1 random walk on $G$ with $q$–matrix $\mathcal{B}$.

Further each particle branches at the local rate $\varrho_t(g)$ according to the (global) offspring law $(q_k)_{k \in \mathbb{N}_0}$ with probability generating function $Q$. Formally the (time-inhomogeneous) process $X^\varrho$ can be defined by its Laplace functionals

$$(3.17) \qquad v_\varphi^\varrho(r, t; g) = \mathbf{E}^\varrho_{r, \delta_g}[\exp(-\langle X_t^\varrho, \varphi \rangle)]$$

that solve the backward equation, $v_\varphi^\varrho(t, t) = \varphi$,

$$(3.18) \qquad -\frac{d}{dr} v_\varrho^\varphi(r, t; g) = \varrho_r(g)\big[Q(v_\varphi^\varrho(r, t; g)) - v_\varphi^\varrho(r, t; g)\big] + (\mathcal{B}v_\varphi^\varrho(r, t))(g).$$

Henceforth, let us restrict to critical binary branching $q_0 = q_2 = \frac{1}{2}$. We are interested in the case where $\varrho$ is itself a sample of critical binary branching random walk (BRW) on $G$ with $q$–matrix $\mathcal{A}$. Compared with the CSBM model, there is an enormous freedom in the choice of the system parameters. For example, one can choose $\mathcal{A}$ transient while $\mathcal{B}$ is recurrent, or one can add a drift to one of the kernels.

Greven, Klenke and Wakolbinger investigate the longtime behaviour of $(\varrho, X^\varrho)$ in the case where $\mathcal{L}(\varrho_0, X_0^\varrho) \in \mathcal{E}_{i_c, i_r}$ is ergodic with intensities $i_c, i_r \in (0, \infty)$. Let $\delta_{\underline{0}}$ be the Dirac measure on the empty configuration. Denote by $\mathcal{H}_i$ the Poisson point process on $G$ with uniform intensity $i \in (0, \infty)$. Finally denote by $\widehat{\mathcal{A}}$ and $\widehat{\mathcal{B}}$ the symmetrization of $\mathcal{A}$ and $\mathcal{B}$ respectively, $\widehat{\mathcal{A}}(g, h) = \frac{1}{2}(\mathcal{A}(g, h) + \mathcal{A}(h, g))$.

Consider first the case where $\widehat{\mathcal{A}}$ is transient so that $\varrho$ is persistent. As discussed earlier the particles of $X^\varrho$ experience an average of the medium $\varrho$. Thus $X^\varrho$ is persistent iff $\widehat{\mathcal{B}}$ is transient (Theorem 1).

The situation is far more delicate if $\widehat{\mathcal{A}}$ is recurrent. The catalyst goes to extinction but there is a subtle difference between a.s. extinction (as for $\mathcal{A}$ Bernoulli on $\mathbb{Z}$) and extinction only in probability (as for $\mathcal{A}$ Bernoulli on $\mathbb{Z}^2$). For this reason only the special cases $G = \mathbb{Z}$ or $G = \mathbb{Z}^2$ are considered.

Assume first $G = \mathbb{Z}$ and additionally $\mathcal{A}$ and $\mathcal{B}$ have the properties $\sum\limits_{x \in \mathbb{Z}} \mathcal{A}(0, x)x = \sum\limits_{x \in \mathbb{Z}} \mathcal{B}(0, x)x = 0$ and

$$\sum_{x \in \mathbb{Z}} \mathcal{A}(0, x)|x|^\alpha < \infty, \qquad \sum_{x \in \mathbb{Z}} \mathcal{B}(0, x)|x|^\beta < \infty, \qquad \text{for some } \alpha > 2 \text{ and } \beta > 1.$$

As with CSBM in $d = 1$ we have (Theorem 2a)

$$(3.19) \qquad \mathcal{L}_{\mathcal{H}_{i_c}, \mathcal{H}_{i_r}}(\varrho, X_t^\varrho) \Longrightarrow \delta_{\underline{0}} \otimes \mathcal{H}_{i_r}, \quad t \to \infty.$$

If $G = \mathbb{Z}^2$ and $\mathcal{A}$ and $\mathcal{B}$ are Bernoulli, then the situation is similar to CSBM in $d = 2$. The law of $X_t^\varrho$ converges to a limit with random homogeneous intensity (Theorem 3):

$$(3.20) \qquad \mathbb{P}_{\mathcal{H}_{i_c}}[\mathbf{P}^\varrho_{\mathcal{H}_{i_r}}[X_t^\varrho \in \bullet] \in \bullet] \Longrightarrow \mathbb{P}_{i_c \ell}[\mathbf{E}^{\widetilde{\varrho}}_{i_r \ell}[\mathcal{H}_{\widetilde{\xi}_1(0)}] \in \bullet],$$

where $\widetilde{\xi}_t(x)$ is the density of the 2-dimensional CSBM $\widetilde{X}^{\widetilde{\varrho}}$. (The assumptions on $\mathcal{A}$ and $\mathcal{B}$ can be weakened to finite variance isotropic random walks with the additional requirement that $\sum_{x \in \mathbb{Z}^2} \mathcal{A}(0, x)|x|^\alpha < \infty$ for some $\alpha > 6$.) The key to this result is a scaling limit of CBRW (Proposition 1.4) and the fact that for large $T$ the support of $\varrho_t$ has large holes (Proposition 1.5) combined with an adaption of a result of Harry Kesten (1995) on the range of branching random walk (Proposition 1.3).

Finally we would like to mention a situation with a striking asymmetry between the catalyst and the reactant. Consider $G = \mathbb{Z}$ and $\mathcal{B}$ Bernoulli with a drift while $\mathcal{A} = 0$ is the random walk that stands still. (Apparently Greven, Klenke and

Wakolbinger would have liked to consider $\mathcal{A}$ Bernoulli but could not overcome technical difficulties.) Obviously $\varrho$ dies out locally a.s. However, the greater mobility of the reactant particles forces them to visit lots of the scattered catalyst clumps and, ironically enough, leads to local extinction (Theorem 2b)

$$(3.21) \qquad \lim_{t \to \infty} \mathbb{E}_{\mathcal{H}_{i_c}} \mathbf{P}^{\varrho}_{\mathcal{H}_{i_r}}[X_t^{\varrho}(g) > 0] = 0, \qquad g \in \mathbb{Z}.$$

## 4. Mutually catalytic branching

Dawson and Perkins (1998) study a model of two spatial branching processes $(u_t)_{t \geq 0}$ and $(v_t)_{t \geq 0}$, where each process acts as the catalyst for the other one. In the continuous space setting $u$ and $v$ can be defined on $\mathbb{R}$ by an SPDE. Let $\gamma > 0$ and $\overset{\bullet}{W}_i(t,x)$ $(i = 1, 2)$ be independent space-time white noises on $\mathbb{R}^+ \times \mathbb{R}$. Consider the SPDE

$$\frac{\partial u}{\partial t}(t,x) = \frac{1}{2}\frac{\partial^2 u}{\partial x^2}(t,x) + (\gamma u(t,x)v(t,x))^{1/2}\,\overset{\bullet}{W}_1(t,x); \quad u(0,x) = u_0(x),$$

(4.1)

$$\frac{\partial v}{\partial t}(t,x) = \frac{1}{2}\frac{\partial^2 v}{\partial x^2}(t,x) + (\gamma u(t,x)v(t,x))^{1/2}\,\overset{\bullet}{W}_2(t,x); \quad v(0,x) = v_0(x).$$

With the aid of a duality going back to Mytnik (1996) it is shown that (for suitable $u_0$ and $v_0$) there exists a unique solution of (4.1). The question of a higher dimensional analogue was left open. There is some recent work (Dawson et al. (1999)) giving an affirmative answer, at least for two dimensions.

On $\mathbb{Z}^d$, however, the model has been constructed by Dawson and Perkins (1998) for any $d \geq 1$. Let $Q$ be the $q$–matrix of a Markov chain on $\mathbb{Z}^d$ with bounded jump rate. Let $(W_i(t,k),\ t \geq 0,\ k \in \mathbb{Z}^d)$, $i = 1, 2$, be independent families of Brownian motions and consider the infinite system of coupled stochastic integral equations

$$u_t(k) = u_0(k) + \int_0^t (u_s Q)(k)ds + \int_0^t (\gamma u_s(k)v_s(k))^{1/2}dW_1(s,k),$$

(4.2)

$$v_t(k) = v_0(k) + \int_0^t (v_s Q)(k)ds + \int_0^t (\gamma u_s(k)v_s(k))^{1/2}dW_2(s,k).$$

Again it is established by means of Mytnik's duality that there exists a unique weak solution to (4.2).

Henceforth let $Q$ be the $q$–matrix of a random walk on $\mathbb{Z}^d$. Dawson and Perkins address the question of co-existence of types. Assume that $\langle u_0, 1 \rangle + \langle v_0, 1 \rangle < \infty$. Then $U_t = \langle u_t, 1 \rangle$ and $V_t = \langle v_t, 1 \rangle$ are nonnegative martingales and hence converge a.s. to some limit $U_\infty$ and $V_\infty$. We say that there is co-existence of types if $\mathbb{P}_{u_0,v_0}[U_\infty > 0 \text{ and } V_\infty > 0] > 0$. This is the case iff $Q$ is transient (Theorem 1.2). Mytnik's duality connects finite initial conditions with infinite initial conditions and converts Theorem 1.2 into a statement on the longtime behaviour of $(u_t, v_t)$ for constant initial condition $u_0 \equiv u > 0$, $v_0 \equiv v > 0$: $\mathbb{P}_{(u,v)}[(u_t, v_t) \in \bullet]$ converges to an equilibrium $\mathbb{P}_{(u,v)}[(u_\infty, v_\infty) \in \bullet]$ (Theorem 1.4) and $\mathbb{P}_{(u,v)}[u_\infty(k)v_\infty(k) > 0] = 1$ if $Q$ is transient (Theorem 1.6). If $Q$ is recurrent then (Theorem 1.5) $k \mapsto u_\infty(k)$ and $k \mapsto v_\infty(k)$ are a.s. constant and $\mathbb{P}_{(u,v)}[u_\infty(0)v_\infty(0) > 0] = 0$. More precisely $\mathbb{P}_{(u,v)}[(u_\infty(k), v_\infty(k)) \in \bullet]$ is the hitting distribution of the set $X := (\{0\} \times [0, \infty)) \cup ([0, \infty) \times \{0\})$ of planar Brownian motion started in $(u, v) \in \mathbb{R}^2$.

Via a duality and comparison argument Cox, Klenke and Perkins (1999) generalised Theorem 1.5 and 1.6 to a class of initial states $u_0, v_0$ that is preserved under time evolution. This result and an abstract new-start argument are employed by Cox and Klenke (1999) to answer the question: "If $Q$ is recurrent, there is (local) extinction of one type. However, is it always (as time evolves) the same type that is locally predominant?" No, it changes infinitely often! In fact for any $x \in X$, $\delta_x$ is a (weak) limit point of $\mathbb{P}_{(u,v)}[(u_t(k), v_t(k)) \in \bullet]$.

## References

[1] Athreya, K.B., and Ney, P. (1972) Branching processes, *Springer Verlag.*

[2] Baillon, J.-B., Clément, Ph., Greven, A., den Hollander, F. (1994) On a variational problem for an infinite particle system in a random medium, *J.reine angew.Math.* **454**, 181–217.

[3] Baillon, J.-B., Clément, Ph., Greven, A., den Hollander, F. (1995) On the attracting orbit of a non-linear transformation arising from renormalization of hierarchically interacting diffusions. Part I: The compact case. *Canadian Journal of Mathematics* **47(1)**, 3–27.

[4] Barlow, M.T., Evans, S.N. and Perkins, E.A. (1991) Collision local times and measure-valued processes. *Can. J. Math.*, **43(5)**, 897–938.

[5] Cox, J.T., and Klenke, A. (1999) Recurrence and Ergodicity of Interacting Particle Systems, *Probab. Th. Rel. Fields* (to appear).

[6] Cox, J.T., Klenke, A., Perkins, E.A. (1999) Convergence to Equilibrium and Linear Systems Duality. in: *Stochastic Models, A Conference in Honour of Professor Don Dawson* (Luis B. Gorostiza and B. Gail Ivanoff, eds.), Conference Proceedings, Canadian Mathematical Society, Amer. Math. Soc., Providence. Conference Proceeding Series of the Canadian Mathematical Society.

[7] Dawson, D.A. (1977) The Critical Measure Diffusion. *Z. Wahr. verw. Geb.* **40**, 125–145.

[8] Dawson, D.A. (1993) Measure-Valued Markov Processes. In: *Ecole d'Eté de Probabilités de St.Flour XXI - 1991, LNM* **1541**, Springer-Verlag.

[9] Dawson, D.A., Etheridge, A.M., Fleischmann, K., Mytnik, L., Perkins, E.A., Xiong, L. (1999) Mutually catalytic super-Brownian motion in $\mathbb{R}^2$. Preprint in preparation (to appear in the preprint series of the Weierstraß Institut, see http://www.wias-berlin.de).

[10] Dawson, D.A., and Fleischmann, K. (1983) On spatially homogeneous branching processes in a random environment, *Math. Nachr.* **113**, 249–257.

[11] Dawson, D.A., and Fleischmann, K. (1985) Critical Dimension for a model of branching in a random medium, *Probab. Th. Rel. Fields* **70**, 315–334.

[12] Dawson, D.A., and Fleischmann, K. (1991) Critical branching in a highly fluctuating random medium, *Probab. Th. Rel. Fields* **90**, 241–274.

[13] Dawson, D.A., and Fleischmann, K. (1994) A super-Brownian motion with a single point catalyst, *Stoch. Proc. Appl.* **49**, 3–40.

[14] Dawson, D.A., and Fleischmann, K. (1995) Super-Brownian motions in higher dimensions with absolutely continuous measure states. *Journ. Theoret. Probab.* **8(1)**, 179–206.

[15] Dawson, D.A., and Fleischmann, K. (1997a) A continuous super-Brownian motion in a super-Brownian medium. *Journ. Theoret. Probab.* **10(1)**, 213–276.

[16] Dawson, D.A., and Fleischmann, K. (1997b) Longtime behavior of a branching process controlled by branching catalysts. *Stoch. Process. Appl.* **71(2)**, 241–257.

[17] Dawson, D.A., Fleischmann, K., and Gorostiza, L. (1989) Stable Hydrodynamic Limit Fluctuations of a Critical Branching Particle System in a random medium, *Ann. Probab.* **17(3)**, 1083–1117.

[18] Dawson, D.A., Fleischmann, K., Li, Y., and Mueller, C. (1995) Singularity of super-Brownian local time at a point catalyst, *Ann. Probab.* **23(1)**, 37–55.

[19] Dawson, D.A., Fleischmann, K., and Mueller, C. (1998) Finite Time Extinction of Super-Brownian Motions with Catalysts, *WIAS Preprint* **431**.

[20] Dawson, D.A., Fleischmann, K., and Roelly, S. (1991) Absolute continuity for the measure states in a branching model with catalysts. In *Stochastic Processes, Proc. Semin. Vancouver/CA 1990*, volume 24 of *Prog. Probab.*, pages 117–160.

[21] Dawson, D.A., Li, Y., and Mueller, C. (1995) The support of measure-valued branching processes in a random environment, *Ann. Probab.* **23(4)**, 1692–1718.

[22] Dawson, D.A., Perkins, E.A. (1998) Long-time behaviour and co-existence in a mutually catalytic branching model, *Ann. Probab.* **26(3)**, 1088–1138.

[23] Delmas, J.-F. (1996) Super-mouvement brownien avec catalyse. *Stochastics and Stochastics Reports* **58**, 303–347.

[24] Dynkin, E.B. (1991) Branching particle systems and super processes. *Ann. Probab.* **19(3)** 1157–1194.

[25] Dynkin, E.B. (1993) Superprocesses and partial differential equations (The 1991 Wald memorial lectures). *Ann. Probab.* **21(3)**, 1185–1262.

[26] Dynkin, E.B. (1994) An Introduction to branching measure-valued processes, *CRM, Amer. Math. Soc., Providence, R.I.*

[27] Etheridge, A., and Fleischmann, K. (1998) Persistence of a two-dimensional super-Brownian motion in a catalytic medium. *Probab. Theory Relat. Fields*, **110(1)**, 1–12.

[28] Evans, S.N., and Perkins, E.A. (1991) Absolute Continuity results for superprocesses with some applications, *Trans. Am. Soc.* **325(2)**, 661–681.

[29] Evans, S.N., and Perkins, E.A. (1994) Measure-valued branching diffusions with singular interactions. *Can. J. Math.* **46(1)**, 120–168.

[30] Evans, S.N., and Perkins, E.A. (1998) Collion local times, historical stochastic calculus, and competing superprocesses, *Electr. J. Probab.* **3**, paper 5, 1–120.

[31] Fleischmann, K. (1994) Superprocesses in catalytic media, In D. Dawson (ed.) *Measure-Valued Processes, Stochastic Partial Differential Equations and Interacting Systems*, CRM Proc. Lecture Notes and Monographs **5**, pp. 119–137. Providence, RI: American Mathematical Society.

[32] Fleischmann, K., and Le Gall, J.-F. (1995) A new approach to the single point catalytic super-Brownian motion. *Probab. Theory Relat. Fields* **102**, 63–82.

[33] Fleischmann, K., and Klenke, A. (1999) Smooth density field of catalytic super-Brownian motion, *Ann. Appl. Probab.*, **9(2)**, 298–318.

[34] Fleischmann, K., and Mueller, C. (1995) A super-Brownian motion with a locally infinite catalytic mass, *Prob. Th. Rel. Fields* **107**, 325-357.

[35] Fleischmann, K., and Vatutin, V.A. (1999) Reduced Supercritical Galton Watson processes in a Random Environment, *Adv. Appl. Probab.*, **31**, 88–111.

[36] Gorostiza, L.G., Wakolbinger, A. (1991) Persistence Criteria for a Class of Critical Branching Particle Systems in Continuous Time. *Ann. Probab.* **19**, 266–288

[37] Gorostiza, L.G., Roelly, S., and Wakolbinger, A. (1992) Persistence of critical multitype particle systems and measure branching processes. *Prob. Th. Rel. Fields* **92**, 313–335.

[38] Gorostiza, L.G., Wakolbinger, A. (1993) Long Time Behaviour of Critical Branching Particle Systems and Applications. In D. Dawson (ed.) *Measure-Valued Processes, Stochastic Partial Differential Equations and Interacting Systems*, CRM Proc. Lecture Notes and Monographs **5**, pp. 119–137. Providence, RI: American Mathematical Society.

[39] Greven, A., and F. den Hollander, F. (1991) Population growth in random media I, Variational formula and phase diagram. *J. Statist. Phys.* **65**, 1123–1146.

[40] Greven, A., and F. den Hollander, F. (1992) Random walk in random environment: Phase transition for local and global growth rates, *Prob. Th. Rel. Fields* **91**, 195–249.

[41] Greven, A., and F. den Hollander, F. (1993) A variational characterization of the speed of a one-dimensional self-repellent random walk, *Ann. Appl. Probab.* **3**, 1067–1099.

[42] Greven, A., Klenke, A., and Wakolbinger, A. (1999) The longtime behaviour of branching random walk in a catalytic medium. *Electron. J. Probab.* **4**, no. 12, 80 pages (electronic).

[43] van der Hofstad, R., den Hollander, F., and König, W. (1997) Central limit theorem for a weakly interacting random polymer, *Markov Processes Relat. Fields* **3**, 1–62.

[44] van der Hofstad, R., and Klenke, A. (1999) Self–attractive Random Polymers, *Preprint.*

[45] Iscoe, I. (1988) On the supports of measure-valued critical branching Brownian motion, *Ann. Probab.* **16(1)**, 200-221.

[46] Jagers, P. (1975) Branching processes with biological applications, *Wiley, London.*

[47] Kallenberg, O. (1983) Random measures, *Akademie Verlag and Academic Press.*

[48] Kesten, H. (1995) Branching Random Walk with a Critical Branching Part, *Journal of Theoretical Probability* Vol **8(4)**, 921–962.

[49] König, W. (1996) A central limit theorem for a one-dimensional polymer measure, *Ann. Probab.* **24**, 1012–1035.

[50] Konno, N., and Shiga, T. (1988) Stochastic partial differential equations for some measure-valued diffusions. *Probab. Theory Relat. Fields* **79**, 201–225.

[51] Maisonneuve, B. (1975) Exit Systems, *Ann. Probab.* **3(3)**, 399-411.

[52] Mytnik, L. (1996) Superprocesses in random environments and related topics, *Ph.D. thesis, Technicon.*

[53] Perkins, E.A. (1989) The Hausdorff measure of the closed support of super-Brownian motion. *Ann. Inst. Henri Poincaré Probab. Statist.* **25**, 205–224.

[54] Perkins, E.A. (1994) Measure-valued branching diffusions with singular interactions, *Can. J. Math.* **46(1)**, 120–168.

[55] D'Souza, J.C. (1994) The rates of growth of the Galton-Watson process in varying environments, *Adv. Appl. Prob.* **26**, 698–714.

[56] D'Souza, J.C., and Biggins, J.D. (1992) The supercritical Galton–Watson process in varying environments, *Stoch. Proc. Appl.* **42**, 39-47.

[57] Sugitani, S. (1989) Some properties for the measure-valued branching diffusion process. *J. Math. Soc. Japan* **41(3)**, 437-462.

UNIVERSITÄT ERLANGEN-NÜRNBERG, MATHEMATISCHES INSTITUT, BISMARCKSTRASSE $1\frac{1}{2}$ 91054 ERLANGEN, GERMANY

*E-mail address*: `klenke@mi.uni-erlangen.de`

*URL*: `http://www.mi.uni-erlangen.de/~klenke`

Canadian Mathematical Society
Conference Proceedings
Volume **26**, 2000

# Exact Infinite Dimensional Filters and Explicit Solutions

## Michael A. Kouritzin

*This work is dedicated to Donald Dawson not only in recognition of his outstanding mathematical career but also for his generosity with his time and advice.*

ABSTRACT. Previously, we defined *infinite dimensional exact filters* as nonlinear filters which can be conveniently reduced without approximation to a single convolution (plus a simple transformation and substitution). We showed that such problems do exist and the observation process can be far more general than those for exact *finite dimensional* filters like the Kalman and Benes filters. Moreover, our infinite dimensional exact filters compare favorably in terms of time efficiency and accuracy to other methods except for the finite dimensional exact filters that have limited utility. Herein, we broaden the realm of applicability for our infinite dimensional exact filters including problems with new nonlinear drifts and nonlinear dispersion coefficients. In particular, we investigate the problem of determining which scalar continuous-discrete filtering problems can be solved with essentially a single convolution with respect to a standard normal distribution. This leads to a particularly simple filtering algorithm because the Fourier transform of the standard normal distribution is known in closed form and very well behaved.

## 1. Introduction

The classical scalar nonlinear filtering problem is concerned with estimating functions of the current state of a scalar *signal* diffusion process

$$(1.1) \qquad X_t = X_0 + \int_0^t \alpha(X_s)ds + \int_0^t \sigma(X_s)dW_s$$

based upon noise-corrupted, distorted observations

$$(1.2) \qquad Y_t = Y_0 + \int_0^t h(X_s)\,ds + B_t,$$

2000 *Mathematics Subject Classification.* Primary 60G35, 93E11, 62M20; Secondary 60H10.
*Key words and phrases.* nonlinear filtering; explicit solutions to stochastic differential equations; exact filters.

Research supported by NSERC, Lockheed Martin Canada, and Lockheed Martin Tactical Defense Systems through a center of excellence in Canadian Mathematics based at the University of Alberta, Edmonton.

where $\{(W_t, B_t),\ t \geq 0\}$ is a standard $\mathbb{R}^2$-valued Brownian motion and $X_0$ and $Y_0$ are random variables such that $\{X_0, Y_0, \{(W_t, B_t),\ t \geq 0\}\}$ are mutually independent. More precisely, one wishes to determine the conditional expectation

$$(1.3) \qquad E\left[\varphi(X_t) | \mathcal{B}(\{Y_s,\ 0 \leq s \leq t\})\right]$$

for a large class of Borel measurable functions $\varphi$ or, equivalently, the conditional probability

$$(1.4) \qquad P\left[X_t \in \Gamma | \mathcal{B}(\{Y_s,\ 0 \leq s \leq t\})\right]$$

for all Borel sets $\Gamma$. This problem is solved mathematically under appropriate regularity conditions by the Kushner-Fujisaki-Kallianpur-Kunita (KFKK), the Duncan-Mortensen-Zakai (DMZ), and the Kallianpur-Striebel (KS) equations. Indeed, many authors have generalized these equations to allow for infinite dimensional signal models in lieu of (1.1) and more general Itô equation models for the observations (1.2). However, regardless of model generality, these (KFKK), (DMZ), and (KS) equations are not readily implementable on computers and further developments have been sought. There are two basic approaches: i) Allow general models within the above Markov framework and introduce approximations to allow computer implementation, or ii) Impose certain restrictions on the model so that the filtering problem degenerates into a computer-convenient algorithm. We are concerned herein with the later approach called *exact filtering*.

Exact filtering refers to the determination of filtering problems that yield a (nearly) readily implementable solution without approximations as well as their implementation schemes. The Kalman filter is the most celebrated exact filter and was thought to be the only exact filter for many years. Here, one considers linear equations of the form

$$(1.5) \qquad X_t = X_0 + \int_0^t A\, X_s ds + \int_0^s \sigma dW_s$$

$$(1.6) \qquad Y_t = Y_0 + \int_0^t C\, X_s ds + B_t,$$

(or a time-inhomogeneous variation) and the idea is to track the conditional mean and error covariance by solving a linear stochastic differential equation and a quadratic ordinary differential equation. However, nowadays, there are other known *finite-dimensional* exact filters where the filtering problem also degenerates to the solution of finite-dimensional stochastic and ordinary differential equations. For examples of such filters the reader is referred to the works of e.g. Benes (1981), Daum (1988) and Leung and Yau (1992).

The finite-dimensionality necessarily imposes stringent conditions on *both* signal and observation processes since there has to be a fixed finite number of sufficient statistics that can be evolved according to a small number of (stochastic and ordinary) differential equations. To avoid some of these restrictions, Kouritzin (1998) introduced the concept of *infinite-dimensional* exact filters, where the filtering problem degenerates into a convolution plus Bayesian update instead of finite dimensional equations. The main benefit of infinite over finite dimensional exact filters initially appeared to be the fact that extremely general observations are allowed at least in the discrete-time observation setting. However, another inherent advantage of this method is it uses highly developed, fast Fourier transform computer algorithms to implement the convolutions. Indeed, in Kouritzin's work all convolution is done with respect to a standard Gaussian distribution so one has control over errors and can save both time and space by using the closed form for the Fourier transform of a standard Gaussian distribution.

Whereas the observations are taken to be a continuous time Itô equation in most mathematical studies, the case of discrete observations may be of more practical importance at least for tracking problems in, for example, the air traffic management and defense industries. Therefore, we return to the general nonlinear diffusion (1.1) but now replace (1.2) with discrete-time measurement $Y_j$ taken at time $t_j$ with $0 < t_1 < t_2 < \dots$ Then, continuous-discrete non-linear filtering is concerned with obtaining the distribution of $X_t$ conditioned on the observations $\{Y_j;\ t_j \leq t\}$ i.e. finding $P(X_t \,|\, \mathcal{B}\{Y_j, t_j \leq t\})$. We let $P_{Y_j|X_{t_j}}(A,x)$ denote a function on $\mathcal{B}(\mathbb{R}) \times \mathbb{R}$ such that (i) for fixed $x$, $A \to P_{Y_j|X_{t_j}}(A,x)$ is a probability measure, (ii) for each $A \in \mathcal{B}(\mathbb{R})$, $x \to P_{Y_j|X_{t_j}}(A,x)$ is measurable, and (iii) for each $A \in \mathcal{B}(\mathbb{R})$, $P_{Y_j|X_{t_j}}(A, X_{t_j}(\omega))$ is a $P$-version of $P\left\{Y_j \in A | \mathcal{B}(X_{t_j})\right\}(\omega)$. (Such a function exists by standard methods see e.g. Theorem 33.3 of Billingsley (1986).) Then, for simplicity, we assume:

**(F1):** The conditional probability measure $A \to P_{Y_j|X_{t_j}}(A,x)$ has a density $p_{Y_j|X_{t_j}}(\cdot|x)$ with respect to Lebesgue measure for almost all $\left[PX_{t_j}^{-1}\right] x$ such that $x \to p_{Y_j|X_{t_j}}(y|x)$ is continuous and bounded for almost all $y \in \mathbb{R}$.

Then, for any $A \in \mathcal{B}(\mathbb{R})$

$$(1.7) \qquad \int_A p_{Y_j|X_{t_j}}(y|X_{t_j})dy = P\left\{Y_j \in A|\mathcal{B}(X_{t_j})\right\} \quad \text{a.s.}$$

We use the boundedness of $x \to p_{Y_j|X_{t_j}}(y|x)$ to ensure that $p_{X_{t_{j+1}}|\mathcal{Y}_{j+1}}$ is bounded in (1.14) below.

Next, we notice that we are not specifying any specific model for our observations. Still, it would be contrary to the theory and practice of filtering to allow these observations to help predict the future. Therefore, it is reasonable to assume that given $\mathcal{X}_{t_j} \doteq \mathcal{B}(X_s, s \leq t_j)$ conditioning on $\mathcal{Y}_j = \mathcal{B}(\{Y_i;\ i \leq j\})$ does not provide any extra information in the sense that:

**(F2):** For all $A \in \mathcal{B}(\mathbb{R})$ and $t \geq t_j$, we have that $E\left[1_{X_t \in A}|\mathcal{X}_{t_j} \vee \mathcal{Y}_j\right] = E\left[1_{X_t \in A}|\mathcal{X}_{t_j}\right]$ almost surely for each $j = 1, 2, \dots$

Moreover, in order to incorporate new observations in a tractable (Bayesian) manner, we assume that *given the current state* the new observation is independent of the past observations in the sense that:

**(F3):** For each $j = 0, 1, 2, \dots$ and $y \in \mathbb{R}$, we have that

$$(1.8) \qquad p_{Y_{j+1}|X_{t_{j+1}}}(y|X_{t_{j+1}}) = p_{Y_{j+1}|(X_{t_{j+1}},\mathcal{Y}_j)}(y|X_{t_{j+1}}) \quad \text{a.s.,}$$

where $p_{Y_{j+1}|(X_{t_{j+1}},\mathcal{Y}_j)}(y|X_{t_{j+1}})$ denotes the conditional density of $Y_{j+1}$ given $\mathcal{B}(X_{t_{j+1}}, \mathcal{Y}_j)$

This type of assumption is quite mild and suitable for filtering theory. We still maintain that our assumptions on the observations are quite general. Indeed, we can have observations of the form

$$(1.9) \qquad Y_k = h(X_{t_k}, V_k),$$

where $h$ is a nonlinear function and $\{V_k, \ k = 1, 2, 3, ...\}$ is any sequence of independent random variables.

Next, for a matter of convenience, we assume that $D$ is an open interval, e.g. $(-\infty, \infty)$, $(0, \infty)$, or $(0, 1)$, in $\mathbb{R}$ and $P(X_t \in D) = 1$ for all $t \geq 0$.

**(F4):** The initial law $\mathcal{L}(X_0)$ of the signal has a density $p_{X_0}$ that is bounded, continuous, and zero off of $D$.

and

**(F5):** The coefficients $\alpha : \mathbb{R} \to \mathbb{R}$ and $\sigma : \mathbb{R} \to [0, \infty)$ are continuous and one-time respectively two-times continuously differentiable on $D$. Moreover, there is a fundamental solution $\Gamma$ to the Cauchy initial data problem for

$$\partial_t u(t, x) = a(x)\partial_x^2 u(t, x) + b(x)\partial_x u(t, x) + c(x)u(t, x),$$

(1.10)

$$u(t)|_{\partial D} = 0, \ u(0, x) = \varphi(x)$$

where

(1.11)

$$a(x) = \frac{\sigma^2(x)}{2}, \ b(x) = 2\sigma(x)\sigma'(x) - \alpha(x),$$

$$c(x) = (\sigma'(x))^2 + \sigma(x)\sigma''(x) - \alpha'(x),$$

such that $E\left[1_{X_t \in A} | \mathcal{X}_{t_j}\right] = \int_A \Gamma(t - t_j; y, X(t_j))dy$ for all $A \in \mathcal{B}(D)$, $t > t_j$ and $(t, y) \to \int_D \Gamma(t - t_j; y, x)\, \varphi(x)\, dx$ is the unique continuous, bounded solution to (1.10) subject to $u(t_j, x) = \varphi(x)$ for any continuous, bounded $\varphi : D \to \mathbb{R}$.

Then, it follows from the tower property of conditional expectation as well as other standard results on conditional probability (see e.g. Billingsley (1986) Theorem 33.3 and 34.5) that the continuous/discrete filtering problem is solved by using the following three steps:

**STEP 1** Set $t_0 = 0$, $\mathcal{Y}_0 = \{\emptyset, \Omega\}$, and

(1.12)
$$p_{X_{t_0}|\mathcal{Y}_0}(x) = p_{X_0}(x),$$

the density of initial state $X_0$;

For all $j = 0, 1, 2, ...,$

**STEP 2** Solve for $p_{X_{t_{j+1}}|\mathcal{Y}_j}(x) = q(t_{j+1}, x)$ from the Cauchy problem

$$\partial_t q_j(t, x) = a(x)\partial_x^2 q_j(t, x) + b(x)\partial_x q_j(t, x) + c(x)q_j(t, x),$$

(1.13)

$$q_j(t)|_{\partial D} = 0, \quad q_j(t_j, x) = p_{X_{t_j}|\mathcal{Y}_j}(x)$$

**STEP 3** Using Bayes' formula with (1.7) at observation time $t_{j+1}$, we have that

(1.14)
$$p_{X_{t_{j+1}}|\mathcal{Y}_{j+1}}(x) = \frac{p_{Y_{j+1}|X_{t_{j+1}}}(Y_{j+1}|x)p_{X_{t_{j+1}}|\mathcal{Y}_j}(x)}{\int_D p_{Y_{j+1}|X_{t_{j+1}}}(Y_{j+1}|\xi)p_{X_{t_{j+1}}|\mathcal{Y}_j}(\xi)d\xi}, \quad \forall x \in D,$$

where $p_{Y_j|X_{t_j}}(\cdot|\cdot)$ and $p_{X_{t_{j+1}}|\mathcal{Y}_j}(\cdot)$ denote the conditional densities of $Y_j$ given $X_{t_j}$ and $X_{t_{j+1}}$ given $\mathcal{Y}_j = \mathcal{B}(\{Y_i; \ i \leq j\})$ respectively.

In particular, (1.14) follows under our conditions from a slight modification of Problem 33.17 in Billingsley (1986) with

$$f(x) = p_{X_{t_{j+1}}|\mathcal{Y}_j}(x), \qquad g_x(y) = p_{Y_{j+1}|X_{t_{j+1}}}(y|x)$$

and

$$p_{Y_{j+1}}(x) = p_{X_{t_{j+1}}|\mathcal{Y}_{j+1}}(x) = p_{X_{t_{j+1}}|\mathcal{B}(\mathcal{Y}_j, Y_{j+1})}(x).$$

The above three-step algorithm is also discussed in Chapter 6 of Jazwinski (1970).

The Kolmogorov equation in Step 2 of the above continuous-discrete filtering algorithm is solved subject to random initial data so it can not be solved off line but rather must be solved in real time (between observations) on a computer. Naturally, this can often be done using time stepping with sophisticated *multi-grid* or *splitting up method* on-line elliptic equation solvers. However, such methods require very complicated computer codes and are usually less efficient than the method presented here, when our method applies. Moreover, many filtering problems result in parabolic equations in Step 2 whose elliptic operators are neither self-adjoint nor diffusive or drift dominant and these features make evaluation through such traditional elliptic equation solvers more difficult. Finally, many signals evolve over infinite domains so (1.13) should be solved over a domain like $\mathbb{R}$, which is impossible for these equation solvers and some artificial boundary conditions must be added.

To motivate our strategy, we let $\Gamma(t - \tau; y, x)$ be a fundamental solution to Kolmogorov's forward equation for $\mathbf{X} = \{X_t; \ 0 \le t < \infty\}$ (i.e. the density for $\mathbf{X}$'s transition probability function) of (1.1). Then,

$$(1.15) \qquad P[X_t^x \in dy] = \Gamma(t; y, x)\, dy,$$

where $X^x$ is defined by

$$(1.16) \qquad X_t^x = x + \int_0^t \alpha(X_s^x)\, ds + \int_0^t \sigma(X_s^x)\, dW_s,$$

and Step 2 above can be rewritten as

$$(1.17) \qquad p_{X_{t_{j+1}}|\mathcal{Y}_j}(\xi) = \int_D \Gamma(t_{j+1} - t_j; \xi, x) p_{X_{t_j}|\mathcal{Y}_j}(x)\, dx.$$

Then, an alternative strategy for solving Step 2 would be to perform the indicated integration in (1.17). However, this would require a separate integration (for each $\xi$) and storage of $\Gamma$ for all $x \in D$ (or subdomain where $X_t^{(x)}$ lives) and $\xi$ "close to $x$" unless convolution can somehow be used.

Herein, we investigate exact infinite dimensional continuous-discrete filters in the single dimensional case further and find explicit representations of (1.16) for which the filtering problem degenerates into a single convolution with respect to a Gaussian kernel. Specifically, we show how to determine scalar (non-linear) Itô equations which have solutions such that (1.17) can be represented by (multiplication, substitution, and) a single convolution with respect to a Gaussian kernel. Our focus is more on exposition than complete generality.

Lockheed Martin is investigating using these methods on their search, surveillance, narcotic smuggling prevention, and military problems. In particular, our convolutional methods are being evaluated for use on long range electro-optical and infrared search and tracking problems by Lockheed Martin Tactical Defense Systems-Eagan. All of these real world problems exhibit discrete-time observations and a continuous-time signal. Also, there are simulations of our infinite-dimensional filters on a problem with a two-dimensional signal in Kouritzin (1998).

The author is grateful to anonymous referee for providing several thoughtful comments.

## 2. Motivation through Gaussian signals

To motivate the method, we take the very simple case where the signal is characterized by affine drift and constant dispersion $\sigma$

$$(2.1) \qquad X_t = X_0 + \int_0^t [\alpha_0 + \alpha_1 X_s]\, ds + \sigma W_t \qquad \forall t \geq 0.$$

Although this signal model does not preclude solution from a Kalman filter our general observation model does since it allows for non-additive, non-Gaussian noise (see e.g. (1.9)). Hence, we use the above continuous-discrete filtering algorithm with $D = \mathbb{R}$ and find that the Kolmogorov equation in Step 2 becomes

(2.2)

$$\partial_t q_j(t,x) = \frac{\sigma^2}{2}\partial_x^2 q_j(t,x) - [\alpha_0 + \alpha_1 x]\partial_x q_j(t,x) - \alpha_1 q_j(t,x) \quad q_j(t_j, x) = p_{X_{t_j}|\mathcal{Y}_j}(x),$$

and we can apply Feynman-Kac's formula (whether $\alpha_1 \geq 0$ or not) in order to find

$$(2.3) \qquad p_{X_{t_{j+1}}|\mathcal{Y}_j}(x) = q_j(t_{j+1}, x) = E\left[ p_{X_{t_j}|\mathcal{Y}_j}(Z_{\delta t_j}^x) \exp\left\{ -\alpha_1 \times \delta t_j \right\} \right],$$

where $\delta t_j = t_{j+1} - t_j$ and

$$(2.4) \qquad Z_t^x = x - \int_0^t [\alpha_0 + \alpha_1 Z_s^x]\, ds + \sigma W_t.$$

However, $Z^x$ has an *explicit solution*

$$(2.5) \quad Z_t^x \;=\; \exp\left[-\alpha_1 t\right]\left\{ x - \alpha_0 \int_0^t \exp\left\{\alpha_1 s\right\} ds + \sigma \int_0^t \exp\left\{\alpha_1 s\right\} dW_s \right\}$$

$$\doteq\; \varphi\left(t, x - \int_0^t f_s dW_s\right) \qquad \forall t \geq 0,$$

where

$$(2.6) \quad \varphi\left(t,u\right) \doteq \exp\left[-\alpha_1 t\right]\left\{ u - \alpha_0 \int_0^t \exp\left\{\alpha_1 s\right\} ds \right\}, \quad f_s = -\sigma \exp\left\{\alpha_1 s\right\}.$$

This means that we can represent $p_{X_{t_{j+1}}|\mathcal{Y}_j}(x)$ in terms of $\delta t_j \doteq t_{j+1} - t_j$ and the standard normal density $\Phi(y) \doteq \frac{1}{\sqrt{2\pi}}\exp\left\{-y^2/2\right\}$ as

$$(2.7) \quad p_{X_{t_{j+1}}|\mathcal{Y}_j}(x) \;=\; \exp\left\{-\alpha_1 \delta t_j\right\} E\left[ p_{X_{t_j}|\mathcal{Y}_j}(\varphi(\delta t_j, x - \int_0^{\delta t_j} f_s dW_s)) \right]$$

$$=\; \exp\left\{-\alpha_1 \delta t_j\right\} \int_{\mathbb{R}} \Upsilon_{\delta t_j}(v_x - \xi)\Phi(\xi)d\xi$$

by a change of variables, where

$$(2.8) \qquad v_x = x / \sqrt{\int_0^{\delta t_j} f_s^2 ds} \text{ and } \Upsilon_{\delta t_j}(y) \doteq p_{X_{t_j}|\mathcal{Y}_j}(\varphi(\delta t_j, \sqrt{\int_0^{\delta t_j} f_s^2 ds} \times y)).$$

This means that $p_{X_{t_{j+1}}|\mathcal{Y}_j}$ can be evaluated by convolution, substitution, and post multiplication as follows

(2.9) $$p_{X_{t_{j+1}}|\mathcal{Y}_j}(x) = \exp\left\{-\alpha_1 \delta t_j\right\} \left[\left(\Upsilon_{\delta t_j} * \Phi\right)(v_x)\right].$$

(Since $\Phi \in \mathcal{S}(\mathbb{R})$, the space of rapidly decreasing functions, it is enough that $\Upsilon_{\delta t_j}$ is a continuous, bounded function whence a tempered distribution for convolution to be defined.) From a computer solvability point of view this convolution representation is a dramatic improvement over the partial differential equation since now we can replace specialized on-line equations solvers and time stepping with two fast Fourier transforms, i.e., we can evaluate

(2.10)

$$p_{X_{t_{j+1}}|\mathcal{Y}_j}(x) = \exp\left\{-\alpha_1 \delta t_j\right\} \mathcal{F}^{-1}\left[\mathcal{F}\Upsilon_{\delta t_j} \cdot \mathcal{F}\Phi\right](v_x), \quad \mathcal{F}\Phi(\xi) = \exp\left\{-\frac{\xi^2}{2}\right\}$$

where $\mathcal{F}$ represents Fourier transform and $\cdot$ denotes pointwise multiplication. The point of this note is that $p_{X_{t_{j+1}}|\mathcal{Y}_j}$ can be calculated in this manner for a class of signal models with *nonlinear* drift and dispersion coefficients. We need only change the functions $\exp\left\{-\alpha_1 \delta t_j\right\}$, $\Upsilon_{\delta t_j}$, and $v_x$.

The case where $X$ has constant dispersion and certain class of non-linear drift has been handled in Kouritzin (1998), even in the vector-valued case. Herein, we consider both nonlinear drift and nonlinear dispersion.

## 3. Explicit Solutions

In our motivation, we used Feynman-Kacs and then an *explicit* solution for a simple Itô equation with affine drift and constant dispersion coefficients. We now return to more general nonlinear models as in (1.1) and recall the definitions in (1.11). We investigate *explicit* solutions of the form

(3.1) $$\zeta_t^x = \varphi\left(x, t, \int_0^t f_s dW_s\right) \qquad \forall t \geq 0$$

for some functions $\varphi \in D \times [0, \infty) \times \mathbb{R} \to \overline{\mathbb{R}}$ (the extended real line) and $f \in \mathcal{C}(\mathbb{R}; (0, \infty))$, where $\zeta^x$ satisfies

(3.2) $$\zeta_t^x = x + \int_0^t \beta(\zeta_s^x) ds + \int_0^t \sigma(\zeta_s^x) dW_s \qquad \forall t \geq 0.$$

Our main contributions of the present section are: (i) to show that any explicit solution to (3.1) with $\sigma$ satisfying (3.16) below is actually a diffeomorphism $\zeta_t^x = \Lambda^{-1}\left(Z^\kappa(t, \Lambda(x))\right)$ of a linear stochastic differential equation $dZ_t^\kappa = (-\kappa Z_t^\kappa + x)dt + dW_t$, $z_0^\kappa = \Lambda(x)$ (the reverse implication that $\Lambda^{-1}\left(Z^\kappa(t, \Lambda(x))\right)$ satisfies (3.1) is obvious), (ii) to characterize the equations (3.2) that yield such explicit solutions, and (iii) state the results that will be required in the next section on infinite dimensional exact filters.

We will not concern ourselves with which explicit solutions can be used immediately within our filtering development or exactly how they would be used in filtering until the next section. Instead, we just fix some open interval $D$, most typically $(-\infty, \infty)$, $(0, \infty)$, or $(0, 1)$, and construct explicit solutions on $D$ up to $\tau = \inf\left\{t > 0 : \varphi(x, t, \int_0^t f_s dW_s) \in D^c\right\}$. We assume that $\frac{d}{dt}\varphi(x, t, u)$, $\frac{d}{du}\varphi(x, t, u)$, and $\frac{d^2}{du^2}\varphi(x, t, u)$ exist and are continuous on $D \times (0, \tau) \times \mathbb{R}$. Moreover, we define

$\tau_\infty = \inf\left\{t > 0 : \varphi(x, t, \int_0^t f_s dW_s) = \pm\infty\right\}$ and assume that $\varphi(x, t, \int_0^t f_s dW_s)$ is almost surely continuous on $D \times (0, \tau_\infty)$. Our first result will establish a condition on $\beta$ and $\sigma$ which will provide a solution to (3.2) in the form (3.1) for each $x \in D$. Throughout this section we assume that $\beta : \mathbb{R} \to \mathbb{R}$ and $\sigma : \mathbb{R} \to [0, \infty)$ are continuous functions which are one-time respectively two-times continuously differentiable on $D$. We *do not* require that $\beta$ or $\sigma$ satisfy a linear growth condition.

LEMMA 3.1. *A necessary and sufficient condition for* (3.2) *to have an explicit (strong) solution of the form* (3.1) *up until* $\tau$ *with* $f$ *continuous and positive is that*

$$(3.3) \qquad \frac{\sigma'(y)\beta(y)}{\sigma(y)} + \frac{\sigma''(y)\sigma(y)}{2} = \beta'(y) + \kappa$$

*for all* $y \in D$ *such that* $\sigma(y) \neq 0$, *any constant* $\kappa \in \mathbb{R}$ *and*

$$(3.4) \qquad \sigma'(y)\beta(y) = 0$$

*for all* $y \in D$ *such that* $\sigma(y) = 0$.

PROOF. We note that $t \to \varphi(x, t, \int_0^t f_s dW_s)$ is a continuous stochastic process on $\{t < \tau_\infty\}$, find $D^c$ to be closed, and use the stopping times (with respect to right continuous filtration $\{\mathcal{F}_{t+}^W, t \geq 0\}$) $\tau_N = \inf\{t > 0 : \varphi(x, t, \int_0^t f_s dW_s) \in D^c \bigcup(-\infty, -N] \bigcup [N, \infty)\}$. Then, we find $\{\tau > t\} = \bigcup_{N=1}^\infty \{\tau_N > t\}$ and $\tau$ is a stopping time. Next, we use Itô's formula on (3.1) to find that

$$(3.5) \quad d\zeta_t^x = \left[\varphi_t\left(x, t, \int_0^t f_s dW_s\right) + \frac{1}{2}f^2(t)\varphi_{uu}\left(x, t, \int_0^t f_s dW_s\right)\right] dt$$
$$+ f(t)\varphi_u\left(x, t, \int_0^t f_s dW_s\right) dW_t,$$

on $\{t < \tau\}$, which matches (3.2) if $\varphi(x, 0, 0) = x$ for all $x \in D$ and

$$(3.6) \qquad \text{(i) } f(t)\frac{\partial\varphi}{\partial u} = \sigma(\varphi) \text{ and (ii) } \frac{\partial\varphi}{\partial t} = \beta(\varphi) - \frac{1}{2}f^2(t)\frac{\partial^2\varphi}{\partial u^2}.$$

Now, suppose $\zeta_{t\wedge\tau}^x = \int_0^{t\wedge\tau} \widehat{\beta}(\zeta_s^x)ds + \int_0^{t\wedge\tau} \widehat{\sigma}(\zeta_s^x)dW_s$ for some continuous $\widehat{\beta}, \widehat{\sigma}$. Then,

$$(3.7) \qquad 0 = \int_0^{t\wedge\tau} [\beta(\zeta_s^x) - \widehat{\beta}(\zeta_s^x)]ds + \int_0^{t\wedge\tau} [\sigma(\zeta_s^x) - \widehat{\sigma}(\zeta_s^x)]dW_s$$

and the first term is a continuous local martingale with zero quadratic variation. Therefore, both terms are zero for all $t > 0$. This with continuity is enough to conclude $\beta = \widehat{\beta}$ and $\sigma = \widehat{\sigma}$ on $\left\{y \in D : P(\bigcup_{s \in Q, s < \tau^+}\{|\zeta_s^x - y| < \frac{1}{n}\}) > 0 \,\forall n = 1, 2, 3, \ldots\right\}$, where $Q^+$ denotes the non-negative rationals. Moreover, we can use (3.6) (i) to find that

$$(3.8) \qquad f^2(t)\frac{\partial^2\varphi}{\partial u^2} = f(t)\frac{\partial\sigma(\varphi)}{\partial\varphi}\frac{\partial\varphi}{\partial u} = \frac{\partial\sigma(\varphi)}{\partial\varphi}\sigma(\varphi),$$

which can be used to reduce (3.6) (ii) and we find that (3.2) has an explicit solution of the form (3.1) if and only if $\varphi(x, 0, 0) = x$ and

$$(3.9) \qquad \text{(i) } \frac{\partial\varphi}{\partial u} = \frac{\sigma(\varphi)}{f(t)} \text{ and (ii) } \frac{\partial\varphi}{\partial t} = \beta(\varphi) - \frac{\sigma'(\varphi)\sigma(\varphi)}{2}.$$

However, this is equivalent to saying that the 1-form

$$(3.10) \qquad d\varphi \doteq \frac{\sigma(\varphi)}{f(t)} du + \left[ \beta(\varphi) - \frac{\sigma'(\varphi)\sigma(\varphi)}{2} \right] dt$$

is exact whence closed over $\mathbb{R} \times (0, \infty)$. Therefore, the (3.2) has an explicit solution of the form (3.1) if and only if $\varphi(x, 0, 0) = x$ and

$$(3.11) \qquad \frac{d}{dt} \frac{\sigma(\varphi(x,t,u))}{f(t)} = \frac{d}{du} \left[ \beta(\varphi(x,t,u)) - \frac{\sigma'(\varphi(x,t,u))\sigma(\varphi(x,t,u))}{2} \right].$$

Yet, the left hand side of (3.11) is easily calculated (with aid of (3.9)) to be

$$(3.12) \qquad \frac{d}{dt} \frac{\sigma(\varphi)}{f(t)} = \frac{\sigma'(\varphi)}{f(t)} \beta(\varphi) - \frac{(\sigma'(\varphi))^2 \sigma(\varphi)}{2f(t)} - \frac{\sigma(\varphi)f'(t)}{f^2(t)}$$

and the right hand side of (3.11) to be

$$(3.13) \qquad \frac{\beta'(\varphi)\sigma(\varphi)}{f(t)} - \frac{\sigma''(\varphi)(\sigma(\varphi))^2}{2f(t)} - \frac{(\sigma'(\varphi))^2 \sigma(\varphi)}{2f(t)}.$$

Comparing these last two equations and making cancellations, we find that

$$(3.14) \qquad \frac{\sigma'(\varphi)}{\sigma(\varphi)} \beta(\varphi) - \frac{f'(t)}{f(t)} = \beta'(\varphi) - \frac{\sigma''(\varphi)\sigma(\varphi)}{2}$$

whenever $\sigma(\varphi) \neq 0$. Since this equation must hold for all $t$ and $\varphi$ with $\sigma(\varphi) \neq 0$ we find that

$$(3.15) \qquad \frac{f'(t)}{f(t)} = \kappa \text{ or } f(t) = f(0) \exp(\kappa t)$$

for some constant $\kappa \in \mathbb{R}$. The case where $\sigma(y) = 0$ follows easily by setting the right hand side of (3.12) to be zero and noting that (3.13) is also zero. □

This lemma allows us to check whether a given SDE has an explicit solution of the form given above. However, from the our filtering point of view, we need to actually construct the explicit solutions and determine if convolution can be used. This problem was already partially solved in Kouritzin and Li (1998) and, some interesting examples are given there. Other work on explicit solutions has been done by Doss (1977) and Sussmann (1978).

Now, we will assume that $\sigma(x) > 0$ almost everywhere on $D$. We know by continuity that our explicit solutions can not explode i.e. hit $\pm\infty$ in finite time unless $\tau_\infty < \infty$. However, our construction does not preclude a possible escape from $D$ when $D$ is a strict subset of $\mathbb{R}$. In this case, our explicit representation may end at the escape time $\tau$.

PROPOSITION 3.2. *Suppose that* $\sigma(\cdot) \in C(\mathbb{R}; [0, \infty)) \cap C^1(D, (0, \infty))$ *is a given dispersion coefficient and* $\lambda \in \overline{D}$ *(the closure of $D$) are such that*

$$(3.16) \qquad \Lambda_\lambda(y) \doteq \int_\lambda^y \frac{dx}{\sigma(x)} < \infty \qquad \forall y \in D.$$

*($\Lambda_\lambda(y)$ is negative when $y < \lambda$ and monotonically increasing.) Then, (3.2) has an explicit solution of form (3.1) up to* $\tau = \inf\left\{ t > 0 : \varphi(x, t, \int_0^t f_s dW_s) \in D^c \right\}$ *if and only if*

$$(3.17) \qquad \beta(x) = \left( \chi - \kappa\Lambda_\lambda(x) + \frac{1}{2}\sigma'(x) \right) \sigma(x) \qquad \forall x \in D,$$

*for some constants* $\chi, \kappa \in \mathbb{R}$. *In this case, we let* $(a, b) \doteq \Lambda_\lambda(D)$ *be the range of* $\Lambda_\lambda$ *for some* $-\infty \le a < b \le \infty$, *set* $f_s = \exp\{\kappa s\}$, *and take*

$$(3.18) \quad \varphi(x, t, u) = \begin{cases} \Lambda_\lambda^{-1}\left(\frac{u + \chi \int_0^t f_s ds + \Lambda_\lambda(x)}{f_t}\right) & a < \frac{u + \chi \int_0^t f_s ds + \Lambda_\lambda(x)}{f_t} < b \\ \inf\left\{\Lambda_\lambda^{-1}(y) : a < y < b\right\} & \frac{u + \chi \int_0^t f_s ds + \Lambda_\lambda(x)}{f_t} \le a \\ \sup\left\{\Lambda_\lambda^{-1}(y) : a < y < b\right\} & \frac{u + \chi \int_0^t f_s ds + \Lambda_\lambda(x)}{f_t} \ge b \end{cases}.$$

PROOF. Inasmuch as the value of $\lambda$ does not effect the proof we will consider the case $\lambda = 0$ and let $\Lambda = \Lambda_0$ to ease the notation. Clearly, $\beta(x)\sigma'(x) = 0$ whenever $x \in D$ and $\sigma(x) = 0$. Moreover, when $\sigma(x) \ne 0$ we find that

$$(3.19) \quad \beta'(x) + \kappa = (\chi - \kappa\Lambda(x) + \frac{1}{2}\sigma'(x))\sigma'(x) + \frac{1}{2}\sigma''(x)\sigma(x)$$
$$= \frac{\sigma'(x)}{\sigma(x)}\beta(x) + \frac{\sigma''(x)\sigma(x)}{2}.$$

Therefore, it follows from the previous lemma that there is an explicit solution. Now, suppose we have such a solution. Then, by the previous result again it must satisfy the overall equality in (3.19), which we can simplify in the case $\sigma(x) > 0$ to

$$(3.20) \quad \frac{d}{dx}\left(\frac{\beta(x)}{\sigma(x)}\right) = \frac{\sigma''(x)}{2} - \frac{\kappa}{\sigma(x)}$$

or taking anti-derivatives to

$$(3.21) \quad \beta(x) = \left[\frac{\sigma'(x)}{2} - \kappa\Lambda(x) + \chi\right]\sigma(x),$$

for any constant $\chi \in \mathbb{R}$. Now, since $\int_0^y \frac{dx}{\sigma(x)} < \infty$ for all $y \in D$ the set $\{x : \sigma(x) = 0\}$ must have measure zero and (3.21) holds for all $x \in D$ by continuity. Then, using the fact $\frac{d\Lambda(x)}{dx} = \frac{1}{\sigma(x)}$, we find that $\left.\frac{d\Lambda^{-1}(y)}{dy}\right|_{y=\Lambda(\varphi)} = \sigma(\varphi)$ and this can be used to show that $\varphi$, as defined in (3.18), is a solution to

$$(3.22) \quad d\varphi = \frac{\sigma(\varphi)}{f(t)}du + \left[\beta(\varphi) - \frac{\sigma'(\varphi)\sigma(\varphi)}{2}\right]dt$$

and wherefore by the proof of the previous result is a valid function to represent our explicit solution as $\zeta_t^x = \varphi\left(x, t, \int_0^t f_s dW_s\right)$. Since $\varphi(x, 0, 0) = x$, $\zeta_t^x$ starts in $D$ and will leave $D$ if $\tau < \infty$. $\qquad \square$

EXAMPLE 3.3. We first take $D = \mathbb{R}$ and consider the case $\sigma(x) = \sigma$ is a constant. Then, we find that (3.2) has an explicit solution of the form (3.1) if and only if

$$(3.23) \quad \beta(x) = \chi\sigma - \kappa x \qquad \forall x \in \mathbb{R}$$

and some constants $\chi, \kappa \in \mathbb{R}$.

Naturally, this just corresponds to the class of Gaussian solution that we discussed in the previous section. However, we can also consider examples with different dispersion coefficients.

EXAMPLE 3.4. We consider the dispersion coefficient of a Feller branching diffusion

$$(3.24) \qquad \sigma(x) = \left\{ \begin{array}{cc} \sqrt{x} & x \geq 0 \\ 0 & x < 0 \end{array} \right.$$

with $D = (0, \infty)$. In this case, $\Lambda_0(x) = 2\sqrt{x}$ for all $x > 0$ and so (3.2) has an explicit solution of the form (3.1) if and only if

$$(3.25) \qquad \beta(x) = \left( \chi \sqrt{x} - 2\kappa x + \frac{1}{4} \right) \qquad \forall x > 0$$

and some constants $\chi, \kappa \in \mathbb{R}$. This positive drift of $\frac{1}{4}$ when $x = 0$ will not allow $0$ to be an absorbing state for (3.2). However, the noise is not "turned off" fast enough to prevent our explicit solutions

$$(3.26) \qquad \zeta_t^x = \frac{1}{4} \left( \int_0^t e^{-\kappa(t-s)}[dW_s + \chi ds] + 2\sqrt{x}e^{-\kappa t} \right)^2$$

from hitting zero.

We can also consider the case of the dispersion coefficient corresponding to geometric Brownian motion. In this case, we can not use $\lambda = 0$.

EXAMPLE 3.5. Suppose that $D = (0, \infty)$ and

$$(3.27) \qquad \sigma(x) = \left\{ \begin{array}{cc} x & x \geq 0 \\ 0 & x < 0 \end{array} \right. .$$

Then, we find that $\Lambda_1(x) = \ln(x)$ for all $x > 0$. Therefore, (3.2) has an explicit solution of the form (3.1) if and only if

$$(3.28) \qquad \beta(x) = \left( \chi x - \kappa x \ln(x) + \frac{x}{2} \right) \qquad \forall x > 0$$

and some constants $\chi, \kappa \in \mathbb{R}$. Our solution $\zeta_t^x = \exp \left[ \int_0^t e^{-\kappa(t-s)}[dW_s + \chi ds] + e^{-\kappa t} \ln(x) \right]$ remains positive and does not escape $D$.

Finally, we can consider an example with a finite interval.

EXAMPLE 3.6. Suppose $D = (0, 1)$ and $\sigma$ is the dispersion coefficient corresponding to the Wright-Fisher models in population genetics:

$$(3.29) \qquad \sigma(x) = \left\{ \begin{array}{cc} \sqrt{x(1-x)} & 0 < x < 1 \\ 0 & x \leq 0, x \geq 1 \end{array} \right.$$

Then,

$$(3.30) \qquad \Lambda_0(x) = \cos^{-1} \left( \frac{1/2 - x}{1/2} \right) \quad \text{and} \quad \lim_{x \nearrow 1} \Lambda_0(x) = \pi.$$

Hence,

$$(3.31) \qquad \beta(x) = \left( \chi - \kappa \cos^{-1} \left( \frac{1/2 - x}{1/2} \right) \right) \sqrt{x(1-x)} + \frac{1}{4}(1 - 2x).$$

Since $\beta(0) = \frac{1}{4}$ and $\beta(1) = -\frac{1}{4}$ there are no absorbing states. However, our explicit solutions $\zeta_t^x = [1 - \cos \left( \int_0^t e^{-\kappa(t-s)}[dW_s + \chi ds] + \cos^{-1} \left( \frac{1/2-x}{1/2} \right) \right)]/2$ will hit $0$ and $1$.

## 4. Filtering algorithms

We will now tie our explicit solution ideas into our filtering interests through the general framework set forth in the Motivation through Gaussian signals section. We will show that there are at least two forms of potential generality for filtering problems that can be solved by exact infinite-dimensional filters, one through parameters in the Riccati equation solutions below and the other through choice of explicit solution. Alternatively, these generalities can be thought of as ways of choosing and decomposing fundamental solutions to Step 2.

REMARK 4.1. Clearly, the most obvious method of obtaining an interesting class of filtering problems from our explicit solutions is through state space diffeomorphism, i.e. to form signals from $\Lambda^{-1}(Z^\kappa(t, \Lambda(x)))$ for some linear stochastic differential equation $Z^\kappa$ and observations $Y_j = h(\Lambda(X_{tj}, V_j)) = \widetilde{h}(Z_{tj}, V_j)$. However, this method would not include analogs to the important Benes filter nor seem to require much discussion. Therefore, we embark along a somewhat different path more aligned with substitution of variables and decomposition of fundamental solutions. Indeed, it is possible by solution of the Riccati equations below to come up with exact infinite dimensional filters with constant dispersion coefficient and nonlinear drift signals. Example 3.3 shows that these can not be handled by diffeomorphism to linear stochastic differential equation alone.

Following the ideas of Section 2, we would be tempted to apply Feynman-Kac's formula immediately on (1.13) to find that

$$(4.1) \quad p_{X_{t_{j+1}}|\mathcal{Y}_j}(x) = q_j(t_{j+1}, x) = E\left[p_{X_{t_j}|\mathcal{Y}_j}(Z^x_{\delta t_j}) \exp\left\{\int_0^{\delta t_j} c(Z^x_s)ds\right\} |\mathcal{Y}_j\right],$$

where $\delta t_j = t_{j+1} - t_j$ and $Z^x$ solves

$$(4.2) \quad Z^x_t = x + \int_0^t b(Z^x_s)ds + \int_0^t \sigma(Z^x_s)dW_s \quad \forall x \in D.$$

REMARK 4.2. A sufficient condition to apply Feynman-Kac's formula in the present setting with $\tau_N \doteq \{t \geq 0 : |Z^x_t| \geq N\}$ is that

$$(4.3) \quad \lim_{M \to \infty} E \sup_{N \geq M}\left[\exp\left\{\int_0^{\tau_N} c(Z^x_s)ds\right\} 1_{\tau_N < \delta t_j}\right] < \infty,$$

which is certainly true when $c : \mathbb{R} \to (-\infty, C]$ for any $C > 0$.

However, returning to (4.1), we see that there now is dependence upon the whole process distribution for $\mathbf{Z}^x$ in (4.1) meaning that convolution will not be immediately possible in Step 2 of the Introduction. Hence, our first task will be to remove the dependence on $Z^x_s$ for $s \neq t$. This is most readily done by changing the potential term in Equation (1.13) of Step 2 into a constant term through a substitution of variables and Bernoulli order reduction:

LEMMA 4.3. *Suppose that $T > 0$, $\eta \in \mathbb{R}$, $m(x) \doteq \frac{b(x)}{2a(x)}$ and $v$ solves the equation*

$$(4.4) \quad \begin{aligned} \partial_t r(t, x) &= a(x)\partial_x^2 r(t, x) + \beta(x)\partial_x r(t, x) + \eta r(t, x), \\ r(0, x) &= \phi(x) \end{aligned}$$

on $D$ or some open interval containing $D$, where $\beta(x) \doteq 2\mu(x)a(x)$ and $\mu$ solves the ordinary differential equation

$$(4.5) \qquad \frac{d\mu(x)}{dx} = \frac{dm(x)}{dx} - \mu^2(x) + m^2(x) + \frac{\eta - c(x)}{a(x)}$$

almost everywhere on $D$. Moreover, assume that $\int_\psi^x |\mu(z) - m(z)|\, dz < \infty$ for some $\psi \in \mathbb{R}$ and all $x \in D$ and

$$(4.6) \qquad \lim_{x \to y} v(t,x) \exp\left\{ \int_\psi^x \mu(z) - m(z)\, dz \right\} = 0$$

for each $y \in \partial D$, $t \in [0,T]$. Then, $u(t,x) = v(t,x) \exp\left\{ \int_\psi^x \mu(z) - m(z)dz \right\}$ solves the equation

$$(4.7) \qquad \partial_t q(t,x) = a(x)\partial_x^2 q(t,x) + b(x)\partial_x q(t,x) + c(x)q(t,x),$$
$$q(0,x) = \phi(x) \exp\left\{ \int_\psi^x \mu(z) - m(z)dz \right\}, \quad q(t)|_{\partial D} = 0$$

on $[0,T] \times D$.

Naturally, this result is proved by substitution and cancellation. Preceding formally for the moment, we know from (F5) that Equation (1.13) in Step 2 of the introduction with the added Dirichlet boundary condition has a unique solution, which must be given by:

$$(4.8) \qquad p_{X_{t_{j+1}}|\mathcal{Y}_j}(x) = v_j(t_{j+1},x) \exp\left\{ \int_\psi^x \mu(z) - m(z)dz \right\},$$

where $m(x) \doteq \frac{b(x)}{2a(x)}$, $\mu$ is a continuous solution to (4.5), $\beta(x) \doteq 2\mu(x)a(x)$, and

$$(4.9) \qquad \partial_t v_j(t,x) = a(x)\partial_x^2 v_j(t,x) + \beta(x)\partial_x v_j(t,x) + \eta v_j(t,x),$$
$$v_j(t_j,x) = p_{X_{t_j}|\mathcal{Y}_j}(x) \exp\left\{ \int_\psi^x m(z) - \mu(z)dz \right\}.$$

Now, we can apply Feynman-Kac's formula to $v_j$ to find that

$$(4.10) \qquad p_{X_{t_{j+1}}|\mathcal{Y}_j}(x) = E\left[ \theta(\zeta_{\delta t_j}^x) \exp\left\{ \eta \times \delta t_j \right\} \right] \exp\left\{ \int_\psi^x \mu(z) - m(z)dz \right\},$$

for each $x \in D$, where

$$(4.11) \qquad \theta(\xi) = p_{X_{t_j}|\mathcal{Y}_j}(\xi) \times \exp\left\{ \int_\psi^\xi m(z) - \mu(z)dz \right\} \quad \forall \xi \in D,$$

$$(4.12) \qquad \zeta_t^x = x + \int_0^t \beta(\zeta_s^x)ds + \int_0^t \sigma(\zeta_s^x)dW_s \quad \forall t \geq 0.$$

However, suppose

$$(4.13) \qquad \frac{\sigma'(y)\beta(y)}{\sigma(y)} + \frac{\sigma''(y)\sigma(y)}{2} = \beta'(y) + \kappa$$

for all $y \in D$ such that $\sigma(y) \neq 0$ and some constant $\kappa \in \mathbb{R}$,

$$(4.14) \qquad \sigma'(y)\beta(y) = 0$$

for all $y$ such that $\sigma(y) = 0$, and $\int_\lambda^x \frac{dy}{\sigma(y)} < \infty$ for all $x \in D$ and some $\lambda \in \mathbb{R}$. Then, $\zeta^x$ has an explicit solution for each $x \in D$ of form

$$(4.15) \quad \zeta_t^x = \Lambda_\lambda^{-1}\left(\frac{\chi \int_0^t f_s ds + \Lambda_\lambda(x) - \int_0^t(-f_s)dW_s}{f_t}\right) \quad \forall 0 \le t \le \tau,$$

$$\tau = \inf\left\{t > 0 : \varphi(x, t, \int_0^t f_s dW_s) \in D^c\right\},$$

where $f_s = \exp\{\kappa s\}$ and $\Lambda_\lambda(y) = \int_\lambda^y \frac{dx}{\sigma(x)}$ for all $y \in D$. Thus, if $P(\tau > \delta t_j) = 1$ we find that $p_{X_{t_{j+1}}|\mathcal{Y}_j}(x)$ can be evaluated by

$(4.16)$

$$p_{X_{t_{j+1}}|\mathcal{Y}_j}(x) = E\left[\theta(\Lambda_\lambda^{-1}\left(\frac{\chi \int_0^{\delta t_j} f_s ds + \Lambda_\lambda(x) - \int_0^{\delta t_j}(-f_s)dW_s}{f_{\delta t_j}}\right))\right] \times H_\psi(\delta t_j, x),$$

where

$$(4.17) \qquad H_\psi(t, x) = \exp\left\{\eta \times t + \int_\psi^x \mu(z) - m(z)dz\right\} \quad \forall t \ge 0, x \in D.$$

However, letting

$$(4.18) \quad \Upsilon_{\delta t_j}(y) = \theta(\Lambda_\lambda^{-1}\left(\frac{\chi \int_0^{\delta t_j} f_s ds + \sqrt{\int_0^{\delta t_j} f_s^2 ds} \times y}{f_{\delta t_j}}\right)), \quad v_x = \frac{\Lambda_\lambda(x)}{\sqrt{\int_0^{\delta t_j} f_s^2 ds}},$$

we find that

$$(4.19) \qquad E\left[\theta(\Lambda_\lambda^{-1}\left(\frac{\chi \int_0^{\delta t_j} f_s ds + \Lambda_\lambda(x) - \int_0^{\delta t_j}(-f_s)dW_s}{f_{\delta t_j}}\right))\right]$$

$$= \int_\mathbb{R} \Upsilon_{\delta t_j}(v_x - \xi)\Phi(\xi)\, d\xi = (\Upsilon_{\delta t_j} * \Phi)(v_x),$$

where $\Phi(y) = \frac{1}{\sqrt{2\pi}}\exp\left\{-\frac{y^2}{2}\right\}$, is again evaluated via convolution.

Now, we suppose again that our signal is

$$(4.20) \qquad X_t = X_0 + \int_0^t \alpha(X_s)ds + \int_0^t \sigma(X_s)dW_s, \ X_t \subset D$$

and determine a class of $\alpha$ corresponding to a particular $\sigma$ for which filtering through convolution with a standard normal distribution is possible. In particular, we assume that Conditions (F1-F5) are true, so we can use the Three-Step Algorithm with

$(4.21)$

$$a(x) = \frac{\sigma^2(x)}{2}, \ b(x) = 2\sigma(x)\sigma'(x) - \alpha(x), \ c(x) = (\sigma'(x))^2 + \sigma(x)\sigma''(x) - \alpha'(x).$$

Then, it follows from (4.21) and $m(x) = b(x)/\sigma^2(x)$ that

$$(4.22) \qquad \frac{dm(x)}{dx} + m^2(x) - \frac{2c(x)}{\sigma^2(x)} = \frac{d}{dx}\left(\frac{\alpha(x)}{\sigma^2(x)}\right) + \frac{\alpha^2(x)}{\sigma^4(x)} \quad \forall x \in D.$$

Moreover, we know from our section on explicit solution that

$$(4.23) \qquad \beta(x) = (\chi - \kappa\Lambda_\lambda(x) + \frac{1}{2}\sigma'(x))\sigma(x), \quad \mu(x) = \beta(x)/\sigma^2(x),$$

on $D$ so

(4.24)

$$\frac{d\mu(x)}{dx} + \mu^2(x) = \frac{1}{\sigma^2(x)}\left[\frac{\sigma''(x)\sigma(x)}{2} - \kappa - \frac{(\sigma'(x))^2}{4} + (\chi - \kappa\Lambda_\lambda(x))^2\right] \quad \forall x \in D.$$

This can be substituted into (4.5) together with (4.22) and the definitions in (4.21) to find that

$$\frac{d}{dx}\frac{\alpha(x)}{\sigma^2(x)} + \left(\frac{\alpha(x)}{\sigma^2(x)}\right)^2 = Q(x),$$

(4.25)

$$Q(x) = \frac{\frac{\sigma''(x)\sigma(x)}{2} - \frac{(\sigma'(x))^2}{4} - 2\eta - \kappa + (\kappa\Lambda_\lambda(x) - \chi)^2}{\sigma^2(x)},$$

which is a Riccati equation in $\frac{\alpha(x)}{\sigma^2(x)}$ to be solved on $D$. Furthermore, (4.23) and $\varphi = p_{X_{t_j}|\mathcal{Y}_j}$ can be substituted into (4.9) to find that

(4.26)

$$\partial_t v_j(t,x) = \frac{\sigma^2(x)}{2}\partial_x^2 v_j(t,x) + (\chi - \kappa\Lambda_\lambda(x) + \frac{\sigma'(x)}{2})\sigma(x)\partial_x v_j(t,x) + \eta v_j(t,x),$$

$$v_j(t_j,x) = \varphi(x)\exp\left\{\int_\psi^x m(z) - \mu(z)dz\right\} \quad \forall x \in D.$$

To implement Step 2 as a convolution, we require a solution for (4.25,4.26) for *some* constants $\kappa, \eta, \chi, \lambda, \psi \in \mathbb{R}$. However, to be a valid signal process we also require that (4.20) have a unique-in-law weak solution. Finally, we also need to be able to apply Feynman-Kac's formula in (4.10), satisfy (4.6) with $v_j$ in place of $v$ and have $P(\tau > \delta t_j) = 1$. These criteria motivate the following definition.

DEFINITION 4.4. Suppose $D$ is some open interval, $\sigma \in C(\mathbb{R}, [0, \infty))$, and $\sigma|_D \in C^2(D, [0, \infty))$. Then, $\alpha : \mathbb{R} \to \mathbb{R}$ corresponds to an infinite-dimensional exact Gaussian filter if: (i) $\frac{\alpha}{\sigma^2(x)}$ is a $C^1(D, \mathbb{R})$-solution to (4.25) for some collection of constants $\kappa, \eta, \chi, \lambda \in \mathbb{R}$, (ii) $\int_\psi^x \left|\frac{\chi - \kappa\Lambda_\lambda(z) - \frac{3}{2}\sigma'(z)}{\sigma(z)} + \frac{\alpha(z)}{\sigma^2(z)}\right| dz < \infty$ for all $x \in D$, the same $\kappa, \chi, \lambda$, and some $\psi \in \mathbb{R}$ (iii) there exists a solution $v_j$ to (4.26) such that

$$\lim_{x \to y} v_j(t,x)\exp\left\{\int_\psi^x \frac{\chi - \kappa\Lambda_\lambda(z) - \frac{3}{2}\sigma'(z)}{\sigma(z)} + \frac{\alpha(z)}{\sigma^2(z)}dz\right\} = 0$$

$$\forall y \in \partial D, t \in [t_j, t_{j+1}]$$

for these constants $\kappa, \chi, \lambda, \psi$ and every continuous, bounded, non-negative $\varphi$ on $D$ with $\varphi|_{\partial D} = 0$, (iv) there is a (unique-in-law) weak solution $X$ to (1.1) such that $X_t \subset D$ and Condition (F5) is true, and (v) our explicit solution from (4.15) satisfies

(4.27) $$\lim_{M \to \infty} E\left[\sup_{N \geq M}\exp\left[2\int_\psi^{\zeta^x(\tau_N)} m(z) - \mu(z)dz\right]1_{\tau_N < \delta t_j}\right] < \infty$$

for each $x \in D$ with $\tau_N^x \doteq \inf\{s \geq t : |\zeta_s^x| > N\}$ in order that the Feynman-Kac formula applies in (4.10) and $P(\tau > t_j) = 1$ with $\tau$ as defined in (4.15).

REMARK 4.5. (ii) is precisely the condition in Lemma 4.3. Suppose $\psi \in D$. Then, $\int_\psi^x \left| \frac{\alpha(z)}{\sigma^2(z)} \right| dz < \infty$ for each $x \in D$ by (i). To utilize (v) in Feynman-Kac's formula at (4.10), we use the fact that $v_j(t,x) \times \exp\left[ \int_\psi^x \mu(z) - m(z)dz \right]$ is bounded. We are investigating new methods that may combine and reduce conditions (ii), (iv), and (v).

This definition allows us to convert the previous formal development into a theorem and a procedure:

THEOREM 4.6. *Suppose Conditions* (F1-F4) *are true,* $\sigma \in C\left(\mathbb{R}, [0, \infty)\right)$, $\sigma|_D \in C^2\left(D, [0, \infty)\right)$, *and* $\alpha$ *corresponds to an infinite-dimensional exact Gaussian filter. Then,* $P_{X_{t_j}|\mathcal{Y}_j}(\cdot)$ *has a density that can be calculated via the following algorithm:*

1. Store the functions

$$(4.28) \qquad F(x) \doteq \exp\left\{ \int_\psi^{\Lambda_\lambda^{-1}(x)} m(z) - \mu(z)dz \right\} \quad \forall x \in \mathbb{R}$$

$$H_\psi(\delta t_j, x) \doteq \exp\left\{ \eta \times \delta t_j + \int_\psi^x \mu(z) - m(z)dz \right\} \quad \forall x \in \mathbb{R}$$

$$v_x \doteq \Lambda_\lambda(x) / \sqrt{\int_0^{\delta t_j} f_s^2 ds} \quad \forall x \in \mathbb{R}$$

2. Store the constants

$$(4.29) \qquad a = \frac{\sqrt{\int_0^{\delta t_j} f_s^2 ds}}{f_{\delta t_j}} \text{ and } b = \frac{\chi \int_0^{\delta t_j} f_s ds}{f_{\delta t_j}}$$

3. $p_{X_{t_0}|\mathcal{Y}_0}(x) = p_{X_0}$      (* Step 1 of Introduction *)
4. Do for $j = 0, 1, 2, ...$
   (a) Form $\Xi(x) = p_{X_{t_j}|\mathcal{Y}_j}(\Lambda_\lambda^{-1}(x)) \times F(x)$
   (b) Take a (Fast) Fourier transform $\widehat{\Xi}$ of $\Xi$
   (c) Pointwise multiply in frequency domain
   $$\widehat{\Psi}(\xi) = \left[ \frac{1}{a} \exp\left( i \frac{b}{a} \xi \right) \widehat{\Xi}(\xi/a) \right] \exp\left[ -\frac{\xi^2}{2} \right]$$
   (d) Take inverse (Fast) Fourier transform $\Psi$ of $\widehat{\Psi}$
   (e) Substitute in and post multiply $p_{X_{t_{j+1}}|\mathcal{Y}_j}(x) = \Psi(v_x) H_\psi(\delta t_j, x)$
   (f) Wait for the next observation $Y_{j+1}(\omega)$.
   (g) Evaluate the Bayes' rule update (* Step 3 of Introduction *)

$$(4.30) \qquad p_{X_{t_{j+1}}|\mathcal{Y}_{j+1}}(x) = \frac{p_{Y_{j+1}|X_{t_{j+1}}}(Y_{j+1}|x) p_{X_{t_{j+1}}|\mathcal{Y}_j}(x)}{\int_\mathbb{R} p_{Y_{j+1}|X_{t_{j+1}}}(Y_{j+1}|z) p_{X_{t_{j+1}}|\mathcal{Y}_j}(z)dz}.$$

In particular, to save time, one can work with unnormalized conditional densities and avoid the calculation of and division by $\int_\mathbb{R} p_{Y_{j+1}|X_{t_{j+1}}}(Y_{j+1}|z) p_{X_{t_{j+1}}|\mathcal{Y}_j}(z)dz$ until one wishes to integrate with the density $p_{X_{t_{j+1}}|\mathcal{Y}_{j+1}}$ to e.g. form the conditional mean. Indeed, it could be avoided entirely if one only wishes to obtain a maximum a posteriori estimator of $X_{t_{j+1}}$.

Finally, we turn to question of finding $\alpha's$ corresponding to infinite dimensional Gaussian filters. We know that (4.25) is a Riccati equation with a continuous right hand side $Q$. Such Riccati equations have been heavily studied. Most solutions on

$D$ are not solvable by quadrature. There are some well known methods to employ in trying to establish solutions. First, if the solution to the linear equation

$$(4.31) \qquad \begin{bmatrix} z' \\ w' \end{bmatrix} = \begin{bmatrix} 0 & Q \\ 1 & 0 \end{bmatrix} \begin{bmatrix} z \\ w \end{bmatrix}, \qquad \begin{bmatrix} z(x_0) \\ w(x_0) \end{bmatrix} = \begin{bmatrix} y(x_0) \\ 1 \end{bmatrix}$$

satisfies $w(x) > 0$ for all $x \in D$. Then, $y = z/w$ is a solution to (4.25). Once, one solution $y_1$ (say) has been found others can be found by solving (the linear equation)

$$(4.32) \qquad\qquad\qquad v' = 2y_1 v + 1$$

and setting $y = y_1 + \frac{1}{v}$ provided $v \neq 0$ on $D$.

Most often one solves (4.25) or (4.31) on a computer. Naturally, in our application this can be done off-line long before the filtering procedure starts. We give a couple of simple examples of solutions:

EXAMPLE 4.7. Suppose that $D = \mathbb{R}$ and $Q(x) = (c_1)^2$ for all $x \in \mathbb{R}$ and some constant $c_1$. Then, $y(x) = c_1 \tanh(c_1 x)$ solves (4.25). Comparing $Q(x) = (c_1)^2$ to (4.25), one finds that this is possible when $\sigma(x) = c_2$ (some constant) as in Example 3.3. This present example is related to the work in Kouritzin (1998) and the original filter of Benes (1981).

EXAMPLE 4.8. Suppose that $D = (0, \infty)$, $0 < x_0 < \infty$, and $Q(x)$ is non-negative and continuous on $D$. Then, there exists a unique continuous solution to

$$(4.33) \qquad \frac{d}{dx}(y(x)) + (y(x))^2 = Q(x), \qquad y(x_0) = 0$$

on $D$. This solution satisfies $y(x) \geq 0$ for all $x \geq x_0$ and $y(x) \leq 0$ for all $x \leq x_0$. The condition $2\eta + \kappa \leq -\frac{1}{4}$ will ensure that $Q$ remains non-negative for $\sigma(x) = x$ as in Example 3.5. There are better results on Riccati equations in many textbooks. We have just given a simple result whose proof is immediate.

In the above example (as well as on different domains), it is well known that there still can be solutions without the constraint that $Q$ is non-negative provided that it does not become "too negative". Still, even when there is a solution $y$ to (4.25) one must check that $\alpha \doteq y \cdot \sigma^2$ corresponds to an infinite-dimensional exact Gaussian filter. This has been done for simple multidimensional examples in Kouritzin (1998) together with simulations and more examples are being worked on.

## References

[1] Benes, V. (1981). Exact finite-dimensional filters for certain diffusions with nonlinear drift. *Stochastics*, **5**:65-92 .

[2] Billingsley, P. (1986). *Probability and Measure, Second Edition*. Wiley, New York.

[3] Daum, F. (1988). New exact nonlinear filters. In: emph Bayesian Analysis of Time Series and Dynamic Models (Ed.: J.C. Spall.) Marcel Dekker, New York.

[4] Doss, H. (1977). Liens entre équations différentielles stochastiques et ordinaires. *Ann. Inst. Henri Poincaré*, **13**:99-125.

[5] Jazwinski, A. (1970). *Stochastic Processes and Filtering Theory*. Academic, New York.

[6] Kouritzin, M. A. (1998). On exact filters for continuous signals with discrete observations. *IEEE Trans. Auto. Control*, **AC-43**:709-715.

[7] Kouritzin, M. A. and Li, D. (1998). On explicit solutions to stochastic differential equations. To appear in *Journal of Stochastic Analysis and Applications*.

[8] Leung, C.W. and Yau, S.S.T. (1992). Recent results on classification of finite dimensional rank estimation algebras with state space dimension 3. In *Proc. IEEE Conf. Decision and Control*, pp.2247–2250.

[9] Sussmann,H. (1978). On the gap between deterministic and stochastic differential equations. *Ann. Probab.*, **6**:19-41.

DEPARTMENT OF MATHEMATICAL SCIENCES, UNIVERSITY OF ALBERTA, EDMONTON, ALBERTA, CANADA T6G 2G1

*E-mail address*: `mkouritz@math.ualberta.ca`

Canadian Mathematical Society
Conference Proceedings
Volume **26**, 2000

# SDE Estimation: Effects of Misspecified Diffusion Functions

## R. J. Kulperger

*This paper is dedicated to Professor Donald Dawson. I also wish to thank Professor Donald Dawson for the help and encouragement he showed when I was a graduate student some years ago, and for his continued support over the years. I am sure that I echo all my fellow students, and other colleagues of Don Dawson in congratulating him on his career achievements up to the present, and to wish him well in the many years yet to come.*

ABSTRACT. Estimation and Euler residuals for discretely observed SDE's are considered. The residuals are also considered under some alternative forms of the diffusion rate function. The alternatives are close to the null process. However their effect is that the Euler residual normal QQ plots look as if they come from a heavy tail noise, even though they are driven by Brownian motion. This property of the Euler residuals makes it difficult to conclude that a heavy tailed noise process drives the SDE.

## 1. Introduction

Consider the stochastic differential equation

$$(1.1) \qquad dX_t = \mu(X_t)dt + \sigma(X_t)dW_t$$

where $W$ is standard Brownian motion, and the mean and standard deviation functions $\mu$ and $\sigma$ are given by

$$
\begin{aligned}
\mu(x) &= a + bx \\
(1.2) \qquad \sigma(x) &= cx^\gamma \, .
\end{aligned}
$$

This model is considered in McLeish and Kolkiewicz [**8**] (hereafter referred to as MK97) generalizing several families of models for interest rates and bonds such as the Cox Ingersoll Ross model (see references in [**8**],the book by Duffie [**3**], or the recent paper [**5**]). MK studied the normality of Euler residuals (described below), and the use of higher ordered Taylor expansions applied to this question.

When applied to 7 years worth of daily rates for 30 years bond MK97 found that the Euler residuals (given by (2.3) as a function of the estimators) looked distinctly non normal in terms of normal QQ plots. They argued that the residuals indicated a stable process might be driving the SDE (1.1) instead of Brownian motion. They

---

2000 *Mathematics Subject Classification.* Primary 62M05; Secondary 60J60, 60J25.
*Key words and phrases.* SDE; discrete observations; estimation, misspecified diffusion.

used a reasonably large sample size of 1818 data points. Note that 7 years times 259 trading days per year yields 1813 trading days over that period.

Several authors have also noted that some financial data exhibits behaviour that is not Gaussian, such as higher kurtosis. Some recent papers discussing these include [2] and [7]. These papers consider a truncated Levy process to account for the non Gaussian behaviour.

The question that we consider in this note is the effect of using an incorrect form of the diffusion function $\sigma$ in (1.2). The process is observed at a sampling interval of length $\Delta = \frac{1}{259}$. It turns out that relatively small perturbations of this function can yield various types of non normal behavior in terms of QQ plots of the Euler residuals. These effects are seen in the sample sizes corresponding to the above. Note that in these cases, if the data come from the SDE model fitted, then the Euler residuals do yield reasonably straight normal QQ plots. We do not consider deviations in the mean function $\mu$ in (1.2) since we do not anticipate that deviations from the mean will not effect the tail behaviour of the residuals as much.

We use the term Euler increment as the related process value of the Euler residual, that is

$$e([k+1]\Delta) = \frac{(X([k+1]\Delta) - X(k\Delta)) - (a + bX(k\Delta))\Delta}{\sigma(X(k\Delta))\sqrt{\Delta}} .$$

## 2. The model, simulation and estimation method

We consider the process (1.1) with rate functions (1.2), and parameter values $a = 3.1827$, $b = -0.3962$, $c = .0075$ and $\gamma = 2.5813$, the values of the estimates in MK97 [8, p 311].

This process was simulated and a discrete observation set was taken. Based on these estimates of the parameters $a, b, c, \gamma$ were obtained. This section describes these in some detail.

The value $X(0)$ was taken to be near the starting value in the graph of the bond data in [8], that is 8. With this starting value, simulate $X^{\Delta,m}$ at discrete intervals $\Delta/m$, and then take every $m$–th value. Simulate $X^{\Delta,m}$ at times $k\Delta/m$, $k = 0, 1, \ldots, [Tm/\Delta]$. Drop the superscript $\Delta, m$ notation. Specifically for given $X(0)$ generate

$$\begin{aligned} &X(\{(k+1)\Delta/m\}) \\ &= \quad X(\{k\Delta/m\}) + \mu(X(\{k\Delta/m\})\Delta/m + \sigma(X(\{k\Delta/m\}))\sqrt{\Delta/m}Z_{k+1} \end{aligned}$$

where $Z$ is a sequence of IID standard normal random variables. The interval spacing parameter is $\Delta = \frac{1}{259}$ corresponding to a unit of one trading day.

The simulation parameters are $T = 7$ years, $\Delta = \frac{1}{259}$ corresponding to one trading day, and $m = 20$ for the tables given in this section. Various values of $m$ were tried, but once $m$ is bigger than about 5 or 10, there were no noticeable changes in the Euler increments. These give rise to a sample size of $7 * 259 = 1813$. The program was written in Fortran.

A typical sample path for Gaussian and a Stable noise are shown in Figures 1 and 2. These were generated using $m = 10$ as the sampling density rate per day. The sample ACF plots usually do not drop to 0 until the range of 200 to 250 days, similar for both noise processes. As expected the stable noise case exhibits more large jumps. Several stable process paths were generated as the first several

**Gaussian Noise, simulated at M*Delta, M = 10**

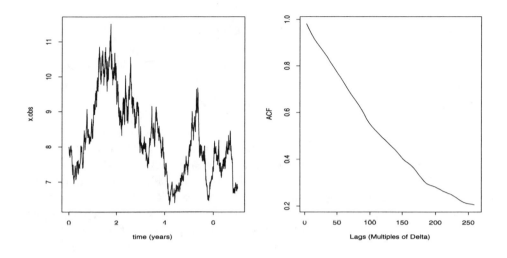

FIGURE 1. A bond SDE path with Gaussian noise.

**Stable 1.7 Noise, simulated at M*Delta, M = 10**

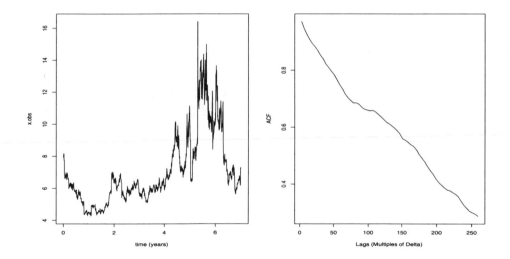

FIGURE 2. A bond SDE path with stable 1.7 noise.

exhibited quite large excursions, and the exhibited one allows finer features to be seen in the plot.

If one observes the process in continuous time on the interval $[0, T]$, and knows the diffusion function $\sigma(x)$, the likelihood function of the drift parameter is given by a Girsanov type formula [**1**, p 219], that is

$$L = \exp\{\int_0^T \frac{(dX_s - \mu(X_s)ds)^2}{\sigma(X_s)^2}\} \ .$$

In the case of the rate functions (1.2) the MLE's satisfy

$$\int_0^T (\hat{a} + \hat{b}X_s)X_s^{-2\hat{g}}ds \ = \ \int_0^T X_s^{-2\hat{g}}dX_s$$

$$\int_0^T (\hat{a} + \hat{b}X_s)X_s^{1-2\hat{g}}ds \ = \ \int_0^T X_s^{1-2\hat{g}}dX_s \ .$$

These equations do not contain $\hat{c}$. If two SDE processes have different diffusion of variance functions then the measures are mutually singular [**1**]. Thus one needs to augment the estimating equations to obtain an estimate of the diffusion function.

MK97 [**8**] considers various methods of using the above likelihood with discrete data $X_{k\Delta}$ based on Ito–Taylor expansions obtained from [**6**]. They indicate that higher order expansions do not make much difference in the scales they consider so we use only the simplest discretization of the MLE's. In addition we obtain $\hat{g}$ iteratively since the likelihood equations are easy to solve for $\hat{a}, \hat{b}$ for a given $\hat{g}$.

In particular this iteration is for a given $\hat{g}_i$,

$$(2.1) \qquad\qquad A(\hat{g}_i)\left(\begin{array}{c} \hat{a} \\ \hat{b} \end{array}\right) = B(\hat{g}_i)$$

where

$$A(g) \ = \ \left(\begin{array}{cc} \sum_{k=0}^{n-1} X(k\Delta)^{-2g}\Delta & \sum_{k=0}^{n-1} X(k\Delta)^{1-2g}\Delta \\ \sum_{k=0}^{n-1} X(k\Delta)^{1-2g}\Delta & \sum_{k=0}^{n-1} X(k\Delta)^{2-2g}\Delta \end{array}\right)$$

$$(2.2) \qquad B(g) \ = \ \left(\begin{array}{c} \sum_{k=0}^{n-1} X(k\Delta)^{-2g}dX([k+1]\Delta) \\ \sum_{k=0}^{n-1} X(k\Delta)^{1-2g}dX([k+1]\Delta) \end{array}\right)$$

and $dX([k+1]\Delta) = X([k+1]\Delta) - X(k\Delta)$.

Let $r_{k+1} = dX([k+1]\Delta) - (\hat{a} + \hat{b}X(k\Delta))\Delta$ be the $(k+1)$–st residual. Since for a small $\Delta$,

$$\text{Var}(dX([k+1]\Delta) \mid X(k\Delta)) \approx \sigma(X(k\Delta))^2\Delta = c^2 X(k\Delta)^{2g}\Delta,$$

$g$ can be estimated by regressing $\log(r_{k+1}^2)$ against $\log(X(k\Delta)^2)$. Use this to then obtain $\hat{g}_{i+1}$ and then iterate (2.1). These iterations stabilized quickly, after only 2 or 3 iterations, based on looking at 20 or 30 preliminary runs. In the tables and residual plots given here a fixed number of 5 iterations were used. It was not sensitive to the initial $\hat{g}_0$, whether it was taken to be 0 or 2.58 the true value.

Finally $\hat{c}$ is obtained from

$$\hat{c}^2 = \frac{\sum_{k=0}^{n-1} r_{k+1}^2}{\Delta \sum_{k=0}^{n-1} X(k\Delta)^{2\hat{g}}} \ .$$

The Euler residuals are then given by

$$(2.3) \qquad r_{k+1} = r([k+1]\Delta) = \frac{(X([k+1]\Delta) - X(k\Delta)) - (\hat{a} + \hat{b}X(k\Delta))\Delta}{\hat{\sigma}(X(k\Delta))\sqrt{\Delta}}$$

where $\hat{\sigma}(x) = \hat{c}x^{\hat{g}}$. The Euler increment is the same object but with the correct rate functions (1.2) used.

## Estimator Property Versus Length of Observations

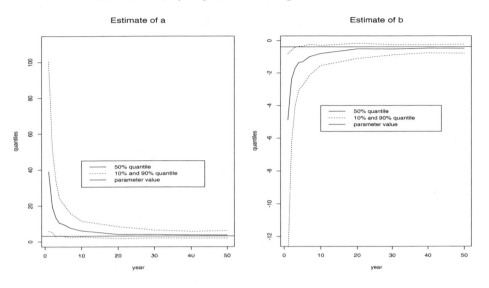

FIGURE 3. Bias of $\hat{a}, \hat{b}$ versus year, with Gaussian noise.

Figures 3 and 4 show results from 200 simulation runs at several different times (years), all for daily readings with $\Delta = \frac{1}{259}$. They give the median and the 10 and 90 percentiles of the four parameter estimates, based on 5 iterations of the estimating equations. These were generated with $m = 10$ giving $m$ generated increments per day. These show that the ad–hoc estimation of $c$ is quite good, and reasonable for $g$. They are median unbiased. This is so even for small $T$ such as 1 year. The big imprecision is in the bias of $a, b$. This is largely due to the long lag dependence of nearly one year in the SDE bond process as seem above. However $a + bx$ is reasonably estimated.

Since we have the simulation data, it is of some interest to report on how fast the central limit theorem takes hold in this setting. It is quite slow, again due in part to the slow rate or decay of the auto correlation function. The estimates are the same data as above. Normal quantile plots are shown for these 200 data points, with the high and low outliers trimmed. This is because of the occasional large value of $\hat{c}$. From Figures 5, 6 and 7 we see that the CLT is slow to take hold. The normal approximation for $\hat{c}$ and $\hat{g}$ are reasonable with 50 years of data (12,950 data points) but is only adequate at best for $\hat{a}$ and $\hat{b}$.

Next we consider the Euler residuals for the simulated null process. Many runs were made over different length of years, with sampling intervals $K * \Delta$, where $\Delta$ corresponds to one day. For example, $K = 5$ corresponds to weekly data, $K = 20$ corresponds to roughly monthly data, and $K = .5$ corresponds to twice daily. Years 1, 3, 5, 7, 10, 20, 30 and 50 were considered. The parameter $K = .5$ was considered for years less than 10, $K = 1$ and 5 for all years, $K = 20$ for years 7 and higher. Ten QQ plots were viewed for years up to 20, and 5 for the higher years. A short summary of the findings is:

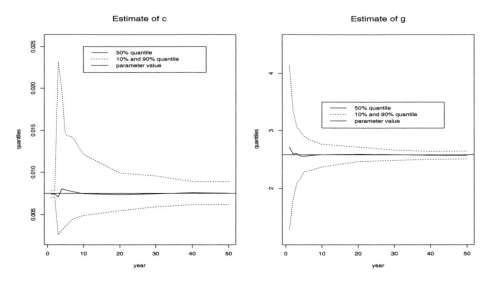

FIGURE 4. Bias of $\hat{c}, \hat{g}$ versus year, with Gaussian noise.

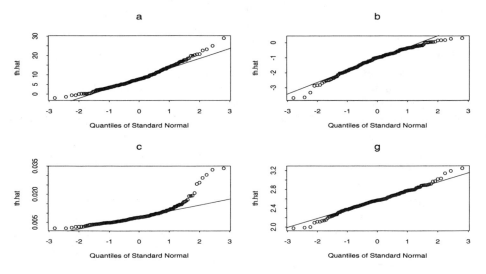

FIGURE 5. Normal QQ plots of estimates for 7 years.

1. QQ plots of the Euler residuals versus the Euler increments is usually near a straight line of slope 1, occasionally with a slope quite different than 1.
2. for sampling rate 1 per $K$ days, and $K \leq 1$, the QQ plots of the Euler residuals is consistent with normal.

QQ Norm of Parameter Estimates, Years = 20, n = 5180

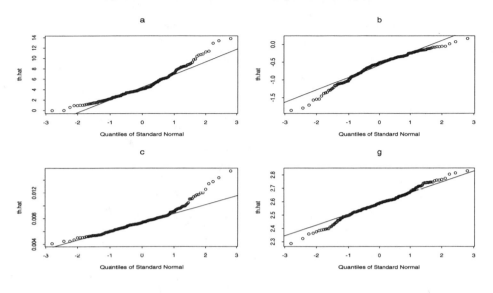

FIGURE 6. Normal QQ plots of estimates for 20 years.

QQ Norm of Parameter Estimates, Years = 50, n = 12950

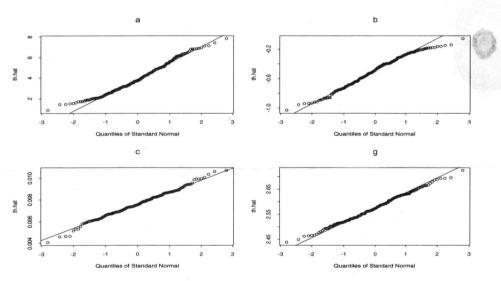

FIGURE 7. Normal QQ plots of estimates for 50 years.

3. for sampling rate 1 per $K$ days, $K = 5$, the Euler residuals often look Normal, but a significant number look non normal, usually more narrow than Gaussian and sometimes with fatter tails than Gaussian.
4. for sampling rate 1 per $K$ days, $K = 20$, the Euler residuals sometimes looked Gaussian but usually not.

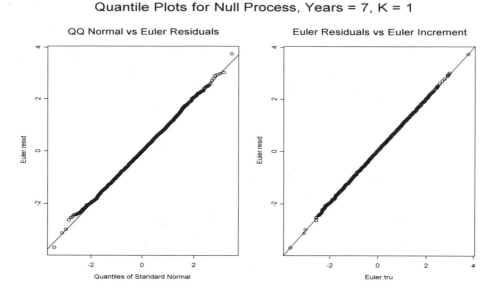

FIGURE 8. Euler residual and increments, years $= 7$, one observation per day, $n = 1814$.

Thus at sampling rate $\Delta$ for the null bond process the Euler residuals should look Gaussian in terms of a QQ plot.

A few QQ plots are shown to indicate how they appeared, see Figures 8, 9 and 10. For year $= 7$, $K = 1$ a typical plot is shown. For (year,$K$) equal to (7,5) and (30,5) the second worst QQ plots are shown.

## 3. Corrected approximation for Euler residuals

Kloeden and Platen [6] describe how to obtain Taylor's expansions for stochastic differential equations. McLeish and Kolkiewicz [8] use this notion to study estimation and the form of residuals from discretely observed SDE's, in particular for the SDE (1.1) and (1.2) which they used as a model for the daily trading values of long term interest bonds. They found heavy tailed Euler residuals, and even using higher order approximations for the residuals were not able to explain the heavy tails. Their conclusion is that the driving term in (1.1) is not Brownian motion, and instead may be a stable process of index $\alpha < 2$. In particular they estimated $\alpha = 1.69$.

The Talyor expansions [6] give a sum in terms of Gaussians, quadratic forms of Gaussians, and higher order polynomial forms. Another way of viewing the Taylor expansion is in this form. Thus one can get a corrected Gaussian approximation for the residuals. When this additional correction was applied, no improvment was obtained in terms of the Euler QQ plots based on the discretized data. A similar observation was made in [8]. This type of correction cannot explain heavy tailed Euler residual plots. These plots are not included in this paper.

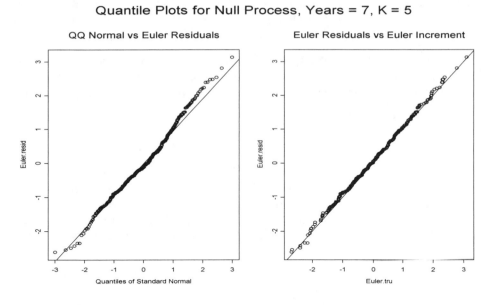

FIGURE 9. Euler residual and increments, years = 7, one observation per 5 days, $n = 363$.

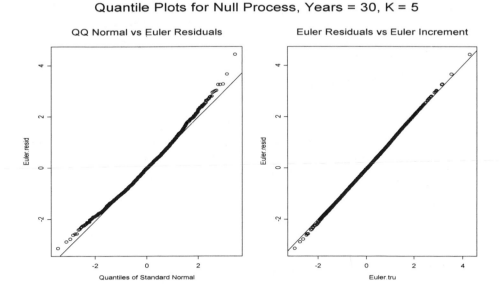

FIGURE 10. Euler residual and increments, years = 30, one observation per 5 days, $n = 1555$.

## 4. Model misspecification and residuals

In this section we consider alternative diffusion rate functions, and study the effect of model misspecification on the Euler residuals. We examine if this could be an explanation of the observed heavy tails in MK97. Before doing so we make a brief excursion into looking at random multipliers of IID normals. This will provide a heuristic for where model misspecification may lead.

**4.1. Mixtures of normals and QQ plots.** Under the alternatives in subsection 4.2, we end up looking at random variables of the form $X = Zh(Y)$, where $Y$ is in a prefield of $Z$ and except for the estimated parameters is independent of $X$, where $Z \sim N(0, 1)$.

Consider a function $h_1$ below

$$h_1(y) = 1 - a_1 * e^{-a_2 * x^2}$$

where $(a_1, a_2) = (.8, .5)$. Suppose also that $Y \sim N(0, 1)$. How will a Normal QQ of $X$ look? Relative to $Z$, $X$ will be shrunk towards 0 if $Y$ is near its centre, and $X$ is approximately $Z$ is $Y$ is not near its centre. In other words $X$ is more tightly packed near 0 than a normal random variable would be. The effect is that mass gets folded into the centre thus giving a density that appears to have fatter than normal tails.

Consider a second function $h_2$

$$h_2(y) = 1 + a_1 * e^{-a_2 * x^2}$$

where $(a_1, a_2) = (2, 1)$. Mass from the centre of $Z$ gets pushed out into the tails, but effect is only important if $Z$ is large and $Y$ is small. It is expected that these tails will be fatter than Gaussian tails, but not as dramatic as $h_1$.

These effects do occur and are illustrated in Figures 11 and 12 with a sample size $n = 400$. Empirically we found that $h_1$ produced fat tail QQ plots very often, while $h_2$ less often, giving normal looking QQ plots about half of the time. Finally we note that the normality of $Y$ is not important, as any other random variable so that $h(Y)$ having a similar effect can be used.

**4.2. Alternative diffusion functions.** The alternatives we consider to the diffusion function $\sigma(x)$ in (1.2) is of the form

$$(4.1) \qquad \sigma_A(x) = \sigma(x)h(x) .$$

The function $h$ will be soup bowl shaped in that it will be constant for $x$ small and large, and smaller and positive for $x$ near 8, roughly the centre of the SDE process $X_t$. These correspond to a process that is slightly less variable near the centre.

These alternatives are in the same spirit and work for a similar reason to that in section 4.1.

For numerical illustration we consider two examples.

$$(4.2) \qquad h_{A,1}(x) = \begin{cases} 2 & \text{if } 0 < x < 6 \\ .75 & \text{if } x = 7 \\ 1 & \text{if } x = 8 \\ 2 & \text{if } 9 < x \end{cases}$$

and linear interpolation in the intervals $[6, 7]$, $[7, 8]$ and $[8, 9]$.

$$(4.3) \qquad h_{A,2}(x) = 1.2 \left( 1 - e^{-.5(x-8)^4} \right) .$$

Can one easily detect these diffusion functions from $\sigma$ based on daily observations for a seven year period? This is examined by looking at empirical sample

**Normal Mixture, h_1**

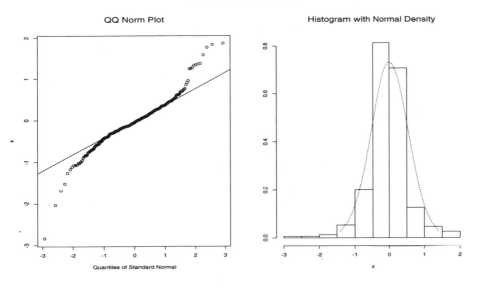

FIGURE 11. Normal plots for $X = Zh_1(Y)$, $n = 400$.

**Normal Mixture, h_2**

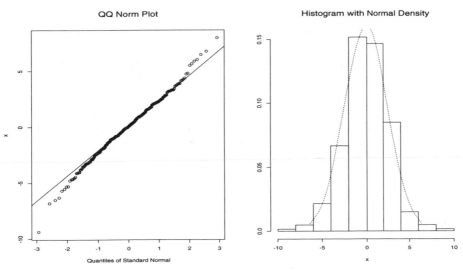

FIGURE 12. Normal plots for $X = Zh_2(Y)$, $n = 400$.

estimates based on bin samples. Specifically an estimate of $\sigma^2(x)$ at a given $x$ is the sample variance of $X([k+1]\Delta)$ over all $k$ such that $\mid X(k\Delta) - x \mid < w$. The bin parameter is taken to be $w = 2\sqrt{\mathrm{Var}(X_\infty)}n^{-.5}$, for the variance taken from the limit distribution under the null process, and $n$ the number of observations over the seven years. It is approximated by simulation runs over 30 year lengths.

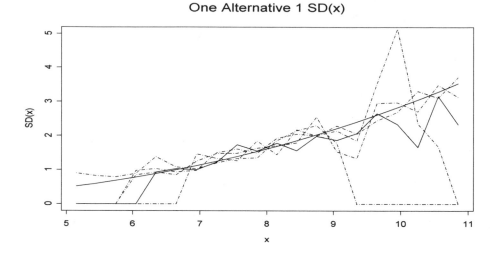

**Sample of 5 SD(x) Estimates For Null Process**

**One Alternative 1 SD(x)**

FIGURE 13. Sample diffusion estimates at selected $x$ for the null and for one alternative type 1. See paragraph after (4.3) for discussion of this plot.

It averages out to be 0.05, and so this value for $w$ is used in the plots here, see Figures 13 and 14. These figures show 5 estimates of the standard deviation for 20 selected $x$. The solid smooth curve is the fitted parametric conditional variance formula. The other solid bumpy line is the nonparametric variance estimate based on one sample path. The 5 bumpy dashed lines are the nonparametric bin variance estimates from 5 sample path realizations from the alternative process. If the bin has fewer than 2 observed data points the estimated SD($x$) is taken to be 0. One sample estimate is also shown for the alternative. Viewing many such plots one sees that the alternative diffusion function sample estimate is quite consistent with the null case. Thus the alternatives considered here are quite close to the null process case.

How do the Euler residuals behave under these alternatives? First consider the heuristics of how these residuals should behave. As seen earlier the Euler increments do behave as standard normals for the sampling intervals considered here. Write $X_k = X(k\Delta)$, and The Euler residuals are

$$
\begin{aligned}
r_{k+1} &= \frac{(X_{k+1} - X_k) - (\hat{a} - \hat{b}X_k)}{\hat{\sigma}(X_k)} \\
&= \frac{(X_{k+1} - X_k) - (\hat{a} - \hat{b}X_k)}{\sigma(X_k)} \frac{\sigma(X_k)}{\hat{\sigma}(X_k)} \\
&\approx \frac{(X_{k+1} - X_k) - (\hat{a} - \hat{b}X_k)}{\sigma(X_k)} h(X_k) \ .
\end{aligned}
$$

This is a product of a $N(0,1)$ variable and the random variable $h(X_k)$. Under the null, $h = 1$, and the result is a standard normal. Under the alternative, $h$ is

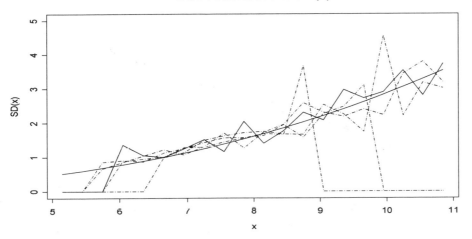

FIGURE 14. Sample diffusion estimates at selected $x$ for the null
and for one alternative type 2.

of differing random values, depending on whether $X_k$ is small, near the centre or
large. This is of the type of effect considered in section 4.1.

This is also studied numerically. A process is simulated under the alternative.
The null process is then fit to these data. The Euler residuals are then obtained.
This is done for sample sizes corresponding to daily observations for 7 years. The
expected heavy tails are seen in the residual plots in Figures 15 and 16. These type
of residual plots came up in one third of the simulation runs. A smaller proportion
of the simulation runs gave relatively normal looking normal QQ plots.

The point of these is that small perturbations in the diffusion function can lead
to QQ plots that look as if a heavy tailed noise process is involved.

## 5. Conclusions

If one fits the correct bond model, for example process (1.1), then the Euler
residuals can and do appear to be Gaussian even for sampling intervals corresponding to daily observations. In some additional numerical work, one needs to take
sampling intervals in the order or 30 or 40 days before the Euler residuals show
definite non Gaussian behavior. Thus one should expect the Euler residuals to
appear to be Gaussian.

Alternative models, with a slightly different diffusion rate function, are considered. The effect of the alternative is to make the Euler residuals appear to have
heavy tails. This makes it very difficult to know if non–Gaussian behavior in the
residuals is due to model misspecification or due to a non Gaussian driving process
in the SDE model.

There are some forms of alternative diffusion rate functions that have little
effect on the Gaussian behavior of the Euler residuals. However if the form is that
a smaller rate appears near the centre of the process, and tails of the rate are similar

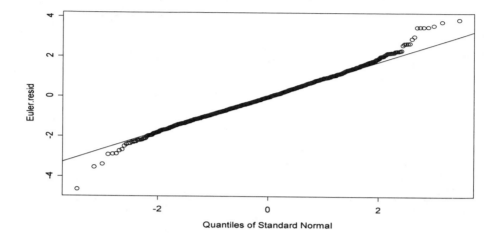

FIGURE 15. Normal QQ plots for Euler residuals under alternative type 1.

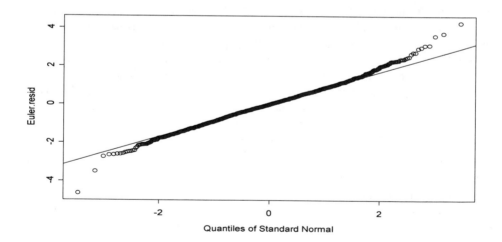

FIGURE 16. Normal QQ Plots for Euler Residuals Under Alternative Type 2

to the null, the effect can be quite big in terms of the Gaussian behavior of the residuals.

Finally we should note that there are many models of bonds. One recent paper by Gallant and Tauchens [4] describes some multifactor models and some estimation methods for these. However the main purpose of this paper is to consider effects of diffusion model misspecification on the Euler residuals. The models in [4] have specific forms of diffusion functions.

**Acknowledgments.** This work has been supported by a grant from the Natural Sciences and Engineering Research Council (NSECRC) of Canada

## References

[1] Basawa, I. V. and B. L. S. Prakasa Rao (1980). Statistical Inference for Stochastic Processes. Academic Press, Toronto.

[2] Bouchard, J.P., Sornette, D. and Potters, M. (1997). Option pricing in the presence of extreme fluctuations. In Proceedings of the Newton Institute session on Mathematical Finance, edited by M. Dempster and S. Pliska. Pages 141-164. Cambridge Unviersity Press.

[3] Duffie, D. (1996). Dynamic Asset Pricing Theory, Second Edition. Princeton University Press.

[4] Gallant, R. and G. Tauchen (1997). Estimation of continuous time models for stock returns and interest rates. Preprint, January 1997, Department of Economics, University of North Carolina, Chapel Hill.

[5] Jiang, G. and J. Knight (1997). A nonparametric approach to the estimation of diffusion processes, with an application to a short–term interest rate model. Econometric Theory, 13, 615–45.

[6] Kloeden, P. and E. Platen (1995). Numerical Solution of Stochastic Differential Equations. Springer, New York.

[7] Matacz, A. (1997). Financial modeling and option theory with the truncated Levy Process. Preprint, Report 97–28, School of Mathematics and Statistics, University of Sydney.

[8] McLeish, D. and A. Kolkiewicz (1997). Fitting diffusion models in finance. in Institute of Mathematical Statistics Lecture Notes, Monograph Series, 32, 327–50.

DEPARTMENT OF STATISTICAL AND ACTUARIAL SCIENCES, THE UNIVERSITY OF WESTERN ONTARIO, LONDON, ONTARIO N6A 5B7, CANADA

*E-mail address*: `rjk@fisher.stats.uwo.ca`

Canadian Mathematical Society
Conference Proceedings
Volume **26**, 2000

# Particle Representations for Measure-Valued Population Processes with Spatially Varying Birth Rates

## Thomas G. Kurtz

ABSTRACT. Representations of measure-valued processes in terms of countable systems of particles are constructed for models with spatially varying birth and death rates. In previous constructions for models with birth and death rates not depending on location or type, the particles were assigned integer-valued "levels", the joint distributions of the particle types were exchangeable, and the measure-valued process $K$ was given by $K(t) = P(t)\bar{Z}(t)$, where $P$ was the "total mass" process and $\bar{Z}(t)$ was the de Finetti measure for the exchangeable particle types at time $t$. In the present construction, particles are assigned real-valued levels and for each time $t$ the joint distribution of locations and levels is conditionally Poisson distributed with mean measure $K(t) \times m$. The representation gives an explicit construction of the boundary measure in Dynkin's probabilistic solution of the nonlinear partial differential equation $\lambda(x)v(x)^{\gamma} - Bv(x) = \rho(x)$, $x \in D$, $v(x) = f(x)$, $x \in \partial D$. The representation also provides a way of generalizing Perkins's models for measure-valued processes in which the individual particle motion depends on the distribution of the population. Questions of uniqueness, however, remain open for most of the models in this larger class.

## 1. Exchangeable population models

We begin by considering a class of finite population models. Let $N(t)$ denote the total population size at time $t$, and let $X(t) = (X_1(t), \dots, X_{N(t)}(t))$ denote the locations of population members in a complete, separable metric space $E$. The state space for the models is then $\hat{E} = \cup_{k=0}^{\infty} E^k$, where $E^0$ denotes the single state in which the population size is zero. If the state $x$ is in $E^k$, we will sometimes write $(x, k)$ to emphasize the length of the vector. We will refer to individual population members as *particles*. The behavior of each particle will depend on the others only through the empirical measure

$$Z(t) = \sum_{i=1}^{N(t)} \delta_{X_i(t)},$$

2000 *Mathematics Subject Classification.* 60J25, 60K35, 60J70, 60J80, 92D10.

*Key words and phrases.* Fleming-Viot process; Dawson-Watanabe process; superprocess; measure-valued diffusion; exchangeability; historical process; nonlinear partial differential equations; conditioning.

Research supported in part by NSF grant DMS 96-26116.

that is, the order of the particles is not significant.

The models we consider will be Markov, specified by their generators, that is, the operators that characterize the processes as solutions of martingale problems. To specify a model, we must define the operator $Af$ for $f$ in an appropriate domain $\mathcal{D}(A)$.

For $k = 1, 2, \ldots$, let $\Gamma_k$ be the collection of all permutations of $\{1, \ldots, k\}$. For $x \in E^k$ and $\sigma \in \Gamma_k$, let $x_\sigma = (x_{\sigma_1}, \ldots, x_{\sigma_k})$. Let $B$ be the generator of a Markov process on $E$. $B$ will determine the individual particle motion. For $1 \leq i \leq k$, $B_i f(x, k)$ will denote the operator $B$ applied to $f$ as a function of $x_i$. For example, for $g_i \in \mathcal{D}(B)$, $i = 1, 2, \ldots$ with $g_i \equiv 1$ for $i$ sufficiently large, define

$$(1.1) \qquad f(x, k) = \prod_{j=1}^{k} g_j(x_j).$$

Then

$$B_i f(x, k) = B g_i(x_i) \prod_{1 \leq j \leq k, j \neq i} g_j(x_j) = \frac{B g_i(x_i)}{g_i(x_i)} f(x, k).$$

Let $\lambda_{-1}(x_i, x)$ denote the "death rate" for the $i$th particle, and for $m \geq 1$, let $\lambda_m(x_i, x)$ be the intensity for a birth event in which the $i$th particle gives birth to $m$ "offspring". We assume that offspring are initially placed at the location of the "parent". For $m = -1$ and $m \geq 1$, we assume that for $x \in E^k$ and $\sigma \in \Gamma_k$, $\lambda_m(x_i, x) = \lambda_m(x_{\sigma_j}, x_\sigma)$, whenever $x_i = x_{\sigma_j}$. In particular, the birth and death rates for a particle depend only on its location and the empirical measure $\sum_{j=1}^{k} \delta_{x_j}$.

In terms of these parameters, we have the generator

$$(1.2) \quad A_0 f(x, k) \;=\; \sum_{i=1}^{k} B_i f(x, k)$$

$$+ \sum_{i=1}^{k} \sum_{m=1}^{\infty} \lambda_m(x_i, x) \frac{1}{\binom{k+m}{m}} \sum_{1 \leq j_1 < \cdots < j_m \leq k+m}$$

$$(f(\theta_{j_1, \ldots, j_m}(x|x_i), k + m) - f(x, k))$$

$$+ \sum_{i=1}^{k} \lambda_{-1}(x_i, x)(f(d_i(x), k - 1) - f(x, k)).$$

For $x \in E^k$, $\theta_{j_1, \ldots, j_m}(x|z)$ is the element $x' \in E^{k+m}$ obtained from $x$ by setting $x'_{j_l} = z$, $l = 1, \ldots, m$, and defining the remaining $k$ components of $x'$ to be the components of $x$, preserving the order, and $d_i(x) = (x_1, \ldots, x_{i-1}, x_{i+1}, \ldots, x_k) \in E^{k-1}$. We take $\mathcal{D}(A_0)$ to be the linear space generated by functions of the form (1.1).

For simplicity, we assume that

$$(1.3) \qquad \sup_k \sup_{x \in E^k} \sum_m m \lambda_m(x_i, x) < \infty.$$

This condition states that the per particle birth rate is uniformly bounded and, in particular, implies that the population size cannot blow up in finite time. If uniqueness holds for the martingale problem of $B$, then this fact implies uniqueness will hold for $A_0$.

The following theorem (Corollary 3.5 from Kurtz (1998)) plays an essential role in our construction. Let $(S, d)$ and $(S_0, d_0)$ be complete, separable metric spaces. An operator $A \subset B(S) \times B(S)$ is *dissipative* if $\|f_1 - f_2 - \epsilon(g_1 - g_2)\| \geq \|f_1 - f_2\|$ for all $(f_1, g_1), (f_2, g_2) \in A$ and $\epsilon > 0$; $A$ is a *pre-generator* if $A$ is dissipative and there are sequences of functions $\mu_n : S \to \mathcal{P}(S)$ and $\lambda_n : S \to [0, \infty)$ such that for each $(f, g) \in A$

$$(1.4) \qquad g(x) = \lim_{n \to \infty} \lambda_n(x) \int_S (f(y) - f(x)) \mu_n(x, dy)$$

for each $x \in S$. $A$ is *graph separable* if there exists a countable subset $\{g_k\} \subset \mathcal{D}(A) \cap \bar{C}(S)$ such that the graph of $A$ is contained in the bounded, pointwise closure of the linear span of $\{(g_k, Ag_k)\}$. (More precisely, we should say that there exists $\{(g_k, h_k)\} \subset A \cap \bar{C}(S) \times B(S)$ such that $A$ is contained in the bounded pointwise closure of $\{(g_k, h_k)\}$, but typically $A$ is single-valued, so we use the more intuitive notation $Ag_k$.) These two conditions are satisfied by essentially all operators $A$ that might reasonably be thought to be generators of Markov processes.

For an $S_0$-valued, measurable process $Y$, $\hat{\mathcal{F}}_t^Y$ will denote the completion of the $\sigma$-algebra $\sigma(\int_0^r h(Y(s))ds, r \leq t, h \in B(S_0))$. For almost every $t$, $Y(t)$ will be $\hat{\mathcal{F}}_t^Y$-measurable, but in general, $\hat{\mathcal{F}}_t^Y$ does not contain $\mathcal{F}_t^Y = \sigma(Y(s) : s \leq t)$. Let $\mathbf{T}^Y = \{t : Y(t) \text{ is } \hat{\mathcal{F}}_t^Y \text{ measurable}\}$. If $Y$ is cadlag and has no fixed points of discontinuity (that is, for every $t$, $Y(t) = Y(t-)$ a.s.), then $\mathbf{T}^Y = [0, \infty)$. $D_S[0, \infty)$ denotes the space of cadlag $S$-valued functions with the Skorohod topology and $M_S[0, \infty)$ denotes the space of Borel measurable functions, $x : [0, \infty) \to S$, topologized by convergence in Lebesgue measure.

THEOREM 1.1. *Let $(S, d)$ and $(S_0, d_0)$ be complete, separable metric spaces. Let $A \subset \bar{C}(S) \times \bar{C}(S)$ be a graph separable, pre-generator, and suppose that $\mathcal{D}(A)$ is closed under multiplication and is separating. Let $\gamma : S \to S_0$ be Borel measurable, and let $\alpha$ be a transition function from $S_0$ into $S$ ($y \in S_0 \to \alpha(y, \cdot) \in \mathcal{P}(S)$ is Borel measurable) satisfying $\int h \circ \gamma(z) \alpha(y, dz) = h(y)$, $y \in S_0$, $h \in B(S_0)$, that is, $\alpha(y, \gamma^{-1}(y)) = 1$. Define*

$$C = \{(\int_S f(z)\alpha(\cdot, dz), \int_S Af(z)\alpha(\cdot, dz)) : f \in \mathcal{D}(A)\}.$$

*Let $\mu_0 \in \mathcal{P}(S_0)$, and define $\nu_0 = \int \alpha(y, \cdot)\mu_0(dy)$.*

a) *If $\tilde{Y}$ is a solution of the martingale problem for $(C, \mu_0)$, then there exists a solution $X$ of the martingale problem for $(A, \nu_0)$ such that $\tilde{Y}$ has the same distribution on $M_{S_0}[0, \infty)$ as $Y = \gamma \circ X$. If $Y$ and $\tilde{Y}$ are cadlag, then $Y$ and $\tilde{Y}$ have the same distribution on $D_{S_0}[0, \infty)$.*

b) *For $t \in \mathbf{T}^Y$,*

$$(1.5) \qquad P\{X(t) \in \Gamma | \hat{\mathcal{F}}_t^Y\} = \alpha(Y(t), \Gamma), \quad \Gamma \in \mathcal{B}(S).$$

c) *If, in addition, uniqueness holds for the martingale problem for $(A, \nu_0)$, then uniqueness holds for the $M_{S_0}[0, \infty)$-martingale problem for $(C, \mu_0)$. If $\tilde{Y}$ has sample paths in $D_{S_0}[0, \infty)$, then uniqueness holds for the $D_{S_0}[0, \infty)$-martingale problem for $(C, \mu_0)$.*

d) *If uniqueness holds for the martingale problem for $(A, \nu_0)$, then $Y$ restricted to $\mathbf{T}^Y$ is a Markov process.*

REMARK 1.2. Theorem 1.1 can be extended to cover a large class of generators whose range contains discontinuous functions. (See Kurtz (1998), Corollary 3.5 and Theorem 2.7.) In particular, suppose $A_1, \ldots, A_m$ satisfy the conditions of Theorem 1.1 for a common domain $\mathcal{D} = \mathcal{D}(A_1) = \cdots = \mathcal{D}(A_m)$ and $\beta_1, \ldots, \beta_m$ are nonnegative functions in $B(S)$. Then the conclusions of Theorem 1.1 hold for

$$Af = \beta_1 A_1 f + \cdots + \beta_m A_m f.$$

**1.1. Example: Empirical measure process.** Let $A = A_0$ defined in (1.2), let $S = \hat{E}$, and let $S_0 = M_c^f(E)$, the space of finite counting measures on $E$. Define $\gamma : S \to S_0$ by

$$\gamma(x, k) = \sum_{i=1}^{k} \delta_{x_i}.$$

Note that each $\mu \in M_c^f(E)$ is of the form $\mu = \sum_{i=1}^{k} \delta_{x_i}$ for some $k$ and $x \in E^k$, and for $\mu$ of this form, define

$$\alpha_0(\mu, \cdot) = \frac{1}{k!} \sum_{\sigma \in E^k} \delta_{x_\sigma}.$$

Define

$$C_0 = \{(\alpha_0 f, \alpha_0 A_0 f) : f \in \mathcal{D}(A_0)\}.$$

We can interpret $\alpha_0 f$ as a function on $M_c^f(E)$ or as a function on $\hat{E}$, $h_f(x, k) = \alpha_0 f(\sum_{i=1}^{k} \delta_{x_i})$, that is symmetric in the sense that $h_f(x, k) = h_f(x_\sigma, k)$ for all $\sigma \in \Gamma_k$. Note that if $f$ is symmetric, then $h_f(x, k) = f(x, k)$ and $\alpha_0 A_0 f(\sum_{i=1}^{k} \delta_{x_i}) = A_0 f(x, k)$. It follows that if $X$ is a solution of the martingale problem for $A_0$, then $\sum_{i=1}^{N(t)} \delta_{X_i(t)}$ is a solution of the martingale problem for $C_0$.

Conversely, if $B \subset \bar{C}(E) \times \bar{C}(E)$ is a graph separable pregenerator and $\mathcal{D}(B)$ is closed under multiplication and separates points and the $\lambda_m$ satisfy (1.3) and are continuous, then $A_0$ satisfies the conditions of Theorem 1.1 and hence any solution of the martingale problem for $C_0$ corresponds to a solution of the martingale problem for $A_0$. Consequently, the two martingale problems are essentially equivalent. (Note that there are variations of Theorem 1.1 that apply under less restrictive conditions. See Kurtz (1998).)

## 2. Marked population models

Next, we introduce a family of "marked" population processes. $F$ will denote the space of marks, so the new state space $S$ will be a subset of $\mathcal{M}_c^f(E \times F)$. In all of the examples $F \subset [0, \infty)$, and with order in mind, we will refer to the marks as "levels".

With reference to Theorem 1.1, let $S_0 = \mathcal{M}_c^f(E)$, and let $\gamma$ be defined by $\gamma(\xi) = \xi(\cdot \times F)$, $\xi \in \mathcal{M}_c^f(E \times F)$. For each $\mu \in \mathcal{M}_c^f(E)$, let $\hat{\alpha}(\mu, \cdot)$ be an exchangeable distribution on $F^{\mu(E)}$. Let $\mu = \sum_{i=1}^{k} \delta_{x_i}$, and define $\alpha_1(\mu, \cdot) \in \mathcal{P}(E \times F)$ by

$$\alpha_1(\mu, G) = \hat{\alpha}(\mu, \{u \in F^k : \sum_{i=1}^{k} \delta_{(x_i, u_i)} \in G\}), \quad G \in \mathcal{B}(\mathcal{M}_c^f(E \times F)).$$

Let $A_1 \subset \bar{C}(S) \times \bar{C}(S)$ and define

$$C_1 = \{(\alpha_1 f, \alpha_1 A_1 f) : f \in \mathcal{D}(A_1)\}.$$

Assuming that $A_1 \subset \bar{C}(S) \times \bar{C}(S)$ is a graph separable, pre-generator, and that $\mathcal{D}(A_1)$ is closed under multiplication and is separating, then Theorem 1.1 applies.

**2.1. Example: Neutral model.** Let $F = \{1, 2, \ldots\}$, and for $\mu = \sum_{i=1}^{k} \delta_{x_i}$, let $\hat{\alpha}(\mu, \cdot)$ satisfy $\hat{\alpha}(\mu, \Gamma_k) = 1$, that is, $\hat{\alpha}(\mu, \cdot)$ is just the distribution of a random permutation. To simplify notation, we identify $\mu$ with $x$ as above and define the generator in terms of functions of $(x, u)$. If we ensure that all functions involved have the property that $h(x, u) = h(x_\sigma, u_\sigma)$ for all permutations $\sigma$, the functions will depend only on the measures $\sum_{i=1}^{k} \delta_{(x_i, u_i)}$.

If $u = (u_1, \ldots, u_k)$ is a permutation of $(1, \ldots, k)$ and $1 \le l < m \le k+1$, define

$$r_i^{lm}(u) = \begin{cases} u_i & u_i < m \\ u_i + 1 & u_i \ge m. \end{cases}$$

$\eta_{lm}(u) = j$ if $u_j = l$, and $\eta_0(u) = j$ where $j$ is the unique index such that $u_j = k$.

Assume that $B \subset \bar{C}(B) \times \bar{C}(E)$, that $\mathcal{D}(B)$ is an algebra (that is, a linear subspace that is closed under multiplication) that separates points and contains the constant functions, and that the martingale problem for $B$ is well posed. Let $\mathcal{D}(A_2)$ be the collection of functions of the form

$$f(x, u, k) = \prod_{i=1}^{k} g(x_i, u_i),$$

where $g(\cdot, j) \in \mathcal{D}(B)$ and for some $j_g \ge 0$, $g(\cdot, j) \equiv 1$ for $j > j_g$. Assume that $\lambda_m : \cup_k E^k \to [0, \infty)$, $m = -1, 1$, satisfy $\lambda_m(x) = \lambda_m(x_\sigma)$ and $\sup_x \lambda_m(x) < \infty$. Define

$$(2.1) \quad A_2 f(x, u, k)$$

$$= \sum_{i=1}^{k} f(x, u, k) \frac{Bg(x_i, u_i)}{g(x_i, u_i)}$$

$$+ \frac{2\lambda_1(x)}{k+1} \sum_{1 \le l < m \le k+1} (g(x_{\eta_{lm}(u)}, m) \prod_{i=1}^{k} g(x_i, r_i^{lm}(u)) - f(x, u, k))$$

$$+ \lambda_{-1}(x) k f(x, u, k)) (\frac{1}{g(x_{\eta_0(u)}, k)} - 1).$$

Note that when there is a "death", it is the particle with the highest level that is eliminated. When there is a "birth", particles with lower levels are more likely to become parents.

If $\alpha(x, k, du) = \frac{1}{k!} \sum_{\sigma \in \Gamma_k} \delta_\sigma(du)$, then defining

$$A_0 f(x, k) = \sum_{i=1}^{k} B_i f(x, k)$$

$$+ \lambda_1(x) \sum_{i=1}^{k} \frac{1}{k+1} \sum_{j=1}^{k+1} (f(\theta_j(x|x_i), k+1) - f(x, k))$$

$$+ \lambda_{-1}(x) \sum_{i=1}^{k} (f(d_i(x), k-1) - f(x, k)),$$

we have $\alpha A_2 f = A_0 \alpha f$. Note that $A_0$ is a special case of $A_0$ defined in (1.2) in Section 1. Here $A_0$ is the generator for a model that is *neutral* in the sense that the birth and death rates are the same for all particles regardless of location.

Let $(X, U)$ be a solution of the martingale problem for $A_2$. Suppose that $U(0)$ is independent of $X(0)$ and is uniformly distributed over all permutations of $(1, \dots, N(0))$. Let $\mu_0$ denote the distribution of $X(0)$. Then, by Theorem 1.1, $X$ is a solution of the martingale problem for $(A_0, \mu_0)$.

Assume that $\lambda_1$ and $\lambda_{-1}$ are bounded and continuous on $S$. It follows that the martingale problem for $A_2$ is well-posed. For $f \in \mathcal{D}(A_2)$, define $\hat{f}(x, k) = f(x_1, \dots, x_k, 1, \dots, k, k)$ and set

$$
\begin{aligned}
A_3 \hat{f}(x, k) \;=\; & \sum_{i=1}^{k} B_i \hat{f}(x, k) \\
& + \lambda_1(x) \sum_{1 \leq l < m \leq k+1} (\hat{f}(\theta_m(x|x_l), k+1) - \hat{f}(x, k)) \\
& + \lambda_{-1}(x)(\hat{f}(d_k(x), k-1)) - \hat{f}(x, u)),
\end{aligned}
$$

where for $1 \leq l < m \leq k+1$ and $x \in E^k$, $x' = \theta_m(x|x_l) \in E^{k+1}$ is given by

$$
x'_i = \begin{cases}
x_i & i < m \\
x_{i-1} & m < i \leq k+1 \\
x_l & i = m.
\end{cases}
$$

Let $(X, U)$ be as above. Let $V_i(t) = j$ if $U_j(t) = i$, that is, $U_{V_i(t)}(t) = i$, and define $(Y_1(t), \dots, Y_{N(t)}(t)) = (X_{V_1(t)}(t), \dots, X_{V_{N(t)}(t)}(t))$. Then $Y$ is a solution of the martingale problem for $A_3$.

Define $\gamma : E^k \to \mathcal{M}(E)$, by $\gamma(x) = \sum_{i=1}^{k} \delta_{x_i}$, and for $\mu = \sum_{i=1}^{k} \delta_{x_i}$, define $\alpha_0(\mu, dx) = \frac{1}{k!} \sum_{\sigma \in \Gamma_k} \delta_{x_\sigma}$. Then for $f \in \mathcal{D}(A_0) = \mathcal{D}(A_3)$,

$$
\alpha_0 A_3 f = \alpha_0 A_0 f.
$$

Let $C = \{(\alpha_0 f, \alpha_0 A_0 f) : f \in \mathcal{D}(A_0)\} = \{(\alpha_0 f, \alpha_0 A_3 f) : f \in \mathcal{D}(A_3)\}$. By the discussion in Section 1, if $X^0$ is a solution of the martingale problem for $A_0$, then $\gamma(X^0)$ is a solution of the martingale problem for $C$. But by Theorem 1.1, any solution of the martingale problem for $C$ corresponds to a solution of the martingale problem for $A_3$. Consequently, for each solution $X^0$ of the martingale problem for $A_0$, there exists a solution $X$ of the martingale problem for $A_3$ such that $\gamma(X^0)$ and $\gamma(X)$ have the same distribution. The process corresponding to $A_3$ is a special case of Model II of [**2**].

## 3. Models with location/type dependent birth and death rates

### 3.1. Critical models. Let $F = [0, n]$. Define

$$(3.1) \quad A^n f(x, u, k)$$

$$= \sum_{i=1}^{k} B_i f(x, u, k)$$

$$+ \sum_{i=1}^{k} 2\lambda(x_i, x) \frac{1}{k+1} \sum_{j=1}^{k+1} \int_{u_i}^{n} (f(\theta_j(x, u|x_i, v), k+1) - f(x, u, k)) dv$$

$$+ \sum_{i=1}^{k} 2\lambda(x_i, x) u_i (f(d_i(x, u), k-1) - f(x, u, k))$$

for $f$ in an appropriate domain. Assume that for $x \in E^k$, and $\sigma \in \Gamma_k$, $k = 1, 2, \ldots$ and $i$ and $j$ such that $x_i = x_{\sigma_j}$, $\lambda(x_i, x) = \lambda(x_{\sigma_j}, x_\sigma)$.

If $\alpha^n(x, k, du) = n^{-k} du_1 \cdots du_k$, then $\alpha^n A^n f = A_0^n \alpha^n f$, where

$$A_0^n f(x, k)$$

$$= \sum_{i=1}^{k} B_i f(x, k)$$

$$+ \sum_{i=1}^{k} n\lambda(x_i, x) \frac{1}{k+1} \sum_{j=1}^{k+1} (f(\theta_j(x|x_i), k+1) - f(x, k))$$

$$+ \sum_{i=1}^{k} n\lambda(x_i, x)(f(d_i(x), k-1) - f(x, k)).$$

Here, $A_0^n$ is again a special case of (1.2). In particular, if $\lambda(x_i, x) \equiv \lambda(x_i)$, then $A_0^n$ is the generator of a critical branching Markov process and the scaling in $n$ is such that a sequence of solutions $X^n$ should satisfy

$$(3.2) \qquad\qquad Z_n = \frac{1}{n} \sum_{i=1}^{N_n(t)} \delta_{X_i^n} \Rightarrow Z,$$

where $Z$ is a Dawson-Watanabe process.

In the remainder of the paper, we concentrate on the Dawson-Watanabe setting, that is, we assume that $\lambda(x_i, x) = \lambda(x_i)$. We will see that this assumption makes existence and uniqueness for the limiting model easy. The relationship between $A^n$ and $A_0^n$, however, insures that the particle representation will be valid for more general models.

With (3.2) in mind, consider (3.1) as $n \to \infty$. To be specific, let $\mathcal{D}(A)$ be the collection of functions of the form

$$f(x, u, k) = \prod_{i=1}^{k} g(x_i, u_i),$$

where $0 < g \leq 1$ is bounded away from zero and there exists $u_g$ such that $g(x_i, u_i) = 1$ if $u_i > u_g$.

If $n > u_g$, then $A^n$ becomes

$$(3.3) \qquad Af(x,u) = \sum_{u_i < u_g} f(x,u) \frac{Bg(x_i, u_i)}{g(x_i, u_i)}$$

$$+ \sum_{u_i < u_g} 2\lambda(x_i) \int_{u_i}^{u_g} f(x,u)(g(x_i, v) - 1) dv$$

$$+ \sum_{u_i < u_g} 2\lambda(x_i) u_i f(x,u) (\frac{1}{g(x_i, u_i)} - 1),$$

and the convergence of $A^n$ is immediate. Assuming that the martingale problem for $B$ is well-posed and (for simplicity) $\lambda$ is bounded and continuous, the martingale problem for $A$ is well-posed. We identify the process with the counting measure

$$\Psi(t) = \sum_i \delta_{(X_i(t), U_i(t))}.$$

Define $\gamma : \mathcal{M}_c(E \times [0,\infty)) \to \mathcal{M}^f(E)$ by

$$\gamma(x,u) = \lim_{r \to \infty} \frac{1}{r} \sum_i I_{[0,r]}(u_i) \delta_{x_i}$$

if the limit exists. Set

$$K(t) = \lim_{r \to \infty} \frac{1}{r} \sum_i I_{[0,r]}(U_i(t)) \delta_{X_i(t)} = \gamma(X(t), U(t)).$$

If $\Psi(0)$ is a Poisson random measure with mean measure $K(0) \times m$, then we claim that, conditioned on $\mathcal{F}_t^K \equiv \sigma(K(s), s \le t)$, $\Psi(t)$ is a Poisson random measure with mean measure $K(t) \times m$.

Let $(S, \mathcal{S})$ be a measurable space, and let $\nu$ be a $\sigma$-finite measure on $\mathcal{S}$. We need the following facts about a *Poisson random measure*, $\xi$, with *mean measure* $\nu$:

a) $\xi$ is a random counting measure on $S$.
b) For each $A \in \mathcal{S}$ with $\nu(A) < \infty$, $\xi(A)$ is Poisson distributed with parameter $\nu(A)$.
c) For $A_1, A_2, \ldots \in \mathcal{S}$ disjoint, $\xi(A_1), \xi(A_2), \ldots$ are independent.

If $\xi$ is a Poisson random measure with mean measure $\nu$

$$E[e^{\int f(z)\xi(dz)}] = e^{\int (e^f - 1)\delta\nu},$$

or letting $\xi = \sum_i \delta_{V_i}$,

$$E[\prod_i g(V_i)] = e^{\int (g-1)d\nu}.$$

Similarly,

$$E[\sum_j h(V_j) \prod_i g(V_i)] = \int hg d\nu e^{\int (g-1)d\nu}.$$

We define $\alpha$ so that if $\mu \in \mathcal{M}^f(E)$, then $\alpha(\mu, \cdot) \in \mathcal{P}(\mathcal{M}_c(E \times [0,\infty))$ is the distribution of a Poisson random measure on $E \times [0, \infty)$ with mean measure $\mu \times m$, that is

$$\int_{\mathcal{M}_c(E \times [0,\infty))} e^{\langle f, z \rangle} \alpha(\mu, dz) = e^{\int (e^{f(x,u)} - 1)\mu(dx)du}.$$

Therefore, if $f(x, u) = \prod g(x_i, u_i)$, then

(3.4) $$\alpha f(\mu) = e^{\int_E \int_0^\infty (g(x,u)-1)du\, \mu(dx)}$$

and

$$\alpha A f(\mu)$$

$$= \alpha f(\mu) \int_E B \int_0^\infty (g(x, u) - 1)du\, \mu(dx)$$

$$+ \alpha f(\mu) \int_E 2\lambda(x) \int_0^\infty \int_u^\infty g(x, u)(g(x, v) - 1)dv\, du\, \mu(dx)$$

$$+ \alpha f(\mu) \int_E 2\lambda(x) \int_0^\infty u(1 - g(x, u))du\, \mu(dx)$$

$$= \alpha f(\mu) \int_E B \int_0^\infty (g(x, u) - 1)du\, \mu(dx)$$

$$+ \alpha f(\mu) \int_E 2\lambda(x) \int_0^\infty \int_u^\infty (g(x, u) - 1)(g(x, v) - 1)dv\, du\, \mu(dx)$$

$$= \alpha f(\mu) \int_E B \int_0^\infty (g(x, u) - 1)du\, \mu(dx)$$

$$+ \alpha f(\mu) \int_E \lambda(x) \left( \int_0^\infty (g(x, u) - 1)du \right)^2 \mu(dx).$$

Consequently, for $f(\mu) = e^{-\langle h, \mu \rangle}$ and $C = \{(\alpha f, \alpha A f) : f \in \mathcal{D}(A)\}$,

(3.5) $$C f(\mu) = f(\mu)\langle -Bh + \lambda h^2, \mu \rangle.$$

But $C$ of this form is the generator for a Dawson-Watanabe process. (See, for example, [**5**], Section 9.4.3.) Since, as defined, $Af$ need not be a bounded function, Theorem 1.1 does not immediately apply; however, Theorem 1.1 can be extended to cover certain operators whose range includes unbounded functions and this extension would apply in the current setting. Alternatively, we could take $F$ to be the space consisting of copies of the closed intervals $[k, k + 1]$, $k = 0, 1, \ldots$, that is, $F$ includes two copies of each integer. Writing $F = \uplus_{k=0}^\infty [k, k + 1]$, assume that the domain of $A$ consists of functions of the form

$$f(x, u) = \prod_i g(x_i, u_i),$$

where $0 \le g(x_i, u_i) \le \rho_g < 1$ for $u_i \in \uplus_{k=0}^{k_g} [k, k + 1]$ and $g(x_i, u_i) = 1$ for $u_i \in \uplus_{k=k_g+1}^\infty [k, k + 1]$. Then, under the assumption that $\lambda$ is bounded, $Af$ is bounded. In any case, we have the following theorem.

THEOREM 3.1. *Suppose that $\tilde{K}$ is a solution of the martingale problem for $C$ given by (3.5), and let $\nu_0 = E[\alpha(Z(0), \cdot)]$. Then there exists a solution*

$$\Psi(t) = \sum_i \delta_{(X_i(t), U_i(t))}$$

*of the martingale problem for $(A, \nu_0)$ such that $K$ defined by*

$$K(t) = \lim_{r \to \infty} \frac{1}{r} \sum_i I_{[0,r]}(U_i(t))\delta_{X_i(t)}$$

*has the same distribution as $\tilde{K}$.*

**3.2. Subcritical models.** The particle construction above can be extended easily to subcritical models of the form

$$A_0^n f(x, k)$$

$$= \sum_{i=1}^{k} B_i f(x, k)$$

$$+ \sum_{i=1}^{k} n\lambda(x_i)(f((x, x_i), k+1) - f(x, k))$$

$$+ \sum_{i=1}^{k} n(\lambda(x_i) + \frac{1}{n}\lambda_0(x_i))(f(\gamma_i(x), k-1) - f(x, k)).$$

For $f(x, u) = \prod g(x_i, u_i)$, the limit for the corresponding marked model is given by

$$Af(x, u)$$

$$= \sum_{u_i < u_g} f(x, u) \frac{Bg(x_i, u_i)}{g(x_i, u_i)}$$

$$+ \sum_{u_i < u_g} 2\lambda(x_i) \int_{u_i}^{u_g} f(x, u)(g(x_i, v) - 1)dv$$

$$+ \sum_{u_i < u_g} (2\lambda(x_i)u_i + \lambda_0(x_i))f(x, u)(\frac{1}{g(x_i, u_i)} - 1),$$

and we have $\alpha f(\mu) = e^{\langle \int_0^\infty (g(\cdot, u) - 1)du, \mu \rangle}$ and

$$\alpha Af(\mu) = \alpha f(\mu) \Big[ \int_E B \int_0^\infty (g(x, u) - 1)du\mu(dx)$$

$$+ \int_E \lambda(x) \left( \int_0^\infty (g(x, u) - 1)du \right)^2 \mu(dx)$$

$$+ \int_E \int_0^\infty \lambda_0(x)(1 - g(x, u))du\mu(dx) \Big].$$

Taking $h = \int_0^\infty (1 - g(\cdot, u))du$ in this formula, for $f(\mu) = e^{-\langle h, \mu \rangle}$, we have

$$Cf(\mu) = f(\mu)\langle -Bh + \lambda h^2 + \lambda_0 h, \mu \rangle.$$

**3.3. Ordered model.** The indexing of the above particle models has no significance. If we order the particles according to increasing level, the generator becomes

$$Af(x, u)$$

$$= \sum_i B_i f(x, u)$$

$$+ \sum_i 2\lambda(x_i) \sum_{j=i}^{\infty} \int_{u_j}^{u_{j+1}} (f(\theta_j(x, u | x_i, v)) - f(x, v))dv$$

$$+ \sum_i (2\lambda(x_i)u_i + \lambda_0(x_i))(f(d_i(x, u)) - f(x, u)).$$

For this ordering, $P(t) = |K(t)|$ is given by

$$P(t) = \lim_{m \to \infty} \frac{m}{u_m},$$

and

$$K(t) = \lim_{m \to \infty} \frac{1}{u_m} \sum_{i=1}^{m} \delta_{X_i(t)}.$$

Assume that $B$ is the generator for a diffusion process satisfying an Itô equation

$$X(t) = X(0) + \int_0^t \sigma(X(s))ds + \int_0^t b(X(s))ds.$$

Then we can write a system of equations for the particle model.

$$
\begin{aligned}
X_k(t) = X_k(0) &+ \int_0^t \sigma(X_k(s))dW_k(s) + \int_0^t b(X_k(s))ds \\
&+ \sum_{1 \le i < j < k} (X_{k-1}(s-) - X_k(s-))dL_{ij}^b(s) \\
&+ \sum_{i<k} (X_i(s-) - X_k(s-))dL_{ik}^b(s) \\
&+ \sum_{j<k} (X_{k+1}(s-) - X_k(s-))dL_j^d(s),
\end{aligned}
$$

$$
\begin{aligned}
U_k(t) = \; &U_k(0) + \sum_{1 \le i < j < k} (U_{k-1}(s-) - U_k(s-))dL_{ij}^b(s) \\
&+ \sum_{i<k} \int_{[0,\infty) \times [0,\infty) \times [0,t]} (u - U_k(s-))I_{[U_{k-1}(s-), U_k(s-))}(u) \\
&\qquad\qquad I_{[0,2\lambda(X_i(s-)))}(v)N_i(du \times dv \times ds) \\
&+ \sum_{j \le k} (U_{k+1}(s-) - U_k(s-))dL_j^d(s),
\end{aligned}
$$

and

$$
\begin{aligned}
L_{ij}^b(t) = \; &\int_{[0,\infty) \times [0,\infty) \times [0,t]} I_{[U_{j-1}(s-), U_j(s-))}(u) \\
&\qquad I_{[0,2\lambda(X_i(s-)))}(v)N_i(du \times dv \times ds) \\
L_i^d(t) = \; &\int_{[0,\infty) \times [0,\infty) \times [0,t]} I_{[0,U_i(s-))}(u) \\
&\qquad I_{[0,2\lambda(X_i(s-)))}(v)N_i(du \times dv \times ds) \\
&+ \int_{[0,\infty) \times [0,t]} I_{[0,\lambda_0(X_i(s))]}(v)N_i^0(dv \times ds).
\end{aligned}
$$

## 3.4. Model with population dependent motion and birth and death rates.
In the system of equations above it is simple to introduce dependence on

the total mass distribution $K$ in the motion and birth and death rates.

$$
\begin{aligned}
X_k(t) \;=\;& X_k(0) + \int_0^t \sigma(X_k(s), K(s))dW_k(s) \\
&+ \int_0^t b(X_k(s), K(s))ds \\
&+ \sum_{1 \le i < j < k} (X_{k-1}(s-) - X_k(s-))dL_{ij}^b(s) \\
&+ \sum_{i < k} (X_i(s-) - X_k(s-))dL_{ik}^b(s) \\
&+ \sum_{j < k} (X_{k+1}(s-) - X_k(s-))dL_j^d(s),
\end{aligned}
$$

$$
\begin{aligned}
U_k(t) \;=\;& U_k(0) + \sum_{1 \le i < j < k} (U_{k-1}(s-) - U_k(s-))dL_{ij}^b(s) \\
&+ \sum_{i < k} \int_{[0,\infty) \times [0,\infty) \times [0,t]} (u - U_k(s-))I_{[U_{k-1}(s-), U_k(s-))}(u) \\
&\qquad\qquad I_{[0, 2\lambda(X_i(s-), K(s-)))}(v) N_i(du \times dv \times ds) \\
&+ \sum_{j \le k} (U_{k+1}(s-) - U_k(s-))dL_j^d(s),
\end{aligned}
$$

and

$$
\begin{aligned}
L_{ij}^b(t) \;=\;& \int_{[0,\infty) \times [0,\infty) \times [0,t]} I_{[U_{j-1}(s-), U_j(s-))}(u) \\
&\qquad I_{[0, 2\lambda(X_i(s-), K(s-)))}(v) N_i(du \times dv \times ds) \\
L_i^d(t) \;=\;& \int_{[0,\infty) \times [0,\infty) \times [0,t]} I_{[0, U_i(s-))}(u) \\
&\qquad I_{[0, 2\lambda(X_i(s-), K(s-)))}(v) N_i(du \times dv \times ds) \\
&+ \int_{[0,\infty) \times [0,t]} I_{[0, \lambda_0(X_i(s), K(s-))]}(v) N_i^0(dv \times ds).
\end{aligned}
$$

The generator for $K$ becomes

$$
(3.6) \qquad Cf(\mu) = f(\mu)\langle -B(K)h + \lambda(\cdot, K)h^2 + \lambda_0(\cdot, K)h, \mu \rangle,
$$

for $f(\mu) = e^{-\langle h, \mu \rangle}$, where

$$
B(\mu)h(z) = \frac{1}{2} \sum_{ij} a_{ij}(z, \mu) \frac{\partial^2}{\partial z_i \partial z_j} h(z) + \sum_i b_i(z, \mu) \frac{\partial}{\partial z_i} h(z)
$$

for $a(z, \mu) = \sigma(z, \mu)\sigma(z, \mu)^T$.

In the case $\lambda$ and $\lambda_0$ constant, models with generators of the form (3.6) were introduced by Perkins [7]. In this setting, the analog of the system was given in Donnelly and Kurtz [2]. For $\lambda$ and $\lambda_0$ constant, uniqueness of the above system can be proved under Lipshitz assumptions on $\sigma$ and $b$. If $\lambda$ and $\lambda_0$ depend on $z$ and/or $\mu$, uniqueness is open.

## 4. Models with simultaneous births

The models above can be extended to allow for multiple simultaneous births. In particular, for $f$ of the form

$$f(x, u, k) = \prod_{i=1}^{k} g(x_i, u_i),$$

let

$$
\begin{aligned}
A^n f(x, u) &= \sum_i f(x, u) \frac{Bg(x_i, u_i)}{g(x_i, u_i)} \\
&\quad + f(x, u) \sum_i \sum_{k=1}^{\infty} (k+1) \lambda_k^n(x_i) n^{-k} \\
&\qquad\qquad \int_{[u_i, n]^k} (\prod_{l=1}^{k} g(x_i, v_l) - 1) dv_1 \cdots dv_k \\
&\quad + f(x, u) \sum_i \sum_{k=1}^{\infty} (k+1) \lambda_k^n(x_i)(1 - (1 - \frac{u_i}{n})^k)(\frac{1}{g(x_i, u_i)} - 1).
\end{aligned}
$$

Note that if $\lambda_1^n(x) = n\lambda_1(x)$ and $\lambda_k^n \equiv 0$ for $k > 1$, then $A^n$ coincides with (3.3). If $\alpha^n(x, k, du) = n^{-k} du_1 \cdots du_k$, then $\alpha^n A^n f = A_0^n \alpha^n f$, where for $f(x, k) = \prod_{i=1}^{k} g(x_i)$,

$$
\begin{aligned}
A_0^n f(x, k) &= \sum_{i=1}^{k} B_i f(x, k) \\
&\quad + f(x, u) \sum_i \sum_{k=1}^{\infty} \lambda_k^n(x_i)(g(x_i)^k - 1) \\
&\quad + f(x, u) \sum_i \sum_{k=1}^{\infty} \lambda_k^n(x_i) k(\frac{1}{g(x_i)} - 1).
\end{aligned}
$$

We see that $\lambda_k^n(x_i)$ is the intensity for the birth of $k$ offspring for a particle located at $x_i$, and setting

$$\lambda_{-1}^n(x_i) = \sum_{k=1}^{\infty} k \lambda_k^n(x_i),$$

$\lambda_{-1}^n(x_i)$ is the death rate that makes the process critical.

Assume that for $u_i > u_g$, $g(x_i, u_i) = 1$, and define $h(x_i, u_i) = \int_{u_i}^{u_g} (1 - g(x_i, v)) dv$. Then

$$\int_{[u_i, n]^k} (\prod_{l=1}^{k} g(x_i, v_l) - 1) dv_1 \cdots dv_k = (n - u_i - h(x_i, u_i))^k - (n - u_i)^k$$

and

$$A^n f(x, u)$$

$$= \sum_i f(x, u) \frac{Bg(x_i, u_i)}{g(x_i, u_i)}$$

$$+ f(x, u) \sum_i \sum_{k=1}^{\infty} (k+1) \lambda_k^n(x_i) \left( (1 - \frac{u_i + h(x_i, u_i)}{n})^k - (1 - \frac{u_i}{n})^k \right)$$

$$+ f(x, u) \sum_i \sum_{k=1}^{\infty} (k+1) \lambda_k^n(x_i)(1 - (1 - \frac{u_i}{n})^k)(\frac{1}{g(x_i, u_i)} - 1).$$

Consequently, if

$$(4.1) \qquad \lim_{n \to \infty} \sum_{k=1}^{\infty} (k+1) \lambda_k^n(x_i)(1 - (1 - \frac{u_i}{n})^k) = \Lambda(x_i, u_i),$$

$A^n f \to A f$ given by

$$Af(x, u) = \sum_i f(x, u) \frac{Bg(x_i, u_i)}{g(x_i, u_i)}$$

$$+ f(x, u) \sum_i (\Lambda(x_i, u_i) - \Lambda(x_i, u_i + h(x_i, u_i)))$$

$$+ f(x, u) \sum_i \Lambda(x_i, u_i)(\frac{1}{g(x_i, u_i)} - 1).$$

We assume that the convergence in (4.1) is uniform in $x_i \in E$ and in $u_i$ on bounded intervals. Assumption (4.1) is essentially equivalent to (9.4.36) of Ethier and Kurtz (1986).

Let $\Lambda^n(x_i, u_i)$, $n = 1, 2, \ldots$ denote the sequence on the left of (4.1), and observe that

$$\frac{\partial^m}{\partial u_i^m} \Lambda^n(x_i, u_i) = (-1)^{m+1} \sum_{k=m}^{\infty} \lambda_k^n(x_i) \frac{(k+1)k \cdots (k-m+1)}{n^m} (1 - \frac{u_i}{n})^{k-m}.$$

The fact that the derivatives alternate in sign and decrease in absolute value implies that

$$\lim_{n \to \infty} \frac{\partial^m}{\partial u_i^m} \Lambda^n(x_i, u_i) = \frac{\partial^m}{\partial u_i^m} \Lambda(x_i, u_i) \equiv \partial^m \Lambda(x_i, u_i)$$

for each $m$, where the convergence is uniform in $u_i$ on bounded intervals that are bounded away from $u_i = 0$. It also follows that $\partial^1 \Lambda(x_i, \cdot)$ is completely monotone and hence must be of the form

$$\partial^1 \Lambda(x_i, v) = \int_0^{\infty} e^{-vz} \hat{\nu}(x_i, dz).$$

Writing $\hat{\nu}(x_i, \cdot) = \lambda(x_i)\delta_0 + \nu(x_i, \cdot)$ where the support of $\nu(x_i, \cdot)$ is in $(0, \infty)$, we have

$$(4.2) \qquad \Lambda(x_i, v) = \lambda(x_i)v + \int_0^{\infty} z^{-1}(1 - e^{-vz})\nu(x_i, dz).$$

Since $\Lambda(x_i, v) < \infty$, we must have

$$\int_0^\infty \frac{1}{1 \vee z} \nu(x_i, dz) < \infty.$$

In terms of $\nu$,

$$\begin{aligned}
Af(x, u) &= \sum_i f(x, u) \frac{Bg(x_i, u_i)}{g(x_i, u_i)} \\
&\quad + f(x, u) \sum_i \lambda(x_i) \int_{u_i}^\infty (g(x_i, v) - 1) dv \\
&\quad + f(x, u) \sum_i \int_0^\infty \left( e^{z \int_{u_i}^\infty (g(x_i, v) - 1) dv} - 1 \right) z^{-1} e^{-u_i z} \nu(x_i, dz) \\
&\quad + f(x, u) \sum_i \Lambda(x_i, u_i)(\frac{1}{g(x_i, u_i)} - 1).
\end{aligned}$$

The fourth term on the right indicates that at rate $\Lambda(x_i, u_i)$, a particle at location $x_i$ and level $u_i$ dies. The second term on the right corresponds to single births. For $u_i \leq a < b$, at rate $\lambda(x_i)(b - a)$, a particle at location $x_i$ and level $u_i$ gives birth to a single particle with level in the interval $(a, b]$. The third term corresponds to multiple births. When such a birth occurs to the particle at level $u_i$, a positive random variable $\zeta$ is generated, and the levels of the offspring form a Poisson process on $[u_i, \infty)$ with intensity $\zeta$. To be precise, suppose that a particle with level $u_i$ lives from time $\tau_i^b$ until time $\tau_i^d$ and that $X_i(t)$ gives the location of the particle for $\tau_i^b \leq t < \tau_i^d$. Then $\nu$ and $X_i$ determine a point process $\xi_i$ on $[\tau_i^b, \tau_i^d) \times [0, \infty)$ through the requirement that

$$\xi_i((\tau_i^b, t] \times G) - \int_{\tau_i^b}^t \int_G z^{-1} e^{-u_i z} \nu(X_i(s), dz) ds, \quad \tau_i^b \leq t < \tau_i^d,$$

is a martingale for each $G \in \mathcal{B}(E)$. Writing

$$\xi_i = \sum_k \delta_{(S_k, \zeta_k)},$$

at time $S_k$, there is a birth event in which new particles are created whose levels form a Poisson process with intensity $\zeta_k$ on $[u_i, \infty)$. Note that

$$\int_0^\infty z^{-1} e^{-u_i z} \nu(x_i, dz)$$

may be infinite, so that a particle may have infinitely many such birth events in a finite amount of time; however, during a finite time interval, only finitely many births will have levels in a bounded interval. In particular, let $u_i \leq a < b$. Noting that for a Poisson process with intensity $z$, $z(b-a)$ is the expected number of points in the interval $(a, b]$,

$$\lambda(x_i)(b - a) + \int_0^\infty z(b - a) z^{-1} e^{-u_i z} \nu(x_i, dz) = (b - a)\partial^1 \Lambda(x_i, u_i) < \infty,$$

is the expected number of births with levels in the interval $(a, b]$ per unit time occuring to a parent at level $u_i$ and location $x_i$.

**4.1. Example: Offspring distribution with finite variance.** Suppose $\lambda_k^n(x) = n\lambda_k(x)$ and

$$\lambda(x) = \sum_{k=1}^{\infty}(k+1)k\lambda_k(x) < \infty.$$

Then

$$\Lambda(x_i, u_i) = \sum_{k=1}^{\infty}(k+1)k\lambda_k(x_i)u_i,$$

and

$$
\begin{aligned}
Af(x,u) \;=\; & \sum_i f(x,u)\frac{Bg(x_i,u_i)}{g(x_i,u_i)} \\
& + f(x,u)\sum_{u_i < u_g}\lambda(x_i)\int_{[u_i,u_g]}(g(x_i,v)-1)dv \\
& + f(x,u)\lambda(x_i)u_i\Big(\frac{1}{g(x_i,u_i)}-1\Big),
\end{aligned}
$$

which is essentially the same as (3.3).

**4.2. Example: Offspring distribution in domain of attraction of stable law.** For $1 < \beta < 2$, let

$$\lambda_k^n(x_i) = \frac{n^{\beta-1}\lambda(x)}{(k+1)^{\beta+1}}.$$

Then

$$\Lambda^n(x_i, u_i) = n^{\beta-1}\sum_{k=1}^{\infty}\frac{\lambda(x_i)}{(k+1)^\beta}\Big(1 - \Big(1 - \frac{u_i}{n}\Big)^k\Big) \to \lambda(x_i)\int_0^{\infty}z^{-\beta}(1 - e^{-u_i z})dz$$

which gives

$$\Lambda(x_i, u_i) = \lambda(x_i)u_i^{\beta-1}\frac{\Gamma(2-\beta)}{\beta-1},$$

$\nu(x_i, dz) = \lambda(x_i)z^{-(\beta-1)}dz$, and

$$
\begin{aligned}
Af(x,u) \;=\; & \sum_i f(x,u)\frac{Bg(x_i,u_i)}{g(x_i,u_i)} \\
& + f(x,u)\frac{\Gamma(2-\beta)}{\beta-1}\sum_i \lambda(x_i)(u_i^{\beta-1} - (u_i + h(x_i,u_i))^{\beta-1}) \\
& + f(x,u)\lambda(x_i)u_i^{\beta-1}\frac{\Gamma(2-\beta)}{\beta-1}\Big(\frac{1}{g(x_i,u_i)}-1\Big).
\end{aligned}
$$

**4.3. Generator for measure-valued process.** Let $h_0(x_i) = h(x_i,0) = \int_0^\infty(1 - g(x_i,v))dv$. With $\alpha$ as in (3.4) and $f(x,u) = \prod_i g(x_i,u_i)$, we have

$$\alpha f(\mu) = e^{-\langle h_0, \mu\rangle}$$

and

$$\alpha A f(\mu)$$

$$= \alpha f(\mu) \int_E B \int_0^\infty (g(x,v) - 1) dv \mu(dx)$$

$$+ \alpha f(\mu) \int_E \int_0^\infty g(x,v)(\Lambda(x,v) - \Lambda(x,v + h(x,v))) dv \mu(dx)$$

$$+ \alpha f(\mu) \int_E \int_0^\infty \Lambda(x,v)(1 - g(x,v)) dv \mu(dx)$$

$$= \alpha f(\mu) \left( -\langle B h_0, \mu \rangle + \int_E \int_0^\infty (\Lambda(x,v) - g(x,v)\Lambda(x,v + h(x,v))) dv \mu(dx) \right)$$

$$= \alpha f(\mu) \left( -\langle B h_0, \mu \rangle + \int_E \int_0^{h_0(x)} \Lambda(x,v) dv \mu(dx) \right),$$

where the last equality follows from the fact that

$$\frac{\partial}{\partial v}(v + h(x,v)) = g(x,v).$$

For the example of Section 4.2, we have

$$\alpha A f(\mu) = \alpha f(\mu) \left( -\langle B h_0, \mu \rangle + \frac{\Gamma(2-\beta)}{\beta(\beta-1)} \langle \lambda h_0^\beta, \mu \rangle \right).$$

## 5. Dynkin's boundary value problem

We now consider a particle model in which the motion process is absorbing on the boundary of an open set $D \subset E$. Let $B_0 \subset \bar{C}(E) \times \bar{C}(E)$ be a graph separable, pre-generator, and suppose that $\mathcal{D}(B_0)$ is closed under multiplication and is separating. (In particular, $B_0$ satisfies the conditions of Theorem 1.1.) Define

$$B f = I_D B_0 f.$$

(Then $B$ satisfies the conditions of Remark 1.2.) If $X$ is a solution of the martingale problem for $B_0$ and $\tau = \inf\{t : X(t) \notin D\}$, then $X(\cdot \wedge \tau)$ is a solution of the martingale problem for $B$. We assume that $\tau < \infty$ a.s. and write $X(\infty)$ for $X(\tau)$.

For $f(x,u) = \prod_i g(x_i, u_i)$, let

$$
\begin{aligned}
A f(x,u) &= \sum_i f(x,u) \frac{B g(x_i, u_i)}{g(x_i, u_i)} \\
&\quad + f(x,u) \sum_i (\Lambda(x_i, u_i) - \Lambda(x_i, u_i + h(x_i, u_i))) \\
&\quad + f(x,u) \sum_i \Lambda(x_i, u_i)(\frac{1}{g(x_i, u_i)} - 1),
\end{aligned}
$$

where $\Lambda$ is as in Section 4, that is, $\Lambda$ is of the form (4.2). We assume that $\Lambda$ is bounded on $D \times [0, a]$ for each $a > 0$ and that $\Lambda(x_i, u_i) = 0$ for $x_i \notin D$. We do not require $\Lambda$ to be continuous; however, $A$ still satisfies the conditions of Remark 1.2. Consequently, each solution of the martingale problem for $C = \{(\alpha f, \alpha A f) : f \in \mathcal{D}(A)\}$ has a particle representation given by a solution of the martingale problem for $A$. Define

$$\psi(x,r) = \int_0^r \Lambda(x,v) dv,$$

so that

$$\alpha A f(\mu) = \alpha f(\mu)\langle -Bh + \psi(\cdot, h), \mu\rangle.$$

Following Dynkin [4], suppose that $V$ satisfies

(5.1) $$-BV(x) + \psi(x, V(x)) = \rho(x), \quad x \in D$$

(5.2) $$V(x) = \varphi(x), \quad x \in \partial D,$$

where $\rho \geq 0$ and $V$ is nonnegative and bounded. We define $\rho(x) = 0$ for $x \notin D$, so (5.1) holds for all $x$. Let $g(x_i, u_i) = 1 - V(x_i)g_0(u_i)$, where $\int_0^\infty g_0(v)dv = 1$ and $0 \leq Vg_0 \leq 1$. Set $f(x, u) = \prod_i g(x_i, u_i)$ and $g_1(u_i) = \int_{u_i}^\infty g_0(v)dv$. Then

$$Af(x, u) = f(x, u)\sum_i \left(\frac{-g_0(u_i)BV(x_i)}{1 - g_0(u_i)V(x_i)} + (\Lambda(x_i, u_i) - \Lambda(x_i, u_i + V(x_i)g_1(u_i)))\right.$$

$$\left. + \Lambda(x_i, u_i)\frac{V(x_i)g_0(u_i)}{1 - V(x_i)g_0(u_i)}\right),$$

$\alpha f(\mu) = e^{-\langle V, \mu\rangle}$ and $\alpha A f(\mu) = \langle \rho, \mu\rangle e^{-\langle V, \mu\rangle}$.

Assume that $X_i(0) = x$ for all $i$ and that $\{U_i(0)\}$ is a Poisson random measure with mean measure $m$. Then

$$
\begin{aligned}
e^{-V(x)} &= E[e^{-\langle V, K(0)\rangle}] \\
&= E[e^{-\langle V, K(t)\rangle - \int_0^t \langle \rho, K(s)\rangle ds}] \\
&= E[\prod_i (1 - V(X_i(t))g_0(U_i(t)))e^{-\int_0^t \langle \rho, K(s)\rangle ds}] \\
&= E[\prod_i (1 - \varphi(X_i(\infty))g_0(U_i(\infty)))e^{-\int_0^\infty \langle \rho, K(s)\rangle ds}] \\
&= E[e^{-\langle \varphi, K(\infty)\rangle - \int_0^\infty \langle \rho, K(s)\rangle ds}],
\end{aligned}
$$

where $\{(X_i(\infty), U_i(\infty))\}$ are the boundary absorption points and the levels for all particles that exit $D$ before dying and

$$K(\infty) = \lim_{r \to \infty} \frac{1}{r}\sum_i I_{[0,r]}(U_i(\infty))\delta_{X_i(\infty)}.$$

The second equality follows from the fact that $e^{-\langle V, K(t)\rangle - \int_0^t \langle \rho, K(s)\rangle ds}$ is a martingale. The third equality follows from (1.5), that is,

$$P\{\Psi(t) \in G | \mathcal{F}_t^K\} = \alpha(K(t), G),$$

where $\alpha(\mu, \cdot)$ is the distribution of a Poisson random measure with mean measure $\mu \times m$. The fourth equality follows by the bounded convergence theorem.

Taking logs, we have

$$
\begin{aligned}
V(x) &= -\log E[e^{-\langle \varphi, K(\infty)\rangle - \int_0^\infty \langle \rho, K(s)\rangle ds}] \\
&= -\log E[\prod_i (1 - \varphi(X_i(\infty))g_0(U_i(\infty)))e^{-\int_0^\infty \langle \rho, K(s)\rangle ds}].
\end{aligned}
$$

The first equality is just (1.11) of Dynkin [4].

# References

[1] Dawson, Donald A. (1993). Measure-valued Markov processes. *Ecole d'Eté de Probabilités de Saint-Flour XXI - 1991. Lect. Notes Math. 1541.* Springer-Verlag, Berlin.

[2] Donnelly, Peter and Kurtz, Thomas G. (1998). Particle representations for measure-valued population models. *Ann. Probab.* (to appear)

[3] Donnelly, Peter and Kurtz, Thomas G. (1999). Genealogical processes for Fleming-Viot models with selection and recombination. (preprint)

[4] Dynkin, E. B. (1991). A probabilistic approach to one class of nonlinear differential equations. *Probab. Th. Rel. Fields* 89, 89-115.

[5] Ethier, Stewart N. and Kurtz, Thomas G. (1986). *Markov Processes: Characterization and Convergence.* Wiley, New York.

[6] Kurtz, Thomas G. (1998). Martingale problems for conditional distributions of Markov processes. *Electronic J. Probab.* **3**, Paper 9.

[7] Perkins, Edwin A. (1992). Measure-valued branching diffusions with spatial interactions. *Probab. Theory Relat. Fields* 94, 189-245.

DEPARTMENTS OF MATHEMATICS AND STATISTICS, UNIVERSITY OF WISCONSIN - MADISON, 480 LINCOLN DRIVE, MADISON, WI 53706-1388 USA

*E-mail address*: kurtz@math.wisc.edu

Canadian Mathematical Society
Conference Proceedings
Volume **26**, 2000

# Minimax Estimation of Exponential Family Means Over $\ell_p$ Bodies Under Quadratic Loss

Brenda MacGibbon, Eric Gourdin, Brigitte Jaumard, and Peter Kempthorne

ABSTRACT. Non-parametric estimation of smooth functions that are members of Sobolev classes is closely related to the problem of estimating the (infinite dimensional) mean of a standard Gaussian shift when the mean is known to lie in an $\ell_p$ body. Donoho, Liu and MacGibbon [**5**] obtained exact results for the comparison between the linear maximum risk and the minimax risk among all estimates for this problem with Gaussian white noise. Here we study minimax estimation of infinite dimensional exponential family mean parameters constrained to belong to $\ell_p$ bodies under the quadratic loss function. The results are illustrated using the Poisson distribution.

## 1. Introduction

Pinsker [**19**] considered the problem of estimating the mean of a certain Gaussian process when the mean is known to lie in an infinite-dimensional "ellipsoid". He found an exact value for the minimax risk of linear estimates and an asymptotic value for the minimax risk among nonlinear estimates. These evaluations allow one to obtain precise constants in the asymptotic minimax risk for certain "real" function estimation problems. See, for example, Efroimovich and Pinsker [**6**, **7**] and Nussbaum [**18**]. A remarkable feature of the Pinsker solution is that it shows the minimax linear estimator to be asymptotically minimax among all estimates. Thus, in normal minimax estimation theory for "ellipsoids" at least, there is little to be gained by nonlinear procedures.

Casella and Strawderman [**4**] were the first to provide analytic and numerical results for the minimax risk in the one dimensional bounded normal mean problem, that is, $\theta \epsilon [-m, m]$ and $X \sim N(\theta, \sigma^2)$ with $\sigma^2$ known. They obtained the minimax estimator for small $m$. Bickel [**2**] obtained second order asymptotically minimax estimates for the normal mean estimation problem given the mean is contained in a finite dimensional sphere.

2000 *Mathematics Subject Classification.* Primary 62C20; Secondary 62F10.

*Key words and phrases.* Poisson; constrained parameters; minimax linear.

Research of the first three authors was supported by NSERC of Canada.

Research of the fourth author was supported by Kempthorne Analytics.

We sincerely wish to thank research assistants Gabriel Lambert and Yvette Nöé for their programming assistance.

Donoho, Liu and MacGibbon [5] considered the problem of estimating the (infinite dimensional) mean of a standard Gaussian shift when the mean is known to lie in an orthosymmetric, convex, quadratically convex set in $\ell_2$. Such sets include ellipsoids, hyperrectangles and $\ell_p$-bodies with $p \geq 2$. They showed that the minimax linear risk is within 25% of the exact minimax risk among all estimates. If the set is not quadratically convex, as in the case of $\ell_1$-bodies, then minimax linear estimators may be outperformed arbitrarily by nonlinear estimates. They used the heuristic that the difficulty of the hardest rectangular subproblem is equal to the difficulty of the full problem and consequently reduced the study to one dimension. They found a bound on the ratio of the linear minimax risk to the minimax risk and called it the Ibragimov-Hasminskii constant since Ibragimov and Hasminskii [12] were the first to study its behaviour. Vidacovic and Dasgupta [21] also studied the efficiency of linear estimators for the bounded normal mean problem.

Johnstone and MacGibbon [13, 14] argued that in many applications the Poisson model is the prototypical one for discrete data and it deserves consideration in its own right. They cited mixture problems, spatial process settings, thinning problems and sparse signal settings as some examples where information about constraints on the Poisson parameters may be available. They studied the problem of estimating a bounded Poisson vector in $\mathbb{R}^p$ using the information normalized quadratic loss function given by $\sum_i (\delta_i - \theta_i)^2 \theta_i^{-1}$ and obtained results analogous to those of Casella and Strawderman [4] and Bickel [2] in the normal case. They also found an upper bound of 1.251 for the Ibragimov-Hasminskii constant for the problem of estimating a Poisson mean vector constrained to lie in a bounded convex $\ell_1$-body, rectangularly convex at 0.

It is our purpose here to study the Poisson parameter estimation problem as the prototype of an exponential family mean estimation problem. The general method given here can be used for any exponential family, although appropriate changes in the form of the estimates and the constants must be made. In this way we extend this study of Johnstone and MacGibbon [13] to the infinite dimensional case and we also use a different loss, the quadratic loss function. We obtain results analogous to those of Casella and Strawderman [4] for the normal problem in Section 2. As linear or affine estimators are often used in practice and often have a closed form expression as compared to the exact minimax estimators, such estimators are also studied on bounded intervals in Section 2 and the associated Ibragimov-Hasminskii constant is obtained for this problem using numerical methods. Section 3 considers the more difficult mathematical problem of comparing linear and nonlinear minimax risks on infinite dimensional subsets. Our main contribution is to show that the minimax linear risk for compact quadratically convex subsets (rectangularly convex at 0) is equal to the minimax linear risk of the hardest rectangular subproblem. The proof of this result is more subtle than the corresponding one for the normal problem in Donoho, Liu and MacGibbon [5] and involves bounding the difference in risks by the corresponding terms in a negative Gâteau derivative.

## 2. Minimax estimation of a bounded Poisson univariate mean under quadratic loss

Let $X$ denote a Poisson random variable with mean $\theta$, that is, the probability that $X = x$ denoted by $P(X = x) = e^{-\theta} \frac{\theta^x}{x!}$ for $x = 0, 1, 2, \cdots$ and $= 0$ otherwise. Let $\mathcal{D}$ represent the space of decision procedures for the problem of estimating

$\theta$ under the quadratic loss function, $L(\delta, \theta) = (\delta - \theta)^2$. Let $R(\delta, \theta)$ denote the associated risk. Under the additional assumption that $\theta$ lies in an interval of the form $[0, m]$, an estimator $\delta_m$ is minimax for the above problem if, for all $\delta \varepsilon \mathcal{D}$,

$$(2.1) \qquad \sup_{0 \leq \theta \leq m} R\left(\delta_m, \theta\right) = \inf_{\delta \varepsilon \mathcal{D}} \sup_{0 \leq \theta \leq m} R(\delta, \theta).$$

Our aim is to compare the efficiency of linear, affine linear and exact minimax estimators where the linear (affine) minimax estimator for this problem is defined to be the linear (affine) estimator $\delta_m^L (\delta_m^A)$ satisfying equation (2.1) with the infimum over $\delta$ referring only to linear (affine) procedures.

Let $\rho(m), \rho_L(m)$ and $\rho_A(m)$ respectively denote the minimax risk, the linear minimax risk and the affine minimax risk for the above problem.

THEOREM 2.1. *The linear minimax estimator for the problem of estimating a Poisson mean $\theta$ that lies in $[0, m]$ under quadratic loss is given by*

$$\delta_m^L(x) = \frac{m}{m+1} x \quad x = 0, 1, 2 \cdots$$

*and the linear minimax risk* $\rho_L(m) = \dfrac{m^2}{m+1}$. *The affine minimax estimator is given by*

$$\delta_m^A(x) = \sqrt{m+1} - 1 + \left(1 - \frac{1}{\sqrt{m+1}}\right) x \quad \text{for} \quad x = 0, 1, 2 \cdots,$$

*and the affine minimax risk* $\rho_A(m) = [\sqrt{m+1} - 1]^2$.

PROOF. Let $\delta(x) = a + bx$ be an affine estimator. Then

$$
\begin{aligned}
R(\delta, \theta) &= \sum_{x=0}^{\infty} [a + bx - \theta]^2 e^{-\theta} \frac{\theta^x}{x!} \\
&= a^2 + 2a(b-1)\theta + (b-1)^2 \theta^2 + b^2 \theta.
\end{aligned}
$$

Since $R(\delta, \theta)$ is a convex function of $\theta$,

$$(2.2) \qquad \sup_{0 \leq \theta \leq m} R(\delta, \theta) = \max[a^2, a^2 + 2a(b-1)m + (b-1)^2 m^2 + b^2 m].$$

If $a = 0$, then $b = \dfrac{m}{m+1}$ minimizes $R(\delta, \theta)$ and the linear minimax risk $= \dfrac{m^2}{m+1}$. If $a$ is not necessarily zero, then the minimum of (2.2) occurs on the boundary

$$2a(b-1)m + (b-1)^2 m^2 + b^2 m = 0;$$

that is, the problem here is reduced to minimizing $\dfrac{1}{4}\left[(1-b)m + \dfrac{b^2}{(1-b)}\right]^2$. The minimum occurs if $b = 1 - \dfrac{1}{\sqrt{m+1}}$ and the minimum value is $\left[\sqrt{m+1} - 1\right]^2 = \dfrac{1}{m+1}[m + 1 - \sqrt{m+1}]^2$. Thus $\delta_m^A(x) = \sqrt{m+1} - 1 + \left(1 - \dfrac{1}{\sqrt{m+1}}\right) x, \quad x = 0, 1, 2 \cdots$ and the affine minimax risk $= \rho_A(m) = [\sqrt{m+1} - 1]^2$.  $\square$

THEOREM 2.2. *An estimator $\delta(x)$ of $\theta$ which is constrained to lie in the interval $[m_1, m_2]$ is said to; be "linear through $m_1$" provided there exists a constant $b$ such* *that $\delta(x) = b(x - m_1)$. The minimax "linear through $m_1$" estimator for this problem of estimating $\theta \in [m_1, m_2]$ is given by*

$$\delta_{m_2}^{L(m_1)}(x) = \frac{(m_2 - m_1)m_2}{m_2 + (m_2 - m_1)^2}(x - m_1)$$

*and the minimax "linear through $m_1$" risk is given by*

$$\rho_{L(m_1)}([m_1, m_2]) = \frac{m_2^3}{(m_2 - m_1)^2 + m_2}.$$

PROOF. Analogous arguments to those used in Theorem 2.1 apply.     □

In order to compare the affine linear and exact minimax risk on infinite dimensional subsets it is also necessary to evaluate $\rho_A([m_1, m_2])$ the affine minimax risk on the interval $[m_1, m_2]$ where $0 \le m_1 < m_2 < \infty$. Let $\delta(x) = a + bx$ then $R(\delta, \theta) = a^2 + \theta^2(b-1)^2 + 2a(b-1)\theta + b^2\theta$, which is a convex function of $\theta$. Thus letting $c = 1 - b$ the affine minimax risk on $[m_1, m_2]$ can be expressed as $g_{m_1, m_2}(a, c) = \inf_{a,c}[f_1(a, c), f_2(a, c)]$ where for $i = 1, 2$, $f_i(a, c) = a^2 + m_i^2 c^2 - 2acm_i + (1-c)^2 m_i$.

A simple proof by contradiction yields $\inf_{a,c} g_{m_1, m_2}(a, c) = \inf_{a \ge 0, c \ge 0} g_{m_1, m_2}(a, c)$. It is possible to solve $f_1(a, c) = f_2(a, c)$ for $a$, giving $a^* = \frac{1}{2}\left[\frac{(1-c)^2}{c} + c(m_1 + m_2)\right]$;

$$g(c) = g_{m_1, m_2}(a^*, c) = \frac{1}{4}\left[c(m_1 + m_2 + 1) - 2 + \frac{1}{c}\right]^2 - m_1 m_2 c^2.$$

Now $g'(c) = \frac{1}{2}\left\{[(m_1 + m_2 + 1)^2 - 4m_1 m_2]c - 2(m_1 + m_2 + 1) + \frac{2}{c^2} - \frac{1}{c^3}\right\}$.

It should be noted that $g''(c) > 0$ for all $c$ and that $g'(c)$ has a discontinuity at $0$. The roots of the equation $g'(c) = 0$ are evaluated numerically and the minimax affine risk obtained.

REMARK 2.3. The above estimators should be compared to the affine minimax estimator for the problem of estimating a Poisson mean $\theta \in [m_1, m_2]$ with the normalized quadratic loss function. This estimator is obtained in Johnstone and MacGibbon [13] and is given by

$$\delta^*(x) = \left\{\frac{(\sqrt{m_2} - \sqrt{m_1})^2 x + \sqrt{m_1 m_2}}{1 + (\sqrt{m_2} - \sqrt{m_1})^2}\right\}$$

and the maximum risk is achieved at $m_2$ and is given by

$$\frac{(\sqrt{m_2} - \sqrt{m_1})^2}{1 + (\sqrt{m_2} - \sqrt{m_1})^2}.$$

Now let us return to the problem of estimating $\theta \in [0, m]$ under quadratic loss. In order to find the minimax solution, the dual problem of finding the least favorable prior distribution for the corresponding Bayes problem is often considered. The duality theory ensures that, if there exist solutions to both problems, then the Bayes procedure with respect to the least favorable prior distribution will be minimax (see Ferguson [8], Kempthorne [15]). In order to determine the minimax risk $\rho(m)$, it is necessary to consider the corresponding Bayes problem. A distribution or prior probability measure $\pi$ is specified on the parameter space $[0, m]$ and a measure of the performance of a procedure $\delta$ is given by its Bayes risk

$$r(\delta, \pi) = \int_0^m R(\delta, \theta)\pi(d\theta),$$

$\delta_\pi$ is called the Bayes procedure with respect to the prior probability measure $\pi$ if $\delta_\pi$ minimizes the Bayes risk, where the Bayes risk $r(\pi)$ is defined as $r(\pi) = r(\delta_\pi, \pi)$. A distribution or prior probability measure is "least favorable" if its Bayes risk is greater than or equal to that of any other distribution. It follows from Ghosh's results [9] that a least favorable prior on $[0, m]$ will put mass on at most a finite number of points. Following the techniques in Casella and Strawderman [4] for the normal case, we will start with a two point prior and determine the interval on which it is minimax. An additional point will be added when two points no longer suffice. This process is continued, adding points one at a time as necessary. For $k$ points $(k < \infty)$, the "least favorable" prior $\pi^*(d\theta)$ is of the form

$$\sum_{i=1}^{k} a_i \epsilon_{\{b_i m\}}$$

where the Dirac measure $\epsilon_{\{b_i m\}}(\theta) = 1$, if $\theta = b_i m$ and $= 0$, otherwise, where $0 \leq b_1 < \cdots < b_k = 1$, and $\sum_{i=1}^{k} a_i = 1$ with $a_i \geq 0$ for all $i$.

The associated Bayes rule $\delta_{\pi^*}$ under quadratic loss is given by

$$\delta_{\pi^*}(x) = \frac{\sum_{i=1}^{k} a_i (b_i m)^{x+1} e^{-b_i m}}{\sum_{i=1}^{k} a_i (b_i m)^x e^{-b_i m}}.$$

The Bayes risk of the least favorable prior $\pi^*$ is defined as

$$\rho(m) = r(\pi^*) = \sum_{i=1}^{k} a_i R(\delta_{\pi^*}, b_i m).$$

In addition, the Bayes rule $\delta_{\pi^*}$ is an "equalizer" rule (see Ferguson [8]); that is $R(\delta_{\pi^*}, b_i m) = R(\delta_{\pi^*}, b_j m) \quad \forall i, j = 1, \cdots k$.

THEOREM 2.4. *If $m \leq m_0 \approx 0.9129569$, that is, $m_0$ is the first positive zero of equation (2.3), the least favorable prior distribution $\pi_m$ is given by $a_m \epsilon_{\{0\}} + (1 - a_m)\epsilon_{\{m\}}$ where $a_m$ satisfies $R(0, \delta_{\pi_m}) = R(m, \delta_{\pi_m})$; that is, $a_m = (1 + e^{m/2})^{-1}$. This gives a minimax estimator: $\delta_m(x) = m$ if $x \geq 1$ and $\delta_m(0) = \frac{(1-a_m)e^{-m}m}{a_m + (1-a_m)e^{-m}}$; and a minimax risk of $\rho(m) = r(\pi_m) = \frac{m^2}{(1+e^{\frac{m}{2}})^2}$, for $0 < m \leq m_0$.*

PROOF. It now suffices by Ferguson [8], Theorem 1, p. 90, to show that there exists $m_0$ such that $r(\pi_m) \geq R(\delta_m, \theta)$ for all $\theta \varepsilon [0, m]$ if $m \leq m_0$.
Let $c_m = m(1 + e^{m/2})^{-2}$. Then $\delta_m(0) = c_m m$.
Now, $R(\delta_m, \theta) = [c_m m - \theta]^2 e^{-\theta} + [m - \theta]^2 [1 - e^{-\theta}]$.
Thus, $\frac{\partial R}{\partial \theta} = e^{-\theta}\left[ 2(m - c_m m) - (c_m m - \theta)^2 + (\theta - m)^2 \right] + 2(\theta - m)$
and $\frac{\partial^2 R}{\partial \theta^2} = 2 + e^{-\theta}\left[ -4(m - c_m m) + (c_m m - \theta)^2 - (\theta - m)^2 \right]$,
$\frac{\partial^3 R}{\partial \theta^3} = e^{-\theta}[m^2(1 - c^2 m) + 2m[1 - c_m](3 - \theta)]$.
Let $m_0$ be the first positive zero of

(2.3) $$\frac{\partial R}{\partial \theta}\bigg|_{\theta=0} = 0$$

Then $m_0 \approx 0.9129569$.

Now

$$\left.\frac{\partial R}{\partial \theta}\right|_{\theta=m} \geq 0 \quad \text{and} \quad \left.\frac{\partial^2 R}{\partial \theta^2}\right|_{\theta=m} \geq 0, \quad \forall m \leq m_0;$$

$$\frac{\partial^3 R}{\partial \theta^3} \geq 0, \quad \forall \theta \leq m, \forall m \leq m_0;$$

which imply that for all $m \leq m_0$, $\max_{0 \leq \theta \leq m} R(\delta_m, \theta) = \max\{R(\delta_m, 0), R(\delta_m, m)\}$

$\Box$

The values of $\rho(m)$ and the Ibragimov-Hasminskii constants $\mu_A(m) = \dfrac{\rho_A(m)}{\rho(m)}$

and $\mu_L(m) = \dfrac{\rho_L(m)}{\rho(m)}$ for selected values of $m < m_0$ are presented in Table 1.

TABLE 1. Minimax estimation of a bounded Poisson mean under quadratic loss for $m \leq m_0$ (with weights $a_m$).

| $m$ | $a_m$ | $\rho(m)$ | $\mu_A(m)$ | $\mu_L(m)$ |
|-----|-------|-----------|------------|------------|
| 0.1 | .487503 | .002377 | 1.002405 | 3.824531 |
| 0.2 | .475021 | .009026 | 1.009304 | 3.693035 |
| 0.3 | .462570 | .019257 | 1.020343 | 3.595096 |
| 0.4 | .450166 | .032424 | 1.035288 | 3.524726 |
| 0.5 | .437823 | .047922 | 1.054002 | 3.477874 |
| 0.6 | .425557 | .065196 | 1.076419 | 3.451132 |
| 0.7 | .413382 | .083734 | 1.102532 | 3.442273 |
| 0.8 | .401312 | .103073 | 1.132386 | 3.449551 |
| 0.9 | .389361 | .122797 | 1.166068 | 3.471712 |

REMARK 2.5. It should be noted, however, that as $m \to \infty$ it is possible to obtain bounds on the ratio of $\rho_A(m)$ to $\rho(m)$ by an appropriate choice of prior in Brown's information inequality for the Bayes risk (Brown and Gajek [3]), that is, if $G$ is a prior with continuous differentiable density $g(\theta)$ on $[0, m]$ and $I(\theta)$ is the Fisher information

$$r(G) \geq \int_0^m I^{-1}(\theta) g(\theta) d\theta - \int_0^m \frac{[\frac{d}{d\theta}(I^{-1}(\theta) g(\theta)]^2}{g(\theta)} d\theta.$$

We can now argue as did Ibragimov and Hasminskii [12] for the bounded normal mean problem that the ratios $\mu_A(m)$ and $\mu_L(m)$ are bounded for all $m$.

For larger values of $m > m_0$ a two-point prior no longer suffices, the problem becomes intractable analytically and numerical methods, to calculate the bounds, must be used. Starting with the three point prior as in Casella and Strawderman [4], a global optimization technique (cf. Gourdin [10] and Gourdin, Jaumard and MacGibbon [11]) is used to solve the optimization problem. This problem involves the maximization of a three variable function, which can be shown to satisfy a Lipschitz property. The problem of maximizing a univariate Lipschitz function defined over a bounded interval was first addressed by Piyavskii [20]. The generalization to the case of multivariate Lipschitz functions was first outlined by Piyavskii himself [20] and later developed by Mladineo [16]. As explained in Gourdin, Jaumard and MacGibbon [11], since the objective function here is concave in some of the variables it is possible to decompose the problem into two parts: one for which the above algorithm is used and one for which efficient constrained concave maximization can be applied (see *e.g.* Avriel [1]).

The numerical results obtained when applying this algorithm to the problem of estimating a bounded Poisson mean under quadratic loss for values of $m$ between 1.0 and 3.0 and with a precision of $\varepsilon = 10^{-4}$ are presented in Table 2. The following notation is used: **J** for the number of function evaluations, and **cpu** for the computation time in seconds.

TABLE 2. Minimax estimation of a bounded Poisson mean under quadratic loss using a global optimization decomposition method with $\varepsilon = 10^{-4}$.

| m | J | CPU | $a_1$ | $a_2$ | $a_3$ | b | $\rho(m)$ | $\mu_A(m)$ | $\mu_L(m)$ |
|---|---|---|---|---|---|---|---|---|---|
| 1.0 | 820579 | 655 | .00000 | .39869 | .60130 | .02414 | .14300 | 1.199810 | 3.49650 |
| 1.1 | 560839 | 451 | .00000 | .40921 | .59079 | .04674 | .16427 | 1.228007 | 3.50758 |
| 1.2 | 407011 | 329 | .00000 | .41789 | .58210 | .06523 | .18655 | 1.251786 | 3.50869 |
| 1.3 | 333471 | 271 | .00000 | .42449 | .57551 | .08009 | .20975 | 1.282228 | 3.50313 |
| 1.4 | 270175 | 222 | .00000 | .43087 | .56913 | .09378 | .23383 | 1.289883 | 3.49256 |
| 1.5 | 203719 | 167 | .00000 | .43556 | .56444 | .10511 | .25871 | 1.305409 | 3.47880 |
| 1.6 | 185427 | 150 | .00000 | .43932 | .56068 | .11511 | .28432 | 1.319277 | 3.46305 |
| 1.7 | 163311 | 134 | .00000 | .44231 | .55768 | .12420 | .31059 | 1.331867 | 3.44625 |
| 1.8 | 142579 | 117 | .00000 | .44457 | .55542 | .13230 | .33746 | 1.343448 | 3.42898 |
| 1.9 | 126611 | 103 | .00000 | .44638 | .55361 | .14003 | .36487 | 1.354243 | 3.41170 |
| 2.0 | 114011 | 93 | .00000 | .44771 | .55229 | .14704 | .39276 | 1.364442 | 3.39478 |
| 2.1 | 135543 | 179 | .00000 | .44884 | .55116 | .15417 | .42110 | 1.374107 | 3.37825 |
| 2.2 | 154331 | 207 | .00000 | .44970 | .55030 | .16085 | .44984 | 1.383361 | 3.36231 |
| 2.3 | 153043 | 208 | .00000 | .45046 | .54954 | .16747 | .47895 | 1.392253 | 3.34697 |
| 2.4 | 148563 | 201 | .00000 | .45113 | .54887 | .17387 | .50840 | 1.400830 | 3.33225 |
| 2.5 | 215115 | 290 | .00000 | .45183 | .54817 | .18040 | .53818 | 1.409087 | 3.31806 |
| 2.6 | 317427 | 428 | .00209 | .45092 | .54700 | .18786 | .56827 | 1.417050 | 3.30438 |
| 2.7 | 422779 | 579 | .01333 | .44328 | .54338 | .20043 | .59875 | 1.424506 | 3.29064 |
| 2.8 | 438543 | 598 | .02175 | .43824 | .54001 | .21170 | .62966 | 1.431379 | 3.27662 |
| 2.9 | 432979 | 589 | .02814 | .43499 | .53688 | .22185 | .66098 | 1.437737 | 3.26244 |
| 3.0 | 426067 | 580 | .03310 | .43295 | .53395 | .23109 | .69270 | 1.443626 | 3.24816 |

Since, in order to determine the Ibragimov-Hasminskii constant, we must study the minimax risk for larger $m$, a modification of Kempthorne's [15] iterative procedure for numerical specification of discrete least favorable priors as described in Nöe [17] is used. Clearly, the Poisson problem studied here satisfies the criteria of Theorems 2.1 and 2.2 of Kempthorne [15] and his algorithm can consequently be applied.

Table 3 contains the results using a modification of Kempthorne's method [15] for larger values of $m$. Note that the results up until $m = 3$ are in agreement with the results using the global optimization procedure where the precision is known to be $10^{-4}$.

If we define

$$\mu^* = \sup_m \frac{\rho_A(m)}{\rho(m)}$$

the Ibragimov-Hasminskii constant for the bounded Poisson mean problem under quadratic loss, then numerically $\mu^* \approx 1.56$.

It should be noted that, for this problem,

$$\mu^{**} = \sup_{m \geq 0} \frac{\rho_L(m)}{\rho(m)} = \frac{\rho_L(0)}{\rho(0)} = 4.$$

Further computations using the iterative procedure were done to determine $\rho[m_1, m_2]$, the minimax risk among all estimates when $\theta \in [m_1, m_2]$. The analogous Ibragimov-Hasminskii constants $\mu^*$ and $\mu^{**}$ for such intervals remained the same.

TABLE 3. Minimax estimation of a bounded Poisson mean under quadratic loss using the iterative algorithm.

| m | $p_1$ | $p_2$ | $p_3$ | $p_4$ | $p_5$ | $b_1$ | $b_2$ | $b_3$ | $b_4$ | $b_5$ | $\rho(m)$ | $\mu_A(m)$ | $\mu_L(m)$ |
|---|---|---|---|---|---|---|---|---|---|---|---|---|---|
| 2.000 | 0.000 | 0.448 | 0.552 | | | 0.000 | 0.147 | 1.000 | | | 0.393 | 1.36444 | 3.39478 |
| 2.100 | 0.000 | 0.449 | 0.551 | | | 0.000 | 0.154 | 1.000 | | | 0.421 | 1.37411 | 3.37825 |
| 2.200 | 0.000 | 0.450 | 0.550 | | | 0.000 | 0.161 | 1.000 | | | 0.450 | 1.38336 | 3.36231 |
| 2.300 | 0.000 | 0.450 | 0.550 | | | 0.000 | 0.167 | 1.000 | | | 0.479 | 1.39225 | 3.34697 |
| 2.400 | 0.000 | 0.451 | 0.549 | | | 0.000 | 0.174 | 1.000 | | | 0.508 | 1.40083 | 3.33225 |
| 2.500 | 0.000 | 0.452 | 0.548 | | | 0.000 | 0.180 | 1.000 | | | 0.538 | 1.40909 | 3.31806 |
| 2.600 | 0.002 | 0.451 | 0.547 | | | 0.000 | 0.189 | 1.000 | | | 0.568 | 1.41705 | 3.30438 |
| 2.700 | 0.013 | 0.443 | 0.543 | | | 0.000 | 0.200 | 1.000 | | | 0.599 | 1.42451 | 3.29064 |
| 2.800 | 0.022 | 0.438 | 0.540 | | | 0.000 | 0.212 | 1.000 | | | 0.630 | 1.43138 | 3.27662 |
| 2.900 | 0.028 | 0.435 | 0.537 | | | 0.000 | 0.221 | 1.000 | | | 0.661 | 1.43774 | 3.26244 |
| 3.000 | 0.033 | 0.433 | 0.532 | | | 0.000 | 0.231 | 1.000 | | | 0.693 | 1.44363 | 3.24816 |
| 3.100 | 0.037 | 0.432 | 0.531 | | | 0.000 | 0.240 | 1.000 | | | 0.725 | 1.44912 | 3.23391 |
| 3.200 | 0.040 | 0.431 | 0.529 | | | 0.000 | 0.247 | 1.000 | | | 0.757 | 1.45424 | 3.21967 |
| 3.300 | 0.042 | 0.431 | 0.526 | | | 0.000 | 0.255 | 1.000 | | | 0.790 | 1.45902 | 3.20553 |
| 3.400 | 0.044 | 0.431 | 0.524 | | | 0.000 | 0.261 | 1.000 | | | 0.823 | 1.46353 | 3.19158 |
| 3.500 | 0.046 | 0.432 | 0.522 | | | 0.000 | 0.267 | 1.000 | | | 0.857 | 1.46778 | 3.17779 |
| 3.600 | 0.047 | 0.433 | 0.520 | | | 0.000 | 0.273 | 1.000 | | | 0.890 | 1.47180 | 3.16422 |
| 3.700 | 0.048 | 0.433 | 0.519 | | | 0.000 | 0.279 | 1.000 | | | 0.924 | 1.47563 | 3.15091 |
| 3.800 | 0.048 | 0.434 | 0.517 | | | 0.000 | 0.284 | 1.000 | | | 0.959 | 1.47928 | 3.13786 |
| 3.900 | 0.049 | 0.435 | 0.516 | | | 0.000 | 0.289 | 1.000 | | | 0.993 | 1.48279 | 3.12511 |
| 4.000 | 0.049 | 0.437 | 0.514 | | | 0.000 | 0.294 | 1.000 | | | 1.028 | 1.48615 | 3.11263 |
| 4.100 | 0.049 | 0.438 | 0.513 | | | 0.000 | 0.300 | 1.000 | | | 1.063 | 1.48938 | 3.10044 |
| 4.200 | 0.049 | 0.439 | 0.512 | | | 0.000 | 0.302 | 1.000 | | | 1.098 | 1.49252 | 3.08858 |
| 4.300 | 0.049 | 0.440 | 0.510 | | | 0.000 | 0.306 | 1.000 | | | 1.133 | 1.49558 | 3.07703 |
| 4.400 | 0.049 | 0.441 | 0.510 | | | 0.000 | 0.309 | 1.000 | | | 1.169 | 1.49854 | 3.06578 |
| 4.500 | 0.000 | 0.052 | 0.440 | 0.508 | | 0.000 | 0.007 | 0.316 | 1.000 | | 1.205 | 1.50141 | 3.05479 |
| 4.600 | 0.000 | 0.055 | 0.438 | 0.506 | | 0.000 | 0.015 | 0.321 | 1.000 | | 1.241 | 1.50414 | 3.04397 |
| 4.700 | 0.000 | 0.059 | 0.436 | 0.505 | | 0.000 | 0.021 | 0.327 | 1.000 | | 1.278 | 1.50674 | 3.03328 |
| 4.800 | 0.000 | 0.062 | 0.435 | 0.503 | | 0.000 | 0.029 | 0.332 | 1.000 | | 1.314 | 1.50920 | 3.02273 |
| 4.900 | 0.000 | 0.065 | 0.433 | 0.502 | | 0.000 | 0.033 | 0.338 | 1.000 | | 1.351 | 1.51156 | 3.01236 |
| 5.000 | 0.000 | 0.068 | 0.432 | 0.500 | | 0.000 | 0.038 | 0.343 | 1.000 | | 1.388 | 1.51381 | 3.00214 |
| 5.500 | 0.000 | 0.081 | 0.427 | 0.492 | | 0.000 | 0.061 | 0.367 | 1.000 | | 1.576 | 1.52358 | 2.95317 |
| 6.000 | 0.000 | 0.091 | 0.424 | 0.485 | | 0.000 | 0.079 | 0.388 | 1.000 | | 1.769 | 1.53132 | 2.90766 |
| 6.500 | 0.000 | 0.099 | 0.423 | 0.478 | | 0.000 | 0.093 | 0.405 | 1.000 | | 1.966 | 1.53753 | 2.86539 |
| 7.000 | 0.000 | 0.105 | 0.422 | 0.473 | | 0.000 | 0.105 | 0.421 | 1.000 | | 2.167 | 1.54257 | 2.82615 |
| 7.500 | 0.000 | 0.111 | 0.422 | 0.467 | | 0.000 | 0.117 | 0.435 | 1.000 | | 2.372 | 1.54669 | 2.78968 |
| 8.000 | 0.000 | 0.116 | 0.422 | 0.462 | | 0.000 | 0.128 | 0.449 | 1.000 | | 2.580 | 1.55011 | 2.75574 |
| 8.500 | 0.002 | 0.121 | 0.420 | 0.457 | | 0.000 | 0.144 | 0.462 | 1.000 | | 2.792 | 1.55288 | 2.72398 |
| 9.000 | 0.003 | 0.126 | 0.418 | 0.452 | | 0.000 | 0.158 | 0.474 | 1.000 | | 3.007 | 1.55507 | 2.69409 |
| 9.500 | 0.004 | 0.131 | 0.417 | 0.448 | | 0.000 | 0.169 | 0.485 | 1.000 | | 3.224 | 1.55680 | 2.66594 |
| 10.000 | 0.004 | 0.135 | 0.417 | 0.444 | | 0.000 | 0.179 | 0.495 | 1.000 | | 3.444 | 1.55816 | 2.63941 |
| 10.500 | 0.000 | 0.005 | 0.139 | 0.418 | 0.440 | 0.000 | 0.005 | 0.189 | 0.505 | 1.000 | 3.700 | 1.55923 | 2.61440 |
| 11.000 | 0.000 | 0.007 | 0.142 | 0.415 | 0.436 | 0.000 | 0.021 | 0.201 | 0.514 | 1.000 | 3.892 | 1.56004 | 2.59072 |
| 11.500 | 0.000 | 0.009 | 0.145 | 0.414 | 0.432 | 0.000 | 0.035 | 0.212 | 0.523 | 1.000 | 4.120 | 1.56059 | 2.56825 |
| 12.000 | 0.000 | 0.011 | 0.413 | 0.148 | 0.428 | 0.000 | 0.046 | 0.531 | 0.223 | 1.000 | 4.349 | 1.56095 | 2.54688 |
| 12.500 | 0.000 | 0.013 | 0.151 | 0.412 | 0.425 | 0.000 | 0.056 | 0.234 | 0.539 | 1.000 | 4.581 | 1.56112 | 2.52652 |
| 13.000 | 0.000 | 0.014 | 0.154 | 0.411 | 0.421 | 0.000 | 0.064 | 0.244 | 0.547 | 1.000 | 4.812 | 1.56113 | 2.50709 |
| 13.500 | 0.000 | 0.016 | 0.156 | 0.410 | 0.418 | 0.000 | 0.072 | 0.253 | 0.554 | 1.000 | 5.051 | 1.56100 | 2.48853 |
| 14.000 | 0.000 | 0.017 | 0.159 | 0.409 | 0.415 | 0.000 | 0.079 | 0.261 | 0.560 | 1.000 | 5.288 | 1.56076 | 2.47078 |
| 14.500 | 0.000 | 0.019 | 0.161 | 0.408 | 0.412 | 0.000 | 0.086 | 0.271 | 0.567 | 1.000 | 5.528 | 1.56042 | 2.45378 |
| 15.000 | 0.000 | 0.020 | 0.164 | 0.407 | 0.409 | 0.000 | 0.092 | 0.279 | 0.573 | 1.000 | 5.769 | 1.55998 | 2.43747 |

## 3. Minimax risk on hyperrectangles and other infinite dimensional subsets

In this section we first consider the problem under quadratic loss of estimating a Poisson mean vector $\theta$ (coordinatewise independent) belonging to a (possibly infinite dimensional) hyperrectangle $\Pi_i[0, m_i]$, which we denote by $[0, m]$.

THEOREM 3.1. *The ratio of minimax linear risk to minimax risk on the infinite dimensional $[0, m]$ under quadratic loss is bounded by the one dimensional Ibragimov-Hasminskii constant $= \mu** = 4$.*

PROOF. Clearly, the minimax risk $\rho([0, m]) = \sum_{i=1}^{\infty} \rho([0, m_i])$, since the least favorable prior is the obvious product measure.

Now let $\delta_L(x)$ be defined coordinatewise by $\delta_L(x) = \left(\frac{m_i x}{m_i + 1}\right)_i$. Now

$$(3.1) \qquad R(\delta_L, m) = \sum \rho_L([0, m_i])$$

and because we are working with linear estimates the *sup* and the *inf* can be interchanged and

$$\rho_L([0, m_i]) = \sup_{\|\theta\| \leq m_i} \inf_{\delta \ linear} R(\delta, \theta);$$

and so $\quad R(\delta_L, m) = \sum \rho_L([0, m_i]) = \sup_{\theta \in [0,m]} \inf_{\delta \ linear} R(\delta, \theta).$

Clearly, $R(\delta_L, m)$ must be greater than or equal to the minimax linear risk. On the other hand

$$\rho_L[0, m] = \inf_{\delta \ linear} \sup_{\theta \in [0,m]} R(\delta, \theta) \leq \sup_{\theta \in [0,m]} \inf_{\delta \ linear} R(\delta, \theta) = R(\delta, \theta)$$

and (3.1) is true.

Since $\rho_L([0, m_i]) \leq \mu^{**} \rho([0, m_i])$ for each $i$ and (3.1) is true, then

$$\rho_L([0, m]) \leq \mu^{**} \rho([0, m]).$$

$\square$

Our next step is to consider other types of infinite dimensional subsets as in Donoho, Liu and MacGibbon [5]. For each $\theta = (\theta_i)_{i=1}^{\infty}$ and $m = (m_i)_{i=1}^{\infty}$ in $\ell_2$, with $\theta_1 \geq m_i$ for each $i$, let $[m, \theta]$ denote the hyperrectangle $= \Pi_i [m_i, \theta_i]$. A set $\Theta$ in $\ell_2$ is said to be rectangularly convex in $\ell_2$ at $m = (m_i)_{i=1}^{\infty}$ if for each $\theta \in \Theta$, the hyperrectangle $[m, \theta]$ is in $\Theta$. $\Theta$ is said to be quadratically convex, if $\{\theta^2 : \theta \in \Theta\}$ is convex.

THEOREM 3.2. *If the mean of an infinite dimensional Poisson vector is known to lie in $\Theta$, a compact convex, rectangularly convex at 0 and quadratically convex set in $\ell_2$, then the minimax risk among linear estimates is within a factor 4 of the minimax risk.*

PROOF. It suffices to show that the difficulty, for linear estimates of the hardest rectangular subproblem is equal to the difficulty, for linear estimates of the full problem. Now for each $\theta \in \Theta$ the minimax risk on the hyperrectangle $[0, \theta]$ is given by

$$J(\theta) = \rho_L([0, \theta]) = \sum_i \frac{\theta_i^2}{\theta_i + 1}.$$

Arguing as is in the proof of Theorem 7 (Donoho, Liu and MacGibbon [5]), we first identify the hardest rectangular subproblem $[0, \tau^{**}]$ of $\Theta$; that is, we show that by $\ell_2$- continuity and compactness $J$ has a maximum $\tau^{**}$ on $\Theta$. Thus

$$(3.2) \qquad J(\theta) \leq J(\tau^{**}) \text{ for all } \theta \in \Theta.$$

Let $\delta_L^*$ represent the corresponding linear estimator which achieves (3.2). In order to complete the proof, it suffices to show that $R(\delta_L^*, \theta) \leq J(\tau^{**})$ for all $\theta \in \Theta$.

Now let $t = (t_i)_i$ be defined by $t_i = \theta_i^2$ for each $i$. We will write $t = \theta^2$. Now
$$T = \{t : t_i = \theta_i^2, \; \theta_i \in \Theta\}$$
is convex, since $\Theta$ is quadratically convex, so let us define $\tilde{J}$ on $T$ by $\tilde{J}(t) = J(\theta)$. Clearly since $\tau^* = \tau^{**2}$ is a maximum of $\tilde{J}$, the Gâteaux derivative of $\tilde{J}$ at $\tau^* = D_{\tau^*}\tilde{J}$ is negative at $\tau^*$.

We shall now proceed to show that

(3.3) $$R(\delta_L^*, \theta) - J(\tau^{**}) \le D_{\tau^*}\tilde{J} \quad \text{for all} \quad \theta \in \Theta.$$

Let $\tau_i^{**}$ denote the $i^{th}$ component of $\tau^{**}$. Then the minimax linear estimator $\delta_L^*$ for $[0, \tau^{**}]$ is of the form $\left(\frac{\tau_i^{**}x}{\tau_i^{**}+1}\right)_{i=1}^{\infty}$. The mean-squared error of this estimator at $\theta$

$$= R(\delta_L^*, \theta) = \sum \frac{\theta_i^2}{(\tau_i^{**}+1)^2} + \frac{(\tau_i^{**})^2}{(\tau_i^{**}+1)^2}\theta_i.$$

By the change of variable $t_i = \theta_i^2$ and $\tau_i^* = \tau_i^{**2}$ for each $i$,

$$R(\delta_L^*, \theta) - J(\tau^{**}) = \sum \frac{(\sqrt{t_i} - \sqrt{\tau_i^*})(\sqrt{t_i} + \sqrt{\tau_i^*} + \tau_i^*)}{(\sqrt{\tau_i^*} + 1)^2}.$$

We also have that

$$\tilde{J}(t) = \sum_i \frac{t_i}{\sqrt{t_i} + 1}.$$

The Gâteau derivative of $\tilde{J}(t)$ at $\tau^*$, $D_{\tau^*}\tilde{J}$ is given by

$$\begin{aligned} D_{\tau^*}\tilde{J} &= \lim_{\varepsilon \to 0} \sum_i \frac{\tilde{J}(\tau_i^* + \varepsilon(t_i - \tau_i^*)) - \tilde{J}(\tau_i^*)}{\varepsilon} \\ &= \sum(\sqrt{t_i} - \sqrt{\tau_i^*})(\tfrac{1}{2}\sqrt{t_i}\sqrt{\tau_i^*} + \tfrac{1}{2}\tau_i^* + \sqrt{t_i} + \sqrt{\tau_i^*}) \end{aligned}$$

We will prove (3.3) by arguing that the result is true term by term. There are two types of terms to consider in the sums involved:

(i) if $(\sqrt{t_i} - \sqrt{\tau_i^*}) > 0$,
   then $(\tfrac{1}{2}\sqrt{t_i}\sqrt{\tau_i^*} + \tfrac{1}{2}\tau_i^* + \sqrt{t_i} + \sqrt{\tau_i^*}) > (\sqrt{t_i} + \sqrt{\tau_i^*} + \tau_i^*)$
   and $(\sqrt{t_i} - \sqrt{\tau_i^*})(\tfrac{1}{2}\sqrt{t_i}\sqrt{\tau_i^*} + \tfrac{1}{2}\tau_i^* + \sqrt{t_i} + \sqrt{\tau_i^*}) > (\sqrt{t_i} - \sqrt{\tau_i^*})(\sqrt{t_i} + \sqrt{\tau_i^*} + \tau_i^*)$

(ii) if $(\sqrt{t_i} - \sqrt{\tau_i^*}) < 0$,
   then $(\tfrac{1}{2}\sqrt{t_i}\sqrt{\tau_i^*} + \tfrac{1}{2}\tau_i^* + \sqrt{t_i} + \sqrt{\tau_i^*}) < (\sqrt{t_i} + \sqrt{\tau_i^*} + \tau_i^*)$
   and $(\sqrt{t_i} - \sqrt{\tau_i^*})(\tfrac{1}{2}\sqrt{t_i}\sqrt{\tau_i^*} + \tfrac{1}{2}\tau_i^* + \sqrt{t_i} + \sqrt{\tau_i^*}) > (\sqrt{t_i} - \sqrt{\tau_i^*})(\sqrt{t_i} + \sqrt{\tau_i^*} + \tau_i^*)$

This clearly implies that (3.3) is true. $\qquad\square$

REMARK 3.3. Arguing as in Section 4 of Donoho, Liu and MacGibbon, the compactness requirement in Theorem 3.2 can be relaxed and it can be shown that the theorem is true for sets of the form:

$$\Theta = \left\{\theta : \sum a_i\|\theta_i\|^p \le 1\right\} \quad \text{for} \quad p \ge 2.$$

REMARK 3.4. It can also be shown as in Donoho, Liu and MacGibbon [5] that linear estimators can be arbitrarily outperformed by nonlinear ones in the sense of minimax risk for $\ell_1-$bodies.

These results for quadratic loss should be compared to the result for $\ell_1$ bodies with the information normalized quadratic loss function in Johnstone and MacGibbon [13]. We give the following infinite dimensional extension of this result.

THEOREM 3.5. *If $\Theta$ is compact, convex and rectangularly convex at 0, then the linear minimax risk on $\Theta$ for information normalized quadratic loss satisfies*

$$\rho_A^n(\Theta) = \sup_{\theta \in \Theta} \rho_A^n([0, \theta]),$$

*where $\rho_A^n([m, \theta])$ denote the affine difficulty of the hyperrectangle $\Pi_i[0, \theta_i] \subset \Theta$ and is equal to*

$$\rho_A^n([0, \theta]) = \sum \frac{\theta_i}{1 + \theta_i}.$$

PROOF. By an $\ell_1$-continuity argument, the same proof as given by Johnstone and MacGibbon [13] in the finite dimensional case works here.  □

In conclusion, as illustrated by the Poisson estimation problem, the form of infinite dimensional subsets $\Theta$ for which the ratio of $\rho_L(\Theta)$ to $\rho(\Theta)$ is bounded by the univariate Ibragimov-Hasminskii constant is very sensitive to the loss function. In general these sets will be quite different for quadratic loss and information normalized quadratic loss. These results can also be extended to other exponential family mean estimation problems.

## References

[1] M. Avriel, *Nonlinear Programming: Analysis and Methods*. Prentice-Hall, Englewood Cliffs, New Jersey, 1976.

[2] P. Bickel, *Minimax estimation of the mean of a normal distribution when the parameter space is restricted*. Annals of Statistics **9** (1981), 1301–1309.

[3] L.D. Brown, and L. Gajek, *Information inequalities for the Bayes risk*. Annals of Statistics **18** (1990), 1578–1594.

[4] G. Casella, and W. Strawderman, *Estimating a bounded normal mean*. Annals of Statistics **9** (1981), 868–876.

[5] D.L. Donoho, R.C. Liu, and B. MacGibbon, *Minimax risk over hyperrectangles and implications*. Annals of Statistics **18** (1990), 1416–1437.

[6] S.H. Efroimovich, and M.S. Pinsker, *Estimation of square-integrable [spectral] density based on a sequence of observations*. Problemy Peredachi Informatsii **17** (1981), 50–68 (in Russian); Problems of Information Transmission (1982), 182–193 (in English).

[7] S.H. Efroimovich, and M.S. Pinsker, *Estimation of square-integrable probability density of a random variable*. Problemy Peredachi Informatsii **18**, (1982), 19–38 (in Russian); Problems of Information Transmission (1983), 175–189 (in English).

[8] T.S. Ferguson, *Mathematical Statistics, A Decision Theoretic Approach*. Academic, New York, 1967.

[9] M.N. Ghosh, *Uniform approximation of minimax point estimates*. Ann. Math. Statist. **35** (1964), 1031–1047.

[10] E. Gourdin, *Global optimization algorithms for the construction of least favorable priors and minimax estimations*. Mémoires d'ingénieur de l'Institut d'Informatique d'Entreprise, 1989, Evry, France.

[11] E. Gourdin, B. Jaumard, and K.B. MacGibbon, *Global optimization decomposition methods for bounded parameter minimax risk evaluation*. SIAM Journal of Scientific Computing **15** (1994), 16–35.

[12] I.A. Ibragimov, and R.Z. Hasminskii, *Non parametric estimation of the value of a linear function in Gaussian white noise*. Theory of Probability and its Applications **29** (1984), 1–32.

[13] I.M. Johnstone, and K.B. MacGibbon, *Minimax estimation of a constrained Poisson vector*. Annals of Statisics **20** (1992), 807–831.

[14]  I.M. Johnstone, and K.B. MacGibbon, *Asymptotically minimax estimation of a constrained Poisson vector via polydisc transforms*. Annals de l'Institut Henri Poincaré, Série B **29** (1993), 289–319.

[15]  P.J. Kempthorne, *Numerical specification of discrete least favorable prior distributions*. SIAM Journal of Scientific and Statistical Computing **8** (1987), 171–184.

[16]  R.H. Mladineo, *An algorithm for finding the global maximum of a multimodal, multivariate function*. Mathematical Programming **34** (1986), 188–200.

[17]  Y. Nöé, *Estimation par le principe du minimax*. Mémoire de l'Université du Québec à Montréal, Montréal, Canada, 1995.

[18]  M. Nussbaum, *Spline smoothing in regression models and asymptotic efficiency in $L_2$*. Annals of Statistics **13** (1985), 984–997.

[19]  M.S. Pinsker, *Optimal filtering of square integrable signals in Gaussian white noise*. Problemy Peredachi Informatisii **16** (1980), 52–68 (in Russian). Problems Inform. Transmission (1980), 120–133 (in English).

[20]  S.A. Piyavskii, *An algorithm for finding the absolute extremum of a function*. USSR Computational Mathematics and Mathematical Physics **12** (1972), 57–67; (Zh. vychisl Mat. mat. Fiz. **12(4)** (1972), 888–896).

[21]  B. Vidacovic, and A. Dasgupta, *Efficiency of linear estimates for estimating a bounded normal mean*. Sankhya A (1996), 81–100.

DÉPARTEMENT DE MATHÉMATIQUES ET D'INFORMATIQUE, UNIVERSITÉ DU QUÉBEC À MONTRÉAL, MONTRÉAL, QUÉBEC CANADA H4B 1X8
*E-mail address*: `macgibbon.brenda@uqam.ca`

GERAD AND DÉPARTEMENT DE MATHÉMATIQUES APPLIQUÉES, ÉCOLE POLYTECHNIQUE
*Current address*: France Telecom — CNET-DAC-OAT, 38–40, Rue du General Leclerc, F–92794 Issy-Les-Moulineaux
*E-mail address*: `eric.gourdin@cnet.francetelecom.fr`

GERAD AND DÉPARTEMENT DE MATHÉMATIQUES APPLIQUÉES, ÉCOLE POLYTECHNIQUE, DÉPARTEMENT DE MATHÉMATIQUES ET DE GENIE INDUSTRIEL, MONTRÉAL, QUÉBEC H3C 3A7, CANADA
*E-mail address*: `brigitt@crt.umontreal.ca`

KEMPTHORNE ANALYTICS, SALEM, MA 01970 U.S.A.
*E-mail address*: `pjk@kemp.com`

Canadian Mathematical Society
Conference Proceedings
Volume 26, 2000

# Steady State Analysis with Heavy Traffic Limits for Semi-Open Networks

William A. Massey and Raj Srinivasan

*Dedicated to Professor Donald A. Dawson*

ABSTRACT. Semi-open queueing networks combine the features of open and closed (or flow controlled) queueing networks, and they belong to the class of non-product form queueing networks. The network we consider in this paper consists of $N+1$ exponential, single-server queues where one queue, the part of the network that is open, has an infinite buffer. The remaining $N$ queues form the closed part of the network, and they are constrained to have collectively no more than $K$ customers between them at all times. The first stage for each customer is to arrive to the open part of the network and receive service there *only* if the number of customers in the closed part of the network is strictly less than $K$. The second stage for service consists of leaving the open part to move through the closed part as if it were in a flow-controlled Jackson network, limited to having no more than $K$ customers. Any number of customers can be in the first open stage, but no more than $K$ customers can be in the second closed stage.

Previous work has treated such models for the special cases of $N = 1$ and $N = 2$. In this paper, we develop a new analysis that allows us to construct the steady state solution for the general $N$-node case where the closed network can have an arbitrary routing topology. We use tensor and Kronecker products to construct the steady state distribution and show that it has a matrix-geometric form. Moreover, we use this structure to show that the heavy traffic limit for the first stage results in an asymptotic product form for the entire network.

## 1. Introduction

Open and closed queueing network models have been found to be very useful in the performance modelling of computer, communication and manufacturing systems. However, many systems that have flow constraints can not be modelled as either open or closed queueing networks. Such systems arise in data networks and manufacturing job shops where parts of the system have limited capacities in terms of buffers or storage space. In this paper, we analyze a class of queueing networks

2000 *Mathematics Subject Classification.* Primary 60K25; Secondary 68M20.
*Key words and phrases.* Jackson Networks; closed Networks; matrix-geometric Solution; heavy-traffic limits; Kronecker products.
The second author was supported in part by a NSERC grant.

that captures such flow constraints in the system. The network we study consists of $N + 1$ exponential single server queues. The first queue is called the feeder queue and it has unlimited buffer space, whereas the remaining $N$ queues are constrained to have collectively no more than $K$ customers. We also assume that at each of the remaining $N$ queues the buffer is size is large enough to handle $K$ customers so that there is no internal blocking. External arrivals to the feeder queue follow a Poisson process. The first stage for the customer is to arrive at this feeder node and receive service only when the total number of customers in the remaining $N$ queues is strictly less than $K$. The second stage for the customer consists of leaving the first queue and traversing through the remaining $N$ queues receiving service and finally leaving the network. Once the customer is in the second stage, it behaves as if it is in a flow controlled open Jackson network, i.e., an open Jackson network with $N$ nodes restricted to having no more than a total of $K$ customers. Since the feeder queue has unlimited buffer space and the remaining queues are restricted to have collectively no more than $K$ customers, we call this network semi-open. One of the special features of this network is that the feeder queue gets blocked whenever the remaining network is full, i.e. when the total number of customers in the remaining queues is equal to $K$. Unlike open and closed queueing networks, even the simplest semi-open queueing network lacks product form solution.

Regarding related work, the literature on *semi-open* networks is very small. Konheim and Reiser [6] [7] developed exact results when the flow controlled part of the *semi-open network* consisted of single node($N = 1$), and a two node cyclic network. Latouche and Neuts [8] studied a two node serial network with blocking and provided algorithmic solutions using the matrix- geometric structure. The work of Kogan and Puhkalski [5] is closely related to this paper. They used the exact results of Konheim and Reiser [6] and showed that when the feeder queue is in heavy traffic, the *semi-open* network decouples into an $M|G|1$ queue, and an independent $M|M|1|K$ queue. In [11], we developed a methodology to obtain the parameters involved in the heavy traffic limit for the general *semi-open* network. See [1], [2], [3] and [12] for other related work on semi-open networks. This paper generalizes the work of Konheim and Reiser [6] [7], Latouche and Neuts [8] and Kogan and Puhkalski [5] simultaneously.

The aim of this paper is to provide exact and approximate solutions for this semi-open network. First, we use tensor and Kronecker products to construct the steady state distribution for this network, and show that it has the matrix-geometric structure. The computation efforts involved in obtaining the matrix-geometric solution are enormous, and are in particular very hard when $N$ and $K$ are large. However, we exploit this matrix-geometric structure to prove that in heavy traffic, the semi-open network decouples into a single server $M|G|1$ queue in heavy traffic and an independent flow controlled Jackson network (see Theorem 3.4 herein). The analysis of the flow controlled Jackson network is straight forward since it has the well known product form solution. However, the determination of the parameters defining the single server $M|G|1$ queue in heavy traffic are more involved. The mean and the variance of the service times in this $M|G|1$ queue are equal to the limiting mean and the variance of the output process from the flow controlled Jackson network with the feeder queue acting as the external Poisson source. The limiting mean of the output process from the flow controlled network can be calculated easily from its equilibrium distribution. In order to obtain the limiting variance of the output process for the flow controlled network, one needs to solve a set of linear equations. These equations are easier to solve than for the general steady state solution of the original semi-open network.

This paper is organized as follows. In Section 2 we describe in detail the semi-open network we analyze in this paper. We also introduce some of the basic tensor and Kronecker product definitions needed to state our results. The main results of this paper are provided in Section 3 where we describe the steady state solution of the semi-open network, and the heavy traffic limit. In Section 4, we derive an explicit expression for the limiting mean and the variance of the effective input process to the flow controlled Jackson network. In Section 5 we prove that the steady state solution of a semi-open network has a matrix-geometric structure and use this structure to prove the heavy traffic limit theorem. In Section 6, we consider a 3-node semi-open network and illustrate how one could obtain the parameters needed for the heavy traffic limit directly from the matrix-geometric solution. In Section 7, we develop hybrid approximation procedures for the average number in the feeder queue based on the heavy and light traffic limits. Numerical results are provided for a 3-node semi-open network when $K = 3$ in order to compare the quality of the heavy traffic approximation. The proof of existence of a solution to the matrix quadratic equation is provided in the appendix (Section 8).

## 2. Semi-open network

The semi-open network we study in this paper consists of two sub systems. The first part is the feeder queue with a single exponential server (service rate $\mu_0$), infinite buffer, and an external Poisson arrival process (rate $\lambda_0$). Let $Q_0(t)$ be the queue length process (including the customer in service) at the feeder queue. The second part of the system consists of $N$ exponential single server queues with service rates $\mu_1$, $\mu_2$, ..., $\mu_N$ respectively at the queues 1, 2, ..., $N$. A customer at the feeder queue receives service only when the total number of customers in the $N$ queues is strictly less than $K$( a pre specified number). Otherwise, the feeder queue is blocked. The service resumes at the feeder queue when there is a departure from one of the $N$ queues. In general, service can proceed uninterrupted when the total number of customers among the $N$ queues is strictly less than $K$. After receiving service at the feeder queue, a customer visits station $i$ with probability $p_{0i}$, $i = 1, 2 \ldots, N$. Let $p_{ij}$, $i, j = 1, 2 \ldots, N$ be the probability that a customer after receiving service at the $i^{th}$ queue visits the $j^{th}$ queue. We also let $q_i = 1 - \sum_{j=1}^{N} p_{ij}$, $i = 1, 2, \ldots, N$ be the probability that a customer leaves the semi-open network after receiving service at the $i^{th}$ node. Let $\mathbf{Q}(t) = (Q_1(t), Q_2(t), \ldots, Q_N(t))$ be the vector queue length process representing the number of customers at each of the $N$ queues. Now consider the queue length process $(Q_0(t), \mathbf{Q}(t))$ of the semi-open network. We are interested in the steady state distribution of this process. First note that if the queue length at the feeder queue becomes infinite, then the remaining network $\mathbf{Q}(t)$ behaves like an flow controlled Jackson network with the feeder queue acting as the Poisson source. A flow controlled Jackson network differs from a Jackson network in that the Poisson arrivals are blocked and lost if there are $K$ customers in the network. Let $\mathbf{Q}^*(t)$ be queue length process of a flow controlled Jackson network with external Poisson arrival with rate $\mu_0$. We will define this flow controlled network $\mathbf{Q}^*(t)$ as the one associated with the semi-open network $(Q_0(t), \mathbf{Q}(t))$. If we let $\delta$ be the equilibrium output rate of the associated flow controlled network $\mathbf{Q}^*(t)$, then one can show that the semi-open network is stable

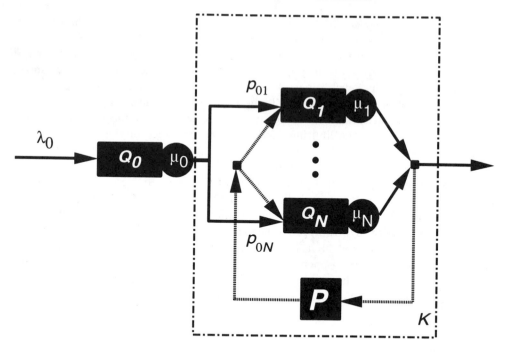

FIGURE 1. Semi-open network.

whenever $\lambda_0 < \delta$. See Massey and Srinivasan[11] for further details. Since $\delta$ is the equilibrium output rate of the flow controlled network, it is also equal to the effective input rate to this network. Note that due to the constraint on the total number of customers in the flow controlled network, the external Poisson input to this network can be interrupted. The calculation of the limiting output rate of the departure process $D(t)$ from the flow controlled network was carried out in [11]. In Section 3, we show how one can alternatively determine $\delta$ from the effective input process $A_e^*(t)$ to the associated network.

Before we state our results, we briefly review some basic tensor and Kronecker product machinery introduced in [9]. These basic definitions will help us to write the generator for the semi-open network in a compact form. If $\mathbf{f}$ and $\mathbf{g}$ both belong to $\ell(Z_+)$, the space of real-valued functions on $Z_+$, let $\mathbf{f} \otimes \mathbf{g}$ be a function in $\ell(Z_+^2)$, where

$$[\mathbf{f} \otimes \mathbf{g}](m, n) \equiv \mathbf{f}(m)\mathbf{g}(n)$$

For example, $\mathbf{f} = \sum_{n \in Z_+} \mathbf{f}(n)\mathbf{e}_n$ for all $\mathbf{f} \in \ell(Z_+)$, where

$$\mathbf{e}_n(m) = \begin{cases} 1 & n = m, \\ 0 & n \neq m. \end{cases}$$

If $\mathbf{h}$ belongs to $\ell(Z_+^2)$, then

$$\mathbf{h} = \sum_{\mathbf{m} \in Z_+^2} \mathbf{h}(\mathbf{m})\mathbf{e}_{m_1} \otimes \mathbf{e}_{m_2}$$

and so $\ell(Z_+^2) = \ell(Z_+)^{(2)} \equiv \ell(Z_+) \otimes \ell(Z_+)$. If $\mathbf{A}$ and $\mathbf{B}$ are both linear operators on $\ell(Z_+)$, then we define $\mathbf{A} \otimes \mathbf{B}$ to be a linear operator on $\ell(Z_+^2)$ where

$$(\mathbf{e}_{m_1} \otimes \mathbf{e}_{m_2})[\mathbf{A} \otimes \mathbf{B}] = (\mathbf{e}_{m_1}\mathbf{A}) \otimes (\mathbf{e}_{m_2}\mathbf{B}).$$

Note that

$$\mathbf{A} \otimes \mathbf{B} = [\mathbf{A} \otimes \mathbf{I}] \cdot [\mathbf{I} \otimes \mathbf{B}].$$

Now define $\mathbf{R}$ and $\mathbf{L}$ respectively be the *right and left shift operators* on $\ell(Z_+)$ such that for all $n \in Z_+ = \{0, 1, \dots\}$,

$$\mathbf{e}_n\mathbf{R} = \mathbf{e}_{n+1}$$

and

$$\mathbf{e}_n\mathbf{L} = \begin{cases} \mathbf{e}_{n-1} & \text{if } n > 0, \\ \mathbf{0} & \text{otherwise.} \end{cases}$$

From the definitions given above it is easy to see that tensor products of unit basis vectors such as $\mathbf{e}_{m_1} \otimes \mathbf{e}_{m_2}$ acts on Kronecker operators such as $\mathbf{R} \otimes \mathbf{I}$ as follows:

$$(\mathbf{e}_{m_1} \otimes \mathbf{e}_{m_2})\,(\mathbf{R} \otimes \mathbf{I}) = \mathbf{e}_{m_1+1} \otimes \mathbf{e}_{n_2}.$$

The above definitions can be easily extended to $N$-fold tensor and Kronecker products. Using the primitive *right and left shift operators* on $\ell(Z_+)$, we define the following Kronecker products. See [9] for further details.

$$\mathbf{R}_i = \mathbf{I} \otimes \cdots \otimes \mathbf{R} \otimes \cdots \otimes \mathbf{I} \quad (i\text{-th position}),$$

$$\mathbf{L}_i = \mathbf{I} \otimes \cdots \otimes \mathbf{L} \otimes \cdots \otimes \mathbf{I} \quad (i\text{-th position}),$$

and

$$(\mathbf{e}_{m_1} \otimes \cdots \otimes \mathbf{e}_{m_N})\mathbf{I}(K) = \begin{cases} \mathbf{e}_{m_1} \otimes \cdots \otimes \mathbf{e}_{m_N} & \text{if } \sum_{i=1}^N m_i < K, \\ \mathbf{0} & \text{otherwise.} \end{cases}$$

Let $\hat{\mathbf{Q}}(t) = (Q_0(t), Q_1(t), \dots, Q_N(t)) = (Q_0(t), \mathbf{Q}(t))$ be the queue length process of the semi-open network. If $\boldsymbol{\pi}[\hat{\mathbf{Q}}]$ is the steady state distribution of $\hat{\mathbf{Q}}(t)$, then it can be expressed in tensor notation as

$$\boldsymbol{\pi}[\hat{\mathbf{Q}}] = \sum_{n \in \mathbf{Z}_+} \sum_{|\mathbf{m}| \le K} \mathrm{P}(Q_0 = n, \mathbf{Q} = \mathbf{m})\mathbf{e}_n \otimes \mathbf{e}_\mathbf{m}$$

which belongs to $\ell(\mathbf{Z}_+)^{(N+1)}$, where

$$\mathbf{e}_\mathbf{m} = \mathbf{e}_{m_1} \otimes \mathbf{e}_{m_2} \cdots \otimes \mathbf{e}_{m_N}.$$

Now the generator for the semi-open network can be written in compact form as

$$\mathbf{A}[\hat{\mathbf{Q}}] = \lambda_0\mathbf{R}_0 + \mu_0\mathbf{L}_0 \cdot \mathbf{I}(K) \cdot \sum_{i=1}^N p_{0i}\mathbf{R}_i + \sum_{i=1}^N \left[ \mu_i q_i\mathbf{L}_i + \sum_{j=1}^N \mu_i p_{ij}\mathbf{L}_i\mathbf{R}_j \right]$$

$$-\lambda_0\mathbf{I} - \mu_0\mathbf{L}_0\mathbf{R}_0 \cdot \mathbf{I}(K) - \sum_{i=1}^N \mu_i\mathbf{L}_i\mathbf{R}_i,$$

where we have slightly abused the notation and let $\mathbf{I}$ represent $\mathbf{I} \otimes \cdots \mathbf{I} \otimes \mathbf{I}$. Given the one to one correspondence between $(n, m_1, m_2 \dots m_N)$ and the basis vector $\mathbf{e}_n \otimes \mathbf{e}_{m_1} \cdots \otimes \mathbf{e}_{m_N}$, we see that $\mathbf{R}_i$ encodes an arrival of a customer to the $i^{\text{th}}$ queue and $\mathbf{L}_i$ encodes the departure of a customer from the $i^{\text{th}}$ queue. In particular, the first term in the generator represents an external arrival to the feeder queue.

The second term represents an arrival of a departing customer from the feeder queue to the $i^{\text{th}}$ queue provided that there are less than $K$ customers in the flow controlled part. Other terms in the generators can be interpreted in a similar way. The generator for the semi-open network can be decomposed further by grouping the terms representing the transitions of a customer who has already joined the controlled part. Let

$$(2.1) \qquad \mathbf{B} = \sum_{i=1}^{N} \left[ \mu_i q_i \mathbf{L}_i + \sum_{j=1}^{N} \mu_i p_{ij} \mathbf{L}_i \mathbf{R}_j - \mu_i \mathbf{L}_i \mathbf{R}_i \right].$$

Now the generator $\mathbf{A}[\hat{\mathbb{Q}}]$ reduces to

$$(2.2) \qquad \mathbf{A}[\mathbb{Q}] \equiv \lambda_0 \mathbf{R}_0 + \mu_0 \mathbf{L}_0 \cdot \mathbf{I}(K) \sum_{i=1}^{N} p_{0i} \mathbf{R}_i - \lambda_0 \mathbf{I} - \mu_0 \mathbf{L}_0 \mathbf{R}_0 \cdot \mathbf{I}(K) + \mathbf{B}$$

## 3. Results

THEOREM 3.1. *If $\lambda_0 < \delta$, then the steady state distribution for the semi-open network has a matrix-geometric solution, or*

$$(3.1) \qquad \mathsf{P}(Q_0 = n, \mathbf{Q} = \mathbf{m}) = \mathbf{g}\boldsymbol{\Omega}^n \cdot \mathbf{e_m^T},$$

*for all $n \in \mathbb{Z}_+$ and $\mathbf{m} \in \mathbb{Z}_+^N$, where $\boldsymbol{\Omega}$ is an operator on tensors of rank $N$ that is the minimal non-negative solution to*

$$(3.2) \qquad \lambda_0 \mathbf{I} + \boldsymbol{\Omega}^2 \mu_0 \mathbf{I}(K) \sum_{i=1}^{N} p_{0i} \mathbf{R}_i + \boldsymbol{\Omega}(\mathbf{B} - \lambda_0 \mathbf{I} - \mu_0 \mathbf{I}(K)) = \mathbf{0},$$

*and the rank $N$ tensor $\mathbf{g}$ is the unique non-negative solution to*

$$(3.3) \qquad \mathbf{g}[\mu_0 \boldsymbol{\Omega} \cdot \mathbf{I}(K) \sum_{i=1}^{N} p_{0i} \mathbf{R}_i + \mathbf{B} - \lambda_0 \mathbf{I}] = \mathbf{0},$$

*such that $\mathbf{g}(\mathbf{I} - \boldsymbol{\Omega})^{-1} \cdot \mathbf{1}^T = 1$, where $\mathbf{1}$ is a vector of ones.*

Below, we list some of the probabilistic properties of $\mathbf{g}$ and $\boldsymbol{\Omega}$.

$$\begin{aligned}
\mathbf{g} \cdot \mathbf{e_m^T} &= \mathsf{P}(Q_0 = 0, \mathbf{Q} = \mathbf{m}) \\
\mathbf{g}\boldsymbol{\Omega}^n \cdot \mathbf{e_m^T} &= \mathsf{P}(Q_0 = n, \mathbf{Q} = \mathbf{m}) \\
\mathbf{g}(\mathbf{I} - \boldsymbol{\Omega})^{-1} \cdot \mathbf{e_m^T} &= \mathsf{P}(\mathbf{Q} = \mathbf{m}) \\
\mathbf{g}\boldsymbol{\Omega}^n \mathbf{I}(K) \cdot \mathbf{1}^T &= \mathsf{P}(Q_0 = n, |\mathbf{Q}| < K) \\
\mathbf{g}(\mathbf{I} - \boldsymbol{\Omega})^{-1} \mathbf{1}^T &= 1.
\end{aligned}$$

COROLLARY 3.2. *The marginal steady state distribution for the scheduler queue $Q_0$ satisfies a quasi local balance condition, or*

$$(3.4) \qquad \lambda_0 \mathsf{P}(Q_0 = n) = \mu_0 \mathsf{P}(Q_0 = n+1, |\mathbf{Q}| < K)$$

*for all $n \in \mathbb{Z}_+$.*

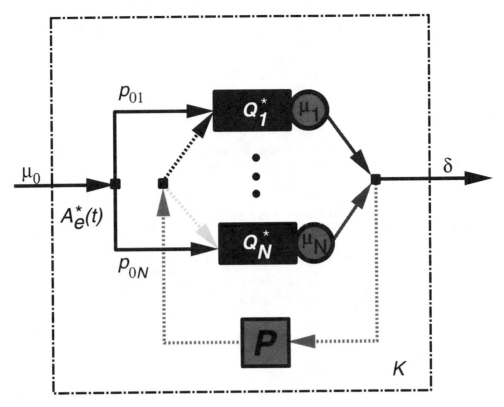

FIGURE 2. Associated window flow controlled Jackson network.

PROPOSITION 3.3. *If* $\mathbf{Q}$ *represents the marginal stationary vector queue length process corresponding to the* $N$ *queues, and* $\mathbf{Q}^*$ *represents the stationary distribution for the associated flow controlled network, then*

$$(3.5) \qquad \lim_{\lambda_0 \uparrow \delta} \mathsf{P}\left(\mathbf{Q} = \mathbf{m}\right) = \mathsf{P}\left(\mathbf{Q}^* = \mathbf{m}\right) = \frac{\prod_{i=1}^{N} \rho_i^{m_i}}{G(N, K)}.$$

*where* $\rho_i = \theta_i/\mu_i$ *and* $(\theta_1, \theta_2, \ldots \theta_n)$ *is the solution to*

$$(3.6) \qquad \theta_i = \mu_0 p_{0i} + \sum_{j=1}^{N} \theta_j p_{ji}$$

*and*

$$(3.7) \qquad G(N, K) = \sum_{|\mathbf{m}| \leq K} \prod_{i=1}^{N} \rho_i^{m_i}.$$

THEOREM 3.4. *The scheduler queue* $Q_0$ *is in heavy traffic when* $\lambda_0$ *approaches* $\delta$, *which gives us*

$$(3.8) \qquad \lim_{\lambda_0 \uparrow \delta} \mathsf{P}\left((\delta - \lambda_0)Q_0 \leq x, \mathbf{Q} = \mathbf{m}\right) = (1 - e^{-x/\sigma'(\delta)}) \cdot \mathsf{P}\left(\mathbf{Q}^* = \mathbf{m}\right),$$

*for all real $x \geq 0$ and all $\mathbf{m} \in \mathbb{Z}_+^N$, where $\sigma(\lambda_0)$, the maximum positive eigenvalue of $\Omega$ is viewed as a differentiable function of $\lambda_0$. Moreover,*

(3.9)   $\dfrac{1}{\sigma'(\delta)} = \lim_{t \to \infty} \dfrac{\mathsf{E}\left[A_e^*(t)\right] + \mathsf{Var}\left[A_e^*(t)\right]}{2t} = \delta + \mu_0 \cdot \lim_{t \to \infty} \mathsf{Cov}\left[A_e^*(t); |\mathbf{Q}(t)| < K\right].$

*Given the hypothesis above, we have*

(3.10)                    $\lim_{\lambda \to \delta}(\delta - \lambda)\mathsf{E}\left[Q_0\right] = \sigma'(\delta).$

The parameter $\delta$ can be calculated explicitly from the steady state distribution of the associated flow control network which has the product form solution (see equation 4.11 herein). In order to calculate the covariance term appearing in Theorem 3.4, one has to solve a system of equations given in Corollary 4.3. However this system of equations are much easier to solve than for the general steady state solution of the original network.

## 4. Analysis of flow controlled Jackson networks

Let $A_e^*(t)$ be the effective input process to the flow controlled network $\mathbf{Q}^*(t)$. Then the transient distribution tensors associated with $A_e^*(t)$ and $\mathbf{Q}^*(t)$ are defined as follows.

(4.1)          $\mathbf{P}\left[\mathbf{Q}^*(t)\right] \equiv \displaystyle\sum_{\mathbf{m} \in \mathbb{Z}_+^N} \Pr\left(\mathbf{Q}^*(t) = \mathbf{m}\right)\mathbf{e_m}$

(4.2)          $\mathbf{P}\left[A_e^*(t)\right] \equiv \displaystyle\sum_{n \in \mathbb{Z}_+} \Pr\left(A_e^*(t) = n\right)\mathbf{e}_n$

(4.3)   $\mathbf{P}\left[A_e^*(t), \mathbf{Q}^*(t)\right] \equiv \displaystyle\sum_{n \in \mathbb{Z}_+}\sum_{\mathbf{m} \in \mathbb{Z}_+^N} \Pr\left(A_e^*(t) = n, \mathbf{Q}^*(t) = \mathbf{m}\right)\mathbf{e}_n \otimes \mathbf{e_m}$

(4.4)   $\boldsymbol{\Gamma}\left[A_e^*(t), \mathbf{Q}^*(t)\right] \equiv \displaystyle\sum_{n \in \mathbb{Z}_+}\sum_{\mathbf{m} \in \mathbb{Z}_+^N} \mathsf{Cov}\left[1_{A_e^*(t)=n}, 1_{\mathbf{Q}^*(t)=\mathbf{m}}\right]\mathbf{e}_n \otimes \mathbf{e_m}$

(4.5)      $\boldsymbol{\Gamma}_A\left[\mathbf{Q}^*(t)\right] \equiv \displaystyle\sum_{\mathbf{m} \in \mathbb{Z}_+^N} \mathsf{Cov}\left[A_e^*(t), 1_{\mathbf{Q}^*(t)=\mathbf{m}}\right]\mathbf{e_m}$

Now observe that

$$\lim_{t \to \infty} \frac{\mathsf{E}[A_e^*(t)]}{t} = \lim_{t \to \infty} \frac{d}{dt}\mathsf{E}[A_e^*(t)]$$

and

$$\lim_{t \to \infty} \frac{\mathsf{Var}[A_e^*(t)]}{t} = \lim_{t \to \infty} \frac{d}{dt}\mathsf{Var}[A_e^*(t)].$$

Also note that

$$\frac{d}{dt}\mathsf{Var}\left[A_e^*(t)\right] = \frac{d}{dt}\mathsf{E}\left[A_e^*(t)^2\right] - 2\mathsf{E}\left[A_e^*(t)\right]\frac{d}{dt}\mathsf{E}\left[A_e^*(t)\right].$$

As the next proposition shows the calculation of the time derivative of the variance of the effective input process involves covariance terms.

PROPOSITION 4.1. *For all $t \geq 0$, we have*

(4.6)                    $\dfrac{d}{dt}\mathsf{E}\left[A_e^*(t)\right] = \mu_0\Pr\left(|\mathbf{Q}^*(t)| < K\right)$

*and*

$$(4.7) \qquad \frac{d}{dt} \mathsf{Var} \left[ A_e^*(t) \right] = 2\mu_0 \mathsf{Cov} \left[ A_e^*(t); |\mathbf{Q}^*(t)| < K \right] + \mu_0 \mathsf{Pr} \left( |\mathbf{Q}^*(t)| < K \right)$$

PROOF. The above formulas follow immediately from the functional version of the forward equations for the Markov process $(A_e^*(t), \mathbf{Q}^*(t))$, i.e.

$$(4.8) \qquad \frac{d}{dt} \mathsf{E} \left[ f(A_e^*(t)) \right] = \mu_0 \mathsf{E} \left[ f(A_e^*(t) + 1) - f(A_e^*(t)); |\mathbf{Q}^*(t)| < K \right].$$

Now apply this formula to the two cases of $f(n) = n$ and $f(n) = n^2$.          □

THEOREM 4.2. *Let* $\mathbf{A}[\mathbf{Q}^*]$ *be the Markov generator for the associated flow controlled network and let*

$$(4.9) \qquad \boldsymbol{\pi}[\mathbf{Q}^*] \equiv \lim_{t \to \infty} \mathbf{P} \left[ \mathbf{Q}^*(t) \right]$$

*Then for all* $t \geq 0$, *we have*

$$
\begin{aligned}
\frac{d}{dt} \boldsymbol{\Gamma}_A \left[ \mathbf{Q}^*(t) \right] &= \boldsymbol{\Gamma}_A \left[ \mathbf{Q}^*(t) \right] \mathbf{A}[\mathbf{Q}^*] \\
(4.10) \qquad &+ \mathbf{P} \left[ \mathbf{Q}^*(t) \right] \left( \sum_{i=1}^N \mu_0 p_{0i} \mathbf{I}(K) \mathbf{R}_i - \mu_0 \mathsf{Pr} \left( |\mathbf{Q}^*(t)| < K \right) \mathbf{I} \right).
\end{aligned}
$$

*and*

$$(4.11) \qquad \delta \equiv \lim_{t \to \infty} \mu_0 \mathsf{Pr} \left( |\mathbf{Q}^*(t)| < K \right) = \boldsymbol{\pi}[\mathbf{Q}^*] \mathbf{I}(K) \cdot \mathbf{1}^\mathsf{T},$$

*where* $\delta$ *equals the limiting effective input rate, which is also the net output rate for the network.*

COROLLARY 4.3. *The limit for* $\boldsymbol{\Gamma}_A[\mathbf{Q}^*(t)]$ *as* $t \to \infty$ *exists and equals* $\boldsymbol{\Gamma}_A[\mathbf{Q}^*]$, *and it satisfies the following system of equations.*

$$(4.12) \qquad \boldsymbol{\Gamma}_A[\mathbf{Q}^*] \cdot \mathbf{A}[\mathbf{Q}^*] = \boldsymbol{\pi}[\mathbf{Q}^*] \left( \delta \mathbf{I} - \sum_{i=1}^N \mu_0 p_{0i} \mathbf{I}(K) \mathbf{R}_i \right).$$

PROOF OF THEOREM 4.2. We use the fact that

$$(4.13) \qquad \frac{d}{dt} \mathbf{P} \left[ A_e^*(t) \right] = \mathbf{P} \left[ A_e^*(t); |\mathbf{Q}^*(t)| < K \right] \cdot \mu_0 (\mathbf{R} - \mathbf{I}),$$

$$(4.14) \qquad \frac{d}{dt} \mathbf{P} \left[ \mathbf{Q}^*(t) \right] = \mathbf{P} \left[ \mathbf{Q}^*(t) \right] \cdot \mathbf{A}[\mathbf{Q}^*],$$

and
(4.15)

$$\frac{d}{dt} \mathbf{P} \left[ A_e^*(t), \mathbf{Q}^*(t) \right] = \mathbf{P} \left[ A_e^*(t), \mathbf{Q}^*(t) \right] \cdot \left( \mathbf{I} \otimes \mathbf{A}[\mathbf{Q}^*] + (\mathbf{R} - \mathbf{I}) \otimes \sum_{i=1}^N \mu_0 p_{0i} \mathbf{I}(K) \mathbf{R}_i \right).$$

Now we combine these results with the identity

$$(4.16) \qquad \boldsymbol{\Gamma} \left[ A_e^*(t), \mathbf{Q}^*(t) \right] = \mathbf{P} \left[ A_e^*(t), \mathbf{Q}^*(t) \right] - \mathbf{P} \left[ A_e^*(t) \right] \otimes \mathbf{P} \left[ \mathbf{Q}^*(t) \right]$$

and apply the operator $\mathbf{f}^\mathsf{T} \otimes \mathbf{I}$, where $\mathbf{f} \equiv [0, 1, 2, \dots]$ so

$$(4.17) \qquad \mathbf{e}_n \otimes \mathbf{e_m} \cdot \mathbf{f}^\mathsf{T} \otimes \mathbf{I} = n \cdot \mathbf{e_m}.$$

Using the additional identities of

(4.18) $\qquad (\mathbf{R} - \mathbf{I}) \cdot \mathbf{f}^{\mathsf{T}} = \mathbf{1}^{\mathsf{T}}$ and $\quad \mathbf{\Gamma}\left[A_e^*(t), \mathbf{Q}^*(t)\right] \cdot \mathbf{f}^{\mathsf{T}} \otimes \mathbf{I} = \mathbf{\Gamma}_A\left[\mathbf{Q}^*(t)\right]$

gives us the theorem. $\hfill\square$

## 5. Proving the results

PROOF OF THEOREM 3.1. Recall that the generator for the semi-open network $\mathbf{A}[\mathbb{Q}]$ is given by

(5.1) $\qquad \mathbf{A}[\mathbb{Q}] \equiv \lambda_0 \mathbf{R}_0 + \mu_0 \mathbf{L}_0 \cdot \mathbf{I}(K) \sum_{i=1}^{N} p_{0i} \mathbf{R}_i - \lambda_0 \mathbf{I} - \mu_0 \mathbf{L}_0 \mathbf{R}_0 \cdot \mathbf{I}(K) + \mathbf{B}$

where

(5.2) $$\mathbf{B} = \sum_{i=1}^{N} \left[ \mu_i q_i \mathbf{L}_i + \sum_{j=1}^{N} \mu_i p_{ij} \mathbf{L}_i \mathbf{R}_j - \mu_i \mathbf{L}_i \mathbf{R}_i \right].$$

Decomposing this generator into $\mathbf{A}[\mathbb{Q}] = \mathbf{A}[\mathbb{Q}] \cdot \mathbf{L}_0 \mathbf{R}_0 + \mathbf{A}[\mathbb{Q}] \cdot (\mathbf{I} - \mathbf{L}_0 \mathbf{R}_0)$, we have

(5.3) $\quad \mathbf{A}[\mathbb{Q}] \cdot \mathbf{L}_0 \mathbf{R}_0 = \left[ \lambda_0 \mathbf{I} + \mu_0 \mathbf{L}_0^2 \cdot \mathbf{I}(K) \sum_{i=1}^{N} p_{0i} \mathbf{R}_i + \mathbf{L}_0 (\mathbf{B} - \lambda_0 \mathbf{I} - \mu_0 \mathbf{I}(K)) \right] \mathbf{R}_0$

and

(5.4) $\qquad \mathbf{A}[\mathbb{Q}] \cdot (\mathbf{I} - \mathbf{L}_0 \mathbf{R}_0) = \left[ \mu_0 \mathbf{L}_0 \cdot \mathbf{I}(K) \sum_{i=1}^{N} p_{0i} \mathbf{R}_i - \lambda_0 \mathbf{I} + \mathbf{B} \right] (\mathbf{I} - \mathbf{L}_0 \mathbf{R}_0).$

Now suppose that $\boldsymbol{\pi}[\mathbb{Q}]$, the steady state distribution tensor of rank $N + 1$ for $\mathbf{A}[\mathbb{Q}]$, has the following matrix-geometric form:

(5.5) $\qquad\qquad\qquad \boldsymbol{\pi}[\mathbb{Q}] \cdot \mathbf{L}_0 = \boldsymbol{\pi}[\mathbb{Q}] \cdot \mathbf{I} \otimes \boldsymbol{\Omega}.$

This is equivalent to having

(5.6) $\qquad\qquad\qquad \boldsymbol{\pi}[\mathbb{Q}] = \sum_{n=0}^{\infty} \mathbf{e}_n \otimes \mathbf{g} \boldsymbol{\Omega}^n.$

Now $\boldsymbol{\pi}[\mathbb{Q}] \cdot \mathbf{A}[\mathbb{Q}] = 0$ implies that $\boldsymbol{\Omega}$ is the minimal non-negative solution to

(5.7) $\qquad \lambda_0 \mathbf{I} + \boldsymbol{\Omega}^2 \mu_0 \mathbf{I}(K) \sum_{i=1}^{N} p_{0i} \mathbf{R}_i + \boldsymbol{\Omega}(\mathbf{B} - \lambda_0 \mathbf{I} - \mu_0 \mathbf{I}(K)) = \mathbf{0},$

and the rank $N$ tensor $\mathbf{g}$ will then be the unique non-negative solution to

(5.8) $\qquad\qquad \mathbf{g} \left[ \mu_0 \boldsymbol{\Omega} \cdot \mathbf{I}(K) \sum_{i=1}^{N} p_{0i} \mathbf{R}_i + \mathbf{B} - \lambda_0 \mathbf{I} \right] = \mathbf{0}.$

Now we want to show that a solution to (5.7) exists. If we let $\mathbf{C} \equiv (\lambda_0 \mathbf{I} + \mu_0 \mathbf{I}(K) - \mathbf{B})^{-1}$, then we can rewrite (5.7) as

(5.9) $\qquad\qquad \boldsymbol{\Omega} = \lambda_0 \mathbf{C} + \boldsymbol{\Omega}^2 \mu_0 \sum_{i=1}^{N} p_{0i} \mathbf{I}(K) \mathbf{R}_i \mathbf{C}$

Since $\mathbf{C}$ is a non-negative matrix, all the matrix coefficients in this reformulated equation are also. This suggests that $\boldsymbol{\Omega}$ can be constructed via the recursion relation

$$(5.10) \qquad \boldsymbol{\Omega}_{n+1} = \lambda_0 \mathbf{C} + \boldsymbol{\Omega}_n^2 \mu_0 \sum_{i=1}^N p_{0i} \mathbf{I}(K) \mathbf{R}_i \mathbf{C}$$

Starting with $\boldsymbol{\Omega}_0 = \mathbf{0}$, the sequence of $\boldsymbol{\Omega}_n$'s will be monotone and its limiting matrix will be $\boldsymbol{\Omega}$, provided that this sequence is bounded above in some operator norm. The proof of boundedness of $\boldsymbol{\Omega}_n$'s is provided in the appendix.

Since $\boldsymbol{\Omega}$ exists, using (5.7), we see that $\boldsymbol{\Omega}$ is an invertible matrix. Moreover since all invertible matrices commute with their inverse, we have

$$(5.11) \qquad \left( \boldsymbol{\Omega}\mu_0 \mathbf{I}(K) \sum_{i=1}^N p_{0i} \mathbf{R}_i + \mathbf{B} - \lambda_0 \mathbf{I} \right) \boldsymbol{\Omega} = \mu_0 \mathbf{I}(K)\boldsymbol{\Omega} - \lambda_0 \mathbf{I}.$$

This shows us that solving for (5.8) is equivalent to solving for

$$(5.12) \qquad \mathbf{g}\mathbf{I}(K)\boldsymbol{\Omega} = \frac{\lambda_0}{\mu_0}\mathbf{g}.$$

Such a $\mathbf{g}$ must exist since from (5.7) we get

$$(5.13) \qquad (\mathbf{I} - \boldsymbol{\Omega})(\lambda_0 \mathbf{1} - \mu_0 \boldsymbol{\Omega}\mathbf{I}(K)\mathbf{1}) = \mathbf{0},$$

and requiring the spectral radius of $\boldsymbol{\Omega}$ to be less than unity, gives us

$$(5.14) \qquad (\lambda_0 \mathbf{1} - \mu_0 \boldsymbol{\Omega} \cdot \mathbf{I}(K)\mathbf{1}) = \mathbf{0}.$$

This means that the column vector of ones serves as a right eigenvector for $\boldsymbol{\Omega}\mathbf{I}(K)$, with eigenvalue $\lambda_0/\mu_0$. Now $\mathbf{I}(K)\boldsymbol{\Omega}$ has the same eigenvalue, and by Perron-Frobenious theory, there must exist a non-negative vector $\mathbf{g}$ that serves as the corresponding *left* eigenvector to $\mathbf{I}(K)\boldsymbol{\Omega}$. Applying $\mathbf{I}(K)$ to both sides, we get

$$(5.15) \qquad \mathbf{g}\mathbf{I}(K) \cdot (\mathbf{I}(K)\boldsymbol{\Omega}\mathbf{I}(K)) = \frac{\lambda_0}{\mu_0}\mathbf{g}\mathbf{I}(K).$$

PROOF OF PROPOSITION 3.3. This follows immediately from the lemma below:

LEMMA 5.1. *If $\mathbf{A}[\mathbf{Q}^*]$ is the generator for the associated flow controlled network, then*

$$(5.16) \qquad \mathbf{g}(\mathbf{I} - \boldsymbol{\Omega})^{-1}\mathbf{A}[\mathbf{Q}^*] = \mu_0\mathbf{g}\mathbf{I}(K) \left( \sum_{i=1}^N p_{0i}\mathbf{R}_i - \mathbf{I} \right)$$

*where*

$$(5.17) \qquad \mathbf{A}[\mathbf{Q}^*] = \mu_0\mathbf{I}(K) \left( \sum_{i=1}^N p_{0i}\mathbf{R}_i - \mathbf{I} \right) + \mathbf{B}.$$

*and*

$$(5.18) \qquad \mathbf{B} = \sum_{i=1}^N \left[ \mu_i q_i \mathbf{L}_i + \sum_{j=1}^N \mu_i p_{ij} \mathbf{L}_i \mathbf{R}_j - \mu_i \mathbf{L}_i \mathbf{R}_i \right].$$

PROOF. Using (5.8), we have

$$(5.19) \qquad \mathbf{g}(\mathbf{I} - \boldsymbol{\Omega})^{-1}(\mathbf{I} - \boldsymbol{\Omega})[\mu_0\boldsymbol{\Omega} \cdot \mathbf{I}(K) \sum_{i=1}^N p_{0i}\mathbf{R}_i + \mathbf{B} - \lambda_0\mathbf{I}] = \mathbf{0}.$$

2 WILLIAM A. MASSEY AND RAJ SRINIVASAN

Since $\boldsymbol{\Omega}$ solves a matrix quadratic equation, we have

$$(\mathbf{I} - \boldsymbol{\Omega})[\mu_0\boldsymbol{\Omega} \cdot \mathbf{I}(K)\sum_{i=1}^{N} p_{0i}\mathbf{R}_i + \mathbf{B} - \lambda_0\mathbf{I}]$$

$$= \mu_0\boldsymbol{\Omega} \cdot \mathbf{I}(K)\sum_{i=1}^{N} p_{0i}\mathbf{R}_i + \mathbf{B} - \lambda_0\mathbf{I}$$

$$-\mu_0\boldsymbol{\Omega}^2 \cdot \mathbf{I}(K)\sum_{i=1}^{N} p_{0i}\mathbf{R}_i - \boldsymbol{\Omega}\mathbf{B} + \lambda_0\boldsymbol{\Omega}$$

$$= \mu_0\boldsymbol{\Omega} \cdot \mathbf{I}(K)\sum_{i=1}^{N} p_{0i}\mathbf{R}_i + \mathbf{B} - \lambda_0\mathbf{I}$$

$$+\lambda_0\mathbf{I} + \boldsymbol{\Omega}(\mathbf{B} - \lambda_0\mathbf{I} - \mu_0\mathbf{I}(K)) - \boldsymbol{\Omega}\mathbf{B} + \lambda_0\boldsymbol{\Omega}$$

$$= \mu_0\boldsymbol{\Omega} \cdot \mathbf{I}(K) \cdot \left[\sum_{i=1}^{N} p_{0i}\mathbf{R}_i - \mathbf{I}\right] + \mathbf{B}.$$

Observing that $(\mathbf{I} - \boldsymbol{\Omega})^{-1}\boldsymbol{\Omega} = (\mathbf{I} - \boldsymbol{\Omega})^{-1} - \mathbf{I}$ completes the proof. $\square$

PROOF OF THEOREM 3.4. Consider the characteristic function

$$(5.20) \qquad \mathsf{E}\left[e^{i\theta(1-\sigma)Q_0}; \mathbf{Q} = \mathbf{m}\right] = \mathbf{g}(\mathbf{I} - e^{i\theta(1-\sigma)}\boldsymbol{\Omega})^{-1} \cdot \mathbf{e_m},$$

where $\sigma$ is the maximum positive eigenvalue of $\boldsymbol{\Omega}$. However, we can rewrite the right hand side as

$$(5.21)\ \mathbf{g}(\mathbf{I} - e^{i\theta(1-\sigma)}\boldsymbol{\Omega})^{-1} = \mathbf{g}(\mathbf{I} - \boldsymbol{\Omega})^{-1}(\mathbf{I} - \boldsymbol{\Omega})(\mathbf{I} - e^{i\theta(1-\sigma)}\boldsymbol{\Omega})^{-1}$$

$$(5.22) \qquad = \mathbf{g}(\mathbf{I} - \boldsymbol{\Omega})^{-1}\left[(\mathbf{I} - \boldsymbol{\Omega})^{-1}(\mathbf{I} - e^{i\theta(1-\sigma)}\boldsymbol{\Omega})\right]^{-1}$$

$$(5.23) \qquad = \mathbf{g}(\mathbf{I} - \boldsymbol{\Omega})^{-1}\left[\mathbf{I} + (1 - e^{i\theta(1-\sigma)})\boldsymbol{\Omega}(\mathbf{I} - \boldsymbol{\Omega})^{-1}\right]^{-1}$$

$$(5.24) \qquad = \mathbf{g}(\mathbf{I} - \boldsymbol{\Omega})^{-1}\left[\mathbf{I} + \frac{1 - e^{i\theta(1-\sigma)}}{\det(\mathbf{I} - \boldsymbol{\Omega})}\boldsymbol{\Omega} \cdot \mathrm{adj}(\mathbf{I} - \boldsymbol{\Omega})\right]^{-1}$$

First note that when $\lambda_0 = \delta$, the maximum eigenvalue $\sigma = 1$, and $\boldsymbol{\pi} = \lim_{\lambda_0 \to \delta} \mathbf{g}(\mathbf{I} - \boldsymbol{\Omega})^{-1}$ which is the steady state distribution of the associated flow control network, and $\boldsymbol{\pi}$ satisfies $\boldsymbol{\pi}\boldsymbol{\Omega} = \boldsymbol{\pi}$. Now observe that $\det(x\mathbf{I} - \boldsymbol{\Omega})$ is a polynomial in $x$ with $x - \sigma$ as a factor, and when $\lambda_0 = \delta$ then $\boldsymbol{\pi}_*$ and $\boldsymbol{\pi}$ are the right and left eigenvectors for $\boldsymbol{\Omega}$. Since $\boldsymbol{\pi} \cdot \mathbf{1}^\mathsf{T} = 1$, we normalize $\boldsymbol{\pi}_*$ such that $\boldsymbol{\pi}_* \cdot \boldsymbol{\pi}^\mathsf{T} = 1$. Unlike the left eigenvector $\boldsymbol{\pi}$, the right eigenvector $\boldsymbol{\pi}_*$ does not have probabilistic meaning. Using Corollary 2 to Theorem 1.1 of Seneta [10] (page 8), we get

$$(5.25) \qquad \lim_{\lambda_0 \uparrow \delta} \mathbf{g}(\mathbf{I} - e^{i\theta(1-\sigma)}\boldsymbol{\Omega})^{-1} = \boldsymbol{\pi}\left[\mathbf{I} - i\theta\boldsymbol{\pi}_*^\mathsf{T} \cdot \boldsymbol{\pi}\right]^{-1} = \frac{1}{1 - i\theta}\boldsymbol{\pi},$$

which is a product of the characteristic function of a exponential random variable and the steady state distribution of the associated flow controlled network. Now to prove (3.9), recall that the matrix quadratic equation can be written as

$$(5.26) \qquad \boldsymbol{\Omega} = \lambda_0\mathbf{C} + \boldsymbol{\Omega}^2\mu_0\sum_{i=1}^{N} p_{0i}\mathbf{I}(K)\mathbf{R}_i\mathbf{C}.$$

If $\mathbf{h}$ is the eigenvector such that $\mathbf{h}(\lambda_0)\boldsymbol{\Omega}(\lambda_0) = \sigma(\lambda_0)\mathbf{h}(\lambda_0)$, then

$$(5.27) \qquad \sigma(\lambda_0)\mathbf{h}(\lambda_0) = \lambda_0\mathbf{h}(\lambda_0)\mathbf{C}(\lambda_0) + \mu_0\sigma(\lambda_0)^2\mathbf{h}(\lambda_0) \cdot \sum_{i=1}^{N} p_{0i}\mathbf{I}(K)\mathbf{R}_i\mathbf{C}(\lambda_0)$$

Multiplying both sides by $\mathbf{C}(\lambda_0)^{-1}$ gives us

(5.28)

$$\sigma(\lambda_0)\mathbf{h}(\lambda_0) \cdot (\lambda_0\mathbf{I} + \mu_0\mathbf{I}(K) - \mathbf{B}) = \lambda_0\mathbf{h}(\lambda_0) + \mu_0\sigma(\lambda_0)^2\mathbf{h}(\lambda_0) \cdot \sum_{i=1}^{N} p_{0i}\mathbf{I}(K)\mathbf{R}_i$$

Differentiating both sides with respect to $\lambda_0$ gives us

$$\sigma'(\lambda_0)\mathbf{h}(\lambda_0) \cdot (\lambda_0\mathbf{I} + \mu_0\mathbf{I}(K) - \mathbf{B})$$
$$+ \sigma(\lambda_0)\mathbf{h}'(\lambda_0)(\lambda_0) \cdot (\lambda_0\mathbf{I} + \mu_0\mathbf{I}(K) - \mathbf{B}) + \sigma(\lambda_0)\mathbf{h}(\lambda_0)$$
$$= \mathbf{h}(\lambda_0) + \lambda_0\mathbf{h}'(\lambda_0) + \mu_0\mathbf{h}'(\lambda_0) \cdot \sum_{i=1}^{N} p_{0i}\mathbf{I}(K)\mathbf{R}_i$$
$$+ 2\mu_0\sigma(\lambda_0)\sigma'(\lambda_0)\mathbf{h}(\lambda_0) \cdot \sum_{i=1}^{N} p_{0i}\mathbf{I}(K)\mathbf{R}_i$$

Setting $\lambda_0 = \delta$, we get $\sigma(\delta) = 1$, $\mathbf{h}(\delta) = \boldsymbol{\pi}[\mathbf{Q}^*]$, and

$$\sigma'(\delta)\boldsymbol{\pi}[\mathbf{Q}^*] \cdot (\delta\mathbf{I} + \mu_0\mathbf{I}(K) - \mathbf{B})$$
$$+ \mathbf{h}'(\delta) \cdot (\delta\mathbf{I} + \mu_0\mathbf{I}(K) - \mathbf{B}) + \boldsymbol{\pi}[\mathbf{Q}^*]$$
$$= \boldsymbol{\pi}[\mathbf{Q}^*] + \delta\mathbf{h}'(\delta) + \mu_0\mathbf{h}'(\delta) \cdot \sum_{i=1}^{N} p_{0i}\mathbf{I}(K)\mathbf{R}_i$$
$$+ 2\mu_0\sigma'(\delta)\boldsymbol{\pi}[\mathbf{Q}^*] \cdot \sum_{i=1}^{N} p_{0i}\mathbf{I}(K)\mathbf{R}_i.$$

Finally, rearranging the terms and using (5.17), and noting that $\boldsymbol{\pi}[\mathbf{Q}^*]\mathbf{A}[\mathbf{Q}^*] = 0$ we obtain

(5.29)

$$\sigma'(\delta)\boldsymbol{\pi}[\mathbf{Q}^*] \cdot \left[\delta\mathbf{I} - \mu_0\sum_{i=1}^{N} p_{0i}\mathbf{I}(K)\mathbf{R}_i\right] = \mathbf{h}'(\delta) \cdot \left[\mu_0\sum_{i=1}^{N} p_{0i}\mathbf{I}(K)\mathbf{R}_i - \mu_0\mathbf{I}(K) + \mathbf{B}\right].$$

By Corollary 4.3, we have

(5.30)

$$\boldsymbol{\pi}[\mathbf{Q}^*] \cdot \left[\delta\mathbf{I} - \mu_0\sum_{i=1}^{N} p_{0i}\mathbf{I}(K)\mathbf{R}_i\right] = \boldsymbol{\Gamma}_A[\mathbf{Q}^*] \cdot \left[\mu_0\sum_{i=1}^{N} p_{0i}\mathbf{I}(K)\mathbf{R}_i - \mu_0\mathbf{I}(K) + \mathbf{B}\right],$$

which gives us

$$(5.31) \qquad \boldsymbol{\Gamma}_A[\mathbf{Q}^*] = \frac{1}{\sigma'(\delta)}\mathbf{h}'(\delta).$$

Using (5.14), we see that

$$(5.32) \qquad \lambda_0\mathbf{h}(\lambda_0) \cdot \mathbf{1}^\mathsf{T} = \mu_0\sigma(\lambda_0)\mathbf{h}(\lambda_0) \cdot \mathbf{I}(K) \cdot \mathbf{1}^\mathsf{T}$$

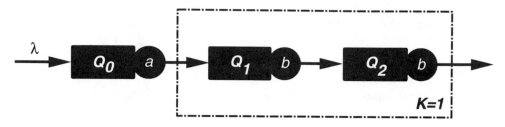

FIGURE 3. 3-node semi-open Nntwork.

Differentiating both sides with respect to $\lambda_0$ gives us

(5.33)
$$\mathbf{h}(\lambda_0) \cdot \mathbf{1}^\mathsf{T} + \lambda_0 \mathbf{h}'(\lambda_0) \cdot \mathbf{1}^\mathsf{T} = \mu_0 \sigma'(\lambda_0) \mathbf{h}(\lambda_0) \cdot \mathbf{I}(K) \cdot \mathbf{1}^\mathsf{T} + \mu_0 \sigma(\lambda_0) \mathbf{h}'(\lambda_0) \cdot \mathbf{I}(K) \cdot \mathbf{1}^\mathsf{T}.$$

Now set $\lambda_0 = \delta$ and observe that if we normalize the eigenvector such that $\mathbf{h}(\lambda_0) \cdot \mathbf{1}^\mathsf{T} = 1$ for all $\lambda_0$, then $\mathbf{h}'(\lambda_0) \cdot \mathbf{1}^\mathsf{T} = 0$. This gives us

(5.34)
$$1 = \mu_0 \sigma'(\delta) \boldsymbol{\pi}[\mathbf{Q}^*] \cdot \mathbf{I}(K) \cdot \mathbf{1}^\mathsf{T} + \mu_0 \mathbf{h}'(\delta) \cdot \mathbf{I}(K) \cdot \mathbf{1}^\mathsf{T}.$$

Dividing both sides by $\sigma'(\delta)$, and noting that

(5.35)
$$\delta \equiv \lim_{t \to \infty} \mu_0 \mathsf{Pr}\left( |\mathbf{Q}^*(t)| < K \right) = \boldsymbol{\pi}[\mathbf{Q}^*] \mathbf{I}(K) \cdot \mathbf{1}^\mathsf{T},$$

we obtain

(5.36)
$$\frac{1}{\sigma'(\delta)} = \delta + \mu_0 \boldsymbol{\Gamma}_A[\mathbf{Q}^*] \cdot \mathbf{I}(K) \cdot \mathbf{1}^\mathsf{T}.$$

Finally using equations (4.6), (4.7) and (4.11) and Corollary 4.3, we have

$$
\begin{aligned}
\lim_{t \to \infty} \frac{\mathsf{E}\left[A_e^*(t)\right] + \mathsf{Var}\left[A_e^*(t)\right]}{2t} &= \lim_{t \to \infty} \left[\mu_0 \mathsf{Pr}\left(|\mathbf{Q}^*(t)| < K\right)\right. \\
&\qquad \left. + \mu_0 \mathsf{Cov}\left[A_e^*(t); |\mathbf{Q}^*(t)| < K\right]\right] \\
&= \delta + \lim_{t \to \infty} \mu_0 \mathsf{Cov}\left[A_e^*(t); |\mathbf{Q}^*(t)| < K\right] \\
\text{(5.37)} \qquad &= \delta + \mu_0 \boldsymbol{\Gamma}_A[\mathbf{Q}^*] \cdot \mathbf{I}(K) \cdot \mathbf{1}^\mathsf{T}.
\end{aligned}
$$

$\square$

## 6. An example

In this section, we consider a 3-node semi-open network (see Figure 3) to illustrate the concepts developed in previous sections. In particular, we explicitly calculate $\delta$, the effective input rate to the flow controlled network, and the limiting variance of the effective input process to obtain the heavy traffic results. We also show how one can obtain the parameters needed to calculate the heavy traffic limit directly from the matrix geometric solution of the semi-open network. The three node semi-open network is illustrated in Figure 3 where we set $N = 2$, $K = 1$. Moreover, we assume that the external arrival rate is $\lambda$, the service rate at the feeder queue is $a$, the service rates of the other two servers equal $b$, and $p_{01} = p_{12} = 1$ and $q_2 = 1$.

First of all in order to calculate the limiting effective input rate $\delta$ and the limiting variance of the effective input process, one needs to calculate the steady state distribution of the flow controlled network (i.e., a two node network with equal service rates $b$ and external Poisson arrival rate $a$). This can be obtained easily using the fact that this network has product form solution and it is given by(the states of the flow controlled network are ordered lexicographically as $(0,0)$, $(0,1)$ and $(1,0)$)

$$(6.1) \qquad \boldsymbol{\pi}[\mathbf{Q}^*] \equiv \lim_{t \to \infty} \mathbf{P}\left[\mathbf{Q}^*(t)\right] = \left[\frac{b}{2a+b} \; \frac{a}{2a+b} \; \frac{a}{2a+b}\right]$$

Using this steady state distribution, the limiting effective input rate $\delta$ is given by

$$(6.2) \qquad \delta \equiv \lim_{t \to \infty} \mu_0 \mathsf{Pr}\left(|\mathbf{Q}^*(t)| < 1\right) = \boldsymbol{\pi}[\mathbf{Q}^*]\mathbf{I}(1) \cdot \mathbf{1}^{\mathsf{T}} = \frac{ab}{2a+b},$$

and via solving the system of equations (4.12) the limiting variance is given by

$$(6.3) \qquad \lim_{t \to \infty} \frac{d}{dt} \mathsf{Var}\left[A_e^*(t)\right] = \frac{ab\left(2a^2 + b^2\right)}{(2a+b)^3}.$$

Note that the semi-open network is stable only when $\lambda < \delta$ and as $\lambda \to \delta$, the semi-open network decouples into an $M|G|1$ queue and a flow controlled network. It follows easily from Theorem 3.4 that in heavy traffic the steady state distribution of the feeder queue is exponentially distributed with mean

$$(6.4) \qquad \lim_{a \to \delta} (a - \delta)\mathsf{E}\left[Q_0\right] = \frac{ab\left(3a^2 + 2ab + b^2\right)}{(2a+b)^3},$$

and that of the remaining two queues is $\boldsymbol{\pi}[\mathbf{Q}^*]$.

For the example considered in this section, one can also calculate $\sigma'(\delta)$, the derivative of the maximum eigenvalue of $\boldsymbol{\Omega}$ evaluated at $\delta$ directly. Solving equations (3.2) and (3.3), we obtain

$$(6.5) \qquad \boldsymbol{\Omega} = \begin{bmatrix} \dfrac{\lambda}{a} & \dfrac{\lambda^2}{ab} & \dfrac{\lambda^2\left(\lambda + b\right)}{ab^2} \\[2mm] \dfrac{\lambda}{a} & \dfrac{\lambda\left(\lambda + a\right)}{ab} & \dfrac{\lambda^2\left(\lambda + a + b\right)}{ab^2} \\[2mm] \dfrac{\lambda}{a} & \dfrac{\lambda\left(\lambda + a\right)}{ab} & \dfrac{\lambda\left(\lambda + a\right)\left(\lambda + b\right)}{ab^2} \end{bmatrix}$$

and

$$(6.6) \qquad \mathbf{g} = \left[\frac{(ab - 2a\lambda - b\lambda)}{ab} \; \frac{\lambda\left(ab - 2a\lambda - b\lambda\right)}{ab^2} \; \frac{\lambda\left(\lambda + b\right)\left(ab - 2a\lambda - b\lambda\right)}{ab^3}\right]$$

Now recalling that $\sigma(\lambda)$, the maximum eigenvalue of $\boldsymbol{\Omega}$ viewed as a function of $\lambda$ can be expressed as

$$(6.7) \qquad \sigma(\lambda) = \mathbf{h}(\lambda)\boldsymbol{\Omega}(\lambda)\mathbf{h}_*^T(\lambda)$$

where $\mathbf{h}(\lambda)$ and $\mathbf{h}_*^T(\lambda)$ are the left and right eigen vectors of $\boldsymbol{\Omega}(\lambda)$ such that $\mathbf{h}(\lambda)\mathbf{h}_*^T(\lambda) = 1$. Now viewing $\sigma(\lambda)$ as a differentiable function of $\lambda$, and noting that when $\lambda = \delta$, $\mathbf{h}(\delta) = \boldsymbol{\pi}[\mathbf{Q}^*]$ and $\mathbf{h}_*^T(\delta) = \boldsymbol{\pi}_*[\mathbf{Q}^*]$, we can express $\sigma'(\delta)$ as

$$(6.8) \qquad \sigma'(\delta) = \boldsymbol{\pi}[\mathbf{Q}^*]\left[\frac{d}{d\lambda}\boldsymbol{\Omega}(\lambda)\right]_{\lambda = \delta} \boldsymbol{\pi}_*[\mathbf{Q}^*]$$

Knowing $\boldsymbol{\pi}[\mathbf{Q}^*]$, one can easily calculate $\boldsymbol{\pi}_*[\mathbf{Q}^*]$, and differentiating the matrix $\boldsymbol{\Omega}$ with respect to $\lambda$ and evaluating at $\delta$, and simplification leads to

$$(6.9) \qquad \sigma'(\delta) = \frac{ab\left(3a^2 + 2ab + b^2\right)}{(2a + b)^3},$$

which is the expected queue length of the feeder queue in heavy traffic. We would like to point out that the calculation of the parameters involved in heavy traffic limit directly from the matrix geometric solution is possible only for small networks such as the one considered here. For large networks, the matrix $\boldsymbol{\Omega}$ is very dense, and even obtaining numerical solutions are tedious. But our procedure only requires solving a systems of equations that are much easier to solve in order to obtain the parameters involved in the heavy traffic limit.

## 7. Numerical results

This section is devoted to developing approximate procedures for the semi-open network. We focus on the average number of customers in the feeder queue. Using the matrix-geometric solution for the semi-open network, the average number of customers in the feeder queue is given by

$$(7.1) \qquad \mathsf{E}[Q_0] = \mathbf{g}(\mathbf{I} - \boldsymbol{\Omega})^{-2}\mathbf{1}^\mathsf{T}$$

As a first approximation, we use the heavy traffic limit (as in Theorem 3.4) and obtain the heavy traffic approximation for the average number of customers in the feeder as

$$(7.2) \qquad \mathsf{E}[Q_0] \approx \frac{\sigma'(\delta)}{\delta(1 - \rho)}$$

where $\rho = \lambda_0/\delta$. Note that $\sigma'(\delta)$ is the expected queue length of an $M|G|1$ queue in heavy traffic. As opposed to the heavy traffic, in light traffic, the feeder queue will behave like an $M|M|1$ queue with arrival rate $\lambda_0$ and service rate $\mu_0$. Therefore, the light traffic approximation for the average number of customers in the feeder queue is given by

$$(7.3) \qquad \mathsf{E}[Q_0] \approx \frac{\lambda_0}{\mu_0 - \lambda_0}.$$

It is expected that the heavy traffic approximation will be good when $\rho$ is close to one and the light traffic approximation will be good when $\rho$ is close to zero. Based on these two approximation, we propose several hybrid approximations for the the average number of customers in the feeder queue. The first hybrid approximation (Hybrid-1) is a simple weighted average of the heavy and light traffic approximations, and it is given by

$$(7.4) \qquad \text{Hybrid } 1 = \rho\left(\frac{\sigma'(\delta)}{\delta(1 - \rho)}\right) + (1 - \rho)\left(\frac{\lambda_0}{\mu_0 - \lambda_0}\right).$$

The second hybrid approximation (Hybrid-2) is the geometric average of the heavy and light traffic approximations and is given by

$$(7.5) \qquad \text{Hybrid } 2 = \left(\frac{\sigma'(\delta)}{\delta(1 - \rho)}\right)^\rho \left(\frac{\lambda_0}{\mu_0 - \lambda_0}\right)^{1 - \rho}.$$

Note that the approximations Hybrid-1 and Hybrid-2 are asymptotically correct.

According to our heavy traffic theorem, the feeder queue behaves like an $M|G|1$ queue only when $\rho \to 1$. As a third approximation, we treat the feeder queue as an $M|G|1$ queue for all values of $\rho$, and calculate the average queue length using the Pollaczek-Khinchin formula

$$(7.6) \qquad \mathsf{E}[Q_0] = \rho + \frac{\lambda^2 \, \mathsf{E}[S^2]}{2(1-\rho)}.$$

where the random variable $S$ denotes the service time of the $M|G|1$ queue. From the limiting mean and the variance of the effective input process to the flow controlled network, it is easily verified that

$$(7.7) \qquad \mathsf{E}[S] = \frac{1}{\delta},$$

and

$$(7.8) \qquad \mathsf{Var}[S] = \frac{v}{\delta^3}.$$

where $v$ is the limiting variance of the effective input process. Using the relation $\sigma'(\delta) = (\delta + v)/2$, the Pollaczek-Khinchin formula reduces to

$$(7.9) \qquad \mathsf{E}[Q_0] = \rho + \frac{\rho^2 \sigma'(\delta)}{\delta(1-\rho)}.$$

Note that as $\rho \to 1$, this formula agrees with the heavy traffic approximation (6.2). The final hybrid approximation (Hybrid-3) is similar to Hybrid-2 approximation. Once again, we consider the geometric average of the $M|G|1$ and the light traffic approximations, and it is given by

$$(7.10) \qquad \text{Hybrid } 3 = \left( \rho + \frac{\rho^2 \sigma'(\delta)}{\delta(1-\rho)} \right)^{\rho} \left( \frac{\lambda_0}{\mu_0 - \lambda_0} \right)^{1-\rho}.$$

In order to compare the performances of the different approximation schemes, we consider the 3-node semi-open network described in the last section. We fix $\mu_1$ and $\mu_2$ (the service rates at the flow controlled part), and vary $\mu_0$ (the service rate of the feeder queue) and $K$ (the maximum number of customers allowed in the flow controlled part). Specifically, when $K = 3$, we consider three different service rates for the feeder queue, $\mu_0 = 0.3$, $\mu_0 = 3$ and $\mu_0 = 30$, and keep $\mu_1 = \mu_2 = 3$. In Figure 4, we plot $\rho$ vs. $E[Q_0]$ when $K = 2$ and $\mu_0 = 0.3$, $\mu_1 = \mu_2 = 3$. To obtain a better understanding of the performance of the different approximation schemes, instead of plotting $\rho$ vs. $E[Q_0]$, we plot $\rho$ vs. $\log(E[Q_0]/M.G)$, where $M.G$ is the exact average number of customers at the feeder queue obtained via the matrix-geometric solution. In these plots, the exact value is always represented by the x-axis, and the approximate values fluctuates around the x-axis. This logarithm of the ratio plots clearly exhibits the disadvantages of the heavy traffic and the light traffic approximations for moderate values of $\rho$, and the advantages of the hybrid approximations. In particular, the approximations Hybrid-2 and Hybrid-3 perform consistently better than other approximations. In particular, for all values of $\mu_0$, as $K$ varies from 1 to 3, the Hybrid-3 approximation becomes very accurate.

E[Q_0]

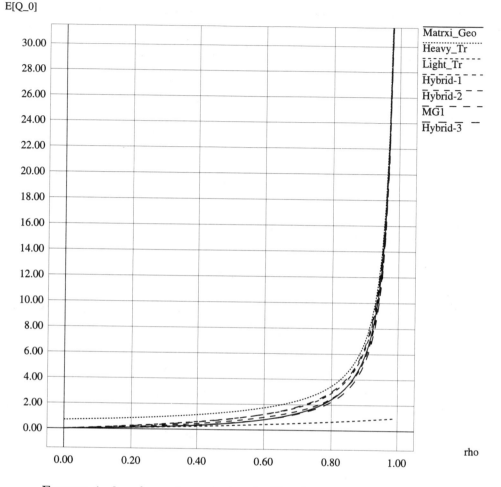

FIGURE 4. 3-node semi-open network, $K = 2$, $\mu_0 = 3$, $\mu_1 = \mu_2 = 3$.

## 8. Appendix

LEMMA 8.1. *Let* $\mathbf{A}$ *be the generator for a continuous time Markov process* $\{X(t), t \geq 0\}$ *on a finite state space* $\{1, 2, \ldots, N\}$. *If* $T_{\mathbf{A}}$ *is the absorption time for this process, and* $T_\lambda$ *is an independent random variable with an exponential distribution of rate* $\lambda$, *then*

$$(8.1) \qquad \left| \lambda \left( \lambda \mathbf{I} - \mathbf{A} \right)^{-1} \right| = \max_{1 \leq i \leq N} \mathsf{P}_i \left( T_{\mathbf{A}} > T_\lambda \right)$$

*where* $| \cdot |$ *is the operator norm induced by the* $\ell_1$*-norm on row vectors.*

We will use this lemma to get a sharp estimate for $|\lambda_0 \mathbf{C}|$. First, if we let $\mathbf{A} = \mathbf{B} - \mu_0 \mathbf{I}(K)$, we can interpret $\mathbf{A}$ as the generator for a modified, flow controlled network of maximum capacity $K$, with no external inputs. In addition, all states with $K - 1$ customers in the network or less have an independent absorption rate of $\mu_0$. If we initialize the network with $K$ customers, then $T_{\mathbf{A}} > T_{\mu_*}$ where $\mu_* = \sum_{i=1}^{N} \mu_i$ and $T_{\mu_*}$ is the time until the first possible network transition. If we initialize with $K - 1$ customers or less, then $T_{\mathbf{A}} = T_{\mu_0}$. Using the former scenario,

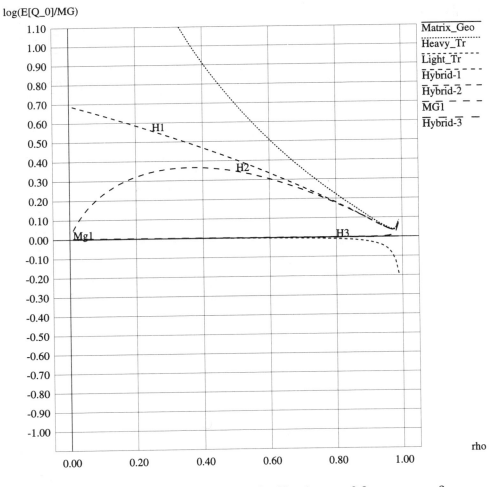

FIGURE 5. 3-node semi-open network, $K = 3$, $\mu_0 = 0.3$, $\mu_1 = \mu_2 = 3$.

we have

$$
\begin{aligned}
\mathsf{P_m}\left(T_{\mathbf{A}} > T_{\lambda_0}\right) &= \mathsf{P_m}\left(T_{\mathbf{A}} > T_{\mu_*} > T_{\lambda_0}\right) + \mathsf{P_m}\left(T_{\mathbf{A}} > T_{\lambda_0} \geq T_{\mu_*}\right) \\
&= \mathsf{P}\left(T_{\mu_*} > T_{\lambda_0}\right) + \sum_{i=1}^{N}\mathsf{P_m}\left(T_{\mathbf{A}} > T_{\lambda_0} \geq T_{\mu_*},\, Q(T_{\mu_*}) = \mathbf{m} - \mathbf{e}_i\right) \\
&\quad + \sum_{i=1}^{N}\sum_{j=1}^{N}\mathsf{P_m}\left(T_{\mathbf{A}} > T_{\lambda_0} \geq T_{\mu_*},\, Q(T_{\mu_*}) = \mathbf{m} - \mathbf{e}_i + \mathbf{e}_j\right)
\end{aligned}
$$

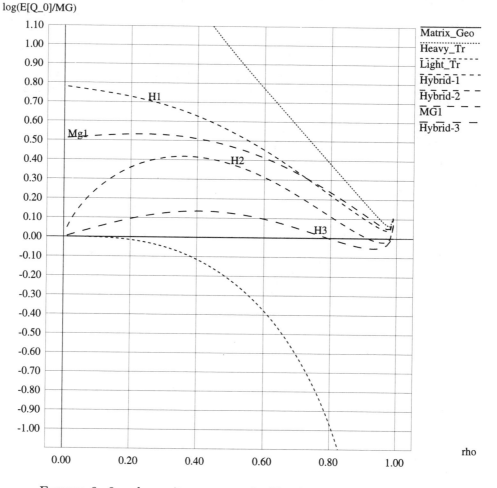

FIGURE 6. 3-node semi-open network, $K = 3$, $\mu_0 = 3$, $\mu_1 = \mu_2 = 3$.

We can obtain the following tight upper bounds for the summation terms since

$$
\begin{aligned}
&\mathsf{P_m}\left(T_{\mathbf{A}} > T_{\lambda_0} \geq T_{\mu_*},\, Q(T_{\mu_*}) = \mathbf{m} - \mathbf{e}_i\right) \\
&= \quad \mathsf{P_m}\left(T_{\mathbf{A}} > T_{\lambda_0} \mid Q(T_{\mu_*}) = \mathbf{m} - \mathbf{e}_i,\, T_{\lambda_0} \geq T_{\mu_*}\right) \\
&\qquad \cdot\ \mathsf{P_m}\left(Q(T_{\mu_*}) = \mathbf{m} - \mathbf{e}_i,\, T_{\lambda_0} \geq T_{\mu_*}\right) \\
&= \quad \mathsf{P_{m-e_i}}\left(T_{\mathbf{A}} > T_{\lambda_0}\right) \cdot \mathsf{P_m}\left(Q(T_{\mu_*}) = \mathbf{m} - \mathbf{e}_i\right) \cdot \mathsf{P}\left(T_{\lambda_0} \geq T_{\mu_*}\right) \\
&= \quad \mathsf{P}\left(T_{\mu_0} > T_{\lambda_0}\right) \cdot \mathsf{P_m}\left(Q(T_{\mu_*}) = \mathbf{m} - \mathbf{e}_i\right) \cdot \mathsf{P}\left(T_{\lambda_0} \geq T_{\mu_*}\right) \\
&\leq \quad \frac{\lambda_0}{\lambda_0 + \mu_0} \cdot \frac{\mu_i q_i}{\mu_*} \cdot \frac{\mu_*}{\lambda_0 + \mu_*} \\
&\leq \quad \frac{\lambda_0}{\lambda_0 + \mu_0} \cdot \frac{\mu_i q_i}{\lambda_0 + \mu_*}
\end{aligned}
$$

and by a similar argument, we also have

$$
\mathsf{P_m}\left(T_{\mathbf{A}} > T_{\lambda_0} \geq T_{\mu_*},\, Q(T_{\mu_*}) = \mathbf{m} - \mathbf{e}_i + \mathbf{e}_j\right) \leq \mathsf{P_{m-e_i+e_j}}\left(T_{\mathbf{A}} > T_{\lambda_0}\right) \cdot \frac{\mu_i p_{ij}}{\lambda_0 + \mu_*}.
$$

log(E[Q_0]/MG)

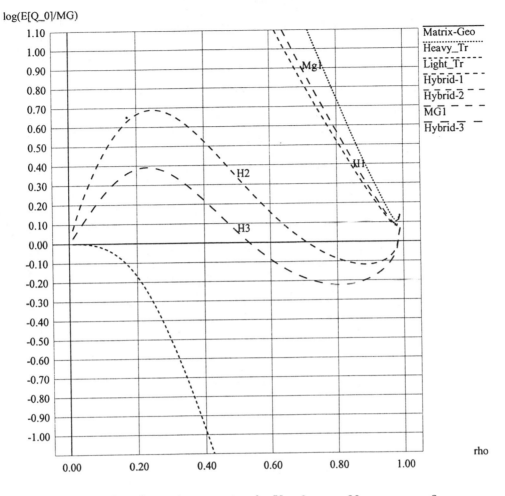

FIGURE 7. 3-node semi-open network, $K = 3$, $\mu_0 = 30$, $\mu_1 = \mu_2 = 3$.

Combining these two inequalities gives us

$$\mathsf{P}_{\mathbf{m}}\left(T_{\mathbf{A}} > T_{\lambda_0}\right) \leq \frac{\lambda_0}{\lambda_0 + \mu_*} + \frac{\lambda_0}{\lambda_0 + \mu_0} \cdot \frac{\sum_{i=1}^{N} \mu_i q_i}{\lambda_0 + \mu_*}$$

$$+ \sum_{i=1}^{N} \sum_{j=1}^{N} \mathsf{P}_{\mathbf{m}-\mathbf{e}_i+\mathbf{e}_j}\left(T_{\mathbf{A}} > T_{\lambda_0}\right) \cdot \frac{\mu_i p_{ij}}{\lambda_0 + \mu_*}.$$

Using Lemma 8.1, we get

$$(8.2) \qquad |\lambda_0 \mathbf{C}| \leq \frac{\lambda_0}{\lambda_0 + \mu_*} + \frac{\lambda_0}{\lambda_0 + \mu_0} \cdot \frac{\sum_{i=1}^{N} \mu_i q_i}{\lambda_0 + \mu_*} + \sum_{i=1}^{N} \sum_{j=1}^{N} |\lambda_0 \mathbf{C}| \cdot \frac{\mu_i p_{ij}}{\lambda_0 + \mu_*}.$$

Since $\sum_{i=1}^{N} \sum_{j=1}^{N} \mu_i p_{ij} = \mu_* - \sum_{i=1}^{N} \mu_i q_i$, then we can simplify this inequality and get

$$(8.3) \qquad |\lambda_0 \mathbf{C}| \leq \frac{\lambda_0}{\lambda_0 + \sum_{i=1}^{N} \mu_i q_i} + \frac{\sum_{i=1}^{N} \mu_i q_i}{\lambda_0 + \sum_{i=1}^{N} \mu_i q_i} \cdot \frac{\lambda_0}{\lambda_0 + \mu_0}.$$

For sufficiently small $\lambda_0$, we can easily show that the $\Omega_n$ are bounded and so $\Omega$ exists.

## References

[1] Avi-Itzhak, B. and Heyman, D. P. *Approximate Queueing Models for Multiprogramming Computer Systems*, Oper. Res., Vol. 21, 1973, 1212–1230.

[2] Buzacott, J. A. and Shanthikumar, J. G. *On approximate Models of Dynamic Job Shops*, Mgmt. Sci., Vol. 31, No. 7, 1985, 870–886.

[3] Dubois, D., *A Mathematical Model of a Flexible Manufacturing System with Limited In-Process Inventory*, Eur. Jnl. of Oper. Res., Vol. 14, No. 1, 1983, 66–78.

[4] Jackson, J. R., *Job-Shop like Queueing Systems*, Mgmt. Sci., Vol. 10, No. 1, 1963, 131–142.

[5] Kogan, Y. A. and Pukhalskii, A. A. *On Tandem Queues with Blocking in Heavy Traffic*, Performance '84, E.Gelenbe (Editor), Elsevier Science Publishers, B.V. (North-Holland), 1984, 549–558.

[6] Konheim, A. G. and Reiser, M. *A Queueing Model with Finite Waiting Room and Blocking*, Journal of A.C.M., Vol. 23, No. 2, 1976, 328–341.

[7] Konheim, A. G. and Reiser, M. *Finite Capacity Queueing Systems with Applications in Computer Modelling*, SIAM. J. Computing, Vol. 7, No. 2, 1978, 210–219.

[8] Latouche, G. and Neuts, R., *Efficient Algorithmic Solutions to Exponential Tandem Queues with blocking*, SIAM. J. Algebra and Discrete Methods, Vol. 1, No. 1, 1980, 93–106.

[9] Massey, W. A. *An Operator Analytic Approach to the Jackson Network*. Journal of Applied Probability, **21**, 1984, 379–393.

[10] Seneta, E., *Non-negative Matrices and Markov Chains*, Second Edition, Spinger-Verlag, 1981.

[11] Massey, W. A. and Srinivasan, R., *A Heavy Traffic Analysis for Semi-Open Networks*, Performance Evaluation, Vol. 13, No. 1, 1991, 59–66.

[12] Srinivasan, R., *Topics in State Dependent Queues and Queueing Networks*, Unpublished Ph.D. thesis, Carleton University, Ottawa, Canada, 1988.

BELL LABORATORIES, LUCENT TECHNOLOGIES, 700 MOUNTAIN AVENUE, MURRAY HILL, NEW JERSEY 07974
*E-mail address*: `will@research.bell-labs.com`

DEPARTMENT OF MATHEMATICS AND STATISTICS, UNIVERSITY OF SASKATCHEWAN, 106 WIGGINS ROAD, SASKATOON, SASKATCHEWAN S7N 5E6
*E-mail address*: `raj@math.usask.ca`

Canadian Mathematical Society
Conference Proceedings
Volume 26, 2000

# Estimating the Orey Index of a Gaussian Stochastic Process with Stationary Increments: An Application to Financial Data Set

R. Norvaiša and D. M. Salopek

*This paper is dedicated to Donald A. Dawson*

ABSTRACT. A new method of time series data analysis is suggested. It is based on a strong limit theorem for a Gaussian stochastic process with stationary increments. The method is used to analyze the local behaviour of a continuous time stochastic process given at finitely many equidistant time moments. In particular, an estimation of the maximal exponent of Hölder's property for sample functions of a stochastic process is the motivation behind the method.

## 1. Introduction

Let $X = \{X(t)\colon t \geq 0\}$ be a real-valued mean zero Gaussian stochastic process with stationary increments and continuous in quadratic mean. Let $\sigma = \sigma_X$ be the incremental variance of $X$ given by

$$\sigma_X(u)^2 := E\big[(X(t+u) - X(t))^2\big]$$

for $t, u \geq 0$. Following Orey [24], we define

$$(1.1) \qquad \gamma_* := \inf\{\gamma > 0\colon u^\gamma/\sigma(u) \to 0 \text{ as } u \downarrow 0\} = \limsup_{u \downarrow 0} \frac{\log \sigma(u)}{\log u}$$

and

$$(1.2) \qquad \gamma^* := \sup\{\gamma > 0\colon u^\gamma/\sigma(u) \to +\infty \text{ as } u \downarrow 0\} = \liminf_{u \downarrow 0} \frac{\log \sigma(u)}{\log u}.$$

The equalities in (1.1) and (1.2) are known representations for limits of the logarithmic ratios (see e.g. Annex A.4 in [29]). By definition, it follows that $0 \leq \gamma^* \leq \gamma_* \leq \infty$. If $\gamma_* = \gamma^*$ then we will say that $X$ has the *Orey index* $\gamma_X := \gamma_* = \gamma^*$. Orey [24] proved that graphs of almost all sample functions of $X$ have Hausdorff-Besicovitch dimension $2 - \gamma_X$ provided $\gamma_X$ exists. If the fractal dimension of a

---

2000 *Mathematics Subject Classification.* Primary 62G05, 62M09, 90A12.

This research is partially supported by NSERC Canada grant number 203232-98 at York University.

graph exists, it is equal to its Hausdorff-Besicovitch dimension. Thus the Orey index is also known as the fractal index or fractional index of $X$ (see e.g. Hall and Roy [14] where this relationship is extended to certain non-Gaussian processes). It is clear that the Orey index of a Brownian motion exists and is equal to $1/2$. Let $B_H = \{B_H(t)\colon t \geq 0\}$ be a fractional Brownian motion with the Hurst exponent $H \in (0,1)$, that is a mean zero Gaussian stochastic process with the covariance function

$$(1.3) \qquad E\big[B_H(t)B_H(s)\big] = \frac{1}{2}\Big[t^{2H} + s^{2H} - |t-s|^{2H}\Big] \qquad \text{for } t, s \geq 0$$

and $B_H(0) = 0$ almost surely. Fractional Brownian motion $B_H$ has stationary increments and its incremental variance $\sigma_H := \sigma_{B_H}$ is given by $\sigma_H(u)^2 = u^{2H}$ for $u \geq 0$. Thus the Orey index $\gamma_{B_H}$ exists and is equal to the Hurst exponent $H$. More generally, if $\sigma_X(u) = u^\gamma \ell(u)$ for $\gamma > 0$ and $\ell$ slowly varying at zero, then $X$ has the Orey index $\gamma_X = \gamma$. Note that the Orey index could be defined for any second order stochastic process with stationary increments.

There are several reasons why the estimation of the Orey index is of interest. For instance, limit theorems for stationary time series which are defined by increments of $X$, depend on the Orey index $\gamma_X$ through the normalizing factor. In addition, the Orey index $\gamma_X$ determines various properties of the sample functions of $X$. For example, almost every sample function of $X$ satisfy Hölder's property with the exponent $\alpha$ for each $\alpha < \gamma_X$. While, if $\alpha > \gamma_X$, then Hölder's property fails to hold almost surely. The behaviour of sample functions determines the meaning of the integral equation used in stochastic modelling. For example, if two Gaussian stochastic processes $X$ and $Y$ have the Orey indices $\gamma_X$ and $\gamma_Y$, respectively, such that $\gamma_X + \gamma_Y > 1$, then the integral $\int_0^T Y(t)\,dX(t)$ exists in the Riemann-Stieltjes sense for almost all sample functions. In particular, $\int_0^T B_H(t)\,dB_H(t)$ exists in the Riemann-Stieltjes sense for almost all sample functions of the fractional Brownian motion $B_H$ if and only if $1/2 < H < 1$. Stochastic modelling based on Riemann-Stieltjes integrals has been used in [22] and [26]. It should be noted that the value $\gamma_X = 1/2$ is critical for giving a suitable meaning to integral equations driven by a stochastic process $X$. Therefore, it is desirable to develop statistical procedures that allow us to determine whether $\gamma_X < 1/2$ or $\gamma_X > 1/2$ is more likely to fit empirical data.

Asymptotic accuracy results for several estimators of the Orey index of a stationary Gaussian stochastic process have been recently established. For example, Hall and Wood [15], Feuerverger, Hall and Wood [11], Constantine and Hall [6], Chan, Hall and Poskitt [5] derived asymptotic properties for the bias and variance of estimators based on box-counting dimension, counting the number of level crossings, the sample variogram and the cosine part of the periodogram, respectively. In the present paper, the estimators of the Orey index are based on the strong limit theorem established by Gladyshev [13] for a Gaussian stochastic process. As compared to the previous estimators, the new estimators are consistent in the stronger sense of almost sure convergence. Furthermore, the present paper focus on analysis of financial data, while the above mentioned works are motivated mainly by problems of natural sciences.

The estimation of the Orey index is related to the estimation of long-range dependence in the following sense. Suppose $X = \{X(t)\colon t \geq 0\}$ is a mean zero second order self-similar stochastic process with stationary increments, and let $\gamma$

be the exponent of self-similarity of $X$. That is, for each $a > 0$, the process $\{a^{-\gamma}X(at)\colon t \geq 0\}$ has the same finite dimensional distributions as $X$. We have then $\sigma_X(u) = u^{\gamma}\sigma_X(1)$ for all $u > 0$, and so the Orey index $\gamma_X$ exists and is equal to the exponent of self-similarity $\gamma$. Furthermore, for each positive integer $n$, let $Y_n := X(n) - X(n-1)$. Then $Y = \{Y_1, Y_2, \dots\}$ is a mean zero second order stationary time series. It can be shown that the autocovariance function $\operatorname{cov}(Y_m, Y_{m+n}) \sim Cn^{2\gamma-2}$ as $n \to \infty$ for some constant $C$, provided $0 < \gamma < 1$ and $\gamma \neq 1/2$. This yields that the time series $Y$ is long-range dependent if $1/2 < \gamma < 1$, and short-range dependent if $0 < \gamma < 1/2$. With the additional assumption that $X$ is Gaussian, $X$ is a fractional Brownian motion and $Y$ is a fractional Gaussian noise, both having the Hurst exponent $\gamma$ which is equal to the Orey index $\gamma_X$. In general, the Orey index $\gamma_X$ reflects the local behaviour of a stochastic process $X$, while long-range dependence is a global property of $X$.

As already noted, the main aim of the present paper is to test empirically two new estimators of the Orey index. These estimators are described in the next section. In Section 3, the two estimators are calculated for a comparison of the estimation of the Hurst exponent from a simulated fractional Brownian motion. Then in Section 4, we calculate the two estimators on financial data. We consider continuous-time financial models based on an integral equation driven by an unknown stochastic process $X$. To estimate the analytical properties of sample functions of $X$ we use and compare two different notions of the return. Conclusions of the present work are stated in Section 5.

## 2. Method of estimation

Let $X = \{X(t)\colon t \geq 0\}$ be a real-valued mean zero Gaussian stochastic process with stationary increments and continuous in quadratic mean. The process will be considered over an arbitrary but fixed time interval $[0, T]$. Given a sequence of increasing positive integers $\{N_m\colon m \geq 1\}$, for each $m \geq 1$, let

$$v_2(X; \lambda(m)) := \sum_{i=1}^{N_m} \left[ X(iT/N_m) - X((i-1)T/N_m) \right]^2,$$

where $\lambda(m) = \lambda(T/N_m) = \{iT/N_m\colon i = 0, \dots, N_m\}$ is a partition of $[0, T]$ into subintervals of length $T/N_m$. Since the increments of $X$ are stationary, by representations (1.1) and (1.2), the Orey index $\gamma_X$ exists if and only if the following limit exists

$$(2.1) \qquad \lim_{m \to \infty} \frac{\log \sqrt{Ev_2(X; \lambda(m))/N_m}}{\log(1/N_m)} = \lim_{m \to \infty} \frac{\log \sigma_X(1/N_m)}{\log(1/N_m)} = \gamma_X$$

for some sequence $N_m \to \infty$.

When $X$ is a fractional Brownian motion $B_H$ with the Hurst exponent $H \in (0, 1)$, Gladyshev [13, Section 3] proved that the relation

$$\lim_{m \to \infty} \frac{\log \sqrt{v_2(B_H; \lambda(1/2^m))/2^m}}{\log(1/2^m)} = H$$

holds for almost all sample functions of $B_H$. In fact, Gladyshev [13, Theorem 3] proved the preceding relation for a more general class of Gaussian stochastic processes than a fractional Brownian motion $B_H$. Gladyshev's result have been extended to non–Gaussian stochastic processes by Pierre [25]. Motivated by these results we suggest the following definition.

DEFINITION 2.1. Let $\{N_m\colon m \geq 1\}$ be a sequence of integers increasing to infinity and let $\lambda = \{\lambda(m)\colon m \geq 1\}$ be a sequence of partitions $\lambda(m) = \lambda(T/N_m) = \{iT/N_m\colon i = 0, \ldots, N_m\}$ of $[0, T]$. We say that a stochastic process $X = \{X(t)\colon t \geq 0\}$ belongs to the Gladyshev class, or class $G^\lambda$ whenever the limit

$$(2.2) \qquad G_\lambda(X) := \lim_{m \to \infty} \frac{\log \sqrt{v_2(X; \lambda(m))/N_m}}{\log(1/N_m)}$$

exists for almost all sample functions of $X$.

For Gaussian stochastic processes, the following can be derived from known results. For each $m \geq 1$ and any positive constant $C_1$, we have the identity

$$(2.3)$$
$$\log \sqrt{\frac{v_2(X; \lambda(m))}{N_m}} = \log \sigma_X(1/N_m) + \log C_1 + \frac{1}{2} \log\left(1 + \frac{v_2(X; \lambda(m))}{C_1^2 N_m \sigma_X(1/N_m)^2} - 1\right).$$

Suppose that the Orey index $\gamma_X$ exists and, for some sequence $\{N_m\colon m \geq 1\}$ of integers increasing to infinity and for some constant $C_1$, there exists the limit

$$(2.4) \qquad \lim_{m \to \infty} \frac{v_2(X; \lambda(m))}{N_m \sigma_X(1/N_m)^2} = C_1^2, \qquad \text{almost surely.}$$

Then by (2.1) and (2.3), it follows that $X$ belongs to the Gladyshev class $\mathcal{G}^\lambda$ and

$$(2.5) \qquad \gamma_X = G_\lambda(X), \qquad \text{almost surely.}$$

For a Brownian motion $X$, relation (2.4) was proved by Lévy [17], and by Cameron and Martin [3]. Baxter [2] and Gladyshev [13] extended (2.4) to a considerably larger class of stochastic processes $X$. Several illuminations and further generalizations of (2.4) have been done by Dudley [8], Fernández de la Vega [10], Giné and Klein [12] and others. For example, by Theorem 1.1 of Shao [27], relation (2.4) with $C_1^2 = T$ holds for any mean zero Gaussian stochastic process $X$ with stationary increments and non-decreasing incremental variance $\sigma_X$ such that either $\sigma_X^2$ is concave on $[0, T]$ and $1/N_m = o(1/\log m)$, or $\sigma_X^2$ is convex on $[0, T + \epsilon]$ for some $\epsilon > 0$ and $(1/N_m)\sigma_X(1/N_m) = o(1/\sqrt{\log m})$. Note that the Orey index $\gamma_X$ need not exist under the stated conditions.

Relation (2.5) can be used to estimate the Orey index from a sample function of a stochastic process $X$ given its values at finitely many points. It provides a naive form of an estimator, to be called the *Gladyshev estimator* $\widehat{\gamma}_X^G$, based directly on the definition (2.2). That is

$$(2.6) \qquad \widehat{\gamma}_X^G(h_m) := \frac{\log \sqrt{v_2(X; \lambda(m))h_m}}{\log h_m},$$

where $h_m := 1/N_m$ and $m$ is the largest integer such that the cardinality of $\cup_{l=1}^m \lambda(l)$ does not exceed the cardinality of available empirical data set. This estimator may not satisfy certain statistical criteria, but it is strongly consistent by its definition. However its statistical properties can be investigated in comparison with other approaches to the estimation problem. Suppose that the incremental variance $\sigma_X$ of $X$ satisfies the condition

$$(2.7) \qquad \log \sigma_X(h) = \gamma \log h + \log C_2 + \epsilon(h) \qquad \text{for } h > 0,$$

where $C_2$ is a positive constant and $\epsilon(h) \to 0$ as $h \downarrow 0$. Then clearly the Orey index $\gamma_X$ exists and equals $\gamma$. Substituting relation (2.7) into (2.3) we deduce that, for each $m \geq 1$,

$$(2.8) \qquad \log \sqrt{v_2(X, \lambda(m))h_m} = \log(C_1 C_2) + \gamma \log h_m + \big[\epsilon(h_m) + \omega(h_m)\big],$$

where

$$\omega(h_m) := \frac{1}{2} \log\left(1 + \frac{h_m v_2(X; \lambda(m))}{C_1^2 \sigma_X(h_m)^2} - 1\right) \to 0, \qquad \text{almost surely as } h_m \downarrow 0$$

provided (2.4) holds. Relation (2.8) is a typical form of a linear regression equation except that $\epsilon + \omega$ are not independent Gaussian random variables. However, one can construct a consistent estimator by the approach of Hall and Wood's [15]. Let $m \geq 2$. For each $j = 1, \dots, m$, let $x_j := \log h_j$ and $\bar{x} := m^{-1} \sum_{j=1}^{m} x_j$. Define the *ordinary least squares estimator* $\widehat{\gamma}_X^{OLS} = \widehat{\gamma}_X^{OLS}(h_1, \dots, h_m)$ by

$$\widehat{\gamma}_X^{OLS}(h_1, \dots, h_m) := \frac{\sum_{j=1}^{m}(x_j - \bar{x})\log \sqrt{v_2(X; \lambda(j))h_j}}{\sum_{j=1}^{m}(x_j - \bar{x})^2}.$$

Then for each fixed $m \geq 2$ and any $\delta > 0$,

$$(2.9) \qquad \lim_{\substack{h_1, \dots, h_m \to 0 \\ \sum\sum |\log h_i - \log h_j| > \delta}} \widehat{\gamma}_X^{OLS}(h_1, \dots, h_m) = \gamma, \qquad \text{almost surely}$$

provided (2.4) holds. Indeed, we can and do assume that $h_j = a_j h$ for $j = 1, \dots, m$, where $a_1, \dots, a_m$ are fixed positive constants and $h \to 0$. Using the relations $\sum_j (x_j - \bar{x}) = 0$ and $\sum_j (x_j - \bar{x})x_j = \sum_j (x_j - \bar{x})^2$, by (2.8), we obtain

$$(2.10) \qquad \widehat{\gamma}_X^{OLS}(h_1, \dots, h_m) - \gamma = \frac{\sum_{j=1}^{m}(x_j - \bar{x})\big[\omega(h_j) + \epsilon(h_j)\big]}{\sum_{j=1}^{m}(x_j - \bar{x})^2}.$$

Since $x_j - \bar{x} = \log a_j - m^{-1} \sum_{i=1}^{m} \log a_i$ do not depend on $h$ for each $j = 1, \dots, m$, the right-hand side of the preceding relation tends to zero almost surely as $h \downarrow 0$ proving (2.9). Therefore under the non-parametric model (2.7), $\widehat{\gamma}_X^{OLS}$ is a strongly consistent estimator of the Orey index $\gamma_X$.

To compare Gladyshev's estimator $\widehat{\gamma}_X^{G}$ versus $\widehat{\gamma}_X^{OLS}$, suppose (2.7) and (2.4) hold. By (2.3), we have

$$(2.11) \qquad \widehat{\gamma}_X^{G}(h) = \gamma + \frac{\log(C_1 C_2) + \omega(h) + \epsilon(h)}{\log h}$$

for each $h > 0$. This relation shows that the leading term in the representation of Gladyshev's estimator has the form $(\text{constant}/\log h)$ as $h \downarrow 0$. The leading term in the representation (2.10) is a linear combination of terms $o(1)/\log h$ as $h \downarrow 0$. A similar, but more precise behaviour, was revealed by Hall and Wood [15] for box-counting estimators of fractal dimension.

## 3. Simulated fractional Brownian motion

In this section, we perform a simulation study of the small-sample properties of the two estimators $\widehat{\gamma}_X^{G}$ and $\widehat{\gamma}_X^{OLS}$. For this we simulate a fractional Brownian motion for several values of the Orey index which is also equal to the Hurst exponent. Using repeated samples we calculate the bias, the standard deviation and the mean square error, and then compare the results for the two estimators.

Table 1: Estimation results for $\widehat{\gamma}_H^G(2^{-14})$ based on 100 repeated samples

| $H$ | 0.1 | 0.2 | 0.3 | 0.4 | 0.5 | 0.6 | 0.7 | 0.8 | 0.9 |
|---|---|---|---|---|---|---|---|---|---|
| $\widehat{H}$ | .1325 | .2320 | .3316 | .4307 | .5300 | .6290 | .7279 | .8267 | .9266 |
| $\widehat{H} - H$ | .0325 | .0320 | .0316 | .0307 | .0300 | .0290 | .0279 | .0267 | .0266 |
| $SD$ | .0007 | .0006 | .0006 | .0007 | .0006 | .0007 | .0010 | .0023 | .0087 |
| $\sqrt{MSE}$ | .0326 | .0320 | .0316 | .0307 | .0300 | .0290 | .0279 | .0268 | .0279 |
| $MSE$ | .0010 | .0010 | .0010 | .0009 | .0009 | .0008 | .0008 | .0007 | .0008 |

Let $B_H = \{B_H(t) : t \geq 0\}$ be a fractional Brownian motion with the Hurst exponent $0 < H < 1$. We used the program of Maeder [19] written in *Mathematica* to simulate a fractional Brownian motion. For each $m \geq 1$, let $\lambda(m) = \lambda(1/2^m) = \{i/2^m : i = 0, \ldots, 2^m\}$ and let

$$u(m)^2 := 2^{-m} v_2(B_H; \lambda(m)) = \frac{1}{2^m} \sum_{i=1}^{2^m} \left[ B_H(i/2^m) - B_H((i-1)/2^m) \right]^2.$$

Then for each $m \geq 1$,

$$\widehat{\gamma}_H^G(2^{-m}) := \widehat{\gamma}_{B_H}^G(2^{-m}) = \frac{\log u(m)}{\log(1/2^m)} = \frac{-\log_2 u(m)}{m}$$

and

$$\widehat{\gamma}_H^{OLS}(2^{-m}) := \widehat{\gamma}_{B_H}^{OLS}(2^{-1}, \ldots, 2^{-m}) = \sum_{j=1}^{m} y_j \log_2 u(j),$$

where $y_j = (x_j - \bar{x})/\sum_{j=1}^{m}(x_j - \bar{x})^2$ and $x_j = \log_2(1/2^j) = -j$ for $j = 1, \ldots, m$. We simulate the vector $\{B_H(i/2^m) : i = 0, \ldots, 2^m\}$ for $m = 14$, $m = 15$ and a chosen values of $H$ and then calculate the two estimators. This gives us four different estimates $\widehat{H}$ of $H$. We repeat this procedure $M = 100$ times to obtain the estimates $\widehat{H}_1, \ldots \widehat{H}_M$ of $H$ for each of the four cases. Then we calculate:

- The estimated expected value $\overline{H} := \left( \sum_{i=1}^{M} \widehat{H}_i \right)/M$;
- The bias $\overline{H} - H$;
- The estimated standard deviation $SD := \sqrt{\sum_{i=1}^{M}(\widehat{H}_i - \overline{H})^2/(M-1)}$;
- The estimated mean square error $MSE := \sum_{i=1}^{M}(\widehat{H}_i - H)^2/M$.

The results are presented in Tables 1 - 4. Note that to calculate $\widehat{\gamma}_H^{OLS}(2^{-m})$ means to find the slope of the least squares line fitted to m points. We also considered fitting the line to fewer number of point. That is, we also considered the estimators $\widehat{\gamma}_{B_H}^{OLS}(2^{-l}, \ldots, 2^{-m})$ for $l = 2, \ldots, m-1$, but the estimation results appear to be increasingly worse as $l$ growth to $m-1$ as compared to the case $l = 1$.

**Bias:** Estimation results show a different behaviour of the bias for the two estimators. The Gladyshev estimator exhibits comparatively positive, constant bias for all values of $H$. While the bias of the OLS estimator monotonically decrease from a positive bias for $H \leq 0.4$ to a negative bias $H \geq 0.5$.

A similar change in the bias for OLS estimator was also illustrated in the simulated results of Bardet [1, Section 5]. To estimate the Hurst exponent Bardet used the aggregated variance method based on the sample variogram.

Table 2: Estimation results for $\widehat{\gamma}_H^G(2^{-15})$ based on 100 repeated samples

| $H$ | 0.1 | 0.2 | 0.3 | 0.4 | 0.5 | 0.6 | 0.7 | 0.8 | 0.9 |
|---|---|---|---|---|---|---|---|---|---|
| $\widehat{H}$ | .1305 | .2299 | .3293 | .4288 | .5280 | .6272 | .7262 | .8250 | .9236 |
| $\widehat{H} - H$ | .0305 | .0299 | .0293 | .0288 | .0280 | .0272 | .0262 | .0250 | .0236 |
| $SD$ | .0005 | .0004 | .0004 | .0004 | .0005 | .0004 | .0006 | .0015 | .0081 |
| $\sqrt{MSE}$ | .0305 | .0299 | .0293 | .0288 | .0280 | .0272 | .0262 | .0250 | .0250 |
| $MSE$ | .0009 | .0009 | .0009 | .0008 | .0008 | .0007 | .0007 | .0006 | .0006 |

Table 3: Estimation results for $\widehat{\gamma}_H^{OLS}(2^{-14})$ based on 100 repeated samples

| $H$ | 0.1 | 0.2 | 0.3 | 0.4 | 0.5 | 0.6 | 0.7 | 0.8 | 0.9 |
|---|---|---|---|---|---|---|---|---|---|
| $\widehat{H}$ | .1599 | .2383 | .3124 | .4029 | .4838 | .5892 | .6816 | .7612 | .8554 |
| $\widehat{H} - H$ | .0599 | .0383 | .0124 | .0029 | -.0162 | -.0108 | -.0184 | -.0388 | -.0446 |
| $SD$ | .0413 | .0371 | .0403 | .0431 | .0490 | .0441 | .0531 | .0988 | .0736 |
| $\sqrt{MSE}$ | .0726 | .0532 | .0420 | .0430 | .0513 | .0452 | .0559 | .1057 | .0857 |
| $MSE$ | .0053 | .0028 | .0018 | .0018 | .0026 | .0020 | .0031 | .0112 | .0073 |

Table 4: Estimation results for $\widehat{\gamma}_H^{OLS}(2^{-15})$ based on 100 repeated samples

| $H$ | 0.1 | 0.2 | 0.3 | 0.4 | 0.5 | 0.6 | 0.7 | 0.8 | 0.9 |
|---|---|---|---|---|---|---|---|---|---|
| $\widehat{H}$ | .1554 | .2276 | .3123 | .4030 | .4916 | .5867 | .6808 | .7755 | .8724 |
| $\widehat{H} - H$ | .0554 | .0276 | .0123 | .0030 | -.0084 | -.0133 | -.0192 | -.0245 | -.0276 |
| $SD$ | .0386 | .0376 | .0393 | .0342 | .0390 | .0388 | .0376 | .0479 | .0495 |
| $\sqrt{MSE}$ | .0674 | .0465 | .0409 | .00342 | .0397 | .0408 | .0420 | .0536 | .0564 |
| $MSE$ | .0045 | .0022 | .0017 | .0012 | .0016 | .0017 | .0018 | .0029 | .0032 |

**SD:** The estimated SD for the Gladyshev estimator is much smaller than the estimated SD for the OLS estimator. The values of SD for $\widehat{\gamma}_H^G$ have the same order for $H \leq 0.7$, while these values steadily increase for $H \geq 0.8$. A similar behaviour of SD is apparent for $\widehat{\gamma}_H^{OLS}$.

A possible explanation for the increase in standard deviation for $H \geq 0.8$ is that different values of $H$ produce different asymptotic behaviours of the variance. A similar difference was discovered in [15], [11] and [6] for several estimators of the Orey index of a stationary Gaussian process $X$. Namely, the variance of the estimators $\widehat{\gamma}_X(1/n)$ considered in the forementioned papers has the order $n^{-1}$ if $0 < \gamma_X < 0.75$ and the order $n^{-4(1-\gamma_X)}$ if $0.75 < \gamma_X < 1$.

An empirical study of the nine estimators of the Hurst exponent was carried out by Taqqu et al. [28]. In particular, they were estimating the Hurst exponent for a simulated fractional Gaussian noise. The estimators in [28] are based on global properties of a fractional Brownian motion $B_H$ and our estimators are concerned with the local properties of $B_H$. However, due to self-similarity of $B_H$, the same Hurst exponent $H$ is estimated in both cases, so some comparisons can be made. Nevertheless some caution should be exercised since we are using a different simulation program and our sample sizes are larger. In general, the performance of $\widehat{\gamma}_H^{OLS}$ is comparable to the estimators studied in [28]. On the other hand, the discrepancy of $\widehat{\gamma}_X^G$ around the estimated expected value $(SD)$ is similar to that of the discrepancy of Whittle's estimator (in [28]) around the actual value of the Hurst exponent $(SD$ and $\sqrt{MSE})$, whose bias is extremely small as compared to the bias of the Gladyshev estimator.

## 4. Financial data

**4.1. Formulation of the estimation problems.** Several financial data sets, representing the price of an asset, will be investigated as values of a sample function of a stochastic process at certain moment of time. A continuous time stochastic process $P = \{P(t): t \geq 0\}$ with almost all sample functions positive, used to model the price, will be called the *price process*. The simplest assumption in *a continuous time financial* model is that $P$ is a geometric Brownian motion. In this case, taking a logarithm of the data set provides a sample which can be used to test unknown parameters of the Brownian motion with a drift. The logarithmic transformation scheme, or the log return, is often used in an econometric analysis of financial data to test even more general assumptions other than the geometric Brownian motion assumption.

In the present paper two types of data transformation will be applied. In addition to the log return, we consider a continuous time modification of a simple net return. First recall the two definitions of returns in the context of discrete time models. Suppose $P = \{P(t): t = 0, 1, \ldots, T\}$ is a discrete time price series, where $P(t)$ denotes the price of an asset at date $t$ which pays no dividends. Then the *simple net return* $\widetilde{R}_{net}$ is defined by $\widetilde{R}_{net}(t) := P(t)/P(t-1) - 1$ for $t = 1, \ldots, T$, and continuously compounded return or the *log return* $\widetilde{R}_{log}$ is defined by $\widetilde{R}_{log}(t) := \log[P(t)/P(t-1)]$ for $t = 1, \ldots, T$. Note that $\widetilde{R}_{net}$ and $\widetilde{R}_{log}$ are interval functions defined on intervals $[t-1, t]$. For detailed comments on returns in econometric analysis we direct the reader to Section 1.4 in [4] and Section 7.6 in [9].

Let $P = \{P(t): t \geq 0\}$ be the price process. In continuous time financial models, the log return $R_{log} = R_{log}(P)$ is a simple extension of its counterpart in a discrete time case, except now it is defined as a point function. That is, we define $R_{log}$ by

$$(4.1) \qquad R_{log}(P)(t) := \log\big[P(t)/P(0)\big], \qquad t \geq 0.$$

On the other hand, any stochastic process $R = \{R(t): t \geq 0\}$ such that $R(0) = 0$ almost surely, determines the price process $P_{exp} = P_{exp}(R)$ by the relation

$$(4.2) \qquad P_{exp}(R)(t) = P(0) \exp\big\{R(t)\big\}, \qquad t \geq 0.$$

We have that $R_{log}(P_{exp}(R)) = R$ and $P_{exp}(R_{log}(P)) = P$, which is an illustration that the price and return processes are in duality to each other. Under the additional assumption that $R$ is a stationary stochastic process, or more generally, a stochastic process with stationary increments, relation (4.2) represents the commonly used model of the price process in insurance and finance. In this paper the log return $R_{log}$ will be assumed to be a Gaussian stochastic process with stationary increments having an Orey index of unknown value. More precisely, we consider the following estimation problem.

ESTIMATION PROBLEM 1. Let $\mathcal{H}_1$ be the class of all mean zero Gaussian stochastic processes $X$ with stationary increments, continuous in quadratic mean and such that the Orey index $\gamma_X$ exists, $X \in \mathcal{G}^\lambda$ for some $\lambda$ and $X(0) = 0$ almost surely. Suppose that the price process $P$ satisfies the relation

$$P(t) = P(0) \exp\{X(t)\}, \quad t \geq 0, \qquad \text{for some unknown } X \in \mathcal{H}_1.$$

Then we consider the problem of estimation of the Orey index $\gamma_X$ from an empirical data representing $P$.

In several areas of financial mathematics, such as option pricing theory, more sophisticated models of the price dynamics of the fundamental asset are considered. For instance, the price process $P$ may be set to be the solution of the Itô stochastic differential equation

$$(4.3) \qquad P(t) = P(0) + \int_0^t \mu P(s)\, ds + (I) \int_0^t \sigma P(s)\, dB(s), \qquad 0 \leq t \leq T,$$

where $B = \{B(t) : t \geq 0\}$ is a standard Brownian motion, $\mu \in \mathbb{R}$, $\sigma > 0$ and $I$ denotes the stochastic integral in the Itô sense. The geometric Brownian motion

$$(4.4) \qquad P_{1/2}(t) := P(0) \exp\left\{\left(\mu - \sigma^2/2\right)t + \sigma B(t)\right\}, \qquad 0 \leq t \leq T,$$

is the unique solution to Eq. (4.3). The statistical analysis of the price dynamic described by either Eq. (4.3), or Eq. (4.4) can be performed in the context of the model (4.2) under the assumption that the log return is an ordinary Brownian motion (see Section 9.3 in [4] for an overview). Furthermore, in financial models based on Itô's stochastic calculus, the Brownian motion in Eq. (4.3) can be replaced by an arbitrary martingale whose coefficients $\mu$, $\sigma$ may depend on $P$. However, if one is concerned with effects of a price dynamics other than assuming the Random Walk Hypothesis or the Efficient Market Hypothesis, then the Brownian motion $B$ in Eq. (4.3) can be replaced by a non-martingale, such as a fractional Brownian motion $B_H$ with the Hurst exponent $H \neq 1/2$. In this case, we consider the Riemann-Stieltjes integral equation

$$(4.5) \qquad P(t) = P(0) + \int_0^t \mu P(s)\, ds + (RS) \int_0^t \sigma P(s)\, dB_H(s), \qquad 0 \leq t \leq T,$$

where $B_H$ is a fractional Brownian motion with $H \in (1/2, 1)$. By Proposition 5.4 of [21], Eq. (4.5) has the unique solution

$$(4.6) \qquad P_H(t) := P(0) \exp\left\{\mu t + \sigma B_H(t)\right\}, \qquad 0 \leq t \leq T.$$

This solution will remain the same if the Riemann-Stieltjes integral in (4.5) is replaced by an extended stochastic integral such as in [18], or solving a non-standard analysis differential equation such as in [7]. The only difference between solutions (4.4) and (4.6) is the quadratic variation term. The quadratic variation of $B_H$ is zero when $1/2 < H < 1$, while the Brownian motion $B = B_{1/2}$ has non-zero quadratic variation. Suppose that $1/2 \leq H < 1$ and that the price process $P$ satisfies either Eq. (4.3) or Eq. (4.5), or $P$ satisfies a more general equation which includes both:

$$(4.7) \qquad P(t) = P(0) + \int_0^t \mu P(s)\, ds + (LC) \int_0^t \sigma P(s)\, d_\lambda B_H(s), \qquad 0 \leq t \leq T,$$

where $LC$-integral is the Left Cauchy integral based on a sequence of partitions $\lambda$ (see e.g. §4 in [22]). That is, $P = P_H$ given by (4.4) if $H = 1/2$ and (4.6) if $1/2 < H < 1$. Then the log return based on $P_H$ gives

$$(4.8) \qquad R_{log}(P_H)(t) = \begin{cases} (\mu - \sigma^2/2)t + \sigma B(t) & \text{if } H = 1/2 \\ \mu t + \sigma B_H(t) & \text{if } 1/2 < H < 1. \end{cases}$$

Thus an estimation of the Hurst exponent $H$ from an empirical data using the log return may contain a systematic error whenever $P$ satisfies equation (4.7). However, this error will not appear if one considers the continuous version of the simple net returns to be defined next.

Let $\mathcal{Q}([0,T])$ be the set of all nested sequences $\lambda = \{\lambda(m)\colon m \geq 1\}$ of partitions $\lambda(m) = \{0 = t_0^m < \cdots < t_{n(m)}^m = T\}$ of $[0,T]$ such that $\cup_m \lambda(m)$ is everywhere dense in $[0,T]$. Given a sample function $P$ on $[0,T]$ and a sequence $\lambda \in \mathcal{Q}([0,T])$, we say that the *net return* $R_{net}^\lambda(P)$ is defined if the limit

$$(4.9) \qquad R_{net}^\lambda(P)(t) := \lim_{m \to \infty} \sum_{i=1}^{n(m)} [P(t_i^m \wedge t) - P(t_{i-1}^m \wedge t)]/P(t_{i-1}^m \wedge t)$$

exists for each $t \in [0,T]$. If $R_{net}^\lambda(P)$ defined for each $\lambda \in \mathcal{Q}([0,T])$ does not depend on $\lambda$, then we write $R_{net}(P)$. Let $1/2 \leq H < 1$ and let $P_H$ be the solution to (4.7) given by (4.4) and (4.6). By Theorem 3.5 (for $1/2 < H < 1$) and by Proposition 3.6 (for $H = 1/2$) of [22], given a sequence $\lambda \in \mathcal{Q}([0,T])$, the net return $R_{net}^\lambda(P_H)$ is defined and the relation

$$(4.10) \qquad R_{net}^\lambda(P_H)(t) = \mu t + \sigma B_H(t), \qquad 0 \leq t \leq T,$$

holds for almost all sample functions of $B_H$. Thus an estimation of the Hurst exponent $H$ from an empirical data using this net return will not contain a systematic error as in (4.8).

To illustrate the difference between the two definitions of returns, we simulated a Brownian motion sample function $B$ with values at each point of the partition $\lambda := \{i/N\colon i = 0,\ldots,N\}$ of $[0,1]$ with $N = 17000$ (see Figure 1). The number 17000 has been chosen because the cardinality of our financial data sets to be considered in the next section has the same order. Then the two returns are applied to the same sample function of the geometric Brownian motion (gBm) $P_{1/2}(t) = \exp\{B(t) - t/2\}$, $t \in [0,1]$. All functions in Figure 1 are determined by their values on the same partition $\lambda$. The quadratic variation was calculated as the vector $[B] = \{\sum_{i=1}^k [B(i/N) - B((i-1)/N)]^2\colon k = 1,\ldots,N\}$. The net return $R_{net}^\lambda(P_{1/2})$ was replaced by its approximation $R[net] := R_{net}^\lambda(\tilde{P}_{1/2})$, where $\tilde{P}_{1/2}(t) := P_{1/2}((i-1)/N)$ for $t \in [(i-1)/N, i/N)$ and $i = 1,\ldots,N$. For the log return $R[log]$ we consider the vector $\{\log P_{1/2}(i/N)\colon i = 0,\ldots,N\}$. The last two pictures in Figure 1 show the difference between the two returns and its comparison to the calculated quadratic variation $[B]/2$, which is relatively small.

The fractional Brownian motion $B_H$ in equations (4.3) and (4.5) can be replaced by a stochastic process $X$ from a larger class of Gaussian stochastic processes with stationary increments. The Orey index $\gamma_X$ of $X$ then plays a role similar to that of the Hurst exponent $H$. Assuming that the price $P$ can be modelled by the solution of the integral equation

$$(4.11) \qquad P(t) = 1 + (LC)\int_0^t P(s)\,d_\lambda X(s), \qquad 0 \leq t \leq T,$$

then one can attempt to estimate $\gamma_X$ from the historical data of $P$, which brings us to our second estimation problem.

ESTIMATION PROBLEM 2. Let $\mathcal{H}_2$ be the subclass of stochastic processes $X \in \mathcal{H}_1$ such that the equation (4.11) has a unique solution. Suppose that a price process $P$ is a solution to the equation (4.11) for some $X \in \mathcal{H}_2$. Then one can consider the problem of estimation of the Orey index $\gamma_X$ from an empirical data of $P$

Estimation Problems 1 and 2 are formulated for a subclass of Gaussian stochastic processes, and can be considered as parameter estimation problem for a non-parametric model. However, we consider these problems as a part of a more general estimation problem. That is, given a function $f$ on $[0, T]$ and a number $0 < p < \infty$, let

$$(4.12) \qquad s_p(f; \lambda) := \sum_{i=1}^{n} |f(t_i) - f(t_{i-1})|^p,$$

where $\lambda = \{0 = t_0 < t_1 < \cdots < t_n = T\}$ is a partition of $[0, T]$. The quantity

$$(4.13) \qquad v_p(f; [0, T]) := \sup \big\{ s_p(f; \lambda) \colon \lambda \text{ is a partition of } [0, T] \big\}$$

may be finite or infinite and is called the *p-variation* of $f$. The function $f$ has bounded $p$-variation if (4.13) is finite. The number

$$v(f) := \inf \big\{ p > 0 \colon v_p(f; [0, T]) < +\infty \big\} = \sup \big\{ p > 0 \colon v_p(f; [0, T]) = +\infty \big\},$$

is called the *p-variation index* of $f$. Let $X$ be a Gaussian stochastic process with stationary increments. Under certain conditions (see Kawada and Kôno [16]), the $p$-variation index $v(X)$ is equal to $1/\gamma_X$ for almost all sample functions of $X$. Thus estimating the Orey index $\gamma_X$ also means estimating the $p$-variation index $v(X)$ of sample functions of $X$. However, the $p$-variation property is different from the properties considered in econometric theory. That is, the $p$-variation index of sample functions of a stochastic process $X$ will not reflect either a type of distribution of $X$ or a type of correlation between increments of $X$. For instance, an $\alpha$-stable Lévy motion with $1 < \alpha < 2$ and a fractional Brownian motion with the Hurst exponent $H \in (1/2, 1)$ are two processes which have opposite statistical properties, but they have the same $p$-variation index when $\alpha = 1/H$. Therefore, the $p$-variation index is an indicator of analytical properties of sample functions such as their degree of non-differentiability, or their integrability properties. The integrability properties are important in the calculus of functions with unbounded variation.

**4.2. Data analysis.** The data set to be analyzed was provided by Olsen & Associates. It is the high frequency data set HFDF96, which consists of 25 different foreign exchange spot rates, 4 spot metal rates, and 2 series of stock indices. This data set was recorded from 1 Jan 1996 GMT to 31 Dec 1996 GMT. Each set has 17568 entries recorded at half hour intervals, and in particular, each record gives a nearest value of an asset both before and after the recording time.

First, we will consider the Estimation Problem 1. That is, the estimation of the Orey index of the log return $R_{log}(P)$ defined by (4.1). Let $\{N_m \colon m \geq 1\}$ be a sequence of increasing integers. We consider two such sequences of integers $N_m = 2^m$ and $N_m = m$ for $m \geq 1$. In the case $N_m = m$ the sequence $\lambda(m)$ is not nested but its mesh $1/m$ tends to zero faster than any power of $1/\log m$ needed for the existence of the limit (2.4), and hence (2.2). Given an empirical data set

$\{P_1, \ldots, P_K\}$ with $K = 17568$ in our case, we find the maximal integer $M$ such that

$$N := \operatorname{card}\left( \cup_{m=1}^{M} \lambda(N_m) \right) \leq K.$$

So that $M = 14$ when $N_m = 2^m$ and $M = 240$ when $N_m = m$. Let $\{t_1, \ldots, t_N\}$ be the ordered set $\cup_{m=1}^{M} \lambda(N_m)$ and let $P$ be the sample function of a price process such that $P(t_j) := P_j$ for each $j = 1, \ldots, N$. In the case $\{\lambda(N_m) : m \geq 1\}$ is a nested sequence of partitions, $N = N_M + 1$ and $t_j = (j-1)/N_M$ for $j = 1, \ldots, N_M + 1$. For each $m \geq 1$, let

$$u(m)^2 := v_2\big(R_{log}(P); \lambda(m)\big)/N_m = \frac{1}{N_m} \sum_{i=1}^{N_m} \Big[ \ln P(i/N_m) - \ln P((i-1)/N_m) \Big]^2,$$

where $\lambda(m) = \lambda(1/N(m)) = \{i/N_m : i = 0, \ldots, N_m\}$ is a partition of $[0, 1]$, and

(4.14) $$h(m) := \frac{\log u(m)}{\log(1/N_m)} = \frac{-\log_2 u(m)}{\log_2 N_m}.$$

Then the Gladyshev estimator will be denoted by

$$\widehat{\gamma}_1^G(1/N_M) := \widehat{\gamma}_{R_{log}(P)}^G(1/N_M) = h(M),$$

and the OLS estimator will be denoted by

$$\widehat{\gamma}_1^{OLS}(1/N_k, \ldots, 1/N_m) := \widehat{\gamma}_{R_{log}(P)}^{OLS}(1/N_k, \ldots, 1/N_m) = \sum_{j=k}^{m} y_j \log_2 u(j),$$

where $y_j = (x_j - \bar{x})/\sum_{j=k}^{m}(x_j - \bar{x})^2$ and $x_j = \log_2(1/N_j)$ for $j = k, \ldots, m$, $1 \leq k < m \leq M$.

Estimation results for the log returns of USD/JPY exchange rate data are given below:

1. Figures 2 and 3 show the values of the function $h$ given by (4.14) for the two sequences of partitions $\lambda(2^{-m})$ and $\lambda(1/m)$. The last value in both cases is the value of the Gladyshev estimator and its numerical value is stated in the caption.

2. Figures 4 and 5 show the results of fitting a line to the sets $\{(\log_2(N_m), -\log_2 u(m)) : m = 1, \ldots, M\}$. The slopes of the fitted lines gives the values of the OLS estimators for the two partitions $\lambda(2^{-m})$ and $\lambda(1/m)$ and their numerical values are stated in the captions.

3. Table 5 shows the results of fitting a line to different sets of points stated in the caption based on the same partition $\lambda(2^{-m})$. The values of the slope provide the corresponding values of the OLS estimator. The values of intercept provide estimates of $\log(C_1 C_2)$ plus the error term in (2.8).

¿From the above results, we conclude that the Orey index of a stochastic process from the class $\mathcal{H}_1$ which "fits best" the estimated log return for USD/JPY exchange rate data is in the interval $[0.4, 0.7]$. The upper bound 0.7 seems to be most reliable because of the monotonicity of the values recorded in Figures 2 and 3. We suspect that the difference of the value 0.976 recorded in Figure 3 from the value 0.737 in Figure 2 is possibly due to the slow convergence rate in (2.4) for the partition $\lambda(1/m)$ as compared to the convergence rate for the partition $\lambda(2^{-m})$. Hence, more partitions from the same amount of data do not seem to improve the preciseness of the estimation. Estimation results for other foreign exchange rates, metal rates, and stock indices are given in Tables 6 - 8.

Table 5: Fitting a line to $\{(j, -\log_2 u(j)): j = k, \ldots, m\}$ for USD/JPY data.

| $(k, m)$ | (1,7) | (2,8) | (3,9) | (4,10) | (5,11) | (6,12) | (7,13) | (8,14) |
|---|---|---|---|---|---|---|---|---|
| Slope | 0.446 | 0.426 | 0.580 | 0.483 | 0.440 | 0.426 | 0.418 | 0.415 |
| Intercept | 4.212 | 4.347 | 4.025 | 3.929 | 4.263 | 4.401 | 4.486 | 4.510 |
| $(k, m)$ | (1,7) | (1,8) | (1,9) | (1,10) | (1,11) | (1,12) | (1,13) | (1,14) |
| Slope | 0.446 | 0.450 | 0.452 | 0.448 | 0.444 | 0.443 | 0.441 | 0.438 |
| Intercept | 4.212 | 4.197 | 4.191 | 4.207 | 4.223 | 4.228 | 4.236 | 4.250 |

Table 6: The Orey index estimates for exchange rates based on a sample of size $2^{14} + 1$.

| Currency | Nearest prior nominal quote | | | | Nearest next nominal quote | | | |
|---|---|---|---|---|---|---|---|---|
| | Bid | | Ask | | Bid | | Ask | |
| | $\widehat{\gamma}^G$ | $\widehat{\gamma}^{OLS}$ | $\widehat{\gamma}^G$ | $\widehat{\gamma}^{OLS}$ | $\widehat{\gamma}^G$ | $\widehat{\gamma}^{OLS}$ | $\widehat{\gamma}^G$ | $\widehat{\gamma}^{OLS}$ |
| AUD/USD | .737 | .439 | .736 | .438 | .736 | .435 | .735 | .433 |
| CAD/USD | .809 | .343 | .808 | .345 | .791 | .327 | .796 | .335 |
| DEM/ESP | .712 | .252 | .717 | .254 | .696 | .220 | .713 | .246 |
| DEM/FIM | .737 | .386 | .734 | .380 | .730 | .380 | .720 | .370 |
| DEM/ITL | .731 | .478 | .731 | .479 | .733 | .473 | .732 | .473 |
| DEM/JPY | .732 | .389 | .732 | .390 | .734 | .398 | .735 | .396 |
| DEM/SEK | .726 | .442 | .724 | .441 | .721 | .444 | .718 | .441 |
| GBP/DEM | .747 | .501 | .748 | .502 | .747 | .500 | .747 | .501 |
| GBP/USD | .750 | .463 | .751 | .464 | .750 | .463 | .750 | .464 |
| USD/BEF | .678 | .385 | .680 | .386 | .680 | .384 | .682 | .385 |
| USD/CHF | .719 | .473 | .718 | .473 | .719 | .474 | .719 | .474 |
| USD/DEM | .745 | .446 | .744 | .445 | .744 | .446 | .743 | .446 |
| USD/DKK | .585 | .291 | .586 | .290 | .585 | .322 | .586 | .317 |
| USD/ESP | .652 | .345 | .673 | .369 | .652 | .338 | .672 | .372 |
| USD/FIM | .689 | .423 | .702 | .433 | .689 | .422 | .703 | .437 |
| USD/FRF | .733 | .434 | .732 | .433 | .733 | .425 | .732 | .424 |
| USD/ITL | .708 | .330 | .712 | .333 | .712 | .325 | .713 | .328 |
| USD/NLG | .724 | .432 | .725 | .432 | .724 | .431 | .721 | .428 |
| USD/SEK | .692 | .355 | .697 | .360 | .690 | .335 | .693 | .355 |
| USD/XEU | .736 | .384 | .736 | .384 | .727 | .370 | .729 | .372 |
| USD/JPY | .737 | .438 | .734 | .436 | .735 | .436 | .733 | .433 |
| USD/MYR | .717 | .298 | .717 | .300 | .729 | .295 | .729 | .295 |
| USD/SGD | .795 | .299 | .787 | .288 | .788 | .303 | .782 | .276 |
| USD/ZAR | .694 | .506 | .693 | .505 | .698 | .514 | .691 | .504 |

Table 7: The Orey index estimates for metal rates based on a sample of size $2^{14}+1$.

| Metal | Nearest prior nominal quote | | | | Nearest next nominal quote | | | |
|---|---|---|---|---|---|---|---|---|
| | Bid | | Ask | | Bid | | Ask | |
| | $\widehat{\gamma}^G$ | $\widehat{\gamma}^{OLS}$ | $\widehat{\gamma}^G$ | $\widehat{\gamma}^{OLS}$ | $\widehat{\gamma}^G$ | $\widehat{\gamma}^{OLS}$ | $\widehat{\gamma}^G$ | $\widehat{\gamma}^{OLS}$ |
| Gold | .741 | .437 | .741 | .436 | .732 | .430 | .729 | .428 |
| Silver | .616 | .382 | .603 | .384 | .603 | .369 | .603 | .369 |
| Palladium | .646 | .427 | .646 | .426 | .637 | .420 | .638 | .420 |
| Platinum | .707 | .437 | .706 | .436 | .702 | .437 | .702 | .437 |

Table 8: The Orey index estimates for stock indices based on a sample of size $2^{14}+1$.

| Index | recorded price prior expiry | | recorded price after expiry | |
|---|---|---|---|---|
| | $\widehat{\gamma}^G$ | $\widehat{\gamma}^{OLS}$ | $\widehat{\gamma}^G$ | $\widehat{\gamma}^{OLS}$ |
| SP 500 | 0.7285 | 0.5216 | 0.7285 | 0.5171 |
| DOW JONES | 0.7174 | 0.5226 | 0.7175 | 0.5229 |

Now consider Estimation Problem 2, that is the estimation of the Orey index of the net return $R_{net}^{\lambda}(P)$ defined by (4.9). In this case, we consider only the nested sequence $\lambda$ of dyadic partitions $\lambda(2^{-m})$ of the interval $[0,1]$. Then the maximal integer $M$ such that $1 + 2^M \leq K = 17568$ again is 14. Let $P$ be the sample function of a price process such that $P(i2^{-M}) := P_{i+1}$ for $i = 0, \ldots, 2^M$. An important difference between the net return $R_{net}^{\lambda}(P)$ and the log return $R_{log}(P)$ is that each value $R_{net}^{\lambda}(P)(t)$ depends on values $P(s)$ for all $s$ from an everywhere dense subset of $[0, t]$. Therefore to calculate the net return $R_{net}^{\lambda}(P)$ we replace $P$ by the sample function $P_M$ such that $P_M(t) := P((i-1)2^{-M})$ if $t \in [(i-1)2^{-M}, i2^{-M})$ for some $i = 1, \ldots, 2^M$ and $P_M(1) := P(1)$. In fact the function $P_M$ is the only empirically available information about the price process $P$. For each $m \geq 1$ and $t \in [0, 1]$, let

$$(4.15) \qquad R_m(t) := \sum_{k=1}^{2^m} \frac{P_M(k2^{-m} \wedge t) - P_M((k-1)2^{-m} \wedge t)}{P_M((k-1)2^{-m} \wedge t)}.$$

Because $R_m \equiv R_M$ for all $m > M$, for each $t \in [0, 1]$, we have

$$(4.16)$$

$$R_{net}^{\lambda}(P_M)(t) \;=\; \lim_{m \to \infty} R_m(t)$$

$$=\; \begin{cases} 0 & \text{if } t \in [0, 2^{-M}), \\ \sum_{i=1}^{j} \frac{P(i2^{-M}) - P((i-1)2^{-M})}{P((i-1)2^{-M})} & \text{if } t \in [j2^{-M}, (j+1)2^{-M}) \\ & \quad \text{for some } j = 1, \ldots, 2^M - 1, \\ \sum_{i=1}^{2^M} \frac{P(i2^{-M}) - P((i-1)2^{-M})}{P((i-1)2^{-M})} & \text{if } t = 1. \end{cases}$$

We look at a difference between the net return and the log return at all points of the partition $\lambda(2^{-M}) = \{i/2^M : i = 0, \ldots, 2^M\}$. Notice that $R_{log}(P)(i/2^M) = R_{log}(P_M)(i/2^M)$ for $i = 0, \ldots, 2^M$. In the preceding subsection this difference is shown by (4.8) and (4.10) when the price process is the solution $P_H$ to the equation (4.7) driven by a fractional Brownian motion $B_H$. In the case $H = 1/2$ the difference between two returns is also illustrated by Figure 1. Here the calculations are done for the sample function $P$ given by USD/JPY exchange rate data and the results are illustrated by Figure 6.

The main problem is to identify the source of the slant in the bottom pictures of Figure 6. If it is the result of rounding and approximation errors, then the net return and the log return are approximately the same. This would support the conclusion that the Orey index of $R_{net}(P) \approx R_{log}(P)$, if it exists, is in the interval $(1/2, 1)$. To evaluate one possible approximation error we compared the functions $R_m$ defined by (4.15) for $m = M$ and $m = M - 1$, and we found that

$$-0.0014 \leq R_{M-1}(t) - R_M(t) \leq 8.2 \cdot 10^{-7} \qquad \text{for all } t \in [0, 1],$$

which indicates that this approximation error is somewhat less than the order of the slant in Figure 6. On the other hand, if the slant is not a result of these errors, then there are several possible reasons of its appearance. One reason is the existence of a non-zero quadratic variation for the processes $X \in \mathcal{H}_2$ (recall that $P$ is assumed to be the solution to (4.11)). Another possible reason for $R_{net}(P) \neq R_{log}(P)$ is that a discontinuous function $P$ may also fit the USD/JPY data. This is possible because $R_{net}(P)$ is defined for a much larger class of functions $P$ than the class of sample functions considered in the present paper. For example, by Proposition 3.2 of [22], if $P$ has bounded $p$-variation (cf. (4.13)) with $p < 2$ and is bounded away from zero then, for each $t \in [0, 1]$,

$$R_{net}^{\lambda}(P)(t) = R_{log}(P)(t) - \sum_{(0,t]} \left[ \log \frac{P}{P_-} - \frac{\Delta^- P}{P_-} \right] - \sum_{[0,t)} \left[ \log \frac{P_+}{P} - \frac{\Delta^+ P}{P} \right],$$

where the two sums converge absolutely, $P_-(t) := P(t-) := \lim_{s \uparrow t} P(s)$, $P_+(t) := P(t+) := \lim_{s \downarrow t} P(s)$, $\Delta^- P(t) := P(t) - P(t-)$ and $\Delta^+ P(t) := P(t+) - P(t)$. However, the last explanation does not allow us to conclude that the Orey index of $P$ if it exists, is equal to $1/2$.

We conclude by applying our two estimators to the net return of USD/JPY exchange rate data. For $m = 1, \ldots, M$, let

$$
\begin{aligned}
w(m)^2 &:= \frac{1}{2^m} v_2(R_{net}^{\lambda}(P_M); \lambda(2^{-m})) \\
&= \frac{1}{2^m} \sum_{k=1}^{2^m} \left[ R_{net}^{\lambda}(P_M)(k 2^{-m}) - R_{net}^{\lambda}(P_M)((k-1)2^{-m}) \right]^2 \\
&= \frac{1}{2^m} \sum_{k=1}^{2^m} \left[ \sum_{i \in J_m(k)} \frac{P(i 2^{-M}) - P((i-1)2^{-M})}{P((i-1)2^{-M})} \right]^2,
\end{aligned}
$$

where $J_m(k) := \{1 + (k-1)2^{M-m}, \ldots, k 2^{M-m}\}$, $k = 1, \ldots, 2^m$, and

$$g(m) := \frac{\log w(m)}{\log(1/2^m)} = \frac{-\log_2 w(m)}{m}.$$

Referring to the Estimation Problem 2 the Gladyshev estimator will be denoted by

$$\widehat{\gamma}_2^G(2^{-M}) := \widehat{\gamma}_{R_{net}(P)}^G(2^{-M}) = g(M),$$

and the OLS estimator will be denoted by

$$\widehat{\gamma}_2^{OLS}(2^{-M}) := \widehat{\gamma}_{R_{net}(P)}^{OLS}(2^{-1}, \ldots, 2^{-M}) = \sum_{j=1}^{M} y_j \log_2 w(j),$$

where $y_j = (x_j - \bar{x})/\sum_{j=1}^{M}(x_j - \bar{x})^2$ and $x_j = \log_2(1/N_j)$ for $j = 1, \ldots, M$.

Estimation results for the net returns of USD/JPY exchange rate data are shown in Figures 7 and 8. These results can be compared with corresponding results for the log returns in Figures 2 and 4. The value of the slope in Figure 8 is 0.443, which is similar to the value of 0.438 obtained using the log returns (see Table 5).

**4.3. Related works.** Mandelbrot [20, Section E6.3] discusses several models of price variations. One of them is the composition process $B_H \circ \Theta = \{B_H(\Theta(t)): t \geq 0\}$, where $B_H$ is a fractional Brownian motion with the Hurst exponent $H$ and $\Theta$ is a multifractal. Also, Mandelbrot [20, pp. 173-176] suggests a new method for estimating the parameters of this model. To describe the idea of the method suppose $\Theta(t) = t$ for $t \geq 0$. Let $\{N_m: m \geq 1\}$ be a sequence of positive integers such that $1/N_m = o(1/(\log m)^a)$ as $m \to \infty$ for each $a > 0$, and let $\lambda(T/N_m) = \{iT/N_m: i = 0, \ldots, N_m\}$ be partitions of $[0, T]$. By Corollary 1.3 of Shao [27], there exists the limit

$$\lim_{m \to \infty} v_{1/H}(B_H; \lambda(T/N_m)) = TE|\eta|^{1/H} \qquad \text{almost surely},$$

where $\eta$ is a standard normal random variable and notation (4.13) is used. Since almost all sample functions of $B_H$ are continuous, it follows that

$$(4.17) \qquad \lim_{m \to \infty} v_q(B_H; \lambda(T/N_m)) = \begin{cases} 0 & \text{if } q > 1/H \\ +\infty & \text{if } q < 1/H \end{cases} \qquad \text{almost surely}.$$

Therefore an estimator of the Hurst exponent $H$ can be defined as the number $\widehat{H}$ with the property $v_q(R_{log}(P); \lambda(1/N_m)) = 0$ for any $q > 1/\widehat{H}$ and $= +\infty$ for any $q < 1/\widehat{H}$. According to Mandelbrot [20, pp. 175, 176], this estimator applied to the USD/DEM exchange rate data from HFDF96 yields the value of $\widehat{H}$ close to $1/2$, but larger than $1/2$. This can be compared with our estimation results in Table 6.

The fractional Brownian motion $B_H$ and its Hurst exponent $H$ in (4.17) can be replaced by a stochastic process $X$ and a number $\gamma$ satisfying conditions of the following statement (and its proof is given in [23]).

THEOREM 4.1. *Let $X = \{X(t): t \geq 0\}$ be a mean zero Gaussian stochastic process with stationary increments and continuous in quadratic mean. Suppose that for some $\gamma \in (0, 1)$, the incremental variance $\sigma_X$ satisfies (2.7) and there exists the limit*

$$(4.18) \qquad \lim_{m \to \infty} v_{1/\gamma}(X; \lambda(T/N_m)) = c \qquad a.s.$$

*for some sequence of integers $\{N_m: m \geq 1\}$ such that $N_m \to \infty$ as $m \to \infty$ and some positive constants $c$, $T$. Then the Orey index $\gamma_X = \gamma$, $X$ belongs to the Gladyshev class $\mathcal{G}^\lambda$ and $G_\lambda(X) = \gamma$ almost surely, where $\lambda = \{\lambda(T/N_m): m \geq 1\}$.*

Therefore the above-defined number $\widehat{H}$ is an estimator of the Orey index $\gamma_X$ for the non-parametric model (2.7) instead of the Hurst exponent of a fractional Brownian motion.

# 5. Conclusions

To the best of our knowledge, this is the first time that a strong limit theorem for Gaussian stochastic processes has been applied to financial data. The results showed that the method can be used to estimate the Orey index. However, to gain a maximal advantage from this estimator, its statistical properties need to be established.

The Monte-Carlo study showed that Gladyshev's estimator (2.6) has a clear bias and a very small standard deviation. Also, calculated for the simulated data as well as for the empirical data, the values of the sequence

$$\left\{ \frac{\log \sqrt{v_2(X; \lambda(m))/N_m}}{\log(1/N_m)} : m = 1, \ldots, M \right\}$$

appear to have a clear monotonicity character (see Figures (2), (3) and (7)). The bias of the OLS estimator also has a predictable character: small for the Orey index values around $1/2$ and opposite signs for the values on both sides of $1/2$. Knowledge of these features can be used in making conclusions about other empirical data sets.

**Acknowledgment.** We thank Thomas Mikosch for pointing out the following papers [5], [6], [11], [14], and [15], and Donald Dawson for his fruitful and insightful discussions.

# References

[1] Bardet, J. M. (1997). Testing self-similarity of processes with stationary increments. Prepublications of Université de Paris-Sud, Mathematiques, 91405 Orsay France, 97.17.

[2] Baxter, G. (1956). Strong limit theorem for Gaussian processes. *Proc. Amer. Math. Soc.*, **7**, 522-527.

[3] Cameron, R. H. and Martin, W. T. (1947). The behavior of measure and measurability under a change of scale in Wiener space. *Bull. Amer. Math. Soc.*, **53**, 130-137.

[4] Campbell, J. Y., Lo, A. W., and MacKinlay, A. C. (1997). *The econometrics of financial markets.* Princeton University Press, Princeton, 1997.

[5] Chan, G., Hall, P., and Poskitt, D. S. (1995). Periodogram-based estimators of fractal properties. *Ann. Statist.*, **23**, 1684-1711.

[6] Constantine, A. G. and Hall, P. (1994). Characterising surface smoothness via estimation of effective fractal dimension. *J. Roy. Statist. Soc. Ser. B*, **56**, 97-113.

[7] Cutland, N. J., Kopp, P. E., and Willinger, W. (1995). Stock price returns and the Joseph effect: a fractional version of the Black-Scholes model. In: *Seminar on stochastic analysis, random fields and application*, Eds. E. Bolthausen et al. *Progress in Probability*, **36**, 327-352; Birkhäuser, Basel.

[8] Dudley, R. M. (1973). Sample functions of the Gaussian process. *Ann. Probab.*, **1**, 66-103.

[9] Embrechts, P., Klüppelberg, C., and Mikosch, T. (1997). *Modelling extremal events for insurance and finance.* Springer, Berlin.

[10] Fernández de la Vega, W. (1974). On almost sure convergence of quadratic Brownian variation. *Ann. Probab.*, **2**, 551-552.

[11] Feuerverger, A., Hall, P., and Wood, A. T. A. (1994). Estimation of fractal index and fractal dimension of a Gaussian process by counting the number of level crossings. *J. Time Ser. Anal.*, **15**, 587-606.

[12] Giné, E. and Klein, R. (1975). On quadratic variation of processes with Gaussian increments. *Ann. Probab.*, **3**, 716-721.

[13] Gladyshev, E. G. (1961). A new limit theorem for stochastic processes with Gaussian increments. *Theor. Probability Appl.*, **6**, 52-61.

[14] Hall, P. and Roy, R. (1994). On the relationship between fractal dimension and fractal index for stationary stochastic processes. *Ann. Appl. Probab*, **4**, 241-253.

[15] Hall, P. and Wood, A. (1993). On the performance of box-counting estimators of fractal dimension. *Biometrika*, **80**, 246-252.

[16] Kawada, T. and Kôno, N. (1973). On the variation of Gaussian processes. In: Maruyama, G., Prokhorov, Yu. V. (eds.); Proc. Second Japan-USSR Symposium on Probability Theory; *Lect. Notes Math.*, **330**, 176-192. Springer, Berlin.

[17] Lévy, P. (1940). Le mouvement brownien plan. *Amer. J. Math.*, **62**, 487-550.

[18] Lin, S. J. (1995). Stochastic analysis of fractional Brownian motions. *Stochastics & Stochastics Rep.*, **55**, 121-140.

[19] Maeder, R. E. (1995). Fractional Brownian motion. *Mathematica J.*, **6**, 38-48.

[20] Mandelbrot, B. B. (1997). *Fractals and Scaling in Finance: Discontinuity, Concentration, Risk.* Selecta volume E. Springer, New York.

[21] Mikosch, T. and Norvaiša, R. (1997). Stochastic integral equations without probability. Preprint.

[22] Norvaiša, R. (1999). Modelling of stock price changes: A real analysis approach. *Finance and Stochastics*, to appear.

[23] Norvaiša, R. (1999). Quadratic variation, $p$-variation and integration with applications to stock price modelling. In preparation.

[24] Orey, S. (1970). Gaussian sample functions and the Hausdorff dimension of level crossings. *Z. Wahrsch. verw. Gebiete*, **15**, 249-256.

[25] Pierre, P. A. (1971). The quadratic variation of random processes. *Z. Wahrsch. verw. Gebiete*, **19**, 291-301.

[26] Salopek, D. M. (1998). Tolerance to arbitrage. *Stoch. Proc. Appl.*, **76**, 217-230.

[27] Shao, Q.-M. (1996). $p$-variation of Gaussian processes with stationary increments. *Studia Sci. Math. Hungar.*, **31**, 237-247.

[28] Taqqu, M. S., Teverovsky, V., and Willinger, W. (1995). Estimators for long-range dependence: an empirical study. *Fractals*, **3**, 785-788.

[29] Tricot, C. (1995). *Curves and fractal dimension.* Springer-Verlag, New York.

INSTITUTE OF MATHEMATICS AND INFORMATICS, AKADEMIJOS 4, VILNIUS 2600, LITHUANIA
*E-mail address*: norvaisa@ktl.mii.lt

YORK UNIVERSITY, DEPARTMENT OF MATHEMATICS AND STATISTICS, 4700 KEELE STREET, TORONTO, ONTARIO M3J 1P3 CANADA
*E-mail address*: dsalopek@mathstat.yorku.ca

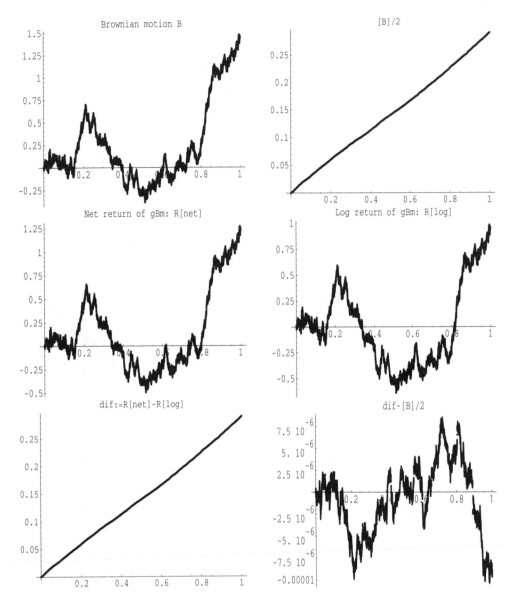

Figure 1: Plot of a Brownian motion sample function $\{B(i/17000)\colon i = 0,\dots,17000\}$ and its transformations.

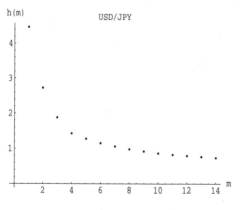

Figure 2: $\widehat{\gamma}_1^G(2^{-14}) = 0.7368.$

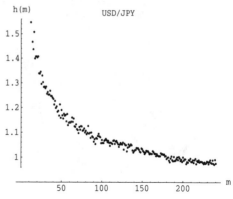

Figure 3: $\widehat{\gamma}_1^G(1/240) = 0.9759.$

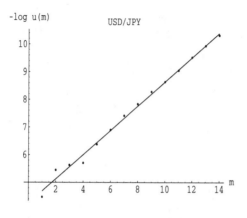

Figure 4: $\widehat{\gamma}_1^{OLS}(2^{-1}, \ldots, 2^{-14}) = 0.438.$

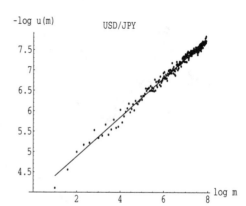

Figure 5: $\widehat{\gamma}_1^{OLS}(1, \ldots, 1/240) = 0.481.$

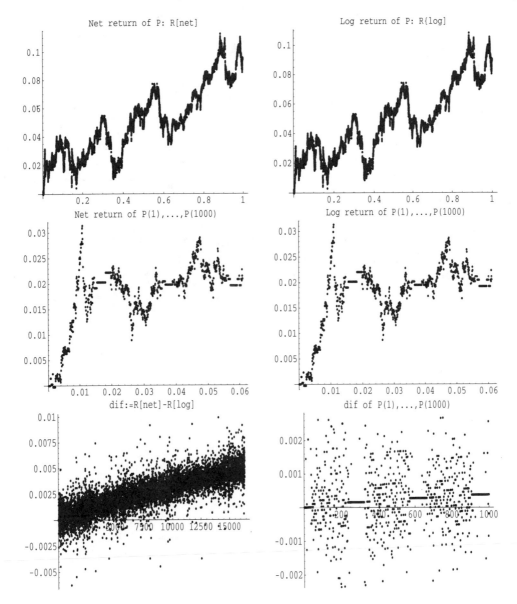

Figure 6: USD/JPY returns.

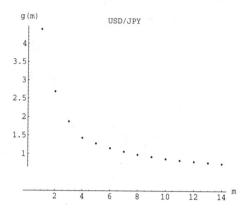

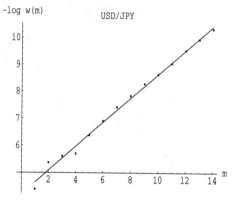

Figure 7: $\widehat{\gamma}_2^G(2^{-14}) = 0.736777$.

Figure 8: $\widehat{\gamma}_2^{OLS}(2^{-14}) = 0.443$.

Canadian Mathematical Society
Conference Proceedings
Volume **26**, 2000

# Large Deviations Estimates for Occupation Time Integrals of Brownian Motion

## Bruno Remillard

*Dedicated to Professor Donald A. Dawson.*

ABSTRACT. Using properties of Dirichlet Laplacian in bounded domains, we derive a large deviation principle for occupation time integrals of Brownian motion in $\mathbb{R}^d$. We also obtain moderate deviations estimates in the case of random potentials.

## 1. Introduction and some results

Let $(x_t)_{t\geq 0}$ be a Brownian motion. In many applications, the asymptotic behaviour of integrals of the form $\dfrac{1}{t}\displaystyle\int_0^t V(x_s)ds$ determines the asymptotic behaviour of many interesting processes. See for example Remillard [**9**]. From a physical point of view such integrals are also important because of the Feynman-Kac formula. More precisely, if we set

$$u(t,x) = E_x\left\{f(x_t)e^{\int_0^t V(x_s)ds}\,\mathbb{I}\{T_D > t\}\right\},$$

where $T_D$ is the exit time of a bounded domain $D$, then $u$ satisfies the following partial differential equation $\dot{u} = -\Delta u + Vu$ and $u(0,\cdot) = f(\cdot)$ in $D$, and $u \equiv 0$ outside $D$. In fact this partial differential equation will play a major role in what follows. Note that $V$ is also called a potential. Occupation time integrals of Brownian motion appears in many physical problems related to Schrödringer and Dirichlet semigroups (e.g. Demuth, Kirsch, and McGillivray [**2**]), random Schrödringer operators (e.g. Carmona [**1**]), and packing dimension estimates (e.g. Gruet and Shi [**5**]).

In this paper, we first prove large deviations result in the natural scaling, namely

$$(1.1) \qquad \frac{1}{\lambda_t}\log P\left\{\frac{1}{t}\int_0^t V(x_s,w)ds \in A\right\} \approx -\inf_{a\in A} I(a)$$

1991 *Mathematics Subject Classification.* Primary 60F10, 60J65; Secondary 81Q10, 60J15.

Work supported by the Natural Sciences and Engineering Research Council of Canada, Grant No. 0GP0042137.

as $t$ tends to infinity, where $\lambda_t = t$. It is interesting to note that the entropy function $I$ is not necessarily convex. Moreover since we are also interested in the case of random $V$'s, this scaling will produce trivial results around the mean, that is the entropy function $I$ will be zero or infinity and the set of zero points is larger than one. Such phenomena could be associated with "phase transitions".

In view of these results, it is natural to consider instead moderate deviations, that is, we consider a scaling $\lambda_t$ in (1.1) satisfying $\lambda_t/t \to 0$ as $t$ goes to infinity.

The paper is organized as follows. In Section 2 we develop some tools in terms of Dirichlet Laplacian in bounded domains. In Section 3, we prove large deviations for the natural scaling. We also find conditions for exponential convergence or the absence of phase transitions and we give some examples when the potential $V$ has compact support, when $V$ is almost periodic and when $V$ is a random scenery.

Section 4 is devoted to the moderate deviations type upper and lower bounds for random $V$'s arising from dynamical systems. Finally Section 5 concerns the asymptotic behaviour of the so-called Brownian motion in a random scenery. For dimension one, we find the appropriate scaling but the upper and lower bounds differ. For dimensions greater than one, the scaling is not the same for the upper and lower bounds.

## 2. Preliminary results

As we shall see, many of our results depend on estimates of

$$I(V, D, B, t) = \int_D P_x \left( \frac{1}{t} \int_0^t V(x_s)ds \in B, \ T_D > t \right) dx,$$

where $B \in \mathcal{B}(\mathbb{R}), V \in \mathcal{B}_b(\mathbb{R}^d)$, D is a bounded domain of $\mathbb{R}^d$ (in what follows, a domain is a nonempty connected open set) and $T_D = \inf\{t > 0; \ x_t \in D^c\}$.

Using Tchebychev inequality, we get

$$(2.1) \quad I(V, D, [\theta, \infty), t) \leq \inf_{\lambda > 0} \exp\{-\lambda \theta t\} \int_D E_x \left( e^{\lambda \int_0^t V(x_u)du}; \ I\{T_D > t\} \right) dx.$$

To get bounds for the integral in (2.1), we will use mainly the spectral theory of self-adjoint operators in Hilbert spaces. Let us now introduce some notation.

Let $C_0^\infty(D) = \{f \in C_0^\infty(\mathbb{R}^d); \text{support}(f) \subset D\}$. For every $f \in C_0^\infty(D)$, we define $\nabla f(x) = \left( \frac{\partial}{\partial x_1} f(x), \cdots, \frac{\partial}{\partial x_d} f(x) \right)$ and $\Delta_0 f(x) = \frac{1}{2} \sum_{k=1}^d \frac{\partial^2}{\partial x_k^2} f(x)$. Further let $H_0^1(D)$ be the completion of $C_0^\infty(D)$ with respect to the norm

$$\|f\|_1 = \left( \int_D |f(x)|^2 dx + \frac{1}{2} \int_D |\nabla f(x)|^2 dx \right)^{1/2}.$$

Since we are dealing with self-adjoint operators, we need only to consider real-valued functions. So, as usual, let $L^2(D) = L^2(D, dx)$ be the real Hilbert space with scalar product $(f,g) = \int_D f(x)g(x)dx, \ f,g \in L^2(D)$.

Next we define the Dirichlet Laplacian $\Delta_D = \Delta$ to be the unique self-adjoint operator on $L^2(D)$ satisfying

(i) $\Delta f = \Delta_0 f$, for all $f \in C_0^\infty(D)$;
(ii) the domain of $(-\Delta)^{1/2}$ is $H_0^1(D)$.

From now on, $\mathcal{D}(D)$ will designate the domain of $\Delta$.

We now state some properties of $\Delta$. For more details, the reader is referred to Reed and Simon [8].

PROPOSITION 2.1. *Let $D$ be a bounded domain and let $V : \mathbb{R}^d \mapsto \mathbb{R}$ be bounded and measurable. Then*

(i) *$\Delta + V$ is self-adjoint with domain $\mathcal{D}(D)$;*

(ii) *for all $f \in L^2(D)$ and for every $x \in D$,*

$$e^{t(\Delta+V)}f(x) = E_x\left(f(x(t))e^{\lambda \int_0^t V(x_u)du}I\{T_D > t\}\right)dx;$$

(iii) *for any $a \in \mathbb{R}$, $\{f \in \mathcal{D}(D); |f| \leq 1, I(f) + (Vf,f) \leq a\}$ is compact in $L^2(D)$, where $I(f) = (-\Delta f, f)$, and $|f|^2 = (f, f)$;*

(iv) *for all $t > 0$, $e^{t(\Delta+V)}$ maps $L^2(D)$ into $C_b(D)$, and if $f \geq 0$ a.e., $f \not\equiv 0$ on $D$, then $e^{t(\Delta+V)}f(x) > 0$, for all $x \in D$;*

(v) *$\Delta + V$ has a discrete spectrum $\{\lambda_k; k \geq 1\}$ where $\lambda_1 > \lambda_2 \geq \lambda_3 \ldots$, and there exists a unique (up to a scalar) $f \in C_b(D) \cap \mathcal{D}(D)$ such that $f(x) > 0$ for all $x \in D$, and $(\Delta + V)f = \lambda_1 f$ on $D$;*

(vi) *if $\mathcal{M}(D) = \{f \in \mathcal{D}(D); (f,f) = 1\}$, and $\mathcal{C}(D) = C_0^\infty(D) \cap \mathcal{M}(D)$, then*

$$\lambda_1 = \sup_{f \in \mathcal{M}(D)} (Vf,f) - I(f) = \sup_{f \in \mathcal{C}(D)} (Vf,f) - I(f);$$

(vii) *for any $f \in L^2(D)$, $\left(e^{t(\Delta+V)}f, f\right) \leq |f|^2 e^{t\lambda_1}$.*

It follows from (2.1) and Proposition 2.1 that

$$(2.2) \qquad \int_D P_x\left(\frac{1}{t}\int_0^t V(x_s)ds \geq \theta, \; T_D > t\right)dx \leq |D|\, e^{-t \sup_{\lambda > 0}\{\lambda\theta - c_D(\lambda)\}},$$

where $c_D(\lambda) = \sup_{f \in \mathcal{M}(D)} \lambda(Vf,f) - I(f)$.

In fact

$$
\begin{aligned}
\frac{I(V,D,t)}{|D|} &= \frac{1}{|D|}\int_D P_x\left(\frac{1}{t}\int_0^t V(x_s)ds \geq \theta, T_D > t\right)dx \\
&\leq \inf_{\lambda > 0}\frac{1}{|D|}e^{-\lambda\theta t}\int_D E_x\left(e^{\lambda\int_0^t V(x_s)ds}I\{T_D > t\}\right)dx \\
&= \inf_{\lambda > 0}\frac{1}{|D|}e^{-\lambda\theta t}\left(e^{t(\Delta+\lambda V)}I_D, I_D\right) \\
&\leq e^{-t\sup_{\lambda > 0}\{\lambda\theta - c_D(\lambda)\}}.
\end{aligned}
$$

Set $J_0(\theta) = \sup_{\lambda \geq 0}\{\lambda\theta - c_D(\lambda)\}$. Using Proposition 2.1 and perturbation theory, we can see that $c_D$ is convex, finite and everywhere differentiable with derivative $c_D'(\lambda) = (Vf_\lambda, f_\lambda)$, where $(\Delta + \lambda V)f_\lambda = c_D(\lambda)f_\lambda$, and $f_\lambda$ satisfies (v) of Proposition 2.1.

Then

$$(2.3) \qquad J_0(\theta) = \begin{cases} I(f_0) & \theta \leq c_D'(0) \\ I(f_\lambda) & \theta = c_D'(\lambda), \ \lambda \geq 0 \\ \lim_{\lambda \uparrow \infty} I(f_\lambda) & \theta = \lim_{\lambda \uparrow \infty} c_D'(\lambda) = \theta_\infty \\ +\infty & \theta > \theta_\infty \end{cases}$$

Next define $J_{00}(\theta) = \inf\limits_{f \in \mathcal{M}(D) \cap \{(Vf,f) \geq \theta\}} I(f)$, where $\inf\limits_{f \in \emptyset} I(f) = +\infty$ .

LEMMA 2.2. $J_0 = J_{00}$ .

PROOF. It follows from the definition of $J_{00}$ and from Proposition 2.1 that $c_D(\lambda) \geq \lambda\theta - J_{00}(\theta)$. Therefore $J_{00}(\theta) \geq J_0(\theta)$, for all $\theta \in \mathbb{R}$, since $J_0$ is the Legendre transform of $c_D$. Next (2.3) yields $J_{00}(\theta) \leq J_0(\theta)$, for all $\theta \in \mathbb{R} \setminus \{\theta_\infty\}$.

If $J_0(\theta_\infty) = +\infty$ we are done. So suppose that $J_0(\theta_\infty) < +\infty$. By Proposition 2.1(iii), one can find $\lambda_n \uparrow +\infty$ so that $f_{\lambda_n} \to f$, $f \in \mathcal{M}(D)$, $(Vf, f) = \theta_\infty$ and $I(f) \leq J_0(\theta_\infty)$. Hence $J_{00}(\theta_\infty) \leq I(f) \leq J_0(\theta_\infty)$ which completes the proof. □

REMARK 2.3. As a by-product of our proof we get

$$(2.4) \qquad J_D(\theta) = \sup_{\lambda \in \mathbb{R}} (\lambda\theta - c_D(\lambda)) = \inf_{f \in \mathcal{M}(D), \ (Vf,f) = \theta} I(f).$$

It follows from Proposition 2.1 that $J_D$ is a decreasing function of D.

We can now state the first general large deviation upper bound.

COROLLARY 2.4. For any $t > 0$ and any closed set $F$,

$$(2.5) \qquad \int_D P_x \left( \frac{1}{t} \int_0^t V(x_s)ds \in F, \ T_D > t \right) dx \leq 2|D| \, e^{-t \inf_{\theta \in F} J_D(\theta)},$$

and

$$(2.6) \qquad \inf_{\theta \in F} J_D(\theta) = \inf_{f \in \mathcal{M}(D), \ (Vf,f) \in F} I(f).$$

PROOF. Combining (2.2) to (2.4) and applying these results to $-V$ as well as $V$, we obtain

$$(2.7) \qquad \int_D P_x \left( \frac{1}{t} \int_0^t V(x_s)ds \geq \theta, \ T_D > t \right) dx \leq |D| \, e^{-t \inf_{\theta_1 \geq \theta} J_D(\theta_1)},$$

for any $\theta \geq \theta_0 = c_D'(0)$, and

$$(2.8) \qquad \int_D P_x \left( \frac{1}{t} \int_0^t V(x_s)ds \leq \theta, \ T_D > t \right) dx \leq |D| \, e^{-t \inf_{\theta_1 \leq \theta} J_D(\theta_1)},$$

for any $\theta \leq \theta_0$. Now if $\theta_0 \in F$, (2.5) holds since $\inf\limits_{\theta \in F} = J_D(\theta_0)$. Next, if $\theta_0 \notin F$, then only one of the following three cases occurs:

- $F \subset [\theta_1, \infty)$ with $\theta_0 < \theta_1 \in F$, so $\inf\limits_{\theta \in F} J_D(\theta) = J_D(\theta_1) = \inf\limits_{\theta \geq \theta_1} J_D(\theta)$, and (2.5) holds because of (2.7);
- $F \subset (-\infty, \theta_2]$ with $\theta_0 > \theta_2 \in F$, so $\inf\limits_{\theta \in F} J_D(\theta) = J_D(\theta_2) = \inf\limits_{\theta \leq \theta_2} J_D(\theta)$, and (2.5) holds because of (2.8);

- $F \subset (-\infty, \theta_2] \cup [\theta_1, \infty)$ with $\theta_2 < \theta_0 < \theta_1 \in F$, so $\inf_{\theta \in F} = \min(J_D(\theta_1), J_D(\theta_2))$, so (2.5) holds because of (2.7) and (2.8) and the following inequality:

$$e^{-J_D(\theta_1)} + e^{-J_D(\theta_2)} \leq e^{-\min\{J_D(\theta_1), J_D(\theta_2)\}}.$$

$\square$

We end this section with two propositions. Set $\mathcal{C} = \{f \in C_0^\infty(\mathbb{R}^d); |f| = 1\}$. Note that $\mathcal{C} = \bigcup_{D \text{ bounded domain}} \mathcal{C}(D)$.

PROPOSITION 2.5. *Suppose* $V \in \mathcal{B}_b(\mathbb{R}^d)$ *and let*

$$c(\lambda) = \sup_{f \in \mathcal{C}} \lambda(Vf, f) - I(f).$$

*Then, if* $\mathcal{C}_\theta = \{f \in \mathcal{C}; I(f) \leq \theta\}$, $\theta \geq 0$, *we have*

(i) *for every* $\lambda, \in \mathbb{R}$ *and any* $y \in \mathbb{R}^d$,

$$\lim_{t \to \infty} t^{-1} \log E_y \left\{ e^{\lambda \int_0^t V(x_s) ds} \right\} = c(\lambda);$$

(ii) $c(\cdot)$ *is convex and* $c(0) = 0$;

(iii)

$$Dc^+(0) = \lim_{\lambda \downarrow 0} c(\lambda)/\lambda = \lim_{\theta \downarrow 0} \sup_{f \in \mathcal{C}_\theta} (Vf, f);$$

(iv)

$$Dc^-(0) = \lim_{\lambda \downarrow 0} c(-\lambda)/\lambda = \lim_{\theta \downarrow 0} \inf_{f \in \mathcal{C}_\theta} (Vf, f);$$

(v) $c(\cdot)$ *is differentiable at the origin if and only if there exists* $h \in \mathbb{R}$ *such that*

(2.9) $$\lim_{\theta \downarrow 0} \sup_{f \in \mathcal{C}_\theta} |(Vf, f) - h| = 0.$$

*Moreover (2.9) holds if and only if* $Dc^+(0) = h = Dc^-(0)$.

PROOF. Because of the translation invariance of Lebesgue measure and the fact that a Brownian motion starting at $y$ is a translate of a Brownian motion starting at the origin $o$, there is no loss of generality in assuming that $y = o$.

Let $D_r = \{x \in \mathbb{R}^d; |x| < r\}$, and set $T_r = T_{D_r}$. We will first prove

(2.10) $$\limsup_{t \to \infty} t^{-1} \log E_o \left( e^{\lambda \int_0^t V(x_s) ds} \right) \leq c(\lambda).$$

Set $K = |\lambda| |V|_\infty$. Then

$$
\begin{aligned}
E_o \left( e^{\lambda \int_0^t V(x_s) ds} \right) &\leq e^{2K} E_o \left( e^{\lambda \int_1^{t+1} V(x_s) ds} \right) \\
&= e^{2K} \int p_1(x) E_x \left( e^{\lambda \int_0^t V(x_s) ds} \right) dx \\
&\leq (2\pi)^{-d/2} e^{2K} \int_{D_r} E_x \left( e^{\lambda \int_0^t V(x_s) ds} \mathbb{I}\{T_r > t\} \right) dx \\
&\quad + (2\pi)^{-d/2} e^{(t+2)K} \int_{D_{r/2}} P_x(T_r \leq t) dx \\
&\quad + e^{(t+2)K} P_0(x_1 \notin D_{r/2}) \\
&= (1) + (2) + (3) \text{ say.}
\end{aligned}
$$

It follows from Proposition 2.1 that (1) is bounded above by

$$(2\pi)^{-d/2}e^{2K}|D_r|e^{tc_{D_r}(\lambda)} \leq (2\pi)^{-d/2}e^{2K}|D_r|e^{tc(\lambda)}.$$

Next, using well-known properties of Brownian motion, we have

$$\int_{D_{r/2}} P_x(T_r \leq t)dx \leq |D_{r/2}|P_0(T_{r/2} \leq t) \leq |D_{r/2}|e^{-\delta r^2/t},$$

for some positive constant $\delta$. Thus the sum of (2) and (3) is bounded above by

$$(1 + |D_{r/2}|)e^{(t+2)K - \delta r^2/t}.$$

Note that the preceding estimates are valid for any $r > 0$. If we choose $r = t^2$, then it is easy to see that (2.10) holds true.

We will now prove that

$$(2.11) \qquad \liminf_{t\to\infty} t^{-1} \log E_o\left(e^{\lambda \int_0^t V(x_s)ds}\right) \geq c(\lambda).$$

To this end, let $f \in \mathcal{C}(D_r)$ be given. Then, using Jensen's inequality and Proposition 2.1, one obtains

$$
\begin{aligned}
e^{t\{\lambda(Vf,f)-I(f)\}} &\leq \left(e^{t(\Delta+\lambda V)}f, f\right)\\
&= \int_{D_r} f(x)E_x\left(f(x_t)e^{\lambda\int_0^t V(x_s)ds}\mathbb{I}\{T_r > t\}\right)dx\\
&\leq (2\pi)^{d/2}|f|_\infty^2 e^{r^2/2}\int_{D_r} p_1(x)E_x\left(e^{\lambda\int_0^t V(x_s)ds}\mathbb{I}\{T_r > t\}\right)dx\\
&\leq (2\pi)^{d/2}|f|_\infty^2 e^{r^2/2}E_o\left(e^{\lambda\int_1^{t+1} V(x_s)ds}\right)\\
&\leq (2\pi)^{d/2}|f|_\infty^2 e^{r^2/2+2K}E_o\left(e^{\lambda\int_0^t V(x_s)ds}\right).
\end{aligned}
$$

Thus

$$
\begin{aligned}
\liminf_{t\to\infty} t^{-1}\log E_o\left(e^{\lambda\int_0^t V(x_s)ds}\right) &\geq \sup_{r>0}\sup_{f\in\mathcal{C}(D_r)}\{\lambda(Vf,f) - I(f)\}\\
&= c(\lambda).
\end{aligned}
$$

This proves (2.11). Now (2.10) together with (2.11) yield (i).

Next (ii) is easy and is left to the reader. It follows from (ii) that $c(\lambda)/\lambda \downarrow Dc^+(0)$ as $\lambda \downarrow 0$. Let $\epsilon > 0$ be given. Then there exists $n_0 = n_0(\epsilon)$ such that $nc(1/n) < \epsilon + Dc^+(0)$, for all $n \geq n_0$. Hence

$$(Vf, f) - nI(f) \leq nc(1/n) < \epsilon + Dc^+(0).$$

Therefore

$$\sup_{f\in\mathcal{C}_\theta}(Vf, f) \leq \theta n + \epsilon + Dc^+(0),$$

so

$$\lim_{\theta\downarrow 0}\sup_{f\in\mathcal{C}_\theta}(Vf, f) \leq \epsilon + Dc^+(0).$$

Since $\epsilon$ is arbitrary, one can conclude that (iii) holds true.

Next (iv) follows from applying (iii) to $-V$. Finally, $c(\cdot)$ is differentiable at 0 if and only if $Dc^+(0) = Dc^-(0)$. The latter is clearly equivalent to (2.9), and the proof is complete. $\qquad\square$

REMARK 2.6. In the literature, $c(\cdot)$ is usually called the free energy function. Its behaviour is very important in proving large deviation results, e.g. Ellis [4].

EXAMPLE 2.7. Suppose $V$ is a stationary and ergodic process of the form $V(x,\omega) = v(\tau_x\omega)$. If $V$ is bounded and independent of the Brownian motion, one can apply Proposition 2.5(i) to see that the free energy is random but almost surely constant. In fact,

$$
\begin{aligned}
c(\lambda, \tau_x\omega) &= \lim_{t\to\infty} t^{-1} \log E_o \left\{ e^{\lambda \int_0^t v(\tau_{x+x_s}\omega)ds} \right\} \\
&= \lim_{t\to\infty} t^{-1} \log E_x \left\{ e^{\lambda \int_0^t v(\tau_{x_s}\omega)ds} \right\} \\
&= c(\lambda, \omega).
\end{aligned}
$$

Because of the ergodicity, $c(\lambda, \omega)$ is almost surely constant in $\omega$.

DEFINITION 2.8. For any $\theta \in \mathbb{R}$, set

$$
J_1(\theta) = \inf_{f \in \mathcal{C},\, (Vf,f)=0} I(f), \quad \text{and} \quad J(\theta) = \sup_{\epsilon > 0} \inf_{|\theta_1 - \theta| < \epsilon} J_1(\theta_1).
$$

PROPOSITION 2.9. *Suppose* $V \in \mathcal{B}_b(\mathbb{R}^d)$. *Then*

(i) $J(\theta) = \sup_{\epsilon>0} \inf_{f \in \mathcal{C},\, |(Vf,f) - \theta| < \epsilon} I(f)$ *and* $\inf_{f \in \mathcal{C},\, |(Vf,f) - \theta| < \epsilon} I(f) = \inf_{|\theta_1 - \theta| < \epsilon} J_1(\theta_1);$

(ii) *for any open set* $O$, $\inf_{\theta \in O} J(\theta) = \inf_{\theta \in O} J_1(\theta);$

(iii) *for any closed set* $F$ *such that* $\partial F$ *is compact,* $\inf_{\theta \in F} J(\theta) = \sup_{\epsilon > 0} \inf_{\theta \in F^\epsilon} J_1(\theta),$

   *where* $F^\epsilon = \displaystyle\bigcup_{\theta \in F} (\theta - \epsilon, \theta + \epsilon);$

(iv) $\{Dc^-(0), Dc^+(0)\} \subset \{\theta; J(\theta) = 0\} \subset [Dc^-(0), Dc^+(0)];$

(v) *if* $F$ *is closed and* $F \cap [Dc^-(0), Dc^+(0)] = \emptyset$, *then* $\inf_{\theta \in F} J(\theta) > 0;$

(vi) $J_2(\theta) = \sup_\lambda \{\lambda\theta - c(\lambda)\} \leq J(\theta).$

PROOF. We first prove (i). Set $\ell = \inf_{f \in \mathcal{C},\, |(Vf,f) - \theta| < \epsilon} I(f)$ and $\ell_1 = \inf_{|\theta_1 - \theta| < \epsilon} J_1(\theta_1)$. To prove (i), it is sufficient to prove that $\ell = \ell_1$.

If both $\ell$ and $\ell_1$ are infinite, there is nothing else to prove. So suppose first that $\ell 1$ is finite. Then for any $n$, one can find $\theta_n$ such that $|\theta_n - \theta| < \epsilon$ and $J_1(\theta_n) < \ell_1 + 1/n$. Hence one can also find $f_n \in \mathcal{C}$ so that $(Vf_n, f_n) = \theta_n$ and $I(f_n) < \ell_1 + 1/n$. Therefore $\ell \leq I(f_n) < \ell_1 + 1/n$, for all $n \geq 1$, proving that $\ell \leq \ell_1$.

Suppose now that $\ell$ is finite. The for any $n \geq 1$, one can find $f_n \in \mathcal{C}$ such that $I(f_n) < \ell + 1/n$ and $\theta_n = (Vf_n, f_n) \in (\theta - \epsilon, \theta + \epsilon)$. It follows from the definition of $J_1$ that $\ell_1 \leq J_1(\theta_n) \leq I(f_n) < \ell + 1/n$, for all $n \geq 1$, proving that $\ell_1 \leq \ell$. Hence $\ell = \ell_1$.

We now prove (ii). Set $\ell = \inf_{\theta \in O} J(\theta)$ and $\ell_1 = \inf_{\theta | \in O} J_1(\theta)$. Again it is sufficient to prove that $\ell = \ell_1$. Since $J \leq J_1$, it follows immediately that $\ell \leq \ell_1$. Suppose that $\ell$ is finite. Then one can find $\theta_n \in O$ so that $J(\theta_n) < \ell + 1/n$. For these $\theta_n$, one can find $\epsilon_n > 0$ such that $(\theta_n - \epsilon_n, \theta_n + \epsilon_n) \subset O$. Using the definition of $J$, one can also find $\theta'_n$ so that $|\theta'_n - \theta_n| < \epsilon_n$ and $J_1(\theta'_n) < \ell + 1/n$. It follows that $\theta'_n \in O$, so $\ell_1 < \ell + 1/n$, for every $n \geq 1$, proving that $\ell_1 \leq \ell$. Hence $\ell = \ell_1$ if at least one of these quantities is finite. This proves (ii).

To prove (iii), note first that it follows from (ii) that

$$\ell_1 = \inf_{\theta \in F} J(\theta) \geq \inf_{\theta \in F^\epsilon} J(\theta) = \inf_{\theta \in F^\epsilon} J_1(\theta).$$

Therefore $\ell = \lim_{\epsilon \to 0} \inf_{\theta \in F^\epsilon} J_1(\theta) \leq \ell_1$. it remains to prove the reverse inequality. This is trivial if $\ell$ is infinite, so suppose it is finite. Then, given $\delta > 0$, one can find a sequence $\theta_n \in F^{1/n}$ such that $J_1(\theta_n) < \ell + \delta$. If $\theta_n \in F$ for some $n$, then $\ell_1 \leq J(\theta_n) \leq J_1(\theta_n) < \ell + \delta$, proving that $\ell_1 \leq \ell$. If $\theta_n \notin F$ for all $n$, then the compactness of $\partial F$ yields the existence of a subsequence $n_k$ such that $\theta_{n_k}$ converges to some $\theta \in \partial F \subset F$. Hence

$$\ell_2 \leq J(\theta) \leq \lim_{k \to \infty} J_1(\theta_{n_k}) < \ell + \delta.$$

Consequently $\ell_1 \leq \ell$.

To prove (iv), note that if $J(\theta) = 0$, then one can find a sequence $f_n \in \mathcal{C}_{1/n}$ such that $(Vf_n, f_n)$ converges to $\theta$. Now, from Proposition 2.5(iii),

$$
\begin{aligned}
Dc^+(0) &= \lim_{n \to \infty} \sup_{f \in \mathcal{C}_{1/n}} (Vf, f) \\
&\geq \lim_{n \to \infty} (Vf_n, f_n) = x \\
&\geq \lim_{n \to \infty} \inf_{f \in \mathcal{C}_{1/n}} (Vf, f) \\
&= Dc^-(0).
\end{aligned}
$$

Therefore $\{x; J(x) = 0\} \subset [Dc^-(0), Dc^+(0)]$. Next, again from Proposition 2.5(iii), one can sequences $f_n$ and $g_n$ such that $I(f_n) \to 0$ and $I(g_n) \to 0$ as $n$ tends to infinity, and

$$Dc^+(0) - 1/n \leq (Vf_n, f_n) \leq Dc^+(0),$$
$$Dc^-(0) \leq (Vg_n, g_n) \leq Dc^-(0) + 1/n.$$

Hence $J(Dc^+(0)) = 0 = J(Dc^-(0))$.

To prove (v), it is enough to show that for any $\theta > Dc^+(0)$, $\inf_{\theta_1 \geq \theta} J(\theta_1) > 0$. For if this is true, then applying it to $-V$ yields $\inf_{\theta_1 \leq \theta} J(\theta_1) > 0$ for any $\theta < Dc^-(0)$. So suppose that $\theta > Dc^+(0)$. Then one can find $a > 0$ such that $\theta - 1/n > \sup_{f \in \mathcal{C}_a} (Vf, f)$ if $n$ is large enough. It follows that if $(Vf, f) > \theta - 1/n$, then $I(f) \geq a$. Thus, using (i) and (iii), we have

$$\inf_{\theta_1 \geq \theta} J(\theta_1) = \lim_{n \to \infty} \inf_{\theta_1 > \theta - 1/n} J_1(x) = \lim_{n \to \infty} \inf_{f \in \mathcal{C},\, (Vf, f) > \theta - 1/n} I(f) \geq a > 0.$$

Finally

$$J_2(\theta) \leq \sup_\lambda \inf_{f \in \mathcal{C}} \{\lambda \theta - \lambda(Vf, f) + I(f)\} \leq J(\theta)$$

since one can find a sequence $f_n \in \mathcal{C}$ such that $(Vf_n, f_n)$ converges to $\theta$ and $I(f_n)$ converges to $J(\theta)$, whenever the latter is finite. When $J(\theta) = \infty$, the inequality holds trivially. This proves (vi). $\qquad \square$

## 3. Large deviation principle

THEOREM 3.1. *Suppose that* $V \in \mathcal{B}_b(\mathbb{R}^d)$. *Then for any open set* $O \subset \mathbb{R}$, *and for any* $y \in \mathbb{R}^d$,

$$(3.1) \qquad \liminf_{t \to \infty} t^{-1} \log P_y \left( \frac{1}{t} \int_0^t V(x_s) ds \in O \right) \geq - \inf_{\theta \in O} J(\theta).$$

PROOF. Without loss of generality, one may suppose that $O \neq \emptyset$. Suppose that one can find $\theta \in O$ such that $\theta = c_{B(y,r)}(\lambda) = c_r(\lambda)$ for some $\lambda \in \mathbb{R}$ and some $r > 0$.

Since $O$ is open and $V$ is bounded, one can find $\epsilon > 0$ such that for $t$ large enough,

$$
\begin{aligned}
P_y\left(\frac{1}{t}\int_0^t V(x_s)ds \in O\right) &\geq P_y\left(\left|\frac{1}{t}\int_1^{t+1} V(x_s)ds - \theta\right| < \epsilon\right) \\
&= \int p_1(x-y)P_x\left(\left|\frac{1}{t}\int_1^{t+1} V(x_s)ds - \theta\right| < \epsilon\right)dx \\
&\geq (2\pi)^{-2d}e^{-r^2/2}\int_{B(y,r)} P_x\left(A_t \cap \{T_r > t\}\right)dx,
\end{aligned}
$$

where $T_r = T_{B(y,r)}$ and $A_t = \left\{\left|\frac{1}{t}\int_0^t V(x_s)ds - \theta\right| < \epsilon\right\}$.

Next, set

$$
\phi_t(\alpha) = \int_{B(y,r)} E_x\left(e^{\alpha\int_0^t V(x_s)ds}\mathbb{I}\{T_r > t\}\right)dx.
$$

From Proposition 2.1 and from the proof of Proposition 2.5, one can conclude that $\lim_{t\to\infty} t^{-1}\log\phi_t(\alpha) = c_r(\alpha)$, for all $\alpha \in \mathbb{R}$.

Next,

$$
\begin{aligned}
\int_{B(y,r)} P_x\left(A_t \cap \{T_r > t\}\right)dx &= e^{-\lambda\theta t}\int_{B(y,r)} E_x\left(e^{\lambda\int_0^t V(x_s)ds}\right. \\
&\qquad \left.\times e^{-\lambda t\left(\frac{1}{t}\int_0^t V(x_s)ds - \theta\right)}\mathbb{I}\{A_t\}\mathbb{I}\{T_r > t\}\right)dx \\
&\geq e^{-\lambda\theta t - |\lambda|\epsilon t}q_t\phi_t(\lambda),
\end{aligned}
$$

where

$$
q_t = \int_{B(y,r)} E_x\left(e^{\lambda\int_0^t V(x_s)ds}\mathbb{I}\{A_t\}\mathbb{I}\{T_r > t\}\right)dx\Big/\phi_t(\lambda).
$$

Now $1 - q_t$ is the sum of

$$
q_{1t} = \int_{B(y,r)} E_x\left(e^{\lambda\int_0^t V(x_s)ds}\mathbb{I}\{T_r > t\}\mathbb{I}\left\{\frac{1}{t}\int_0^t V(x_s)ds \geq \theta + \epsilon\right\}\right)dx\Big/\phi_t(\lambda),
$$

and

$$
q_{2t} = \int_{B(y,r)} E_x\left(e^{\lambda\int_0^t V(x_s)ds}\mathbb{I}\{T_r > t\}\mathbb{I}\left\{\frac{1}{t}\int_0^t V(x_s)ds \leq \theta - \epsilon\right\}\right)dx\Big/\phi_t(\lambda).
$$

By Tchebychev's inequality, for any $\alpha, \beta > 0$,

$$
q_{1t} \leq e^{-\alpha t(\theta+\epsilon)}\phi_t(\lambda + \alpha)/\phi_t(\lambda),
$$

and

$$
q_{2t} \leq e^{\beta t(\theta-\epsilon)}\phi_t(\lambda - \beta)/\phi_t(\lambda).
$$

Therefore

$$
\limsup_{t\to\infty} t^{-1}\log q_{1t} \leq -\sup_{\alpha>0}\{\alpha(x+\epsilon) - c_r(\lambda+\alpha) + c_r(\lambda)\} < 0,
$$

and

$$
\limsup_{t\to\infty} t^{-1}\log q_{2t} \leq -\sup_{\beta>0}\{-\beta(x-\epsilon) - c_r(\lambda-\beta) + c_r(\lambda)\} < 0,
$$

since $\theta = c'_r(\lambda)$. It follows that $q_t \to 1$ as $t$ tends to infinity, so

$$(3.2) \qquad \liminf_{t\to\infty} t^{-1} \log P_y \left( \frac{1}{t} \int_0^t V(x_s) ds \in O \right) \geq -|\lambda|\epsilon - \{\lambda\theta - c_r(\lambda)\}.$$

Since $\theta = c'_r(\lambda)$, we have $\lambda\theta - c_r(\lambda) = J_r(\theta)$, where $J_r = J_{B(y,r)}$ and $J_{B(y,r)}$ is defined in (2.4).

Because (3.2) holds for any $\epsilon > 0$, one obtains

$$(3.3) \qquad\qquad \liminf_{t\to\infty} t^{-1} \log P_y \left( \frac{1}{t} \int_0^t V(x_s) ds \in O \right) \geq -J_r(\theta),$$

for any $\theta \in O$ such that $\theta = c'_r(\lambda)$ for some $\lambda$.

Suppose now that $J_1(\theta) < \infty$. The for any $\epsilon > 0$, one can find $r_\epsilon > 0$ and $f_\epsilon \in C_{r_\epsilon}$ such that $(Vf_\epsilon, f_\epsilon) = \theta$ and $I(f_\epsilon) < \epsilon + J_1(\theta)$. It follows that $J_{r_\epsilon}(\theta) < \epsilon + J_1(\theta)$. Since $J_{r_\epsilon}(\theta)$ is finite, it follows from (2.3) that either $\theta = c'_{r_\epsilon}(\theta)$ for some $\lambda$, or $\theta = \lim_{\lambda\uparrow\infty} c'_{r_\epsilon}(\theta)$, or $\theta = \lim_{\lambda\downarrow\infty} c'_{r_\epsilon}(\theta)$. If the second possibility holds true, then $\theta_\lambda = c'_{r_\epsilon}(\theta) \in O$ for $\lambda$ large enough since $O$ is open. Applying (3.3), one obtains

$$\liminf_{t\to\infty} t^{-1} \log P_y \left( \frac{1}{t} \int_0^t V(x_s) ds \in O \right) \geq -J_{r_\epsilon}(\theta_\lambda),$$

for all $\lambda$ large enough. Since $J_{r_\epsilon}(\theta_\lambda)$ tends to $J_{r_\epsilon}(\theta) < \epsilon + J_1(\theta)$, it follows that for any $\theta \in O$ with $J_1(\theta) < \infty$,

$$\liminf_{t\to\infty} t^{-1} \log P_y \left( \frac{1}{t} \int_0^t V(x_s) ds \in O \right) \geq -J_1(\theta) - \epsilon,$$

for any $\epsilon > 0$. Letting $\epsilon$ tends to zero in the last inequality yields

$$\liminf_{t\to\infty} t^{-1} \log P_y \left( \frac{1}{t} \int_0^t V(x_s) ds \in O \right) \geq - \inf_{x\in O} J_1(\theta).$$

Finally (3.1) follows by using Proposition 2.9(ii).                                    □

THEOREM 3.2. *Suppose that $V \in \mathcal{B}_b(\mathbb{R}^d)$. Let $F$ be a closed subset of $\mathbb{R}$. Then for any $y \in \mathbb{R}^d$,*

$$(3.4) \qquad\qquad \limsup_{t\to\infty} t^{-1} \log P_y \left( \frac{1}{t} \int_0^t V(x_s) ds \in F \right) \leq - \inf_{\theta\in F} J(\theta).$$

PROOF. Since $V$ is bounded, we can restrict our attention to compact sets. In view of Proposition 2.9(iii), it is enough to prove that for any compact set $F$, any $\delta > 0$ and $y \in \mathbb{R}^d$,

$$(3.5) \qquad\qquad \limsup_{t\to\infty} t^{-1} \log P_y \left( \frac{1}{t} \int_0^t V(x_s) ds \in F \right) \leq - \sup_{\delta>0} \inf_{\theta\in F^\delta} J_1(\theta).$$

Let $\delta > 0$ be given. Set $T_{t^2} = T_{B(y,t^2)}$. For any $t > 2|V|_\infty/\delta$,

$$P_y\left(\frac{1}{t}\int_0^t V(x_s)ds \in F\right)$$

$$\leq P_y\left(\frac{1}{t}\int_1^{t+1} V(x_s)ds \in \overline{F^\delta}\right)$$

$$= \int p_1(x-y)P_x\left(\frac{1}{t}\int_0^t V(x_s)ds \in \overline{F^\delta}\right)dx$$

$$\leq (2\pi)^{-d/2}\int_{B(y,t^2)} P_x\left(\frac{1}{t}\int_0^t V(x_s)ds \in \overline{F^\delta},\ T_{t^2} > t\right)dx$$

$$+ P_o(x_1 \in B(y,t^2)) + (2\pi)^{-d/2}|B(y,t^2)|P_0(T_{t^2} > t)$$

$$\leq (2\pi)^{-d/2}|B(y,t^2)|e^{-t\{I(f);f\in\mathcal{M}(B(y,t^2)),\ (Vf,f)\in\overline{F^\delta}\}}$$

$$+(1+|B(y,t^2)|)e^{-\alpha t^3},$$

for some $\alpha > 0$, using Corollary 2.4 and the same technique as in the proof of Proposition 2.5.

Next, using (2.6), one obtains

$$\inf_{f\in\mathcal{M}(B(y,t^2)),\ (Vf,f)\in\overline{F^\delta}}I(f) = \inf_{\theta\in\overline{F^\delta}}J_{B(y,t^2)}(\theta)$$

$$\geq \inf_{\theta\in F^{2\delta}}J_{B(y,t^2)}(\theta)$$

$$= \inf_{f\in\mathcal{C}(B(y,t^2)),\ (Vf,f)\in F^{2\delta}}I(f)$$

$$\geq \inf_{f\in\mathcal{C},\ (Vf,f)\in F^{2\delta}}I(f)$$

$$= \inf_{\theta\in F^{2\delta}}J_1(\theta),$$

since $C_0^\infty(D)$ is dense in $H_0^1(D)$. Combining the last results, one obtains (3.5). $\square$

We are now in a position to characterize exponential convergence of $\frac{1}{t}\int_0^t V(x_s)ds$. Recall that $\frac{1}{t}\int_0^t V(x_s)ds$ converges exponentially fast to $h \in \mathbb{R}$ if and only if for any $\delta > 0$,

$$(3.6) \qquad \limsup_{t\to\infty} t^{-1}\log P_y\left(\left|\frac{1}{t}\int_0^t V(x_s)ds - h\right| \geq \delta\right) < 0,$$

The result is the following.

THEOREM 3.3. *Suppose $V \in \mathcal{B}_b(\mathbb{R}^d)$. Then $\frac{1}{t}\int_0^t V(x_s)ds$ converges exponentially fast to $h \in \mathbb{R}$ if and only if the free energy $c$ is differentiable at $0$ and $c'(0) = h$.*

REMARK 3.4. It is well-known that the differentiability of the free energy at $0$ is sufficient for exponential convergence but it is not obvious that this is necessary because large deviation lower bounds are difficult to obtain under this condition.

PROOF. Suppose that $c'$ exists and is equal to $h$. Then from Theorem 3.2, it follows that

$$\limsup_{t\to\infty} t^{-1}\log P_y\left(\left|\frac{1}{t}\int_0^t V(x_s)ds - h\right| \geq \delta\right) \leq -\inf_{|\theta-h|\geq\delta}J(\theta) < 0$$

because of Proposition 2.9(v), since $\{\theta; J(\theta) = 0\} = \{h\}$.

Now suppose that $Dc^-(0) < Dc^+(0)$. It follows from Theorem 3.1 that for any $\delta > 0$, and any $h \in \mathbb{R}$,

$$0 \geq \liminf_{t \to \infty} t^{-1} \log P_y \left( \left| \frac{1}{t} \int_0^t V(x_s)ds - h \right| \geq \delta \right) \geq - \inf_{|\theta - h| > \delta} J(\theta) = 0,$$

because $\{Dc^-(0), Dc^+(0)\} \cap [h - \delta, h + \delta]^c \neq \emptyset$ and because $J(Dc^-(0)) = J(Dc^+(0)) = 0$, by Proposition 2.9(iv).                                                                          $\square$

EXAMPLE 3.5. Suppose $V \in \mathcal{B}_b(\mathbb{R}^d)$ has compact support. In that case, $c$ is differentiable at 0 and $c'(0) = 0$, meaning that $\dfrac{1}{t} \displaystyle\int_0^t V(x_s)ds$ converges exponentially fast to zero. As a by-product, the strong law of large numbers holds true.

To check the differentiability, set $G(x) = \displaystyle\int_0^{x_1} V(s, x_2, \dots, x_d)ds$. Since $V$ has compact support, $G$ is bounded. Next, integrating by parts, we get

$$(Vf, f) = \int V(x)f^2(x)dx = -2 \int G(x)f(x)\frac{\partial}{\partial x_1}f(x)dx.$$

Using Cauchy-Schwartz inequality, we obtain

$$|(Vf, f)| \leq 2|G|_\infty |f||\frac{\partial}{\partial x_1}f| \leq 2|G|_\infty |f||\nabla f| \leq 2\sqrt{2a}|G|_\infty,$$

if $f \in \mathcal{C}_a$. Hence $\lim_{a \downarrow 0} \sup_{f \in \mathcal{C}_a} |(Vf, f)| = 0$ and we conclude from Proposition 2.5(v) that $c$ is differentiable at 0 and $c'(0) = 0$.

EXAMPLE 3.6. Suppose $V \in \mathcal{B}_b(\mathbb{R}^d)$ is almost periodic. In that case, $c$ is differentiable at 0 and

$$c'(0) = \lim_{T \to \infty} \frac{1}{T^d} \int_{[-T/2, T/2^d]} V(x)dx.$$

To prove this, we just have to prove it for trigonometric polynomials $p$ of the form

$$p(x) = a_0 + \sum_{k=1}^n (a_k \cos <x, c_k> + b_k \sin <x, c_k>),$$

where $c_k \neq 0$ for any $1 \leq k \leq n$, since by definition, almost periodic functions are the closure of the trigonometric polynomials under the supremum norm on $\mathbb{R}^d$.

Now if $V = p$, for a trigonometric polynomial $p$, it is easy to see that

$$a_0 = \lim_{T \to \infty} \frac{1}{T^d} \int_{[-T/2, T/2]^d} p(x)dx,$$

so we only have to prove that $c'(0) = a_0$. If $(c_k)_i \neq 0$, then for any $f \in \mathcal{C}_a$,

$$\left| \int \cos <x, c_k> f^2(x)dx \right| = 2 \left| \int \frac{\sin <x, c_k>}{(c_k)_i} f(x) \frac{\partial}{\partial x_i} f(x)dx \right| \leq \frac{2\sqrt{2ad}}{|c_k|}.$$

Therefore

$$\sup_{f \in \mathcal{C}_a} |(Vf, f) - a_0| \leq 2\sqrt{2ad} \left( \sum_{k=1}^n \frac{|a_k| + |b_k|}{|c_k|} \right),$$

and it follows that $\lim\sup\limits_{a\downarrow 0}\limits_{f\in\mathcal{C}_a} |(Vf,f)-a_0| = 0$. We conclude from Proposition 2.5(v) that $c$ is differentiable at 0 and $c'(0) = a_0$.

EXAMPLE 3.7. Suppose $V(x) = \xi_{[x+U]}$, where the $\xi_k$'s, $k \in \mathbb{Z}^d$ are independent and identically distributed copies of a random variable $\xi$, and where $U$ is uniformly distributed over $[0,1)^d$ and $[\cdot]$ denotes the multidimensional integer part of $x \in \mathbb{R}^d$. It is easy to see that $V$ is a stationary and ergodic process, so using Example 2.7, it follows that the corresponding free energy is almost surely constant. We will prove that almost surely

$$Dc^+(0) = a_+ = \sup\{a; P(\xi > a) > 0\},$$

and

$$Dc^-(0) = a_- = \inf\{a; P(\xi < a) > 0\}.$$

If $\lambda \geq 0$, then

$$c(\lambda) = \sup_{f\in\mathcal{C}}\{\lambda(Vf,f) - I(f)\} \leq \sup_{f\in\mathcal{C}}\{\lambda a_+ - I(f)\} = \lambda a_+,$$

since $P(\xi \leq a_+) = 1$ and $\inf\limits_{f\in\mathcal{C}} I(f) = 0$.

Now for any $\epsilon > 0$, one can find $\delta > 0$ and $f \in \mathcal{C}$ such that $\lambda a_+ - \epsilon < \lambda(a_+ - \delta) - I(f)$.

Set $p_k(u) = \int_{[x]=k} f^2(x-u)dx$. Then $\sum\limits_{k\in\mathbb{Z}^d} p_k(u) = 1$ and $A_f = \{k; p_k(u) > 0$ for some $u \in [0,1)^d\}$ is finite because $f$ has compact support. Hence

$$P((Vf,f) > a_+ - \delta) = P\left(\sum_{k\in A_f}\xi_k p_k(U) > a_+ - \delta\right) \geq \{P(\xi > a_+ - \delta)\}^{\text{card}(A_f)} > 0.$$

It follows that

$$P(c(\lambda) > \lambda a_+ - \epsilon) \geq P(\lambda(Vf,f) - I(f) > \lambda a_+ - \epsilon) \geq P((Vf,f) > a_+ - \delta) > 0.$$

Since $c$ is almost surely constant, it follows that $P(c(\lambda) = \lambda a_+) = 1$, for all $\lambda \geq 0$. Similarly, $P(c(\lambda) = \lambda a_-) = 1$, for all $\lambda \leq 0$. Therefore with probability one $Dc^\pm(0) = \pm\lim_{n\to\infty} nc(\pm 1/n) = a_\pm$, with probability one. Therefore unless the $\xi_k$'s are constant, $c$ is not differentiable. A more detailed study will be done in section 5. This example shows that for random potentials, exponential convergence does not hold.

Note also that for any $n$, $U_\epsilon = \inf\limits_{f\in\mathcal{C},\ |(Vf,f)-\theta|<1/n} I(f)$ is measurable and almost surely constant. If $\theta$ is in the support of $\xi$, that is $P(|\xi - \theta| < \epsilon) > 0$ for all $\epsilon > 0$, then $\theta$ is in the support of $(Vf,f)$. So choose a sequence $f_n \in \mathcal{C}_{1/n}$. It follows that for any $n$,

$$P(U_\epsilon \leq 1/n) \geq P(|(Vf_n,f_n) - \theta| < \epsilon) > 0.$$

This proves that $P(U_\epsilon = 0) = 1$. Hence $P(J(\theta) = 0) = 1$ for all $\theta$ in the support of $\xi$. Consequently, with probability one

$$J(\theta) = \begin{cases} 0 & \text{if } \theta \in \text{support}(\xi); \\ +\infty & \text{if } \theta \notin \text{support}(\xi). \end{cases}$$

## 4. Random potentials arising from dynamical systems

From now on, we will consider potentials $V(\cdot)$ of the form

$$V(x,\omega) = v(\tau_x \omega),$$

where $(\Omega, \mathcal{B}, \{\tau_x\}_{x \in \mathbb{R}^d}, \mu)$ is a dynamical system and $v$ is bounded and measurable. Our aim is to find upper and lower bounds for

$$(4.1) \qquad \lambda_t^{-1} \log P \otimes \mu \left( \frac{1}{t} \int_0^t V(x_s) \, ds \in A \right), \quad A \in \mathcal{B}(\mathbb{R}),$$

and $\lambda_t$ is an increasing positive function such that $\lim_{t \to \infty} \lambda_t/t = 0$. Here $P = P_o$.

To motivate the study of the asymptotic behavior of (4.1), recall that in section 3, it was shown that the exponential convergence of $\frac{1}{t} \int_0^t V(x_s) \, ds$ was related to the differentiability of the free energy function $c(\cdot)$ at the origin. For most random potentials, we believe that $c(\cdot)$ will not be differentiable at the origin. This is why we consider scaling functions $\lambda(\cdot)$ satisfying $\lim_{t \to \infty} \lambda_t/t = 0$.

Next, it follows from Lemma 4.2.1 in Remillard and Dawson [11] that if $\int_1^\infty \frac{e^{-\theta \lambda_t}}{t} \, dt < \infty$ for all $\theta > \theta_0 \geq 0$, then there exists a measurable set $N$ such that $\mu(N) = 0$ and for every $\omega \notin N$, the following hold true: for any closed set $F$,

$$\limsup_{t \to \infty} \lambda_t^{-1} \log P \left( \frac{1}{t} \int_0^t V(x_s, \omega) \, ds \in F \right) \leq$$

$$\theta_0 + \inf_{\delta > 0} \left( \limsup_{t \to \infty} \lambda_t^{-1} \log P \otimes \mu \left( \frac{1}{t} \int_0^t V(x_s, \omega) \, ds \in F^\delta \right) \right).$$

for almost every $\omega$, with respect to $\mu$.

We first state the upper bound results.

THEOREM 4.1. *Suppose that for any open set $O$*

$$\limsup_{t \to \infty} \lambda_t^{-d} \log \left( \sup_{f \in \mathcal{C}(B(0, R\sqrt{t\lambda_t}), I(f) \leq a\lambda_t/t} \mu\{(Vf, f) \in O\} \right) \leq -\psi_1(a, O, R),$$

*for some non negative function $\psi_1$. Let $F$ be a closed subset of $\mathbb{R}$.*

(i) *If $d = 1$ and $\psi_2$ is defined by (4.3), then*

$$\limsup_{t \to \infty} \lambda_t^{-1} \log P \otimes \mu \left( \frac{1}{t} \int_0^t V(x_s) \, ds \in F \right)$$

$$\leq - \sup_{r > 0} \min \left\{ r^2/2, \inf_{a > 0} (a + \psi_2(a, F, r)) \right\}.$$

(ii) *If $d \geq 2$ and $\psi_3$ is defined by (4.4), then*

$$\limsup_{t \to \infty} \lambda_t^{-1} \log P \otimes \mu \left( \frac{1}{t} \int_0^t V(x_s) \, ds \in F \right) \leq - \sup_{r > 0} \min \left\{ k_d r^2/2, \psi_3(F, r) \right\}.$$

PROOF. Set $a_t = \sqrt{t\lambda_t}$. For $0 < r_1 < r$, set $G_t = B(0, r_1 a_t)$ and $D_t = B(0, ra_t)$. For a closed set $F$, set

$$p(t, F) = P \otimes \mu \left( \frac{1}{t} \int_0^t V(x_s)\, ds \right)$$

Because of the stationarity of the potential $V$, and using (2.5) one obtains

$$
\begin{aligned}
p(t, F) &= E^\mu \left\{ \frac{1}{|D_t|} \int_{D_t} P_x \left( \frac{1}{t} \int_0^t V(x_s)\, ds \in F \right) dx \right\} \\
&\leq \frac{|D_t|}{|G_t|} E^\mu \left\{ \frac{1}{|D_t|} \int_{D_t} P_x \left( \frac{1}{t} \int_0^t V(x_s)\, ds \in F, T_{D_t} > t \right) dx \right\} \\
&\quad + \frac{1}{|G_t|} \int_{G_t} P_x \left( T_{D_t} \leq t \right) dx \\
&\leq P \left( \sup_{s \leq t} |x_s| \geq a_t (r - r_1) \right) + 2 \frac{r^d}{r_1^d} e^{-\lambda_t J_t},
\end{aligned}
$$

where $J_t = \dfrac{t}{\lambda_t} \inf_{\theta \in F} J_{D_t}(\theta)$.

Next, it is well known that there exits a constant $k_d$ such that

$$\lim_{t \to \infty} t^{-2} \log P \left( \sup_{s \leq 1} |x_s| \geq t \right) = -k_d/2.$$

Hence

$$\lim_{t \to \infty} \frac{1}{\lambda_t} \log P \left( \sup_{s \leq t} |x_s| \geq a_t (r - r_1) \right) = -k_d (r - r_1)^2 / 2.$$

Finally, from a standard argument,

$$\limsup_{t \to \infty} \frac{1}{\lambda_t} \log \mu(J_t \leq a) \leq -\psi(a),$$

yields

$$\limsup_{t \to \infty} \frac{1}{\lambda_t} \log E^\mu (e^{-\lambda_t J_t}) \leq - \inf_{a > 0} (a + \psi(a)) = -l.$$

Therefore, letting $r_1$ go to zero, one obtains

$$(4.2) \qquad \limsup_{t \to \infty} \frac{1}{\lambda_t} \log p(t, F) \leq - \min(k_d r^2/2, l)$$

Now $J_t \leq a$ if and only if there exists $f \in K_{a\lambda_t/t}(D_t)$ such that $(Vf, f) \in F$, where $K_a(D') = \{f \in \mathcal{M}(D'); I(f) \leq a\}$, for any bounded domain $D'$. It follows from Proposition 2.1 that $K_a(D')$ is compact.

Let $0 < \lambda_1 < \lambda_2 \leq \lambda_3 \leq \cdots$ be the eigenvalues of $-\Delta = -\Delta_D$ associated with the eigenvectors $f_1, f_2, f_3, \ldots$. Further let $N(\lambda) = \text{card}\{\lambda_k; \lambda_k \leq \lambda\}$ and $P_\lambda f = \sum_{k: \lambda_k \leq \lambda} (f, f_k) f_k$, $f \in L^2(D)$.

Since $|(Vf, f) - (Vg, g)| \leq \|V\|_\infty |f - g|(|f| + |g|)$ for any $f, g \in L^2(D)$, and $|f - P_\lambda f|^2 \leq I(f)/\lambda$ for any $f \in \mathcal{D}(D)$, it follows that $(Vf, f) \in F$ for some $f \in K_a(D)$ implies that $(Vf, f) \in F^{2\epsilon\|V\|_\infty}$ for some $f \in P_{a/\epsilon^2}(K_a(D))$. Note that $P_\lambda$ being a projection, $P_{a/\epsilon^2}(K_a(D))$ is isometrically isomorphic to a compact subset of the closed unit ball in $\mathbb{R}^{N(a/\epsilon^2)}$.

For any integer $m$, let $n(\epsilon, m)$ be the minimal number of balls of radius $\epsilon$ required to cover the closed unit ball of $\mathbb{R}^m$. Then $\epsilon^{-m} \leq n(\epsilon, m) \leq 2m \left( \frac{1+2\epsilon}{\epsilon} \right)^m$.

It follows that $P_{a/\epsilon^2}(K_a(D))$ can be covered by $n(\epsilon, N(a/\epsilon^2))$ closed balls of radius $3\epsilon$ with centers in $K_a(D)$. Since $C_0^\infty(D)$ is dense in $H_0^1(D)$, we see that for any $a' > a$, one can find $m = n(\epsilon, N(a/\epsilon^2))$ functions $f_i \in \mathcal{C}(D)$ such that $I(f_i) < a'$ and $(V f_i, f_i) \in F^{9\epsilon\|V\|_\infty}$ for some $1 \le i \le m$, if $(Vf, f) \in F^{2\epsilon\|V\|_\infty}$ for some $f \in P_{a/\epsilon^2}(K_a(D'))$. Note that the choice of the $f_i$'s is independent of $V$. Finally, since $\{\lambda_k/\theta^2\}_{k\ge1}$ are the eigenvalues of $-\Delta(\theta D)$, we obtain

$$\mu(J_t \le a) \le 2N(a\lambda_t^2/\epsilon^2)\left(\frac{1+2\epsilon}{\epsilon}\right)^{N(a\lambda_t^2/\epsilon^2)}$$

$$\times \inf_{a'>a} \sup_{f\in\mathcal{C}(D_t), I(f)\le a'\lambda_t/t} \mu\left\{(Vf, f) \in F^{9\epsilon\|V\|_\infty}\right\}.$$

Next, it follows from Weyl's result on the asymptotic distribution of eigenvalues of $-2\Delta$ (e.g. Reed and Simon [8][ Theorem XIII.78], that

$$\lim_{\lambda\to\infty} N(\lambda)/\lambda^{d/2} = (r/\sqrt{2})^d \left(\Gamma\left(\frac{d+2}{2}\right)\right)^{-2}.$$

In particular

$$\lim_{t\to\infty} N(a\lambda_t^2/\epsilon^2)/\lambda_t^d = \left(\frac{ar^2}{2\,\epsilon^2}\right)^{d/2} \left(\Gamma\left(\frac{d+2}{2}\right)\right)^{-2}.$$

It follows from the above calculations that

$$\limsup_{t\to\infty} \frac{1}{\lambda_t^d} \log\mu(J_t \le a) \le -\psi_2(a, F, r),$$

where

$$(4.3) \quad \psi_2(a, F, r) = \sup_{\epsilon>0}\left\{\sup_{a'>a} \psi_1(a', F^{9\epsilon\|V\|_\infty}, r)\right.$$

$$\left. -\left(\frac{ar^2}{2\,\epsilon^2}\right)^{d/2} \left(\Gamma\left(\frac{d+2}{2}\right)\right)^{-2} \log\left(\frac{1+2\epsilon}{\epsilon}\right)\right\}.$$

This proves the case $d = 1$. When $d \ge 2$, note that

$$\limsup_{t\to\infty} \frac{1}{\lambda_t} \log\mu(J_t \le a) = -\infty,$$

whenever $\psi_2(a, F, r) > 0$. Therefore

$$\limsup_{t\to\infty} \frac{1}{\lambda_t} \log p(t, F) \le -\psi_3(F, r),$$

where

$$(4.4) \qquad \psi_3(F, r) = \inf\{a > 0; \psi_2(a, F, r) = 0\}.$$

$\square$

The rate we found for the lower bound is the same as to the one obtained in Donsker and Varadhan [3] for the Wiener sausage.

THEOREM 4.2. *Let $O$ be an open subset of $\mathbb{R}$ and let $D$ be a bounded domain of $\mathbb{R}^d$. For any bounded domain $D'$, let $M(D')$ denotes the set of all probability measures on $D'$, and set $\mathcal{F}_{D'} = \sigma\{V(x); x \in D'\}$. Suppose there exists a probability measure $\nu$ on $(\Omega, \mathcal{B})$ satisfying the following conditions:*

$$(4.5) \qquad\qquad \lim_{a\to\infty} \nu(B_\delta(aD)) = 1, \text{ for all } \delta > 0,$$

where $B_\delta(aD) = \left\{ a^{-d} \log \left. \dfrac{d\nu}{d\mu} \right|_{\mathcal{F}_{aD}} \leq h + \delta \right\}$ for some $h \geq 0$;

(4.6)
$$\liminf_{a \to \infty} \inf_{R \in M(aD)} \nu(A(R)) > 0,$$

where $A(R) = \left\{ \displaystyle\int_{\mathbb{R}^d} V(x) \, R(dx) \in O \right\}$. Then

$$\liminf_{t \to \infty} t^{-d/(d+2)} \log P \otimes \mu \left( \frac{1}{t} \int_0^t V(x_s) \, ds \in O \right)$$
$$\geq -\left( \frac{d+2}{2} \right) \left( \frac{2\lambda_1}{d} \right)^{d/(d+2)} h^{2/(d+2)},$$

where $\lambda_1 = \displaystyle\inf_{f \in \mathcal{C}(D)} I(f)$.

PROOF. Let $a_t = \lambda t^{1/(d+2)}$, $\lambda > 0$, and let $\delta > 0$ be given. Further let $L_t$ be the occupation time measure of the Wiener process. It follows that $A(L_t) = \left\{ \dfrac{1}{t} \displaystyle\int_0^t V(x_s) \, ds \in O \right\}$.
Then

$$
\begin{aligned}
P \otimes \mu(A(L_t)) &\geq P \otimes \mu\left( A(L_t) \cap \{T_{a_t D} > t\} \right) \\
&= E^P \left[ I\!\!I\{T_{a_t D} > t\} E^\nu \left\{ e^{-\log \frac{d\nu}{d\mu} \big|_{\mathcal{F}_{a_t D}}} I\!\!I\{A(L_t) \cap B_\delta(a_t D)\} \right\} \right] \\
&\geq e^{-(h+\delta)a_t^d} P(T_{a_t D} > t) \left\{ \inf_{R \in M(a_t D)} \nu(A(R)) - \nu(B_\delta(a_t D)^c) \right\}.
\end{aligned}
$$

Therefore, using conditions (4.5) and (4.6), one obtains

$$\liminf_{t \to \infty} t^{-d/(d+2)} \log P \otimes \mu \left( \frac{1}{t} \int_0^t V(x_s) \, ds \in O \right) \geq$$

$$-(h+\delta)\lambda^d + \liminf_{t \to \infty} t^{-d/(d+2)} \log P(T_{a_t D} > t) \geq -(h+\delta)\lambda^d - \lambda_1/\lambda^2,$$

where the last inequality follows from Donsker and Varadhan [3]. Finally,

$$\sup_{\delta > 0} \sup_{\lambda > 0} \left( -(h+\delta)\lambda^d - \lambda_1/\lambda^2 \right) = -\left( \frac{d+2}{2} \right) \left( \frac{2\lambda_1}{d} \right)^{d/(d+2)} h^{2/(d+2)}.$$

Hence the result. $\qquad\square$

## 5. Illustration: Brownian motion in a random scenery

In this section, we find upper and lower bounds for

$$\lambda_t^{-1} \log P(W_t \geq a),$$

where $\lambda_t$ is an appropriate increasing sequence and $W_t = \dfrac{1}{t} \displaystyle\int_0^t \xi_{[x(s)+U]} ds$ where $\{\xi_k\}_{k \in \mathbb{Z}^d}$ are i.i.d. bounded random variables and $U$ is uniformly distributed over $T_d = [0,1)^d$ and is independent of the $\xi_k$'s.

The analog discrete process, called random walk in a random scenery in Kesten and Spitzer [6], is defined by $W_n = \displaystyle\sum_{k=1}^n \xi_{S_k}$ where $\{S_k\}_{k \geq 0}$ is a symmetric simple random walk on $\mathbb{Z}^d$. In Remillard and Dawson [12], the authors proved a conjecture of Kesten and Spitzer for both Brownian motion and random walk in a random scenery.

Let $\Phi(\lambda) = \log E\left(e^{\lambda\xi_0}\right)$ and let $\Phi^*(\cdot)$ be the Legendre transform of $\Phi(\cdot)$, that is, for any real number $a$,

$$\Phi^*(a) = \sup_\lambda\{\lambda a - \Phi(\lambda)\}.$$

The main results are stated in the following two theorems.

THEOREM 5.1. *Suppose $F$ is a closed subset of $\mathbb{R}$. Then*

(i)

$$\limsup_{t\to\infty} \frac{1}{t^{1/3}} \log P(W_t \in F) \le -\frac{3}{2} \inf_{a\in F} (\Phi^*(a))^{2/3},$$

*if $d = 1$;*

(ii)

$$\limsup_{t\to\infty} \frac{1}{\sqrt{t/\log t}} \log P(W_t \in F) \le -2\sqrt{2\pi} \inf_{a\in F} (\Phi^*(a))^{1/2},$$

*if $d = 2$;*

(iii)

$$\limsup_{t\to\infty} \frac{1}{t^{1/2}} \log P(W_t \in F) \le -\frac{2}{\sqrt{c_d}} \inf_{a\in F} (\Phi^*(a))^{1/2},$$

*when $d \ge 3$, where $c_d = \displaystyle\int_0^\infty h(s)ds$ and $h(s) = \displaystyle\sup_{y\in T_d} \int_{T_d} p(s, x - y)dx$.*

THEOREM 5.2. *Suppose that $\lambda_1 = \displaystyle\inf_{f\in\mathcal{C}(B_1)} I(f)$, where $B_1$ is the ball centered at the origin with volume 1. Then for any $d \ge 1$, one has*

$$\liminf_{t\to\infty} \frac{1}{t^{d/(d+2)}} \log P(W_t > a) \ge -\left(\frac{d+2}{2}\right)\left(\frac{2\lambda_1}{d}\right)^{d/(d+2)} (\Phi^*(a))^{2/(d+2)},$$

*for $a \ge E(\xi_0)$, and*

$$\liminf_{t\to\infty} \frac{1}{t^{d/(d+2)}} \log P(W_t < a) \ge -\left(\frac{d+2}{2}\right)\left(\frac{2\lambda_1}{d}\right)^{d/(d+2)} (\Phi^*(a))^{2/(d+2)},$$

*for $a \le E(\xi_0)$. In particular, when $d = 1$, one obtains*

$$\liminf_{t\to\infty} \frac{1}{t^{1/3}} \log P(W_t \in O) \ge -\frac{3}{2} \inf_{a\in O} (\pi\Phi^*(a))^{2/3},$$

*for any infinite open interval $O$.*

To prove Theorem 5.1, we will need the following proposition.

PROPOSITION 5.3. *Set $p_*(t) = \displaystyle\sup_k p_k(t)$, where $p_k(t) = \dfrac{1}{t}\displaystyle\int_0^t 1_{\{[x(s)+U]=k\}}ds$, $k \in \mathbb{Z}^d$. Then for any closed set $F$, one gets*

$$P(W_t \ge a) \le 2E\left(e^{-\inf_{a\in F} \Phi^*(a)/p_*(t)}\right).$$

PROOF. Let $\tilde{\mathcal{F}}_t = \sigma\{x[s] + U; s \leq t\}$. Then for any $a \geq E(\xi_0)$, one has

$$P(W_t \geq a|\tilde{\mathcal{F}}_t) \leq \inf_{\lambda > 0} e^{-\lambda a} E\left(e^{\lambda W_t}|\tilde{\mathcal{F}}_t\right) \quad \text{a. s.}$$

Now

$$
\begin{aligned}
E(\exp\{\lambda W_t\}|\tilde{\mathcal{F}}_t) &= E(\exp\{\lambda \sum_k \xi_k p_k(t)\}|\tilde{\mathcal{F}}_t) \\
&= \prod_k E(\exp\{\lambda \xi_0 p_k(t)\}|\tilde{\mathcal{F}}_t) \\
&= \exp\{\sum_k \Phi(\lambda p_k(t))\}.
\end{aligned}
$$

Since $\Phi(\cdot)$ is convex and $\Phi(0) = 0$, it follows that $\Phi(\lambda)/\lambda$ is increasing for $\lambda > 0$. Therefore

$$\Phi(\lambda p_k(t)) \leq \frac{p_k(t)}{p_*(t)} \Phi(\lambda p_*(t)),$$

yielding

$$\sum_k \Phi(\lambda p_k(t)) \leq \sum_k \frac{p_k(t)}{p_*(t)} \Phi(\lambda p_*(t)) = \Phi(\lambda p_*(t))/p_*(t).$$

Next

$$\sup_{\lambda > 0} \left\{ \lambda a - \frac{\Phi(\lambda p_*(t))}{p_*(t)} \right\} = \sup_{\lambda > 0} \left\{ \frac{\lambda a - \Phi(\lambda)}{p_*(t)} \right\} = \frac{\Phi^*(a)}{p_*(t)}.$$

Hence

$$P(W_t \geq a|\tilde{\mathcal{F}}_t) \leq e^{-\Phi^*(a)/p_*(t)} \quad \text{a. s.}$$

Similarly, if $a \leq \xi_0$,

$$P(W_t \leq a|\tilde{\mathcal{F}}_t) \leq e^{-\Phi^*(a)/p_*(t)} \quad \text{a. s.}$$

Combining these two inequalities and using a standard argument as in the proof of (2.5), one obtains

$$P(W_t \in F|\tilde{\mathcal{F}}_t) \leq 2e^{-\inf_{a \in F} \Phi^*(a)/p_*(t)} \quad \text{a. s.}$$

Thus

$$
\begin{aligned}
P(W_t \in F) &= E\left\{ P(W_t \in F|\tilde{\mathcal{F}}_t) \right\} \\
&\leq 2E\left\{ 2e^{-\inf_{a \in F} \Phi^*(a)/p_*(t)} \right\}.
\end{aligned}
$$

This completes the proof of the proposition. $\qquad\square$

The next proposition is a well-known result of large deviations theory.

PROPOSITION 5.4. *Suppose that* $\limsup_{t \to \infty} \lambda_t^{-1} \log P(\lambda_t p_*(t) > a) \leq -\phi(a)$, *for some increasing functions* $\lambda(\cdot)$ *and* $\phi(\cdot)$ *on* $(0, \infty)$. *Then, for any* $\lambda > 0$, *one obtains*

$$\limsup_{t \to \infty} \lambda_t^{-1} \log E\left(e^{-\lambda/p_*(t)}\right) \leq -\inf_{a > 0}\left(\frac{\lambda}{a} + \phi(a)\right).$$

PROPOSITION 5.5. *For any* $a > 0$, *we have*

(i)

$$\lim_{t \to \infty} t^{-1/3} \log P(t^{1/3} p_*(t) > a) = -a^2/2$$

when $d = 1$;

(ii)

$$\limsup_{t \to \infty} (t/\log t)^{-1/2} \log P\left((t/\log t)^{1/2} p_*(t) > a\right) \leq -2\pi a$$

if $d = 2$;

(iii)

$$\limsup_{t \to \infty} t^{-1/2} \log P(t^{1/2} p_*(t) > a) \leq -a/c_d,$$

when $d \geq 3$.

PROOF. We first find a general upper bound for $P(\lambda_t p_*(t) > a)$ and then we will take into account the dimension $d$. Let $S_n(t)$ be the set of all $(t_1, \ldots, t_n)$ such that $0 < t_1 < \cdots < t_n < t$.

For $n \geq 2$, we have

$$P(\lambda_t p_*(t) > a) \leq \sum_k P(\lambda_t p_k(t) > a) \leq \left(\frac{\lambda_t}{a}\right)^n \sum_k E(\{p_k(t)\}^n).$$

Now

$$E(\{p_k(t)\}^n) =$$

$$\frac{n!}{t^n} \int_{S_n(t)} \int_{T_d} \int_{(T_d+k)^n} p(t_1; x_1-u) p(t_2-t_1; x_2-x_1) \cdots p(t_n-t_{n-1}; x_n-x_{n-1}) \, dx \, du \, dt.$$

Making the change of variables $y_i = x_i - k$, $1 \leq i \leq n$, $v = u - k$, and summing over $k$, we obtain

$$\sum_k E(\{p_k(t)\}^n) =$$

$$\frac{n!}{t^n} \int_{S_n(t)} \int_{\mathbb{R}^d} \int_{T_d^n} p(t_1; y_1 - v) p(t_2 - t_1; y_2 - y_1) \cdots p(t_n - t_{n-1}; y_n - y_{n-1}) \, dy \, dv \, dt.$$

Making the change of variables $s_i = t_{i+1} - t_1$, $1 \leq i \leq n-1$ and $s_n = t - t_1$, one finally gets

$$\sum_k E(\{p_k(t)\}^n) = \frac{n!}{t^n} \int_{S_n(t)} \int_{T_d^n} p(s_1; y_2 - y_1) p(s_2 - s_1; y_3 - y_2) \cdots$$

(5.1)
$$\cdots p(s_{n-1} - s_{n-2}; y_n - y_{n-1}) \, dy \, ds.$$

Suppose that $d = 1$. Then $p(s; y) \leq (2\pi s)^{-1/2}$, for any $y \in \mathbb{R}$. Hence

$$\sum_k E(\{p_k(t)\}^n) \leq \left(\frac{n!}{t^n}\right) \int_{S_n(t)} s_1^{-1/2} \prod_{i=2}^{n-1} (s_i - s_{i-1})^{-1/2} \, ds$$

$$= n! (2t)^{-\left(\frac{n-1}{2}\right)} / \Gamma\left(\frac{n+3}{2}\right).$$

Therefore

(5.2)
$$P(t^{1/3} p_*(t) > a) \leq \left(\frac{t^{1/3}}{a}\right)^n n! (2t)^{-\left(\frac{n-1}{2}\right)} / \Gamma\left(\frac{n+3}{2}\right).$$

Setting $n = 2[\lambda t^{1/3}] - 1$, $\lambda > 0$ and letting $t$ go to infinity, in (5.2), one gets

(5.3)
$$\limsup_{t \to \infty} t^{-1/3} \log P(t^{1/3} p_*(t) > a) \leq \lambda \log \lambda - \lambda - \lambda \log(a^2/2),$$

where we used the following equality borrowed from Rainville [7][p. 30]: for any $x > 0$,

$$\log \Gamma(x) = \left( x - \frac{1}{2} \right) \log x - x + \frac{1}{2} \log(2\pi) - \int_0^\infty \frac{u - [u] - \frac{1}{2}}{x + u} \, du.$$

Taking the infimum over $\lambda > 0$ in (5.3), one finally obtains

$$\limsup_{t \to \infty} t^{-1/3} \log P(t^{1/3} p_*(t) > a) \le -a^2/2.$$

On the other hand, it follows from Theorem 4.2 in Remillard [10] that for any $\lambda > 0$,

$$\lim_{t \to \infty} t^{-1/3} \log E \left( e^{\lambda t^{2/3} p_0(t)} \right) = \lambda^2/2.$$

Therefore, using Gärtner-Ellis Theorem, one finds

$$\liminf_{t \to \infty} t^{-1/3} \log P(t^{1/3} p_*(t) > a) \ge \liminf_{t \to \infty} t^{-1/3} \log P(t^{1/3} p_0(t) > a) \ge -a^2/2.$$

This completes the proof of (i).

If $d \ge 2$, set $h(t) = \sup_{y \in T_d} \int_{T_d} p(t; x - y) \, dx$, and let $H(t) = \int_0^t h(s) \, ds$. Then it follows (5.1) that

$$\sum_k E(\{p_k(t)\}^n) \le n! \, (H(t)/t)^{n-1} .$$

Therefore, if $0 < \frac{\lambda \lambda_t^2 H(t)}{at} < 1$,

$$\frac{(\lambda_t \lambda)^n}{n!} P(\lambda_t p_*(t) > a) \le \left( \frac{t}{H(t)} \right) \left( \frac{\lambda \lambda_t^2 H(t)}{at} \right)^n ,$$

for any $n \ge 0$. So summing over $n \ge 0$, one obtains

(5.4) $$P(\lambda_t p_*(t) > a) \le \left( \frac{t}{H(t)} \right) e^{-\lambda \lambda_t} \left( 1 - \frac{\lambda \lambda_t^2 H(t)}{at} \right)^{-1} ,$$

whenever $0 < \frac{\lambda \lambda_t^2 H(t)}{at} < 1$.

When $d = 2$, it is easy to see that $H(t)/\log t$ converges to $1/2\pi$ as $t$ tends to infinity. Therefore

$$\limsup_{t \to \infty} (t/\log t)^{-1/2} \log P \left( (t/logt)^{1/2} p_*(t) > a \right) \le -\lambda,$$

for any $0 < \lambda < 2\pi a$. Taking the infimum of the right-hand side over $\lambda$ yields (ii). The proof is (iii) is similar. When $d \ge 3$, $H(t)$ converges to $c_d$ as $t$ tends to infinity. By taking $\lambda_t = t^{1/2}$, (5.4) yields

$$\limsup_{t \to \infty} t^{-1/2} \log P(t^{1/2} p_*(t) > a) \le \inf_{0 < \lambda < a/c_d} (-\lambda) = -a/c_d.$$

$\square$

We are now in a position to prove Theorems 5.1 and 5.2.

PROOF OF THEOREM 5.1. It follows from Proposition 5.3 that for any closed set $F$,

$$P(W_t \in F) \le 2E \left( e^{-l/p_*(t)} \right),$$

where $l = \inf_{a \in F} \Phi^*(a)$.

When $d = 1$, one can apply Proposition 5.4 and Proposition 5.5(i) to obtain

$$\limsup_{t \to \infty} t^{-1/3} \log P(W_t \in F) \le -\inf_{x>0} \left( \frac{l}{x} + \frac{x^2}{2} \right) = -\frac{3}{2} l^{2/3}.$$

Simarly, when $d = 2$, one obtains

$$\limsup_{t \to \infty} (t/\log t)^{-1/2} \log P(W_t \in F) \le -\inf_{x>0} \left( \frac{l}{x} + 2\pi x \right) = -2\sqrt{2\pi l}.$$

Finally, when $d \ge 3$, one gets

$$\limsup_{t \to \infty} t^{-1/2} \log P(W_t \in F) \le -\inf_{x>0} \left( \frac{l}{x} + \frac{x}{c_d} \right) = -2\sqrt{l/c_d}.$$

□

PROOF OF THEOREM 5.2. We only prove the case $a \ge E(\xi_0)$, since the other case can be deduced from the other one. Since the result is obviously true if $\Phi^*(a) = \infty$, one may suppose that $l = \Phi^*(a) < \infty$. Because $\Phi(b) = \infty$ whenever $b > \lim_{\lambda \to \infty} \Phi'(\lambda)$, one can also suppose that the set of all $\lambda > 0$ such $\Phi'(\lambda) > a$ is not empty. Choose such a $\lambda$.

Let $D = B_1$ be the ball centered at the origin with volume 1. and let $\nu$ be the unique product measure satisfying

$$\log E^\nu \left( e^{\theta \xi_k} \right) = \Phi(\theta + \lambda) - \Phi(\lambda),$$

for every $\theta \in \mathbb{R}$. Note that for this measure, $E^\nu(\xi_0) = \Phi'(\lambda)$.

It is easy to see that

$$\lim_{b \to \infty} b^{-d} \log \left. \frac{d\nu}{d\mu} \right|_{\mathcal{F}(bB_1)} = \lambda \Phi'(\lambda) - \Phi(\lambda) = \Phi^*(\Phi'(\lambda)) = l, \quad \nu \text{ almost surely,}$$

where $\mathcal{F}(bB_1)$ is the $\sigma$-algebra generated by $\{\xi_{[x+U]}; x \in bB_1\}$.

Thus condition (4.5) holds with $h = l$. Next, condition (4.6) holds if

$$(5.5) \qquad \inf_{b>0} \inf_{R \in M(bB_1)} \nu \left( \int V(x) R(dx) > a \right) > 0.$$

It is easy to see that (5.5) holds if there exists $\delta > 0$ such that for all finite set $A \subset \mathbb{Z}^d$ and for any set $\{p_k\}_{k \in A}$ such that $p_k > 0$ and $\sum_{k \in A} p_k = 1$,

$$(5.6) \qquad \nu \left( \sum_{k \in A} p_k \xi_k \le a \right) < 1.$$

Since $\Phi$ is convex, $\left( \Phi(\lambda - \theta) - \Phi(\lambda) \right) \big/ \theta$ is non decreasing in $\theta > 0$. Therefore

$$\Phi(\lambda - \theta p_k) - \Phi(\lambda) \le p_k \left( \Phi(\lambda - \theta) - \Phi(\lambda) \right),$$

so

$$\sum_{k \in A} \Phi(\lambda - \theta p_k) - \Phi(\lambda) \le \Phi(\lambda - \theta) - \Phi(\lambda).$$

Therefore

$$\nu\left(\sum_{k\in\mathcal{A}} p_k\xi_k \le a\right) \le \inf_{\theta>0} e^{\theta a} E^\nu\left(e^{-\theta\sum_{k\in\mathcal{A}} p_k\xi_k}\right)$$

$$= \inf_{\theta>0} e^{\theta a+\sum_{k\in\mathcal{A}}\{\Phi(\lambda-\theta p_k)-\Phi(\lambda)\}}$$

$$\le \inf_{\theta>0} e^{\theta a+\Phi(\lambda-\theta)-\Phi(\lambda)}$$

$$\le e^{-\Phi^*(a)+\lambda a-\Phi(\lambda)}$$

$$< 1,$$

since $\Phi^*(a) > \lambda a - \Phi(\lambda)$.

Hence (5.6) holds true and it follows that from Theorem 4.2 that

$$\liminf_{t\to\infty} t^{-d/(d+2)} \log P\otimes\mu\left\{\frac{1}{t}\int_0^t V(x_s)ds) > a\right\} \ge$$

$$-\left(\frac{d+2}{2}\right)\left(\frac{2\lambda_1}{d}\right)^{d/(d+2)}(\Phi^*(a))^{2/(d+2)}$$

for all $\lambda$ so that $\Phi'(\lambda) > a$.

The proof is completed if one observes that $\inf_{\lambda;\Phi'(\lambda)>a} \Phi^*(\Phi'(\lambda)) = \Phi^*(a)$ and $\lambda_1 = \pi^2/2$ when $d = 1$. $\qquad\square$

## Acknowledgment

The author wishes to thank Don Dawson for many helpful discussions on this subject that lead to considerable improvement of the results. The author also wishes to thank the referees for their helpful comments.

## References

[1] Carmona, R. (1986) Random Schrödinger operators. *Ecole d'ete de probabilites de Saint-Flour XIV - 1984*, Lect. Notes Math. **1180**, 1-124.

[2] Demuth, M., Kirsch, W., and McGillivray, I. (1995). Schrödinger operators – geometric estimates in terms of the occupation time. *Commun. Partial Differ. Equations* **20**, No. 1-2, 37-57.

[3] Donsker, M.D. and Varadhan, S.R.S. (1975). Asymptotics for the Wiener sausage. *Comm. Pure Appl. Math.* **28** 525-565.

[4] Ellis, R.S. (1985). *Entropy, Large Deviations, and Statistical Mechanics*. Grundlehren der mathematischen Wissenschaften **271**, Springer-Verlag, New York.

[5] Gruet, J.-C. and Shi, Z. (1996). The occupation time of Brownian motion in a ball. *J. Theor. Probab.* **9**, No. 2, 429-445.

[6] Kesten, H. and Spitzer, F. (1979). A limit theorem related to a new class of self similar processes. *Z. Wahrsch. Verw. Gebiete* **50** 5-25.

[7] Rainville, E.D. (1971). *Special Functions*. Chelsea Publishing Company, Bronx, New York.

[8] Reed, M. and Simon, B. (1972). *Methods of Modern Mathematical Physics, IV*. Academic Press, New York.

[9] Remillard, B. (1987). *Large deviations and laws of the iterated logarithm for multidimensional diffusion processes with applications to diffusion processes with random coefficients*. PhD thesis, Carleton University.

[10] Remillard, B. (1990). Asymptotic behaviour of the Laplace transform of weighted occupation times of random walks and applications. *Diffusion Processes and Related Problems in Analysis*, Volume 1, Mark A. Pinsky ed. Progress in Probability, **22**, Birkäuser, Boston, 497-519.

[11] Remillard, B. and Dawson, D.A. (1989). Laws of the iterated logarithm and large deviations for a class of diffusion processes. *Canad. J. Statist.* **17** 349-376.

[12] Remillard, B and Dawson, D.A. (1991). A limit theorem for Brownian motion in a random scenery. *Canad. Math. Bull.* **34** 385-391.

DÉPARTEMENT DE MATHÉMATIQUES ET INFORMATIQUE, UNIVERSITÉ DU QUÉBEC À TROIS-RIVIÈRES, TROIS-RIVIÈRES (QUÉBEC), CANADA G9H 5H7

*E-mail address*: `remillar@uqtr.uquebec.ca`

Canadian Mathematical Society
Conference Proceedings
Volume **26**, 2000

# Comparative Genomics Via Phylogenetic Invariants For Jukes-Cantor Semigroups

David Sankoff and Mathieu Blanchette

*Dedicated to Professor Donald A. Dawson*

ABSTRACT. We review the theory of invariants as it has been developed for comparing the DNA sequences of homologous genes from phylogenetically related species, with particular attention to the semigroups used to model sequence evolution. We also outline the computational theory of genome rearrangements, including the optimization problems in calculating edit distances between genomes and the simpler notion of breakpoint distance. The combinatorics of rearrangements, involving non-local changes in the relative order of genes in the genome, are more complex than the base substitutions responsible for gene sequence evolution. Nevertheless we can construct models of gene order evolution through symmetry assumptions about disruptions of gene adjacencies. Based on the extended Jukes-Cantor semigroup that emerges from this modeling, we derive a complete set of linear phylogenetic invariants. We use these invariants to relate mitochondrial genomes from a number of animal phyla and compare the results to parsimony trees also based on gene adjacencies and to minimal breakpoint trees.

## 1. Invariants for models of sequence evolution

Consider the aligned DNA sequences: $X_1^{(1)} \cdots X_n^{(1)}, \cdots, X_1^{(N)} \cdots X_n^{(N)}$, all of length $n$, representing $N$ species whose history of evolutionary divergence, or *phylogeny*, is represented by a tree **T** with vertex set $V$ and edge set $E$, as in Figure 1. The terminal vertices represent observed, or present-day, species. The non-terminal vertices represent hypothetical ancestral species. For each $i$, the $X_i^{(J)}$ are the terminal points of a trajectory indexed by **T**, taking on values in the alphabet of bases $\{A, C, G, T\}$. This trajectory is a sample from a process described by $|E|$ $4 \times 4$ Markov matrices with positive determinant all belonging to the same semigroup, one matrix associated to each of the edges in $E$. Such semigroups have been proposed by Jukes and Cantor [**46**], Kimura [**51, 52**], Tajima and Nei [**75**], Hasegawa *et al.* [**43**], Cavender [**17**], Jin and Nei [**45**], Tamura [**76**], Nguyen and Speed [**57**], Tamura and Nei [**77**], Steel [**70**] and Ferretti and Sankoff [**32**].

2000 *Mathematics Subject Classification.* Primary 92D15, 60J27; Secondary 68Q25, 13P10.

*Key words and phrases.* phylogenetic trees, linear invariants; Jukes-Cantor semigroup; genomics; sorting by reversals; breakpoints; Metazoa; mitochondria.

Aside from the fact that it has $N$ terminal vertices, the tree $\mathbf{T}$ is unknown. In particular, the $|E|$ matrices associated with the edges are unknown, though the common semigroup from which they are drawn is given. The central problem of phylogenetic inference is to estimate $\mathbf{T} = (V, E)$, given only $n$ data vectors each consisting of the values at the $N$ terminal vertices of the trajectory, of form $(X_i^{(1)}, \cdots, X_i^{(N)})$, where $X_i^{(J)}$ is the $i$-th base in the $J$-th DNA sequence.

In DNA evolution, rates of change between any two elements of $\{A, C, G, T\}$ tend to be symmetric. With this type of data, it is usually preferable not to try to locate a root, or earliest ancestor node, in the tree. Thus in this paper we will make the simplifying assumption that $\mathbf{T}$ is an unrooted binary branching tree (all non-terminal vertices of degree 3), hence $|V| = 2N - 2$, $|E| = 2N - 3$, and will confine ourselves to symmetric transition matrices.

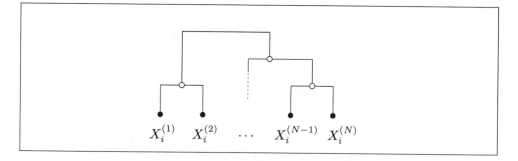

FIGURE 1. Sample trajectory $X_i^{(\cdot)}$. Indexing tree $\mathbf{T}$ is unknown, but the same for all $i = 1, \cdots, n$. Filled dots at terminal vertices indicate $N$ present-day species at which values of the process can be observed, unfilled dots represent unobservable ancestral species.

Phylogenetic invariants are predetermined functions of the probabilities of the observable $N$-tuples. These functions are identically zero only for $\mathbf{T}$ (and possibly a limited number of other trees), no matter which $|E|$ matrices are chosen from the semigroup. Evaluating the invariants associated with all possible trees, using observed $N$-tuple frequencies as estimates of the probabilities, enables the rapid inference of the (presumably unique) tree $\mathbf{T}$ for which all the invariants are zero or vanishingly small.

The chief virtue of the method of invariants is that it is not sensitive to "branch length", i.e. to which $|E|$ matrices are chosen from the semigroup; for a matrix $M$, this length may be taken to be $-\log \det M$. Methods of phylogenetic reconstruction which do not take account of the model used to generate the data may be susceptible to an artifact which tends to group long lineages together and short lineages together.

Lake [53] introduced linear invariants in 1987, studying the case $N = 4$ for a 2-parameter semigroup originally suggested by Kimura [51]. In the same year, Cavender and Felsenstein [18] published quadratic invariants for a 1-parameter semigroup of $2 \times 2$ matrices. Subsequently a great deal of research has been carried out into both linear invariants, by Cavender [17], Fu [34], Nguyen and Speed [57], Steel and Fu [71], Hendy and Penny [44], and polynomial invariants, by Drolet and Sankoff [23], Sankoff [61], Felsenstein [28], Ferretti *et al.* [31, 29, 32, 33], Evans and Speed [25], Steel *et al.* [72], Szekeley *et al.* [74], Steel [70], Evans and Zhou [26], Hagedorn [37] and Hagedorn and Landweber [38].

## 2. The study of gene order evolution.

The aligned DNA sequences discussed in Section 1 represent the internal structure of the same gene in $N$ related species or, more correctly, homologous genes (i.e., diverging from the same ancestral gene) in these species. For the rest of this paper, we turn to a higher level of analysis, the *genome*, comparing the linear ordering of all the different genes in the chromosomes of these species. For these purposes, we do not take into account variation in the DNA sequences among a set of $N$ homologous genes, but simply consider them identical.

As individual genes evolve through the Markovian base substitution discussed in Section 1, as well as other *local* processes such as base deletion or insertion, several additional, *non-local*, evolutionary mechanisms also operate, at the genomic level.

The study of comparative genomics has focused on inferring the most economical explanation for observed differences in gene orders in two or more genomes in terms of certain *rearrangement* processes. For single-chromosome genomes, this has been formulated as the problem of calculating an edit distance between two linear orders on the same set of objects, representing the ordering of homologous genes in two genomes. In the most realistic version of the problem, a sign (plus or minus) is associated with each object in the linear order, representing the direction of transcription of the corresponding gene, which depends on which of the two complementary DNA strands the gene is located. The elementary edit operations (Figure 2) may include one or more of the processes:

1) **inversion**, or **reversal**, of any number of consecutive terms in the ordered set, which, in the case of signed orders, also reverses the sign of each term within the scope of the inversion. Kececioglu and Sankoff [50] re-introduced the problem—earlier posed by Waterson *et al.*[80], and even earlier in the genetics literature, e.g. [73]—of computing the minimum reversal distance between two given permutations in the unsigned case, and gave approximation algorithms and an exact algorithm feasible for moderately long permutations. Bafna and Pevzner [4] gave improved approximation algorithms and Caprara [14] showed this problem to be NP-complete. Kececioglu and Sankoff [49] also found tight lower and upper bounds for the signed case and implemented an exact algorithm which worked rapidly for long permutations. Indeed, Hannenhalli and Pevzner [41] showed that the signed problem is only of polynomial complexity, improvements to their algorithm were given by Berman and Hannenhalli [6] and by Kaplan *et al.* [47].

2) **transposition** of any number of consecutive terms from their position in the order to a new position between any other pair of consecutive terms. This may or may not also involve an inversion. Computation of the transposition distance between two permutations was considered by Bafna and Pevzner [5], but its NP-completeness has not yet been confirmed. An edit distance which is a weighted combination of inversions, transpositions and deletions has been studied by Sankoff [62] and Blanchette *et al.* [8].

There is no difficulty in reformulating these notions to apply to circular rather than linear genomes as is often the case of unichromosomal genomes (e.g. bacteria or

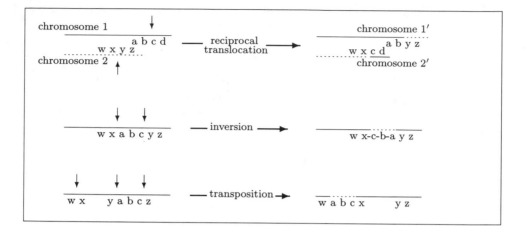

FIGURE 2. Schematic view of genome rearrangement processes. Letters represent positions of genes. Vertical arrows at left indicate breakpoints introduced into original genome. Reciprocal translocation exchanges end segments of two chromosomes. Inversion reverses the order and sign of genes between two breakpoints (dotted segment). Transposition removes a segment defined by two breakpoints and inserts it at another breakpoint (dotted segment), in the same chromosome or another.

mitochondria). For multichromosomal genomes, the most important role is played by:

3) **reciprocal translocation.** Kececioglu and Ravi [48] began the investigation of translocation distances, and Hannenhalli [39, 42] has shown that two formulations of the problem are of polynomial complexity. Ferretti *et al.* [30] proposed a relaxed form of translocation distance applicable when chromosomal assignment of genes, but not their order, is known. The complexity of its calculation was shown to be NP-complete by DasGupta *et al.* [22] and its structure was further investigated by Liben-Nowell [54].

Note that although our discussion in this paper is phrased in terms of the order of genes along a chromosome, the key aspect for mathematical purposes is the order and not the fact that the entities in the order are genes. They could as well be blocks of genes contiguous in all $N$ species, conserved chromosomal segments in comparative genetic maps (cf. Nadeau and Sankoff [56]) or, indeed, the results of any decomposition of the chromosome into disjoint ordered fragments, each identifiable in all $N$ genomes.

## 3. Phylogeny based on gene order.

The extension of edit-distances for gene order data to finding globally optimal phylogenetic trees is inherently difficult. Not only are some of the measures of genomic edit-distance in Section 2 computationally complex, but the extension of any of them, even the reversals-distance for signed genomes (itself only of quadratic complexity), to three or more genomes — multiple genome rearrangement — is NP-hard [15]. An example is the "median" problem: find the "ancestor" genome which is closest to three given genomes. Heuristics are available [**40, 69**], but they are only feasible for small genomes.

The breakpoint distance between two genomes containing the same genes, is the number of pairs of adjacent genes in one genome which are not adjacent in the other [**80**]. (Figure 2 illustrates how breakpoints are created.) This is not an edit distance, but tends to be highly correlated with such distances and has the advantage of being computable in linear time. Nevertheless its extension to three or more genomes is also NP-hard [**58, 13**]. It does have a simple reduction to the Traveling Salesman Problem [**64**] and can thus benefit from relatively efficient software available for the latter to solve examples on three genomes with moderate-sized $n$. This can then be extended to the optimization of fixed-topology phylogenies [**7, 65**], and ultimately to the search for optimal topologies [**9**].

In this kind of phylogenetic inference, breakpoint distance is used as a *parsimony* criterion. And parsimony methods are among those which, under the simplest probabilistic models of mutation, may sometimes reconstruct trees incorrectly when there are some very short and some very long branches [**27**]. This problem, together with the computational complexity of all versions of the multiple genome rearrangement problem, leads us to investigate the potential of phylogenetic invariants for studying gene order evolution. Not only do they avoid branch length problems, but they require negligible calculation time.

But can we find invariants for gene order data? After all, the various sets of breakpoints in a multi-genome comparison do not resemble a multiple alignment of sequences in any way, so that the phylogenetic invariants developed in the context of DNA base sequence data are not applicable. In Section 4 we will present models for genome rearrangement processes analogous to the base substitution models for gene sequence evolution, and examine the evolution of the adjacencies of pairs of genes over time. In one case, that of inversions on unsigned genomes, we obtain a matrix semigroup of transition probabilities among these adjacencies. In the other cases, the time-indexed matrices of transition probabilities do not form a semigroup. Nevertheless, in Section 5, we propose a simpler model for the evolution of breakpoints, not based on any assumptions about the rearrangement processes responsible for them, and use this to calculate a complete set of linear invariants for the fifteen binary unrooted trees where $N = 5$.

## 4. Probability models for breakpoint distances

We will propose models for inversion on unsigned and signed circular genomes, as well as for transpositions on unsigned genomes. We will assume in all three models that all pairs of adjacent genes $fg$ are equally likely to be disrupted, though this is a simplification of biological reality [**9, 63**]. Recall that, as in Figure 2, two different pairs must be disrupted for each inversion, and three for each transposition.

First we will provide detailed notation for the operations described in Section 2 in the case of circular genomes. Consider such a genome with gene order $\gamma_1 \cdots \gamma_n$. The origin is arbitrary so that the genome could also be written $\gamma_{i+1} \cdots \gamma_n \gamma_1 \cdots \gamma_i$. Label the genes found on one of the two complementary strands of the genome with a plus sign and those on the other with a minus, resulting in $g_1 \cdots g_n$. ($g_i = \gamma_i$ or $g_i = -\gamma_i$.) By convention, we "view" the circle from the side which ensures that the positively labeled strand is the one read in a clockwise manner, the other counterclockwise. Changing the sign on all genes is equivalent to viewing the circle from the "flip" side, and does not change the identity of the genome.

Consider any two pairs of adjacent genes $ab$ and $cd$ (possibly $b = c$ or $d = a$). An example of an inversion is the operation which takes $g_1 \cdots ab \cdots cd \cdots g_n$ to $g_1 \cdots a$ -$c \cdots$ -$bd \cdots g_n$ (or, equivalently, to -$g_n \cdots$ -$db \cdots c$ -$a \cdots$ -$g_1$, as illustrated in Figure 3).

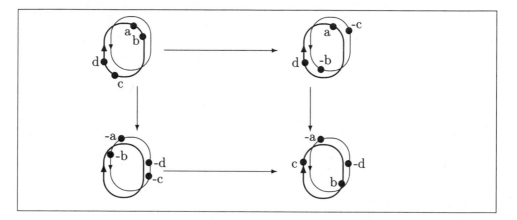

FIGURE 3. Reading direction, sign assignment to genes, and inversion. Reading direction is indicated by arrowheads on each DNA strand. The two genomes on the left are biologically identical; one view can be derived from the other by flipping the genome over and assigning signs to each gene according to whether it is on the "front", (i.e. read clockwise) or the "back" (read counterclockwise) strand. The two views of the genome on the right result from inverting the segment from gene $b$ to gene $c$, inclusive. The commutivity of flipping and inversion accords with the fact that it does not matter biologically from which side we view the genome.

We may also consider *unsigned* genomes where the reading direction (or strand) of each gene is unknown. In Figure 3, we may imagine the two strands superimposed and ignore the signs on the genes. In this case, the inversion transforms $g_1 \cdots ab \cdots cd \cdots g_n$ to $g_1 \cdots ac \cdots bd \cdots g_n$ or, equivalently, to $g_n \cdots db \cdots ca \cdots g_1$; in the former representation, reading clockwise at the top right of Figure 3, genes $b \cdots c$ were in the *scope* of the inversion; in the latter, reading clockwise at the bottom right, these genes were not in the scope of the inversion. Though flipping the genome does not change its identity, considering the two representations separately will be important for probabilistic modeling in the next section.

Consider any three pairs of adjacent genes $ab, cd$ and $fg$, where $fg$ occurs in the interval $d \cdots a$. The operation which takes $g_1 \cdots ab \cdots cd \cdots fg \cdots g_n$ to $g_1 \cdots ad \cdots fb \cdots cg \cdots g_n$ is a transposition. In some models, $g_1 \cdots ad \cdots f$ -$c \cdots$ -$bg \cdots g_n$ can also be produced by a single transposition; in other models, it requires an inversion as well.

**4.1. Inversions, unsigned case.** For an unsigned circular genome, consider a continuous time process with rate $\lambda = 1$. Each change of state involves an inversion, where any two pairs of adjacent genes $fg$ and $hk$ are is equally likely to be disrupted. We focus on a particular gene $f$ and, after each inversion, choose the representation of the resultant genome where $f$ has *not* been in the scope of the inversion. If $fg$ is one of the two pairs chosen (probability $1/n$), any of the $n-1$ genes other than $f$ is equally likely to replace $g$. In the case $g = h$, gene $g$ replaces itself and the inversion is "invisible". The matrix of transition probabilities for the occupant, at a specific time $t$, of the slot in the genome originally occupied by $g$, whose columns and rows are labeled by the $n-1$ candidate genes, is of form $(1 - (n-1)\alpha)I + \alpha J$, where $I$ is the identity and $J$ the matrix of 1's, and $1 - (n-1)\alpha = e^{-t/(n-2)}$. This is a generalization to $n-1 \times n-1$ matrices of the Jukes-Cantor [**46**] semigroup of $4 \times 4$ matrices.

**4.2. Inversions, signed case.** A model consisting only of random inversions on signed genomes, however, is quite different. Suppose all genes are on the same strand and have positive sign. Then if $fg$ is disrupted by an inversion, the new successor to $f$ will necessarily have negative sign. All negatively signed genes (other than $-f$) will have probability $1/(n-1)$ of replacing the successor to $f$. All positively signed genes will have probability zero. So the Jukes-Cantor equiprobability among the $2n-3$ possible new successors definitely does not hold. Moreover, after the first inversion, the strandedness of some genes will have changed, so that for the next inversion, some of the transition probabilities for successors will change. In other words, the process cannot be modeled by a semigroup of matrices as in the unsigned case.

Without loss of generality, we label the genes from 1 to $n$, and after each inversion we flip the genome if necessary so as to ensure gene 1 always has positive sign. In addition, we designate the position occupied by gene 1 to be position 1, the position occupied by its successor to be position 2, and so on. Let $x_i$ be the occupant of the $i$-th position. After $k$ inversions, the probability that the $i$-th position will be occupied by gene $h$ is

$$
\begin{aligned}
P_k(x_i = h) &= P_{k-1}(x_i = h)\Pr[h \text{ not in scope of } k\text{--th inversion}] \\
&\quad + \sum_{j=2}^{n} P_{k-1}(x_j = -h)Pr[k\text{--th inversion moves } h \text{ from } j \text{ to } i] \\
&= P_{k-1}(x_i = h)\left(1 - \binom{n}{2}^{-1}(i-1)(n+1-i)\right) \\
&\quad + \binom{n}{2}^{-1}\sum_{j=2}^{n} P_{k-1}(x_j = -h)\min\left\{ \begin{array}{ll} i-1, & n+1-j, \\ j-1, & n+1-i \end{array} \right\}
\end{aligned}
$$

For $n = 4$, this recurrence produces the pattern in Table 1.

It can be seen that it takes a relatively large number of inversions to "scramble" the genome enough so that the successor to gene 1 is equally likely to be any other gene, with either sign.

To compare this rearrangement process to the one generated by the Jukes-Cantor semigroup, we define $P_t(x_i = h)$ as the probability that the $i$-th position will be occupied by gene $h$ at time $t$. Then

$$
P_t(x_2 = h) = \sum_{k=0}^{\infty} \frac{e^{-t}t^k}{k!} P_k(x_2 = h)
$$

TABLE 1. Approach of $P_k(x_2 = h)$ to equiprobability.

| $h$ \\ $k$ | 2 | 3 | 4 | -2 | -3 | -4 |
|---|---|---|---|---|---|---|
| 1 | 0.500 | 0 | 0 | 0.167 | 0.167 | 0.167 |
| 2 | 0.333 | 0.111 | 0.083 | .167 | 0.139 | 0.167 |
| 4 | 0.205 | 0.154 | 0.143 | 0.170 | 0.158 | 0.170 |
| 8 | 0.169 | 0.166 | 0.165 | 0.167 | 0.166 | 0.167 |

Table 2 illustrates the approach to Jukes-Cantor probabilities of the inversion on signed genomes model for $n = 4$.

TABLE 2. Approach of $P_t(x_2 = h)$ to Jukes-Cantor probabilities.

| $h$ \\ $t$ | random inversions | | | | | | Jukes-Cantor | |
|---|---|---|---|---|---|---|---|---|
| | 2 | 3 | 4 | -2 | -3 | -4 | 2 | others |
| 1 | 0.632 | 0.032 | 0.025 | 0.106 | 0.099 | 0.106 | 0.672 | 0.066 |
| 2 | 0.433 | 0.078 | 0.065 | 0.145 | 0.133 | 0.145 | 0.473 | 0.105 |
| 4 | 0.258 | 0.134 | 0.123 | 0.166 | 0.154 | 0.166 | 0.279 | 0.144 |
| 8 | 0.178 | 0.163 | 0.160 | 0.168 | 0.164 | 0.168 | 0.182 | 0.164 |

This table shows that the transition probabilities remain rather inhomogeneous for a considerable time, even for $n$ as small as 4. For $t = 4$, there have been about 8 opportunities on the average for each of the four adjacencies to be disrupted (two per inversion); nonetheless the probabilities are decidedly non-uniform, even among the genes where $h \neq 2$. For larger $n$, such as the case $n = 37$ of interest in Section 7 below, the situation is analogous. Even after all the original adjacencies have had ample opportunity to be disrupted, the transition probabilities remain quite different from Jukes-Cantor, especially for low or high values of $h$, e.g. $\pm h = 2, 3, 36$ or 37. But the values of $t$ of biological interest will be those during which a fair proportion of the original adjacencies will be conserved. In other words, for those lengths of time for which we wish to apply these methods, the Jukes-Cantor semigroup is not a good approximation for the random inversions model.

**4.3. Transpositions.** Finally, consider transpositions on unsigned circular genomes. Again, we assume a uniform probability rate $\lambda = 1$ of such events occurring. At each event, any choice of three different pairs of adjacent genes $ab, cd$ and $fg$ is equally likely to be disrupted. Any of the $n - 2$ genes other than $f$ or $g$ is equally likely to play the role of $b$ in replacing $g$ as the neighbour of $f$. But the fact that $g$ cannot replace itself as it could in the unsigned inversion model, leads to the same sort of difficulty as with signed inversion. A Jukes-Cantor model cannot be formulated.

## 5. Extended Jukes-Cantor model for breakpoints

In this section, we construct a model for signed genomes. We will not assume that inversion, or any other particular process, is the only mechanism of genome rearrangement. Inversion, transposition or single-gene movement could also play a role, in unknown proportions. Thus, we will not assume that only -$h$ can replace $g$, where $h$ and not -$h$ appears in the original genome, as in the pure inversions case. Indeed, inspired by Jukes-Cantor, we assume that for any gene $f$, whose successor is $g$, the probability $\alpha$ that, over a given time interval, the successor to $f$ will have changed from $g$ to $h$, is the same for all pairs of genes $f$ and $g$, and for all $h \neq g$. Note that $h = $ -$g$ is not excluded. There are $2n - 3$ such changes possible. The probability that $g$ will remain the successor is then $1 - (2n - 3)\alpha$. Note that $1 - (2n - 3)\alpha > \alpha$ since, for consistency's sake, this event, including both no change and reversed changes, is at least as likely as any other particular change.

We have in effect defined a $2n - 2 \times 2n - 2$ Jukes-Cantor matrix $M(\alpha)$, where the rows and columns are indexed by the $2n - 2$ possible signed genes different from $f$ and -$f$. The entries are all $\alpha$ except for $1 - (2n - 3)\alpha$ on the diagonal. The model defines a semigroup which determines (stochastically) the trajectory of the occupant of the "successor to $f$" slot across a phylogeny. From it, if we were given the branch lengths, we could calculate the probabilities of all possible $N$-tuples at the terminal vertices.

We are not, however, given the branch lengths, nor are we directly interested in these lengths, since our goal is to find the correct tree topology in a way which is *insensitive* to them.

For a given $f$, and there are $2n$ of them, since we analyze $f$ and -$f$ separately, the $(2n - 2)^N$ different $N$-tuples in the successor slot may be summarized by far fewer patterns. The 5-tuple $gghhh$ has the same probability as $gg$-$h$-$h$-$h$ or $hhkkk$, because of the symmetries in the model. We identify these configurations as follows: The first component of the $N$-tuple is labeled $x$, the second — if it is not also labeled $x$ by virtue of being identical to the first — is labeled $y$. The label $z$ is reserved for the third different gene name in the $N$-tuple, if there is one, and so on. If $g$ and -$g$ occur in the same $N$-tuple, they require two distinct labels.

In the case of 37 genes (74 distinct gene names), instead of more than a billion 5-tuples there are only 52 distinct configurations. In effect, this is the fifth term in the Bell series:

$$a(N) = 1 + \sum_{i=1}^{N-1} a(i) \binom{N}{i} = 1, 2, 5, 15, 52, 203, \cdots,$$

which is the number of ways of distributing five indistinguishable objects into five labeled boxes.

## 6. The invariants

Using the algorithm of Fu [**34**], we find the following complete set of phylogenetic linear invariants for the $k \times k$ Jukes-Cantor semigroup on the unrooted binary tree ((AB)C(DE)), as in Figure 4.

The term "complete" is used in the sense that these eleven invariants below form a basis for the ideal of invariants. We use the configuration label, e.g. $xyzxw$, as a shorthand for the configuration probability normalized by the number of $N$-tuples it represents, or for simply the probability of any one of these $N$-tuples, e.g. Prob($hg$-$ghk$).

$$xyzyx - xyzyw - xyzzx + xyzzw$$

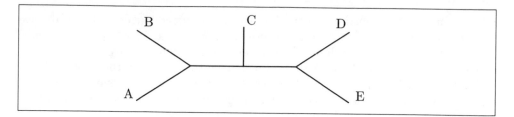

FIGURE 4. Unrooted binary tree ((AB)C(DE)). The other 14 trees are obtained by permuting the 5 labels

$$xyzyz - xyzyx - xyzwz + xyzwx$$

$$xyzxy - xyzxw - xyzzy + xyzzw$$

$$xyzxz - xyzxy - xyzwz + xyzwy$$

$$xyzzx - xyzzy - xyzwx + xyzwy$$

$$xxyxy - xxyyx + xxyyz - xxyxz - xxyzy + xxyzx$$

$$xyyxy - xyyyx + xyyyz - xyyzy - xyyxz + xyyzx$$

$$xyxxy - xyxyx + xyxyz - xyxxz - xyxzy + xyxzx$$

$$xyxzy - xyxyz + xyyxz - xyyzx + xyzyw - xyzxw - xyzwy + xyzwx$$

$$xyxxy - xyxxz - xyxyy + xyxzz - xyyyx + xyyyz$$

$$+xyyxx - xyyzz + xyzyy - xyzxx - xyzwy + xyzwx$$

$$xyxyy - xyxzz - xyyxx + xyyzz - xyzyy + xyzxx$$

$$+k(xyxzy - xyxzw - xyyzx + xyyzw - xyzwy + xyzwx)$$

In our context, $k = 2n - 2 = 72$. There are other invariants, but they are not *phylogenetic*, i.e. they are zero for all trees. For the unsigned inversion model in Section 4.1, $k = n - 1$.

**6.1. Remarks on the invariants.** In examining the eleven invariants, we observe that 14 of the 52 possible configurations enter into no invariant. Seven of these contain no information on the branching structure of the tree:

$$xxxxx \qquad xyyyy, xyxxx, xxyxx, xxxyx, xxxxy \qquad xyzwu$$

and so it is not surprising that they play no role here. The other seven are:

$$xxyyy, xxyzw, xyzzz \qquad xxxyy, xxxyz, xyzww \qquad xxyzz$$

These are precisely the configurations which characterize one (top three configurations), the other (second three configurations), or both (bottom configuration) of the two internal edges of the tree ((AB)C(DE)). They could be expected to be among the most frequent configurations (along with the seven non-informative configurations above and other configurations requiring no "extra steps" such as $xyyzw$ or $xyyyz$). Were all the data concentrated on these fourteen configurations, then all the eleven invariant functions would be exactly zero.

**6.2. Evaluating the invariants.** To estimate the configuration probabilities, we analyze the successor slot for each of the $2n$ gene names, treating $f$ and $-f$ separately, and calculating the relative frequency of each configuration, normalized by the number of different $N$-tuples which it contains. Though the configurations for different genes are not statistically independent, the expected value of a relative frequency is nonetheless the probability that generated it. By the linearity of the invariant functions, the expected value of each of the invariants evaluated using the relative frequencies is zero for ((AB)C(DE)) and non-zero for some other trees.

Note that with 37 genes as in the application of Section 7 below, or 74 data points, the 52 configurations will not all be estimated with any degree of accuracy. Neither will the invariant functions, especially since much of the data will be concentrated on the configurations that do not even appear in the invariant formulae. The situation would be much worse for $N = 6$ with 203 configurations, one of the reasons for not proceeding beyond $N = 5$ here.

# 7. An application to metazoan phylogeny

The mitochondrion is an "organelle" occurring in profusion in animal, plant, fungal and most other eukaryotic cells. It has its own genome with a small number ($< 100$) of genes, usually organized as a single circular chromosome. The mitochondrial genome of many metazoan animals has been completely sequenced and the genes they contain identified. The breakpoints in comparisons among the gene orders of these genomes have proven to contain much information pertinent to the inference of metazoan phylogeny [9]. The conservatism of certain genomes, such as human, *Drosophila* and *Katharina tunicata* (a chiton), versus the extreme divergence of related lineages, such as echinoderms or snails, i.e. the presence of both short and long branches, is the chief difficulty in the reconstruction of this phylogeny. In the next sections we apply our theory of breakpoint invariants to explore three problems in the phylogeny of higher metazoans, the true coelomates, based on the species in Table 3. These problems pertain to the protostome-deuterostome split, the internal structure of the protostomes, and the internal branching order of the deuterostomes.

We will evaluate the eleven invariant relations, substituting the observed $N$-tuple frequencies for their probabilities; with larger genomes these frequencies should satisfy the invariant relations more closely, but with just 37 genes in the mitochondrial genome, we can only hope that the invariants associated with the true tree **T** are better satisfied than are those which are not associated with it. We carry out extensive simulations to assess to what extent the trees we infer are likely to be the correct ones.

TABLE 3. Coelomate mitochondrial genomes compared in this investigation, with higher taxonomic levels. Citations: HU [2], SS [3], BA [16], DR [20], KT [10], LU [11].

| | ORGANISM | | PHYLUM |
|---|---|---|---|
| HU | Human | CHO | chordate (deuterostome) |
| SS | *Asterina pectinifera* (sea star) | ECH | echinoderm (deuterostome) |
| BA | *Balanoglossus carnosus* (acorn worm) | HEM | hemichordate (deuterostome) |
| DR | *Drosophila yakuba* (insect) | ART | arthropod (protostome) |
| KT | *Katharina tunicata* (chiton) | MOL | mollusc (protostome) |
| LU | *Lumbricus terrestris* (earthworm) | ANN | annelid (protostome) |

## 8. Metazoan phylogeny

Aspects of coelomate metazoan phylogeny are controversial, e.g. [1, 19]; among the groupings in Table 3, only the split between deuterostomes and protostomes seems undisputed. Eernisse *et al.* [24], Giribet and Ribera [36] and most others would group annelids and molluscs as sister groups, with arthropods related to these at a deeper level. But there are still proponents, e.g. [59], of a traditional grouping (*Articulata*) of annelids and arthropods as sister taxa. Hemichordates have been grouped with the chordates as in Brusca and Brusca [12] or in the "Tree of Life" [55], but evidence by Wada and Satoh [79] has led many to group them closer to the echinoderms, e.g. [60, 78].

Aside from these unsettled questions, efforts to infer phylogeny based on distances between mitochondrial gene orders have tended to group *Drosophila* closer to human than the echinoderms are, e.g. [68] and [9], Figures 4a and 4b, an artifact of the mitochondrial genome of the latter being highly divergent, the former two relatively conservative.

Figure 5 contrasts three phylogenies, one representing the "Tree of Life" [55], another the summary phylogeny by Valentine [78] on the University of California Museum of Paleontology website, and the third the *Drosophila*-human artifact.

## 9. Test procedures

Different invariants contain different numbers of configurations and, when evaluated with frequency data on the correct and incorrect trees, have different ranges, so that it may be misleading to compare trees on the basis of how close they are to zero with respect to all the invariants. To standardize the comparisons, we simulated 10,000 trees of form ((AB)C(DE)) on 37-gene genomes, with all branches disrupted by $R$ random inversions, and compiled the distribution of each the 11 invariants evaluated using the sample configuration frequencies. The value of $R$ is determined by counting the number of breakpoints on a minimum breakpoint tree [65] and dividing by $2\theta(2N-3)$, each inversion contributing up to two breakpoints, and there being $2N-3$ branches on an unrooted binary tree. The parameter $\theta$ corrects for "multiple hits" — we used $\theta = 0.75$. This only approximates the situation with the mitochondrial data (some lineages are clearly much longer than others), nonetheless the 11 test distributions constructed this way can serve as comparable scales to judge the fit of each of the 15 possible trees.

The score for each combination of tree and invariant can thus be transformed into a significance level. (Highly significant implies a poor fit.) A summary score

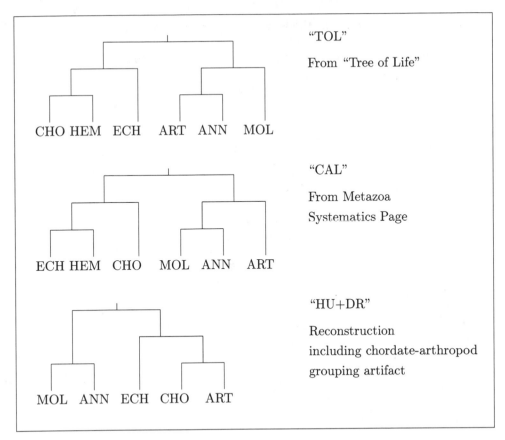

FIGURE 5. Three alternative views of coelomate evolution.

for each tree can then be produced by taking the product of the 11 significance levels.

A more clear-cut result of our method would see the tree **T** emerge with no invariant scoring less than $\Psi$ and all other trees scoring less than $\Psi$ (i.e. "significant") on at least one invariant, for some threshold $\Psi$. Simulations show, however, that for genomes with only 37 genes and the degree of divergence present on the coelomate data, this criterion will select the true tree at most 40% of the time, using the most favorable choice of $\Psi$ [**66**]. Thus we rely here on the summary score based on statistical significance.

## 10. Results

In this section, we will first present and comment on the trees selected by the use of invariants. We then compare these to the most parsimonious trees based on the same data, but simply minimizing the total number of changes (in terms of presence versus absence) over the tree for each possible adjacency of two genes. (See [**35**] for an approach based on three-gene adjacencies, also applied to mitochondrial gene orders of invertebrates.) These will both be compared to the minimum breakpoint trees in the sense of Section 3, i.e. minimizing the sum over all branches of the number of breakpoints between the genomes at each endpoint.

Note that the latter two methods, while both relying on a parsimony criterion, are quite distinct, and may have different results. It is not hard to show that if $A$ is the cost of the most parsimonious tree, using presence versus absence of all possible gene adjacencies as characters, then $A \leq 2B$, where $B$ is the minimum sum of the breakpoint costs over all branches of the same tree. Consider the three (unsigned) circular genomes in the top of Table 4. In calculating the breakpoint distance,

TABLE 4. Data sets contrasting adjacency parsimony and breakpoint distance.

| 1 | 2 | 5 | 6 | 4 | 3 |
|---|---|---|---|---|---|
| 1 | 3 | 2 | 5 | 4 | 6 |
| 1 | 2 | 3 | 4 | 5 | 6 |
| 1 | 2 | 3 | 4 | 5 | 6 |
| 1 | 2 | 3 | 4 | 5 | 6 |
| 1 | 3 | 5 | 2 | 6 | 4 |

an optimal median genome is found to be 1 2 3 4 5 6, with breakpoint distance 3 from each of the first two data genomes and zero from the third, for a total tree distance of 6. The 18 gene adjacencies in the data, however, consist of 9 pairs, each occurring twice and absent once, for a total cost of 9, so that $A < 2B$.

In contrast, for the three genomes in the bottom of Table 4, we again have $B = 6$ based on median genome 1 2 3 4 5 6, but $A = 12 = 2B$. The variability in the relation between $A$ and $B$ implies that the optimal breakpoint tree does not need coincide with the most parsimonious tree in terms of adjacencies, and indeed it generally does not.

**10.1. Deuterostomes and protostomes.** The first subset of the data to be examined includes HU, SS, DR, KT and LU, in order to compare the results with those of [68] and [9]. In this case $R = 10$.

The best three trees manifested scores of $2 \times 10^{-12}, 6 \times 10^{-15}, 7 \times 10^{-17}$. The first of these was consistent with the CAL tree in Figure 5, and the third was the artifactual tree in that figure. The second also contained the HU+DR artifact.

Nevertheless, according to the best tree, our method succeeded in correctly grouping CHO and ECH, despite the discordance of branch lengths which defeat distance-matrix-based attempts. And it also confirmed the ANN+MOL grouping in CAL versus the TOL grouping of ANN+ART.

**10.2. The Balanoglossus data.** The recently sequenced mitochondrial genome of *Balanoglossus carnosa* allows a more detailed investigation of deuterostome-protostome branching. Here we focus on the deuterostome-arthropod relationship, retaining *Katharina* as a second protostome, but dropping *Lumbricus* from the analysis. The simulations for constructing the statistical tests were redone with $R = 6$. The results in this analysis clearly confirm the deuterostome grouping. The three best trees, with summary scores $10^{-7}, 10^{-7}, 6 \times 10^{-8}$, all group the deuterostomes together and no other tree scores better than $3 \times 10^{-15}$ (which is the score when DR groups more closely with HU and BA than SS does). In this analysis the best tree is consistent with the TOL tree in Figure 5, while the CAL tree is third best.

**10.3. A comparison of methods.** Table 5 shows how the candidate phylogenies fare under the method of invariants compared to two parsimony approaches. Both the methods of invariants and adjacency parsimony operate on the configuration frequencies in the gene-successor data described in Section 6.2, in the former case as detailed in that section, and in the latter by counting total "extra steps" required by all data configurations on each tree. The third method minimizes the sum of the breakpoint distances over all branches of the tree, involving the optimization of ancestral genomes [9]. It can be seen that the two methods operating on gene-

TABLE 5. Comparison of three methods on three data sets (top: without BA, middle: without LU, bottom: all six species). Figures indicate rank of trees built using the same method on the same data sets. Asterisks indicate ranks which are tied with at least one other. Parentheses indicate 6-species supertree based on best tree in 5-species analysis without BA, combined with three best trees in 5-species analysis without LU. Note that for 5-species analyses, ranks are out of 15, while for 6 species, they are out of 105.

|     | Tree | invariants | adjacency parsimony | breakpoint distance |
|-----|------|-----------|---------------------|---------------------|
| TOL | (HU SS)((DR LU) KT) | 5* | 5 | 3* |
| CAL | (HU SS)(DR (LU KT)) | 1 | 1* | 1 |
| HU | ((HU DR) SS)(LU KT) | 3 | 1* | 5* |
| + | ((HU DR) LU)(SS KT) | 4 | 6* | 3* |
| DR | ((HU DR) KT)(LU SS) | 2 | 3 | 2 |
| TOL | ((HU BA) SS)(DR KT) | 1 | 1 | 1* |
| CAL | (HU (BA SS))(DR KT) | 3 | 3 | 1* |
|     | ((HU SS) BA)(DR KT) | 2 | 2 | 1* |
| HU+DR | ((HU BA) DR)(SS KT) | 4 | 4 | 4* |
| CAL | (HU (BA SS))(DR (LU KT)) | (3) | 5 | 1* |
|     | ((HU BA) SS)(DR (LU KT)) | (1) | 1 | 3 |
|     | ((HU SS) BA)(DR (LU KT)) | (2) | 4 | 1* |
|     | ((HU BA) SS)(LU (DR KT)) |  | 2* | 13* |
| HU+DR | ((HU BA) DR)(SS (LU KT)) |  | 2* | 13* |

successor configuration frequencies tend to agree as to the best tree, although the method of invariants seems slightly less susceptible to the HU+DR artifact. The breakpoint distance method does not resolve among the best trees for two of the data sets, but whether this is a virtue or a shortcoming remains to be seen.

All of the analyses support the protostome-deuterostome split, and they all support the annelid-mollusc grouping as a sister group to the arthropods. On the other hand, they do not agree on the internal grouping of the deuterostomes. The breakpoint distance gives equal support to a ECH+CHO grouping (which is of little credibility) as to the ECH+HEM analysis, whereas the other methods favour a more traditional CHO+HEM grouping.

## Further work

Though much probabilistic modeling of gene sequence changes has been incorporated into phylogenetic analysis, very little research has gone into mathematical approaches to phylogenetics based on gene order, and even less, previous to the present undertaking, into probability models for the evolution of gene order. (See, however,[21]).

Of both mathematical and biological interest is whether this theory can be developed in the direction of other semigroups. Linear invariant theory is well-developed, for the Kimura models (e.g. [72]) and others, and biological interpretation in the breakpoint context is possible. Even though an *exact* representation of models such as random inversions only on signed data (Section 4) in terms of semigroups of matrices is not possible, significantly better approximations than Jukes-Cantor may well be feasible.

In comparing the invariant method to the two parsimony methods, we cannot do more based on these small applications than note that in one example, in Section 10.1, adjacency parsimony did not discriminate against a branch-length artifact, while the invariants, based on the same data, did.

Perhaps the most promising direction for the method of invariants lies towards larger genome size — plastids, prokaryotes and, when more eukaryotes are completely sequenced, nuclear genomes. Multichromosomal genomes are handled as easily as single-chromosome ones, since the model pertains to single breakpoints and not to whole fragments, which behave differently in inversions, transpositions and reciprocal translocations. Increasing $n$ only linearly increases the time to compute configuration frequencies, which is negligible. Our simulations [66] indicate that the method should be able to identify the true tree with a high degree of accuracy for large genomes. Note that heterogeneity of rates is not a problem with this approach, either from lineage to lineage, nor from gene to gene in their quantitative susceptibility to be adjacent to breakpoints; this stems from the linearity of the invariants. Thus the fact that tRNA genes may be more mobile [9], either because they tend to be at the end of rearranged fragments or because they may be individually transposed in the genome, does not affect the results.

Enlarging the method to handle six species and perhaps more is quite feasible, though the book-keeping involved with hundreds of invariants is considerable. Beyond this, some way of handling decomposition of the problem, such as we used in Sections 10.1 and 10.2, might be systematized.

The biological results obtained here include the relatively early branching of arthropods within the protostomes, and the grouping of the hemichordates with the chordates, though the latter is equivocal. Our method clearly distinguishes between deuterostomes and protostomes, which is not always the case with other approaches using rearrangement data.

## Acknowledgments

Research supported by grants from NSERC and the Canadian Genome Analysis and Technology program to DS and an NSERC graduate scholarship to MB. DS is a Fellow of the Canadian Institute for Advanced Research. We are grateful to the referees for their comments, and for advice and discussion to Jeffrey Boore, Kenneth Halanych, Franz Lang, Mitchell Sogin, Veronique Barriel, Gonzalo Giribet, Martin Christofferson, Mary Mickevich and Takashi Kunisawa, though we take full responsibility for all shortcomings remaining in this article. Much of this material also appears in [**67**].

# References

[1] Aguinaldo, A.M.A., Turbeville, J.M., Linford, L.S., Rivera, M.C., Garey, J.R., Raff, R.A. and Lake, J.A., *Evidence for a clade of nematodes, arthropods and other moulting animals*, Nature **387** (1997), 489-493.

[2] Anderson, S., Bankier, A.T., Barrell, B.G., de Bruijn, M.H.L., Coulson, A.R., Drouin, J., Eperon, I.C., Nierlich, D.P., Roe, B.A., Sanger, F., Schreier, P.H., Smith, A.J.H., Staden, R. and Young, I.G., *Sequence and organization of the human mitochondrial genome*, Nature **290** (1981), 457-465.

[3] Asakawa, S., Himeno, H., Miura, K. and Watanabe, K., *Nucleotide sequence and gene organization of the starfish Asterna pectinifera mitochondrial genome* (1993), unpublished.

[4] Bafna, V. and Pevzner, P.A., *Genome rearrangements and sorting by reversals*, SIAM Journal of Computing **25** (1996), 272-289.

[5] Bafna, V. and Pevzner, P.A., *Sorting by transpositions*, In: Combinatorial Pattern Matching. 6th Annual Symposium, Z.Galil and E. Ukkonen, eds., Lecture Notes in Computer Science 937, Springer Verlag, New York (1995), 614-623.

[6] Berman, P. and Hannenhalli, S., *Fast sorting by reversal*, In: Combinatorial Pattern Matching. 7th Annual Symposium, D. Hirschberg and G. Myers, eds., Lecture Notes in Computer Science 1075, Springer Verlag, New York (1996), 168-185.

[7] Blanchette, M., Bourque, G. and Sankoff, D. *Breakpoint phylogenies*. In: Genome Informatics 1997, S. Miyano and T. Takagi, eds., Universal Academy Press, Tokyo (1997), 25-34.

[8] Blanchette, M., Kunisawa, T. and Sankoff, D., *Parametric genome rearrangement*, Gene **172** (1996), GC 11-17.

[9] Blanchette, M., Kunisawa, T., Sankoff, D., *Gene order breakpoint evidence in animal mitochondrial phylogeny*, Journal of Molecular Evolution **49** (1999), 193-203.

[10] Boore, J.L. and Brown, W.M., *Complete DNA sequence of the mitochondrial genome of the black chiton, Katharina tunicata*, Genetics **138** (1994), 423-443.

[11] Boore, J.L. and Brown, W.M., *Complete sequence of the mitochondrial DNA of the annelid worm Lumbricus terrestris*, Genetics **141** (1995), 305-319.

[12] Brusca, R.C. and Brusca, G.J., *Invertebrates*, Sinauer, Sunderland, MA, (1990).

[13] Bryant, D., *Complexity of the breakpoint median problem*, manuscript, Centre de recherches mathématiques (1998).

[14] Caprara, A., *Sorting by reversals is difficult*, In: Proceeedings of the First Annual International Conference on Computational Molecular Biology (RECOMB 97), ACM, New York (1997), 75-83.

[15] Caprara, A., *Formulations and hardness of multiple sorting by reversals*, In: Proceeedings of the Third Annual International Conference on Computational Molecular Biology (RECOMB 99), S. Istrail, P. Pevzner and M. Waterman, eds., ACM, New York (1999), 84-93.

[16] Castresana, J., Feldmaier-Fuchs, G. and Paabo, S., *Codon reassignment and amino acid composition in hemichordate mitochondria*, Proceedings of the National Academy of Sciences (U.S.A.) **95** (1998), 3703-3707.

[17] Cavender, J.A., *Mechanized derivation of linear invariants*, Molecular Biology and Evolution **6** (1989), 301-316.

[18] Cavender, J.A. and Felsenstein, J., *Invariants of phylogenies: Simple case with discrete states*, Journal of Classification **4** (1987), 57-71.

[19] Christofferson, M.L. and Araújo-de-Almeida, E.A., *A phylogenetic framework of the enterocoela (Metameria: Coelomata)*, Revista Norestina de Biologia (Brazil) **9** (1994), 172-208.

[20] Clary, D.O. and Wolstenholme, D.R., *The mitochondrial DNA molecular of Drosophila yakuba: nucleotide sequence, gene organization, and genetic code*, Journal of Molecular Evolution **22** (1985), 252-271.

[21] Dalkie, K.-S., *Analysis of breakpoints for genome rearrangement*, Honours essay, Department of Mathematics, University of Canterbury, New Zealand (1998).

[22] DasGupta, B., Jiang, T., Kannan, S., Li, M. and Sweedyk, Z., *On the complexity and approximation of syntenic distance*, In: Proceedings of the First Annual International Conference on Computational Molecular Biology (RECOMB 97), ACM, New York, (1997), 99-108.

[23] Drolet, S. and Sankoff, D., *Quadratic invariants for multivalued characters*, Journal of Theoretical Biology **144** (1990), 117-129.

[24] Eernisse, D.J., Albert, J.S. and Anderson, F.E., *Annelida and Arthropoda are not sister taxa. A phylogenetic analysis of spiralian metazoan morphology*, Systematic Biology **41** (1992), 305-330.

[25] Evans, S.N. and Speed, T.P., *Invariants of some probability models used in phylogenetic inference*, Annals of Statistics **21** (1993), 355-377.

[26] Evans, S.N. and Zhou, X., *Constructing and counting phylogenetic invariants*, Journal of Computational Biology **5** (1998), 713-724.

[27] Felsenstein, J., *Cases in which parsimony or compatibility methods will be positively misleading*, Systematic Zoology **27** (1978), 401-410.

[28] Felsenstein, J., *Counting phylogenetic invariants in some simple cases*, Journal of Theoretical Biology **152** (1991), 357-376.

[29] Ferretti, V., Lang, B.F. and Sankoff, D., *Skewed base compositions, asymmetric transition matrices and phylogenetic invariants*, Journal of Computational Biology **1** (1994), 77-92.

[30] Ferretti, V., Nadeau, J.H. and Sankoff, D., *Original synteny*, In: Combinatorial Pattern Matching. 7th Annual Symposium, D. Hirschberg and G. Myers, eds., Lecture Notes in Computer Science 1075, Springer Verlag, New York (1996), 159-167.

[31] Ferretti, V. and Sankoff, D., *The empirical discovery of phylogenetic invariants*, Advances in Applied Probability **25** (1993), 290-302.

[32] Ferretti V. and Sankoff D., *Phylogenetic invariants for more general evolutionary models*, Journal of Theoretical Biology **173** (1995), 147-162.

[33] Ferretti, V. and Sankoff, D., *A remarkable nonlinear invariant for evolution with heterogeneous rates*, Mathematical Biosciences **134** (1996), 71-83.

[34] Fu Y. X., *Linear invariants under Jukes' and Cantor's one-parameter model*, Journal of Theoretical Biology **173** (1995), 339-352.

[35] Gallut, C., *Codage de l'ordre des génes du génome mitochondrial animal en vue d'une analyse phylogénétique*, Mémoire de DEA, Université Paris VI Pierre et Marie Curie (1998).

[36] Giribet, G. and Ribera, C., *The position of Arthropods in the animal kingdom: A search for as reliable outgroup for internal arthropod phylogeny*, Molecular Phylogenetics and Evolution **9** (1998), 481-488.

[37] Hagedorn, T.R., *On the number and structure of phylogenetic invariants*, Advances in Applied Mathematics (in press).

[38] Hagedorn, T.R. and Landweber, L.F., *Phylogenetic invariants and geometry*, manuscript, College of New Jersey and Princeton University (1999).

[39] Hannenhalli, S., *Polynomial algorithm for computing translocation distance between genomes.* In: Combinatorial Pattern Matching. 6th Annual Symposium, Z.Galil and E. Ukkonen, eds., Lecture Notes in Computer Science 937, Springer Verlag, New York (1995), 162-176.

[40] Hannenhalli, S., Chappey, C., Koonin, E.V. and Pevzner, P.A., *Genome sequence comparison and scenarios for gene rearrangements: a test case*, Genomics **30** (1995), 299-311.

[41] Hannenhalli, S. and Pevzner, P.A., *Transforming cabbage into turnip. (polynomial algorithm for sorting signed permutations by reversals)*, In: Proceedings of the 27th Annual ACM-SIAM Symposium on the Theory of Computing (1995), 178-189.

[42] Hannenhalli, S. and Pevzner, P.A., *Transforming men into mice (polynomial algorithm for genomic distance problem)*, In: Proceedings of the IEEE 36th Annual Symposium on Foundations of Computer Science (1995), 581-592

[43] Hasegawa, M., Kishino, H. and Yano, T., *Dating of the human-ape splitting by a molecular clock of mitochondrial DNA*, Journal of Molecular Evolution **22** (1985), 160-174.

[44] Hendy, M.D. and Penny, D., *Complete families of linear invariants for some stochastic models of sequence evolution, with and without the molecular clock assumption*, Journal of Computational Biology **3** (1996), 19-31.

[45] Jin, L. and Nei, M., *Limitations of the evolutionary parsimony method of phylogenetic analysis*, Molecular Biology and Evolution **7** (1990), 82-102.

[46] Jukes, T.H. and Cantor C.R., *Evolution of protein molecules*, In: Mammalian Protein Metabolism, H.N. Munro, ed., Academic Press, New York (1969), 21-132.

[47] Kaplan, H., Shamir, R. and Tarjan, R.E., *Faster and simpler algorithm for sorting signed permutations by reversals*, In: Proceedings of the 8th Annual ACM-SIAM Symposium on Discrete Algorithms, ACM, New York (1997), 344-351.

[48] Kececioglu, J. and Ravi, R., *Of mice and men. Evolutionary distances between genomes under translocation*, In: Proceedings of the 6th Annual ACM-SIAM Symposium on Discrete Algorithms, (1995), 604-613.

[49] Kececioglu, J. and Sankoff, D., *Efficient bounds for oriented chromosome inversion distance*, In: Combinatorial Pattern Matching. 5th Annual Symposium, M. Crochemore and D. Gusfield, eds., Lecture Notes in Computer Science, 807, Springer Verlag, New York (1994), 307-325.

[50] Kececioglu, J. and Sankoff, D., *Exact and approximation algorithms for sorting by reversals, with application to genome rearrangement*, Algorithmica **13** (1995),180-210.

[51] Kimura, M., *A simple method for estimating evolutionary rate of base substitutions through comparative studies of nucleotide sequences*, Journal of Molecular Evolution **16** (1980), 111-120.

[52] Kimura, M., *Estimation of evolutionary sequences between homologous nucleotide sequences*, Proceedings of the National Academy of Sciences (U.S.A.) **78** (1981), 454-458.

[53] Lake, J.A., *A rate-independent technique for analysis of nucleic acid sequences: Evolutionary parsimony*, Molecular Biology and Evolution **4** (1987), 167-191.

[54] Liben-Nowell, D., *On the structure of syntenic distance*, In: Combinatorial Pattern Matching. 10th Annual Symposium, M. Crochemore and M. Paterson, eds., Lecture Notes in Computer Science 1645, Springer Verlag, New York (1999), 50-65.

[55] Maddison, D. and Maddison, W., *Tree of Life metazoa page* (1995), http://phylogeny.arizona.edu/tree/eukaryotes/animals/animals.html

[56] Nadeau, J.H. and Sankoff, D., *Counting on comparative maps*, Trends in Genetics **14** (1998), 495-501.

[57] Nguyen, T and Speed, T.P., *A derivation of all linear invariants for a nonbalanced transversion model*, Journal of Molecular Evolution **35** (1992), 60-76.

[58] Pe'er, I. and Shamir, R., *The median problems for breakpoints are NP-complete*, Electronic Colloquium on Computational Complexity Technical Report 98-071 (1998), http://www.eccc.uni-trier.de/eccc

[59] Rouse, G.W. and Fauchald, K., *The articulation of annelids*, Zoologica Scripta **24** (1995), 269-301

[60] Ruppert, E.E. and Barnes, B.D., *Invertebrate Zoology*, Saunders, Philadelphia (1994).

[61] Sankoff, D., *Designer invariants for large phylogenies*, Molecular Biology and Evolution **7** (1990), 255-269.

[62] Sankoff, D., *Edit distance for genome comparison based on non-local operations*, In: Combinatorial Pattern Matching. 3rd Annual Symposium, A. Apostolico, M. Crochemore, Z. Galil and U. Manber, eds., Lecture Notes in Computer Science 644, Springer Verlag, New York (1992), 121-135.

[63] Sankoff, D., *Comparative mapping and genome rearrangement*, In: From Jay Lush to Genomics: Visions for Animal Breeding and Genetics, J.C.M. Dekkers, S.J. Lamont and M.F. Rothschild, eds., Ames, Iowa, Iowa State University (1999), 124-134, http://agbio.cabweb.org/conference/index.html

[64] Sankoff, D. and Blanchette, M., *The median problem for breakpoints in comparative genomics*, In: Computing and Combinatorics, Proceedings of COCOON '97, T. Jiang and D.T. Lee, eds., Lecture Notes in Computer Science 1276, Springer Verlag, New York (1997), 251-263.

[65] Sankoff, D. and Blanchette, M., *Multiple genome rearrangement and breakpoint phylogeny*, Journal of Computational Biology **5** (1998), 555-570.

[66] Sankoff, D. and Blanchette, M., *Phylogenetic invariants for metazoan mitochondrial genome evolution*, In: Genome Informatics 1998, S. Miyano and T. Takagi, eds., Universal Academy Press, Tokyo (1998) 22-31.

[67] Sankoff, D. and Blanchette, M., *Probability models for genome rearrangement and linear invariants for phylogenetic inference*, In: Proceedings of the Third Annual International Conference on Computational Molecular Biology (RECOMB 99), S.Istrail, P.Pevzner and M.Waterman, eds., ACM, New York (1999), 302-309.

[68] Sankoff, D., Leduc, G., Antoine, N., Paquin, B., Lang, B.F and Cedergren, R.J., *Gene order comparisons for phylogenetic inference: Evolution of the mitochondrial genome*, Proceedings of the National Academy of Sciences (U.S.A.) **89** (1992), 6575-6579.

[69] Sankoff, D., Sundaram, G. and Kececioglu, J., *Steiner points in the space of genome rearrangements*, International Journal of the Foundations of Computer Science **7** (1996), 1-9.

[70] Steel, M.A., *Recovering a tree from the leaf colorations it generates under a Markov model*, Applied Mathematics Letters **7** (1994), 19-23.

[71] Steel, M. A. and Fu, Y.X., *Classifying and counting linear phylogenetic invariants for the Jukes-Cantor model*, Journal of Computational Biology **2** (1995), 39-47.

[72] Steel, M. A., Szekeley, L.A., Erdos, P.L. and Waddell, P., *A complete family of phylogenetic invariants for any number of taxa under Kimura's 3ST model*, New Zealand Journal of Botany **31** (1993), 289-296.

[73] Sturtevant, A.H. and Novitski, E., *The homologies of chromosome elements in the genus Drosophila*, Genetics **26** (1941), 517-541.

[74] Szekeley, L.A., Steel, M. A. and Erdos, P.L., *Fourier calculus on evolutionary trees*, Advances in Applied Mathematics **14** (1993), 200-216.

[75] Tajima, F. and Nei, M., *Estimation of evolutionary distance between nucleotide sequences*, Molecular Biology and Evolution **1** (1984), 269-285.

[76] Tamura, K., *Estimation of the number of nucleotide substitutions when there are strong transition-transversion and $G + C$-content biases*, Molecular Biology and Evolution **9** (1992), 678-687.

[77] Tamura, K. and Nei, M., *Estimation of the number of nucleotide substitutions in the control region of mitochondrial DNA in humans and chimpanzees*, Molecular Biology and Evolution **10** (1993), 512-526.

[78] Valentine, J.W. *University of California Museum of Paleontology Metazoa Systematics Page* (no date), http://www.ucmp.berkeley.edu/phyla/metazoasy.html.

[79] Wada, H., and Satoh, N., *Details of the evolutionary history from invertebrates to vertebrates, as deduced from the sequences of 18S rDNA*, Proceedings of the National Academy of Sciences (U.S.A.) **91** (1994), 1801-1804.

[80] Watterson, G.A., Ewens, W.J., Hall, T.E. and Morgan A., *The chromosome inversion problem*, Journal of Theoretical Biology **99** (1982), 1-7.

CENTRE DE RECHERCHES MATHÉMATIQUES, UNIVERSITÉ DE MONTRÉAL, CP 6128 SUCCURSALE CENTRE-VILLE, MONTRÉAL, QUÉBEC H3C 3J7
*E-mail address*: `sankoff@ere.umontreal.ca`

DEPARTMENT OF COMPUTER SCIENCE & ENGINEERING, UNIVERSITY OF WASHINGTON, SEATTLE, WASHINGTON 98195-2350
*E-mail address*: `blanchem@cs.washington.edu`

Canadian Mathematical Society
Conference Proceedings
Volume **26**, 2000

# Some Exceptional Configurations

## Byron Schmuland

ABSTRACT. The Dirichlet form given by the intrinsic gradient on Poisson space
is associated with a Markov process consisting of a countable family of interacting diffusions. By considering each diffusion as a particle with unit mass,
the randomly evolving configuration can be thought of as a Radon measure
valued diffusion.

The quasi-sure analysis of Dirichlet forms is used to find exceptional sets
for this Markov process. We show that the process never hits certain unusual
configurations, such as those with more than unit mass at some position, or
those that violate the law of large numbers. Some of these results also hold
for Gibbs measures with superstable interactions.

## 1. Introduction

In recent work [**1, 2, 3, 4, 10, 11, 12, 14, 17**] the theory of Dirichlet forms has
been used to construct and study Markov processes that take values in the space
$\Gamma_X$ of locally finite configurations on a Riemannian manifold $X$. The configuration
space is defined by

$$\Gamma_X := \{\gamma \subset X : |\gamma \cap K| < \infty \text{ for every compact } K\}.$$

A configuration, then, is simply a collection $\gamma$ of points in $X$ with the property that
only finitely many points inhabit any compact set.

A probability measure on $\Gamma_X$ models a randomly chosen configuration, that is,
a point process on $X$. The most well-known point process, the Poisson process,
corresponds to the Poisson measure $\pi_\sigma$ on $\Gamma_X$ (see section 2.1 below). A diffusion
process $((X_t)_{t\geq 0}, (P_\gamma)_{\gamma\in\Gamma_X})$ with values in $\Gamma_X$ describes the evolution in time of a
system of point processes on $X$, in other words, a countable family of $X$-diffusions.

The purpose of this paper is to examine the sample path properties of such a
$\Gamma_X$-valued diffusion, and in particular, to find exceptional sets for this process, that
is, subsets $N \subset \Gamma_X$ such that

$$\int_{\Gamma_X} P_\gamma\left(X_t \notin N \text{ for all } t \geq 0\right) \mu(d\gamma) = 1,$$

for a suitable probability measure $\mu$ on $\Gamma_X$.

For simplicity, in this paper we will always take the manifold $X$ to be the
Euclidean space $\mathbb{R}^d$. The space $\Gamma_{\mathbb{R}^d}$ will be given the topology of vague convergence

2000 *Mathematics Subject Classification.* Primary 60H07, 31C25, 60G57, 60G60.

of measures, and measures on $\Gamma_{\mathbb{R}^d}$ are defined on the corresponding Borel sets $\mathcal{B}(\Gamma_{\mathbb{R}^d})$. We reserve the notation $m$ for Lebesgue measure on $\mathbb{R}^d$.

Every configuration $\gamma$ can be identified with the Radon measure $\sum_{x \in \gamma} \varepsilon_x$, and we will make this identification without comment. For $f \in C_0(\mathbb{R}^d)$ we let $\langle f, \gamma \rangle$ be the integral of $f$ with respect to the measure $\gamma$, that is, $\langle f, \gamma \rangle = \sum_{x \in \gamma} f(x)$. Also the symbol $\emptyset$ will refer to the empty set or the zero measure as needed.

## 2. Dirichlet form

**2.1. Poisson measures and Gibbs measures.** In this section we describe the two families of probability measures that will serve as the invariant measure for our configuration-valued diffusion $(X_t)_{t \geq 0}$. In the free case, a Poisson measure is used to model random particles that act independently; while in the Gibbs case, a Gibbs measure is used to model random particles that interact via a potential function. Although we are mainly interested in the mathematically more challenging Gibbs case, analysis of the free case often serves as a useful guideline.

DEFINITION 2.1. For any non-atomic Radon measure $\sigma$ on $\mathbb{R}^d$, the *Poisson measure* $\pi_\sigma$ with intensity $\sigma$ is the probability measure on $\Gamma_{\mathbb{R}^d}$ characterized by the formula

$$(2.1) \qquad \int_{\Gamma_{\mathbb{R}^d}} \exp(\langle f, \gamma \rangle) \, \pi_\sigma(d\gamma) = \exp\left( \int_{\mathbb{R}^d} (e^{f(x)} - 1) \, \sigma(dx) \right),$$

for $f \in C_0(\mathbb{R}^d)$.

If $A$ and $B$ are disjoint Borel subsets of $\mathbb{R}^d$, then under the measure $\pi_\sigma$, $\gamma(A) := \langle 1_A, \gamma \rangle$ and $\gamma(B) := \langle 1_B, \gamma \rangle$ are independent Poisson random variables with means $\sigma(A)$ and $\sigma(B)$. Two other useful formulas that can be derived from equation (2.1) are

$$(2.2) \qquad \int_{\Gamma_{\mathbb{R}^d}} \langle f, \gamma \rangle \, \pi_\sigma(d\gamma) = \int_{\mathbb{R}^d} f(x) \, \sigma(dx),$$

$$(2.3) \quad \int_{\Gamma_{\mathbb{R}^d}} (\langle f, \gamma \rangle \langle g, \gamma \rangle - \langle fg, \gamma \rangle) \, \pi_\sigma(d\gamma) = \int_{\mathbb{R}^d} \int_{\mathbb{R}^d} f(x) g(y) \, \sigma(dx) \sigma(dy).$$

Equation (2.2) shows us that the intensity measure $\sigma$ is the mean of the distribution $\pi_\sigma$.

Following [16], we now describe measures that correspond to interacting systems. These measures are examples of grand canonical Gibbs measures that appear in the classical statistical mechanics of continuous systems. More details on this class of measures can be found in [4, 15, 16].

A *pair potential* is any measurable function $\phi : \mathbb{R}^d \to \mathbb{R} \cup \{+\infty\}$ such that $\phi(-x) = \phi(x)$. For a pair potential $\phi$, a bounded measurable subset $\Lambda$ in $\mathbb{R}^d$, and a configuration $\gamma \in \Gamma_{\mathbb{R}^d}$, the *conditional energy* of $\gamma$ in $\Lambda$ is given by the formula

$$E_\Lambda^\phi(\gamma) := \begin{cases} 0 & \text{if } \gamma(\Lambda) = 0, \\ \sum \phi(x - y) & \text{if } \sum |\phi(x - y)| < \infty, \\ +\infty & \text{otherwise}, \end{cases}$$

where the summation is taken over all pairs $\{x, y\} \subset \gamma$ such that $\{x, y\} \cap \Lambda \neq \emptyset$. The term $\phi(x - y)$ is meant to represent the repulsive energy between a pair of particles located at $x$ and $y$. Of course, when $\phi(x - y) < 0$ the particles are attracted rather than repulsed. We adopt the convention that a sum over the empty set is zero so that $E_\Lambda^\phi(\gamma) = 0$ if either $\gamma(\mathbb{R}^d) = 1$ or $\gamma(\Lambda) = 0$. We also define

$$Z_\Lambda^{z,\phi}(\gamma) := \int_{\Gamma_{\mathbb{R}^d}} \exp\left[-E_\Lambda^\phi(\gamma_{\Lambda^c} + \omega_\Lambda)\right]\,\pi_{z\sigma}(d\omega).$$

Here $\gamma_{\Lambda^c} + \omega_\Lambda$ is the configuration formed by combining the part of $\gamma$ outside $\Lambda$ with the part of $\omega$ inside $\Lambda$. The parameter $z > 0$ is called the *activity* and $\pi_{z\sigma}$ is the Poisson measure as in Definition 1. Notice that $Z_\Lambda^{z,\phi}(\gamma)$ is always strictly positive since it is greater than or equal to $\pi_{z\sigma}(\omega_\Lambda = 0) = \exp(-z\sigma(\Lambda)) > 0$.

DEFINITION 2.2. A probability measure $\mu$ on $\Gamma_{\mathbb{R}^d}$ is called a *Gibbs measure* with activity $z$, pair potential $\phi$, and intensity measure $\sigma$ if, for every bounded measurable $\Lambda \subset \mathbb{R}^d$ we have $Z_\Lambda^{z,\phi}(\gamma) < \infty$ for $\mu$-almost every $\gamma \in \Gamma_{\mathbb{R}^d}$ and for every $\Delta \in \mathcal{B}(\Gamma_{\mathbb{R}^d})$,

$$(2.4) \qquad \mu(\Delta) = \iint_{\Gamma_{\mathbb{R}^d}\Gamma_{\mathbb{R}^d}} 1_\Delta(\gamma_{\Lambda^c} + \omega_\Lambda) \frac{\exp\left[-E_\Lambda^\phi(\gamma_{\Lambda^c} + \omega_\Lambda)\right]}{Z_\Lambda^{z,\phi}(\gamma)}\,\pi_{z\sigma}(d\omega)\,\mu(d\gamma).$$

Gibbs measures do not always exist. Apart from the obvious case when $\phi \equiv 0$, which yields the Poisson measure $\pi_{z\sigma}$, there is no guarantee that we can always find a measure satisfying the conditions of Definition 2. Looking at (2.4), we see that the exponential term encourages configurations of low energy and discourages those of high energy. The pair potential of a Gibbs measure must not force the particles to diverge to infinity or allow them to converge in a bounded region of space. For instance, when $z = 1$ and $\sigma$ is Lebesgue measure, here are two pair potentials whose corresponding Gibbs measure fails to exist, for opposite reasons;

$$\phi(x) \equiv \infty, \qquad \phi(x) = \begin{cases} -1 & \text{if } |x| \leq 1, \\ 0 & \text{otherwise.} \end{cases}$$

In [**15, 16**], Ruelle studies pair potentials that avoid these problems. He begins by defining a partition of $\mathbb{R}^d$ into cubes. For every $r = (r_1, \ldots, r_d) \in \mathbb{Z}^d$ let

$$(2.5) \qquad Q_r = \left\{x \in \mathbb{R}^d : \left(r_i - \frac{1}{2}\right) \leq x_i < \left(r_i - \frac{1}{2}\right)\right\}.$$

DEFINITION 2.3.

(SS) A pair potential $\phi$ is called *superstable* if there exist $A > 0$ and $B \geq 0$ so that if $\Lambda = \cup_{r \in R} Q_r$ is a finite union of cubes, then

$$E_\Lambda^\phi(\gamma_\Lambda) \geq \sum_{r \in R}\left[A\gamma(Q_r)^2 - B\gamma(Q_r)\right].$$

(LR) A pair potential $\phi$ is called *lower regular* if there exists a decreasing positive function $\Psi : \mathbb{N} \to [0, \infty)$ such that $\sum_{r \in \mathbb{Z}^d} \Psi(|r|_\infty) < \infty$, and for any disjoint $\Lambda'$ and $\Lambda''$ that are finite unions of cubes as in (2.5), then we have

$$\iint_{\Lambda'\Lambda''} \phi(x - y)\,\gamma(dx)\gamma(dy) \geq -\sum_{r',r'' \in \mathbb{Z}^d} \Psi(|r' - r''|_\infty)\gamma_{\Lambda'}(Q_{r'})\gamma_{\Lambda''}(Q_{r''}),$$

for all $\gamma \in \Gamma_{\mathbb{R}^d}$. Here $|\cdot|_\infty$ refers to the maximum norm on $\mathbb{R}^d$.

(I) A pair potential $\phi$ is called *integrable* if $\int_{\mathbb{R}^d} |\exp(-\phi(x)) - 1|\,dx < \infty$.

Ruelle argues that a physically realistic model for the atoms of a rare gas, for instance, ought to use a spherically symmetric pair potential that is very repulsive at short distances and that dies out at long distances. Using Lebesgue measure as the intensity $\sigma$, he shows [**16**, Theorem 5.5] that if $\phi$ satisfies (SS), (LR), and (I), then there exists a Gibbs measure for $\phi$ and any $z > 0$. In fact he shows that there is always a Gibbs measure $\mu$ that is tempered in the following sense.

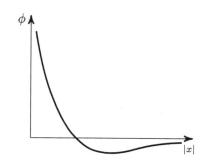

*A typical pair potential*

DEFINITION 2.4. A measure $\mu$ on $\Gamma_{\mathbb{R}^d}$ is called *tempered* if

$$\limsup_{l \to \infty} \frac{\sum_{|r| \le l} \gamma(Q_r)^2}{(2l+1)^d} < \infty \text{ for } \mu\text{-almost every } \gamma \in \Gamma_{\mathbb{R}^d}.$$

DEFINITION 2.5. A probability measure $\mu$ on $\Gamma_{\mathbb{R}^d}$ is called a *Ruelle measure* if $\mu$ is a tempered Gibbs measure with activity parameter $z > 0$, intensity $\sigma$ equal to Lebesgue measure, and a pair potential $\phi$ that is superstable, lower regular, and integrable.

Here are a couple of examples of pair potentials that appear often in the literature and satisfy the conditions of the above definition.

1. The *Lennard-Jones* potential given, for some $a, b > 0$, by

$$\phi_{a,b}(x) := \frac{a}{|x|^{12}} - \frac{b}{|x|^6}, \quad x \in \mathbb{R}^3 \setminus \{0\},$$

is especially important in atomic and molecular physics.

2. The potential for the *hard-core model* is given, for some radius $R > 0$, by

$$\phi_R(x) = \begin{cases} +\infty & \text{if } |x| \le R, \\ 0 & \text{otherwise.} \end{cases}$$

Suppose that $\mu$ is a Gibbs measure and $\Lambda$ a bounded measurable subset of $\mathbb{R}^d$. Let $\mathcal{F}(\Lambda)$ be the $\sigma$-algebra of events $\Delta \in \mathcal{B}(\Gamma_{\mathbb{R}^d})$ that only depend on the part of the configuration in $\Lambda$, that is, $1_\Delta(\gamma) = 1_\Delta(\gamma_\Lambda)$ for every $\gamma \in \Gamma_{\mathbb{R}^d}$. Exchanging the order of integration in (2.4), we find that $\mu|_{\mathcal{F}(\Lambda)}$ is absolutely continuous with respect to $\pi_{z\sigma}|_{\mathcal{F}(\Lambda)}$ with density

$$(2.6) \qquad \omega \mapsto \int_{\Gamma_{\mathbb{R}^d}} [Z_\Lambda^{z,\phi}(\gamma)]^{-1} \exp\left[-E_\Lambda^\phi(\gamma_{\Lambda^c} + \omega_\Lambda)\right] \mu(d\gamma).$$

In other words, a Gibbs measure is always locally absolutely continuous with respect to its corresponding Poisson measure. In general we have very little information about the density (2.6), but for Ruelle measures it is known to be bounded, with a bound that depends on $\Lambda$.

Another important tool in studying Ruelle measures are the (infinite-volume) correlation functions $\rho_m : (\mathbb{R}^d)^m \to \mathbb{R}$ which can be expressed as

$$\rho_m(x_1, \ldots, x_m)$$
$$= z^m \exp(-E_\Lambda^\phi(\{x_1, \ldots, x_m\})) \int_{\Gamma_{\mathbb{R}^d}} \exp\left(-\sum_{i=1}^m \langle \phi(x_i - \cdot), \gamma \rangle\right) \mu(d\gamma).$$

These provide us with the analogues of (2.2) and (2.3) for Ruelle measures:

$$(2.7) \qquad \int_{\Gamma_{\mathbb{R}^d}} \langle f, \gamma \rangle \, \mu(d\gamma) = \int_{\mathbb{R}^d} f(x)\rho_1(x) \, dx,$$

$$(2.8) \quad \int_{\Gamma_{\mathbb{R}^d}} (\langle f, \gamma \rangle \langle g, \gamma \rangle - \langle fg, \gamma \rangle) \, \mu(d\gamma) = \int_{\mathbb{R}^d}\int_{\mathbb{R}^d} f(x) \, g(y)\rho_2(x, y) \, dx \, dy.$$

**2.2. Gradient.** We define a linear space $\mathcal{F}C_b^\infty$ of functions on $\Gamma_{\mathbb{R}^d}$ by taking all smooth cylinder functions. That is,

$$\mathcal{F}C_b^\infty := \{ u : u(\gamma) = g(\langle f_1, \gamma \rangle, \langle f_2, \gamma \rangle, \ldots, \langle f_n, \gamma \rangle)$$
$$\text{for some } f_i \in C_0^\infty(\mathbb{R}^d) \text{ and } g \in C_b^\infty(\mathbb{R}^n) \}.$$

For $u \in \mathcal{F}C_b^\infty$, we define the gradient $\nabla^\Gamma u$ at the point $\gamma \in \Gamma_{\mathbb{R}^d}$ as an element of the "tangent space" $T_\gamma(\Gamma_{\mathbb{R}^d}) := L^2(\mathbb{R}^d \to \mathbb{R}^d; \gamma)$ by the formula

$$\left(\nabla^\Gamma u\right)(\gamma; x) := \sum_{i=1}^n \frac{\partial g}{\partial x_i}(\langle f_1, \gamma \rangle, \langle f_2, \gamma \rangle, \ldots, \langle f_n, \gamma \rangle)\nabla f_i(x).$$

Here $\nabla$ refers to the usual gradient on $\mathbb{R}^d$. It is not hard to prove that $\nabla^\Gamma u$ is well-defined, even though the representation of $u$ as a cylinder function is not unique.

**2.3. Dirichlet form.**

DEFINITION 2.6. For $u, v \in \mathcal{F}C_b^\infty$ define the *square field* $⽥(u, v)$[1] as the real-valued function on $\Gamma_{\mathbb{R}^d}$ given by

$$⽥(u, v)(\gamma) := \langle \nabla^\Gamma u, \nabla^\Gamma v \rangle_{T_\gamma(\Gamma_{\mathbb{R}^d})}$$
$$= \int_{\mathbb{R}^d} \langle (\nabla^\Gamma u)(\gamma; x), (\nabla^\Gamma v)(\gamma; x) \rangle_{\mathbb{R}^d} \, \gamma(dx).$$

We will often use the abbreviation $⽥(u) := ⽥(u, u)$.

DEFINITION 2.7. For $u, v \in \mathcal{F}C_b^\infty$ define the *pre-Dirichlet form* by

$$(2.9) \qquad \mathcal{E}(u, v) := \int_{\Gamma_{\mathbb{R}^d}} ⽥(u, v)(\gamma) \, \mu(d\gamma).$$

We would like to study the form in (2.9), where $\mu$ is a Gibbs measure or a mixture of Poisson measures. A basic requirement is that $(\mathcal{E}, \mathcal{F}C_b^\infty)$ should be a well-defined and closable bilinear form on $L^2(\Gamma_{\mathbb{R}^d}, \mu)$. In general, proving closability is difficult and the closability problem is still an active area of research. In the free case, the specific conditions in Definition 2.8 (1.) guarantee closability, but in the Ruelle case we will simply assume that the pair potential $\phi$ satisfies the (mild) additional smoothness and integrability assumptions to ensure that (2.9) is well-defined and closable. In particular, the Lennard-Jones and hard-core potentials yield closable forms. Proofs of closability can be found in [**3**, Corollary 4.1 and

---

[1]This notation for square field is based on the Chinese character Tián, which means 'field'.

Remark 4.3] for the free case, and in [**4**, Proposition 5.1], [**11**, Section 6.3], and [**10**] in the Ruelle case.

DEFINITION 2.8.

1. Let $\sigma$ be a measure on $\mathbb{R}^d$ that has a density $\varrho$ with respect to Lebesgue measure satisfying $\varrho > 0$ almost everywhere, and $\varrho^{1/2} \in H^{1,2}_{\text{loc}}(\mathbb{R}^d)$. Here $H^{1,2}_{\text{loc}}(\mathbb{R}^d)$ denotes the local Sobolev space of order 1 in $L^2_{\text{loc}}(\mathbb{R}^d; dx)$. We are in the *free case* when $\mu$ is a mixture of $\pi_{z\sigma}$, that is,

$$(2.10) \qquad \mu := \int_{\mathbb{R}_+} \pi_{z\sigma}\, \lambda(dz),$$

where $\lambda$ is a probability measure on $\mathbb{R}_+$ with $\int_{\mathbb{R}_+} z\, \lambda(dz) < \infty$.

2. We are in the *Ruelle case* when $\mu$ is a Ruelle measure such that $(\mathcal{E}, \mathcal{F}C_b^\infty)$ is a well-defined and closable form in $L^2(\Gamma_{\mathbb{R}^d}, \mu)$.

From now on, we assume that the measure $\mu$ is in one of the two categories described in Definition 2.8, and we will let $(\mathcal{E}, D(\mathcal{E}))$ denote the closure of $(\mathcal{E}, \mathcal{F}C_b^\infty)$ in $L^2(\Gamma_{\mathbb{R}^d}, \mu)$. Standard Dirichlet form theory shows that the closure $(\mathcal{E}, D(\mathcal{E}))$ is a symmetric, local, Dirichlet form. In particular, the space $D(\mathcal{E})$ is complete with respect to the norm

$$|u|_1 := \big(\mathcal{E}(u, u) + (u, u)_{L^2(\mu)}\big)^{1/2}.$$

The map $(u, v) \to \boxplus(u, v)$ is continuous from $\mathcal{F}C_b^\infty \times \mathcal{F}C_b^\infty$ into $L^1(\Gamma_{\mathbb{R}^d}; \mu)$ when $\mathcal{F}C_b^\infty$ is equipped with the $|\ |_1$-norm, and so the square field $\boxplus$ extends to the full domain $D(\mathcal{E})$ in such a way that formula (2.9) continues to hold.

The usual functional calculus for Dirichlet forms ensures (eg. [**13**, Lemma 3.2]) that if $u, v \in D(\mathcal{E})$ and $\psi$ is a smooth function on $\mathbb{R}$ that vanishes at the origin and has bounded derivative, then $\psi(u)$ belongs to $D(\mathcal{E})$ and

$$(2.11) \qquad \boxplus(\psi(u)) = (\psi'(u))^2 \boxplus(u).$$

In the same vein, you can show that $u \vee v$ and $u \wedge v$ belong to $D(\mathcal{E})$ and

$$(2.12) \qquad \boxplus(u \vee v) \le \boxplus(u) \vee \boxplus(v) \quad \text{and} \quad \boxplus(u \wedge v) \le \boxplus(u) \vee \boxplus(v).$$

Let us do a sample calculation with the easiest kind of cylinder function: $u(\gamma) = \langle f, \gamma \rangle$ where $f \in C_0^\infty(\mathbb{R}^d)$. Although $u$ doesn't belong to $\mathcal{F}C_b^\infty$, it is easy to show that it belongs to $D(\mathcal{E})$ and that the following calculations are valid. For this function $u$, the gradient $\nabla^\Gamma u$ is equal to $\nabla f$ at all points $\gamma \in \Gamma_{\mathbb{R}^d}$, and so the square field is $\boxplus(u)(\gamma) = \int_{\mathbb{R}^d} |\nabla f|^2(x)\gamma(dx)$. In the free case, it follows from equation (2.2) and (2.10) that the Dirichlet form applied to such a function $u$ gives

$$\mathcal{E}(u, u) = \int_{\Gamma_{\mathbb{R}^d}} \langle |\nabla f|^2, \gamma \rangle\, \mu(d\gamma) = \Big( \int_{\mathbb{R}_+} z\, \lambda(dz) \Big) \Big( \int_{\mathbb{R}^d} |\nabla f|^2(x)\sigma(dx) \Big),$$

In the Ruelle case, (2.7) shows us that

$$\mathcal{E}(u, u) = \int_{\mathbb{R}^d} |\nabla f|^2(x)\, \rho_1(x)\, dx.$$

**2.4. Stochastic process.** In order to prove that the Dirichlet form $(\mathcal{E}, D(\mathcal{E}))$ has an associated Markov process we need to show that $(\mathcal{E}, D(\mathcal{E}))$ is quasi-regular [**7**, Chapter IV, Theorem 3.5]. The quasi-regularity and locality of $(\mathcal{E}, D(\mathcal{E}))$ has been proven for certain cases by Yoshida [**17**], and in general by Ma and Röckner [**8**] but since $\Gamma_{\mathbb{R}^d}$ is not complete with respect to the vague topology it is necessary to use the completed state space

$$\ddot{\Gamma}_{\mathbb{R}^d} := \{\mathbb{Z}_+ \cup \{+\infty\}\text{-valued Radon measures on } \mathbb{R}^d\}.$$

Since $\Gamma_{\mathbb{R}^d} \subset \ddot{\Gamma}_{\mathbb{R}^d}$ and $\mathcal{B}(\ddot{\Gamma}_{\mathbb{R}^d}) \cap \Gamma_{\mathbb{R}^d} = \mathcal{B}(\Gamma_{\mathbb{R}^d})$, we can consider $\mu$ as a measure on $(\ddot{\Gamma}_{\mathbb{R}^d}, \mathcal{B}(\ddot{\Gamma}_{\mathbb{R}^d}))$ and correspondingly $(\mathcal{E}, D(\mathcal{E}))$ as a Dirichlet form on $L^2(\ddot{\Gamma}_{\mathbb{R}^d}; \mu)$.

The associated Markov process $((X_t)_{t \geq 0}, (P_\gamma)_{\gamma \in \ddot{\Gamma}_{\mathbb{R}^d}})$ has vaguely continuous sample paths since $(\mathcal{E}, D(\mathcal{E}))$ is a local form [**7**, Chapter V, Theorem 1.11].

**2.5. Exceptional sets.** Let $(\mathcal{E}, D(\mathcal{E}))$ be the local, quasi-regular Dirichlet form given by the closure of the pre-Dirichlet form in (2.9), and $(X_t)_{t \geq 0}$ the associated $\ddot{\Gamma}_{\mathbb{R}^d}$-valued diffusion process.

Exceptional sets and quasi-continuous functions are important tools for understanding the diffusion process corresponding to a Dirichlet form. Exceptional sets are "almost empty", and quasi-continuous functions are "almost continuous" in a sense appropriate for Dirichlet forms. We first give the definition of these objects in terms of the form $(\mathcal{E}, D(\mathcal{E}))$, and then describe their interpretation using the process $(X_t)_{t \geq 0}$.

DEFINITION 2.9. Let $(\mathcal{E}, D(\mathcal{E}))$ be a Dirichlet form on $L^2(\ddot{\Gamma}_{\mathbb{R}^d}; \mu)$.

(a) For a closed subset $F \subseteq \ddot{\Gamma}_{\mathbb{R}^d}$ we define a closed subspace of $D(\mathcal{E})$ by

$$D(\mathcal{E})_F := \{u \in D(\mathcal{E}) : u = 0 \ \mu\text{-a.e. on } \ddot{\Gamma}_{\mathbb{R}^d} \setminus F\}.$$

(b) An increasing sequence $(F_k)_{k \in \mathbb{N}}$ of closed subsets of $\ddot{\Gamma}_{\mathbb{R}^d}$ is called an $\mathcal{E}$-nest if $\cup_{k \geq 1} D(\mathcal{E})_{F_k}$ is $|\ |_1$-dense in $D(\mathcal{E})$.

(c) A function $u : \ddot{\Gamma}_{\mathbb{R}^d} \to \mathbb{R}$ is called $\mathcal{E}$-quasi-continuous if there exists an $\mathcal{E}$-nest $(F_k)_{k \in \mathbb{N}}$ so that $u|_{F_k}$ is continuous for each $k \in \mathbb{N}$.

(d) A subset $N \in \mathcal{B}(\ddot{\Gamma}_{\mathbb{R}^d})$ is called $\mathcal{E}$-exceptional if $1_N$ is $\mathcal{E}$-quasi-continuous and $\mu(N) = 0$.

We shall use the following result throughout this paper (see [**7**, Chapter III, Proposition 3.5], [**7**, Chapter IV, Lemma 4.5], and [**6**]).

LEMMA 2.10. *Let* $u_n \in D(\mathcal{E})$ *be a sequence of* $\mathcal{E}$-*quasi-continuous functions with* $\sup_n \mathcal{E}(u_n, u_n) < \infty$ *and* $u_n \to u$ *pointwise. Then* $u$ *is an* $\mathcal{E}$-*quasi-continuous function, in particular, for* $\mu$-*almost every* $\gamma \in \ddot{\Gamma}_{\mathbb{R}^d}$,

(2.13) $$P_\gamma (t \to u(X_t) \text{ is continuous }) = 1.$$

*If* $u$ *is* $\mu$-*square integrable, then* $u \in D(\mathcal{E})$.

It is also worthwhile noting the interpretation of an $\mathcal{E}$-exceptional set in terms of the associated diffusion [**7**, Chapter IV, Proposition 5.30].

LEMMA 2.11. *A set* $N \in \mathcal{B}(\ddot{\Gamma}_{\mathbb{R}^d})$ *is* $\mathcal{E}$-*exceptional if and only if, for* $\mu$-*almost every* $\gamma \in \ddot{\Gamma}_{\mathbb{R}^d}$,

$$P_\gamma (X_t \in N \text{ for some } 0 \leq t < \infty) = 0.$$

## 3. Support properties for the process

Throughout this section, we will assume that $(\mathcal{E}, D(\mathcal{E}))$ is the closure of the pre-Dirichlet form (2.9) and that $\mu$ is in one of the two categories of measures spelled out in Definition 2.8. From the results in [8] we know that $(\mathcal{E}, D(\mathcal{E}))$ is a local, quasi-regular Dirichlet form on the space $L^2(\ddot{\Gamma}_{\mathbb{R}^d}, \mu)$. We are interested in the sample path properties of the associated $\ddot{\Gamma}_{\mathbb{R}^d}$-valued diffusion process $X_t$.

**3.1. Local properties.** The space $\ddot{\Gamma}_{\mathbb{R}^d}$ is the completion of the space $\Gamma_{\mathbb{R}^d}$ in the vague topology. Unlike the measures $\sum_{x \in \gamma} \varepsilon_x$ in the space $\Gamma_{\mathbb{R}^d}$, the measures in $\ddot{\Gamma}_{\mathbb{R}^d}$ allow for the possibility that two (or more) particles could occupy the same position in $\mathbb{R}^d$, resulting in a point with mass of two (or more). The following proposition gives conditions so that, with probability one, the process $X_t$ will not hit the set of such measures, so the completion of the state space was unnecessary after all.

The results in the first two propositions in this section have been announced in [14].

PROPOSITION 3.1.   **(Free case)**   *If* $d \geq 2$, $\varrho \in L^2_{\mathrm{loc}}(dx)$, *and* $\int_{\mathbb{R}_+} z^2 \lambda(dz) < \infty$, *then the set* $\ddot{\Gamma}_{\mathbb{R}^d} \setminus \Gamma_{\mathbb{R}^d}$ *is* $\mathcal{E}$*-exceptional.*

PROOF.   Our goal is to show that the set of measures $\gamma$ that take values greater than one is $\mathcal{E}$-exceptional. It clearly suffices to prove this locally, that is, to show that for every positive integer $a$, the function $u := 1_N$ is $\mathcal{E}$-quasi-continuous, where

$$(3.1) \qquad N := \{\gamma : \sup(\gamma(\{x\}) : x \in [-a, a]^d) \geq 2\}.$$

We first note that if $\mu$ is a mixed Poisson measure, then $\sup_x \gamma(\{x\}) = 1$ $\mu$-almost everywhere, and so $\mu(N) = 0$.

Our analysis begins with a smooth partition of $\mathbb{R}^d$ into small pieces. Let $f$ be a $C_0^\infty(\mathbb{R})$ function satisfying $1_{[0,1]} \leq f \leq 1_{[-1/2, 3/2)}$ and $|f'| \leq 3 \times 1_{[-1/2, 3/2)}$, and for any $n \in \mathbb{N}$ and $i = (i_1, \ldots, i_d) \in \mathbb{Z}^d$, define a $C_0^\infty(\mathbb{R}^d)$ function by

$$f_i(x) := \prod_{k=1}^{d} f(nx_k - i_k).$$

We also let $I_i(x) := \prod_{k=1}^{d} 1_{[-1/2, 3/2)}(nx_k - i_k)$ and note that $f_i \leq I_i$. Taking the $j$th partial derivative of $f_i$ gives

$$\partial_j f_i(x) = nf'(nx_j - i_j) \prod_{k \neq j} f(nx_k - i_k),$$

and so $(\partial_j f_i(x))^2 \leq 9n^2 I_i(x)$. Adding over $j$ from 1 to $d$ gives us

$$(3.2) \qquad |\nabla f_i(x)|^2 \leq 9n^2 d I_i(x).$$

Let $\psi$ be a smooth function on $\mathbb{R}$ satisfying $1_{[2,\infty)} \leq \psi \leq 1_{[1,\infty)}$ and $|\psi'| \leq 2 \times 1_{(1,\infty)}$. Choosing $A := \mathbb{Z}^d \cap [-na, na]^d$, define a continuous element of $D(\mathcal{E})$ by

$$(3.3) \qquad u_n(\gamma) := \psi\big(\sup_{i \in A} \langle f_i, \gamma \rangle\big).$$

Then $u_n \to u$ pointwise as $n \to \infty$, so to apply Lemma 2.10 we must prove that $\sup_n \mathcal{E}(u_n, u_n) < \infty$. We begin by bounding $\mathbb{H}(u_n)$, the square field applied to $u_n$.

First note that

$$(3.4) \qquad \left(\psi'\big(\sup_{i\in A}\langle f_i,\gamma\rangle\big)\right)^2 \le 4\times 1_{(\sup_{i\in A}\langle f_i,\gamma\rangle>1)} \le 4\times 1_{(\sup_{i\in A}\langle I_i,\gamma\rangle\ge 2)},$$

where for the final inequality we use the fact that $\langle I_i,\gamma\rangle$ is an integer. Therefore, using (2.12) along with the inequalities in (3.2) and (3.4), we get

$$
\begin{aligned}
\text{Ⅲ}(u_n)(\gamma) &= \left(\psi'\big(\sup_{i\in A}\langle f_i,\gamma\rangle\big)\right)^2 \text{Ⅲ}\big(\sup_{i\in A}\langle f_i,\cdot\rangle\big)(\gamma)\\
&\le \left(\psi'\big(\sup_{i\in A}\langle f_i,\gamma\rangle\big)\right)^2 \sup_{i\in A}\text{Ⅲ}(\langle f_i,\cdot\rangle)(\gamma)\\
(3.5)\qquad &= \left(\psi'\big(\sup_{i\in A}\langle f_i,\gamma\rangle\big)\right)^2 \sup_{i\in A}\int|\nabla f_i(x)|^2\,\gamma(dx)\\
&\le 4\times 1_{(\sup_{i\in A}\langle I_i,\gamma\rangle\ge 2)}\,9\,n^2\,d\,\sup_{i\in A}\langle I_i,\gamma\rangle\\
&\le 36n^2 d\sum_{i\in A}1_{(\langle I_i,\gamma\rangle\ge 2)}\langle I_i,\gamma\rangle.
\end{aligned}
$$

From equation (2.10) we get

$$
\begin{aligned}
(3.6)\qquad \int_{(\langle I_i,\gamma\rangle\ge 2)}\langle I_i,\gamma\rangle\mu(d\gamma) &= \int_{\mathbb{R}_+} z\langle I_i,\sigma\rangle\big(1-e^{-z\langle I_i,\sigma\rangle}\big)\,\lambda(dz)\\
&\le \langle I_i,\sigma\rangle^2\int_{\mathbb{R}_+}z^2\,\lambda(dz)
\end{aligned}
$$

and combined with (3.5) this gives

$$(3.7)\qquad \mathcal{E}(u_n,u_n)\le c\,n^2\sum_{i\in A}\langle I_i,\sigma\rangle^2.$$

Although the supports of the indicator functions $I_i$ are not disjoint, each point belongs to at most $2^d$ of the sets $\{I_i=1\}$ for $i\in A$. Therefore the Cauchy-Schwarz inequality gives us

$$
\begin{aligned}
(3.8)\qquad \sum_{i\in A}\langle I_i,\sigma\rangle^2 &= \sum_{i\in A}\left(\int I_i(x)\,\varrho(x)\,dx\right)^2\\
&\le \sum_{i\in A}\left(\int I_i(x)\,\varrho(x)^2\,dx\right)\left(\int I_i(x)\,dx\right)\\
&\le 2^d\int_{[-(a+1),a+1]^d}\varrho(x)^2\,dx\,(2/n)^d,
\end{aligned}
$$

and combining this with (3.7) we find that

$$\mathcal{E}(u_n,u_n)\le cn^{2-d}.$$

Since we have assumed that $d\ge 2$ we see that $\sup_n \mathcal{E}(u_n,u_n)<\infty$. We conclude that $N$ is $\mathcal{E}$-exceptional. $\qquad\square$

**Note** Corollaries 1 and 2 in the next section tell us when we can drop the condition $\int_{\mathbb{R}_+}z^2\,dz<\infty$.

PROPOSITION 3.2. **(Ruelle case)** *The set $\ddot{\Gamma}_{\mathbb{R}^d}\setminus\Gamma_{\mathbb{R}^d}$ is $\mathcal{E}$-exceptional.*

PROOF. To compare with the free case we first set $\varrho=z$ so that $\sigma$ is $z$ times Lebesgue measure, and put $\lambda=\varepsilon_z$. The inequality (5.14) in [**16**] guarantees that the Radon-Nikodym derivative in (2.6) is bounded, for any bounded $\Lambda$. Putting $\Lambda=[-a,a]^d$ and $A=\mathbb{Z}^d\cap[-na,na]^d$ we see that the inequalities (3.6), (3.7), and (3.8) are valid (up to a constant) for a Ruelle measure. Therefore, the function $u$ is $\mathcal{E}$-quasi-continuous and $\mu(N)=0$ in the Ruelle case as well. $\qquad\square$

**3.2. Global properties.** In this section we consider properties of the random configuration $X_t$ that depend on the whole configuration. Here, the local absolute continuity of a Ruelle measure with respect to Poisson measure is not helpful, so it is harder to get results for Ruelle measures.

The next result shows that, in the free case, under the exponential growth condition in (3.9), the total number of particles $X_t(\mathbb{R}^d)$ in the random configuration $X_t$ remains constant in time. It leaves open the possibility that if (3.9) is violated, then $X_t$ may hit the empty configuration $\emptyset$ at exceptional times, that is, the particles may explode to infinity simultaneously. Note that (3.9) is trivially satisfied if $\sigma(\mathbb{R}^d) < \infty$, and that in this case $X_t(\mathbb{R}^d) < \infty$ also. In all of the calculations below, $c$ is a constant that does not depend on the index $n$, but whose value may change from line to line. Also, $S_r$ refers to the sphere in $\mathbb{R}^d$ with radius $r$, centered at the origin.

PROPOSITION 3.3.  **(Free case)**  *If there exist $a, b > 0$ so that*

(3.9)
$$\sigma(S_r) \le a \exp(br),$$

*then*

$$\int_{\ddot{\Gamma}_{\mathbb{R}^d}} P_\gamma \left( t \to X_t(\mathbb{R}^d) \text{ is constant } \right) \mu(d\gamma) = 1.$$

PROOF. Let $\psi_k$ be a smooth function on $\mathbb{R}$ that vanishes outside of the interval $(k-1, k+1)$ and so that $\psi_k(k) = 1$. Also let $f_r$ be a smooth function on $\mathbb{R}^d$ such that $1_{S_r} \le f_r \le 1_{S_{r+1}}$ and $|\nabla f_r|^2 \le c$. Let

(3.10)
$$u_r(\gamma) := \psi_k \left( \langle f_r, \gamma \rangle \right).$$

In order to estimate $\mathcal{E}(u_r, u_r)$ we first note that

$$\nabla^\Gamma u_r(\gamma; x) = \psi_k' \left( \langle f_r, \gamma \rangle \right) \nabla f_r(x),$$

so that the square field satisfies

(3.11)
$$\begin{aligned}
\boxplus(u_r)(\gamma) &\le \left( \psi_k' \left( \langle f_r, \gamma \rangle \right) \right)^2 \int_{\mathbb{R}^d} |\nabla f_r(x)|^2 \, \gamma(dx) \\
&\le c 1_{\{\gamma(S_r) \le k\}} \gamma(S_{r+1} \setminus S_r).
\end{aligned}$$

Therefore it follows that

$$\begin{aligned}
\mathcal{E}(u_r, u_r) &\le c \int_{\mathbb{R}_+} \int_{(\gamma(S_r) \le k)} \gamma(S_{r+1} \setminus S_r) \, \pi_{z\sigma}(d\gamma) \, \lambda(dz) \\
&\le c \int_{\mathbb{R}_+} \left[ \sum_{j=0}^{k} e^{-z\sigma(S_r)} z^j \sigma(S_r)^j \right] z\sigma(S_{r+1} \setminus S_r) \, \lambda(dz) \\
&\le c \int_{\mathbb{R}_+} \left[ \sum_{j=1}^{k+1} e^{-z\sigma(S_r)} z^j \sigma(S_r)^j \right] \sigma(S_{r+1})/\sigma(S_r) \, \lambda(dz) \\
&\le c \sigma(S_{r+1})/\sigma(S_r).
\end{aligned}$$

We used the fact that $S_r$ and $S_{r+1} \setminus S_r$ are disjoint so the random variables $\gamma(S_r)$ and $\gamma(S_{r+1} \setminus S_r)$ are independent on $(\ddot{\Gamma}_{\mathbb{R}^d}, \pi_{z\sigma})$. By the assumption in (3.9), we can find a sequence $r_n \to \infty$ so that $\sup_n \sigma(S_{r_n+1})/\sigma(S_{r_n}) < \infty$ and therefore

$\sup_n \mathcal{E}(u_{r_n}, u_{r_n}) < \infty$. As $n \to \infty$ the sequence $(u_{r_n})_{n \in \mathbb{N}}$ converges pointwise to the function $\psi_k(\gamma(\mathbb{R}^d)) = 1_{\{\gamma(\mathbb{R}^d)=k\}}$. By Lemma 2.10, we find that

$$\int_{\ddot{\Gamma}_{\mathbb{R}^d}} P_\gamma \left( t \to 1_{\{\gamma(\mathbb{R}^d)=k\}}(X_t) \text{ is continuous } \right) \mu(d\gamma) = 1,$$

and since $k$ is arbitrary, we obtain our result. □

COROLLARY 3.4. *The condition $\int_{\mathbb{R}_+} z^2 \lambda(dz) < \infty$ in Proposition 1 can be dropped if $\sigma(\mathbb{R}^d) < \infty$.*

PROOF. The proof depends on a truncation argument based on the fact that if $u, v \in D(\mathcal{E})$ are bounded functions, then the product $uv$ belongs to $D(\mathcal{E})$ and

(3.12) $$\boxplus(uv) = u^2 \boxplus(v) + v^2 \boxplus(u).$$

If $\sigma(\mathbb{R}^d) < \infty$, then the mixed measure $\mu$ is absolutely continuous with respect to $\pi_\sigma$ and the density is given by

$$\frac{d\mu}{d\pi_\sigma}(\gamma) = \int_{\mathbb{R}_+} e^{(1-z)\sigma(\mathbb{R}^d)} z^{\gamma(\mathbb{R}^d)} \lambda(dz).$$

In particular, $\gamma(\mathbb{R}^d) < \infty$ for $\mu$-almost every $\gamma \in \ddot{\Gamma}_{\mathbb{R}^d}$ and $\mu$ is just a multiple of $\pi_\sigma$ on each of the sets $\{\gamma : \gamma(\mathbb{R}^d) = k\}$. In Proposition 3.3 we showed that $v(\gamma) := 1_{\{\gamma(\mathbb{R}^d)=k\}}$ is an $\mathcal{E}$-quasi-continuous element of $D(\mathcal{E})$ and by letting $r \to \infty$ in (3.11) we see that $\boxplus(v) = 0$. In particular, (3.12) shows that for any bounded $u \in D(\mathcal{E})$, we have $u1_{\{\gamma(\mathbb{R}^d)=k\}} \in D(\mathcal{E})$ and $\boxplus(u1_{\{\gamma(\mathbb{R}^d)=k\}}) = 1_{\{\gamma(\mathbb{R}^d)=k\}} \boxplus(u)$.

Now apply this formula to the functions $(u_n)_{n \in \mathbb{N}}$ defined in (3.3). Using the density $d\mu/d\pi_\sigma$ and (3.5), we find that

$$\begin{aligned}
\int_{\ddot{\Gamma}_{\mathbb{R}^d}} \boxplus(u_n 1_{\{\gamma(\mathbb{R}^d)=k\}})(\gamma) \, \mu(d\gamma) &= \int_{\{\gamma(\mathbb{R}^d)=k\}} \boxplus(u_n)(\gamma) \, \mu(d\gamma) \\
&= c \int_{\{\gamma(\mathbb{R}^d)=k\}} \boxplus(u_n)(\gamma) \, \pi_\sigma(d\gamma) \\
&\leq cn^2 \sum_{i \in A} \langle I_i, \sigma \rangle^2.
\end{aligned}$$

As in the proof of Proposition 1, since $d \geq 2$, the expression above is bounded in $n$ and so we conclude that $u1_{\{\gamma(\mathbb{R}^d)=k\}}$ is an $\mathcal{E}$-quasi-continuous element of $D(\mathcal{E})$, where $u$ is the limit of the sequence $(u_n)_{n \in \mathbb{N}}$. This means that the set

$$\{\gamma : \gamma(\mathbb{R}^d) = k\} \cap N$$

is $\mathcal{E}$-exceptional, where $N$ is defined in (3.1). This gives the result, since $k$ and $a$ are arbitrary, and since Proposition 3.3 shows that $\{\gamma : \gamma(\mathbb{R}^d) = \infty\}$ is $\mathcal{E}$-exceptional. □

We would now like to prove the result corresponding to Proposition 3.3 in the Ruelle case, that is, show that $X_t(\mathbb{R}^d) = \infty$ for all times $t \geq 0$. We begin by proving the fixed time result which says that $\mu(\gamma(\mathbb{R}^d) < \infty) = 0$. Here $\mu$ need not even be a Ruelle measure, any Gibbs measure with intensity satisfying $\sigma(\mathbb{R}^d) = \infty$ works.

LEMMA 3.5. *A Gibbs measure $\mu$ with $\sigma(\mathbb{R}^d) = \infty$ does not charge the set of finite configurations.*

PROOF. First consider the case where $Z_\Lambda^\phi(\emptyset) < \infty$ for all bounded $\Lambda$. Then

$$
\begin{aligned}
\mu(\{\emptyset\}) &= \int \frac{1}{Z_\Lambda^\phi(\gamma)} \int_{\{\omega_\Lambda + \gamma_{\Lambda^c} = \emptyset\}} \exp\left[-E_\Lambda^\phi(\omega_\Lambda + \gamma_{\Lambda^c})\right] \pi_{z\sigma}(d\omega)\, \mu(d\gamma) \\
\text{(3.13)} \qquad &= \int_{\{\gamma_{\Lambda^c} = \emptyset\}} \frac{1}{Z_\Lambda^\phi(\gamma)} \exp\left[-z\sigma(\Lambda)\right] \mu(d\gamma) \\
&= \frac{1}{Z_\Lambda^\phi(\emptyset)} \exp\left[-z\sigma(\Lambda)\right] \mu(\gamma_{\Lambda^c} = 0).
\end{aligned}
$$

Now $Z_\Lambda^\phi(\emptyset) \geq \exp[-z\sigma(\Lambda)](1 + z\sigma(\Lambda))$, because $E_\Lambda^\phi(\omega_\Lambda) = 0$ if $\omega(\Lambda)$ is either 0 or 1. Combined with (3.13) this yields $\mu(\{\emptyset\}) \leq (1 + z\sigma(\Lambda))^{-1}$ which goes to zero as $\Lambda \nearrow \mathbb{R}^d$. Plugging this back into the equation (3.13) gives $\mu(\gamma_{\Lambda^c} = 0) = 0$ for all $\Lambda$, which in turn implies that $\mu(\gamma(\mathbb{R}^d) < \infty) = 0$.

On the other hand, if $Z_\Lambda^\phi(\emptyset) = \infty$ for large $\Lambda$, then setting $\Delta$ in (2.4) to be the set of finite configurations, and bearing in mind that $\gamma_\Lambda(\mathbb{R}^d) < \infty$ and $\omega_\Lambda(\mathbb{R}^d) < \infty$ anyway, we get

$$
\mu(\gamma(\mathbb{R}^d) < \infty) = \iint_{\{\gamma(\mathbb{R}^d) < \infty\}} \frac{1}{Z_\Lambda^\phi(\gamma)} \exp\left[-E_\Lambda^\phi(\omega_\Lambda + \gamma_{\Lambda^c})\right] \pi_{z\sigma}(d\omega)\, \mu(d\gamma).
$$

However, if $\gamma(\mathbb{R}^d) < \infty$, then $Z_\Lambda^\phi(\gamma) \to Z_\Lambda^\phi(\emptyset) = \infty$ as $\Lambda \nearrow \mathbb{R}^d$ and by bounded convergence we obtain $\mu(\gamma(\mathbb{R}^d) < \infty) = 0$. $\qquad\square$

PROPOSITION 3.6. **(Ruelle case)** *For sufficiently small activity $z$, the set* $\{\gamma : \gamma(\mathbb{R}^d) < \infty\}$ *is $\mathcal{E}$-exceptional.*

PROOF. Mimicking the proof of Proposition 3.3, we define $u_r$ as in (3.10). Use Cauchy-Schwarz with $q$ defined by $1/q + 1/d = 1$ to integrate the bound from (3.11) and obtain

$$
\text{(3.14)} \qquad \mathcal{E}(u_r, u_r) \leq c\, \mu\left(\gamma(S_r) \leq k\right)^{1/q} E_\mu\left(\gamma(S_{r+1} \setminus S_r)^d\right)^{1/d}.
$$

The proof can be completed along the lines of Proposition 3.3 provided we can show that $\mathcal{E}(u_r, u_r)$ is bounded in $r$. This will require analysis of the random variables $\gamma(\Lambda)$ on $(\ddot{\Gamma}_{\mathbb{R}^d}, \mu)$. Roughly speaking, we must show that $\gamma(\Lambda)$ is of the order $m(\Lambda)$ as $\Lambda \nearrow \mathbb{R}^d$, where $m$ is Lebesgue measure.

For $k \in \mathbb{Z}_+$ the $k$th falling power of $x$ is $x^{\underline{k}} = x(x-1)\cdots(x-(k-1))$, and the correlation function $\rho_k$ gives

$$
\text{(3.15)} \qquad E_\mu\left(\gamma(\Lambda)^{\underline{k}}\right) = \int \cdots \int_{\Lambda^k} \rho_k(x_1, \ldots, x_k)\, dx_1 \ldots dx_k.
$$

Now Ruelle [**16**, Proposition 2.6] shows that there exists a constant $\xi$ so that $\rho_k \leq \xi^k$. This means that the random variable $\gamma(\Lambda)$, which has a Poisson distribution under $\pi_{zm}$, is comparable to a Poisson random variable under $\mu$. To be precise, let's rewrite $x^d$ in terms of falling powers of $x$ as in

$$
\text{(3.16)} \qquad x^d = \sum_{k=1}^{d} \left\{\begin{matrix} d \\ k \end{matrix}\right\} x^{\underline{k}},
$$

where $\left\{^d_k\right\}$ are Stirling numbers of the second kind [**5**, (6.10)]. Combining (3.15) and (3.16) we get

$$E_\mu\left(\gamma(\Lambda)^d\right) = \sum_{k=1}^d \left\{^d_k\right\} E_\mu\left(\gamma(\Lambda)^{\underline{k}}\right) \leq \sum_{k=1}^d \left\{^d_k\right\} (\xi m(\Lambda))^k$$

$$= E(Z^d_{\xi m(\Lambda)}) \leq c\, E(Z^d_1)(\xi m(\Lambda))^d,$$

where $Z_\nu$ denotes a Poisson random variable with mean $\nu$, and where we must assume that $m(\Lambda)$ is bounded away from zero for the final inequality. In particular, this establishes the bound, for $m(\Lambda)$ bounded away from zero,

$$(3.17) \qquad E_\mu\left(\gamma(\Lambda)^d\right)^{1/d} \leq c\, \xi\, m(\Lambda).$$

The most difficult part of the proof is to establish bounds for the probabilities $\mu(\gamma(\Lambda) \leq k)$, in other words, to show that Ruelle measures do not allow too few particles per unit volume in $\mathbb{R}^d$. If we assume that $z$ is sufficiently small, then by [**16**, Theorems 5.7 and 5.8], $\mu$ is the unique Gibbs measure for $\phi$ and $z$, and is translation invariant. In particular, the first two correlation functions satisfy $\rho_1(x) = \rho$ for some constant, and $\rho_2(x, y) = \varphi(x - y)$ for some function $\varphi : \mathbb{R}^d \to \mathbb{R}$. First of all, (2.7) shows that the mean of $\mu$ is $\rho$ times Lebesgue measure, so that $E_\mu(\gamma(\Lambda)) = \rho m(\Lambda)$. In addition, [**16**, Theorem 4.4.8] shows that $\varphi$ is absolutely integrable on $\mathbb{R}^d$, so that using (2.8) we obtain

$$E_\mu(\gamma(\Lambda)^2 - \gamma(\Lambda)) = \iint_{\Lambda^2} \varphi(x - y)\, dx\, dy \leq c m(\Lambda).$$

This gives $\mathrm{Var}_\mu\left(\gamma(\Lambda)\right) \leq cm(\Lambda)$, and now Chebyshev's inequality implies

$$(3.18) \qquad \mu(\gamma(\Lambda) \leq k) \leq \frac{c}{m(\Lambda)}.$$

Combining (3.18) and (3.17) with the inequality (3.14) shows that $(u_r)_{r \geq 0}$ is $\mathcal{E}$-bounded, and the conclusion follows. $\qquad\square$

In the free case, under suitable growth conditions on $\sigma$, we have seen that if $\sigma(\mathbb{R}^d) = \infty$, then the number of particles in the random configuration $X_t$ remains infinite at all times. The following proposition gives more detailed information on the growth of $X_t$.

LEMMA 3.7. **(Free case)** *If* $\sigma(\mathbb{R}^d) = \infty$, *then* $\lim_{r \to \infty} \gamma(S_r)/\sigma(S_r) = z$, *for* $\pi_{z\sigma}$-*almost every* $\gamma \in \ddot{\Gamma}_{\mathbb{R}^d}$.

PROOF. Choose radii $(r_n)_{n \in \mathbb{N}}$ so that $\sigma(S_{r_n}) = n$. Then under the measure $\pi_{z\sigma}$, the function $\gamma(S_{r_n}) = \gamma(S_1) + \gamma(S_2 \setminus S_1) + \cdots + \gamma(S_{r_n} \setminus S_{r_{n-1}})$ is the sum of $n$ independent Poisson($z$) random variables. By the law of large numbers, $\gamma(S_{r_n})/n \to z$ $\pi_{z\sigma}$-almost surely. Now for $r_n \leq r \leq r_{n+1}$ we have

$$(3.19) \qquad \frac{n}{n+1}\, \frac{\gamma(S_{r_n})}{n} \leq \frac{\gamma(S_r)}{\sigma(S_r)} \leq \frac{\gamma(S_{r_{n+1}})}{n+1}\, \frac{n+1}{n}$$

and so the limit is attained over the continuous index $r$ as well. $\qquad\square$

As the following result shows, under certain conditions, the result is also true $\mathcal{E}$-quasi-everywhere; that is, the "intensity" of $X_t$ is constant along sample paths.

PROPOSITION 3.8. **(Free case)** *Suppose that* $\sigma(\mathbb{R}^d) = \infty$,

(3.20)
$$\lim_{\epsilon \to 0} \liminf_{r \to \infty} \sigma(S_{r-\epsilon})/\sigma(S_r) = 1,$$

*and*

(3.21)
$$\int_{\mathbb{R}_+} z \log^+(z)\, \lambda(dz) < \infty.$$

*Then, for $\mu$-almost every $\gamma \in \ddot{\Gamma}_{\mathbb{R}^d}$,*

$$P_\gamma \left( \lim_{r \to \infty} \frac{X_t(S_r)}{\sigma(S_r)} = \lim_{r \to \infty} \frac{\gamma(S_r)}{\sigma(S_r)} \text{ for all } t \geq 0 \right) = 1.$$

PROOF. For every $r > 0$ and $\epsilon > 0$ let $\psi_{r,\epsilon}$ be a smooth function satisfying $1_{(-\infty, r-\epsilon]} \leq \psi_{r,\epsilon} \leq 1_{(-\infty, r]}$, and $|\psi'_{r,\epsilon}| \leq c/\epsilon$. Define a continuous element of $D(\mathcal{E})$ by

(3.22)
$$u_{r,\epsilon}(\gamma) := \langle \psi_{r,\epsilon}(| \cdot |), \gamma \rangle / \sigma(S_r).$$

Bounding the square field gives

(3.23)
$$\boxplus(u_{r,\epsilon})(\gamma) \leq \frac{c^2}{\epsilon^2} \frac{\gamma(S_r)}{\sigma(S_r)^2}.$$

As in the previous lemma, define radii $(r_n)_{n \in \mathbb{N}}$ so that $\sigma(S_{r_n}) = n$. It is known [9] that therefore,

$$\begin{aligned}
\int_{\ddot{\Gamma}_{\mathbb{R}^d}} \sup_{n \in \mathbb{N}} \frac{\gamma(S_{r_n})}{n}\, \pi_{z\sigma}(d\gamma) &\leq c + c \int_{\ddot{\Gamma}_{\mathbb{R}^d}} \gamma(S_{r_1}) \log^+(\gamma(S_{r_1}))\, \pi_{z\sigma}(d\gamma) \\
&\leq c + c\left( z \log^+(z) \right).
\end{aligned}$$

From the inequality (3.19), we get

$$\int_{\ddot{\Gamma}_{\mathbb{R}^d}} \sup_{r \geq r_1} \frac{\gamma(S_r)}{\sigma(S_r)}\, \pi_{z\sigma}(d\gamma) \leq c + c\left( z \log^+(z) \right),$$

and hence, integrating with respect to $\lambda(dz)$ and using (3.21), it follows that

$$\int_{\ddot{\Gamma}_{\mathbb{R}^d}} \sup_{r \geq r_1} \frac{\gamma(S_r)}{\sigma(S_r)}\, \mu(d\gamma) < \infty.$$

Let's denote the random variable $X^*(\gamma) := \sup_{r \geq r_1} \gamma(S_r)/\sigma(S_r)$.

For fixed $n \geq r_1$, let $(A_j)_{j \in \mathbb{N}}$ be an increasing sequence of finite subsets of $[n, \infty)$ so that $\cup_j A_j$ is dense in $[n, \infty)$. For fixed $\gamma$ and $\epsilon$, the function $r \mapsto u_{r,\epsilon}(\gamma)$ is continuous on $[r_1, \infty)$ and so

$$\sup_{r \geq n} u_{r,\epsilon}(\gamma) = \sup_j \sup_{r \in A_j} u_{r,\epsilon}(\gamma).$$

Now for each $j \in \mathbb{N}$, $\sup_{r \in A_j} u_{r,\epsilon}(\gamma) \in D(\mathcal{E})$ and is $\mathcal{E}$-quasi-continuous. Repeated use of the inequality (2.12) combined with the bound (3.23) gives

$$\boxplus\left( \sup_{r \in A_j} u_{r,\epsilon} \right) \leq c^2 X^*/\epsilon^2 \sigma(S_n),$$

and so

$$\sup_j \mathcal{E}\left( \sup_{r \in A_j} u_{r,\epsilon}, \sup_{r \in A_j} u_{r,\epsilon} \right) \leq \int_{\ddot{\Gamma}_{\mathbb{R}^d}} \frac{c^2 X^*(\gamma)}{\epsilon^2 \sigma(S_n)}\, \mu(d\gamma) < \infty.$$

Applying Lemma 1, we see that the pointwise limit $\sup_{r \geq n} u_{r,\epsilon}$ belongs to $D(\mathcal{E})$ and is $\mathcal{E}$-quasi-continuous. In addition, the bound for the square field also carries over; $\boxplus(\sup_{r \geq n} u_{r,\epsilon}) \leq c^2 X^*/\epsilon^2 \sigma(S_n)$. Applying the same argument to the decreasing sequence $(\sup_{r \geq n} u_{r,\epsilon})_{n \in \mathbb{N}}$, we find that the pointwise limit $u_\epsilon := \limsup_{r \to \infty} u_{r,\epsilon}$ belongs to $D(\mathcal{E})$, is $\mathcal{E}$-quasi-continuous, and has $\boxplus(u_\epsilon) = 0$. The extra factor of $\sigma(S_n)$ in the denominator accounts for the fact that the square field is zero in the limit. Since $\mathcal{E}(u_\epsilon, u_\epsilon) = 0$ is bounded in $\epsilon$, we may apply Lemma 1 to conclude that $u := \lim_{\epsilon \to 0} u_\epsilon$ belongs to $D(\mathcal{E})$, is $\mathcal{E}$-quasi-continuous, and has $\boxplus(u) = 0$. The regularity assumption (3.20) on the measure $\sigma$ means that

$$u(\gamma) = \limsup_{r \to \infty} \gamma(S_r)/\sigma(S_r).$$

For any two rational numbers $0 < a < b$, we let $(\psi_n)_{n \in \mathbb{N}}$ be a sequence of smooth, compactly supported functions that vanish at the origin, decreasing pointwise to the indicator function $1_{[a,b]}$. The bound (2.11) shows us that $\psi_n(u)$ belongs to $D(\mathcal{E})$, is $\mathcal{E}$-quasi-continuous, and has $\boxplus(\psi_n(u)) = 0$. Letting $n \to \infty$ and applying Lemma 2.10 once more, we find that $1_{[a,b]}(u)$ belongs to $D(\mathcal{E})$ and is $\mathcal{E}$-quasi-continuous.

Applying the continuity result (2.13) simultaneously to the countable set of functions $\{1_{[a,b]}(u) : 0 < a < b \in \mathbb{Q}\}$, we conclude that the value of $\limsup_{r \to \infty} X_t(S_r)/\sigma(S_r)$ is almost surely constant in $t$.

A parallel argument shows that $\liminf_{r \to \infty} X_t(S_r)/\sigma(S_r)$ is also almost surely constant in $t$, and this gives the result. $\qquad \square$

COROLLARY 3.9. *The condition $\int_{\mathbb{R}_+} z^2 \lambda(dz) < \infty$ in Proposition 1 can be dropped if $\limsup_r \sigma(S_{r-\epsilon})/\sigma(S_r) > 0$ for some $\epsilon > 0$.*

PROOF. Since the case $\sigma(\mathbb{R}^d) < \infty$ was covered in Corollary 1, we will assume that $\sigma(\mathbb{R}^d) = \infty$. Choose $\epsilon > 0$ and $(r_k)_{k \in \mathbb{N}}$ so that $r_k \uparrow \infty$, $\sum_k \sigma(S_{r_k})^{-1} < \infty$, and $\lim_{k \to \infty} \sigma(S_{r_k - \epsilon})/\sigma(S_{r_k}) = c > 0$. Define $v_k$ to be the function $u_{r_k, \epsilon}$ as defined in (3.22). We have

$$\boxplus\left(\sup_k v_k\right) \leq \sup_k \boxplus(v_k) \leq \sum_k \boxplus(v_k),$$

and using the bound (3.23) and the condition $\sum_k \sigma(S_{r_k})^{-1} < \infty$, we find that the right-hand side is $\mu$-integrable. As in the proof of Proposition 3.8, we conclude that $\limsup_k v_k$ belongs to $D(\mathcal{E})$ and $\boxplus(\limsup_k v_k) = 0$, so that for any $b > 0$, $v := 1_{[0,b]}(\limsup_k v_k)$ belongs to $D(\mathcal{E})$ and $\boxplus(v) = 0$.

Now apply the product formula (3.12) to $u_n v$ where the functions $(u_n)_{n \in \mathbb{N}}$ are defined in (3.3). Since

$$\limsup_k v_k(\gamma) \geq c \limsup_k \gamma(S_{r_k - \epsilon})/\sigma(S_{r_k - \epsilon}),$$

we know from Lemma 4 that $v$ vanishes $\pi_{z\sigma}$ almost every for $z > b/c$. In particular,

$$
\begin{aligned}
\mathcal{E}(u_n v, u_n v) &= \int_{\mathbb{R}_+} \int_{\ddot{\Gamma}_{\mathbb{R}^d}} \boxplus(u_n v)(\gamma)\, \pi_{z\sigma}(d\gamma)\lambda(dz) \\
&= \int_{\mathbb{R}_+} \int_{\{\liminf_k v_k \leq b\}} \boxplus(u_n)(\gamma)\, \pi_{z\sigma}(d\gamma)\lambda(dz) \\
&\leq \int_0^{b/c} \int_{\ddot{\Gamma}_{\mathbb{R}^d}} \boxplus(u_n)(\gamma)\, \pi_{z\sigma}(d\gamma)\lambda(dz).
\end{aligned}
$$

So as in the proof of Proposition 1, with the measure $\lambda$ effectively truncated at the value $b/c$, we see that $(u_n\, v)_{n \in \mathbb{N}}$ is $\mathcal{E}$-bounded and so conclude that the set $N \cap \{\limsup_k v_k \le b\}$ is $\mathcal{E}$-exceptional. Since $b$ is arbitrary, this yields the desired result. $\qquad\square$

**Acknowledgements.** This work was completed during visits to the Mathematical Sciences Research Institute at Berkeley and to the Universität Bielefeld. I thank both institutes for inspiring working conditions. Special thanks to the whole stochastics group at Bielefeld, in particular, to my host Michael Röckner.

## References

[1] S. Albeverio, Yu. G. Kondratiev, and M. Röckner: Differential geometry of Poisson spaces, *Comptes Rendus de L'Académie des Sciences Paris* **323**, 1129–1134 (1996).

[2] S. Albeverio, Yu. G. Kondratiev, and M. Röckner: Canonical Dirichlet operator and distorted Brownian motion on Poisson spaces, *Comptes Rendus de L'Académie des Sciences Paris* **323**, 1179–1184 (1996).

[3] S. Albeverio, Yu. G. Kondratiev, and M. Röckner: Analysis and geometry on configuration spaces, *Journal of Functional Analysis* **154**, 444–500 (1998).

[4] S. Albeverio, Yu. G. Kondratiev, and M. Röckner: Analysis and geometry on configuration spaces: The Gibbsian case, SFB 343 (Bielefeld) Preprint 97-091. To appear in the *Journal of Functional Analysis*.

[5] R.L. Graham, D.E. Knuth, and O. Patashnik: *Concrete Mathematics (Second edition)*. Addison-Wesley: Reading, Massachusetts, 1994.

[6] K. Kuwae: Functional calculus for Dirichlet forms, preprint 1997.

[7] Z.M. Ma and M. Röckner: *Introduction to the Theory of (Non-Symmetric) Dirichlet Forms*. Springer: Berlin, 1992.

[8] Z.M. Ma and M. Röckner: Construction of diffusion processes on configuration spaces, SFB 343 (Bielefeld) Preprint.

[9] J. Marcinkiewicz and A. Zygmund: Sur les fonctions indépendantes, *Fundamenta Mathematicae* **29**, 60–90 (1937).

[10] H. Osada: Dirichlet form approach to infinite-dimensional Wiener processes with singular interactions, *Communications in Mathematical Physics* **176**, 117–131 (1996).

[11] M. Röckner: Stochastic analysis on configuration spaces: Basic ideas and recent results, SFB 343 (Bielefeld) Preprint 98-031.

[12] M. Röckner and A. Schied: Rademacher's theorem on configuration spaces and applications, Mathematical Sciences Research Institute (Berkeley) Preprint 1998-011.

[13] M. Röckner and B. Schmuland: Quasi-regular Dirichlet forms: Examples and counterexamples, *Canadian Journal of Mathematics* **47** (1), 165–200 (1995).

[14] M. Röckner and B. Schmuland: A support property for infinite-dimensional interacting diffusion processes, *Comptes Rendus de L'Académie des Sciences Paris* **326**, 359–364 (1998).

[15] D. Ruelle: *Statistical Mechanics: Rigorous Results*. W.A. Benjamin: New York, 1969.

[16] D. Ruelle: Superstable interactions in classical statistical mechanics, *Communications in Mathematical Physics* **18**, 127–159 (1970).

[17] M. W. Yoshida: Construction of infinite dimensional interacting diffusion processes through Dirichlet forms, *Probability Theory and Related Fields* **106**, 265–297 (1996).

DEPARTMENT OF MATHEMATICAL SCIENCES UNIVERSITY OF ALBERTA EDMONTON, ALBERTA, CANADA T6G 2G1

*E-mail address*: schmu@stat.ualberta.ca

Canadian Mathematical Society
Conference Proceedings
Volume **26**, 2000

# On a Conjecture of B. Jørgensen and A. D. Wentzell: From Extreme Stable Laws to Tweedie Exponential Dispersion Models

Vladimir Vinogradov

*I am pleased to dedicate this paper to Don Dawson. In fact, Don suggested to write a paper on Tweedie models for the probability audience and advised on its contents after my lecture at his seminar at the Fields Institute.*

ABSTRACT. A quite heterogeneous class of Tweedie exponential dispersion models was introduced in statistics by Tweedie mainly because of the simple form of their unit variance function. It is known that some of the Tweedie models are derived by exponential tilting of extreme stable laws. Also, Tweedie models possess scaling properties similar to those of stable laws, and they emerge as weak limits of appropriately scaled natural exponential families. Domains of attraction to Tweedie models are described as well. Recently, Jørgensen and Wentzell independently conjectured that the classical limit theorems on weak convergence to stable laws and those on weak convergence to Tweedie models should be related.

In this work, it is shown how the theorems on weak convergence to Tweedie models with index $p > 2$ can be derived from those on weak convergence to positive stable laws.

## 1. Introduction

The class of Tweedie exponential dispersion models generalizes, in a certain way, the class of extreme stable laws. This class of models was introduced in statistics by Tweedie (1984), mainly because of the simple form of their *unit variance function* $V(\mu)$ which is defined below by (1.1). They may be characterized as the only exponential dispersion models closed either under scale transformations, or under location changes. Convergence results to the Tweedie models were established in Jørgensen, Martínez and Tsao (1994) and Jørgensen, Martínez and Vinogradov (1999). See, e.g., Theorem 4.1 of Jørgensen (1997) and also Theorems 3.1 and 3.2 of Jørgensen, Martínez and Vinogradov (1999) for details.

In the theory of exponential dispersion models, one usually employs the techniques of *unit variance functions* and/or the *generating measures*. Below, we give

2000 *Mathematics Subject Classification.* Primary 62J12, 60E07, 60F05.
*Key words and phrases.* exponential dispersion models; stable laws; weak convergence.
Research partially supported by an NSERC grant.

the relevant definitions and terminology required for the sequel. A more comprehensive treatment of the properties of the exponential dispersion models can be found in Jørgensen (1997).

We first introduce the concept of *unit deviance* which generalizes the concept of squared distance. Namely, the bivariate function $d(\cdot, \cdot)$ is called a unit deviance if

$$d(y, y) = 0$$

and

$$d(y, \mu) > 0$$

whenever $y \neq \mu$. Here, $\mu$ stands for a location (or position) parameter, which is usually the mean. The unit deviance $d(y, \mu)$ is called *regular* if

$$d(y, \mu) \in C^2$$

with respect to $(y, \mu)$ such that

$$\frac{\partial^2 d}{\partial \mu^2}(\mu, \mu) > 0.$$

Under this regularity condition, it is possible to define the *unit variance function*

$$(1.1) \qquad V(\mu) := \frac{2}{\frac{\partial^2 d}{\partial \mu^2}(\mu, \mu)}.$$

It is clear that the squared distance $(y - \mu)^2$ is a special case of a unit deviance corresponding to unit variance function $V(\mu) = 1$. In fact, the latter property that $V(\mu)$ is equal to one identically is a characteristic property of the class of normal distributions which is a subclass of Tweedie models corresponding to the value of parameter $p = 0$ (see formula (1.16) below).

At this stage, we use the concept of the unit deviance to characterize a special class of the *natural exponential families*. The latter ones are described in terms of their probability densities having the following form:

$$(1.2) \qquad c(y) \cdot \exp\left\{-\frac{1}{2}d(y, \mu)\right\},$$

where $c(y)$ is a sufficiently smooth function that does not depend on location parameter $\mu$ (see, e.g., Jørgensen (1997, p. 6)). However, a more general class of the natural exponential families is derived by the use of the concept of the *generating measure*. Namely, a particular natural exponential family can be defined as the family of probability densities $p(y, \theta)dy$ of a certain random variable $Z$ which have the following form:

$$(1.3) \qquad p(y, \theta)dy = \frac{1}{b(\theta)}e^{\theta y}\nu(dy).$$

Here, $\nu(dy)$ is the generating measure, $\theta$ is the canonical parameter, and $b(\theta)$ is the normalizing constant which is expressed in terms of the Laplace transform of the generating measure. The latter definition is probably more appealing to probabilists than (1.2). In fact, some notation and concepts of the theory of natural exponential families were developed in parallel to those of the theory of large deviations. In particular, two representatives of family (1.3) are obtained from each other by the corresponding Cramér's transforms.

It should be pointed out that the above-mentioned location parameter $\mu$ and the *canonical parameter* $\theta$ which emerged in formulas (1.2) and (1.3), respectively, are related by means of formula (1.7) given below which involves the concept of the *mean value mapping* $\tau(\cdot)$. The latter function $\tau(\theta)$ is just the first derivative of cumulant $G(\theta)$ of the generating measure $\nu(dy)$. Here,

$$(1.4) \qquad G(\theta) := \log \int_{-\infty}^{\infty} e^{\theta y} \nu(dy).$$

Now, let us denote the domain of the canonical parameter by

$$(1.5) \qquad \Theta := \left\{ \theta \in \mathbb{R}^1 : G(\theta) < \infty \right\}.$$

Following Jørgensen (1997), assume that the least upper bound of $\Theta$ is zero provided it is finite. Clearly, this can be easily achieved by reparametrization. Let

$$(1.6) \qquad \tau(\theta) := G'(\theta).$$

It is relatively easy to see that $\tau(\theta)$ is the mean of r.v. $Z$ with respect to the probability measure $p(y, \theta)$ which is defined by (1.3). In the sequel, we will need to solve the equation

$$(1.7) \qquad \tau(\theta) = \mu$$

for $\theta$, in order to reparametrize a given natural exponential family. It is relatively easy to show (see, e.g., formula (2.18) of Jørgensen (1997)) that the solution $\theta = \tau^{-1}(\mu)$ of equation (1.7) and the unit variance function $V(\mu)$ are related as follows:

$$(1.8) \qquad \frac{\partial \tau^{-1}}{\partial \mu}(\mu) = \frac{1}{V(\mu)}.$$

Also, it is interesting to observe that starting from definition (1.3) of a natural exponential family, one can express the unit deviance $d(y, \mu)$ in the following form:

$$(1.9) \qquad d(y, \mu) = 2 \left\{ \sup_{\theta \in \Theta} (y\theta - G(\theta)) - (y\tau^{-1}(\mu) - G(\tau^{-1}(\mu))) \right\}$$

(see, e.g., formula (2.20) of Jørgensen (1997)). In particular, the first term within the braces is the rate function that corresponds to cumulant $G(\theta)$, whereas the subtrahend is the log-likelihood for location parameter $\mu$.

At this stage, note that representation (1.3) of a natural exponential family can be rewritten as follows in terms of the above-introduced notation:

$$(1.10) \qquad p(y, \theta)dy = \exp\{\theta y - G(\theta)\}\nu(dy),$$

where $\theta \in \Theta$. Now, in addition to the domain $\Theta$ of the canonical (or the *exponential tilting*) parameter $\theta$, we introduce the domain $\Lambda \subset (0, \infty)$ of the *scaling parameter* $\lambda$ as follows. By definition, $\lambda \in \Lambda$ iff there is a measure $\nu_\lambda$ such that

$$(1.11) \qquad\qquad \lambda G(\theta) := \log \int_{-\infty}^{\infty} e^{\theta y} \nu_\lambda(dy)$$

(see., e.g., Section 3.1 of Jørgensen (1997)). Obviously , $1 \in \Lambda$.

It is relatively easy to see that for each $\lambda \in \Lambda$, measure $\nu_\lambda$ generates a certain natural exponential family

$$(1.12) \qquad\qquad p_\lambda(y,\theta)dy := \exp\{\theta y - \lambda G(\theta)\}\nu_\lambda(dy)$$

with the same domain $\Theta$ of the canonical parameter $\theta$.

Following Jørgensen (1997, definition 3.1), hereinafter we refer to this family of densities (1.12) of r.v. $Z$ parametrized by $\theta$ and $\lambda$ as the *additive exponential dispersion model generated by* $\nu(dy)$. Also, we will use the following notation in the case of *additive* model:

$$Z \sim ED^*(\theta, \lambda),$$

with $(\theta, \lambda) \in \Theta \times \Lambda$.

At the same time, it is sometimes more convenient to reparametrize this bi-parametric family (1.12) in terms of parameters $(\mu, \sigma^2)$. Here,

$$\sigma^2 := 1/\lambda$$

is a scaling parameter rather than variance, whereas parameters $\mu$ and $\theta$ are related by (1.7). In particular, one can consider the *reproductive form*

$$(1.13) \qquad\qquad Y := Z/\lambda = ED(\mu, \sigma^2)$$

of the exponential dispersion model (1.12). The term *reproductive* is used in the sense that the sample mean of the observations from this model belongs to the same model. Moreover, we have that for r.v. $Y$ defined by (1.13),

$$(1.14) \qquad\qquad EY = \mu,$$

and

$$(1.15) \qquad\qquad \mathrm{Var}\, Y = \sigma^2 V(\mu),$$

where the unit variance function $V(\mu)$ is defined by (1.1).

By definition, the *Tweedie exponential dispersion models* hereinafter denoted by $Tw_p(\mu, \sigma^2)$ are those characterized by the following simple form of the unit variance function $V(\mu)$:

$$(1.16) \qquad\qquad V(\mu) = \mu^p.$$

It is known (see, e.g., Chapter 4 of Jørgensen (1997)) that such models exist iff $p \in \mathbb{R}^1 \backslash (0,1)$.

## 2. Tweedie models and extreme stable laws

It is known (see, e.g., Chapter 4 of Jørgensen (1997)) that Tweedie models and stable laws share many common features. In particular, those Tweedie models which are closed with respect to scale transformations possess the following scaling property similar to that for the stable laws:

$$(2.1) \qquad cTw_p(\mu, \sigma^2) \stackrel{\mathrm{d}}{=} Tw_p(c\mu, c^{2-p}\sigma^2)$$

for each $c > 0$.

The class of Tweedie models is completely described (see, e.g., Chapter 4 of Jørgensen (1997)). Here, we note that Tweedie models which correspond to $p < 0$ and $p > 2$ are obtained by exponential tilting (or considering Cramér's conjugate measure) of extreme stable distributions with index $1 < \alpha < 2$ and positive stable distributions with index $0 < \alpha < 1$, respectively (see, e.g., Theorem 4.3 of Jørgensen (1997)). Hereinafter, index $\alpha$ of the stable law and parameter $p$ of the Tweedie model are related as follows:

$$(\alpha - 1)(p - 1) = -1,$$

which implies that $\alpha = (p - 2)/(p - 1)$. Note that an important special case of the positive stable distribution with index $\alpha = 1/2$ corresponds to the *inverse Gaussian family*, which is a subclass of Tweedie models corresponding to the value of parameter $p = 3$. Also, Hougaard (1986) used the class of exponentially tilted stable laws for fitting medical data.

The other common feature shared by Tweedie models and stable laws is that they both emerge as weak limits of certain families of distributions. Various results on weak convergence to Tweedie models were derived in Jørgensen, Martínez and Tsao (1994), Jørgensen (1997, Chapter 4), and Jørgensen, Martínez and Vinogradov (1999). For the sake of simplicity, here we quote only a very special case of a result of such type.

THEOREM 2.1. *(see, e.g., Theorem 4.5 of Jørgensen (1997)). Assume that $p > 2$, and that the unit variance function $V(\mu)$ of the reproductive exponential dispersion model $Y = Z/\lambda = ED(\mu, \sigma^2)$ defined by (1.13) is regular of order $p$ at infinity, i.e.,*

$$(2.2) \qquad V(m) \sim c_0 \cdot m^p$$

*as $m \to \infty$, where $c_0 > 0$ is a certain constant.*

   *Then for each fixed $\mu$ and $\sigma^2 > 0$,*

$$(2.3) \qquad \frac{1}{c} ED(c\mu, \sigma^2/c^{p-2}) \stackrel{\mathrm{weakly}}{\longrightarrow} Tw_p(\mu, c_0\sigma^2)$$

*as $c \to \infty$ provided the values of $c\mu$ and $\sigma^2/c^{p-2}$ belong to the domains of the corresponding parameters.*

REMARK 2.2. The results of the same type as (2.3) can be used in statistics for approximating various exponential families.

At the same time, Jørgensen (1997, p. 150) conjectured that 'If a distribution belongs to the domain of attraction of a positive stable or extreme stable distribution, then probably the exponential dispersion model generated from it converges according to the Tweedie convergence theorem.' Subsequently, Wentzell (1998, personal communication) conjectured that it is likely that the classical limit theorems on weak convergence to stable laws imply those on weak convergence to Tweedie models.

The following result demonstrates that this conjecture due to Jørgensen and Wentzell is correct. For simplicity, we only formulate and prove our result for the case of convergence to a positive stable law with index $0 < \alpha < 1$.

THEOREM 2.3. *The limit theorems on weak convergence to Tweedie models with parameter $p > 2$ follow from the classical limit theorems on weak convergence to positive stable laws with index $0 < \alpha < 1$.*

REMARK 2.4. It seems likely that this approach of obtaining limit theorems for the Tweedie models from the classical ones could also be used for a relatively simple derivation of (unknown) analogs of the Berry–Essen estimates and the Edgeworth expansions in the case of convergence to the Tweedie models. Such hypothetical results could be useful for approximating various exponential famlies.

## 3. Proof of Theorem 2.2

For simplicity of notation, assume that our exponential dispersion model that satisfies (2.2) is non-negative and absolutely continuous.

First, it is interesting to note that condition (2.2) on regularity of $V(\mu)$ at infinity is equivalent to the condition on the regular variation of the generating measure $\nu(dy)$ at infinity such that the (non-tilted) distribution of an appropriately scaled sum of i.i.d.r.v.'s $\{Y_n, n \geq 1\}$ would belong to the domain of attraction of a certain positive stable law with index

$$\alpha = (p-2)/(p-1)$$

(see, e.g., Jørgensen, Martínez and Vinogradov (1999)). However, this fact is not directly used in our proof. Instead, we directly determine the asymptotic behavior of function $\tau^{-1}(u)$ as $u \to \infty$ under fulfilment of condition (2.2). Recall that $\tau^{-1}(\infty) = 0$. Then it follows from (1.8) that

$$(3.1) \qquad \tau^{-1}(u) = -\int_u^\infty \frac{dx}{V(x)} \sim -\frac{1}{c_0(p-1)} u^{1-p} = -\frac{1-\alpha}{c_0} u^{-1/(1-\alpha)}$$

as $u \to \infty$, by (2.2). In particular, starting from the corresponding Tweedie model with index $p > 2$ which trivially belongs to its own domain of attraction, one easily gets that

$$(3.1') \qquad\qquad \tau^{-1}\left(n^{1/(p-2)}\mu\right) = -\frac{1-\alpha}{n^{1/\alpha}\mu^{1/(1-\alpha)}},$$

since in this case, cumulant $G(\theta)$ defined by (1.4) acquires the following form:

$$G(\theta) = -\frac{(1-\alpha)^{1-\alpha}(-\theta)^\alpha}{\alpha},$$

where $\theta \leq 0$ (see, e.g., Chapter 4 of Jørgensen (1997)).

Now, assume for simplicity that $n := c^{p-2}$ is an integer. Clearly, $n \to \infty$ since $p > 2$. Then it follows from Jørgensen (1997, pp. 81–82) that the expression on the left–hand side of (2.3) can be rewritten as

$$(3.2) \qquad \frac{1}{c} \cdot \frac{1}{c^{p-2}} \cdot \sum_{i=1}^{c^{p-2}} ED_i(c\mu, \sigma^2),$$

where $\{ED_i(c\mu, \sigma^2),\ 1 \le i \le n\}$ are the independent copies $\{Y_i, 1 \le i \le n\}$ of r.v. $Y$ in the reproductive form (see (1.13)) with unit variance function satisfying (2.2). Clearly, the value of location parameter $\mu = \infty$ corresponds to the value of tilting parameter $\theta = 0$. Also, the Cramér's conjugate distribution of $ED_i(c\mu, \sigma^2)$ corresponds to the value of the tilting parameter $\theta = \tau^{-1}(n^{1/(p-2)}\mu)$, where $n^{1/(p-2)}\mu = c$. Hence, expression (3.2) can be rewritten as

$$(3.3) \qquad \frac{1}{n^{1+1/(p-2)}} \left( Y_1^{\tau^{-1}(n^{1/(p-2)}\mu)} + \cdots + Y_n^{\tau^{-1}(n^{1/(p-2)}\mu)} \right),$$

where superscript $\tau^{-1}(n^{1/(p-2)}\mu)$ just indicates that Cramér's conjugate distribution of r.v.'s $\left\{ Y_i^{\tau^{-1}(n^{1/(p-2)}\mu)},\ 1 \le i \le n \right\}$ is considered, with the value of the exponential tilting parameter equal to $\tau^{-1}(n^{1/(p-2)}\mu)$. Also, recall that the expectation $\tau(\tau^{-1}(n^{1/(p-2)}\mu))$ of each of $Y_i$'s with respect to this measure is equal to $n^{1/(p-2)}\mu = c\mu$.

At this stage, note that $1 + 1/(p-2) = 1/\alpha$ and set

$$S_n(\tau^{-1}(n^{1/(p-2)}\mu)) := Y_1^{\tau^{-1}(n^{1/(p-2)}\mu)} + \cdots + Y_n^{\tau^{-1}(n^{1/(p-2)}\mu)}.$$

Then expression (3.3) can be rewritten as

$$(3.4) \qquad S_n(\tau^{-1}(n^{1/(p-2)}\mu))/n^{1/\alpha}.$$

Recall that our goal is to show that this exponentially tilted and scaled sum $S_n(\tau^{-1}(n^{1/(p-2)}\mu))/n^{1/\alpha}$ converges weakly to the corresponding tilted version of the positive stable law with index $\alpha = (p-2)/(p-1)$ as $n \to \infty$ provided that non-tilted and scaled sum $S_n(0)/n^{1/\alpha}$ converges weakly to the non-tilted positive stable law with the same index $\alpha$ as $n \to \infty$. To this end, fix an arbitrary $x > 0$ and consider

$$(3.5) \qquad \mathbb{P}\left\{ S_n(\tau^{-1}(n^{1/(p-2)}\mu))/n^{1/\alpha} \le x \right\} = \int_0^{xn^{1/\alpha}} p_{S_n}^{\tau^{-1}(n^{1/(p-2)}\mu)}(v)dv,$$

where

$$p_{S_n}^{\tau^{-1}(n^{1/(p-2)}\mu)}(\cdot)$$

is the density of $S_n$ with respect to Cramér's conjugate measure that corresponds to exponential tilting parameter $\theta$ equal to $\tau^{-1}(n^{1/(p-2)}\mu)$. It is well known from the theory of Cramér's conjugates that

$$p_{S_n}^{\tau^{-1}(n^{1/(p-2)}\mu)}(v) = \exp\left\{ \tau^{-1}(n^{1/(p-2)}\mu)v \right\} p_{S_n}^0(v)/m_{\alpha,\sigma}(\tau^{-1}(n^{1/(p-2)}\mu))^n,$$

where $m_{\alpha,\sigma}(\theta)$ is common Laplace transform of r.v.'s $\{Y_i^0,\ 1 \le i \le n\}$. Hence, the expression on the right-hand side of (3.5) can be rewritten as

$$(3.6) \qquad \frac{1}{m_{\alpha,\sigma}\left(\tau^{-1}\left(n^{1/(p-2)}\mu\right)\right)^n} \int_0^{xn^{1/\alpha}} \exp\left\{ \tau^{-1}\left(n^{1/(p-2)}\mu\right)v \right\} p_{S_n}^0(v)dv.$$

Now, note that by (3.1),

$$\tau^{-1}\left(n^{1/(p-2)}\mu\right) \sim -\frac{1-\alpha}{c_0}\mu^{-1/(1-\alpha)}n^{-(p-1)/(p-2)} = -\frac{1-\alpha}{c_0}\mu^{-1/(1-\alpha)}n^{-1/\alpha}$$

as $n \to \infty$ (compare to (3.1')). Therefore, expression (3.6) is equivalent to the following one as $n \to \infty$:

$$m_{\alpha,\sigma}\left(-\frac{1-\alpha}{c_0}\mu^{-1/(1-\alpha)}n^{-1/\alpha}\right)^{-n}$$

$$\cdot \int_0^{xn^{1/\alpha}} \exp\left\{-\frac{1-\alpha}{c_0}\mu^{-1/(1-\alpha)}n^{-1/\alpha}v\right\} p_{S_n}^0(v)dv.$$

Making an obvious change of variables $z = vn^{-1/\alpha}$ and using the relationship

$$p_{S_n/n^{1/\alpha}}^0(z) = n^{1/\alpha}p_{S_n}^0(v),$$

one gets that the latter expression is equal to

(3.7)
$$m_{\alpha,\sigma}\left(-\frac{1-\alpha}{c_0}\mu^{-1/(1-\alpha)}n^{-1/\alpha}\right)^{-n}$$

$$\cdot \int_0^x \exp\left\{-\frac{1-\alpha}{c_0}\mu^{-1/(1-\alpha)}z\right\} p_{S_n/n^{1/\alpha}}^0(z)dz.$$

Now, recall that $m_{\alpha,\sigma}(\cdot)$ stands for common Laplace transform of r.v.'s $\{Y_i^0, 1 \le i \le n\}$ which belong to the domain of attraction of the corresponding positive stable law with index $\alpha$. Use a well–known relationship between Laplace transforms $f_{S_n}(\cdot)$ and $f_{S_n/n^{1/\alpha}}(\cdot)$ of r.v.'s $S_n$ and $S_n/n^{1/\alpha}$, respectively, namely,

$$f_{S_n/n^{1/\alpha}}\left(n^{1/\alpha}t\right) = f_{S_n}(t) = f_{X_i}(t)^n.$$

Choosing

$$t = -\frac{1-\alpha}{c_0}\mu^{-1/(1-\alpha)}\,n^{-1/\alpha},$$

one can ascertain that expression (3.7) is equal to

$$f_{S_n/n^{1/\alpha}}\left(-\frac{1-\alpha}{c_0}\mu^{-1/(1-\alpha)}\right)^{-1}\int_0^x \exp\left\{-\frac{1-\alpha}{c_0}\mu^{-1/(1-\alpha)}z\right\}p_{S_n/n^{1/\alpha}}^0(z)dz.$$

Finally, provided that the random sequence $S_n/n^{1/\alpha}$ converges to the corresponding positive stable r.v. $Y_\infty$ with index $\alpha$, the latter expression converges to that obtained from it by replacing $S_n/n^{1/\alpha}$ by r.v. $Y_\infty$. This was shown by the author using straightforward methods. However, it could also be derived from more general convergence results recently mentioned to me by S. R. S. Varadhan. These arguments conclude the proof of the fact that convergence theorems to Tweedie models are derived in some cases from the classical limit theorems on weak convergence to stable laws.

## Acknowledgment

I thank B. Jørgensen, A. D. Wentzell, and S. R. S. Varadhan for their advice. Special thanks to A. Feuerverger for permitting to use some material from the introduction to our unpublished work [1].

# References

[1] Feuerverger, A. and Vinogradov, V. (1999). The joy of Tweedie models (in progress).

[2] Hougaard, P. (1986). Survival models for heterogeneous populations derived from stable distributions. *Biometrika* **73** 387–396.

[3] Jørgensen, B. (1997). *The Theory of Dispersion Models*, Chapman & Hall, London.

[4] Jørgensen, B. Martínez, J.R. and Tsao, M. (1994). Asymptotic behaviour of the variance function. *Scand. J. Statist.* **21** 223–243.

[5] Jørgensen, B. Martínez, J.R. and Vinogradov, V. (1999). On weak convergence to Tweedie laws and regular variation of natural exponential families (in progress).

[6] Tweedie, M.C.K. (1984). An index which distinguishes between some important exponential families. In: Statistics: Applications and New Directions. Proceedings of the Indian Statistical Institute Golden Jubilee International Conf. (Eds.: J.K. Ghosh and J. Roy) Indian Statistical Institute, Calcutta, pp. 579–604.

DEPARTMENT OF MATHEMATICS, 321 MORTON HALL, OHIO UNIVERSITY, ATHENS, OH 45701 USA

*E-mail address*: vlavin@math.ohiou.edu

Canadian Mathematical Society
Conference Proceedings
Volume **26**, 2000

# Valuation of a Barrier European Option on Jump-Diffusion Underlying Stock Price

## H. Wang

*Dedicated to Professor Donald A. Dawson*

ABSTRACT. In this paper, a formula for the fair price of a barrier European option on jump-diffusion underlying stock price is given. The model for the discontinuous underlying stock price is assumed to be the same as that introduced by Merton [**8**] (1976). A dynamic programming method is used in the derivation of the fair price formula.

## 1. Introduction

In this paper, based on the same model and assumptions as in Merton [**8**] on jump-diffusion stock price, an analytic option pricing formula for an European barrier option is derived. It is difficult to derive an analytic formula for this option because the distribution of the lifetime maximum of the underlying jump-diffusion stock price is not available. To overcome this, a dynamic programming method for the conditional expectations of Markov processes is introduced.

In his classic paper on option pricing on discontinuous underlying stock price, Merton [**8**] introduced a model for jump-diffusion stock price and presented an analytic pricing formula for an European option on the jump-diffusion stock price. This formula has no closed-form solution and is very complicated. However, in recent years, there have been considerable number of papers on discontinuous price models (e.g. Zhou [**12**], [**11**], Cox-Ross [**3**], Ahn-Thompson [**1**], Amin [**2**], Das-Tufano [**4**], Duffie-Singleton [**5**], Jarrow-Turnbull [**7**], Jarrow-Lando-Turnbull [**6**], Page-Sander [**10**], ⋯). Many of them showed by market data that in some situations the diffusion approach produces very disappointing results, while the jump-diffusion approach gives results which are consistent with many empirical facts.

2000 *Mathematics Subject Classification.* Primary 60H30, 62P05; Secondary 91B70.
*Key words and phrases.* barrier option; jump-diffusion stock price; dynamic programming.

## 2. Stock price dynamic and preliminary

The dynamic of the underlying stock price is assumed to be driven by the following stochastic differential equation

(2.1) $$dS_t = (\alpha - \lambda k)S_t dt + \sigma S_t dW_t + (Y_t - 1)S_t dq_t.$$

For above equation, we have following assumptions.

(1) $k, \alpha, \lambda, \sigma$ are positive fixed numbers.
(2) For any different $t, s > 0$, $Y_t$ and $Y_s$ are independent. $Y_t > 0$ is a random variable with distribution $F$ and expectation $k + 1$.
(3) $\{q_t\}$ is a Poisson process with intensity $\lambda$.
(4) $\{W_t\}$ is a standard Brownian motion.
(5) Here $\{W_t\}, \{q_t\}$, and $\{Y_t\}$ are assumed to be independent.

Then, an explicit solution to the above equation is

(2.2) $$S_t = S_0 Y(n_t) \exp\{(\alpha - \frac{\sigma^2}{2} - \lambda k)t + \sigma W_t\},$$

where $n_t$ is the number of jumps of the Poisson process until time $t$ and $Y(n_t)$ is defined by

$$Y(n_t) \equiv \prod_{i=1}^{n_t} Y_i,$$

where $\{Y_i\}$ are i.i.d random variables with distribution $F$.

To derive the fair pricing formula for the barrier European option on the jump-diffusion stock price, we have to find distributions of certain functionals of a Brownian motion. First, let us cite following basic result.

LEMMA 2.1. *Let $\{B_t\}$ be a standard Brownian motion. Define $M(T) \equiv \max_{0 \le t \le T} B_t$. Then*

$$\mathbb{P}\{M(T) \in dm, B(T) \in db\} = \frac{2(2m - b)}{T\sqrt{2\pi T}} \exp\left\{-\frac{(2m - b)^2}{2T}\right\} dm db$$

(2.3)                                      *for $m > 0$, $b < m$.*

PROOF. The result follows from the reflection principle of Brownian motion. $\square$

For a more general process – namely, a Brownian motion with non-zero drift. Consider $X_t = \sigma B_t + \nu t$, where $\{B_t\}$ is a standard Brownian motion, $\sigma > 0$, $\nu$ are real numbers. Denote $M^X(T) \equiv \max_{0 \le t \le T} X_t$ and $m^X(T) \equiv \min_{0 \le t \le T} X_t$. By virtue of Girsanov's theorem, we have the following lemma.

LEMMA 2.2. *For every $t > 0$, the joint distribution of $X_t$ and $M^X(t)$ is given by the formula*

(2.4) $$\mathbb{P}\{M^X(t) \in dy, X_t \in dx\} = f_t(x, y, \sigma, \nu) dx dy,$$

*where*

$$f_t(x, y, \sigma, \nu) \equiv \left[-\frac{2(x - 2y - \nu t)}{\sigma^3 t\sqrt{2\pi t}} - \frac{2\nu}{\sigma^3 \sqrt{2\pi t}}\right] \exp\left\{2\nu y\sigma^{-2} - \frac{(2y - x + \nu t)^2}{2t\sigma^2}\right\}$$

$$\text{for } y > 0, \ x < y.$$

*The joint distribution of $X_t$ and $m^X(t)$ is given by the formula*

(2.5) $$\mathbb{P}\{m^X(t) \in dy, X_t \in dx\} = \tilde{f}_t(x, y, \sigma, \nu) dx dy,$$

*where*

$$\tilde{f}_t(x, y, \sigma, \nu) \equiv \left[\frac{2(x - 2y - \nu t)}{\sigma^3 t\sqrt{2\pi t}} + \frac{2\nu}{\sigma^3\sqrt{2\pi t}}\right] \exp\left\{2\nu y\sigma^{-2} - \frac{(2y - x + \nu t)^2}{2t\sigma^2}\right\}$$
$$\text{for } y \leq 0, \; y \leq x.$$

PROOF. See the proof of Lemma B.3.2 and Corollary B.3.3 of Musiela and Rutkowski [**9**]. □

LEMMA 2.3. *Let* $\{B_t\}$ *be a Brownian motion starting at* $B_0 = x \neq 0$. *Define* $M(T) \equiv \max_{0 \leq t \leq T} B_t$. *Then*

$$(2.6) \qquad \mathbb{P}\{M(T) \in dm, B(T) \in db\} = g_T^x(m, b)dmdb,$$

*where*

$$g_T^x(m, b)$$
$$(2.7) \quad \equiv \quad \begin{cases} \dfrac{2(2m - b - x)}{T\sqrt{2\pi T}} \exp\left\{-\dfrac{(2m - b - x)^2}{2T}\right\} & \text{for } m > x, \; b < m \\ 0 & \text{otherwise.} \end{cases}$$

PROOF. This can be directly derived from Lemma 2.1. □

Barrier options have payoffs that depend on whether the underlying stock price reaches a barrier, $L$ before expiry $T$. Barrier options can be classified as either knock-out options or knock-in options. Knock-out options are options that become worthless if the stock price reaches a barrier but otherwise have payoffs identical to a standard option. Knock-in options are options that become standard options if the stock price reaches a barrier but otherwise are worthless. A down-and-out call option is a knock-out option that becomes worthless if the stock price falls below the barrier $L$, but otherwise is the same as a call option with payoff $[S_T - K]^+$ at expiry, where $K$ is the strike price. The "down" means that the barrier level $L$ is below the initial stock price. A down-and-out put option has payoff $[K - S_T]^+$ provided the underlying stock price does not fall below the barrier $L$, otherwise it becomes worthless. An up-and-out call is a regular call option that ceases to exist if the stock price reaches a barrier $L$ which is higher than the current stock price. According to the probability complementary relationship, the price of a regular call option equals the value of a down-and-in call plus the value of a down-and-out call. That is

$$(2.8) \qquad c = c_{di} + c_{do},$$

and

$$(2.9) \qquad c = c_{ui} + c_{uo}.$$

for up-and-in call and up-and-out call, respectively. Similarly, For put options, we have

$$(2.10) \qquad p = p_{ui} + p_{uo},$$

$$(2.11) \qquad p = p_{di} + p_{do}.$$

Since the formulae for regular call or put options are already given by Merton's paper [**8**], from the above identities (2.8) - (2.11), we only need to give formulae for $c_{uo}, c_{do}, p_{uo}, p_{do}$. The derivation of these formulae follows from Lemma 2.2. Hence it is sufficient to only calculate $c_{uo}$ for this paper.

In the remaining section, we propose our method to derive the desired fair up-and-down call option price formula. This method is similar to Dynamic Programming in Stochastic Optimization and Control Theory, and hence we call it, the dynamic programming method. Roughly speaking, the method is as follows. For a given option maturity time $T$ and the stock price $S_t$ given by (2.2), it is difficult to directly find the joint distribution function of $\max_{0 \le t \le T} S_t$ and $W_T$ due to the jump perturbation. However we can divide the time interval $[0, T]$ into $n$ subintervals $[0, \tau_1), [\tau_1, \tau_2), \ldots, [\tau_n, T]$ if there are $n$ jumps exactly before $T$, where $\tau_i$ denotes the $i^{th}$ jump time. If $W_{\tau_i}$ already known, then from Lemma 2.2 it is not very difficult to find the joint distribution function of $\max_{\tau_i \le t < \tau_{i+1}} S_t$ and $W_{\tau_{i+1}}$. In short, the idea is to transform an evaluation of an expectation into an evaluation of a sequence of one-period conditional expectations, working forward in time in a recursive manner.

Another concept we need to introduce is the risk-neutral measure or martingale measure. That is, a risk-neutral measure (martingale measure) is any probability measure, equivalent to the market measure, which makes all discounted stock prices martingales.

To simplify our notation, we assume that $\mathbb{P}$ is the risk-neutral measure. Then

$$(2.12) \qquad S_t = S_0 Y(n_t) \exp\left\{ (r - \frac{\sigma^2}{2} - \lambda k)t + \sigma W_t \right\},$$

where $r$ is the risk-free rate and $\{W_t\}$ is a standard Brownian motion. Since $\{W_t\}, \{q_t\}$, and $\{Y_t\}$ are assumed to be independent, we can use Girsanov's theorem again to define

$$(2.13) \qquad \tilde{\mathbb{P}}(A) \equiv \int_A Z(T) d\mathbb{P},$$

where $A \in \mathcal{F}_T$, $\{\mathcal{F}_t\}$ is the $\sigma$-field generated by $\{W_s, s \le t\}$,

$$(2.14) \qquad Z(t) = \exp\left\{ -\theta W_t - \frac{\theta^2 t}{2} \right\},$$

and $\theta = (r - \lambda k - \sigma^2/2)/\sigma$. Therefore, $\tilde{W}_t = \theta t + W_t$ is a standard Brownian motion under $\tilde{\mathbb{P}}$ and

$$(2.15) \qquad S_t = S_0 Y(n_t) \exp\{\sigma \tilde{W}_t\}.$$

## 3. Up-and-out European call

Before deriving the barrier option formula, we should mention again that this paper is based on the same assumptions as that given in Merton's paper [8]. Given $0 < K < L$ and assume $S_0 < L$, the central aim of this section is to evaluate the following present value of an up-and-out European call option:

$$(3.1) \qquad v(0, S_0) = e^{-rT} \mathbb{E}[(S_T - K)^+ 1_{\{S_T^* < L\}}],$$

where $S_t$ is defined by (2.12) and $S_T^* \equiv \max_{0 \le t \le T} S_t$. Recall that $\tau_i$ is the $i^{th}$ jump time of the Poisson process with $\tau_0 \equiv 0$. By the independence of the Poisson process and the Brownian motion, we have

$$(3.2)\, v(0, S_0) = e^{-(r+\lambda)T} \sum_{n \ge 0} \frac{(\lambda T)^n}{n!} \mathbb{E}[(S_T - K)^+ 1_{\{S_T^* < L\}} | \tau_n \le T < \tau_{n+1}].$$

Since $Y_t > 0$, the joint distribution of $(Y(1), Y(2), \cdots, Y(n))$ can be derived by a random vector transformation. Denote its joint distribution by $G(y_1, y_2, \cdots, y_n)$. Define $Y(0) \equiv y_0 \equiv 1$ and

$$
\begin{aligned}
R(x, y, z) \quad &\equiv \quad (zS_0 \exp\{\sigma x\} - K) \exp\left\{\theta x - \frac{\theta^2 T}{2}\right\} \\
&\times 1_{\{y < \frac{\ln(L/(zS_0))}{\sigma}, x > \frac{\ln(K/(zS_0))}{\sigma}\}},
\end{aligned}
$$
(3.3)

$$
h(y, z) \equiv 1_{\{y < \frac{\ln(L/(zS_0))}{\sigma}\}}.
$$
(3.4)

Then, for $n = 0$

$$
\mathbb{E}[(S_T - K)^+ 1_{\{S_T^* < L\}} | \tau_n \leq T < \tau_{n+1}] = \tilde{\mathbb{E}}[R(\tilde{W}_T, \max_{t \in [0,T]} \tilde{W}_t, y_n)],
$$

and for $n \geq 1$

$$
\mathbb{E}[(S_T - K)^+ 1_{\{S_T^* < L\}} | \tau_n \leq T < \tau_{n+1}]
$$
$$
= \int_0^T \int_{t_1}^T \cdots \int_{t_{n-1}}^T \int_{\mathbb{R}^n} D \, dG(y_1, y_2, \cdots, y_n) \lambda^n e^{-\lambda t_n} dt_1 \cdots dt_n,
$$
(3.5)

where

$$
\begin{aligned}
D \quad &\equiv \quad \tilde{\mathbb{E}}[R(\tilde{W}_T, \max_{t \in [t_n, T]} \tilde{W}_t, y_n) h(\max_{t \in [t_{n-1}, t_n)} \tilde{W}_t, y_{n-1}) \\
&\times h(\max_{t \in [t_1, t_2)} \tilde{W}_t, y_1) \cdots h(\max_{t \in [0, t_1)} \tilde{W}_t, 1)].
\end{aligned}
$$
(3.6)

Above $\tilde{\mathbb{E}}$ is the expectation with respect to the probability measure $\tilde{\mathbb{P}}$. Recall that $\{\tilde{W}_t\}$ is a standard Brownian motion under the probability $\tilde{\mathbb{P}}$. Using the Markov property, we have

$$
\begin{aligned}
D \quad = \quad &\tilde{\mathbb{E}} \underbrace{\{}_{(0)} h(\max_{0 \leq t < t_1} \tilde{W}_t, 1) \tilde{\mathbb{E}}^{\tilde{W}_{t_1}} \underbrace{\{}_{(1)} h(\max_{0 \leq t < t_2 - t_1} \tilde{W}_t, y_1) \tilde{\mathbb{E}}^{\tilde{W}_{t_2 - t_1}} \underbrace{\{}_{(2)} \cdots \\
&\tilde{\mathbb{E}}^{\tilde{W}_{t_{n-1} - t_{n-2}}} \underbrace{\{}_{(n-1)} h(\max_{0 \leq t < t_n - t_{n-1}} \tilde{W}_t, y_{n-1}) \\
&\times \tilde{\mathbb{E}}^{\tilde{W}_{t_n - t_{n-1}}} \underbrace{\{}_{(n)} R(\tilde{W}_{T - t_n}, \max_{0 \leq t \leq T - t_n} \tilde{W}_t, y_n) \underbrace{\}}_{(n)} \cdots \underbrace{\}}_{(0)} \\
= \quad &\int_{\mathbb{R}} \int_{\mathbb{R}} h(z_1, y_0) g_{t_1}^0(z_1, x_1) dz_1 dx_1 \int_{\mathbb{R}} \int_{\mathbb{R}} h(z_2, y_1) g_{t_2 - t_1}^{x_1}(z_2, x_2) dz_2 dx_2 \cdots \\
&\int_{\mathbb{R}} \int_{\mathbb{R}} h(z_n, y_{n-1}) g_{t_n - t_{n-1}}^{x_{n-1}}(z_n, x_n) dz_n dx_n \times \\
&\int_{\mathbb{R}} \int_{\mathbb{R}} R(x_{n+1}, z_{n+1}, y_n) g_{T - t_n}^{x_n}(z_{n+1}, x_{n+1}) dz_{n+1} dx_{n+1}.
\end{aligned}
$$

**Comments:** In spite of concise definitions of the density functions in above pricing formula, it is not easy to evaluate these multiple integrals. Therefore, in the future we need to find a simpler, approximate formula for the barrier option on jump-diffusion underlying stock prices.

**Acknowledgments.** The author thanks Dr. Ian Iscoe for helpful comments.

# References

[1] Ahn, Chang M. and Howard E. Thompson (1988). Jump-diffussion processes and the term structure of interest rates. *The Journal of Finance*, XLIII, No 1:155–174, 1988.

[2] Amin, Kaushik I. (1993). Jump diffusion option valuation in discrete time. *The Journal of Finance*, XLVIII, No 5:1833–1863, 1993.

[3] Cox, John C. and Stephen A. Ross (1976). The valuation of options for alternative stochastic processes. *Journal of Financial Economics*, 3:145–166, 1976.

[4] Das, Sanjis R. and Peter Tufano (1995). Pricing credit-sensitive debt when interest rates, credit ratings and credit spreads are stochastic. *The Journal of Financial Engineering*, Vol. 5 No. 2:161–198, 1995.

[5] Duffie, Darrell and Kenneth J. Singleton (1998). Modeling term structures of defaultable bonds. *Working paper of Graduate School of Business, Stanford University*, 1998.

[6] Jarrow, R. A., D. Lando, and S. M. Turnbull (1997). A markov model for the term structure of credit risk spreads. *The Review of Financial Studies*, Vol. 10, No. 2:481–523, 1997.

[7] Jarrow, Robert A. and Stuart M. Turnbull (1995). Pricing derivatives on financial securities subject to credit risk. *The Journal of Finance*, L, No 1:53–85, 1995.

[8] Merton, Robert C. (1976). Option pricing when underlying stock returns are discontinuous. *Journal of Financial Economics*, 3:125–144, 1976.

[9] Musiela, Marek and Marek Rutkowski (1997). *Martingale Methods in Financial Modelling*. Springer, 1997.

[10] Page,Jr., Frank H. and Anthony. B. Sanders (1986). A general derivation of the jump process option pricing formula. *Journal of Financial and Quantitative Analysis*, 21, No 4:437–446, 1986.

[11] Zhou, Chunsheng (1997). A jump-diffusion approach to modeling credit risk and valuing defautable securities. *Research papers of Federal Reserve Board*, :1–47, 1997.

[12] Zhou, Chunsheng (1997). Path-dependent option valuation when the underlying path is discontinuous. *Research papers of Federal Reserve Board*, :1–19, 1997.

ALGORITHMICS INC., 185 SPADINA AVENUE, TORONTO, ON M5T 2C6 CANADA
*E-mail address*: hwang@algorithmics.com